116. Gylbert L, Asplund O, Berggren A, Jurell G, Ramj U, Ostrup I. Preoperative antibiotics and capsular contracture in augmentation mammaplasty. Plast Reconstr Surg 1990; 86:260.
117. Mladick RA. Prevention of capsular contracture. Plast Reconstr Surg 1999;103:1773–1774.
118. Mladick RA. "No-touch" submuscular breast augmentation techniques. Aesthetic Plast Surg 1993;17:183.
119. Collins N, Mirza S, Stanley PR, Cambell L, Sharpe DT. Reduction of potential contamination of breast implants by the use of "nipple shields." Br J Plast Surg 1999;52:445.
120. Baker JL Jr., Handler ML, LeVier RR. Occurrence and activity of myofibroblasts in human capsular contracture tissue surrounding mammary implants. Plast Reconstr Surg 1981;68:905.
121. Burkhardt BR, Eades E. The effect of Biocell texturing and povidone-iodine irrigation on capsular contracture around saline-inflatable breast implants. Plast Reconstr Surg 1995; 96:1317–1325.
122. McGregor RR. Silicones and their uses. New York, NY: McGraw Hill; 1954, pp. 9–10, 255–256.
123. Adams WP Jr., Conner WC, Barton FE Jr., Rohrich RJ. Optimizing breast-pocket rrigation: the post-betadine era. Plast Reconstr Surg 2001;107:1596–1601.
124. Virden CP, et. al. Subclinical infection of the silicone breast implant surface as a possible cause of capsular contracture. Aesthetic Plast Surg 1992;16:173–179.
125. Jennings DA, Morykwas MJ, Burns WW, Crook ME, Hudson WP, Argenta LC. In vitro adhesion of endogenous skin microorganisms to breastprostheses. Ann Plast Surg 1991; 27:216–220.
126. Waldrogel FA, Bisno AC. Infections associated with indwelling medical devices. American Society For Microbiology. Portland, OR: Book News, 2000.
127. Freedman AM, Jackson IT. Infections in breast implants. Infect Dis Clin North Am 1989;3:275.
128. Ahn CY, Clifford YK, Wagar EA, Wong RS, Shaw WW. Microbial evaluation: 139 implants removed from symptomatic patients. Plast Reconstr Surg 1996;98:1225.
129. Netscher DT, Weizer G, Wigoda P, Walker LE, Thornby J, Bowen D. Clinical relevance of positive breast periprosthetic cultures without overt infection. Plast Reconstr Surg 1995; 96:1125–1129.
130. Burkhart B. Fibrous capsular contracture around breast implants: The role of subclinical infection. Infectious Surgery 1985;4:469.
131. Sanger JR, Sheth NK, Franson TR. Adherence of microorganisms to breast prostheses: an in vitro study. Ann Plast Surg 1989;22:337–342.
132. Walker LE, Breiner MJ, Goodman CM. Toxic shock syndrome after explantationof breast implants: a case report and review of the literature. Plast Reconstr Surg 1997;99:875–879.
133. Poblete JVP, Rodgers JA, Wolfort FG. Toxic shock syndrome as a complication ofbreast prostheses. Plast Reconstr Surg 1995;96:1702–1708.
134. Olesen LL, Ejlertsen T, Nielsen T. Toxic shock syndrome following insertion of breast prosthesis. Br J Surg 1991;78:585.
135. Bartlett P, Reingold AL, Graham DR, et al. Toxic shock syndrome associated with surgical wound infections. JAMA 1982;247:1448.
136. Centers for Disease Control. Case definitions for public health surveillance. Morb Mortal Wkly Rep 1990;39:38.
137. Lauter C. Recent advances in toxic shock syndrome. Contemp Int Med 1994;6:11.
138. Broome CV. Epidemiology of toxic shock syndrome in the United States: overview. Rev Infect Dis 1989;11:S14.
139. Gabriel SE, Woods JE, O'Fallon WM, Beard CM, Kurland LT, Melton LJ. Complications leading to surgery after breast implantation. N Engl J Med 1997;336:677–682.

93. Gumucio CA, Pin P, Young VL, et al. The effect of breast implant on the radiographic detection of microcalcification and soft-tissue masses. Plast Reconstr Surg 1989;84:772.
94. Hayes H, Vandergrift J, Diner WC. Mammography and breast implants. Plast Reconstr Surg 1988;82.1.
95. Spear SL, Mardini S. Alternative filler materials and new implant designs: what's available and what's on the horizon? Clin Plast Surg 2001;28:435–443.
96. Gorczyca DP, Brenner RJ. The Augmented Breast: radiologic and clinical perspectives. New York, NY: Thieme, 1997.
97. Paletta C, Paletta FX, Paletta FX. Squamous cell carcinoma following breast augmentation. Ann Plast Surg 1992;29:425–432.
98. Rohrich RJ, Adams WP Jr, Beran SJ, et al. An analysis of silicone gel-filled breast implants: diagnosis and failure rates. Plast Reconstr Surg 1998;102:2304–2308.
99. Theophelis LG, Stevenson TR. Radiographic evidence of breast implant rupture. Plast Reconstr Surg 1986;78:673.
100. Samuels JB, Rohrich RJ, Weatherall PT, Ho AMW, Goldberg KL. Radiographic diagnosis of breast implant rupture: current status and comparison of techniques. Plast Reconstr Surg 1995;96:865–877.
101. Silfen R, amir A, Hauben DJ. Plastic surgeon's responsibility following breast reconstruction. Plast Reconstr Surg 2001;108:275–276.
102. Venta LA, Salomon CG, Flisak ME, Venta ER, Izquierdo R, Angelats J. Sonographic signs of breast implant rupture. Am J Roentgenol 1996;166:1413.
103. Beekman WH, Feitz R, Hage JJ, Mulder JW. Life span of silicone gel-filled mammary prostheses. Plast Reconstr Surg 1997;100:1723.
104. DeBruhl ND, Gorczyca DP, Ahn CY, Shaw WW, Bassett LW. Silicone breast implants: US evaluation. Radiology 1993;189:95–98.
105. Caskey CI, Berg WA, Anderson N, et al. Evaluation of breast implant rupture: diagnosis by ultrasonography. Radiology 1994;190:819.
106. Beekman WH, van Straalen WR, Hage JJ, Taets can Amerongen M, Mulder JW. Imaging signs and radiologists' jargon of ruptured breast implants. Plast Reconstr Surg 1998;102:1281–1289.
107. Berg WA, Caskey CI, Hamper UM, et al. Diagnosing breast implant rupture with MR imaging, US, and mammography. Radiography 1993;13:1323.
108. Berg WA, Anderson ND, Zerhouni EA, Chang BW, Juhlman JE. MR imaging of the breast in patients with silicone breast implants: normal postoperative variants and diagnostic pitfalls. Am J Roentgenol 1994;163:575.
108. Mund DF, Farria DM, Gorczyca DP. MR imaging of the breast in patients with silicone-gel implants: spectrum of findings. Am J Roentgenol 1993;161:773.
110. Soo MS, Kornguth PJ, Walsh R, Elenberger CD, Georgiade GS. Complex radial folds versus subtle signs of intracapsular rupture of breast implants: MR findings with surgical correlation. M J Roentgenol 1996;166:1421.
111. Levine RA, Collins TL. Definitive diagnosis of breast implant rupture by ultrasonography. Plast Reconstr Surg 1991;87:1126.
112. Venta LA, Salomon CG, Flisak ME, Venta ER, Izquierdo R, Angelats J. Sonographic signs of breast implant rupture. Am J Roentgenol 1996;166:1413.
113. Reynolds HE, Buckwalter KA, Jackson VP, Siwy BK, Alexander SG. Comparison of mammography, sonography, and magnetic resonance imaging in the detection of silicone gel breast implant rupture. Ann Plast Surg 1994;33:247–255.
114. Lossing C, Hansson H. Peptide growth factors and myofibroblasts in capsules around human breast implants. Plast Reconstr Surg 1993;91:1277–1285.
115. Handel N, Jensen A, Black Q, et al. The fate of breast implants: a critical analysis of complications and outcomes. Plast Reconstr Surg 195;96:7.

69. Spear SL, Baker JL. Classification of capsular contracture after prosthetic breast reconstruction. Plast Reconstr Surg 96;1119–1129:1995.
70. Eklund GW, Busby RC, Miller SH, Job J. Improved imaging of the augmented breast. Am J Roentgenol 1988;151:469.
71. Fodor PB, Isse NG. Endoscopically assisted aesthetic plastic surgery. St. Louis, MO: Mosby-Year Book, Inc., 1996, pp. 123–192.
72. Brody GS. On the safety of breast implants Plast Reconstr Surg 1997;100:1314–1321.
73. Levine JJ. Sclerodermalike esophageal disease in children breast-fed by mothers with silicone breast implants. JAMA 1994;271:213–216.
74. Foster RS. Follow-up after mastectomy with or without breast reconstruction. In: Spear SL. Surgery of the breast: principles and art. Philadelphia, PA:Lippincott-Raven Publishers, 1998.
75. Rosselli Del Turco M, Palli D, Cariddi A, et al. Intensive diagnostic follow-up after treatment of primary breast cancer: a randomized trial. JAMA 1994;271:1903.
76. GIVIO Investigators. Impact of follow-up testing on survival and health-related quality of life in breast cancer patients: a multicenter randomized controlled trial. JAMA 1994;271:1587.
77. Cher DJ, Conwell JA, Mandel JS. MRI for detecting silicone breast implant rupture: meta-analysis and implications Ann Plast Surg 2001;47:367–380.
78. Gorczyca DP. MR imaging of breast implants. Magn Reson Imaging Clin N Am 1994;2:659–672.
79. O'Tolle M, Caskey CI. Imaging spectrum of breast implant complications: mammography, ultrasound, and magnetic resonance imaging. Semin Ultrasound CT MR 2000;21:351–361.
80. Mitnick, JS, Vazqyes, MF, Plesser K, et al. Fine needle aspiration biopsy in patients with augmentation mammoplasty. Plast Reconstr Surg 1993;91:241–244.
81. Simon MS, Hoff M, Hessein M, et al. An evaluation of clinical follow-up in women with early stage breast cancer among physician members of the American Society of Clinical Oncology. Breast Cancer Res Treat 1993;27:211.
82. Loomer L, Brockschmidt JK, Muss HB, et al. Postoperative follow-up of patients with early breast cancer. Cancer 1991;76:55.
83. Coleman J, Bostwick J. Breast cancer. In: Bostwick J, ed. Plastic and reconstructive breast surgery. St. Louis, MO: Quality Medical Publishing, Inc., 2000.
84. Yu LT, Latorre G, Marotta J, et al. In vitro measurement of silicone bleed from breast implants. Paper presented at the 27th Annual Meeting of the Aesthetic Society, Dallas, TX, April 15, 1994.
85. Brown SL, Silverman BG, Berg WA. Rupture of silicone-gel breast implants: causes, sequelae, and diagnosis. Lancet 1997;350:1531–1537.
86. Gorczyca DP, Sinha S, Ahn C, et al. Silicone breast implants in vivo: MR imaging. Radiology 1992;185:407–410.
87. Harris KM, Ganott MA, Shestak KC, Losken HW, Tobon H. Silicone implant rupture detection with US. Radiology 1993;187:761–768.
88. Gorczyca DP, DeBruhl ND, Ahn CY, et al. Silicone breast implant ruptures in an animal model: comparison of mammography, MR imaging, US, and CT. Radiology 1994;190:227–232.
89. Ramirez OM, Daniel RK. Endoscopic plastic surgery. New York, NY: Springer-Verlag, Inc., 1996, pp. 185–193.
90. Codner MA, Cohen AT, Hester TR. Complications in breast augmentation: prevention and correction. Clin Plast Surg 2001;28:587–596.
91. Mentor data product insert: saline-filled and spectrum mammary prostheses, May 2000.
92. Carlson GW, Curley SA, Martin JE, Fornage BD, Ames FC. The detection of breast cancer after augmentation mammaplasty. Plast Reconstr Surg 1993;91:837.

42. Fisher JC. The silicone controversy: When will science prevail? New Engl J Med 1992; 326:1696.
43. Kessler DA. The basis of the FDA's decision on breast implants. New Engl J Med 1992; 326:1713.
44. Chan SC, Birdsell DC, Gradeen CY. Detection of toluene-diamines in the urine of a patient with polyurethane-covered breast implants. Clin Chem 1991;37:756.
45. Rohrich RJ, Kenkel JM, Adams WP. Preventing capsular contracture in breast augmentation: in search of the holy grail. Plast Reconstr Surg 1999:103;1759–1760.
46. Independent Review Group: Silicone gel breast implants: The report of the Independent Review Group (www.silicone-review.gov.uk).
47. Angell M. Shattuck Lecture—Evaluating the health risks of breast implants: the interplay of medical science, the law, and public opinion. N Engl J Med 1996;334:1513–1518.
48. Janowsky EC, Kupper LL, Hulka BS. Meta-analyses of the relation between silicone breast implants and the risk of connective-tissue diseases. N Engl J Med 2000;342:781.
49. Bondurant S, Ernster V, Herdman R, eds. Safety of silicone breast implants. Washington, DC: Institute of Medicine, National Academy Press, 1999.
50. Brinton LA, Brown SL. Breast implants and cancer. J Natl Cancer Inst 1997;89:1341–1349.
51. Herdman RC, Fahey TJ. Silicone breast implants and cancer. Cancer Invest 2001;19:821–832.
52. Deapen DM, Bernstein L, Brody GS. Are breast implants anticarcinogenic? A 14-year follow-up of the Los Angeles study. Plast Reconstr Surg 1997;99:1346–1353.
53. Park AJ, Black RJ, Sarhadi NS, Chetty U, Watson ACH. Silicone gel-filled breast implants and connective disease. Plast Reconstr Surg 1998;101:261–267.
54. Tebbetts, JB. Alternatives and trade-offs in breast augmentation. Clin Plast Surg 2001; 28:485–500.
55. Hamas RS. The comparative dimensions of round and anatomical saline-filled breast implants. Aesthetic Surg J 2000;20:281–290.
56. Camirand A, Doucet J, Harris J. Breast augmentation: compression—a very important factor in preventing capsular contracture. Plast Reconstr Surg 1999;104:529–538.
57. Pound EC, Pound EC. Transumbilical breast augmentation (TUBA). Clin Plast Surg 2001; 28:597–605.
58. Biggs TM, Yarish RS. Augmentation mammaplasty: a comparative analysis. Plast Reconstr Surg 1990;85:368.
59. Becker H, Springer R. Prevention of capsular contracture. Plast Reconstr Surg 1999;103: 1766–1774.
60. Kulber DA, Mackenzie D, Steiner JH, et al. Monitoring the axilla in patients with silicone gel implants. Ann Plast Surg 1995;35:580–584.
61. Biggs TM. Augmentation mammaplasty: a comparative analysis. Plast Reconstr Surg 1999;103:1761–1765.
62. Cook RR, Delongchamp RR, Woodbury M, Perkins LL, Harrison MC. The prevalence of women with breast implants in the United States-1989. J Clin Epidemiol 1995; 48:519.
63. Bright RA, Moore RM. Estimating the prevalence of women with breast implants (Letter). Am J Public Health 1996;86:891.
64. Brody, Garry S. Safety and effectiveness of breast implants. Surgery of the breast: principles and art. Philadelphia, PA: Lippincott-Raven, 1998:335–345.
65. Goldberg EP, Habal MB. Future directions in breast implant surgery. Clin Plast Surg 2001;28:687–702.
66. American Society of Plastic Surgeons: 2000 Breast Surgery Statistics. (http://www.plasticsurgery.org/mediactr/stats-11.pdf).
67. Netscher DT, Clamon J. Smoking: adverse effects on outcomes from plastic surgical patients. Plast Surg Nurs 1994;14:205.
68. Becker II, Prysi MF. Quantitative assessment of postoperative breast massage. Plast Reconstr Surg 1990;86:355.

16. Regnault P, Daniel RK, eds. Aesthetic plastic surgery: principles and techniques. Boston, MA: Little, Brown, 1984.
17. Cronin TD, Gerow FJ. Augmentation mammaplasty: a new "natural feel" prosthesis. Transactions of the Third International Congress of Plastic and Reconstructive Surgery. Amsterdam: Excerpta Medica, 1964.
18. Young VL, Watson ME. Breast implant research: where we have been, where we are, where we need to go. Clin Plast Surg 2001;28:451–464.
19. Angell M. Breast implants—protection or paternalism? New Engl J Med 1992;326:1694.
20. Miller TA. Capsulectomy. Plast Reconstr Surg 1998;102:882.
21. McKinney P, Tresley G. Long-term comparison of patients with gel and saline mammary implants. Plast Reconstr Surg 1983;72:27.
22. Baker JL. Augmentation mammaplasty: a comparative analysis. Plast Reconstr Surg 1999;103:1763–1764.
23. Lavine DM. Saline inflatable prostheses: 14 years' experience. Aesthetic Plastic Surgery 1993;17:325.
24. Tebbetts JB. Alternative and trade-offs in breast augmentation. Clin Plast Surg 2001;28:485–500.
25. Worton EW, Seifert LN, Sherwood R. Late leakage of inflatable silicone breast prostheses. Plast Reconstr Surg 1980;65:302–306.
26. Weiner DL, Aiache AE, Silver L. A new soft, round, silicone gel breast implant. Plast Reconstr Surg 1974;53:174–178.
27. Collis N, Coleman D, Foo ITH, Sharpe DT. Ten-year review of a prospective randomized controlled trail of textured versus smooth subglandular silicone gel breast implants. Plast Reconstr Surg 2000;106:786–791.
28. Hester TR, Nahai F, Bostwick J, et al. A 5-year experience with polyurethane-covered mammary prostheses for treatment of capsular contracture, primary augmentation mammaplasty and breast reconstruction. Clin Plast Surg 1988;15:569.
29. Hakelius L, Ohlsen L. Tendency to capsular contracture around smooth and textured gel-filled silicone mammary implants: a 5-year follow-up. Plast Reconstr Surg 1997;100:1566.
30. Spear SL, Elmaraghy M, Hess C. Textured-surface saline-filled implants for augmentation mammaplasty. Plast Reconstr Surg 2000;105:1542–1552.
31. Shaw WW. Guidelines and indications for breast implant capsulectomy. Plast Reconstr Surg 1988;102:893.
32. Kjøller, K, Hölmich LR, Jacobsen PH, et al. Capsular contracture after cosmetic breast implant surgery in Denmark. Ann Plast Surg 2001;47:359–66.
33. Peters W, Smith D, Fornasier V, et al. An outcome analysis of 100 women after explantation of silicone gel breast implants. Ann Plast Surg 1997;39:9–19.
34. Tebbetts JB. A surgical perspective from two decades of breast augmentation: toward state of the art in 2001. Clin Plast Surg 2001;28:425–434.
35. Tarpila E, Ghassemifar R, Fagrell D, et al. Capsular contracture with textured versus smooth saline-filled implant for breast augmentation: a prospective clinical study. Plast Reconstr Surg 1997;99:1934–1939.
36. Worseg A, Kuzbari R, Tairych G, et al. Long-term results of inflatable mammary implants. Br J Plast Surg 1995;48:183–188.
37. Levine JJ. Trachtman H. Gold DM. Pettei MJ. Esophageal dysmotility in children breast-fed by mothers with silicone breast implants. Dig Dis Sci 1996;41:1600–1603.
38. Rascoe DS, Greene WB, Greene AS, McGown ST. The absence of esophageal lesions in maternal progeny of silicone injected rats. Plast Reconstr Surg 1997;99:1784–1785.
39. Mentor. Breast augmentation—is it right for you? Santa Barbara, CA: Mentor, 2001.
40. McGhan Medical. Saline-filled breast implant surgery: making an informed decision [patient brochure]. Santa Barbara, CA: McGhan Medical, 2000.
41. Hölmich LR, Kjøller K, Vejborg I, et. al. Prevalence of silicone breast implant rupture among Danish women. Plast Reconstr Surg 2001;108:848–863.

APPENDIX: SUPPORT GROUPS (CONTINUED)

National Alliance of Breast Cancer Organizations 10th Floor (NABCO)	9 East 37th Street New York, NY 10016	(888) 80-NABCO www.nabco.org
National Breast Cancer Coalition	1707 L Street, NW Suite 1060 Washington, D.C. 20036	(202) 296-7477 www.natlbcc.org
National Coalition for Cancer Survivorship Silver Spring, CO	1010 Wayne Avenue Suite 770	(877) 622-7937 www.cansearch.org
National Cancer Institute	NCI Public Inquiries Office Bldg. 31, Rm. 10A31 31 Center Drive, MSC 2580 Bethesda, MD 20892-2580	(800) 4-CANCER www.cancernet.nci.nih.gov
The Wellness Community	1320 Centre Street, Suite 305 Newton Centre, MA 02459	(617) 332-1919 www.wellnesscommunity.org
Women's Information Network Against Breast Cancer (WINABC)	536 S. Second Avenue, Suite K Covina, CA 91723-3043	(626) 332-2255 www.winabc.com
Y-ME National Organization for Breast Cancer Information and Support	212 West Van Buren Chicago, IL 60607-3908	(800) 221-2141 www.y-me.org

REFERENCES

1. Grigg M. Institute of Medicine. In: Bondurant S, Ernster VL, Herdman R, eds. Information for women about the safety of silicone breast implants. Washington, DC: National Academy of Sciences, 2000.
2. LaTrenta GS. Breast augmentation. In: Saunders WB, ed. Breast and body contouring. Philadelphia, PA: W.B. Saunders, 1994.
3. Beekman WH, Hage JJ, Jorna LB, Mulder JW. Augmentation mammaplasty: the story before the silicone bag prosthesis. Ann Plast Surg 1999;43:446–451.
4. Lexer E. Jahre Transplantationsforschung. Munch Med Wochenschr 1925;72:830.
5. Passot R. Atrophie Mammaire: reflection aesthetique par las greffe graisseuseepiploique pure. Presse Med 1930;37:627–629.
6. Berson M. Dermo-fat transplants used in building up the breast. Surgery 1945;15:451–456.
7. Bames HO. Augmentation mammaplasty by lipo-transplant. Plast Reconstr Surg 1953; 11:404–456.
8. Milward TM. Calcification in dermofat grafts. Br J Plast Surg 1973;26:179–180.
9. Peer LA. Transplantation of fat. In: Converse JM, ed. Reconstructive plastic surgery. Philadelphia, PA: W.B. Saunders, 1965;35:51–59.
10. Gersuny R. Harte und wieche paraffinprothesen. Zentralbl Chir 1903;30:1–5.
11. Uchida J. Clinical application of cross-linked: restoration of breasts, cheeks, atrophy of infantile paralysis, funnel shape chest, ect. Jpn J Plast Reconstr Surg 1961;4:303–309.
12. Arons MS, Sabesin SM, Sith RR. Experimental studies with Etheron sponge. Plast Reconstr Surg 1961;28:72–80.
13. Conway H, Dietz GH. Augmentation mammaplasty. Surg Gynecol Obstet 1962;114: 579–577.
14. Williams F, Roaf R. Implants in surgery. London, Philadelphia, PA: W.B. Saunders, 1973.
15. Edwards B. Teflon-silicone breast implants. Plast Reconstr Surg 1963;32:519–526.

implants), wrinkling, malposition, irritation/inflammation, keloid or hypertrophic scar formation, rash, pneumothorax, filler port malfunction, and chest wall deformity *(39,40,139)*.

ACKNOWLEDGMENTS

Special thanks to Dianne Hart and the Mentor Corporation for their contributions to this chapter.

APPENDIX: SUPPORT GROUPS

Support group	Address	Contact information
American Academy of Cosmetic Surgery (AACS)	401 N. Michigan Avenue Chicago, IL 60611-4267	(312) 527-6713 ww.cosmeticsurgery.org
American Cancer Society (Reach to Recovery)	1599 Clifton Road NE Atlanta, GA 30329	(800) ACS-2345 www.cancer.org
American College of Rheumatology	1800 Century Place, Suite 250 Atlanta, GA 30345	(404) 633-3777 www.rheumatology.org
American Medical Association (AMA)	515 N. State Street Chicago, IL 60610	(312) 464-5000 www.ama-assn.org
American Society of Plastic Surgeons (ASPS)	444 East Algonquin Road Arlington Heights, IL 60005	(888) 4-PLASTIC www.plasticsurgery.org
The American Society for Aesthetic Plastic Surgery (ASAPS)	11081 Winners Circle Los Alamitos, California 90720	(800) 364-2147 asaps@surgery.org
Breast Augmentation & Breast Implants Information Web - by Nicole	542 Hypoluxo Road Suite 339 Lake Worth, FL 33467	(800) 260-9497 www.implantinfo.com
BreastCancerinfo.com/ Susan G. Komen Alliance	5005 LBJ Freeway Suite 250 Dallas, TX 75244	(800) 462-9273
Cancer Care, Inc.	275 Seventh Avenue New York, NY 10001	(800) 813-HOPE www.cancercare.org
Cancer Connection	1500 E. Duarte Road Duarte, CA 91010	(626) 359-1111 www.cityofhope.org
Cancer News on the Net	4184 Shore Crest Drive West Bloomfield, MI 48303	www.cancernews.com
Food and Drug Administration (FDA)	5600 Fishers Lane RM. 16-59 Rockville, MD 20857	(888) INFO-FDA www.fda.gov/cdrh/ breastimplants
Institute of Medicine	2101 Constitution Ave., NW Washington, D.C. 20418	(202) 334-3300 www.iom.edu
La Leche League International	1400 N. Meacham Road Schaumburg, IL 60173	(847) 519-7730 www.lalecheleague.org
McGhan Medical Corp.	700 Ward Drive Santa Barbara, CA 93111-2936	(800) 862-4426 (800) 624-4261 www.mcgahan.com
Mentor Corporation	5425 Hollister Avenue Santa Barbara, CA 93111	(800) MENTOR-8 www.mentorcorp.com

(continued)

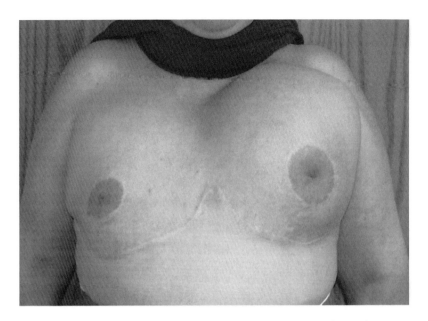

Fig. 5. Ruptured left breast implant and hematoma formation.

patients and 40% of reconstruction patients have a complication within the first year of implantation *(1,39)*. Complications are more prevalent in women with silicone-filled implants than in women with saline implants *(1)*. Breast prostheses are not considered lifetime devices and most women will experience one or more complications within a 3-year period.

Other complications are also more prevalent in women who have had breast reconstruction, prior radiation, implant replacement, and tissue expansion than in patients who have undergone augmentation mammaplasty *(1)*. Breast pain, tenderness, and hardness are common after breast implantation and are usually caused by a capsular contracture, bacterial infection, or rupture. Seroma formation occurs in 3–6% of women after 3 years, especially in patients with textured implants when an implant does not bond with the surrounding capsule *(39,40)*. Sensory complications, such as insensate nipple and hyperesthesia, are typically related to the size of the pocket and implant, and less frequently associated with the extent of surgical dissection *(53)*. The perioperative incidence of hematoma is 0.5–3% and may be minimized by preoperative medical treatment of hypertension. Hematomas usually occur within the first day after surgery and frequently require surgical drainage (Fig. 5) *(39,40,90)*. Calcification of the fibrous capsule is not common but noteworthy because it can mimic carcinoma on mammogram. Skin necrosis is associated with infection, excessive heat or cold therapy, steroid use in the pocket, smoking, chemotherapy, and radiation. The inability to breastfeed has been reported in up to 64% of patients with breast implants vs 7% of women without implants. Mondor's disease is a rare complication caused by thrombosed and inflamed thoracoepigastric veins. These tender cords extend from the inframammary fold inferiorly and result in an indurated string-like band. Fortunately, this condition is self-limited and resolution can be expected within a few months. Other less common complications after implantation include synmastia (medial communication between

antibiotic treatment beyond that which is used to treat the primary infection is beneficial in preventing seeding of the breast implant. There is no evidence that continuing antibiotic therapy in such patients for a longer duration than in patients without prostheses is valuable nor that higher doses should be used. Other techniques used to prevent capsular contracture include excellent hemostasis, implant and pocket lavage with Betadine, Bacitracin or the like, use of large volumes of dilute local anesthetic, minimal to no-touch technique in handling implants (implants placed inside a sterile sleeve during insertion), nipple shields, postoperative oral/topical vitamin E, sterile saline infusion via a closed system, surgical drains, and talc-free gloves *(45,59,117–123)*. Treatment for severe capsular contracture may require capsulotomy, capsulectomy, or removal of the implant *(90)*.

The adherence of bacteria to breast implants is independent of surface texture *(124–126)*. The incidence of infection is 1–3% after augmentation mammaplasty and 6% after reconstructive breast surgery during the first year, according to the McGhan and Mentor implant manufacturers *(39,40)*. Clinically obvious abscess formation or cellulitis rarely occurs *(127)*. However, when it does, the implant must be removed until the infection has resolved. It is best to wait for 3 to 6 months after resolution of the clinical infection before placing another implant. Infection may also occur if an implant becomes exposed. Occasionally, a surgeon can salvage an implant exposure assuming that an overt infection is not present. Otherwise, explantation should be performed without delay.

According to a study of 139 implants removed from symptomatic patients, axillary pain and skin rash were significantly more common in women with culture-positive explants than culture-negative patients *(128)*. However, there were no significant differences in other local symptoms. Constitutional symptoms reported in women with culture-positive implants include fatigue, weakness, intermittent fever, and myalgia, but systemic symptoms and implant integrity are not associated with positive bacterial cultures *(20,129)*.

Staphylococcus epidermis, Staphylococcus aureus, and *Propionibacterium acnes* are the most commonly isolated organisms from the surface of breast implants. These Gram-positive bacteria are isolated in greater than 90% of surgically removed implants *(130)*. *Escheria coli* is cultured in 1.5% of explanted prostheses *(128)*. Brief exposure to organisms, especially Gram-positive bacteria, is sufficient to inoculate breast prostheses *(131)*.

Nonmenstrual toxic shock syndrome (TSS) is a rare, life-threatening illness caused by an *S. aureus* infection *(132,133)*. Most commonly, the focus of infection is a packing, foreign body, or infected wound *(132)*. Although the pathogenesis remains unclear, it is believed that certain strains of *S. aureus* produce an exotoxin called TSS toxin 1 that results in disseminated toxic injury, release of inflammatory mediators, shock, and multisystem failure *(133)*. This disorder has been reported after breast prosthesis implantation *(134,135)*. In 1990, the Centers for Disease Control described criteria for the diagnosis of TSS *(136)*. The mortality rate is 3 to 50% *(137,138)*. This disease and other serious infections can occasionally be detected during routine follow-up of patients with breast prostheses and require prompt medical treatment *(132)*.

Very large implants are particularly prone to complications such as extrusion through the overlying soft tissue and shifts in position. Approximately 20% of augmentation

Table 6
Rosenberg Endoscopic Grading System of Breast Implants

Grade	Description
I	Intact
II	Consistent with silicone-gel bleed
IIIA	Implant has a small rupture
IIIB	Gross perforation with silicone contained within the capsule
IV	Gross perforation with silicone outside of the capsule or disintegrated implant

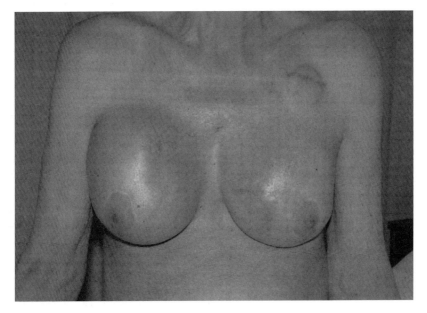

Fig. 4. Baker's grade III capsular contracture and intracapsular rupture of the right submuscular silicone breast implant.

generally higher in women having repeat implantation and in women having secondary reconstruction vs reconstruction performed at the time of mastectomy.

Capsular contracture can occur following subclinical infection on the surface of breast implants or systemic infection *(115)*. However, the incidence of infection is much lower than the incidence of capsular contracture, suggesting that the relationship is not causal *(1)*. Subclinical infection (colonization or biofilm formation) more likely exacerbates rather than causes capsular contracture. Nevertheless, we recommend prophylactic antibiotics for patients with breast prostheses who are at risk for transient bacteremia from surgery on the gastrointestinal or genitourinary tract, dental procedures, and the like. This is analogous to recommendations for patients with prosthetic cardiac valves or joint prostheses *(114)*. One dose of an intravenous antibiotic (most commonly a cephalosporin) 30–60 minutes prior to induction of anesthesia is usually sufficient *(59,116)*. In patients with established systemic infections such as pneumonia, urinary tract infection, or abscess with septicemia, there is no evidence that extending

nation to decrease respiratory motion *(96)*. Electrocardiographic gating and respiratory compensation techniques further attenuate motion artifact.

The most reliable criterion to make the definitive diagnosis of intracapsular rupture of a silicone-filled implant is the linguini or free-floating loose-thread sign *(86,106)*. Also known as internal membranes, these hypointense lines represent pieces of collapsed implant envelope that clearly invaginate from the implant surface *(107,108)*. Their parallel orientation in relation to the fibrous capsule clearly differentiates them from normal radial folds that are found in a more perpendicular orientation. This classic sign of intracapsular rupture is not easily identified by mammographic examination *(79)*. The inverted teardrop, noose, or keyhole sign represents silicone that has entered a radial fold on the exterior of the implant *(78,109)*. Subscapular lines or the pull-away sign suggest gel bleed or minor leakage that results in displacement of the implant from the fibrous capsule *(100,110)*. Rupture of the inner lumen in double-lumen implants is often detected by fluid droplets in the silicone compartment. There are many other signs of implant rupture *(100,111–113)*.

COMPUTED TOMOGRAPHY

CT is not frequently performed to evaluate the status of breast prostheses because of the radiation that is required for this examination. However, capsular rupture and other disorders can be detected using this technique *(46)*.

ENDOSCOPY

The evaluation of breast implants by endoscopy was first described by Dowden and Anain in 1993 *(89)*. A small surgical window allows the integrity of the entire implant to be assessed without directly examining the entire implant surface. If a leak is present, a thin coating of silicone forms over the entire implant surface regardless of the size of the leak or distance from the endoscope *(86)*. In 1994, Rosenberg described a grading system for endoscopic assessment of breast implants for failure (Table 6).

CAPSULAR CONTRACTURE

The most common complication in patients with breast prostheses is capsular contracture (Fig. 4). Partially in response to silicone bleed, a macrophage-mediated foreign body reaction occurs and results in periprosthetic fibroplasia and capsule formation. Other factors that may be involved in the pathophysiology of capsule formation include autoimmune/connective tissue disorders, genetic predisposition, hematoma, infection, necrotic tissue, seroma, smoking, surface characteristics of the implant, and talcum powder *(114)*. Most contractures occur within 2 to 3 months after surgery *(56)*. Manufacturers report a 5 to 7% risk rate of severe contracture (Baker's grade III–IV) in the first year and 9% 3 years after augmentation with saline-filled implants *(39,40)*. In patients who have had breast reconstruction with saline-filled implants, severe contracture risk rates are significantly higher; 13% after 1 year and 25% after 3 years *(39,40)*. Some experts believe that severe contracture in women with silicone implants approaches 100% after 25 years *(1,18)*. In our own experience, Baker's grade III–IV contracture occurs in approximately 10–15% of patients with saline implants and 40% of patients with silicone implants. Contracture rates are also

is otherwise clinically insignificant *(79)*. Mammography in conjunction with sonography is more sensitive than MRI in detecting extracapsular silicone implant ruptures, identified as a radio-opaque mass, linear streak, or lymph node opacification *(96,99,100)*.

Patients with severe contracture or painful breasts may not be able to tolerate a mammographic examination. Furthermore, the examination itself may result in rupture, especially of older implants. Nevertheless, the American College of Radiology does not recommend that consent forms be obtained during routine mammography.

Approximately 20% of women who have had a mastectomy elect to have breast reconstruction with an implant. Although most patients who have undergone mastectomy are seen for follow-up by their oncologist and general surgeon, the plastic surgeon generally assumes full responsibility for all breast reconstruction patients with respect to the status of their implants *(101)*.

ULTRASOUND

Ultrasound is 70% sensitive to changes in the integrity of silicone implants *(46)*. Examination is operator- and equipment-dependent and the best results are obtained when the radiologist is present during the evaluation. Ultrasound may be used as a screening tool for implants that have been implanted for more than 10 years *(102,103)*.

A characteristic echogenic snowstorm of silicone mixed with breast tissues is the most specific sign of extracapsular implant rupture *(87)*. The second most useful sign of implant rupture is the accumulation of low-level homogenous echoes, or coarse aggregates within the silicone gel. These were reported in 54% of ruptures by one investigator *(104,105)*. These are analogous to the droplets observed during MRI examination. Intracapsular rupture may be identified by a series of horizontal echogenic, discontinuous, straight or curvilinear lines known as the stepladder sign *(104)*. The stepladder sign is analogous to the linguini sign found during MRI examination *(106)*. Hyper- or hypoechoic masses, dispersion of the beam, discontinuity of the implant membrane, and heterogeneous echoes inside of the implant are other findings of implant rupture.

Herniation is easily differentiated from extracapsular rupture during ultrasound examination *(28)*. Images of the axilla may detect adenopathy or free silicone. Although implant margins are well defined by sonography, ultrasound has limited value in the evaluation of the posterior implant and breast tissue due to attenuation of the ultrasound beam by silicone.

MAGNETIC RESONANCE IMAGING

MRI has a sensitivity of 96% and specificity of 76% in the detection of intra- and extracapsular rupture *(86)*. It is not operator-dependent, requires no compression of the breast, provides a view of the entire implant and surrounding tissues in various planes, and uses no ionizing radiation. However, MRI is the most expensive of the radiographic modalities and is impractical or contraindicated in patients with claustrophobia, marked obesity, cardiac pacemakers, aneurysm clips, and other ferromagnetic devices. MRI examination with a dedicated breast coil can distinguish between normal folds and ruptured implants. Patients should be placed in the prone position during MRI exami-

than 10–15 years *(89)*. The International Review Group recommends that all adverse incidents be reported to the Medical Devices Agency Adverse Incident Center *(46)*.

The two most common causes of early implant failure are faulty valves or rupture of the implant shell *(90)*. Other causes include underfilling saline implants and rupture occurring during closed capsulotomy *(91)*. Closed capsulotomy is not recommended by the FDA or the McGhan or Mentor companies as a treatment for capsular contracture. Fortunately, for women with breast implants, the health risks after implant failure are negligible.

MAMMOGRAPHY

Both silicone-gel and saline-filled implants are radio-opaque and obscure the appearance of breast tissue during radiographic examination. Furthermore, all breast implants compress breast tissue and may mask fine details that otherwise might have been identified *(92)*. Silicone gel-filled implants have been reported to obscure from 22 to 83% of breast tissue. Saline-filled implants obscure breast tissue to a lesser degree *(93,94)*. Although saline-filled silicone-envelope implants are the gold standard, there is ongoing research to develop better shapes, shells, and fillers. Bio-oncotic gel polymer matrix, dextran, hyaluronic acid, hydrogel, lipids, methylcellulose, peanut oil, polyethylene glycol, polyvinylpyrolidone, seaweed, soybean and triglyceride oil are examples of materials currently under investigation *(95)*. Additional implant envelope and filler materials are available outside of the United States. McGhan has two implants it sells internationally. The 410 model is a shaped, textured cohesive gel implant that is believed to prevent silicone migration owing to an advanced cross-linkage design. The 150 model is a silicone-gel filled implant with an adjustable saline-filler inner lumen and fixed-volume silicone gel-filled outer lumen. A polyvinylpyrolidone-hydrogel implant called NovaGold is also available outside the United States *(95)*. Hydrogel is a mixture of polysaccharide and water that is produced by Arion in France *(95)*. However, it has a propensity to cause an anaphylactic reaction and its hypertonicity can result in swelling over time. Polyurethane-textured implants and implants with partial texturing are available from other manufacturers.

Compression mammography is more difficult to perform in patients with capsular contracture and the distortion of breast tissue during the examination may reduce the detection of microcalcification and soft-tissue masses *(92)*. Despite these implications, mammographic and sonographic features of malignancy in patients with implants are no different from those without *(96)*. More importantly, no studies have shown an increased incidence of cancer in patients with implants or a higher stage of cancer at the time of diagnosis. In fact, only one case study to date has described a breast implant as directly associated with a carcinoma *(97)*.

Mammographic findings have poor statistical correlation with actual implant integrity. The sensitivity and specificity of mammography in detecting implant failure are 55% and 69%, respectively *(98)*. Because of the high radiodensity of silicone gel and the implant envelope, an implant with an intracapsular leak is difficult to distinguish from an intact implant. Ill-defined borders and irregular density of the implant contents may indicate intracapsular leak and a contour bulge may suggest intracapsular rupture *(79)*. Capsule weakness may also result in herniation, mimicking implant rupture, but in a contiguous location *(60)*. Patients with axillary masses may require both mammography and MRI to rule out underlying malignancy.

Women who have had a mastectomy and subsequent breast reconstruction with an implant require follow-up surveillance to detect recurrence, metastasis, or new primary breast cancer. Our own recommended surveillance schedule entails monthly breast self-examination, a history and physical examination every 4 to 6 months for the first 5 years, and annual mammography (Table 5) *(81,82)*. Physical examination should focus on the reconstructed breast and chest wall, the contralateral breast, the axilla, supraclavicular, and cervical nodal areas *(74)*. Visual inspection for edema, rash, and estimation of range of shoulder motion should also be performed. A mammogram of the reconstructed breast is not required if a complete mastectomy has been performed because all breast tissue is removed during this procedure. However, women who have had a mastectomy for cancer should be informed that they are at increased risk of breast cancer in the contralateral breast and should have annual mammograms of the contralateral breast performed indefinitely. Women 50 years of age and older who have been diagnosed with breast cancer have a 4% lifetime risk of cancer in the contralateral breast; women under 50 have a 14% lifetime risk *(83)*.

Various terms are used in the literature to describe breast implant failure. Bleed denotes the microscopic transudation of silicone across an intact implant shell and occurs at a rate of approximately 1 g over 5 years *(84)*. Gel bleed is not a complication and occurs in all women with breast implants *(46)*. Leak refers to the drainage of silicone or saline through a tiny hole in the implant envelope or valve. Rupture is the term for a macroscopic tear in the implant. If the fibrous capsule also tears and silicone leaks through the tear, this is called extrusion of silicone.

Approximately 80–90% of implant ruptures are intracapsular and 10–20% are extracapsular *(78,85,86)*. In intracapsular failures, the silicone is contained within a fibrous envelope of scar tissue. This may result in a slight change in the shape or firmness of the breast. However, more than 50% of intracapsular ruptures are undetected by physical examination because a normal breast appearance is maintained by the fibrous capsule *(87)*. Intracapsular rupture can be detected equally well by ultrasound (if the radiologist is present), MRI, and computed tomography (CT) *(88)*.

Most women with extracapsular failure notice tearing of the capsule as the sudden appearance of pain or other unpleasant sensation and a new mass *(64)*. Other signs include change in breast size, appearance, and/or consistency. Should gel escape through the scar envelope into adjacent breast tissue, all involved breast tissue containing extravasated silicone should be removed to avoid masking or mimicking breast cancer. If silicone gel disseminates beyond breast tissue, and the patient is asymptomatic, there are no disease prevention-related concerns that necessitate excision of the material. In general, saline implant failure usually results in deflation, whereas silicone-gel implant failure more commonly remains undetected.

Physical findings of implant rupture require further investigation and radiographic confirmation. The imaging modality of choice for evaluation of implant rupture is MRI *(23)*. Mammography, ultrasound, or MRI findings consistent with rupture require direct visualization of the prosthesis by open or endoscopic technique. Endoscopic examination may also be useful for evaluation of prostheses that have been in place for more

Table 4
Recommended Surveillance After Breast Augmentation With an Implant for Asymptomatic Patients at Saint Louis University Hospital

	Postoperative year					
	1	2	3	4	5	10
Office visit	3	1	1	1	1	1
Mammogram[a]	1	1	1	1	1	1

[a] The number of times per year mammography is recommended only pertains to women ≥40 years of age. Additional tests are indicated only if signs or symptoms warrant them.

Table 5
Recommended Surveillance After Breast Reconstruction With an Implant for Asymptomatic Patients at Saint Louis University Hospital

	Postoperative year					
	1	2	3	4	5	10
Office visit	3	2	2	2	2	1
Mammogram[a]	1	1	1	1	1	1

[a] The number of times per year mammography is recommended only pertains to women ≥40 years of age. Additional tests are indicated only if signs or symptoms warrant them.

cal Oncology discourages the use of routine biochemical or imaging studies for asymptomatic patients as they do not affect patient survival or quality of life *(74–76)*.

Some patients warrant additional medical attention. Patients with capsular contracture or fibrous breasts should have biannual physical examinations in addition to their annual mammograms. If the implant is not mobile and displacement is not adequate for physical examination, a straight lateral view is added to image the posterior breast tissue. Magnetic resonance imaging (MRI) examination of the breast is still considered a confirmatory diagnostic test and is not recommended for routine screening of asymptomatic women *(77–79)*.

After augmentation mammaplasty, 55–95% of patients who return to their physician for a nonroutine visit have a palpable mass *(60)*. Implant valves are frequently the source. Free silicone may also mimic disease processes and should be distinguished from fibroadenoma and carcinoma by mammographic or ultrasound examination. Fine needle aspiration of palpable masses is feasible in patients with implants *(80)*. This must be done carefully in order to avoid damage to an intact implant. Open biopsy of the mass is sometimes required. A biopsy positive for silicone does not rule out cancer

Table 3
Modified Baker Classification
of Capsular Contracture After Prosthetic Breast Reconstruction

Grade	Description
IA	Absolutely natural, cannot tell breast was reconstructed
IB	Soft, but the implant is detectable by physical examination or inspection because of mastectomy.
II	Mildly firm reconstructed breast with an implant that may be visible and detectable by physical examination.
III	Moderately firm reconstructed breast. The implant is readily detectable, but the result may still be acceptable.
IV	Severe capsular contracture with an unacceptable aesthetic outcome and/or significant patient symptoms requiring surgical intervention

A history and physical examination is performed during follow-up office visits at 1 and 4 weeks after surgery. This includes inspection for symmetry, signs of infection, degree of ptosis (Table 1), and class of capsular contracture (Tables 2 and 3) *(69)*. Palpation of the breast, axilla, and nipple-areola complex is essential to rule out rupture, hematoma, seroma, or mass.

The Society for Breast Imaging, the American Society of Plastic and Reconstructive Surgeons, the American College of Surgeons, the American College of Radiology, and the American Cancer Society concur that women with breast implants should be on the same schedule of routine mammography as all other women according to their age and risk group *(60)*. Our current protocol for average-risk patients entails a baseline mammogram at age 40 and annually thereafter (Tables 4 and 5). Women with breast implants having screening or diagnostic mammography should have craniocaudal and mediolateral oblique views as well as displaced views, known as the Eklund technique *(70)*. Eklund views (four push-back or displacement views) are superior in sensitivity and specificity to standard mammography views in the detection of breast pathology. Using this technique, the total breast volume visualized by mammography is increased from 51 to 61% in patients with subglandular implants and from 72 to 91% with submuscular implants. Mammographic units not accredited by the American College of Radiology may not have technicians trained in this technique. The patient should return to the same mammography unit for future mammograms and inform the technician that she has breast implants prior to each examination *(64)*.

Follow-up should be continued on an annual basis for life. Patients do not need additional follow-up if they are asymptomatic, have no physical problems, and no clinical or radiographic signs of a ruptured implant (Tables 4 and 5) *(71)*. Breastfeeding should be encouraged in mothers with breast implants. It confers no known risk to the infant *(72)*. Anecdotal reports of esophageal motility disorders have been reported in children who have breastfed from mothers with silicone breast implants. However, no scientific studies have demonstrated a causal relationship between silicone breast implants and esophageal motility disorders *(37,38,73)*. The American Society of Clini-

Table 1
Regnault Classification of Breast Ptosis

Degree	Nipple-areolar complex
1st	At or above the inframammary fold
2nd	Below the inframammary fold
3rd	On dependent aspect of inferior convexity

Table 2
**Original Baker Classification
of Capsular Contracture After Augmentation Mammaplasty**

Grade	Description
I	Breast absolutely natural; no one could tell breast was augmented
II	Minimal contracture; the surgeon can tell surgery was performed, but patient has no complaint
III	Moderate contracture; patient feels some firmness
IV	Severe contracture; obvious just from observation

effect and interaction with anesthetics. Cessation of tobacco use should begin 2 weeks prior to and extend for a minimum of 2 weeks after the operation to prevent impaired wound healing and risk of postoperative infection (67). Routine laboratory studies should be obtained prior to surgery as indicated by risk factors, including pregnancy testing in women of childbearing years. No laboratory tests are routinely ordered postoperatively.

After breast prosthesis implantation, women should receive instructions on pain management, limitations of physical activity, time off work, how to perform monthly breast self-examinations, methods to detect implant rupture and contracture, and when to seek medical attention. Patients should also be informed of the type of prosthesis and the manufacturer of each implant and a letter should be sent to the patient's primary care physician regarding their prostheses and the surgical procedure.

A postoperative brassiere or compression bandage is worn postoperatively for 1 to 2 days to minimize hematoma formation. A sports brassiere is recommended for an additional 3 months to reduce the incidence of contracture (56). If the implants are high after surgery, an elastic band may be placed over the top half of the breast to encourage a more caudal position. Drains are rarely used. If drains are used, they are usually removed between postoperative days 1 and 9 depending on output. Implant movement exercises are initiated 2 to 3 weeks after surgery in all breast quadrants three times per day and continued indefinitely (68). Normal activity may be resumed after 24 hours, as tolerated. More vigorous physical activity, such as running and aerobics, should be avoided for 4 weeks. Proper support garments should be worn once these activities are resumed. Owing to the resilience and durability of the current generation of breast implants, unrestricted activity is usually allowed after 4 weeks.

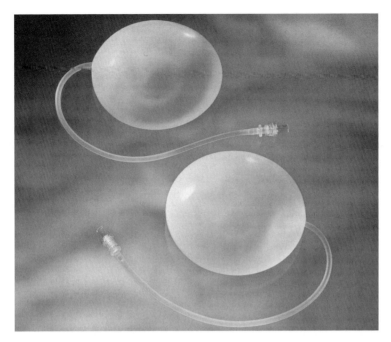

Fig. 3. Mentor saline-filled breast implants. Round smooth tissue expanders.

graphic examination and lessen the chance for local complications and repeat operations *(1,60)*. For women with minimal breast tissue, the submuscular implant location is preferred. The implant outline tends to be less obvious because the pectoral muscle fills the space cephalad to the implant *(61)*. The incidence of areolar anesthesia is also decreased. However, submuscular implantation usually entails longer surgery, longer recovery, and more pain postoperatively.

Although the precise number is unknown, it is estimated that more than 1.5 million American women currently have breast prostheses; 75–80% for breast augmentation and 20–25% for breast reconstruction after mastectomy *(1,46,62–64)*. More women are having elective breast surgery every year. Between 1992 and 2000, the number of breast augmentation procedures in the United States increased by 476% and the number of breast reconstruction procedures increased by 166%. The number of breast implant procedures is continuing to increase at a rate of 13% to 15% per year *(65)*.

According to Breast Surgery Statistics, published annually by the American Society of Plastic Surgeons, 272,891 devices were implanted during the year 2000. There were 187,755 patients who had breast augmentation; 13,401 patients had breast reconstruction with implant placement; 25,226 patients had breast reconstruction with tissue expander placement (Fig. 3). An additional 46,509 patients had breast implant removal with subsequent replacement *(66)*.

As with all women, those having breast implantation surgery should be encouraged to maintain a healthy lifestyle, which includes a balanced diet, sufficient rest, and exercise. Patients are instructed to stop taking aspirin, aspirin products, nonsteroidal anti-inflammatory agents, and vitamin E supplements for 2 weeks prior to surgery. Abstinence from alcohol 5 days prior to surgery is advised because of its vasodilatory

patients and have a low incidence of malposition over time *(34)*. However, when the patient is in a recumbent position, contoured implants typically do not spread out as well as round implants or natural breasts *(55)*.

There are four different incision sites currently described for clinical practice—inframammary, transaxillary, periareolar, and periumbilical. The inframammary incision is the standard for breast implantation procedures. According to one source, 68% of patients choose an inframammary incision vs 22% for transaxillary, 9% for periareolar, and 1% for umbilical locations *(34)*. Surgical exposure is excellent and an inconspicuous scar location can be achieved with an inframammary approach. Furthermore, preservation of the ductal elements of the breast results in less bacterial contamination as well as a decreased incidence of unsuccessful future breast feeding *(56)*.

The transaxillary approach is especially appealing to patients who do not desire a visible scar on the aesthetic unit of the breast. This technique also allows for access to both subglandular and submuscular locations. However, the brachial plexus and axillary vasculature are close to the incision site and plane of dissection and are vulnerable to injury. The axillary approach is not recommended for secondary or reoperations other than replacement of a deflated smooth-shelled saline implant *(54)*.

The periareolar incision provides good access for either subglandular or submuscular placement as well as superior positioning. This technique also results in minimal scarring and superior exposure of the inframammary ligament. In patients requiring concurrent areolar correction or repositioning, this is the incision of choice. However, there is an increased incidence of inability to breast feed owing to transection of ductal tissue. Intercostal nerve injury and subsequent nipple paresthesia is an uncommon complication.

Like the transaxillary technique, the transumbilical breast augmentation approach does not create a visible scar on the aesthetic unit of the breast. However, exposure is less than ideal and independent control of the anatomical elements of the breast is not possible *(54)*. Because of the small caliber of the endoscopic instruments required for this procedure, only inflatable implants can be used. Patients should consent to conversion to a standard breast incision should technical difficulties arise *(57)*. Transumbilical and endoscopic techniques of breast implantation are not recommended by either Mentor or McGhan manufacturers.

Implants may be placed in a subglandular or submuscular position. Approximately one-third of implants are placed in a subglandular location during breast augmentation. This is technically less difficult and may result in superior cosmetic results in patients with mammary ptosis. Surgery and recovery are generally shorter in duration, access for future reoperation is easier, and most patients report less postoperative pain. However, such implants are generally more palpable, have a greater incidence of contracture, and tend to obscure breast tissue during mammographic examination.

Submuscular implants (deep to the pectoralis major muscle) are generally less obvious and have a more natural appearance and feel. Owing to the constant massage between muscle and chest wall, there is decreased potential for contracture, undesirable breast shape and/or nipple position, and hematoma or seroma formation *(56,58)*. Furthermore, because of decreased contamination from ductal tissue and the rich vascular bed, the postoperative infection rate may be lower *(59)*. The Institute of Medicine recommends submuscular implant placement to improve the sensitivity of mammo-

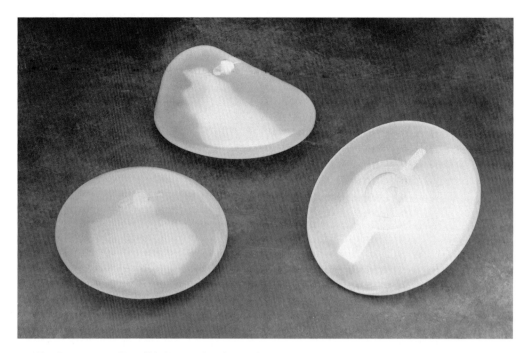

Fig. 2. Mentor saline-filled breast implants. Clockwise from the top: contoured smooth implant, posterior aspects of round smooth implant, and anterior aspect of round smooth implant.

ease, and impaired mammographic detection of cancer *(42,43,45)*. Despite this moratorium, there has yet to be any epidemiological data confirming a direct association between silicone or saline breast implants and an increased incidence of breast cancer or autoimmune disease *(46–51)*. On the contrary, studies have shown that silicone breast implants may significantly decrease the risk of breast cancer and that women having reconstruction with implants after mastectomy may have a significant reduction in death caused by breast cancer *(52,53)*. The reason for this is unknown.

Prior to the moratorium, the silicone-filled silicone-shell prosthesis was the gold standard for breast augmentation. Today, they are only available for patients undergoing reconstructive surgery and those enrolled in clinical trials studying the safety of the next generation of silicone devices for use in breast augmentation. Saline-filled silicone-shell implants are the current industry standard and are the only FDA-approved breast prostheses available for elective breast augmentation.

Tebbetts states that "any highly skilled surgeon can achieve acceptable results in breast augmentation with a wide variety of implant types and sizes in any type of breast." Each selection of implant alternative, however, involves different trade-offs *(54)*. We endorse this view. The round breast implant is the traditional shape and most frequently used design. These are especially useful when greater projection is desired. The contoured implant was designed to more closely resemble the natural female breast (Fig. 2). On its anterior surface, it is concave superiorly and convex inferiorly with greater fullness at its lower pole. Also known as anatomical or teardrop implants, these prostheses provide a more natural and aesthetically pleasing appearance for many

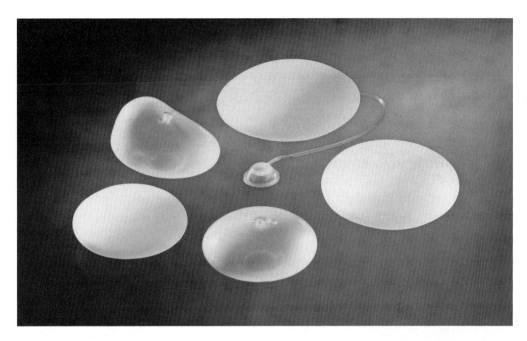

Fig. 1. Mentor saline-filled breast implants. Clockwise from top: round textured implant with filler port, round textured implant, round smooth implant, round textured implant, contoured smooth implant.

structive purposes, and on the duration of implantation *(37,38)*. Third-generation saline devices have a first-year failure rate of 1–4% after augmentation and 3% after reconstruction *(39,40)*. At 3 years, the incidence of failure is 3–5% after augmentation and 6–9% after reconstruction *(39,40)*. Thereafter, a cumulative rupture rate of 1–2% per year after 5 years can be expected. Approximately 5–10% of saline implants rupture after 10 years. Silicone-gel implants have a higher rupture rate than saline implants. The rupture rate of silicone implants is approximately 26% after an average of 12 years according to one study *(41)*. This provides a compelling reason for lifelong follow-up of patients with implants.

In 1991, concerns about the carcinogenicity of polyurethane-coated implants surfaced. Sarcoma formation observed in rats was linked to two carcinogenic metabolites of polyurethane: toluene 2,3-diisocyanate and toluene 2,6-diisocyanate diamine *(42,43)*. Laboratory studies confirmed that polyurethane does break down into toxic byproducts in vivo and 2,4-toluene diamine was found in the urine of women with these implants *(44)*. After the release of this information, Bristol Meyers/Surgitek decided not to apply to the Food and Drug Administration (FDA) for premarket approval of polyurethane coating. Later that year, the FDA banned all polyurethane implants. In the more than 100,000 women who received these implants between the years 1968 and 1991, no clinical health problems have since been linked to polyurethane *(18)*.

In 1992, the FDA announced a moratorium on the use of all silicone-gel implants owing to concerns about their carcinogenicity, an association with autoimmune dis-

In the early 1970s, second-generation devices were introduced. They featured thinner silastic walls, less viscous gels, and optional teardrop or double-lumen designs *(1,26)*. These seamless implants resulted in a much more natural feel and less trouble with capsular contracture. However, second-generation silicone implants had a higher rupture rate (50–95% at 12 years) than either first- or third-generation implants because of their relatively thin silicone envelopes *(1)*. A polyurethane coating was added to these silicone implants in 1972 in an attempt to prevent capsular contracture *(27)*. Breast softness and shape were well preserved after 5 years *(28)*. However, as the polyurethane coating eventually wore off, the contracture rate increased.

Another second-generation innovation, the textured surface, was designed to reduce encapsulation and contracture *(1)*. Organized capsule formation around traditional smooth-surfaced implants caused a distinct collagen layer to form over the surface of the implant and this constricted over time. It was felt that, by altering the pattern of organized collagen around the implant, the incidence of capsular contracture would be reduced and that the collagen would fix the implants in position, limiting displacement or migration *(18)*. Several investigators have since reported that textured implants do in fact have a decreased incidence of encapsulation and contracture compared to smooth implants in both the subglandular and submuscular locations *(20,29–32)*. In 1997, Peters reported findings of severe capsular contracture in 81% of women with smooth silicone implants placed in a subglandular position vs 21% of women with smooth implants in a submuscular position *(33)*. In a later study by Collis of 53 women with subglandular implants followed over a 10-year period, 65% of patients with smooth implants had severe capsular contracture vs 11% in patients with textured implants *(27)*. However, the pathogenesis of capsular contracture is multifactorial and not entirely owing to surface characteristics and implant location. Other putative variables include implant size, envelope tightness, and pressure on the implant, all of which increase the adherence of surrounding tissues to the implant surface *(34)*. In 1997, Tarpila reported on 21 women who had a smooth saline implant in one breast and a textured implant in the other *(35)*. The contracture rates were 38% for the smooth implants and 29% for the textured ones. These rates were not significantly different. In 1995, Worseg described 167 women with smooth saline implants who had a mean capsular contracture rate of approximately 35%, irrespective of anatomical location *(36)*. It is unknown whether anatomic location or surface characteristics will affect the incidence of contracture in long-term follow-up. Smooth-surfaced implants remain the most frequently used design of breast prostheses owing to less long-term palpability and visibility in both submuscular and subglandular positions *(34)*.

In a further attempt to combat contracture, rupture, and leakage, third-generation implants were introduced in the 1980s and are the current generation (Fig. 1) *(1)*. They have thin walls (similar to second-generation implants) but greater durability and an inner lining of fluorosilicone or modified elastomer to reduce the rate of silicone bleed *(1)*. In 1987, the Becker expansion prosthesis was introduced. It has a removable valve and subcutaneous port that allows for serial saline infusion and size adjustments for up to 6 months after implantation. The valve can then be permanently removed after all postoperative volume adjustments have been made *(2)*.

As with first- and second-generation implants, the frequency of failure for third-generation implants depends on whether the implants are placed for cosmetic or recon-

uid silicone injections were introduced in Japan and later described by Uchida in 1961 *(11)*. However, owing to silicone migration, pain, inflammation, recurrent infection, chronic drainage, ulceration, granuloma formation, calcification, breast nodules, cysts, skin discoloration, tissue necrosis, pulmonary embolism, coma, and death, this procedure has since been condemned *(1,2)*. Previous silicone injections may also compromise radiographic evaluation of implant integrity.

The use of glass balls as an implant material was first described by Schwarzmann in 1930 and was the first "alloplastic implant" *(3)*. For obvious reasons, their use as implants was extremely limited. Other materials included ivory, ground rubber, ox cartilage, sponges, sacs, and tapes *(1)*. Polystan, a polyethylene sponge implant, was introduced in 1951 and used for many years with few complications *(3)*. Ivalon, also a polyvinyl sponge-like material, was first reported for breast implantation by Pangman in 1953 *(2)*. However, Ivalon prostheses became very hard over time because of a fibrotic reaction *(3)*. Other undesirable side effects included infection and chronic drainage. A compound implant designed by Pangman and Wallace in 1955, consisting of an Ivalon insert and an outer shell of polyethylene, elicited less fibrosis and had a more natural feel *(2)*. Concerns about carcinogenicity eventually led to the removal of Ivalon implants from the market. Plastic implants, such as the Etheron sponge and the polyurethane implant, were described in 1961 and 1962, respectively, but were abandoned because of their potential carcinogenicity *(12,13)*. In 1964, Malbec designed an air-filled prosthesis using a Polystan shell. Despite its ability to withstand extreme pressures, its rounded shape and lightness limited its popularity *(14)*.

The first clinical trial using a recognizably modern silicone prosthesis was reported by Cronin and Gerow in 1962 *(2)*. In 1963, Edwards developed a silicone foam prosthesis covered with a silicone plate and a layer of Teflon felt for fixation *(15)*. These first-generation silicone devices decreased the incidence of severe periprosthetic fibrosis, later called capsular contracture, from 100 to 40% *(16)*. That same year, Cronin and Gerow described their prosthesis with a silicone rubber shell and silicone gel filler at the Third International Congress of Plastic Surgery *(17)*. On the posterior surface of the implant was a patch of Dacron for fixation, which was removed in later versions owing to an increased incidence of contracture. The major advantage of silicone implants was their natural feel *(18)*. Unfortunately, these implants were heavy and had deflation rates as high as 76% *(1)*.

In 1964, Arion introduced the first saline-filled implant marketed as the Simaplast device *(19)*. Like silicone-filled implants, it had an exterior shell of semipermeable silicone and a filling valve that allowed for volume adjustments at the time of placement. It was easy to insert immediately after mastectomy through a small incision and available in a wide range of sizes. Incisions could be made in more remote locations and the incidence of capsular contracture was lower than that observed with silicone-filled prostheses *(18,20–22)*. However, saline-filled implants had a less natural feel, contour irregularities, rounded shape, poor fixation posteriorly, and deflation rates as high as 84% *(14,23,24)*. Early failures were caused by valve and valve stem problems, whereas later failures were to the result of internal abrasions or "fold flaws" at the end of a wrinkle in the silicone shell *(25)*. In 1968, Dempsey and Latham described a technique whereby implants were inserted deep rather than superficial to the pectoralis major muscle. This decreased the incidence of prosthesis exposure and of palpable prosthetic margins *(14)*.

11
Breast Prostheses

Current Recommendations for Care of Patients After Implantation of Breast Prostheses

Forrest S. Roth, David J. Gray, and Christian E. Paletta

INTRODUCTION

The history of breast augmentation predates the field of modern plastic surgery *(1)*. The first reported augmentation mammoplasty was performed by Czerny in 1895. He resected a benign breast adenoma from a young actress and then transplanted a lipoma from her back to fill the defect *(2)*. In 1917, Barlett described free tissue transfer of subcutaneous fat excisions from the abdominal wall, thighs, and buttocks to reconstruct breasts *(3)*. This technique required 50% more allograft tissue than the volume of excised breast tissue to compensate for postoperative shrinkage. A breast reconstructed in this manner was usually similar in size to the original breast after 6 months, but was firmer and less pendulous. Lexer in 1925 *(4)* and Passot in 1930 *(5)* published their techniques of fat transplantation with as little as 25% total weight and volume losses in their fat grafts. In 1943, May reported a technique that resulted in even less postoperative fat absorption *(3)*. Later called the derma-fat-fascia graft by Berson in 1945, May's fat graft was transferred along with the underlying fascia; it seemed to be more resistant to infection as well *(6)*. In 1953, Bames described a 90% survival rate for derma-fat-fascia grafts vs 60% for traditional fat grafts *(7)*. In 1959, Watson published his experience and reported no postoperative shrinkage *(2)*. However, unsightly donor scars, a high incidence of fat necrosis, calcification, and chronic drainage resulted in generally poor long-term results *(8)*. In 1964, Peer described a two-stage fat transplant procedure *(9)*. However, this technique resulted in unnatural breast textures, misleading mammographic findings, and was followed by a sharp decline in the use of autogenous tissue for breast augmentation.

These difficulties with autogenous tissue transfer led others to implant or inject foreign substances. In 1899, Gersuny was the first to describe subcutaneous paraffin injections *(10)*. This technique was quickly abandoned because of breast ulceration, sinus formation, inflammation, necrosis, and embolization of the paraffin. In the 1950s, liq-

25. van der Graaf Y, de Waard F, van Herwerden LA, Defauw J. Risk of strut fracture of Bjork-Shiley valves. Lancet 1992;339:257–261.
26. van der Maas PJ, de Koning HJ, van Ineveld BM, et al. The cost-effectiveness of breast cancer screening. Int J Cancer 1989;43:1055–1060.
27. Loeve F, Boer R, van Oortmarssen GJ, van Ballegooijen M, Habbema JD. The MISCAN-COLON simulation model for the evaluation of colorectal cancer screening. Comput Biomed Res 1999;32:13–33.
28. Law AM, Kelton WD. Simulation modeling and analysis. McGraw-Hill series in industrial engineering and management science. Boston, MA: McGraw-Hill; 2000.
29. Gold MR, Siegel JE, Russell LB, Weinstein MC. Cost-effectiveness in health and medicine. New York, NY: Oxford University Press; 1996.
30. Drummond MF, Jefferson TO. Guidelines for authors and peer reviewers of economic submissions to the BMJ. BMJ 1996;313:275–283.
31. Hunink MGM. Decision making in health and medicine : integrating evidence and values. Cambridge, UK ; New York, NY: Cambridge University Press, 2001.
32. Sox HC. Medical decision making. Boston, MA: Butterworths, 1988.
33. Sackett DL, Haynes RB, Tugwell P. Clinical epidemiology: a basic science for clinical medicine. Boston, MA: Little Brown , 1985.

2. Karlson EW, Hankinson SE, Liang MH, et al. Association of silicone breast implants with immunologic abnormalities: a prospective study. Am J Med 1999;106:11–19.
3. Stein ZA. Silicone breast implants: epidemiological evidence of sequelae. Am J Public Health 1999;89:484–487.
4. Ledley RS, Lusted LB. Reasoning foundations of medical diagnosis. Science 1959;130:9–21.
5. Lusted LB. Decision-making studies in patient management. N Engl J Med 1971;284:416–424.
6. Kassirer JP, Moskowitz AJ, Lau J, Pauker SG. Decision analysis: a progress report. Ann Intern Med 1987;106:275–291.
7. Habbema JD, Bossuyt PM, Dippel DW, Marshall S, Hilden J. Analysing clinical decision analyses. Stat Med 1990;9:1229–2242.
8. Weinstein MC, Fineberg HV. Clinical decision analysis. Philadelphia: Saunders, 1980.
9. Fryback DG, Dasbach EJ, Klein R, et al. The Beaver Dam Health Outcomes Study: initial catalog of health-state quality factors. Med Decis Making 1993;13:89–102.
10. Essink-Bot ML, Krabbe PF, Bonsel GJ, Aaronson NK. An empirical comparison of four generic health status measures: the Nottingham Health Profile, the Medical Outcomes Study 36-item Short- Form Health Survey, the COOP/WONCA charts, and the EuroQol instrument. Med Care 1997; 35:522–537.
11. Chung KC, Greenfield ML, Walters M. Decision-analysis methodology in the work-up of women with suspected silicone breast implant rupture. Plast Reconstr Surg 1998;102:689–695.
12. Peters W, Pugash R. Ultrasound analysis of 150 patients with silicone gel breast implants. Ann Plast Surg 1993;31:7–9.
13. Harris KM, Ganott MA, Shestak KC, Losken HW, Tobon H. Silicone implant rupture: detection with US. Radiology 1993;187:761–768.
14. Netscher DT, Weizer G, Malone RS, Walker LE, Thornby J, Patten BM. Diagnostic value of clinical examination and various imaging techniques for breast implant rupture as determined in 81 patients having implant removal. South Med J 1996;89:397–404.
15. Ahn CY, DeBruhl ND, Gorczyca DP, Shaw WW, Bassett LW. Comparative silicone breast implant evaluation using mammography, sonography, and magnetic resonance imaging: experience with 59 implants. Plast Reconstr Surg 1994;94:620–627.
16. Pauker SG, Kassirer JP. Therapeutic decision making: a cost–benefit analysis. N Engl J Med 1975;293:229–234.
17. Pauker SG, Kassirer JP. The threshold approach to clinical decision making. N Engl J Med 1980;302:1109–1117.
18. Beck JR, Pauker SG. The Markov process in medical prognosis. Med Decis Making 1983; 3:419–458.
19. Sonnenberg FA, Beck JR. Markov models in medical decision making: a practical guide. Med Decis Making 1993;13:322–338.
20. Hunink MG, Bult JR, de Vries J, Weinstein MC. Uncertainty in decision models analyzing cost–effectiveness: the joint distribution of incremental costs and effectiveness evaluated with a nonparametric bootstrap method. Med Decis Making 1998;18:337–346.
21. Chessa AG, Dekker R, van Vliet B, Steyerberg EW, Habbema JD. Correlations in uncertainty analysis for medical decision making: an application to heart-valve replacement. Med Decis Making 1999;19:276–286.
22. Birkmeyer JD, Marrin CA, O'Connor GT. Should patients with Bjork-Shiley valves undergo prophylactic replacement? Lancet 1992;340:520–523.
23. van der Meulen JH, Steyerberg EW, van der Graaf Y, et al. Age thresholds for prophylactic replacement of Bjork-Shiley convexo- concave heart valves: a clinical and economic evaluation. Circulation 1993;88:156–164.
24. Steyerberg EW, Kallewaard M, van der Graaf Y, van Herwerden LA, Habbema JD. Decision analyses for prophylactic replacement of the Bjork-Shiley convexo-concave heart valve: an evaluation of assumptions and estimates. Med Decis Making 2000;20:20–32.

information from tests performed in a patient. Since time aspects are less relevant, a static and relatively simple decision tree can often be constructed.

State-transition models are useful for long-term evaluations of prognosis, including estimates of life expectancy and time spent in certain health states. A simple state-transition model was presented, based on the life table for the general population, with an extension incorporating health states other than "alive" and "dead" in a multistate life table. More complex state-transition models may be required to compare follow-up policies. The ingredients of such models include baseline prognostic estimates (e.g., the incidence of a problem with a device when no intervention takes place), diagnostic estimates (e.g., the sensitivity and specificity of a test), and therapeutic effects (e.g., the effectiveness of antibiotics to treat a certain type of infection). Note that quantitative empirical evaluations of alternative follow-up policies are scarce.

For all types of models it is important to structure the problem well before implementation in computer code is considered, to quantify the model using empirical data whenever possible, and to consider medically relevant outcomes. In addition to medical effects, costs can be included. Sensitivity analysis is a central tool in both the development of the model and the presentation of results from the model.

Development of a valid and credible model is far from easy, and requires both computer expertise and medical expertise. Both structural and parameter uncertainty can usually be recognized in a model, leading to subjective aspects in a model.

The reward for all this can be a computer model that is a powerful tool to study a difficult medical question. The application of computer models for patients with devices may well increase in the future, when more data from well-performed, large-scale studies will be available to structure and quantify the models. Models may potentially play a large role in the treatment of patients, either by the development of guidelines or by individual decision support. For all applications of models, credibility is important, which may be achieved with the active participation of the users of the models and thorough attempts of validation.

FURTHER READING

Several excellent texts exist that discuss issues on medical decision making and modeling in more detail than presented here. Texts on clinical decision analysis give a theoretical background to performing decision analyses and the construction of decision trees *(8,31)*. Somewhat more practical texts in the same area are also available *(32,33)*. An authoritative reference on cost-effectiveness analysis has appeared some years ago, including a thorough discussion on the development of models *(29)*. The development of simulation models is discussed in depth by others *(28)*. Applications of decision trees and state-transition models are becoming more frequent in the medical literature, and some journals such as *Medical Decision Making* focus on methodological issues in model development and application.

REFERENCES

1. Edworthy SM, Martin L, Barr SG, Birdsell DC, Brant RF, Fritzler MJ. A clinical study of the relationship between silicone breast implants and connective tissue disease. J Rheumatol 1998;25:254–260.

comparison of the total model output with empirical data. When discrepancies are noted between predicted and observed outcomes, these can sometimes be used for *calibration* of the model (i.e., the adjustment of certain model parameters to decrease discrepancies). Preferably, the adjustment should be justifiable by experts on more grounds than just that the model fit improves, and the adjusted model should be validated using other relevant data.

Measures to increase validity may also positively influence *credibility* (i.e., whether the model results will be accepted as "correct"). At least, the output of the model should be consistent with that anticipated by experts ("face validity").

Uncertainty

Computer models generally suffer from two types of uncertainty: structural and parameter uncertainty. Structural uncertainty is caused by assumptions about the design of the model. Examples include the independence of test results, similarity of test characteristics among different patient groups, stability of risks over time, influence of an invasive procedure on mortality, whether a disease has an additive or multiplicative effect on the mortality hazards observed in the general population *(24)*, and the shape of dose–response relationships *(29)*. Structural uncertainty is hard to investigate, and is often ignored. Developing separate models with different assumptions (e.g., a simple version of a model and a more complex version) or using different groups of analysts to develop separate models can be valuable.

Parameter uncertainty is caused by limitations in empirical data that are used to quantify a model. Some events may be so rare that reliable estimation of their incidence is difficult. However, accurate estimation is crucial when the events are serious threats to the patient's health (e.g., fatal complications). An example is the estimation of fracture risk of mechanical heart valves in various subgroups. Some notable factor in the model may not have been studied often, or even be impossible to study empirically, such as a problem in a device that can only be detected after explantation of the device. Finally, predictions relate to the future and various unpredictable developments may influence actual patient outcomes.

Structural and parameter uncertainty together determine the validity of a model. Evaluation of the joint effect of both types of uncertainty is difficult and may not readily be possible. Ideally, a model would be constructed and quantified with detailed information from large-scale empirical studies. However, in practice, such information is often not available and expert knowledge will need to be incorporated. This inevitably leads to subjectivity in a model and hence to a potential for questioning the validity of a model. On the other hand, a model often provides the best summary of the current state of knowledge on a certain medical problem for a particular application such as deciding about diagnostic testing and treatment.

SUMMARY AND CONCLUSION

In this chapter, various issues regarding the potential roles of computer modeling have been discussed with a focus on patients with prosthetic devices. Three types of problems were discussed: diagnostic, prognostic, and follow-up. A model for a diagnostic problem typically includes Bayes' rule to update probabilities with diagnostic

model is a simplification of reality, but it should not be an oversimplification. This judgment depends on the objective; a model may be valid for one type of evaluation but not for another. A model may be considered valid if the decisions that are based on it are similar to those that would be made if it were feasible to perform empirical experiments.

As examples, we may consider validation of the models discussed before. For the example of rupture of breast implants (example 1), an empirical study should show similar probabilities of rupture to those predicted by the decision tree. For example 2 (defective heart valve), an empirical study should show fracture occurring over time as predicted by the model, with fracture rates depending on age as predicted. For example 3 (evaluation of follow-up schemes), a valid model should provide an estimate of the cost-effectiveness of a particular follow-up scheme compared to an alternative that is similar to that observed in an empirical study of two cohorts of patients who each receive follow-up with one of these schemes.

Some of these empirical studies are hypothetical since their limited feasibility is the very reason to develop a computer model. Thinking about empirical studies does, however, help in defining "validity." Practical obstacles include the fact that substantial numbers of patients would need to be included to limit random noise and that detailed information on every patient would need to be available to enable persuasive comparisons of model predictions and empirical observations.

The level of simplicity in a model depends on factors such as the purpose of the model (28). This is essential because it suggests which aspects need to be modeled in detail and which do not. Sensitivity analysis may be used for this purpose also; factors with a large influence need to be modeled carefully and quantified as accurately as possible. In practice, the initial model is usually "moderately detailed" and then modified through discussions with experts. The availability of data, time, and money limit the amount of model detail.

Various practical suggestions have been made to increase model validity (28). Subject-matter experts (e.g., experienced clinicians) need to be involved in the model development. These may indicate the plausibility of simplifying assumptions and relevant outcomes of a model. Also, empirical data are often not available for all relevant parameters. Experts in the clinical field may then provide subjective probability estimates. In principle, such estimates satisfy the same properties as probabilities from empirical data. Getting reliable estimates from experts can be quite challenging. How to maximize reliability is a topic of ongoing research.

The analyst should document the "conceptual model," especially when it is complex. This documentation should be accessible to clinicians, such that model elements, underlying assumptions, and quantifications are made clear. Such a document is also useful for a "structured walk-through" of the model, where consensus is sought among all persons involved in the development of the model. Sensitivity analyses may be used to obtain insight in the relevance of particular factors in the model. Aspects to vary in such analyses may include the values of a parameter, the choice of the distribution for a parameter, and the level of detail for a subsystem.

Finally, components of the models may separately be validated with available data, such that the validity of the total model is supported. The strongest test of validity is a

considered modeling of medical effects, such as survival. Costs can be considered in these models as well.

Costs are commonly divided into "direct" and "indirect" costs. Direct costs generally refer to changes in resource use directly attributable to the follow-up and include the value of goods, services, and other resources that are consumed in or outside the medical setting. Often, direct medical costs are the most important cost component, including costs of tests, drugs, supplies, health care personnel, and medical facilities. Direct nonmedical costs include time spent by patients and their family in relation to the disease and transportation costs to and from medical facilities. Indirect costs generally refer to productivity losses related to illness or death ("productivity costs") *(29)*.

Cost-effectiveness analyses can be undertaken from different perspectives. The broadest is the societal perspective, which incorporates all costs and health effects regardless of who incurs them. Alternative perspectives include that of the hospital, the insurer, or the patient. From a societal perspective, opportunity costs of resources that are valuable to society should be measured. A detailed method of estimating costs has been referred to as microcosting. With this method, cost items are measured in detail, and unit costs are valued separately, for example, based on market prices *(29)*.

Costs can be related to medical effects in the cost-effectiveness ratio. Note that both costs and effects should be calculated as the difference among competing choices. Incremental costs and incremental effects should be studied, rather than average costs and average effects. The medical effects should be relevant to the patients involved, for example, detection of a problem with a device (e.g., rupture of a breast prosthesis) or survival. Survival can combined with health-related quality of life in quality-adjusted life years (QALYs). QALYs are currently the preferred measure of effectiveness *(29)* and are often estimated with some type of state-transition model. The comparison of competing strategies may then be summarized in one single number: the costs per QALY of one strategy over the other. However, it is also very important to aim for transparency (e.g., by providing insight in the important components of the analysis) such as the number of tests performed in either strategy, the number of complications, the number of treatments given, and so on *(30)*. Furthermore, note that some speak about a cost–utility analysis when costs are related to QALYs and about a cost–benefit analysis when medical effects or QALYs are valued in monetary terms (e.g., by "willingness-to-pay" studies). More details on cost-effectiveness can be found elsewhere *(29)*.

VALIDITY AND CREDIBILITY

A difficult problem in the development of a computer model is that of trying to determine whether the model is an accurate representation of the actual problem being studied (i.e., whether the model is valid). On a technical level, validity depends on whether the conceptual model has correctly been translated into a computer program (i.e., that no programming errors have been made). This has been referred to as model *verification (28)*. In a large and complex model, verification is a difficult task despite the various advances in modern computer software.

On a conceptual level, validation is the process of determining whether a model is an accurate representation of the problem for the particular objectives of the study. A

made about the results of treatment of patients in different health states, that is, shortly or later after failure.

Utilities and Evaluation

Deciding which outcomes to consider for follow-up schemes is not straightforward. When lethal events are considered, life expectancy would be a natural choice, possibly extended with measures like event-free life expectancy or quality-adjusted life expectancy. For nonfatal events, a summary might be considered of all relevant events in the model under alternative follow-up schemes. Possibly a weight might be attached to each event, enabling a summary evaluation with a single number for each policy.

Technically, the evaluation of a complex model is often done with microsimulation techniques. Microsimulation is a type of Monte Carlo (or "random") simulation. Individual life histories are generated for large numbers of patients under each follow-up policy. Conclusions may then be based on the averages over simulations. The same patient histories should be used for all follow-up policies ("common random numbers") *(28)*. There are specialized software programs that perform such evaluations efficiently.

Sensitivity Analysis

The more complex the problem addressed, the larger the number of assumptions that usually have to be made and the greater the need for sensitivity analyses. Sensitivity analyses may first be used to check model consistency and correct programming errors. For example, when extreme values are assumed for some parameters, the results from the model can be predicted and model output can be compared with the expected results.

Furthermore, one might vary all important parameters over their plausible ranges. However, when the number of parameters gets large, this leads to a confusing overview. A full analysis of the uncertainty in the model can be obtained by defining distributions of the parameter values and drawing parameter estimates from this distribution. This is sometimes referred to as second-order Monte Carlo simulation, in contrast to a first-order simulation, where probability distributions are fixed *(29)*.

Furthermore, one might first perform analyses for an idealized world, assuming perfect follow-up attendance or 100% compliance with antibiotic treatment, for example. Such parameter estimates are not realistic and might be varied in scenario analyses where sets of parameter values are used that correspond to certain expected or observed settings in the real world.

GENERAL MODELING ISSUES

Some further important issues are addressed here, including cost-effectiveness analysis and aspects of computer modeling such as validity, credibility, and uncertainty of a model.

Cost-Effectiveness Analysis

In a cost-effectiveness analysis, costs of an intervention or procedure are related to their medical effects. This enables a judgment on the desirability of a medical intervention when a limited budget is available. In the previous sections we predominantly

management surrounding implantation of a device, with follow-up schedules aiming at late effects. Also note that we do not consider follow-up for the purpose of registration of possible complications ("postmarketing surveillance") or fear of lawsuits for missed complications. Instead we focus on the benefit of follow-up to individual patients, which is determined by the balance between positive and negative effects.

Next, what type of tests would be used during follow-up? Taking a history and some simple, noninvasive tests are obvious initial steps, but when are more advanced diagnostics indicated? Furthermore, what would the clinical policy be after an adverse event? Examples include medical treatment (e.g., antibiotics for infected prosthesis), device replacement for mechanical failure, or observation (e.g., for hip pain after hip replacement).

Is it necessary to model the natural course of the disease? When the natural course is defined, alternative follow-up schedules can systematically be evaluated and compared to each other. However, empirical data regarding the natural course are often difficult to obtain and analyze. The natural course has often been modeled in cancer, where the disease process is related both to the sensitivity of tests (e.g., detection of metastases) and to the prognosis (poor when advanced disease has developed) (26,27).

Quantification

The model needs to be quantified. Empirical data from clinical studies are preferable. Prognostic estimates might be obtained from a follow-up study in a well-defined cohort. For example, the risk of prosthesis failure can be estimated by analyzing the number of failures during follow-up of a series of patients with a prosthesis implanted in a certain time frame in one or more medical centers. The analysis of such data needs to account for potential differences in surveillance intensity among patients. Examples include patient dropout and the end of the study period on a certain date. Patients without complete follow-up are considered as censored observations in statistical survival analysis (also referred to as "time-to-event" or "failure-time analysis"). When different outcomes are considered (e.g., infection and disease progression), these may initially be analyzed separately, but there are often questions about the independence of these outcomes. Such issues require the involvement of knowledgeable clinicians and statistically skilled researchers in the development of the model.

Diagnostic estimates are required for all tests used during follow-up and for all outcomes considered. A problem arises when test characteristics depend on the timing of follow-up visits. For example, when annual visits are compared with visits every two years, the sensitivity for detection of prosthesis failure is likely higher with the latter schedule since failure often causes more obvious abnormalities that can be detected more easily with a certain test. For example, a rupture in a breast prosthesis will increase in time and hence become better recognizable.

Therapeutic effects are ideally determined from randomized clinical trials, which enable unbiased comparisons between alternative treatments. For evaluation of follow-up regimens, the timing of treatment is an essential element. How important is it if diagnosis and treatment of prosthesis failure is delayed, for example? Trials that address this type of question are rare. In the absence of empirical data, assumptions have to be

valued as 60% of the present year. The discounted life expectancy was reduced from 10.3 to 9.4 years by the assumed risk of valve failure, or by 8.8%, which still exceeds the estimate of 4% surgical risk.

Lessons From the Example

A number of lessons can be learned from the example:

1. As the complexity of a problem increases, more assumptions must be made to construct a manageable model. Some assumptions can be addressed in a sensitivity analysis. Uncertainty in baseline survival estimates or the lethality of fracture had small influences on the model outcome, but uncertainty concerning the risk of strut fracture and the surgical risk each had larger influence. Furthermore, some structural assumptions were made in the modeling process. These include the assumption that the replacement valve is functionally identical to the Björk-Shiley valve, the only difference being a zero risk of failure in the replacement valve. Mechanical valves that are currently being used in cardiothoracic centers may have improved hemodynamic properties and reduced risks of complications, which would make replacement more attractive. The reoperation was assumed not to have adverse effects on long-term survival, which may not be true.
2. When sensible assumptions are made, a relatively simple state-transition problem emerges, which can be programmed in a spreadsheet program (*see*, e.g., http://www.eur.nl/fgg/mgz/software.html#bscc). Alternatives include a Markov model or stochastic programming with simulation software.
3. A state-transition model can be used to provide individualized decision support. However, the decision to perform a prophylactic explantation of a device is very difficult because the risks of surgery are evident in the short term. The benefits of avoiding valve failure are evident in the future. A rather fundamental issue concerns the dimension of time (patient age and follow-up duration). Based on the negative association between the risk of valve dysfunction and time since implantation, a decrease in risk is assumed over time. However, the future risks of valve failure are by definition extrapolations from current observations, while these future risks are the reason to consider explantation now. This all implies that decision making cannot be based solely on the model but requires careful discussion among physicians (e.g., cardiologist and thoracic surgeon) and involvement of the patient.

EXAMPLE 3: EVALUATION OF FOLLOW-UP SCHEMES

The systematic evaluation of alternative follow-up schemes might be addressed with a state-transition model. The question might be which follow-up scheme is most cost-effective for a certain problem in which various tests can be used at various times during follow-up. In theory, an optimal scheme might be identified by examining a huge set of combinations of tests and timings of follow-up. In practice, a limited number of alternative policies may be compared that appear sensible from a clinical perspective. A number of steps can be considered.

Structuring the Problem

The problem needs to be defined clearly. The purpose of follow-up should be defined. This includes the identification of long-term complications of prosthesis placement (e.g., failure, infection), detection of progression of the disorder for which the prosthesis was placed, and detection of other conditions of medical significance. Note that short-term complications are probably best considered as part of the initial clinical

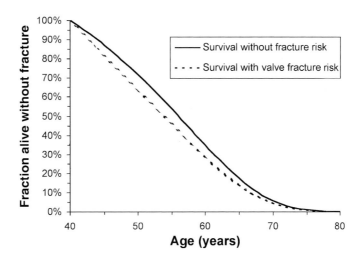

Fig. 12. Survival for a 40-year-old male with a Björk-Shiley Convexo-Concave valve according to the annual risks shown in Fig. 11 (with or without fracture risk).

successful replacement have a survival that is identical to that of patients with heart valve that lack the risk of mechanical failure. If the risk of replacement is lower than 10.8%, replacement would be the preferable strategy for this patient. For example, if the surgical risk is estimated as 4%, a life expectancy of 96% × 16.0 = 15.4 years is expected, which is 15.4 – 14.3 = 1.1 years longer than with expectant management.

Sensitivity Analysis

Several implicit and explicit assumptions were made in the analysis. Some of these can be assessed in a sensitivity analysis. The most important issue is the risk of strut fracture; the estimate of 1.6% at age 40 may be too low or too high, which directly affects the desirability of replacement. If the risk is less than 0.6%, the decrease in life expectancy is less than 4%. Furthermore, a 100% lethality of fracture was assumed; if 40% of the patients survive fracture, the decrease in life expectancy is 6.7% instead of 10.8%.

Baseline survival estimates (i.e., without risk of valve failure) are important but less crucial because survival is a component in both strategies (surgery or expectant management). If survival were much worse (i.e., a life expectancy without valve failure of 3.5 instead of 16.0 years), the loss in life expectancy would be less than 4%.

The surgical risk may be lower or higher than 4%, which directly affects the desirability of replacement. In practice, an individualized estimate of surgical mortality would have to be made, incorporating the presence of individual risk factors. These evaluations might be restricted to patients with a substantial loss of life expectancy, according to the model.

An important issue is the outcome considered. In our illustration, we simply compared the number of years of life with either strategy. In practice, patients may be risk-averse and prefer years in the present to years in the far future. In decision analysis, this preference is often quantified by discounting years further away (e.g., by 5% per year). This means that the next year is valued as 95% of the present year, and year 10 is

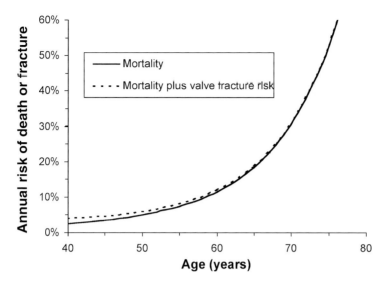

Fig. 11. Annual risks for a 40-year-old male with a Björk-Shiley Convexo-Concave valve: mortality, excluding the fracture risk, and the combination of mortality or fracture.

order) characteristics. For a mitral valve in a 40-year-old patient, we might know that the valve had a large diameter (29 mm), was produced in a large shop order (>175 valves), but that no device failure had been reported in this shop order. The risk in the first year of follow-up is then estimated as 1.6%, which is substantially higher than the average risk of 0.1%. A negative association was found between mechanical failure risk and age, which makes the expected risk decrease to 0.9% at age 50, 0.6% at age 60, and 0.3% at age 70. The combined risk of mortality or failure is shown in Fig. 11.

The lethality after fracture is very high in aortic valve patients (>90%) and also substantial in mitral valve patients (>60%) *(24)*. For the present illustration we assume 100% lethality or, equivalently, analyze survival time until fracture.

Prophylactic surgery has morbidity and mortality risks that are difficult to quantify with empirical data because reoperations are generally done for indications other than prophylactic valve replacement. We might assume that the mortality risk is similar to that of primary implantation and that it depends on several risk factors. A comparison of several published studies indicates that the risk of prophylactic replacement might be between 1% and 4% for 40-year-old patients without clear risk factors *(24)*. We assume a 4% 30-day mortality risk for our evaluation.

Utilities and Evaluation

The most relevant outcome in this illustration is survival: valve replacement should be considered if a higher survival is expected than with expectant management. The expected survival for our patient is shown in Fig. 12, both for the situation in which the valve is present and the situation in which the risk of failure is zero. The areas under the curves are 14.3 and 16.0 years. This means that the risk of strut failure reduces life expectancy by 1.7 years or 10.8%. Interestingly, this percentage equals the risk of prophylactic surgery for which valve replacement and expectant management strategies yield identical life expectancies. This follows from the assumption that patients with a

A stochastic model implies that Monte Carlo simulation or micro-simulation is used. Life histories of individuals in a population are simulated many times. If large numbers are generated, the average becomes stable and approaches that of deterministic calculations. However, stochastic modeling offers more flexibility, which is required for more complex models than simple state-transition models. Also, probabilistic sensitivity analyses can more easily be performed with stochastic modeling, which provide insight in the joint uncertainty caused by all parameter values *(20)*. In these analyses, correlations between probabilities can potentially be incorporated *(21)*.

EXAMPLE 2: BJÖRK-SHILEY CONVEXO-CONCAVE HEART VALVE PATIENTS

Patients with this particular mechanical heart valve are at risk of suffering outlet strut fracture, which leads to dislodgment and embolization of the disk. This is often lethal. Therefore, prophylactic replacement may be considered to avert fracture. We consider decision making for a 40-year-old male patient with such a heart valve implanted in the mitral position 20 years ago.

Structuring the Model

Without replacement, the structure of the computer model is like that shown in Fig. 7 *(22–24)*. After failure (strut fracture), a substantial mortality must be assumed. Survivors are in the state "alive after failure," which implies that another heart valve was implanted. With replacement, elective surgical mortality would need to be taken into account. After successful replacement, the expected survival is assumed to be identical to that of patients with another mechanical heart valve (without the risk of strut fracture).

Transition Probabilities

Many of the required transition probabilities can be derived from a large-scale follow-up study in Björk-Shiley Convexo-Concave valve recipients in the Netherlands *(25)*. The annual mortality risk of patients with mechanical valves was analyzed with exclusion of the risk of strut fracture by considering fracture as a censoring event in the statistical analysis. Various risk factors were statistically taken into account in regression models, including demographics (age and gender), cardiac risk factors (previous valve surgery, previous myocardial infarction, left ventricular function, coronary artery disease), and concomitant surgery performed at the time of implantation of the valve (bypass surgery, tricuspid valve surgery, ascending aorta surgery) *(24)*. Figure 11 shows the estimated annual risk of mortality for our 40-year-old male with a mitral valve, assuming absence of risk factors. Comparison with Fig. 5 shows a similar steep increase in risk with age. However, the risks of heart valve patients (Fig. 11) are substantially higher than those of males in the general population at every age (Fig. 5). This reflects the severity of the underlying cardiovascular disease and specific events that may jeopardize the patient's health (e.g., thromboembolic events, bleeding, endocarditis). Annual risks exceeding 10% are seen from age 60, whereas such high risks are only seen from age 80 in the general population.

The average risk of strut fracture is low (around 0.1% per year). However, regression analysis has identified subgroups of patients with a considerably higher risk *(24)*, based on patient (age), valve (size, position, type), and production (e.g., size of shop

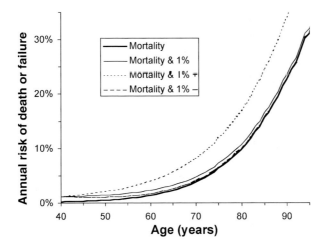

Fig. 9. Annual risks of mortality (as in Fig. 5), and the combination of mortality or failure (as in Fig. 8).

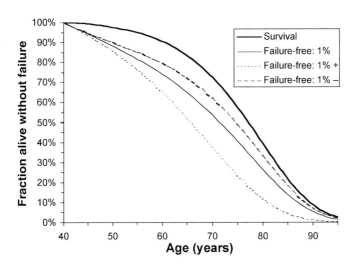

Fig. 10. Survival curve without failure for a 40-year-old Dutch male according to the annual risks shown in Fig. 9. The top line ("survival," i.e., no failure risk) is identical to that shown in Fig. 6. The areas under the curves represent the (failure-free) life expectancies.

of 1%). This assumes that failures are not treated (e.g., replacement of the device) and that failure does not lead to excess mortality. Treatment and risks of failure can readily be taken into account in state-transition models by inclusion of the relevant transition probabilities.

Calculations with these types of multistate life tables can be performed with deterministic or stochastic models. With a deterministic model, average numbers of events in a population are used. This can be achieved with a relatively simple spreadsheet program or a "Markov model" *(18)*. A Markov model assumes that the probability of an event in a certain state is independent from the previous state. Markov models are available through "Markov nodes" in most modern software for decision analysis *(19)*.

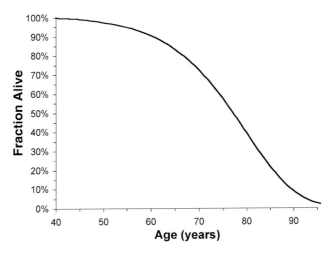

Fig. 6. Survival curve for a 40-year-old Dutch male according to the annual mortality risks shown in Fig. 5. The area under the curve represents the life expectancy.

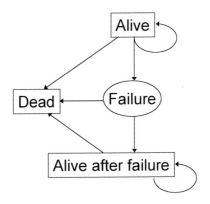

Fig. 7. State-transition model for a person with a failure-prone device. Three states are considered: "alive," "alive after failure," and "dead." "Failure" is an acute event.

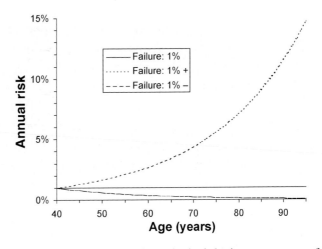

Fig. 8. Annual risks of failure for three hypothetical devices: a constant failure risk of 1% per year, an increasing risk, starting at 1% and increasing by 5% per year, and a decreasing risk, starting at 1% and decreasing by 5% per year.

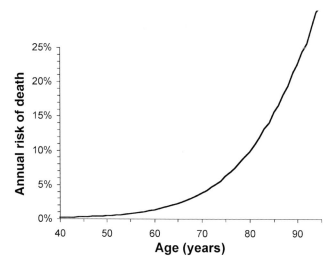

Fig. 5. Annual mortality risks for Dutch males in the general population.

typical of countries in the developed Western world (Fig. 5). The expected survival is 99.84% after 1 year, 99.66% after 2 years, and so on (Fig. 6). The expected survival of a 40-year-old male can be summarized by the life expectancy, which is the area under the survival curve (35.9 years). So, a 40-year-old male may expect to live nearly 36 years more on average. Note that he may actually die much earlier or later.

A Multistate Life Table

A multistate life table contains more states than just alive and dead. For example, consider patients with a device that may suffer from failure. This is assumed in Fig. 7, where the model contains a state "alive after failure," which is reached through "failure" (an acute event). The risk of failure may be constant, decrease, or increase over time. For illustration we consider three failure risks: constant at 1% per year, starting at 1% and proportionally decreasing by 5% per year, and starting at 1% and proportionally increasing by 5% per year (Fig. 8). The proportional decrease or increase in risk is fully hypothetical; in practice, many other patterns of decreasing or increasing risk can be imagined. If these failure risks apply to an otherwise healthy 40-year-old male, the previously shown mortality risks may be assumed (Fig. 5). The risk of mortality plus failure and the corresponding probability of remaining free from failure is shown in Figs. 9 and 10. With decreasing risk, the probability of being failure-free stays closest to the survival curve. At age 60, the probability of being alive without failure then is 79.6%, whereas the survival estimate is 90.5%. If failure has no consequences for survival, 90.5–79.6 = 10.9% of the patients are expected to be alive after failure. The expected percentage of patients remaining failure-free is 74.1% (constant 1% annual risk) or 65.0% (1% annual risk, increasing by 5% per year) at age 60. The failure-free life expectancies are 32.1, 29.7, and 25.0 years for the curves shown in Fig. 10. These life expectancies are calculated as the areas under the curves. Other statistics can be calculated relatively easily. For example, the expected time alive after failure is the area between the survival and failure-free curves (6.2 years assuming a constant annual risk

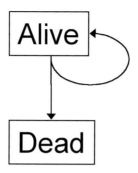

Fig. 4. State-transition model for a healthy person. Only two states are considered: "alive" and "dead."

Lessons From the Example

A number of lessons can be learned from the example.

1. Even for a relatively simple diagnostic decision problem, assumptions are being made to construct a manageable model. These may not be realistic. An example is the assumption of independence between US and MRI test results *(14)*.
2. Probability estimates are of primary importance. After filling in prevalence, probability of equivocal test results, sensitivity and specificity, we could estimate the probability of rupture using Bayes' rule. The posttest probability provides the primary input for the decision-making process *(16,17)*.
3. The valuation of outcomes is important but comes after considerations of probability estimates for actual decision making. A transparent overview of the likelihood of each outcome is more important than a summary measure such as the weighted misclassification rate.
4. Sensitivity analyses should be performed to study the effect of uncertainty in assumptions on the results of the analysis.

STATE-TRANSITION MODELS

Decision trees are not well suited to represent events that repeat over time. For example, a patient with a device may die or encounter a problem with that device during the next year, but also during another year. The modeling of events like device failure and mortality can more efficiently be represented with a state-transition model.

State-transition models allocate members of a population into one of several health states. Transitions occur from one state to another at certain time intervals, according to transition probabilities.

A Simple Life Table

A simple state-transition model is based on the life table for the general population. For example, consider the survival of a 40-year-old male in the general Dutch population. The states are simply "alive" or "dead" (Fig. 4). Over time, he may die of any cause, as indicated by the arrow to "dead." Note that "dead" is an absorbing state. The probability of death between age 40 and 41 is very low: 0.16%. One year later, when he is 41 years old, the transition probability from "alive" to "dead" is 0.18%. The annual risk of mortality increases gradually with age to more han 20% at age 90, which is

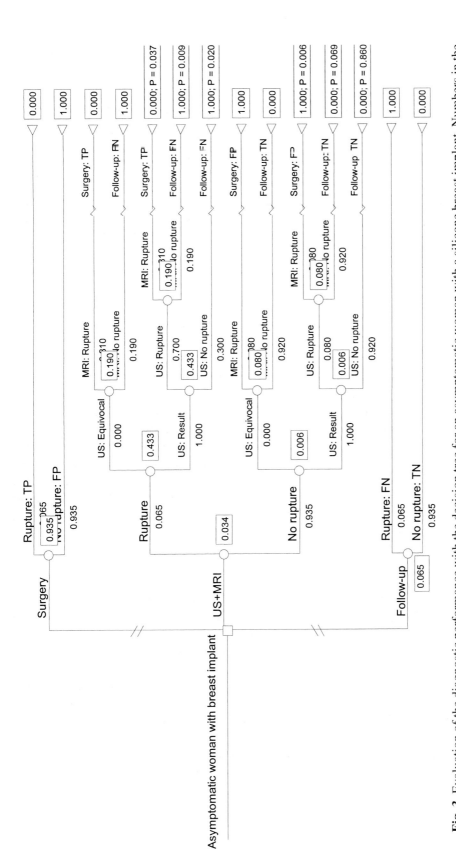

Fig. 3. Evaluation of the diagnostic performance with the decision tree for an asymptomatic woman with a silicone breast implant. Numbers in the tree replace the abbreviations shown in Fig. 2. The strategy "US + MRI" has the lowest misclassification error (3.4%), which is calculated as the sum of the probabilities ("P") of the false-positive (FP) and false-negative (FN) classifications.

Table 1
Overview of the Diagnostic Performance of Three Strategies Shown in Figs. 1 to 3

Strategy	TP	FP	TN	FN	Error
Surgery	6.5%	93.5%	0%	0%	93.5%
Follow-up	0%	0%	93.5%	6.5%	6.5%
US + MRI	3.7%	0.6%	92.9%	2.9%	3.5%

TP and FP denote the true- and false-positive rates (i.e., surgery in ruptured or intact prostheses, respectively). TN and FN denote the true- and false-negative rates (i.e., follow-up in intact or ruptured prostheses, respectively). The classification error is the sum of FP and TN rates. US, ultrasound; MRI, magnetic resonance imaging.

The misclassifications are directly available for the surgery and follow-up strategies (Fig. 3 and Table 1). For surgery, a 6.5% true-positive rate is achieved (all ruptured prostheses are explanted), but false-positive decisions equal 93.5% (explantation of intact prostheses). For follow-up, the true-negative rate is 93.5% and false-negative rate is 6.5% (missed ruptured prostheses). For the US + MRI strategy, the misclassification is calculated as 3.5%, based on the weighted average of the false-positive and false-negative classifications and assuming no equivocal US results (Fig 2: p_equivocal=0). For example, the branch Rupture | US:Result | US:Rupture | MRI: No rupture results in a 0.9% probability of false-negative misclassification: $0.065 \times 1.0 \times 0.70 \times 0.19 = 0.008645$. The classifications of the US + MRI test strategy are 3.7% true-positive, 0.6% false-positive, 92.9% true-negative, and 2.9% false-negative (Table 1). Approximately a halving of the misclassification rate is achieved compared to the follow-up strategy (3.5% vs 6.5%). So, although tests may help to distinguish ruptured from intact prostheses, they are far from perfect.

Sensitivity Analysis

Performing a sensitivity analysis is illustrated by varying a probability (of equivocal test results) and a utility (of false-positive decisions). We assumed that US would not give equivocal test results (Figs. 2 and 3, p_equivocal=0). However, if we suppose that equivocal results occur in 10% of the women, the misclassification probability would increase to 3.9%.

We assumed that false-negative and false-positive decisions were equally important. If false-positive decisions are considered more important than false-negative, e.g., 10 times, a weighted misclassification rate would be estimated as 6.5% for follow-up, and 8.8% for the test strategy. So, with this weighting, follow-up would be preferable to testing, which is explained by the highly undesirable risk of false-positive decisions (unnecessary surgery). A full sensitivity analysis would include varying all relevant probabilities (prevalence of rupture, sensitivities and specificities) and structural assumptions (independence of US and MRI test results, no MRI after negative US, equivocal test results independent of rupture).

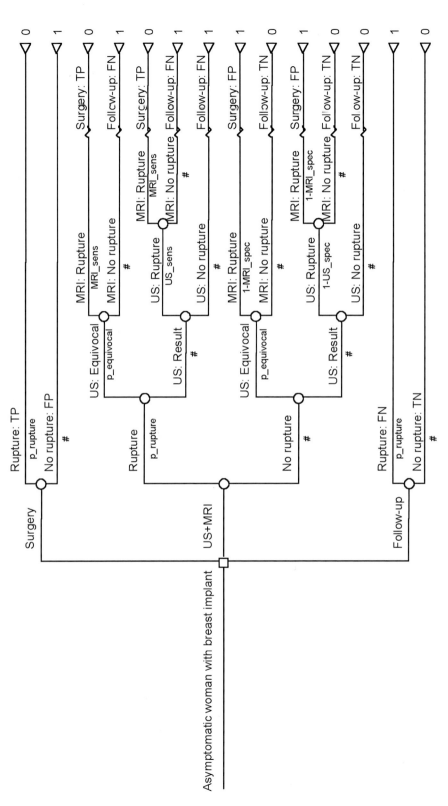

Fig.2. Decision tree for an asymptomatic woman with a silicone breast implant, with insertion of nodes that denote the presence or absence of rupture. p_rupture, probability of rupture; p_equivocal, probability of equivocal test results; US, ultrasound; MRI, magnetic resonance imaging; sens, sensitivity; spec, specificity; TP, true-positive; FP, false-positive; TN, true-negative; FN, false-negative. The "#" sign denotes the complement (1-value in other arm of the tree). Misclassifications (FN or FP) are denoted with the value 1 at the right end of the tree.

If US results are equivocal, we assume no change in the probability of rupture. If US shows no rupture, the probability of 6.5% decreases to 2.2%, since the LR = 0.33, the prior odds 0.0695, and the posterior odds 0.023 (0.33 × 0.0695).

After US, MRI follows, with assumed sensitivity of 81% and specificity of 92% *(15)*. If a rupture on US is confirmed on MRI, the probability that the prosthesis has truly ruptured increases to 86%. In this case, the prior odds is the posterior odds after a positive US (0.608). The post-MRI odds is LR × odds = 10.1 × 0.608 = 6.2, or a probability of 86%. If MRI shows no rupture, the probability becomes 11% (post odds = LR × odds = 0.21 × 0.608 = 0.13). After equivocal US results, the probability becomes 41% or 1.4% with a positive or negative MRI result respectively (post odds = LR × odds = 10.1 × 0.695 = 0.70 and 0.21 × 0.695 = 0.014, respectively).

In this calculation, an important assumption is made, namely independence of US and MRI test results (technically speaking: independence given the disease, or "conditionally independence"). We assume that the sensitivity and specificity of MRI are identical in patients before and after performing US. If certain ruptures are difficult to diagnose on both US and MRI, independence is violated. This would imply that the additional diagnostic information from MRI after US is smaller than when MRI is considered without US *(14)*.

Furthermore, we assume that no MRI is performed after a negative US. The probability of rupture would decrease from 2.2% to 0.5% if MRI was also negative (post odds = LR × odds = 0.21 × 0.023 = 0.005), or increase from 2.2% to 18.7% with a positive MRI (post odds = LR × odds = 10.1 × 0.023 = 0.23). Both probabilities are so low that surgery would not be considered indicated by most surgeons. Hence, MRI is not necessary after a negative US in an asymptomatic woman.

Similar calculations can be performed for women with another risk profile. For example, women with symptoms like breast asymmetry or capsular contracture and implants over 10 years old may have a prior probability of rupture of 64% *(12)*. When rupture is that likely before testing, a positive US would need no confirmation by MRI to recommend implant removal *(11)*.

Utilities and Evaluation

Figure 1 illustrates how the probabilities of rupture change after testing with US and MRI. We now indicate the fraction of patients in each branch, and evaluate how much is gained by testing. A simple criterion is the misclassification rate, assuming that false-positive decisions (explantations of intact prostheses) are as bad as false-negative decisions (no explantation of ruptured prostheses). Misclassification rate or classification error is complementary to accuracy, which is the rate of correct classifications. To enable this type of evaluation, the decision tree is restructured (Fig. 2). For every strategy (surgery, US + MRI, follow-up), a chance node is inserted indicating the underlying true state of the prosthesis (rupture/no rupture). Figure 2 looks more complex than Fig. 1, but it has an advantage over Fig. 1. Now we can work directly with sensitivity and specificity in the US + MRI strategy, rather than use Bayes' rule. For US, we make binary splits: first whether the US was equivocal, and if not, whether it showed rupture. Initially, a zero probability of equivocal test results is assumed. For the misclassification rate, we count all false-positive and false-negative classifications and a value of 1 is used for both classifications at the end of the tree ("value node").

are negative test results in women with ruptured prostheses. False-positive and false-negative findings occur because tests usually are not perfect.

The sensitivity of US is about 70% and specificity is about 92% *(11,14,15)*. The sensitivity of MRI is about 80% and specificity is about 92% *(14,15)*. This means that the diagnosis of prosthesis rupture is missed by US in 30% of the cases and by MRI in 20% of the cases. Both tests misclassify 8% of women with an intact prosthesis.

Bayes' Rule

Sensitivity and specificity are conditional probabilities: given that a patient does (not) have the disease, what is the probability of a positive (negative) test result: p(T|D)? In clinical practice we are interested in the probability of disease given the test result: p(D|T). This requires a reversal of conditional probabilities, which is achieved with Bayes' rule.

A rather standard formulation of Bayes' rule is as follows. For formula notation, p(T+|D+) is the sensitivity, p(T–|D–) is the specificity, p(D+) is the prevalence of the disease before the test is done ("prior probability"), p(D+|T) is the probability of disease after the test ("posterior probability"). The probability of disease after a positive test result, p(D+|T+), is:

$$p(D+|T+) = p(T+|D+) \times p(D+) / [\, p(T+|D+) \times p(D+) + p(T+|D-) \times p(D-)\,]$$

In words: the probability of disease after a positive test is equal to the likelihood of finding a positive test result in those with disease (true-positives) divided by the total likelihood of finding a positive test result (true-positives and false-positives).

For a negative test result, the formula is:

$$p(D+|T-) = p(T-|D+) \times p(D+) / [\, p(T-|D+) \times p(D+) + p(T-|D-) \times p(D-)\,]$$

A more general formulation of Bayes' rule incorporates the likelihood ratio (LR). The LR is defined as the probability of a test result in those with disease divided by the probability of this test result in those without disease:

$$LR = p(T|D+) / p(T|D-)$$

This definition applies to tests with two results (positive/negative), but also when more detailed test results are available (e.g., a continuous laboratory measurement). Bayes' rule is simplified further when we consider odds rather than probabilities, with odds = p / (1-p) and p = odds / (1+odds):

$$Odds(D+|T) = LR \times Odds(D+)$$

In words: the odds of disease after a test result is the product of the LR of that test result and the odds of disease before testing.

To illustrate the use of Bayes' rule, we show the probability of rupture of the prosthesis in Fig. 1, starting with a probability of 6.5% *(13)*. For US, we assume a sensitivity of 70% and a specificity of 92% *(15)*, leading to an LR of 70%/(1–92%) = 8.75 for a positive test result and an LR of (1–70%)/92% = 0.33 for a negative test result. If US is indicative for rupture, the probability is increased to 38%. The underlying calculation is based on multiplication of the LR and the prior odds. The LR = 8.75, the prior odds is p/(1-p) = 6.5%/93.5% = 0.0695. The resulting posterior odds is LR × odds = 8.75 × 0.0695 = 0.608. The probability corresponding to an odds is calculated as odds/(1+odds), in this case 0.608 / (1+0.608) = 38%.

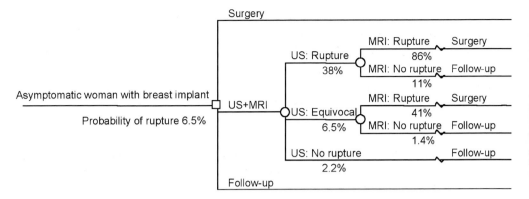

Fig. 1. Schematic decision tree for an asymptomatic woman with a silicone breast implant who is concerned about rupture of the prosthesis. Rupture is assumed to be present in 6.5% of asymptomatic women. The probability is changed according to ultrasound (US) and magnetic resonance imaging (MRI) test results, e.g., to 2.2% after a negative US.

in asymptomatic patients). A "test" strategy is then designed, which falls between the other two: a patient is treated if the test is positive and not if the test is negative.

For the evaluation of breast prostheses, two tests may be considered: ultrasonography (US) and magnetic resonance imaging (MRI) *(11)*. Both are well-accepted tests and one might wonder about how to combine these tests. In principle, many permutations of the tests are possible: one can choose to use either test, or both. A more complicated strategy is to let the use of MRI depend on the US results (e.g., no further testing when US is negative). US may not always give a positive or negative result so the decision tree must accommodate the possibility of an equivocal result.

The proposed structure of diagnosing rupture in asymptomatic women with silicone breast implants is shown in Fig. 1. The tree is read from left to right and starts with a decision node (a square denoting the point where a decision is made), while chance nodes are denoted with circles for events that are determined by chance. At the top, the "direct treatment" strategy is shown (i.e., surgery without testing). At the bottom, the "no test, no treatment" strategy is shown (i.e., follow-up without further testing). The proposed testing strategy starts with US because this is the cheaper test, with MRI if US shows rupture or equivocal results.

Probabilities

After defining a structure for the tree, we need to fill in the relevant probabilities. The first issue is the likelihood of rupture in asymptomatic patients. A literature review *(11)* identified studies that estimated the prevalence of rupture in asymptomatic patients at 8% *(12)* and 6.5% *(13)*.

Next we need to know the diagnostic quality of the tests. Commonly used test characteristics are sensitivity and specificity. Sensitivity is the probability of a positive test result in patients with a condition. In this example, sensitivity is the probability that a test shows rupture in a woman with a ruptured prosthesis. Such a test result may be considered a true-positive. Specificity is the probability of a negative test result in patients without a condition. In this example, specificity is the probability that a test shows no rupture in a woman with an intact prosthesis, a true-negative. False-positive findings are positive test results in women with intact prostheses, and false-negatives

1. Definition of the clinical problem. This should include a description of the type of patient for whom the decision is analyzed, possible diagnostic and therapeutic actions, and possible short- and long-term outcomes. The clinical problem should then be structured as a decision tree, which needs subsequently to be quantified. A decision tree typically includes decision nodes and chance nodes *(8)*.
2. Quantification of the decision tree. This relates to two aspects: probabilities and utilities (or relative value judgments). Probabilities may relate to tests (e.g., a positive result when the disease is present, or sensitivity), therapies (e.g., the risk of short-term complications, the probability of long-term success), and prognosis (e.g., outcome without treatment). Utilities are used to indicate the relative desirability of outcomes (e.g., disability or malfunctioning on a scale that ranges from dead (valued as 0) to well (valued as 1)). Utility estimates of health states are frequently based on specialized assessments (e.g., "standard gamble" or "time trade-off") or quality-of-life questionnaires (e.g., SF-36, or EUROQOL) *(8–10)*.
3. Calculation of the preferred clinical strategy. This can easily be achieved with modern computer programs, which "fold back" utilities. The expected utility is calculated for each strategy, which means that the utilities in each possible arm of the tree are weighted by their probabilities. The preferred action has the highest expected utility.
4. Evaluation of uncertainty with a sensitivity analysis. Considerable uncertainty may be present in the preferred clinical strategy because of modeling assumptions and limited available data. The term *sensitivity analysis* refers to a technique to estimate the influence of changes in assumptions concerning the preferred strategy. In a sensitivity analysis, the value of one or more parameters in a model is varied to study the effect on the outcome. If the variation has small effects, the uncertainty in the value of the parameter is predicted to be relatively unimportant. If such variation causes large changes in the results from the analysis, the parameter is predicted to be important and needs to be considered particularly carefully.
5. Presentation in a clinically useful way. The decision tree may need simplification for presentation, and qualitative and quantitative results should be presented such that they are easily understandable.

Steps 1 to 5 need not be followed in strict order. For example, quantification of a highly realistic, detailed tree may be problematic (step 2), leading to a revision with simplifying assumptions (step 1). Also, sensitivity analysis may reveal a large influence of variation in some parameters (step 4) which hence need further study (step 2).

EXAMPLE 1: BREAST IMPLANTS

Here we discuss considerations concerning a typical decision tree dealing with diagnosis of rupture in breast implants.

Structuring the Decision Tree

The diagnosis of rupture of a silicone breast prosthesis cannot be made with certainty based on clinical presentation and symptoms, and available diagnostic tests are not perfect. This uncertainty makes a decision analysis relevant. It is customary in diagnostic problems to consider two reference strategies for comparison: a "no test" strategy, which would imply no testing or treatment, and a "direct treatment" strategy, which would imply surgical removal of the implant without further testing. The "no test" strategy might be considered if the risk of rupture is considered low, whereas direct treatment would be considered when rupture is very likely (which is not the case

EXAMPLE 1: A DECISION TREE FOR A DIAGNOSTIC PROBLEM

Women with silicone breast implants may be worried about the integrity of their implant because leakage of silicone has been associated with connective tissue diseases in the lay press. Note that epidemiological evidence suggests no association or at most a weak one *(1–3)*. A plastic surgeon may be confronted with an asymptomatic woman asking about the odds that her implants are ruptured. The surgeon may reply that the likelihood of rupture of an implant is generally low in asymptomatic women but that diagnostic tests are available to exclude or confirm rupture. What diagnostic strategy should the surgeon advise, and what is the probability of rupture before and after doing a test? How much improvement is obtained with respect to the selection of women for surgery when one or more tests are performed? These questions are addressed with a decision tree.

EXAMPLE 2: A STATE-TRANSITION MODEL FOR A PROGNOSTIC PROBLEM

In the early 1980s, mechanical valves of the Björk-Shiley Convexo-Concave type were implanted in about 80,000 patients. This valve was withdrawn from the market after repeated reports of acute mechanical failure that were caused by fracture of the outer strut. By the year 2000, more than 30,000 patients were still alive with this type of valve. Should prophylactic explantation be considered in all such individuals or only in subgroups? How can these subgroups be identified? What is the risk that patients currently alive with this valve will suffer from a fracture of the valve in their remaining lifetime? These questions are addressed with a state-transition model.

EXAMPLE 3: A MODEL FOR A FOLLOW-UP SCHEME

After implantation of a device, some form of follow-up is generally undertaken. Examples of follow-up schemes from various institutions are presented in this book for a variety of implants. A follow-up scheme may be seen as a screening policy that aims to detect or prevent problems earlier than without active follow-up. Various issues may come up, for example, what type of tests to include, and at what intervals follow-up visits have to take place. On the one hand, problems with a device should be detected in a timely fashion whereas, on the other hand, follow-up should not be too burdensome for patients, physicians, and the health care system. Some of the relevant issues may be treated with complex state-transition models.

DECISION TREES

Decision trees are the central element in decision analysis, a multidisciplinary science that emerged from operations research and game theory in the 1950s. Applications in medicine were limited until the 1970s *(4,5)*, but decision analysis has since been applied in a variety of medical problems *(6)*. Decision analysis enables a systematic and explicit comparison of clinical strategies with respect to their benefits for the patient.

A decision tree represents the sequence of chance events and decisions over time. Its construction entails a number of steps *(7)*.

10
Computer Modeling

Ewout W. Steyerberg

INTRODUCTION

With the increase in computational power, computer models can be constructed that address complex problems related to the clinical management of the bionic patient. Models may be used for a variety of assessments related to the occurrence of short- and long-term complications, expected survival, choice of follow-up scheme, and more. In this chapter, computer modeling is discussed with a focus on its potential role for the clinical management of patients with prosthetic devices.

Models generally become relevant when uncertainty exists about what decision is optimal and when empirical studies in patients are not easily possible. Models may also be extensions of empirical studies (e.g., to enable the estimation of long-term outcome after the end of follow-up in a study). Advantages of a model derive from the fact that it represents a systematic and internally consistent approach to a problem. The process of explicit formulation of a problem may be valuable in itself to enhance understanding of the problem. Furthermore, parameter values can easily be changed in a model to study their effects on the expected outcomes for a patient or group of patients that requires much less effort (once a model is developed) than setting up a new empirical study. Disadvantages are related to the fact that a model is only an approximation of reality, and various simplifications have to be made. Also, even with simplifications, quantitative estimates for the model may be hard to obtain.

We might define a model as an explicit and quantified version of a theory about reality. The fact that assumptions have to be made in the development of a model is inherent to the process. Assumptions can be assessed explicitly in a model and hence be discussed critically. Without a model, the same assumptions may underlie a proposed solution for a problem, but the assumptions are implicit and less open to discussion.

Three types of questions are discussed in this chapter, relating to diagnosis, prognosis, and follow-up schedules. Each question is illustrated with an example and a type of model: a decision tree, a state-transition model, and a more dynamic state-transition model. Pros and cons are discussed as well as general characteristics of computer modeling, but no attempt is made to be comprehensive.

From: *The Bionic Human: Health Promotion for People With Implanted Prosthetic Devices*
Edited by: F. E. Johnson and K. S. Virgo © Humana Press Inc., Totowa, N

37. Wooster PL, Louch RE, Krajden S. Intraoperative bacterial contamination of vascular grafts. Can J Surg 1985;28:407–409.
38. Kaiser AB, Clayson KR, Mulherin JL Jr, et sl. Antibiotic prophylaxis in vascular surgery. Ann Surg 1978;188:283–289
39. Vicaretti M, Hawthorne W, Ao PY, Fletcher JP. Does in situ replacement of a staphylococcal infected vascular graft with a rifampicin impregnated gelatin sealed Dacron graft reduce the incidence of subsequent infection? Int Angiol 2000;19:158–65.
40. Vandone PL, Tolva V, Trimarchi S, Rampaldi V, Giuffrida GF, Bortolani EM. Infection prophylaxis in cardiovascular surgery. Minerva Chir 1998;53:397–403.
41. Burkhardt BR, Fried M, Schnur PL, Tofield JJ. Capsules, infection, and intraluminal antibiotics. Plast Reconstr Surg 1981;68:43–49.
42. Virden C, Dobke MK, Stein P, Parsons CL, Frank DH. Subclinical infection of the silicone breast implant surface as a possible cause of capsular contracture. Aesth Plast Surg 1992; 16:172–179.
43. Petit F, Maladry D, Werther JR, Mitz V. Late infection of breast implant, complication of colonic perforation. Review of the literature. Role of preventive treatment. Ann Chir Plast Esthet 1998;43:559–562.
44. Baker KA. Antibiotic prophylaxis for selected implants and devices. J Calif Dent Assoc 2000;28:620–626.
45. Brand KG. Infection of mammary prostheses: a survey and the question of prevention: Ann Plast Surg 1993;30:289–295.
46. Brown SL, Hefflin B, Woo EK, Parmentier CM. Infections related to breast implants reported to the Food and Drug Administration, 1977–1997. J Long Term Eff Med Implants 2001;11(1–2):1–12.
47. Adams WP Jr, Conner WC, Barton FE Jr, Rohrich RJ. Optimising breast pocket irrigation: an in vitro study and clinical implications. Plast Reconstr Surg 2000;105:334–338.
48. Burkhardt BR, Demas CP. The effect of siltex texturing and povidone–iodine irrigation on capsular contracture around saline inflatable breast implants. Plast Reconstr Surg 1992; 93:123–128.
49. Darouiche RO, Meade R, Mansouri MD, Netscher DT. In vivo efficacy of antimicrobe-impregnated saline-field silicone implants. Plast Reconstr Surg 2002;109:1352–1357.
50. Hortskotte D, Piper C, Niehues R, Wiemer M, Schultheiss HP. Late prosthetic valve endocarditis. Eur Heart J 1995;16 Suppl B:39–47.
51. Hyde JA, Darouiche RO, Costerton JW. Strategies for prophylaxis against prosthetic valve endocarditis: a review article. J Heart Valve Dis 1998;7:316–326.
52. Tozzi P, Al-Darweesh A, Vogt P, Stumpe F. Silver-coated prosthetic heart valve: a double-bladed weapon. Eur J Cardiothorac Surg 2002;19:729–731.
53. Oxford Textbook of Medicine, Online Vol 2: Prosthetic Valve Endocarditis.
54. British Society for Antimicrobial Chemotherapy. Recommendations for endocarditis prophylaxis in adults.
55. Babighian G. Problems in cochlear implant surgery: Adv Otorhinolaryngol 1993;48:65–69.
56. Donnelly MJ, Pyman BC, Clark GM. Chronic middle ear disease and cochlear implantation. Ann Otol Rhinol Laryngol Suppl 1995;166:406–408.
57. Jackler RK, O'Donoghue GM, Schindler RA. Cochlear implantation: strategies to protect the implanted cochlea from middle ear infection. Ann Otol Rhinol Laryngol 1986; 95:66–70.
58. Robinson PJ, Chopra S. Antibiotic prophylaxis in cochlear implantation: current practice. J Laryngol Otol Suppl 1989;18:20–21.

17. Thomalla JV, Thompson ST, Rowland RG, Mulcahy JJ. Infectious complications of penile prosthetic implants. J Urol 1987;138:65–67.
18. Maffezzini M, Capone M, Ciampalini S, De Stefani S, Simonato A, Carmignani G. Antibiotic prophylaxis in prosthetic penile surgery: critical assessment of results in 75 consecutive patients. Int J Impot Res 1996;8:87–89.
19. Culley C, Carson III. Management of penile prosthesis infection. Prob Urol 1993;7: 368–380.
20. Toktas G, Uysal Z, Erkan I, Remzi D. Fournier's gangrene: report of 6 cases. Mikrobyiol Bul 1988;22:72–75.
21. Gross AJ, Sauerwein DH, Kutzenberger J, Ringert RH. Penile prostheses in paraplegic men. Br J Urol 1996;78:262–264.
22. Atkins BL, Athanasou N, Deeks JJ. Prospective evaluation of criteria for microbiological diagnosis of prosthetic joint infection at revision arthroplasty. J Clin Microbiol 1998;36: 2932–2939.
23. Glenny AM, Song F. Antimicrobial prophylaxis in total hip replacement: a systematic review: Health Technol Assess 1999;3:1–57.
24. Borderon E, Sillion D, Delplace J, Klifa G. Treatment of bone infections with gentamicin-containing acrylic cement beads. Prognostic value of the study of the suction drainage fluid. Pathol Biol (Paris) 1984;32:455–458.
25. Armstrong MS, Spencer RF, Cunningham JL, Gheduzzi S, Miles AW, Learmonth ID. Mechanical characteristics of antibiotic-laden bone cement. Acta Orthop Scand 2002;73:688–690.
26. Cerretani D, Giorgi G, Fornara P, et al. The in vitro elution characteristics of vancomycin combined with imipenem-cilastatin in acrylic bone-cements: a pharmacokinetic study. J Arthroplasty. 2002;17:619–626
27. Newman GR, Hobot JA. Resins for combined light and electron microscopy: a half century of development. Histochem J 1999;31:495–505.
28. Liem MSL, Van der Graaf Y, Van Steensel CJ, et al. Comparison of conventional anterior surgery and laparoscopic surgery for inguinal-hernia repair. NEJM 1997;336:1541–1547.
29. Wellwood J, Sculpher MJ, Stoker D, Nicholls GJ, Geddes C, Whitehead A, Singh R, Spiegelhalter D. Randomised controlled trial of laparoscopic vs open mesh repair for inguinal hernia: outcome and cost. BMJ 1998;317:103–110.
30. Barreca M, Stipa F, Cardi E, Bianchini L, Lucandri G, Randoni B. Antibiotic prophylaxis in the surgical treatment of inguinal hernia: need or habit? Minerva Chir 2000;55: 599–605.
31. Taylor SG, O'Dwyer PJ. Chronic groin sepsis following tension-free inguinal hernioplasty. Br J Surg 1999;86:562–565.
32. Gilbert AI, Felton LL. Infection in inguinal hernia repair considering biomaterials and antibiotics. Surg Gynecol Obstet 1993;177:126–130.
33. Taylor EW, Byrne DJ, Leaper DJ, Karran SJ, Browne MK, Mitchell KJ. Antibiotic prophylaxis and open groin hernia repair. World J Surg 1997;21:811–814; discussion: 814–815.
34. Yerdel MA, Akin EB, Dolalan S, et al. Effect of single-dose prophylactic ampicillin and sulbactam on wound infection after tension-free inguinal hernia repair with polypropylene mesh: the randomised, double-blind, prospective trial. Ann Surg 2001;233:26–33.
35. Javier Sanchez-Manuel F, Luis Seco-Gil J, Lozano-Garcia J. Antibiotic prophylaxis and hernia repair. Systematic quantitative review results. Enferm Infecc Microbiol Clin 2001; 19:107–113.
36. Schwetling R, Barlehner E. Is there an indication for general perioperative antibiotic prophylaxis in laparoscopic plastic hernia repair with implantation of alloplastic tissue? Zentralbl Chir 1998;123:193–195.

a significant effect in reducing prosthetic infections, but there is a distinct lack of proof. Large multicenter trials on the efficacy of specific operative measures or antimicrobial prophylaxis have the potential to rationalize and improve surgical practice.

There is little data on the general topic of how to care for patients with most types of prosthetic implants. Most surgeons rely on tradition, anectodal experiences, and other forms of low quality evidence. The authors are not aware of any guideline at a regional, national, or international level. Surprisingly, even the manufacturers have not come up with clear and homogeneous guidelines. Theoretically, there is a possibility of infection in a prosthesis for as long as the host lives. However, lifelong follow-up is costly and generally has a very low yield. There is a need for studies in this area to determine the optimum intensity of follow-up. Increased awareness on the part of primary care services, bolstered with adequate training to diagnose infections at an early stage, and a quick and easily accessible referral system to the relevant surgical specialty, might provide the most viable answer to this problem.

REFERENCES

1. Dougherty, SH. Infections in the bionic man. Pharmacotherapy 1988;8:7S–10S.
2. Sugarman B, Young EJ. Infections associated with prosthetic devices: magnitude of the problem. Inf Dis Clin North Am 1989;3:187–198.
3. Gristina AG, Costerton JW, Leake E., et al. Bacterial colonization of biomaterials: clinical and laboratory studies. Orthop Trans 1980;4:355.
4. Segreti J, Levin S. The role of prophylactic antibiotics in the prevention of prosthetic device infections. Inf Dis Clin North Am 1989;3;2:357–370.
5. Scheibel JH, Jensen I, Pedersen S. Bacterial contamination of air and surgical wounds during joint replacement operations. Comparison of two types of staffing clothing. J Hosp Infect 1991;19:167–174.
6. Lidwell OM, Lowbury EJ, Whyte W, Blowers R, Stanley SJ, Lowe D. Effect of ultra-clean air in operating rooms on deep sepsis in the joint after total hip or knee replacement: a randomized trial. Br Med J 1982;285:10–14.
7. Burke JF. The effective period of preventive antibiotic action in experimental incision and dermal lesions. Surgery 1961;50:161–168
8. Platt R, Zaleznik DF, Hopkins CC, et al. Perioperative antibiotic prophylaxis for herniorrhaphy and breast surgery. NEJM 1990;322(3):153–160.
9. Polk HC, Lopez-Mayor JF. Postoperative wound infection: a prospective study of determinant factors and prevention. Surgery 1969;66:97–103.
10. Hill C, Flamant R, Mazas F, Evrard J. Prophylactic cefazolin versus placebo in total hip replacement. Report of a multicentric double blind randomized trial. Lancet 1981;1: 795–797.
11. An YH, Friedman RJ. Prevention of sepsis in total joint arthroplasty. J Hosp Infect 1996; 33:93–108.
12. Dougherty SH, Simmons RL. Endogenous factors contributing to prosthetic device infections. Inf Dis Clin North Am 1989;3:199–209.
13. Haines SJ, Walters BC. Antibiotics Prophylaxis for cerebrospinal fluid shunts: a metanalysis. Neurosurgery 1994;34:87–92.
14. Choux M, Genitori L, Lang D, Lena G. Shunt implantation: reducing the incidence of shunt infection. J Neurosurg 1992;77:875–880.
15. O'Brien M, Parent A, Davis B. Management of ventricular shunt infections. Child's Brain 1979;5:304–309.
16. Quigley MR, Reigel DH, Kortyna R. Cerebrospinal fluid shunt infections. Report of 41 cases and a critical review of the literature. Pediatr Neurosci 1989;15:111–120.

Table 1 *(Continued)*
**Current Prophylactic Antibiotic Regimen Advocated
by the British Society for Antimicrobial Chemotherapy**

Procedure
3. Obstetric and gynecology procedures   Cover is suggested only for patients with prosthetic valves or a previous attack of endocarditis and is as for 1d, vancomycin, or teicoplanin, directed against fecal *streptococci*
4. Gastrointestinal procedures   Cover is suggested only for patients with prosthetic valves or a previous attack of endocarditis and is as for vancomycin, teicoplanin, or clindamycin, directed against fecal *streptococci*

[a] Includes patients who should be referred to hospital. iv, intravenously; im, intramuscularly. (From ref. *42*.)

mon indications for cochlear implant surgery, especially in the pediatric age group. These patients are particularly prone to subsequent infection of their implants *(50)*. Thus, careful patient selection, preoperative optimization and timing seem to be the keys to successful cochlear implant surgery. Otitis media must be eradicated. Most authors are of the opinion that conservative surgery is probably inadequate, and radical mastoidectomy, with obliteration of the middle ear cleft with blind pit closure of the middle ear canal, appears to provide the best conditions for subsequent implantation. This is normally carried out after a period of 3 to 6 months *(56)*.

Other local preoperative measures include eradication of associated nose or scalp infection, clearing the external ear (hair clipping, cerumen removal, sterilization of the canal), pre- and postoperative antiseptic shampoos, and so on *(55,56)*.

Some promising intraoperative measures are being tried to minimize the migration of pathogens from the middle to the inner ear. The most effective such measure attempts to re-establish a separation of the cochlea from the middle ear by developing a seal around the implant at the level of the round window, using a cuff of bioactive ceramic material. Initial animal experiments have found this to be effective in preventing or minimizing the incidence of implant infection *(57)*. Further studies are awaited.

Most surgeons use perioperative antibiotic prophylaxis during cochlear implant surgery, although some studies have, paradoxically, found higher rates of implant removal in patients treated with antibiotics *(58)*. Most cases of infection appear to be associated with technical problems in the design or the positioning of the device *(58)*. Also, although few surgeons prefer to cover their patients with antibiotics during minor ear, nose, and throat procedures, there is no evidence to support this practice. As with most other cases of implant surgery, there is a need for prospective randomized controlled trials to establish the efficacy of perioperative antibiotic prophylaxis in cochlear implant surgery.

COMMENTS

There are few evidence-based guidelines concerning prevention of infection in patients with an implanted device. Many pre-, per-, and postoperative measures probably have

Table 1
Current Prophylactic Antibiotic Regimen Advocated by the British Society for Antimicrobial Chemotherapy

Procedure

1. Dental surgery (includes dental extractions, scaling, or periodontal surgery; surgery or instrumentation of the upper respiratory tract).

Type of anesthesia
    Local or no anesthesia
      a. For patients not allergic to penicillin and not prescribed penicillin more than once in the previous month
      b. For patients allergic to penicillin

Dosages
    Amoxycillin 3 g as a single dose 1 hour before the procedure
    Clindamycin 600 mg as a single oral dose 1 hour before the procedure
    General anesthesia
      c. For patients not allergic to penicillin and not given penicillin more than once in the previous month

Amoxycillin 1 g iv or 1 g im in 2.5 mL of 1% lignocaine hydrochloride just before induction plus 0.5 g amoxycillin by mouth 6 hours later

Or

Amoxycillin 3-g oral dose 4 hours before anaesthesia followed by a further 3-g dose by mouth as soon as possible after the procedure

Or

Amoxycillin 3 g together with probenecid 1 g orally 4 hours before the procedure

Special risk patients[a]
    Patients with prosthetic valves who are to have a general anaesthetic

Or

Patients who are to have a general anaesthetic and who are allergic to penicillin or have had a penicillin more than once in the previous month

Or

Patients who have had a previous attack of endocarditis
      d. For those patients not allergic to penicillin and who have not had penicillin more than once in the previous month
      e. For patients allergic to penicillin or who have had penicillin more than once in the previous month

Amoxycillin 1 g iv or 1 g im in 2.5 mL of 1% lignocaine hydrochloride plus 120 mg gentamicin iv or im at the time of induction, then 0.5 g amoxycillin orally 6 hours later

Vancomycin 1 g by slow iv infusion over at least 100 minutes followed by gentamicin 120 mg iv at the time of induction or 15 minutes before the surgical procedure

Or

    Teicoplanin 400 mg iv plus gentamicin 120 mg iv at the time of induction or 15 minutes before the surgical procedure

Or

    Clindamycin 300 mg iv in 50 mL diluent over 10 minutes at the time of induction or 15 minutes before the surgical procedure, followed by 150 mg orally or iv in 25 mL diluent over 10 minutes 6 hours later

2. Genitourinary surgery or instrumentation
    As for 1d or 1e, directed against fecal *streptococci*

(continued)

cated in the development of infection *(51)*. Patient optimization and treatment of remote infections are also crucial. Infection rates in patients with prosthetic valves have been found to be higher in the presence of native valve endocarditis or a history of native valve endocarditis, and such infections should be eradicated before valve replacement surgery is undertaken, if possible. Most surgeons use broad-spectrum antibiotic for perioperative prophylaxis. This is commonly a second-generation cephalosporin, although there are no prospective randomized controlled trials demonstrating a clear benefit *(51)*.

Interestingly, some authors have explored the effectiveness of local antimicrobial measures to prevent prosthetic valve endocarditis. Most commonly, this has been in the form of either a silver thread that has been used to suture the tissue cuff along the perimeter of the valve or a silver-coated sewing cuff *(52)*. A prospective study in Russia using silver-thread sutures for cardiac valve replacement found no early or late infection over a follow-up period of 2 years. However, subsequent case reports in which similar implants were used have described acute cardiac failure owing to partial detachment of the prosthetic valve. This was found to be a result of chronic inflammation and multiple erosions in the myocardial tissue, as a toxic reaction to silver *(52)*. Such silver-impregnated valves were subsequently withdrawn from the market in January 2000. Despite this problem, the idea of using a prosthetic cardiac valve impregnated with an antimicrobial agent to prevent infection in the region is a promising one and needs to be explored further.

Given the clear relationship between late prosthetic valve endocarditis and transient bacteremias, and a current mortality rate of 20–30% in late prosthetic valve endocarditis despite surgical and medical treatment, most authorities recommend adequate antibiotic prophylaxis for any manipulation or instrumentation likely to give rise to a bacteremia, as well as for any infection in which a bacterial cause can be assumed or proved *(53)*. It is interesting to note that antibiotic prophylaxis is also recommended for prosthetic valve recipients having a normal vaginal delivery, an event that does not necessitate antibiotics under normal circumstances. The British Society for Antimicrobial Chemotherapy *(54)* developed guidelines summarized in Table 1.

COCHLEAR IMPLANTS

Implanting an electronic device in the totally deaf ear relies on three key technical points: (a) the introduction along the scala tympani of an electrode array and its stabilization, (b) the creation of a seat for the internal receiver/stimulator into the squama temporalis, and (c) the elevation of a wide, well-vascularized skin flap, adequate to cover without tension the internal package of the implanted device *(55)*. With all the devices, the electrode is inserted into the cochlea via an opening at or near the round window.

The unique anatomic features of the middle and inner ear make infection of a cochlear implant a potentially life-threatening complication. Infection of the middle ear can easily travel, via the implant, into the inner ear and thereby into the meninges. Meningitis in a patient with a cochlear implant generally necessitates prompt removal of the device *(55)*.

Cochlear implant surgery is unsafe in the presence of otitis media, yet complicated chronic suppurative otitis media, with resulting total deafness, is one of the most com-

ideal and suggests that use of this combination may reduce the incidence of clinical capsular contracture (47). However, some pharmaceutical companies advise against immersion of silicone implants in antiseptic solutions, especially povidone-iodine because this seems to interfere with some of the surface properties of the implant. Intraluminal impregnation of breast implants with antibiotics is another option that has been investigated. This seems particularly suited for silicone gel prostheses because the unique permeability characteristics of the silicone shell appears to permit sustained antimicrobial activity at the surface of the prosthesis. Reports using animal models and in vitro studies are available (48). Darouiche et al. found that implants impregnated with minocycline/rifampin were 12 times less likely to be colonized than nonimpregnated implants in a rabbit model of S. aureus colonization and infection of subcutaneously placed implants. The results indicate that minocycline/rifampin-impregnated implants can significantly decrease the rate of bacterial colonization, implant-related infection, and implant-related abscess, as well as the likelihood of capsular contracture (49). Burkhardt et al. found that the use of intraluminal cephalothin and gentamicin significantly reduces the incidence of capsular contracture following both primary mammary augmentation and secondary open capsulotomy (48). These studies are promising and dictate the need for more exhaustive clinical trials if they are to be accepted as standard practice. Until such time, careful patient selection and optimization, meticulous, and atraumatic surgical technique with minimal tissue handling, maintenance of sterile technique, judicious use of appropriate systemic antibiotic prophylaxis intraoperatively, and adequate follow-up are the major available methods of preventing and detecting infection in patients with breast implants.

PROSTHETIC CARDIAC VALVES

Infection is an important cause of the morbidity and mortality associated with heart valve replacement surgery, the overall risk being 2 to 4% over the life span of the valve. Once established, it carries a mortality rate as high as 70% (4). Although some infected prostheses may be salvaged by rigorous antimicrobial therapy, in general this has extremely limited success and the majority of cases require surgical removal and replacement of the infected prostheses. Prevention of prosthetic valve infection, both early and late, is, therefore, crucial to the long-term success of prosthetic heart valves.

Prosthetic valve endocarditis is normally classified as early or late, depending on whether infection occurs within or after 60 days. It appears that early valve endocarditis is usually secondary to infection introduced at the time of surgery, although late infections occur as result of transient bacteremia owing to various procedures (50). A probable source of infection can be found in 25 to 80% of patients, the most frequent being dental procedures, urological infections and interventions, and in-dwelling catheters. The microbiological pattern also seems to be different in the two cases. Although early endocarditis is commonly associated with *staphylococci*, Gram-negative bacteria, and fungi, late infections are commonly caused by *S. epidermidis*, *S. aureus*, *viridans streptococci,* and *enterococci* (51).

At present, the only means of preventing prosthetic valve endocarditis are scrupulous asepsis and prophylactic perioperative antibiotic therapy. Maintenance of sterile technique is essential. Inadequate scrubbing and glove perforation have been impli-

1. Antibiotic therapy.
2. Operative time reduction.
3. Limited vascular surgery in the presence of concomitant gastrointestinal surgical disease.
4. Use of alloplastic vascular grafts with high biological compliance.
5. Reduction of invasive diagnostic techniques *(40)*.

BREAST IMPLANTS

Given the huge number of breast prostheses implanted worldwide, it is somewhat surprising to note that infection is actually uncommon, especially because the human breast is not sterile. *S. epidermidis* can be isolated from nipple secretions in up to 67% of women, and has been cultured from 55% of surgical mammary pockets prior to implant insertion. *S. epidermidis* and *S. aureus* are, in fact, the commonest organisms isolated from breast implant infections. Rarely, *Mycobacterium fortuitum* has been implicated as the causative organism *(41)*. Chronic, ongoing, subclinical infections with *S. epidermidis* have also been implicated in the pathogenesis of capsular contracture, although a recent review by an Independent Review Group suggests that *S. epidermidis* infection might be an aggravating rather than a causative factor of this complication *(36)*. The nature of the implants used may have a bearing on infection as well. The time interval between implantation and infection seems to be longer for silicone implants than for saline breast implants *(42)*. Late infections from a remote source have also been known to occur. Petit et al. published a case report of late implant infection 40 years after an augmentation mammoplasty following colonic perforation and faecal peritonitis. *Bacteroides fragilis* was found to be the causative organism *(43)*. Some authors have reported that a preceding event such as laryngitis or a flu-like illness can result in implant infection months or years after the initial operation *(41)*. Following a review of the incidence of infection after dental procedures, Baker concluded that dental treatment-related bacteremias are a very rare cause of metastatic infections in breast implants or other implants *(44)*. Thus, although some authors advocate Penicillin V for breast-implant patients undergoing dental procedures, routine antibiotic prophylaxis for breast implants is not given during dental procedures in most centers. Undoubtedly, in the vast majority of breast-implant infections, bacterial inoculation occurs at the time of surgery. These organisms can thrive for an indefinite period of time in the periprosthetic biofilm and produce infections at an opportune moment. Presence of a peri-implant hematoma or seroma obviously encourages this *(45)*.

Patient factors also need to be considered in the prevention of breast-implant infection. Preoperative chemotherapy often makes these patients more susceptible to implant infection, and allowing adequate recovery time before such operations are undertaken is probably a wise policy. The issue of whether postmastectomy breast reconstructions are more prone to get infected than elective breast augmentation procedures is a controversial one. Although some researchers have found an increased incidence in breast reconstructions, the difference in incidence is not statistically significant *(46)*.

Various intraoperative measures have been practiced to reduce breast-implant infection, the most important of which seems to be the irrigation of the breast pocket with an antiseptic solution. In a study employing povidone-iodine and two double antibiotic solutions, Adams et al. concluded that 10% povidone-iodine in gentamicin/cefazolin is

abscess following laparoscopic hernia repair is probably lower than after open repair, irrespective of whether antibiotics are used or not. In fact, these studies have found no added benefit of prophylactic antibiotics in preventing wound infections in laparoscopic hernia repairs *(28,29,36)*. There have been no reports to the contrary. However, further studies are needed to resolve this issue satisfactorily.

In the final analysis, it must be left to the surgeon's discretion whether or not to use antibiotics prophylaxis for hernia repairs. Surgeons who carry out a large number of these procedures are best advised to audit their own work and formulate their own protocols based on such results as well as local bacterial flora and antibiotic resistance.

VASCULAR GRAFTS

Vascular reconstruction using prosthetic grafts is one of the most common vascular surgical operations. The very nature of the underlying disease makes vascular graft implantation particularly prone to infection. Grafts often have to be tunnelled through tissues that are chronically ischemic and hence liable to get infected *(3)*.

It is believed that the vast majority of vascular grafts get infected perioperatively. *Staphylococcus epidermidis* is the most common organism to infect grafts: however, mixed flora are more frequent in patients with rest pain and ulceration or gangrene. In a prospective study investigating intraoperative bacterial contamination of vascular grafts, Wooster et al. found that 56% of grafts became contaminated during surgery: this was lowered to 35% when the surgeon changed gloves before preclotting the graft *(37)*. Thorough skin antisepsis is also of utmost importance in preventing graft infection. Povidone-iodine skin preparation has been found to be superior to hexachlorophene-ethanol skin preparation in preventing infection *(38)*. Careful patient selection and preoperative optimization, scrupulous attention to antisepsis, and adequate surgical technique with minimal tissue injury are essential in preventing graft infection. Because these grafts are often tunnelled through ischemic tissues, it is vital that they are laid on a well vascularized muscle bed or covered with well-vascularized flaps if subsequent infection is to be prevented.

Both systemic and local antibiotic prophylaxis have been used to prevent vascular graft infections. Animal experiments from Australia have shown a significant reduction in the incidence of graft infection when grafts were soaked in a rifampicin solution and the animals were given prophylactic antibiotics at the same time *(39)*. This combination seems to produce better results than either local antibiotics or systemic antibiotics used separately. Another animal experiment found that replacement of a vascular graft infected with *S. epidermidis* with a rifampicin-soaked Gelsoft graft is effective in reducing the rate of subsequent *S. epidermidis* infection *(33)*. However, there are no published reports of any clinical trials using antibiotic-impregnated vascular grafts.

All large studies on this subject indicate that perioperative antibiotic prophylaxis brings about a significant reduction in the incidence of vascular graft infection. Vandone et al., in a retrospective study of 2950 patients, showed a significant reduction of vascular graft infection over a 12-year period using second- and third-generation cephalosporins *(40)*. Aortobifemoral bypass grafts appear to be most prone to infection. He also suggested the following measures as gold standards in the reduction of cardiovascular surgical infections, and it seems logical to conclude this section with a reminder of these:

MESH REPAIRS OF HERNIAS

Lichtenstein's tension-free hernioplasty using a polypropylene mesh is the most common form of groin hernia repair. Although laparoscopic hernia repair using an onlay polypropylene mesh is gaining widespread acceptance, open repair still constitutes the bulk of hernia surgery. Polypropylene mesh is also used as an onlay for large incisional hernias with wide defects, using the same principles of tension-free repair *(28)*.

Given the sheer numbers of hernia repairs carried out worldwide, it is not surprising that a large volume of published literature is available on this subject. Most studies are concerned with the use and efficacy of antibiotic prophylaxis in hernia repairs. However, it must be stressed that all other factors that are important in any prosthesis surgery should also be carefully considered when carrying out these kinds of surgery. It is essential to ensure that any underlying medical conditions are treated or optimized before hernia surgery. Control of local and remote infections is also important. In patients who have features of prostatism, or a history of urinary retention and catheterization, it is vital to culture urine and treat any urinary tract infection before hernia surgery is undertaken. Hospital stay should be at a minimum and, as far as possible, these procedures should be carried out on a day-case basis. Operative technique is also important, and maintenance of a sterile field, adequate haemostasis and minimal soft tissue injury is of utmost importance *(28,29)*.

Surprisingly, there does not seem to be a definite consensus on whether prophylactic antibiotics should be used for mesh hernioplasties, despite the large number of studies on the subject *(30–32)*. A multicenter prospective study of 2493 inguinal hernial repairs, both primary and recurrent, in Miami, Florida, found an overall infection rate of 1%, which was unaffected by the use of antibiotics. Interestingly, more than 70% of wound infections occurred in patients 60 years of age or older. The authors concluded that expense of routine prophylactic antibiotics treatment in inguinal hernia operations could not be reconciled by any benefits obtained *(32)*. A similar study in Scotland produced similar results *(33)*. However, a double-blind, prospective, randomized trial comparing a single dose of intravenous prophylactic ampicillin and sulbactam with placebo during open prosthetic inguinal hernia repairs in Turkey found an infection rate of 0.7% in the antibiotic group and 9% in the placebo group, a significant difference *(34)*. Deep infections and wound infection-related re-admissions were also reduced by the use of antibiotics. The study concluded that proponents of mesh repairs may therefore be advised to use prophylactic single-dose intravenous antibiotic coverage during surgery *(34)*. A meta-analysis of eight studies found that antibiotic prophylaxis in hernia repair, whether prosthetic material was used or not, diminished the rate of infection by 42, 61, and 48% in herniorrhaphies, hernioplasties, and the two combined, respectively *(35)*. It also concludes that antibiotic prophylaxis in hernia repair is useful in preventing wound infection. The study does not endorse indiscriminate administration; rather, the decision should be based on the local rate of wound infection and on each patient's risk factors *(35)*. Some reports suggest that the incidence of wound infection and groin

mechanism of action, and dose-related bacterial killing effect than the systemic regimen, it is unwise to conclude that local antibiotic therapy is superior to systemic antibiotic prophylaxis. Most surgeons prefer systemic antibiotic therapy to gentamicin-impregnated cement.

In 1999, Glenny and Song published their systematic review on antimicrobial prophylaxis in total hip replacement *(22)*, carried out as part of the NHS Research and Development Health Technology Assessment Programme. This was a detailed and critically evaluated review of 25 randomized controlled trials carried out over a period of 20 years (1977–1997). The study proved, beyond any reasonable doubt, that administration of systemic antibiotics can in fact reduce infection rates in joint replacement surgery to an extent which is statistically significant. In fact, they stressed that given the disastrous effect of an infection of a joint prosthesis, causing increased patient morbidity and increased costs, antimicrobial prophylaxis is necessary for joint replacement surgery. Cephalosporins (first- and second-generation) are the most commonly used antibiotics. They cover a broad spectrum of bacteria and are relatively nontoxic. Third-generation cephalosporins have similar antimicrobial activity, but are generally more expensive and hence not as cost-effective. The chosen agent should penetrate the tissues and, if possible, the postoperative hematoma that is commonly present. Ultimately, however, the exact type of antibiotic to be used will be determined by the spectrum of bacterial flora in a particular center, and their antibiotic resistance profiles. The use of combination antibiotic regimens and the use of new, more expensive, broad spectrum agents is to be discouraged, unless justified by local conditions or unless they are part of a study protocol. The first dose of the antibiotic should be administered within 1 hour prior to anesthesia. There is no evidence to suggest that administering prophylactic antibiotics for more than 1 day postoperatively reduces the number of infections following total hip replacement surgery. Extending the duration of the regimen for longer than 24 hours may not only be wasteful but is potentially hazardous in terms of toxicity and the increased risk of developing bacterial resistance. Using these principles of antibiotic prophylaxis it is possible to keep the overall wound infection rate for both total hip and knee replacements to less than 2.5% *(23)*.

Directional or laminar flow may be vertical or horizontal. Here, in addition to normal turbulent airflow through the operating room, which is necessary to maintain humidity, temperature, and air circulation, an increased rate of air change is employed to reduce the number of organisms near the patient. Air is pumped into the room through filters, passed out of vents in the periphery of the operating room, and does not return into the operating suite. This helps reduce contamination of operating room air and also reduces eddies that pick up bacteria from the floor. Most operating rooms have 20 to 40 air changes per hour, but this rate may increase to 400 per hour in the vertical laminar flow system of a Charnley tent.

Some studies indicate that the use of these types of special hypersterile surgical suites with laminar airflow may lower the infection rate even without antibiotic prophylaxis. From other studies there is evidence, although not compelling, that the benefits of prophylactic antibiotics are additive to those of filtered air surgical suites. In their review, Glenny and Song found that, with laminar airflow and the use of ultraclean operating rooms, it is possible to reduce the number of total joint infections from 1.5% (conventional operating rooms) to 0.6% *(23)*.

JOINT PROSTHESES

Total hip replacements and total knee replacements have become some of the most common and successful operations ever introduced. Total hip replacements has been practised in the United Kingdom for more than 25 years. More than 42,000 total hip replacements were performed in the United Kingdom alone during the period from 1994 to 1995. The majority of patients who undergo these procedures are frail and elderly, often with underlying serious medical conditions and deficient immune systems, and thereby prone to infections, both local and remote. It is not surprising, then, that prosthesis infection is a common and devastating problem in this group, resulting in significant morbidity in terms of pain, disability, prolonged hospitalization, and so on, necessitating revision arthroplasties. Infection is, in fact, the second most common indication for revision prosthetic surgery after mechanical failure caused by loosening, breaking, or wearing of the prosthesis. The National Health Service (NHS) spends in excess of £170 million per year in treating these infections. The costs are going to keep rising as life expectancy increases and the incidence of such age-related conditions as osteoarthritis increases with it. The magnitude of the problem is truly enormous *(21)*. *S. aureus* and *S. epidermidis* are the most frequently isolated pathogens in most series. Other commonly identified bacteria include *aerobic streptococci* and *anaerobic cocci (22)*.

All general measures that are essential for any prosthetic surgery should be followed rigorously in all patients and are not repeated here. Two issues that are central to the whole argument of prevention of infection in joint prosthesis are highlighted: the use of prophylactic antibiotics, both systemic and local, and the use of ultraclean operating theaters with laminar airflow *(23)*.

One approach to preventing postoperative infections following total hip arthroplasty is the use of antibiotic-impregnated bone cement. Antibiotics mixed with bone cement used in total joint replacement leach from the hardened cement by diffusion. This leaching varies among the types of bone cement and antibiotics available. Gentamicin-impregnated cement allows a high rate of initial release as well as protracted release of the antibiotic into the surrounding tissues. Gentamicin is found in wound secretions in trace amounts of up to 7 weeks after surgery *(24)*. Medullary bone contains low levels of antibiotics for weeks and, in occasional cases, years. Cortical bone has lower levels of antibiotics. β-Lactam antibiotics also have been combined with bone cement and clinically evaluated *(25,26)*. However, most of the studies are in animals. Although β-lactams are heat labile, these drugs retain substantial antimicrobial activity despite exposure to the heat generated by polymerization of the bone cement *(27)*. The theoretical difficulties in the clinical application of antibiotic-impregnated bone cement include possible loss of structural integrity of the cement, systemic toxicity of diffusing antibiotics, and hypersensitivity reactions to the sustained release of the antibiotic. However, none of these have been proven in clinical studies *(21)*.

Multicentric trials comparing infection rates in arthroplasties involving prostheses treated with gentamicin-impregnated cement or systemic cloxacillin, dicloxacillin, or a cephalosporin have shown a lower infection rate in the patients treated with the gentamicin-impregnated cement. However, the superficial wound infection rate was higher in the local therapy group when compared with the systemic antibiotic group. Additionally, because the local antibiotic (gentamicin) has a different antibiotic spectrum,

surgical procedures are combined with penile prosthesis implantation. These include simultaneous inguinal hernia repair, circumcision, penile straightening, and so on. A history of infectious osteitis pubis may also be a risk factor for prosthesis infection. Patients with paraplegia or spinal cord injury also appear to be at an increased risk of prosthesis infections, and an increased incidence of infection from 8 to 33% has been recorded by some investigators (21). And, finally, it is imperative to document sterile urine before any surgery is undertaken since there is a well-documented relationship between infection of genitourinary prostheses and urinary tract infections (18,19).

Short preoperative hospital stays are important in order to prevent the dramatic changes in skin flora that take place with hospitalization. Within 24 hours of hospital admission, normal skin flora consisting of susceptible Gram-positive organisms are replaced by nosocomial Gram-negative, anaerobic, or resistant Gram-positive organisms, all of which may be resistant to perioperative preventative antibiotics. Preoperative preparation must include elimination of possible remote infection sites that can subsequently produce contamination or hematogenous spread of infection (19).

The surgical environment is critical to the prevention of prosthesis infection. Limitation of operating room traffic, adequate sterile technique, minimal tissue devitalization, short surgical duration, and effective wound closure and management are well-known variables affecting the risk of perisurgical infections. Patient shaving and hair removal should be performed in the operating room with minimal trauma. Effective skin disinfectant programs are also critical in decreasing perioperative infections. Although the prosthesis is in the operating room environment before implantation, it should be submerged in an antibacterial agent to prevent airborne bacteria and debris from attaching to the prosthetic device. Use of laminar airflow, protective bubbles, or vented shields continues to be controversial. Bacterial inoculation from remote infections such as dental abscesses, skin ulcers, and activation of Crohn's disease with intra-abdominal fecal spillage have all been documented to produce late penile prosthesis infections. These infections may occur many years after initial implantation and it is prudent to cover such patients with antibiotics during genitourinary, dental, or other surgical manipulations (17).

Use of prophylactic perioperative antibiotics for implantation of penile prosthesis is a subject of controversy. Some studies have demonstrated a reduction of postoperative infections from 15 to less than 1% when perioperative antibiotics were administered (4,18). However, these studies do not stand up to careful statistical scrutiny, and well-designed prospective randomized controlled studies are needed to determine the optimal antibiotics and duration of therapy. Presently, most surgeons use a single preoperative dose of a cephalosporin and base subsequent antibiotic use on individual surgical judgment (19).

It is also important to diagnose penile prosthesis infections at an early stage and treat them vigorously if explantation and revision surgery is to be avoided. Local signs such as redness, tenderness, and induration tend to predominate over systemic signs of infection like pyrexia and a raised white cell count in penile prosthesis infection and, if caught at this stage, it might be possible to control the infection with intravenous antibiotics, thereby avoiding explantation (17).

important in premature babies and neonates. The patient's hair is washed the morning after surgery and again before discharge using aqueous povidone-iodine. Dressings are changed every 2 days using strictly aseptic techniques (14). Using this protocol, Choux et al. managed to reduce the rate of shunt infection from the existing 7.75% to an astonishing 0.17% (14).

There is no clearly defined role for prophylactic antibiotic medications in the prevention of shunt infections. A large number of studies has been carried out, but quite a few of them do not hold up to statistical scrutiny. Immersing the shunt material in gentamicin solution seems to be a common practice. This is supposed to reduce the electrostatic properties of the silicone tubing and reduce bacterial adherence. Meta-analysis, using pooled data from existing publications on this subject, has shown that antibiotics have a beneficial effect only where a high baseline rate of infection exists. Although the exact level of the rate where this happens is not known, it is advised that surgeons experiencing shunt infection rates above 5% are better off using a short course of antibiotic prophylaxis (13). A large variety of antibiotics, including gentamicin, cloxacillin, trimethoprim, sulphamethoxazole, rifampin, and second-generation cephalosporins, have been used (15,16). The exact group of antibiotics to be used depends on the bacterial flora and antibiotic protocol unique to a particular center.

PENILE PROSTHESIS

Although mechanical failure or malfunctioning seem to be the most common complications of penile prostheses, infection can be the most disastrous, often resulting in severe disability, loss of function, or loss of tissue so extensive that subsequent reimplantation becomes impossible (17).

What makes penile prostheses particularly prone to infection is the fact that bacteria seem to have a high propensity to adhere to the silicone elastomers used in making the most common types of these prostheses. This is most common at the time of implantation. However, bacterial seeding from a hematogenous or lymphatic spread has also been known to occur, particularly around the periprostatic space, and may account for the late infections. Some investigators are of the opinion that the periprosthetic biofilm may harbor infections for a long period of time and that these might be activated after long periods of quiescence to produce late infections. One half of penile prosthetic infections are reported to occur within 7 months of implantation, and 2.6% occur after 5 years (17–19).

S. epidermidis is most commonly implicated, accounting for 35 to 56% of infections, followed by Gram-negative enteric bacteria (23%), including P. mirabilis, P. aeruginosa, E. coli, and S. marcesen. Fungal infection with C. albicans is a rare but deadly occurrence and can result in corporal abscess and extensive tissue loss (19).

All the risk factors to be considered before any prosthetic operation obviously hold true in penile prosthetic surgery as well. These include the general medical condition of the patient, conditions that result in immunocompromise, systemic disorders like diabetes mellitus, skin infections, and so on. Fournier's gangrene of the penis and scrotum has been reported following penile prosthesis insertion in transplant patients (20). Some risk factors are, however, unique to penile prosthesis surgery. There seems to be a significantly increased rate of infection when any other

sis can occur in up to 20% of cases *(13)*. The most common organisms are the *Staphylococci*, especially coagulase-negative. Gram-negative bacilli are less common. Mortality from an infected shunt ranges from 30 to 40% and those who survive risk major neurological deficits. The age of the patient, the etiology of the hydrocephalus, and the type of shunt used correlate with the risk of shunt infection *(13,14)*.

One of the most definitive studies on shunt infection was published from Hospital des Enfants in La Timone, Marseilles, France, by Choux et al. in 1992 *(14)*. This was a clear, concise study in which definite problems were identified, a standard protocol regarding pre-, peri-, and postoperative management instituted, the results audited and impressive improvement in infection rates achieved. It was concluded that children admitted for either shunt insertion or revision should be carefully assessed from a general medical point of view. If an intercurrent infection is diagnosed, or if there is a localized skin problem, shunt insertion should be deferred until the patient's condition has improved. In patients where urgent treatment of hydrocephalus is required and the patient's general medical condition contraindicates shunt placement, a period of temporary external ventricular drainage is preferred over definitive shunt implantation until the problem has been rectified.

All patients admitted for shunt insertion or revision should undergo a hair shampoo the evening before and on the morning of the operation, using an aqueous povidone-iodine preparation. Neonates and infants should not be routinely shaved regardless of whether the operation is for implantation of a new shunt or revision of an existing shunt. In older children admitted for a shunt revision, a limited area of the head should be shaved by the surgeon in the anesthesia room immediately before the patient is transferred to the operating room. The scalp and the skin should then be prepared with aqueous povidone-iodine or chlorhexidine in neonates while the surgeon scrubs.

Some intraoperative precautions against shunt infections were also implemented by Choux and co-authors *(14)* such as carrying out shunt implantation early in the day prior to other neurosurgical procedures. Neonates, infants, and young children were operated on before older children, regardless of whether the procedure was a primary shunt insertion or a revision. The number of people in the operating room was limited to four: the surgeon and the assistant, the anesthesiologist, and a circulating nurse. To avoid possible contamination of the instrumentation and materials, there was no scrub nurse and the sterile equipment was opened at the last moment by the surgeon and the assistant. It is recommended that the shunt implantation be carried out by a neurosurgeon with considerable experience with both the selection and use of shunt materials and the operative techniques required for shunt insertion. After the packaging is opened, and before insertion, the shunt is immersed in a gentamicin bath. Formal testing of the valve for patency or verification of the opening pressure is avoided to reduce the risk of introducing infection.

During surgery, the abdominal incision is made first to minimize the length of time the cranial wound is open. The shunt valve should be carefully positioned in the subgaleal space to avoid skin damage from the shunt material. Throughout the operation, the wounds are irrigated with a dilute solution of aqueous povidone-iodine, and all shunt materials are rinsed with gentamicin prior to implantation.

Postoperatively, the nursing staff ensure that the patient's head is positioned correctly to avoid pressure on the cranial wound and shunt valve. This is particularly

Because almost all bacteria causing clinically significant deep prosthetic infections are likely to have been introduced at surgery, antibiotics started after the incision is closed are not really prophylactic and should be considered early therapy. If the length of the operation exceeds the effective duration of the antimicrobial agent, another intraoperative dose should be given to maintain adequate tissue levels *(4)*.

Although there seems to be general agreement that intraoperative antibiotic use does reduce the incidence of prosthetic infection in most cases, it is not clear that antibiotics are useful and warranted in preventing remote infections occurring late after prosthetic insertion. Late prosthetic infection as a result of the hematogenous spread of an organism from a remote site with the seeding of that organism on the prosthetic interface is a well-documented event. However, the value of antibiotic prophylaxis in such cases is controversial *(2)*. The disadvantages might in fact outweigh the benefits in terms of complications and cost. Some studies have found that the incidence of bacteremia and line infection increases if prophylactic antibiotics are continued for more than 4 days *(12)*, resulting in an exponential increase in health care costs. *Clostridium difficile* infections alone result in a mean lifetime cost of nearly $11,000 per patient *(13)*. These factors will obviously need to be considered when prescribing prophylactic antibiotics for these patients.

Nearly all implant procedures are examples of clean surgery. In other words, there is no break in sterile technique, there is no entry to contaminated body areas, and the wound is electively created. The anticipated wound infection rate should be approximately 2% or less. Traditionally, this would not merit prophylactic antibiotics as the consequences of wound infection in most instances do not outweigh the cost and potential toxicity of prophylactic antibiotics. With many permanent prostheses, however, postoperative infection is associated with such excessive morbidity and hospital cost that therapies appear to outweigh the cost and toxicity of prophylactic antibiotics in this patient subset *(4)*.

PATIENT FACTORS THAT CONTRIBUTE TO INFECTION OF PROSTHESES

Any condition or procedure that encourages transient bacteremia or dampens the host immune response predisposes to prosthesis infection. This includes a wide variety of skin conditions, minor surgical procedures such as dental extraction, invasive diagnostic procedures such as barium enema, as well as a host of systemic illnesses such as diabetes mellitus, cirrhosis, uremia, and the like *(4)*.

Even old age and obesity are adverse factors. Intravascular prosthetic devices are particularly prone to get infected from transient bacteremia and any infection in these patients should be treated aggressively. It has become standard practice to cover this group of patients with suitable antibiotics before minor surgery (e.g., dental extraction or invasive diagnostic procedures). Recently, it has been found that breast prostheses are also prone to get infected during these procedures as well and might benefit from antibiotic cover during such procedures *(4,5)*.

VENTRICULOPERITONEAL SHUNTS

Ventriculoperitoneal (VP) shunts are used in conditions where there is an impairment of cerebrospinal fluid (CSF) circulation. Bacterial contamination of the prosthe-

Prosthetic devices also promote bacterial growth by providing a sanctuary from the host defense mechanisms, and there seems to be a crucial period of time before graft healing, hence the increased susceptibility of recently implanted prostheses to infection in comparison to established implants. An implanted prosthesis gets covered by a layer of fibrin clot as early as 30 minutes after surgery, and this layer adsorbs bacteria and promotes their growth, safe from the immune mechanisms *(1)*.

Some bacteria have the ability to adhere to prostheses without the benefit of a fibrin layer. The classic example is *Staphylcoccus epidermidis*, which produces a glycocalyx layer ("slime") that helps it to adhere to the prosthetic surface and also offers protection against antibodies and, more importantly, against antibiotics. However, slime is also produced by a large number of other bacteria implicated in prosthesis infection, notably *S. aureus* and *Pseudomonas*. Finding a means for drugs to penetrate slime would presumably be useful in decreasing the rate of infections in prostheses *(3)*.

THE ROLE OF PROPHYLACTIC ANTIBIOTICS

It may be useful to distinguish between acute, often postoperative, hospital-acquired infections commonly associated with temporary prosthetic devices like urinary catheters, peripheral venous access lines, and the like, and the subacute and chronic infections associated with permanent devices that arise months to years after implantation. The former are usually caused by the patient's indigenous bacterial flora and are controlled relatively easily by removing the device. The majority of deep, late-appearing infections, on the other hand, result from low-grade pathogenic bacteria that may originate from the patient but are also present in the ambient air or on personnel involved in the surgical procedure. It is believed that these are almost always introduced at the time of surgery and contaminate surgical wounds even when the utmost care is taken to maintain a sterile field at the time of surgery *(2–4)*. Peroperative use of polypropylene coveralls *(5)* and whole-body exhaust-ventilated suits in an ultraclean system *(6)* can effectively reduce the bacterial contamination rate by 50%; unfortunately, contamination cannot be fully avoided. The administration of prophylactic antibiotics can prevent wound infection if an appropriate antimicrobial is present in the tissues in sufficient concentration during surgical exposure *(4)*. Several experimental and clinical studies undertaken during the last half century have proven the efficacy of prophylactic systemic antibiotics in preventing infection *(7–9)* but there is a shortage of such data when dealing with surgical implants *(10)*.

It is impossible and probably risky to categorically declare certain antibiotics as the gold standard for specific procedures as bacterial resistance can occur after continuous use of a particular antibiotic for any length of time. Every center will have its own unique spectrum of bacterial resistance and antibiotic protocols. However, it is important to highlight certain key features of prophylactic antibiotic use in order to make the best choice. Therapy should be instituted within 15 to 60 minutes of the initial surgical incision *(11)*. Starting sooner (6 to 24 hours before surgery) may result in the selection of resistant bacteria at the time of incision. Starting antibiotics too soon may also result in intraoperative antibiotic blood and tissue levels that are too low. Starting antibiotic therapy later (more than 2 hours) diminishes the extent of infection but not the incidence. The continued administration of antibiotics for more than 6 to 48 hours after surgery has not been proven advantageous, with the exception of intravascular devices.

European Counterpoint to Chapter 9

Soumen Ghosh and Riccardo A. Audisio

INTRODUCTION

Surgical insertion of prosthetic devices into the human body has become a common practice worldwide. Prosthetic hip and knee replacements, prosthetic cardiac valve insertion, mesh repair of hernias, insertion of breast implants, to name but a few, are carried out in enormous numbers all over the world at astronomical health care costs in terms of equipment, personnel, hospital stay, and management of complications *(1)*. Infection of prosthetic devices is one of the most common complications and it has huge implications as measured by patient morbidity, mortality, and cost. It has been estimated that the cost of managing an infected hip joint prosthesis, including removal, exceeds the cost of insertion by a factor of six *(2)*.

Tissue injury is inevitable with surgery. The combination of tissue injury and bacteria introduced at the time of surgery triggers a cascade of events. If the bacteria are cleared, the injured tissue is repaired but the presence of a foreign body affects the cellular and humoral immunity responsible for this cascade and also provides direct protection to bacteria, helping them to grow and colonize. As a consequence, a much smaller bacterial inoculum is required to produce a wound infection if an implanted prosthesis is present than in a prosthesis-free environment *(1)*.

It has been found that the extent to which such colonization is favored depends also on several properties of prostheses that influence tissue reactivity. Some of these are chemical composition (protein-based materials are more reactive than nonprotein-based materials), shape (sharp edges are more reactive than smooth contours), surface characteristics (abraded surfaces are more reactive than smooth surfaces), particle size (particulate material is more reactive than solid material), and so on *(1)*.

In the presence of a prosthesis, the normal inflammatory process is exaggerated and persistent, thereby producing more tissue damage. The phagocytic and bactericidal properties of granulocytes are decreased in the presence of a foreign body, probably because direct contact between the two results in a premature lysosomal discharge, thereby exhausting the leucocyte metabolically and influencing its phagocytic ability *(2)*.

Foreign bodies embedded in biological systems also exert adverse effects on humoral immunity, activating the complement and clotting cascades, among others. The resultant microvascular thrombosis decreases tissue perfusion, causing more tissue damage *(3)*.

From: *The Bionic Human: Health Promotion for People With Implanted Prosthetic Devices*
Edited by: F. E. Johnson and K. S. Virgo © Humana Press Inc., Totowa, NJ

81. Mohammad SF. Enhanced risk of infection with device-associated thrombi. ASAIO J 2000;46:S63–S68.
82. Dieter RS. Coronary artery stent infection. Clin Cardiol 2000;23:808–810.
83. Bouchart F, Dubar A, Bessou JP et al. Pseudomonas aeruginosa coronary stent infection. Ann Thorac Surg 1997;64:1810–1813.
84. Peyton JW, Hylemon MB, Greenfield LJ, Crute SL, Sugerman HJ, Quershi GD. Comparison of Greenfield filter and vena caval ligation for experimental septic thromboembolism. Surgery 1983;93:533–537.
85. Robinson PJ, Chopra S. Antibiotic prophylaxis in cochlear implantation: current practice. J Laryngol Otol Suppl 1989;18:20–21.
86. Clark GM, Pyman BC, Pavillard RE. A protocol for the prevention of infection in cochlear implant surgery. J Laryngol Otol 1980;94:1377–1386.
87. Luntz M, Hodges AV, Balkany T, Dolan-Ash S, Schloffman J. Otitis media in children with cochlear implants. Laryngoscope 1996;106:1403–1405.
88. Mills RP. The use of cortical bone grafts in ossiculoplasty. I: surgical techniques and hearing results. J Laryngol Otol 1993;107:686–689.
89. Nikolaou A, Bourikas Z, Maltas V, Aidonis A. Ossiculoplasty with the use of autografts and synthetic prosthetic materials: a comparison of results in 165 cases. J Laryngol Otol 1992;106:692–694.
90. Antimicrobial prophylaxis in surgery. Med Lett Drugs Ther 2001;43:93–98.

60. Braithwaite BD, Davies B, Heather BP, Earnshaw JJ. Early results of rifampicin-bonded Dacron grafts for extra-anatomic vascular reconstruction. Br J Surg 1998; 85:1378–1381.
61. D'Addato M, Curti T, Freyrie A. Prophylaxis of graft infection with rifampicin-bonded Gelseal graft: 2-year follow-up of a prospective clinical trial. Cardiovasc Surg 1996;4: 200–204.
62. Healy DA, Keyser J III, Holcomb GW III, Dean RH, Smith BM. Prophylactic closed suction drainage of femoral wounds in patients undergoing vascular reconstruction. J Vasc Surg 1989;10:166–168.
63. Hall EH, Sherman RG, Emmons WW III, Naylor GD. Antibacterial prophylaxis. Dent Clin North Am 1994;38:707–718.
64. Yogev R, Bisno AL. Infections of the central nervous system shunts. In: Waldvogel FA, Bisno AL, eds. Infections associated with indwelling medical devices (3rd ed.). Washington DC: ASM Press, 2000, pp. 231–246.
65. Little JW. Prosthetic implants: risk of infection from transient dental bacteremias. Compendium 1991;12:160–168.
66. Bayston R, Bannister C, Boston V et al. A prospective randomized controlled trial of antimicrobial prophylaxis in hydrocephalus shunt surgery. Z Kinderchir 1990;45 Suppl 1:5–7.
67. Choux M, Genitori L, Lang D, Lena G. Shunt implantation: reducing the incidence of shunt infection. J Neurosurg 1992;77:875–880.
68. Eggimann P, Waldvogel F. Pacemaker and defibrillator infections. In: Waldvogel FA, Bisno AL, eds. Infections associated with indwelling medical devices (3rd ed.). Washington DC: ASM Press, 2000, pp. 247–264.
69. Bluhm G, Jacobson B, Julander I, Levander-Lindgren M, Olin C. Antibiotic prophylaxis in pacemaker surgery—a prospective study. Scand J Thorac Cardiovasc Surg 1984;18: 227–234.
70. Bluhm G, Nordlander R, Ransjö U. Antibiotic prophylaxis in pacemaker surgery: a prospective double blind trial with systemic administration of antibiotic versus placebo at implantation of cardiac pacemakers. Pacing Clin Electrophysiol 1986; 9:720–726.
71. Stillman N, Douglass CW. Developing market for dental implants. J Am Dent Assoc 1993; 124:51–56.
72. Tanner A, Maiden MFJ, Lee K, Weber HP. Dental implant infections. Clin Inf Dis 1997;25: S213–S217.
73. Belser UC, Meyer JM. Dental implants. In: Walgvogel FA, Bisno AL, eds. Infections associated with indwelling medical devices (3rd ed.). Washington DC: ASM Press, 2000, pp. 373–393.
74. Lambert PM, Morris HF, Ochi S. The influence of 0.12% chlorhexidine digluconate rinses on the incidence of infectious complications and implant success. J Oral Maxillofac Surg 1997;55:25–30.
75. Dent CD, Olson JW, Farish SE et al. The influence of preoperative antibiotics on success of endosseous implants up to and including stage II surgery: a study of 2,641 implants. J Oral Maxillofac Surg 1997;55:19–24.
76. Reiser GM, Nevins M. The implant periapical lesion: etiology, prevention, and treatment. Compend Contin Educ Dent 1995;16:768,770,772.
77. Myles O, Thomas WJ, Daniels JT, Aronson N. Infected endovascular stents managed with medical therapy alone. Catheter Cardiovasc Interv 2000;51:471–476.
78. James E, Broadhurst P, Simpson A, Das S. Bacteremia complicating coronary artery stenting. J Hosp Inf 1998;38:154–155.
79. Hearn AT, James KV, Lohr JM, Thibodeaux LC, Roberts WH, Welling RE. Endovascular stent infection with delayed bacterial challenge. Am J Surg 1997;174:157–159.
80. Paget DS, Bukhari RH, Zayyat EJ, Lohr JM, Roberts WH, Welling RE. Infectibility of endovascular stents following antibiotic prophylaxis or after arterial wall incorporation. Am J Surg 1999;178:219–224.

37. Osher RH, Amdahl LD, Cheetam JK. Antimicrobial efficacy and aqueous humor concentration of preoperative and postoperative topical trimethoprim/polymyxin B sulfate versus tobramycin. J Cataract Refract Surg 1994;20:3–8.
38. Stewart RH, Kimbrough RL, Smith JP, deFaller JM. Use of steroid/antibiotic prophylaxis in intraocular lens implantation: a double-masked study v placebo. Ann Ophthal 1983;15: 24–28.
39. De Gevigney G, Pop C, Delahaye JP. The risk of infective endocarditis after cardiac surgical and interventional procedures. Eur Heart J 1995;16 Suppl B:7–14.
40. Yagiela JA. Prophylactic antibiotics: cardiac and prosthetic considerations. J Calif Dent Assoc 1995;23:29–40.
41. Durack DT, Lukes AS, Bright DK. New criteria for diagnosis of infective endocarditis: utilization of specific echocardiographic findings. Am J Med 1994; 96:200–209.
42. Goldmann DA, Hopkins CC, Karchmer AW et al. Cephalothin prophylaxis in cardiac valve surgery. A prospective, double blind comparison of two-day and six-day regimens. J Thorac Cardiovasc Surg 1977;73:470–479.
43. Karchmer AW. Infections of prosthetic heart valves. In: Waldvogel FA, Bisno AL, eds. Infections associated with indwelling medical devices (3rd ed.). Washington DC: ASM Press, 2000, pp. 145–172.
44. Little JW. Antibiotic prophylaxis for prevention of bacterial endocarditis and infectious major joint prostheses. Curr Opin Dent 1992;2:93–101.
45. Wahl MJ. Clinical issues in the prevention of dental-induced endocarditis and prosthetic joint infection. Pract Periodontics Aesthet Dent 1994;6:25–32.
46. Dajani AS, Taubert KA, Wilson W et al. Prevention of bacterial endocarditis. Recommendations by the American Heart Association. JAMA 1997;277:1794–1801.
47. Bengston S. Prosthetic osteomyelitis with special reference to the knee: risks, treatment, and costs. Ann Med 1993;25:523–529.
48. Steckelberg JM, Osmon DR. Prosthetic joint infections. In: Waldvogel, Bisno eds. Infections associated with indwelling medical devices (3rd ed.). Washington DC: ASM Press, 2000, pp. 173–209.
49. Maguire JH. Advances in the control of perioperative sepsis in total joint replacement. Rheum Dis Clin North Am 1988;14:519–535.
50. Friedman RJ, Friederich LV, White RL, Kays MB, Brundage DM, Graham J. Antibiotic prophylaxis and tourniquet inflation in total knee arthroplasty. Clin Orthop 1990;260:17–23.
51. Josefsson G, Kolmert L. Prophylaxis with systemic antibiotics versus gentamicin bone cement in total hip arthroplasty. A ten-year survey of 1,688 hips. Clin Orthop 1993;292: 210–214.
52. McQueen MM, Hughes SP, May P, Verity L. Cefuroxime in total joint arthroplasty. Intravenous or in bone cement. J Arthroplasty 1990;5:169–172.
53. Hasselgren PO, Ivarsson L, Risberg B, Seeman T. Effects of prophylactic antibiotics in vascular surgery. A prospective, randomized, double blind study. Ann Surg 1984; 200:86–92.
54. Yeager RA, Moneta GL, Taylor LM Jr, McConnell DB, Porter JM. Can prosthetic graft infection be avoided? If not, how do we treat it? Acta Chir Scand Suppl 1990;555:155–163.
55. Jensen LJ, Aagaard MT, Schifter S. Prophylactic vancomycin versus placebo in arterial prosthetic reconstructions. Thorac Cardiovsc Surg 1985;33:300–303.
56. Salzmann G. Perioperative infection prophylaxis in vascular surgery—a randomized prospective study. Thorac Cardiovasc Surg 1983;31:239–242.
57. Pitt HA, Postier RG, MacGowan WA et al. Prophylactic antibiotics in vascular surgery. Topical systemic, or both? Ann Surg 1980;192:356–364.
58. Walker M, Litherland HK, Murphy J, Smith JA. Comparison of prophylactic antibiotic regimens in patients undergoing vascular surgery. J Hosp Infect 1984;5 Suppl A:101–106.
59. Lalka SG, Malone JM, Fisher DF et al. Efficacy of prophylactic antibiotics in vascular surgery: an arterial wall microbiologic and pharmacokinetic perspective. J Vasc Surg 1989; 10:501–509.

15. Polk HC, Lopez-Mayor JF. Postoperative wound infection: a prospective study of determinant factors and prevention. Surgery 1969;66:97–103.
16. Segreti J, Levin S. The role of prophylactic antibiotics in the prevention of prosthetic device infections. Infect Dis Clin North Am 1989;3:357–370.
17. Schaff H, Carrel T, Steckelberg JM, Grunkemeier GL, Holubkov R. Artificial Valve Endocarditis Reduction Trial (AVERT): protocol of a multicenter randomized trial. J Heart Valve Dis 1999;8:131–139.
18. Hyde JA, Darouiche RO, Costerton JW. Strategies for prophylaxis against prosthetic valve endocarditis: a review article. J Heart Valve Dis 1998;7:316–326.
19. Willemen D, Paul J, White SH, Crook DW. Closed suction drainage following knee arthroplasty. Effectiveness and risks. Clin Orthop 1991;264:232–234.
20. Drinkwater CJ, Neil MJ. Optimal timing of wound drain removal following total joint arthroplasty. J Arthroplasty 1995;10:185–189.
21. Blomgren G. Hematogenous infection of total joint replacement: an experimental study in the rabbit. Acta Orthop Scand 1981;52:1–64.
22. Little JW. Patients with prosthetic joints: are they at risk when receiving invasive dental procedures? Spec Care Dentist 1997;17:153–160.
23. Averns HL, Kerry R. Role of prophylactic antibiotics in the prevention of late infection of prosthetic joints. Results of a questionnaire and review of the literature. Br J Rheumatol 1995;34:380–382.
24. Pennisi VR. Long-term use of polyurethane breast prostheses: a 14-year experience. Plast Reconstr Surg 1990;86:368–371.
25. Gabriel SE, Woods JE, O'Fallon WM, Beard CM, Kurland LT, Melton JL. Complications leading to surgery after breast implantation. N Engl J Med 1997 6;336:718–719.
26. Brown SL, Hefflin B, Woo EK, Parmentier CM. Infections related to breast implants reported to the Food and Drug Administration, 1977–1997. J Long Term Eff Med Implants 2001;11:1–12.
27. Ahn CY, Ko CY, Wagar EA, Wong RS, Shaw WW. Microbial evaluation: 139 implants removed from symptomatic patients. Plast Reconstr Surg 1996;98:1225–1229.
28. Adams WP, Conner WC, Barton FE, Rohrich RJ. Optimizing breast pocket irrigation: an in vitro study and clinical implications. Plast Reconstr Surg 2000; 105:334–338.
29. Kernodle DS, Kaiser AB. Postoperative infections and antimicrobial prophylaxis. In: Mandell D, Bennett JE, eds. Principles and practices of infectious diseases (5th ed., Vol 2). Philadelphia, PA: Churchill Livingston, 2000, pp. 3177–3191.
30. Hunter JG, Padilla M, Cooper-Vastola S. Late Clostridium perfringens breast implant infection after dental treatment. Ann Plast Surg 1996;36:309–312.
31. Hessen MT, Zuckerman JM, Kaye D. Infections associated with foreign bodies in the urinary tract. In: Waldvogel FA, Bisno AL, eds. Infections associated with indwelling medical devices (3rd ed.). Washington DC: ASM Press, 2000, pp. 325–344.
32. Fishman IJ, Scott FB, Selim AM. Rescue procedure: An alternative to complete removal for treatment of infected penile prosthesis. J Urol 1987;137:202A.
33. Schwartz BF, Swanzy S, Thrasher JB. A randomized prospective comparison of antibiotic tissue levels in the corpora cavernosa of patients undergoing penile prosthesis implantation using gentamicin plus cefazolin versus an oral fluoroquinolone for prophylaxis. J Urol 1996;156:991–994.
34. Licht MR, Montague DK, Angermeier KW, Lakin MM. Cultures from genitourinary prostheses at reoperation: questioning the role of Stapylococcus epidermidis in periprosthetic infection. J Urol 1995;154:387–390.
35. Martins FE, Boyd SD. Post-operative risk factors associated with artificial urinary sphincter infection-erosion. Br J Urol 1995;75:354–358.
36. Barequet IS, Baker AS, Schein OD. Endophtalmitis associated with cataract surgery. In: Waldvogel FA, Bisno AL, eds. Infections associated with indwelling medical devices (3rd ed.). Washington DC: ASM Press, 2000, pp. 294–301.

tions, infection is particularly significant because of its morbidity and potential mortality. Prevention of infection through meticulous postimplant care is widely believed to be valuable, but this has not been demonstrated by prospective, randomized trials. In general, the risk of adverse effects from antibiotics is small and is outweighed by the significance of prosthetic infections. Table 7 summarizes perioperative antibiotic prophylaxis for patients undergoing implantation of prosthetic devices.

When published guidelines are not available to help the decision-making process, consultation with the appropriate specialist is probably in the best interest of the patient who will also need to be fully informed of the risks and benefits of any approach. Large, prospective human trials are needed to identify the answers to these questions, but they are difficult and expensive to perform because of the large number of patients and the long duration of follow-up required. There are also major ethical issues because prophylaxis is already commonly used in device implantation surgery.

Although progress is being made in developing new types of prosthetic materials that will reduce bacterial adherence to the foreign material, their implementation will take time.

REFERENCES

1. Maderazo EG, Judson S, Pasternak H. Late infections of total joint prostheses. A review and recommendations for prevention. Clin Orthop 1988;229:131–142.
2. Cioffi GA, Terezhalmy GT, Taybos GM. Total joint replacement: a consideration for antimicrobial prophylaxis. Oral Surg Oral Med Oral Pathol 1988;66:124–129.
3. Reid G. Bacterial colonization of prosthetic devices and measures to prevent infection. New Horiz 1998;6:58–63.
4. An YH, Friedman RJ. Prevention of sepsis in total joint arthroplasty. J Hosp Infect 1996;33: 93–108.
5. Young EJ, Sugarman B. Infections in prosthetic devices. Surg Clin North Am 1988;68: 167–180.
6. Carson CC. Management of prosthesis infections in urologic surgery. Urol Clin North Am 1999;26:829–839.
7. Bell SM, Gatus BJ, Shepherd BD. Antibiotic prophylaxis for the prevention of late infections of prosthetic joints. Aust N Z J Surg 1990;60:177–181.
8. Little JW. Managing dental patients with joint prostheses. J Am Dent Assoc 1994;125: 1374–1378.
9. Williams DN, Gustilo RB. The use of preventive antibiotics in orthopedic surgery. Clin Orthop 1984;190:83–88.
10. Lidwell OM, Lowbury EJ, Whyte W, Blowers R, Stanley SJ, Lowe D. Effect of ultra-clean air in operating rooms on deep sepsis in the joint after total hip or knee replacement: a randomized study. Br Med J 1982;285:10–14.
11. Hill C, Flamant R, Mazas F, Evrard J. Prophylactic cefazolin versus placebo in total hip replacement. Report of a multicentre double blind randomized trial. Lancet 1981;1: 795–797.
12. Scheibel JH, Jensen I, Pedersen S. Bacterial contamination of air and surgical wounds during joint replacement operations. Comparison of two different types of staff clothing. J Hosp Infect 1991;19:167–174.
13. Norden W. Antibiotic prophylaxis in orthopedic surgery. Rev Infect Dis 1991; Suppl 10: S842—S846.
14. Burke JF. The effective period of preventive antibiotic action in experimental incisions and dermal lesions. Surgery 1961;50:161–168.

Table 7 (*Continued*)
Perioperative Antibiotic Prophylaxis in Surgical Patients With Implantable Devices

Type of prosthesis	Likely organisms	Perioperative antibiotic	Dose for adult patient
Penile and urinary sphincter prostheses	streptococci, enteric Gram-negative bacilli Enteric Gram-negative bacilli, enterococci	Ciprofloxacin	500 mg orally or 400 mg iv
Intraocular lenses	*Staphylococcus epidermidis*, *Staphylococcus aureus*, streptococci, enteric Gram-negative bacilli, *Pseudomonas* spp.	Gentamicin, tobramycin, ciprofloxacin, ofloxacin or neomycin-gramicidin-polymyxin B, cefazolin	Multiple drops topically over 2 to 24 hours 100 mg subconjunctivally 6000 mg iv or 1.5 mg/kg iv 1–2 g iv
Ossicular and cochlear implants	*Staphylococcus aureus*, *Staphylococcus epidermidis*, enteric Gram-negative organisms, anaerobes	Clindamycin+ gentamicin Cefazolin	
Ventriculoperitoneal shunts	*Staphylococcus aureus*, *Staphylococcus epidermidis*	Cefazolin or vancomycin[a]	1–2 g iv 1 g iv
Dental implants	Anaerobes, enteric Gram-negative bacilli, *Staphylococcus aureus*	Clindamycin + gentamicin or cefazolin	600 mg iv 1.5 mg/kg iv 1–2 g iv

[a]Vancomycin should be used only in patients with severe allergy to penicillin or if the institution rate of methicillin-resistant *Staphylococcus aureus* or *Staphylococcus epidermidis* is high. iv, intravenously. (Adapted from ref. 90.)

Table 7
Perioperative Antibiotic Prophylaxis in Surgical Patients With Implantable Devices

Type of prosthesis	Likely organisms	Perioperative antibiotic	Dose for adult patient
Prosthetic heart valve	Staphylococcus aureus, Staphylococcus epidermidis, corynebacteria, enteric Gram-negative bacilli	Cefazolin or Cefuroxime or vancomycin[a]	1–2 g iv 1.5 g iv 1 g iv
Vascular graft	Staphylococcus aureus, Staphylococcus epidermidis, enteric Gram-negative bacilli	Cefazolin or vancomycin[a]	1–2 g iv 1 g iv
Pacemaker and implantable cardioverter-defibrillators	Staphylococcus aureus, Staphylococcus epidermidis, corynebacteria, enteric Gram-negative bacilli	Cefazolin or Cefuroxime or vancomycin[a]	1–2 g iv 1.5 g iv 1 g iv
Endovascular stents and filters	Staphylococcus aureus, Staphylococcus epidermidis, corynebacteria, enteric Gram-negative bacilli	No recommendations	
Joint prostheses	Staphylococcus aureus, Staphylococcus epidermidis	Cefazolin or vancomycin[a]	1–2 g iv 1 g iv
Breast implants	Staphylococcus aureus, Staphylococcus epidermidis	Cefazolin or Cefuroxime or vancomycin[a] (*)	1–2 g iv 1.5 g iv 1 g iv

(continued)

untreated dogs and the treated dog that died were positive. The authors concluded that the presence of sepsis does not preclude the use of the Greenfield filter *(84)*. The AHA does not recommend prophylaxis during a procedure that might induce bacteremia in patients with Greenfield filters.

Cochlear and Ossicular Implants

Cochlear implant surgery consists of insertion of a receiver-stimulator unit and involves opening of the labyrinth, which always carries a risk of meningitis *(85)*. The middle ear is not normally sterile and most otolaryngologists use prophylactic antibiotics in cochlear implant operations. There are no prospective, randomized studies to evaluate the benefit of prophylactic antibiotics. In a multicenter (United States, Europe, Australia) retrospective study, antibiotic prophylaxis was employed in 56.4% of cochlear implant procedures. Analysis of the data was unable to demonstrate a benefit from the use of antibiotics for this procedure. However, the overall rate of infection was low (2.9%) *(85)*.

Clark et al. developed a protocol for prevention of infection in cochlear implant surgery, but the impact on infection rate was not studied. Some of the measures instituted by these investigators overlap the general surgical principles and others are specific to otological surgery. They recommend that the patient undergoing cochlear implant surgery should be screened 1 week prior to surgery with nasal swabs for *Staphylococcus aureus,* and that therapy be given in those with positive cultures. The external auditory canal should be cleaned using the operating microscope (hair clipped, wax removed, the canal cleaned with swabs soaked in 70% alcohol). The authors recommend a hair shampoo with cetrimide solution the night before surgery, and on the morning of surgery a shower with 4% solution of chlorhexidine *(86)*. The same authors recommend sealing the space in the round window along the electrode of the device with a collar of Teflon felt in order to prevent possible bacterial penetration from the middle ear.

In a retrospective study in children with cochlear implants, Luntz et al. did not find an increased prevalence of acute otitis media; routine antibiotics were effective in treating postimplantation acute otitis media *(87)*. They concluded that a history of recurrent acute otitis media is not an absolute contraindication for cochlear implants.

Concern about transmitting Creutzfeld-Jacob disease or HIV in ossicular reconstruction by using homograft dura or bone has caused most surgeons to use cortical bone autografts *(88)* or synthetic prostheses *(89)*. In a series of 165 patients who underwent ossiculoplasty with synthetic material, only one extrusion of the ossicular replacement prosthesis secondary to infection was reported *(89)*.

There are no published recommendations based on high-quality evidence favoring prophylaxis during procedures associated with high risk of bacteremia, although some authors recommend antibiotic prophylaxis during these procedures *(86)*. Middle ear infections need prompt antibiotic therapy in patients with cochlear or ossicular implants to minimize the risk of device infection.

SUMMARY

Implantable prosthetic devices are increasingly being used both in the United States and the rest of the world. Most procedures are elective. Among the possible complica-

Table 6
Risk Factors of Infection After Angioplasty

1. A repeat percutaneous puncture of the same femoral artery
2. Indwelling arterial sheath that remains for several days with repeated catheterizations performed through the same sheath
3. Hematoma around the sheath
4. Diabetes mellitus

Adapted from ref. 83.

Paget et al. demonstrated that giving prophylactic antibiotics (cefazolin) at the time of stent deployment significantly reduced the rate of stent or artery infection after both immediate and delayed (4 weeks) *Staphylococcus aureus* intra-aortic challenge. Seventeen percent of stent/artery complexes in the antibiotic group with immediate bacterial challenge had positive cultures compared with 70% in the control group. In the group with delayed bacterial challenge the stent infection rate was 10% in the antibiotic group and 50% in the control group *(80)*. Another experiment conducted in vitro using stents and ^{111}In labeled *Staphylocooccus epidermidis* demonstrated inhibition of bacterial adhesion to the stent with the addition of IIb/IIIa platelet receptor antagonist and heparin, but not with rifampicin. *Staphylococcus epidermidis* did not adhere to the stent in the absence of thrombosis *(81)*. Signs and symptoms of stent infection (fever, bacteremia, thrombosis, mycotic aneurysms, pericarditis) typically occur a few days to a few weeks after the initial intervention *(82)*. Infections developing later are uncommon. The major risk factors of infection after angioplasty are shown in Table 6.

Cases of infected coronary stents have been reported in the recent literature. Coronary stents were introduced in 1987 and are currently placed in 40 to 60% of all interventional coronary artery procedures. Although there are a large number of stents placed, there were only four cases of coronary artery stent infections reported in a recent review. Two were infected with *Staphylococcus aureus* and two with *Pseudomonas aeruginosa (82)*.

There are no published recommendations regarding the use of prophylactic antibiotics prior to cardiac catheterization and angioplasty with stenting. Nevertheless, assuring a surgically sterile procedure, minimizing the time the arterial sheath is in place and avoiding repeated catheterization through the same sheath appear to be important steps in avoiding endovascular stent infection. The AHA does not recommend antibiotic prophylaxis prior to procedures associated with high risk of bacteremia (dental or surgical) in patients with endovascular stents *(46)*.

The literature regarding Greenfield filter infection is relatively sparse. In an experimental study done by Peyton in dogs with Greenfield filters and introduction of infected autologous thrombi, treatment for 5 days with antibiotics resulted in negative Greenfield filter and thrombus cultures 2 weeks later when the dogs were sacrificed. In a similar study performed in 12 dogs with Greenfield filters and sterile thrombus, a remote abscess was created by implantation of an infected gauze sponge in one of the extremities. All dogs became toxic and 6 died. The remainder was subsequently treated with removal of the sponge, drainage and antibiotics. One treated dog died within 24 hours and 5 survived 8 days until sacrifice. Cultures of the filters and emboli in the

prophylactic antibiotics, and the fact that these patients are dependent on their devices for survival in most cases, antimicrobial prophylaxis seems to be a reasonable strategy. Cefazolin (1 g) preoperatively and every 6 hours postoperatively for 24 hours is the preferred choice *(29)*. The AHA does not recommend antibiotic prophylaxis with dental or surgical procedures for these patients.

Dental Implants

In 1992, 300,000 dental implants were placed in the United States and that number is growing *(71)*. These impants replace single or multiple missing teeth. The dental implants are mainly made of titanium and surgically implanted in alveolar bone. A process of osteointegration takes place over a 3- to 6-month period *(72)*. The implant survival rate is 98.9% at 15 years *(73)*. The two main causes for failure are mechanical stress and infection. Formation of a biofilm on the implant is followed by bacterial plaque deposition, which generally elicits a clinical inflammatory reaction *(73)*.

The peri-implant flora includes bacterial species present in healthy gingival sites (*Streptococcus sanguis, Actinomyces viscosus, Actinomyces odontolyticus*) as well as putative periodontal pathogens (*Porphyromonas gingivalis, Prevotella intermedia, Prevotella melaninogenica* and *Fusobacterium* spp.) *(72)*. Implants failing because of infection are likely to be colonized by putative periodontal pathogens *(72)*.

The peri-implant infection can involve the gingival margins, the periodontal deeper tissues or the bone adjacent to the implant *(72)*. Clinically, the probing depth of the peri-implant sulcus in peri-implant infection is increased and associated with bleeding; suppuration can also be present. Another feature is crestal bone loss detected radiologically *(73)*.

The use of perioperative chlorhexidine (0.12%) in dental implant surgery was studied by Lambert et al. in a prospective study. There was a significant reduction in the risk of infectious complications from 8.7% compared with 4.1% in the group in which chlorhexidine was used *(74)*. The results of a prospective study evaluating the role of perioperative antibiotics in dental implants found fewer failures with use of antibiotics *(75)*. Bone necrosis caused by high bone temperature during implant preparation is thought to play a role in development of peri-implant infection *(76)*.

The bacterial plaque is a key player in the development of peri-implant infection; therefore, the goal is to maintain the dental implants plaque-free *(73)*. These patients should have regular (every 6 months) professional teeth cleaning and checkups and also be instructed on rigorous home care. Twice-daily mouth rinse with an antiseptic has been shown to reduce the plaque formation in patients with dental implants *(73)*.

Intravascular Stents and Filters

More than 400,000 endovascular stents are placed annually in the United States for the treatment of stenosed arteries or veins (coronary, renal, carotid, subclavian, and others) *(77)*. Once inserted, they cannot be removed percutaneously and local bacteriological sampling is not possible *(78)*. Experimental iliac artery stent infection was induced in a swine model by bacterial challenge with *Staphylococcus aureus,* both immediately after stent placement and 4 weeks later (80% of the stent cultures were positive in the first case and 50% in the second scenario, the swine being sacrificed at 72 hours after the bacterial challenge in both experiments) *(79)*. Using the same model,

Unfortunately, not enough participants could be enrolled to provide meaningful statistical analysis *(66)*.

Two meta-analyses found that antibiotic prophylaxis is effective if the expected rate of shunt infection is high *(64)*. In an infectious disease textbook, perioperative antibiotics were judged to be not indicated in institutions with low-shunt infection rates (less than 10% overall) *(29)*. In one study, trimethoprim (160 mg)/sulfamethoxazole (800 mg) intravenously preoperatively and every 12 hours for three doses proved beneficial in a center in which the infection rate was above 20% *(29)*. Choux et al. emphasized that the shunt implantation should be brief ($\leq$40 minutes) and the number of people allowed in the operating room minimized. Shunt material should not be opened until it is ready to be implanted and the valve should not be tested. Some authors recommend rinsing the implant in gentamicin before implantation *(67)*.

Cerebrospinal fluid shunts are rarely infected during routine procedures (dental, endoscopic) associated with transient bacteremia. Therefore, antibiotic prophylaxis for such procedures is usually not necessary. However, consultation with the patient's neurosurgeon may be appropriate in specific circumstances *(65)*. In ventriculoatrial shunts, which are less commonly used, the risk of infection associated with transient bacteremia may be higher than with ventriculoperitoneal shunts, particularly in the first 6 months. Therefore antibiotic prophylaxis before invasive procedures should be considered.

Cardiac Pacemakers and Implantable Cardioverter Defibrillators

Approximately 200,000 cardiac pacemakers are implanted in the United States annually and there are more than 200,000 patients with implantable cardioverter-defibrillators (ICDs) worldwide *(68)*. The 10-year infection rate with cardiac pacemakers varies between 1.4 and 3.3 % *(68)*. The 5-year infection rate in ICDs is reported to be 1.9 to 3.8% *(68)*.

Most factors predisposing to infection of these devices are the same as for other prosthetic devices, but some are specific for these procedures. The presence of a hematoma at the implantation pocket site has been reported to be a risk factor for both early (2 months from procedure) and late infection *(68)*. The use of temporary pacing preceding the permanently implanted device is associated with a higher infection rate (2.9 vs 0.4%) *(68)*. As in other types of prostheses implanted during clean surgical procedures, the main organisms responsible are *Staphylococcus aureus* and coagulase-negative *staphylococci*, with Gram-negative bacteria and anaerobes identified in a smaller proportion of cases *(68)*. Analysis of the reasons for infection suggests that most are acquired during the implantation procedure.

The clinical presentation of infection depends on whether the event is close to or remote from the date of surgery. Persistent fever without any other source of infection in a patient with a pacemaker or ICD is highly suggestive of device-related infection. Erosion of the pacing system through the skin almost always indicates infection *(68)*. Lead-related endocarditis, with or without septic pulmonary emboli, is another potential infectious complication.

Two separate randomized, controlled trials of patients undergoing pacemaker implantation surgery using either flucloxacillin or cloxacillin vs placebo gave contrary results *(69,70)*. Because of the lack of definitive data in the literature on the need for

However the rate of infection in this population was low (<2.5% at 2 years). The use of prophylactic suction drainage in groin wounds resulting from elective vascular surgery does not result in a decrease in perioperative infection rate *(62)*. Patients with synthetic vascular grafts are at higher risk of developing secondary bacterial endarteritis for about 6 months postoperatively. After this time, most grafts are typically covered with endothelium. After incorporation of the graft into local tissue there is no need for antibiotic prophylaxis in the event of surgical/dental procedures *(63)*. The AHA does not recommend antibiotic prophylaxis for vascular grafts before dental/surgical procedures. Nevertheless, if this is deemed necessary in an individual case, the regimens recommended for prosthetic valves could be used.

Cerebrospinal Fluid Shunts

These shunts are used for diversion of the cerebrospinal fluid in conditions associated with chronic hydrocephalus. Approximately 75,000 are placed each year in the United States. Of the various complications associated with shunt placement, infections are particularly worrisome because of the deleterious effects on the patient's cerebral function and the difficulty of treatment.

The perioperative and long-term infection rate over the past decade has decreased to less than 10% *(64)*. Most infections occur within the first 2 months after implantation *(65)*. The factors that increase the risk of infection are similar to those for other types of surgery and are host-related (extremes of age, etiology of hydrocephalus, previous shunt infection, concurrent infection, etc.) and procedure-related (duration of surgery, surgeon's experience, operating room traffic, technical errors, etc.).

The responsible microorganisms are primarily skin flora. The most common are coagulase-negative staphylococcus, followed by *Staphylococcus aureus, viridans streptococci, Streptococcus pyogenes, Enterococcus, Corynebacterium, Propionibacterium* and Gram-negative bacilli *(64)*. Most shunt infections are probably caused by organisms introduced at the time of the surgical procedure. Shunts can also be infected at the distal end by bowel perforation, skin infection adjacent to the shunt, or less commonly owing to bacteremia *(64)*.

Diagnosis can be subtle and provides a major justification for routine postoperative evaluation. It is based on the clinical findings of infection, such as fever and malaise. Nausea, vomiting, headache, and altered mental status may result from shunt malfunction. The classical signs of meningitis are present in less than one-third of the patients. Infection of the surgical site can be an early warning of deeper shunt infection. Sometimes peritoneal exudates can give rise to a peritoneal pseudocyst detected by computed tomography scan, or frank peritonitis can be the defining symptom.

Cerebrospinal fluid should be obtained from the shunt reservoir and not a lumbar puncture site and sent for glucose, protein, gram stain, cell count, and culture (both aerobic and anaerobic). Blood cultures are rarely positive except in ventriculoatrial shunts *(64)*.

There are no prospective, randomized studies evaluating the utility of antibiotics in preventing perioperative infections in patients undergoing cerebrospinal fluid shunt implantation. A tentative study to address this issue randomized patients undergoing insertion or revision of ventriculoperitoneal shunts to receive intraventricular vancomycin vs controls (usual antimicrobial prophylaxis used in the eligible centers, if any).

Table 4
Prophylactic Antibiotic Regimens Recommended
in Patients With Prosthetic Joints Undergoing Dental Procedures

Cephalexin	2 g orally, 1 hour before the procedure
Cephradine	2 g orally, 1 hour before the procedure
Amoxicillin	2 g orally, 1 hour before the procedure
Clindamycin (for the penicillin-allergic patient)	600 mg orally, 1 hour before surgery

Adapted from ref. 22.

Table 5
Factors Associated With Increased Risk of Vascular Graft Infection

1. Type, size, and anatomic location of the prosthesis
2. Condition of the patient (diabetic, uremic, obese, or immunosuppressed patients have impaired wound healing and higher risk of prosthetic graft infection)
3. Preoperative percutaneous arteriography (because of tissue trauma and hematoma)
4. Pre-existing gangrene and limb infections
5. Revision surgery
6. Emergency surgery

Adapted from refs. 5 and 54.

The use of perioperative prophylaxis has been studied in several controlled trials. In a prospective, double-blind study, the effects of 1-day and 3-day courses of cefuroxime vs placebo on the infection rate in patients undergoing peripheral vascular surgery were studied by Hasselgren et al. (53). There were statistically fewer surgical site infections in both groups given antibiotics than the placebo group. One graft infection occurred in the placebo group and none in either treatment group. Prophylaxis of 3 days duration was no better than 1 day. Jensen et al. (55) compared vancomycin with placebo in 128 vascular graft operations. Fourteen wound infections (3 of which were graft infections) occurred in the placebo group and only 1 wound infection (no graft infection) in the vancomycin group.

In 300 reconstructive arterial operations of the abdominal aorta and lower extremities, Salzman found a vascular prosthesis infection rate of 0.8% in a group receiving antibiotics and 2.4% in the group without prophylaxis (56). Pitt et al. randomized patients undergoing a vascular procedure with a groin incision to receive no antibiotics, topical cephradine in the incision before closure, 24 hours of intravenous cephradine or both topical and intravenous cephradine. The wound infection rate was significantly reduced by cephradine administered by either topical or systemic route with no advantage for the combination treatment (57). In another study no difference in infection rate was observed with 1 vs 2 g of cefazolin given preoperatively (58). Cefazolin had significantly higher serum and tissue concentrations as compared to cefamandole in patients undergoing aortofemoral reconstruction (59).

The use of vascular grafts immersed in rifampin solution before surgery, or rifampin-bonded Dacron grafts, did not result in a significant decrease in the incidence of vascular graft infection when compared to the use of standard untreated grafts (60,61).

temic antibiotics vs gentamicin-bone cement was studied in a muticenter, prospective, randomized trial in patients undergoing total hip arthroplasty in Sweden. The gentamicin-cement group had a significant advantage at 1 year, but this was not maintained at 5 or 10 years *(51)*. Other investigators studied cefuroxime-impregnated cement vs systemic cefuroxime; there was no difference in outcomes between groups *(52)*. At the present time, antibiotic-impregnated cement cannot be recommended as the only prophylaxis for patients undergoing total joint replacement surgery *(13)*.

The technical skills of the operating team are paramount in reducing the risk of infection *(49)*. The use of closed suction drainage for 48 hours following total knee arthroplasty led to colonization of 25% of the drain tips with coagulase-negative *staphylococcus* or *Staphylococcus aureus (19)*. Whether or not this correlates with early postoperative infections is not known, but there seems to be no benefit in continuing the drainage beyond 24 hours *(19)*. The value of prophylaxis against prosthetic joint infection in situations associated with possible bacteremia is unclear. Infections at any site, including the skin, urinary tract, and periodontium should be aggressively treated in all patients in general and especially in those with prosthetic devices.

In a review of 1000 patients with total joint replacement followed for 6 years, none of 224 patients who later underwent dental or surgical procedures developed hematogenous infections *(13)*. The very few reported cases of joint prostheses infected with dental organisms were in patients who had severe periodontal disease *(13)*. Therefore, prompt identification and control of periodontal disease and infection appears to be more important than routine antibiotics during dental procedures.

The American Dental Association and American Academy of Orthopedic Surgeons have recommended consideration of antibiotics in high-risk patients (rheumatoid arthritis, systemic lupus erythematosus, hemophilia, insulin-dependent diabetes mellitus, malnourishment, immunosuppression, previous prosthesis infection, the first 2 years following prosthetic insertion) undergoing procedures with high rates of bacteremia (dental extractions, periodontal procedures, dental implants, endodontic instrumentation /surgery beyond the apex, intraligamentary local anesthetic, placement of orthodontic bands, cleaning of teeth or implants). The recommended regimens are listed in the Table 4. The American Society of Colon and Rectal Surgeons and the American Society of Gastrointestinal Endoscopy do not recommend antibiotic prophylaxis for their procedures in patients with joint prostheses *(48)*.

Vascular Grafts

The incidence of vascular graft infection in a series of 134 patients with lower extremity vascular reconstruction followed for 30 months was 0.9% *(53)*. Increased risk of infection is associated with factors listed in Table 5.

An infected vascular graft is suggested by the presence of drainage from the surgical site, failure of the wound to heal, systemic sepsis, and graft thrombosis or fistula formation. Arteriography, coupled with standard blood cultures, is often used if the diagnosis is suspected but not obvious. The organisms involved, *Staphylococcus* spp., and Gram-negative enteric organisms, are believed to be commonly introduced at the time of surgery or later from bacteremia from various sources (skin, urinary tract, oral cavity, etc.). It is common to find graft infections in patients at routine follow-up visits. Because this is a life-threatening and limb-threatening complication, there is a rationale for routine follow-up of these patients.

Table 3
Risk Factors for Infection of Total Joint Prostheses

1. Rheumatoid arthritis
2. Severe type 1 diabetes mellitus
3. Corticosteroid therapy
4. Hemophilia
5. Immunosuppression
6. Revision surgery
7. Previous infection of the same joint
8. Distant focus of infection
9. Large prostheses
10. Hinged knee and total elbow prostheses

Adapted from refs. 22,47,48,50.

The diagnosis of an infected joint prosthesis is based on the presence of pain, spontaneous wound drainage, or failure of the wound to heal in the early postoperative period. Culture of a sinus tract, if present, is not reliable and radioisotope scans have limited value for at least 1 year postoperatively (48). Because the best chance for cure of the infection is a combination of device removal and antibiotics, Reid recommends cutting sections of the device at explantation, placing them in enriched medium and sonication (7 minutes) to remove the adherent organisms and thus increasing the yield of the culture (3).

Several studies have been conducted using perioperative antibiotics in joint replacement surgery. Very few of these studies that compared antibiotic to placebo were well designed (13). In a double-blind, randomized, placebo-controlled study conducted in nine centers on 2137 patients undergoing hip replacement surgery, the cefazolin group had a perioperative site infection rate of 0.9% as compared with 3.3% in the placebo group (49). The most commonly recommended regimen is 1 g of cefazolin before surgery (30–60 minutes) and then no more than two or three doses postoperatively (48). In patients undergoing total knee arthroplasty, maintaining the interval between the cefazolin administration and tourniquet inflation at 5 minutes resulted in better soft tissue and bone antibiotic concentrations (50). Unless the patient is allergic to penicillin, the routine use of vancomycin for prophylaxis is not justified because its efficacy in prophylaxis has not been well studied or established (49).

The effect of ultra-clean air (defined arbitrarily as air containing less than 10 bacterial particles/m^3) has been addressed in several studies. Laminar airflow operating rooms reduced the incidence of joint prosthesis infection in a study by Lidwell et al. (10), but this was not confirmed by data from the Mayo Clinic (both the laminar airflow and conventional operating room groups were given cefazolin perioperatively) (48). The cost of providing operating rooms with ultra-clean air is an issue that may be prohibitive for some centers.

Buchholz and Engelbrecht developed antibiotic-incorporated cement more than two decades ago (49). The antibiotics are slowly released from the cement and can achieve high local concentrations with nontoxic serum concentrations. Prophylaxis with sys-

Table 2
Procedures Associated With Bacteremia in Which Antibiotic Prophylaxis is Warranted in High-Risk Patients

Dental Procedures	Dental extractions
	Periodontal procedures (surgery, scaling and root planing, probing, and recall maintenance)
	Dental implant placement and reimplantation of avulsed teeth
	Endodontic (root canal) instrumentation or surgery only beyond the apex
	Subgingival placement of antibiotic fibers or strips
	Initial placement of orthodontic bands but not brackets
	Intraligamentary local anesthetic injections
	Prophylactic cleaning of teeth or implants when bleeding is anticipated
Respiratory tract procedures	Tonsillectomy and/or adenoidectomy
	Surgical operations that involve respiratory mucosa
	Bronchoscopy with a rigid bronchoscope
Gastrointestinal tract procedures	Sclerotherapy for esophageal varices
	Esophageal stricture dilatation
	Endoscopic retrograde cholangiography with biliary obstruction
	Biliary tract surgery
	Surgical operations that involve intestinal mucosa
Genitourinary tract procedures	Prostatic surgery
	Cystoscopy
	Urethral dilatation

Adapted from recommendations by the American Heart Association (1997) *(46)*.

tis Working Party for Antimicrobial Chemotherapy from the United Kingdom limits the use of prophylaxis to dental extractions, scaling, and periodontal surgery *(40)*.

Patients with dentures should be encouraged to be seen periodically or if denture-associated irritation occurs. Of paramount importance is the maintenance of an excellent oral hygiene and good oral health. Because most cases of endocarditis related to invasive procedures are recognized within 2 to 4 weeks following the procedure, it is important that the patients be instructed to report any symptoms suggestive of infection and have a prompt evaluation for this complication.

Prosthetic Joints

Approximately 600,000 joint prostheses are implanted each year in the United States. The incidence of infection of total joint prostheses implanted over 10 years at one institution was 1.7% *(1)*, similar to rates reported in other studies *(4)*. The most common pathogens are the staphylococcus species. These represent 50 to 60% of the cases with a high incidence of coagulase-negative *staphylococci*, which appear to be increasing lately. Less common organisms are Gram-negative bacilli, *peptostreptococci, streptococci,* and *anaerobes (4)*. The clinical conditions associated with an increased risk of infection are listed in Table 3.

Antibiotic prophylaxis in patients with prosthetic valves before dental procedures is a well-accepted practice but this is not based on a randomized, prospective trial. Such a trial would likely need to enroll 6000 patients to demonstrate the effectiveness of antibiotic prophylaxis in endocarditis prevention *(44)*. Infective endocarditis has been demonstrated after tooth extraction in rats with periodontal disease and catheter-induced aortic valve vegetations *(40)*. A combination of antibiotics (penicillin and streptomycin) was shown to prevent infection in all rabbits with experimentally induced sterile valve vegetations when inoculated with 10^8 *Streptococcus sanguis*, which was significantly better than in groups treated with penicillin G or streptomycin alone *(40)*. It is estimated that no more than 10% of all cases of native valve-infective endocarditis are related to medical or dental procedures and there are no good data for that risk in prosthetic valve endocarditis *(45)*. The protective effect of antibiotics in procedures associated with bacteremia is not 100%. The antibiotics may exert their protective effect by decreasing adherence of bacteria to the heart valve, thus avoiding subsequent clinical infection *(45)*. In a survey of health care providers, only 30% complied with American Heart Association (AHA) guidelines *(18)*. Problems included lack of awareness of guidelines, administration of inappropriate antibiotics, and administration of antibiotics when not recommended.

According to the AHA guidelines, individuals with prosthetic valves are judged to be at highest risk for infective endocarditis and antibiotic prophylaxis for them is recommended for both dental and other procedures likely to be associated with bacteremia (Table 2) *(46)*. The AHA guidelines state that prophylaxis is optional in the following circumstances: bronchoscopy with a flexible bronchoscope, with or without biopsy; transesophageal echocardiography; vaginal hysterectomy or delivery.

The recommended antibiotic regimen for dental, oral, respiratory, and esophageal procedures is amoxicillin 2 g 1 hour before procedure. For patients unable to take oral medications, 2 g of ampicillin intramuscularly or intravenously 30 minutes before the procedure is the recommended choice. In penicillin-allergic patients the options are clindamycin (600 mg orally, intramuscularly or intravenously); a cephalosporin (if the allergy to penicillin is not the immediate type); cephalexin or cefadroxil (2 g orally); cefazolin (1 g intramuscularly or intravenously) or azithromycin or clarithromycin (500 mg orally), also shortly before surgery. Parenteral antibiotics have not been shown to be more effective than oral antibiotics *(45)*. When multiple dental procedures are necessary, the AHA recommends that treatments be done at an interval of at least seven days in order to prevent the emergence of resistant flora.

The recommended prophylactic regimens for genitourinary and nonesophageal gastrointestinal (GI) procedures are ampicillin (2 g) plus gentamicin (1.5 mg/kg intramuscularly or intravenously), 30 minutes before procedure. For the penicillin-allergic patient, 1 g of vancomycin intravenously over 1.5 hours plus 1.5 mg/kg gentamicin intramuscularly or intravenously is the recommended alternative. The *ad hoc* committee recommends parenteral antibiotics for genitourinary or GI procedures, particularly in patients at high risk for infective endocarditis (which includes those with prosthetic valves). Patients on anticoagulant treatment should not receive an intramuscular regimen but rather an intravenous or oral one *(46)*. In patients who are already taking antibiotics for other conditions, it is recommended that a drug from another class be selected because of the potential of a relatively resistant oral flora *(46)*. The Endocardi-

The most common microorganisms are *Staphylococcus* spp., with coagulase-negative *staphylococcus* more common than *Staphylococcus aureus*. *Staphylococcus aureus* is the main pathogen in intravenous drug users *(18)*. The microbiology of late prosthetic valve endocarditis is similar to that of native valve endocarditis and includes *viridians streptococci: Streptococcus sanguis, mutans, milleri; staphylococci* and the HACEK (*Haemophylus* spp., *Actinobacillus, Cardiobacterium, Eikenella* and *Kingella*) group of organisms *(40)*.

Diagnosis is based on the presence of continuous bacteremia regardless of fever, which is not always present. A new or changing murmur is found in about half of the patients *(5)*. Echocardiographic interpretation is made difficult by the presence of echoes generated by the prosthesis. If there is a suspicion that slow-growing organisms or fastidious organisms are involved, or if initial cultures are negative, the microbiology laboratory should be notified and requested to incubate the blood cultures longer (3 to 4 weeks). For a review of the diagnostic criteria for infective endocarditis the reader is referred to a recent review *(41)*.

Because the main etiological agents of early prosthetic valve infection are associated with surgical site and catheter-related infections *(18)*, special attention should be paid to preventing them by good aseptic technique and early removal of the catheters. Hyde et al. found that inadequate scrubbing and glove perforation were two common intraoperative factors responsible for development of infection *(18)*.

A computer literature review revealed no prospective, controlled trials demonstrating a clear benefit of perioperative antibiotics in prosthetic valve surgery. The only such study was discontinued early because of two episodes of fatal pneumococcal sepsis in the placebo group *(16)*. Perioperative antibiotics are routinely used in valve replacement surgery, primarily because of the disastrous complications of valve infection and the relatively low toxicity of commonly used antibiotics. When selecting an antibiotic for prophylaxis, one must take into consideration the potential infecting organism, the spectrum of antibiotic activity, and the tissue penetration of the agent. The currently preferred regimen is cefazolin 1 g intravenously preoperatively and then every 6 to 8 hours for 24 hours. In a study by Goldmann et al. *(42)*, cephalothin given for 6 days was compared prospectively with a 2-day regimen in 200 patients undergoing prosthetic valve replacement. No case of endocarditis developed in either group during the 2-month follow-up. Of 11 patients with no detectable cephalothin in their sera at the close of the surgery, 3 developed staphylococcal surgical site infections, compared with 2 of 175 patients whose sera contained cephalothin ($p = 0.002$). Although no infective endocarditis developed in the mentioned follow-up time, the surgical site infections, which increase the risk for developing early prosthetic valve endocarditis, were identified when inadequate levels of antibiotics were present at the operative site. The authors also concluded that 6 days of prophylaxis was not justified. Recently, vancomycin has been increasingly used in centers where there is a high incidence of methicillin-resistant *Staphylococcus* spp. The superiority of vancomycin to cefazolin or cefamandole in preventing prosthetic valve endocarditis caused by coagulase-negative staphylococci has not been established *(43)*. In patients allergic to, or intolerant of, penicillin it is justified to select vancomycin as a prophylactic agent *(29)*. Doses should be sufficient to achieve acceptable levels throughout the procedure and for 24 hours afterward.

dental, or other) in patients with urinary sphincter prostheses. Perioperative antibiotics are routinely used, though, when artificial urinary sphincters are implanted *(35)*.

Intraocular Lens Implants

The incidence of endophthalmitis after lens implantation has been reported to be around 0.1% in the first month after surgery, which is when most of the infections occur *(36)*. The most common organisms involved are coagulase-negative *staphylococci* and *Staphylococcus aureus*, but other pathogens (*Proteus* spp., *Pseudomonas aeruginosa*, *Propionibacterium acnes*) are not rare *(36)*. The presenting symptoms are ocular pain, decreased vision, headache, and photophobia; fever is not a common feature *(36)*. For the diagnosis of pseudophakic endophthalmitis, vitreous cultures are very important.

In a study of 99 patients undergoing intraocular lens implantation, the preoperative ocular surface disinfection was done in a randomized fashion either with trimethoprim sulfate/polymyxin B sulfate or tobramycin *(37)*. The baseline cultures from the conjunctiva yielded *Staphylococcus epidermidis* (66%), *Corynebacterium* spp. (15%), and *Staphylococcus aureus* (8%). All organisms except *Staphylococcus epidermidis* were eradicated in both groups on the day of surgery. *Staphylococcus epidermidis* was eradicated on the day of surgery in 58% of patients in trimethoprim/polymyxin group and 68% in the tobramycin group.

A prospective, randomized, double-blind study compared the efficacy of steroid/antibiotic (neomycin/polymyxin B) treatment to placebo in reducing postoperative ocular inflammation in patients with lens implantation *(38)*. Thirty-eight percent of the patients in the placebo group had significant iritis postoperatively but none in the treatment group. It is impossible to distinguish the effect of steroid from the effect of antibiotic in this study.

Preoperative disinfection of the ocular surface with a 5% povidone-iodine solution and 10% iodine wash of the lids are commonly recommended *(36)*. Although there are no controlled studies to support their use, many surgeons use subconjunctival injections of antibiotics at the end of the surgery *(36)*. Systemic antibiotics have no role *(36)*.

There have been no identified reports of intraocular lens infections caused by remote infections and bacteremia. Intraocular lenses differ from other prostheses in that they are not in contact with the vascular compartment *(36)*. The aqueous humor–blood barrier is temporarily disrupted by the surgical procedure, but it is generally restored 12 weeks later, reducing the likelihood of hematogenous seeding from a remote focus *(31)*. After 12 weeks, systemic antibiotic prophylaxis before dental or medical procedures would not achieve meaningful intraocular levels.

Prosthetic Heart Valves

The incidence of early prosthetic valve endocarditis (defined as infection identified within 60 days postoperatively) has decreased from 1.14% patients per year between 1965 and 1976 to 0.3% patients per year between 1977 and 1982 *(39)*. This is approximately four times more frequent than late prosthetic valve endocarditis (developing after 60 days). The overall mortality rate of prosthetic valve is higher than that of native valve endocarditis *(39)*. Risk factors associated with prosthetic valve endocarditis are previous history of infective endocarditis and more than one valve replaced *(39)*.

Table 1
Risk Factors for Infection in Penile Prostheses

1. Prolonged hospitalization prior to surgery
2. Diabetes mellitus
3. Repeat implantation
4. Concomitant surgical procedures at the time of implantation (circumcision, hernia repair, reconstructive procedures)
5. Remote infection
6. Spinal cord injury
7. Immunocompromised host

Adapted from ref. 6.

undergoing penile prosthesis surgery who were randomized to receive either ofloxacin orally or gentamicin and cefazolin parenterally *(33)*. In both regimens, the antibiotic concentrations were equal to or exceeded the minimal inhibitory concentrations for *Staphylococcus epidermidis* and *Escherichia coli* in 80% of the patients. The authors thus concluded that penile implant surgery could be done on an outpatient basis using oral ofloxacin with significant economic advantages compared to the parenteral regimen. The rate of infection was zero in both groups. There was an insufficient number of patients to determine clinical equivalence, however.

Carson states that routine prophylaxis for penile implants prior to dental procedures does not seem to be cost-effective but may be appropriate in immunocompromised patients *(6)*. The evidence on this topic is quite sparse.

Artificial urinary sphincter prosthesis features a periurethral cuff that either exerts a fixed or variable periurethral pressure depending on device design *(31)*. In a series of 109 artificial urinary sphincters followed up postoperatively for 1 to 32 months, the infection rate was 2.7% (6). Licht et al. *(34)* conducted a study of 65 penile prostheses and 22 artificial urinary sphincter implantation procedures in which cultures were obtained prospectively at reoperation for reasons other than infection. Patients were followed for 12 months or until development of infection. *Staphylococcus epidermidis* was isolated from 40% of clinically uninfected penile prostheses and 36% of clinically uninfected artificial urinary sphincters, but only three of these devices later became infected and in all cases another organism was found at explantation *(34)*. The authors concluded that the role of *Staphylococcus epidermidis* in prosthetic infection may be overestimated.

Documentation of a sterile urine before implantation is critical in prevention of early postoperative infections *(6)*. Because most infections in the artificial urinary sphincters involve the pump device, close monitoring of this portion of the device for signs of infection has been recommended *(6)*. In a review of 145 patients with an artificial urinary sphincter, Martins et al. found that the risk of infection/erosion was increased with improper urethral catheterization or endoscopic manipulation and with exposure to radiation for cancer treatment *(35)*.

As with penile implants, there are no randomized, prospective studies regarding the use of perioperative antibiotics or prophylaxis before minor procedures (endoscopic,

A significant complication of breast implant surgery remains the capsular contracture. It has been hypothesized that the etiology of this might be a subclinical infection with *Staphylococcus epidermidis* (28). Surgeons frequently irrigate implant pockets with diluted solutions of povidone-iodine or antibiotic solutions before implantation. Povidone-iodine solutions have been shown to be cytotoxic to fibroblasts in vitro (28). Adams et al. studied effects of various dilutions of povidone-iodine in saline or double antibiotic solutions on *Staphylococcus epidermidis, Staphylococcus aureus, Propionibacterium acnes, Pseudomonas*, and *E. coli*, in vitro (28). They found that 10% betadine in gentamicin/cefazolin was effective at killing the bacteria tested.

We found no randomized, controlled studies of antimicrobial prophylaxis in breast implant surgery in a computer literature search. General principles of asepsis discussed in the introductory part of this chapter are valid for breast implant surgery.

The routine use of antibiotic prophylaxis in total or partial mastectomy is not recommended in medical textbooks of infectious disease, but the data are controversial (29). Even though breast implant surgery is a clean procedure, most surgeons recommend the use of perioperative prophylactic antibiotics owing to the large surgical dead space that hinders proper phagocytic bacterial killing (28). Late infections after breast implantation are rare; nevertheless prompt diagnosis and treatment of all local and distant bacterial infection is essential (30). Although there are no published guidelines regarding antibiotic prophylaxis before dental or surgical procedures in patients with breast implants, prophylaxis is rational and should be considered, based on individual circumstances.

Penile Prostheses and Prosthetic Urinary Sphincters

More than 20,000 penile prostheses for the treatment of impotence are inserted annually in the United States (6). Artificial urinary sphincters have been used for the treatment of urinary incontinence since the 1970s (31). In a meta-analysis of 39 studies involving 9361 patients with penile prostheses, the infection rate was 2.7% per year (6). A retrospective review of 1337 penile prostheses implanted over 7 years found an infection rate of 18% in diabetic patients requiring revision for mechanical failure, iatrogenic causes or patient dissatisfaction (3). In a series published by Fishman et al., 56% of penile prosthesis infections occurred within 7 months of implantation, 36% between 7 and 12 months, and 2.6% after 5 years (32). The responsible organisms are *Staphylococcus epidermidis* (35–56%), Gram-negative enteric organisms (20%), and mixed flora of Gram-negative aerobic and anaerobic bacteria (6).

Clinically apparent infections present with new-onset penile pain, erythema, and induration at the site, fever, drainage, and even device extrusion (6). Pain at the site or device migration is highly suspicious for prosthetic infection. Risk factors are summarized in Table 1 and are similar to other recognized risks for surgical site infection.

Risk factors do not contraindicate penile prosthesis implantation, but increase the likelihood of infectious complications. These patients warrant increased education regarding infection prevention in the postoperative period (6). This is one of the major reasons that postimplantation management is important.

Despite continued controversy, most surgeons use perioperative antibiotics for implantation of genitourinary devices (6), most commonly second- or third-generation cephalosporins. In a prospective trial, Schwartz et al. studied 20 consecutive patients

temic lupus erythematosus; disease, drug- or radiation-induced immunosuppression; insulin-dependent diabetes mellitus; previous prosthetic joint infection; first two years following joint placement; malnourishment and hemophilia) may be at higher risk for hematogenous infections. Antibiotic prophylaxis for such patients undergoing dental procedures with a higher bacteremic incidence (dental extractions, periodontal procedures, dental implants, endodontic instrumentation or surgery beyond the apex, intraligamentary local anesthetic, placement of orthodontic bands, and prophylactic cleaning of teeth or implants) should be considered (22).

Although only 47% of the orthopedic surgeons in a national survey published in 1985 believed that dental bacteremias were associated with infection of the prosthetic joints, 93% recommended antibiotic prophylaxis for their patients undergoing invasive dental procedures (22). This contrasts with a British survey that found that only 10% of the orthopedic surgeons, general practitioners, and rheumatologists thought that dental extraction poses a risk of hematogenous infection of a prosthetic joint, such that antibiotic prophylaxis should be considered (23). The explanation for these differences may lie in the fact that the United States is a more litigious society. The Working Party of the British Society for Antimicrobials and Chemotherapy did not consider that antibiotics were indicated for dental procedures for patients with prosthetic joints (23). The presence of obvious dental infection should be appropriately treated with both antibiotics and dental care.

The literature regarding perioperative antibiotic prophylaxis at the time of implantation of prostheses and that preceding other procedures is scarce for the other prostheses treated in this text. This chapter does not address this general topic further. At present, there are no firm recommendations regarding differences in dose, route, or duration of the treatment of infections remote from the prosthetic device. Any infection with the potential for bacteremia should be promptly recognized and treated.

INDIVIDUAL CHARACTERISTICS OF THE MOST COMMONLY USED PROSTHETIC DEVICES

Breast Implants

Breast implants, usually textured-surface silicone, gel- or saline-filled, are used for breast reconstruction after subcutaneous mastectomy, cancer ablation, or for breast augmentation (24). Fifty-thousand implants for breast augmentation and 40,000 after cancer surgery are inserted each year in the United States. The incidence of clinical wound infections after placement of breast implants followed for a mean of 7.8 years was 2.5% in 749 women who had breast implants at Mayo Clinic between 1964 and 1991 (25).

The most common organisms for breast implants reported to the US Food and Drug Administration surveillance system for monitoring adverse events related to medical devices were staphylococcus species (26). In a study by Ahn et al., 47% of 139 breast implants removed from symptomatic patients for any reason were culture positive (27). Propionibacterium acnes were isolated in 57.5%, *Staphylococcus epidermidis* in 41% and *Escherichia coli* in 1.5% of cases (27). It is likely that an infected breast implant is contaminated at the time of insertion, with bacteria remaining dormant for months or years in the biofilm layer before producing any symptoms (28).

Local Use of Antibiotics or antimicrobials

A large, multicenter, international, randomized trial is presently ongoing to assess the efficacy of Silzone™ coating (a silver-containing substance) in the sewing cuff of St. Jude Medical heart valves compared to the conventional cuff in preventing infection *(17)*. Cement impregnated with antibiotics during prosthetic joint surgery has been evaluated in several studies. This is discussed later in the section on joint prostheses. Many surgeons practice pre-immersion of valve, prosthetic joint, or other prosthetic devices in antibiotic solutions, but the benefit is unclear *(18)*.

Postoperative Measures

Overt infections in the surgical patient remote from the operative site can cause bacteremia and can be the source of prosthetic device infections. These infections should be aggressively treated or preferably prevented altogether. In the early postoperative period, the most common sources are surgical site infections, vascular catheters, and drains *(18)*.It cannot be stressed enough to promptly remove the urinary and intravascular catheters or surgical drains as soon as the clinical condition allows. Both Willemen et al. *(19)*and Drinkwater et al. *(20)* have demonstrated that there is no measurable gain by continuing drainage beyond 24 hours in total knee arthroplasties and other total joint arthroplasties.

In the late postoperative period, sustained or transient bacteremia could conceivably seed the prosthetic devices. The most common case scenarios are urinary tract, respiratory, skin, or dental infections. Blomren evaluated the effect of transient bacteremia on finger joint prostheses implanted in New Zealand white rabbits by injection of bacteria intravenously and found that hematogenous spread is capable of producing infection similar to local inoculation of bacteria in this model *(21)*. The greatest risk of infection posed by this was in the early postoperative period. This experimental model has been criticized because the inoculum size of *Staphylococcus aureus* was so large that 56% of the rabbits died from the intravenous challenge of bacteria *(22)*.

The only published guidelines to reduce infection of the prosthetic devices by antibiotic prophylaxis are those related to prosthetic heart valves. These recommendations are directed to dental and other invasive procedures in the patient with prosthetic heart valves. This includes both mechanical and bioprosthetic valves. The microbiological etiology of late prosthetic valve endocarditis overlaps that of native valve endocarditis with a high rate of organisms found in the normal mouth flora. There are few reports of oral organisms causing late total arthroplasty infections, and therefore there is no consensus on whether antibiotic prophylaxis should be used for dental procedures in these patients.

In 1990, the Council on Dental Therapeutics released a statement concluding "data are insufficient at the present time to support the need for, or the effectiveness of antibiotic prophylaxis for the dental patient who has a prosthetic joint" *(22)*. In 1997, the American Dental Association and American Academy of Orthopedic Surgeons issued an advisory statement emphasizing that antibiotic prophylaxis is not indicated for patients with pins, plates or screws, nor is it routinely indicated for most dental patients with total joint replacement. There is limited evidence that some immunocompromised and other patients with total joint replacement (rheumatoid arthritis; sys-

hip and knee replacement was evaluated in a randomized controlled trial in 1982 by Lidwell et al. *(10)*. The incidence of joint infections (confirmed at reoperation) in patients whose prostheses were inserted in an operating room supplied by ultra-clean air was one-half that of patients who were operated on in a conventionally ventilated room. In a study done in Denmark, the use of polypropylene coveralls during joint replacement surgeries reduced the bacterial contamination of the air by 62% but surgical site infection rate was unaffected *(12)*. When the operating room personnel wore a whole-body exhaust-ventilated suit in the ultra-clean system, the incidence of joint infections was 25% of that found with conventional air supply *(10)*. This study did not include controls for the effect of prophylactic antibiotics, but Hill et al., in another prospective trial, found that when an ultra-clean operating room was used, the rates of infection in total hip replacement were the same for antibiotic and for placebo group *(11)*. Operating rooms with ultra-clean air may not be cost-effective for departments performing less than 100 arthroplasties a year *(13)*. Airborne contamination with bacteria can also be reduced by limiting operating room traffic, reduction of conversation *(9)*, and optimizing the operative times *(2)*.

Perioperative antibiotic prophylaxis has been the object of numerous studies and debates. As early as 1961, it was demonstrated experimentally that antimicrobial agents given immediately before or immediately after contamination of an incision could prevent the development of soft tissue infections *(14)*. The role of prophylactic antibiotics in general surgery was studied by Polk et al. in a prospective study that demonstrated that systemic antibiotics administered before incision was made could prevent subsequent infections in operations associated with bacterial contamination *(15)*. This data was subsequently extrapolated to surgery for prosthetic device implantation even though the overwhelming majority of these operations are clean procedures in which the role of prophylactic antibiotics is more difficult to demonstrate *(16)*. Nevertheless, because the consequences of infection of the prosthetic device during insertion are severe, antibiotic prophylaxis is now routine practice.

Few randomized, controlled trials have compared antibiotics to placebo in prosthetic device surgery *(11)*. Several studies compared either various antibiotics one to another, different durations of antibiotics postoperatively, or the same antibiotics given parenterally vs cement-impregnated in the case of joint prostheses. The particular studies are discussed with the individual prostheses.

The following general principles of antibiotic prophylaxis to prevent surgical site infections should be followed in prosthetic device surgery:

1. The antibiotics administered should cover the likely bacteria involved in the subsequent wound or prosthetic infections.
2. The antibiotics should be administered in adequate doses for good tissue penetration.
3. The antibiotics should be relatively nontoxic and inexpensive.
4. The timing of administration is very important: the first dose should be given 15 to 60 minutes before incision *(4)*.
5. If the surgery is of long duration, another intraoperative dose of antibiotics should be given to maintain adequate tissue levels *(16)*. This depends on the half-life of the drug.
6. Administration of the antibiotic beyond 24 hours has not been found to have additional benefit.

GENERAL PRINCIPLES OF INFECTION PREVENTION IN PROSTHETIC DEVICES

Preoperative Measures

Before a prosthetic device is placed, the patient's overall medical condition should be at its optimum. Any acute or chronic physical condition should have been treated or controlled *(4)*. Optimizing the control of diabetes mellitus, rheumatoid arthritis, heart failure, and chronic lung disease should be part of the preoperative management. This includes a careful evaluation for the presence of any active infection that may seed the prosthetic device during insertion. Elimination of infection is not always possible. An example is the insertion of a prosthetic heart valve in a patient with infective endocarditis. Attention needs to be given to skin conditions: eczema, folliculitis, and chronic skin ulcers *(7)*, which can be a source of the infection. Little *(8)* recommends a thorough dental evaluation and treatment of any active infection before joint prosthesis surgery. This is reasonable for most types of surgery involving prosthetic material. The patient should be educated on good oral hygiene and continued care to avoid future dental problems *(8)*. Although a sterile urine is mandatory for penile prosthesis implantation *(6)*, it should be included in the evaluation for other devices as well. The best approach to prevent patient colonization with hospital flora is to minimize the duration of hospitalization prior to surgery *(6)*.

Intraoperative Measures

An important first step in the colonization and subsequent infection of a prosthetic device is adherence of bacteria. It therefore makes sense to prevent bacteria from reaching biomaterial surfaces or at least minimize the number of them at the time of implantation. It is well established that the risk of clinical infection is directly related to inoculum size. The inoculum size needed to reach a particular risk level of infection is several orders of magnitude greater in surgical sites without prosthetic material compared to sites with such implants. The infection rate in prosthetic joint implantation decreased from 8.2% in 1972 to 1.1% in 1993; this has been attributed mainly to improved operating room techniques and the use of prophylactic antibiotics *(4)*.

Skin is the major source for staphylococcus species that are responsible for more than 60% of all prosthetic infections. Careful skin preparation at the operative site with standard betadine-alcohol is an important first step. Use of water-impermeable drapes may further reduce contamination of the operative field *(4)*. For patients in whom urological prostheses are planned, shaving and hair removal with minimal trauma should be performed in the operating room *(6)*. Conventional operating room policies and procedures, including proper scrubbing, the use of good-quality surgical attire, double gloving *(9)*, and the like, is important.

Wound irrigation with saline, or saline-antibiotic solution, removes blood clots and other debris from the wound and reduces the amount of bacteria, all of which have been repeatedly demonstrated to be of value in reducing surgical site infection rates. The mechanical action and dilution of irrigation is probably more important than the activity of the various antibiotics used.

The effect of ultra-clean air (arbitrarily defined as air containing less than 10 bacterial particles/m^3) in the operating rooms on the incidence of deep infections after total

of the antibiotics directed against various bacteria under experimental conditions *(3)*. *Staphylococcus aureus* produces fibronectin-binding protein A and B *(3)* that facilitates bacterial adherence. Sometimes the implantation process itself facilitates subsequent infection. Two examples illustrate this. The inflammatory reaction around the foreign material may render leukocytes unable to mount a phagocytic response *(3)*. Implantation of a prosthetic joint utilizes polymethylmethacrylate cement, which polymerizes *in situ* by an exothermic reaction, resulting in temperatures exceeding 100°C *(5)*. This heat may cause local tissue necrosis, which strongly favors bacterial growth.

The sources of bacteria responsible for infection early after implantation are the patient's skin flora, the environment, the operating room personnel, or the device itself. The device can also be seeded by hematogenous or lymphatic routes from a remote site of active infection (cutaneous, urinary, dental, or other) after implantation.

In prosthetic valve endocarditis that occurs early (within 60 days of surgery), *Staphylococcus* species (*epidermidis* and *aureus*) are the most commonly recovered organisms. If it develops later (more than 60 days), the responsible agents resemble native valve endocarditis and the microorganisms primarily responsible are *Streptococcus viridans*, Group D *Streptococcus* and *Staphylococcus* spp. Because most reported cases of prosthetic joint infections, both early and late, involve coagulase-negative *staphylococcus,* it is felt that these bacteria may be introduced at the time of surgery and infection is not manifested clinically for a long time because of their low virulence *(5)*. This information, and the observation that few cases of infected arthroplasties are caused by common dental pathogens, cast doubt on the value of antibiotic prophylaxis during routine dental procedures.

Staphylococcus epidermidis is the most common organism in infected penile prosthesis but Gram-negative organisms, and occasionally mixed Gram-negative organisms and anaerobes, are involved in some cases *(6)*. In ventriculo-peritoneal or ventriculo-atrial shunt infections, the most commonly encountered organisms are staphylococci, especially coagulase-negative species.

The signs and symptoms of active infection in patients with infected prosthetic material, such as fever, elevated white blood cell count, and local signs of inflammation, are often surprisingly inconspicuous. This is an important reason why careful follow-up can be so valuable. Awareness of the insidious nature of the clinical presentation and serious consequences of these infections can facilitate early diagnosis and prompt treatment. The involved organisms must be identified in order to target them with appropriate antimicrobial treatment. The best chance for eradication of infection involves removal of the infected prosthesis, but this frequently leads to inherent morbidity and patient dissatisfaction. Therefore, accurate diagnosis of the presence of infection, the etiological agents involved, their microbiological susceptibility and the appropriate surgical intervention must be coordinated to produce the best outcome. Realistic specific goals with respect to eradication or suppression of infection, as well as overall function, must be defined, and early diagnosis is helpful in these decisions. Currently, no follow-up schedule designed to detect or prevent infection in patients with implantable devices has been evaluated by properly controlled clinical trials, but follow-up is routinely performed, as documented in subsequent chapters. Clearly, it is important that patients be educated to seek prompt medical care in the event of signs or symptoms suggestive of infection close to or remote from the implanted prosthesis.

9
Prevention of Infection in Prosthetic Devices

Ramona E. Simionescu and Donald J. Kennedy

INTRODUCTION

Each year in the United States millions of implantable prosthetic devices are utilized. They are designed to replace lost or deteriorated function of organs such as joints, heart valves, lenses, and a variety of other body parts. All have shortcomings, primarily mechanical failure and infectious complications. Infection is often devastating and frequently entails significant morbidity, both immediate and long term, and appreciable mortality in selected devices. It has significant economic impact (1).

The general aspects of pathogenesis of infection of bioprosthetic devices, incidence of infection, and the accepted prophylaxis measures, are discussed including the various controversies that persist on this important topic. Because there are no well-defined evidence-based guidelines for preventing prosthetic device infections, with the exception of prosthetic heart valves, this chapter presents the existing data and the opinions of various authorities.

Principles for the prevention of such infection are widely accepted, but there is no consensus regarding specific practices. The principal data from which current practices have originated are animal studies. Prospective, randomized, double-blind studies in humans evaluating the efficacy of antimicrobial prophylaxis to prevent infection in prosthetic devices are unlikely to be done because of ethical issues (2) and because an extraordinarily large number of study participants would be required in order to have sufficient statistical power.

This chapter summarizes general measures to reduce the risk of infection during and after insertion of prosthetic devices, followed by data pertaining to each type of prosthetic device. In order to understand preventive measures, the pathogenic mechanisms of infection and the most commonly encountered pathogens are reviewed.

Once a foreign object is implanted, complex physical and biological reactions take place that ultimately result in adherence of cells and proteins to its surface (3). A concept termed "the race for the surface" postulates that there is competition between microorganisms and human cells for attachment to the biomaterial introduced into the body. It is attractive but not universally accepted (3). Once organisms have attached to the surface they can form a biofilm containing bacterial glycocalyx (slime) in 6 to 12 hours. This promotes the growth of additional bacteria, hinders recovery of organisms during diagnostic procedures (4), and increases the minimal bactericidal concentration

From: *The Bionic Human: Health Promotion for People With Implanted Prosthetic Devices*
Edited by: F. E. Johnson and K. S. Virgo © Humana Press Inc., Totowa, NJ

94. Absolom DR, Lamberti FV, Policova Z, Zingg W, van Oss CJ, Neumann AW. Surface thermodymanics of bacterial adhesion. Appl Environ Micrbiol 1983;46:90–97.
95. Busscher HJ, Weerkamp AH, van der MeiHC, van Pelt AWJ, de Jong HP, Arends J. Measurements of the surface free energy of bacterial cell surfaces and its relevance in adhesion. Appl Environ Microbiol 1984;48:980–983.
96. Udipi K, Ornberg RL, Thurmond KB, Settle SL, Forster D, Riley D. Modification of inflammatory response to implanted biomedical materials in vivo by surface bound superoxide dismutase mimics. J Biomed Mater Res 2000;51:549–60.

71. Tang L, Jennings TA, Eaton JW. Mast cells mediate acute inflammatory responses to implanted biomaterials. Proc Natl Acad Sci USA 1998;95:8841–8846.
72. Tang L, Eaton JW. Tissue engineering of vascular grafts. Zilla PP, Greisler HP, eds. Georgetown, TX: Landes; 1998.
73. Audran R, Lesimlple T, Delamaire M, Picot C, Van Damme J, Toujas L. Adhesion molecule expression and response to chemotactic agents of human monocyte-derived macrophages. Clin Exp Immunol 1996;103:155–160.
74. Naif HM, Li S, HoShon M, Mathijs J, Williamson P, Cunningham AL. The state of maturation of monocytes into macrophages determines the effects of IL-4 and IL-13 on HIV replication. J Immunol 1997;158:501–511.
75. Athanasou NA, Quinn J, Bulstode DJK. Resorption of bone by inflammatory cells derived from the joint capsule of hip arthroplasties. J Bone Joint Surg 1992;74:57–62.
76. Holtrop ME, Cox KA, Glowacki J. Cells of the mononuclear phagocyte system resorb implanted bone matrix: a histological and ultrastructural study. Calcif Tissue Int 1982;34: 488–494.
77. Murray DW, Rushton N. Macrophages stimulate bone resorption when they phagocytose particles. J Bone Joint Surg 1990;72:988–992.
78. Sutherland K, Mahoney JR, Coury AJ, Eaton JW. Degradation of biomaterials by phagocyte-derived oxidants. J Clin Invest 1993;92:2360–2367.
79. Chan SC, Birdsell DC, Gradeen CY. Urinary excretion of free toluene diamines in a patient with polyurethane-covered breast implants. Clin Chem 1991;37:756–758.
80. Picha GJ, Goldstein JA, Store E. Analysis of the soft-tissue response to components used in the manufacture of breast implants. Plast Reconstr Surg 1990;85:903–916.
81. Stark GB, Gobel M, Jaeger K. Intraluminal cyclosporin A reduces capsular thickness around silicone implants. Ann Plast Surg 1990;24:156–161.
82. McNally AK, Anderson JM. Interleukin-4 induces foreign body giant cells from human monocytes/macrophages. Am J Pathol 1995;147:1487–1499.
83. Most J, Neumayer HP, Dierich MP. Cytokine-induced generation of multinucleated giant cells in vitro requires interferon and expression of LFA-1. Eur J Immunol 1990;20: 1661–1667.
84. Kazazi F, Chang J, Lopez A, Vadas M, Cunningham AL. Interleukin-4 and immunodeficiency virus stimulate LFA-1-ICAM-1-mediated aggregation of monocytes and subsequent giant cell formation. J Gen Virol 1994;75:2795–2802.
85. Tananka H, Shinki T, Ttakito J, Jim CH and Suda T. Trans-glutaminase is involved in the fusion of mouse alveolar macrophages induced by 1,25-dihydroxyvitamin D3. Exp Cell Res 1991;192:165–172.
86. Smenta K, Jr., Multinucleate foreign-body giant cell formation. Exp Mol Pathol 1987;46: 258–265.
87. Kao WJ. Evaluation of protein-modulated macrophage behavior on biomaterials: designing biomimetic materials for cellular engineering. Biomaterials 1999;20:2213–2221.
88. Flory PJ. Principles of polymer chemistry. New York, NY: Cornell University Press; 1953.
89. Lappin-Scott HM, Costerton JW, eds. Microbial Biofilms. New York, NY: Cambridge University Press; 1995.
90. Habash M, Reid, G. Therapeutic Review. Microbial biofilms: Their development and significance for medical device-related infections. J Clin Pharmacol 1999;39:887–898.
91. Costerton JW, Stewart PS. Battling biofilms. Sci Am 2001;285:74–81.
92. Derjaguin BV, Landau L. Theory of the stability of strongly charged lyophobic soils on the adhesion of strongly charged particles in solutions of electrolytes. Acta Physicochim 1941;14:633–662.
93. Verwey EWJ, Overbeek JTG. Theory of the solubility of lyophilic colloids. Amsterdam: Elsiever, 1948.

49. Bostman OM, Philajamaki HK. Adverse tissue reactions to bioabsorbable fixation devices. Clin Orthop 2000;(371):217–227.
50. Bostman O, Pihlajanmaki H. Clinical biocompatibility of biodegradable orthopaedic implants for internal fixation: a review. Biomaterials 2000;21:2615–2621.
51. Neale SD, Athanasou NA. Cytokine receptor profile of arthroplasty macrophages, foreign body giant cells and mature osteoclasts. Acta Orthop Scand 1999;70:452–458.
52. Bauer JJ, Harris MT, Kreel I, Gelernt IM. Twelve-year experience with expanded polytetrafluoroethylene in the repair of abdominal wall defects. Mt Sinai J Med 1999;66:20–25.
53. Kennedy GM, Matyas JA. Use of expanded polytetrafluoroethylene prosthetic patches in repair of the difficult hernia. Am J Surg 1994;168:304–306.
54. Horikoshi M, Macaulay W, Booth RE, Crossett LS, Rubash HE. Comparison of interface membranes obtained from failed cemented and cement less hip and knee prostheses. Clin Orthop 1994;309:69–87.
55. Tucci M, Tsao A, Hughes J, Jr. Analysis of capsular tissue from patients undergoing primary and revision total hip arthroplasty. Biomed Sci Instrum 1996;32:119–125.
56. El-Seifi A, Fouad B. Long-term fate of Plastipore in the middle ear. ORL J Otorhinolaryngol Relat Spec 1998;60:198–201.
57. Kerr AG, Riley DN. Disintegration of porous polyethylene prostheses. Clin Otolaryngol 1999;24:168–170.
58. Soliman HE, Milad MF, Ayyat FM, Zein TA, Hussein ES. Penile implants in the treatment of organic impotence. Saudi Med J 2001;22:30–33.
59. Greenland S, Finkle WD. A retrospective cohort study of implanted medical devices and selected chronic diseases in Medicare claims data. Ann Epidemiol 2000;10:205–213.
60. Mulcahy JJ. Long-term experience with salvage of infected penile implants. J Urol 2000;163:481–482.
61. Anderson JM. Inflammatory responses to implants. Trans Am Soc Artif Organs 1988;34:101–107.
62. Anderson JM. Inflammation and the foreign body response. Problems Gen Surg 1994;11:147–160.
63. Jenny CR, DeFife KM, Colton E, Anderson JM. Human monocyte/macrophage adhesion, macrophage motility, and IL-4-induced foreign body giant cell formation on silane-modified surfaces in vitro. J Biomat Sci 1998;171–184.
64. Jenny CR, Anderson JM. Effects of surface-coupled polyethylene oxide on human macrophage adhesion and foreign body giant cell formation in vitro. J Biomed Mater Res 1999;44:206–216.
65. Kao WJ, Zhao QH, Hiltner A, Anderson JM. Theoretical analysis of in vivo macrophage adhesion and foreign body giant cell formation on polydimethylsiloxane, low density polyethylene and polyetherurethanes. J Biomed Mater Res 1994;28:73–79.
66. Zhao Q, Topham N, Anderson JM, Hiltner A, Lodoen G and Payet CR. Foreign-body giant cells and polyurethane biostability; in vivo correlation of cell adhesion and surface cracking. J Biomed Mater Res 1991;25:177–183.
67. Rosales C, Julia RL. Signal transduction by cell adhesion receptors in leukocytes. J Leukocyte Biol 1995;57:189–198.
68. Ruoslahti E and Pierschbacher MD. New perpectives in cell adhesion: RGD and intergrins. Science 1987:238;491–497.
69. Jenny CR, Anderson JM. Adsorbed serum proteins responsible for surface dependent human macrophage behavior. J Biomed Mater Res 2000;49:435–447.
70. Jenny CR, Anderson JM. Adsorbed IgG: A potent adhesive substrate for human macrophages. J Biomed Mater Res 2000;50:281–290.

27. Mullner-Eidenbock A, Amon M, Schauersberger J, et al. Cellular reaction on the anterior surface of four types of intraocular lenses. J Cataract Refract Surg 2001;27:734–740.
28. D'Hermies F, Korobelnik JF, Chauvaud D, Pouliquen Y, Parel JM, Renard G. Scleral and episcleral histological changes related to encircling explants in 20 eyes. Acta Ophthalmol Scand 1999;77:279–285.
29. Butnay J, de Sa Mauro, Feindel M, David TE. The Toronto SPV Bioprosthesis: Review of morphological findings in eight valves. Semin Thorac Cardiovasc Surg 1999:11:157–162.
30. Grabenwoger M, Grimm M, Eybl E, et al. New aspects of the degeneration of bioprosthetic heart valves after long-term implantation. J Thorac Cardiovac Surg 1992;104:14–21.
31. Grabenwoger M, Fitzal F, Gross C, et al. Different modes of degeneration in autologous and heterologous heart valve prostheses. J Heart Valve Dis 2000;9(1):104–111.
32. Edelman ER, Campbell R, Pathobiologic responses to stenting. Am J Cardiol 1998;81: 4E–6E.
33. Virmani R, Farb A. Pathology of in-stent restenosis. Curr Opin Lipid 1999;10:499–506.
34. Von Segesser LK, Olah A, Leskosek B, et al. Coagulation patterns in bovine left heart bypass with phospholipid versus heparin surface coating. ASAIO J 1993;39:43–46.
35. Campbell EJ, O'Byrne V, Stratford PW, et al. Biocompatible surfaces using methacrylolphosphorylcholine laurylmethacrylate copolymer. ASAIO J 1994;40:M853–M857.
36. Whelan DM, van der Giessen WJ, Krabbendam SC, et al. Biocompatibility of phosphorylcholine coated stents in normal procine coronary arteries. Heart 2000;83:338–345.
37. Thomas WO, Harper LL, Wong SW, et al. Explantation of silicone breast implants. Am Surg 1997;63:421–429.
38. Hameed MR, Erlandson R, Rosen PP. Capsular synovial-like hyperplasia around mammary implants similar to detritic synovitis. A morphologic and immunohistochemical study of 15 cases. Am J Surg Pathol 1995;19:433–438.
39. Rosa DS, Greene WB. Silicone breast implants: Pathology. Ultrastructural Pathology 1997;21:263–271.
40. Mena EA, Kossovsky N, Chu C, Hu C. Inflammatory intermediates produced by tissue encasing silicone breast prosthesis. J Invest Surg 1995;8:31–42.
41. Abbondanzo SL, Young VL, Wei MQ, Miller FW. Silicone gel-filled breast and testicular implant capsules: a histologic and immunophenotypic study. Mod Pathol 1999;12: 706–713.
42. Boynton EL, Henry M, Morton J, Waddell JP. The inflammatory response to particulate wear debris in total hip arthroplasty. Can J Surg 1995;38:507–515.
43. Sanatavirta S, Gristina, A, Knottinen YT. Cemented versus cementless hip arthroplasty. A review of prosthetic biocompatibility. Acta Orthop Scand 1992;63:225–232.
44. Korovessis P, Repanti M. Evolution of aggressive granulomatous periprosthetic lesions in cemented hip arthroplasties. Clin Orthop 1994;300:155–161.
45. Van de Belt H, Neut D, Schenk W, van Horn JR, van der Mei HC, Busscher HJ. Infection of orthopedic implants and the use of antibiotic-loaded bone cements. A review. Acta Orthop Scand 2001;72:557–571.
46. Chiba J, Schwendeman LJ, Booth RE Jr, Crossett LS, Rubash HE. A biochemical, histologic, and immunohistologic analysis of membranes obtained from failed cemented and cementless total knee arthroplasty. Clin Orthop 1994;299:114–124.
47. Urban RM, Jacobs JJ, Gilbert JL, Galante JO. Migration of corrosion products from modular hip prostheses. Particle microanalysis and histopathological findings. J Bone Joint Surg Am 1994;76:1345–1359.
48. Trindade MC, Schurman DJ, Maloney WJ, Goodman SB, Smith RL. G-protein activity requirement for olymethylmethacrylate and titanium particle-induced fibroblast interleukin-6 and monocyte chemoattractant protein-1 release in vitro. J Biomed Mater Res 2000;51(3):360–368.

3. Jozefowicz J, Jozefowicz M. Interactions of biospecific functional polymers with blood proteins and cells. J Biomater Sci Polym Ed 1990;1:147–165.
4. Andrade JD. Needs, problems, and opportunities in biomaterials and biocompatibility. Clinical Materials 1992;11:19–23.
5. Pourbaix M. Electrochemical corrosion of metallic. Biomaterials 1984;5:122–134.
6. Zimmerman MC, Alexander H, Parsons JR, Bajpai PK. The design and analysis of laminated degradable composite bone plates for fracture fixation. In: TL Vigo and AF Turbak, eds. High-tech textiles. ACS Symposium Series 457. American Chemical Society, Washington DC. 1999:132–148.
7. Aklonis, JJ. Introduction to polymer viscoelasticity. New York, NY: Wiley, 1983.
8. Ferry JD. Viscoelastic properties of polymers. New York, NY: Wiley, 1980.
9. Fung YC. Biomechanics. Mechanical properties of living tissues. New York, NY: Springer, 1993.
10. Goodwin JW, Hughes RW. Rheology for chemists: an introduction. Royal Society of Chemists, UK, 2000.
11. Riande E, Diaz-Calleja R, Drolongo MG, Masegosa C, Salom C. Polymer viscoelasticity: stress and strain in practice. New York, NY: Marcel Dekker, 2000.
12. Yang M, Taber LA. The possible role of poroelasticity in the apparent viscoelastic behavior of passive cardiac muscle. J Biomech 1991;24:587–597
13. Simon BR, Kaufmann MV, McAfee MA, Baldwin AL. Finite element models for arterial wall mechanics. J Biomech Eng 1993;115:489–496.
14. Konofagou EE, Harrigan TP, Ophir J, Krouskop TA.Poroelastography: imaging the poroelastic properties of tissues. Ultrasound Med Biol 2001;27:1387–1397.
15. Scott GC, Korostoff E. Oscillatory and step response electromechanical phenomena in human and bovine bone. J Biomech 1990;23:127–143.
16. Roitt I, Brostoff J, Male D. Immunology (5th ed.). St. Louis, MO: Mosby, 1998.
17. Cotran RS, Kumar V, Collins T. Robbins pathological basis of diseases. Philadelphia, PA: W. B. Saunders, 1999.
18. Piattelli A, Scarano A, Piattelle M. Histological observations on 230 retrieved dental implants: 8 years experience (1989–1996). J Periodontol 1998;69:178–184.
19. Tal H, Dayan D. Spontaneous early exposure of submerged implants: III. Histopathology of perforated mucosa covering submerged implants. J Periodontol 2000;71:1231–1235.
20. Schliephake H, Schmelzeisen R, Maschek H, Haese M. Int J Oral Maxillofac Surg 1999; 28:323–329.
21. National Eye Institute Summary http://www.nei.nih.gov/resources/strategicplans/neiplan/frm./.5Flens.htm
22. Beasley AN, Auffarth GU, Von Recum AF. Intraocular lens implants: A biocompatibility review. J Inves Surg 1996;9:399–413.
23. Hollick EJ, Spalton DJ, Ursell PG, Pande MV. Biocompatibility of poly(methyl methacrylates), silicone, and AcrySof intraocular lenses; randomized comparison of the cellular reaction on the anterior lens surface. J Cataract Refract Surg 1998;24:361–366.
24. Saika S, Miyamoto T, Yamanaka A, et al. Immunohistochemical evaluation of cellular deposits on posterior chamber intraocular lenses. Graefes Arch Clin Exp Ophthalmol 1998;236:758–765.
24a. Abbas AK, Lichtman AH. Basic immunology: the functions and disorders of the immune system (2nd ed.). Philadelphia, PA: W. B. Saunders, 2004.
25. Amon M, Menapace R. In vivo observation of surface precipitates of 200 consecutive hydrogel intraocular lenses. Ophthalmologica 1992;204:13–18.
26. Amon M, Menapace R, Radax U, Freyler H. In vivo study of cell reactions on poly(methyl methacrylate) intraocular lenses with different surface properties. J Cataract Refract Surg 1996;22 Suppl 1:825–829.

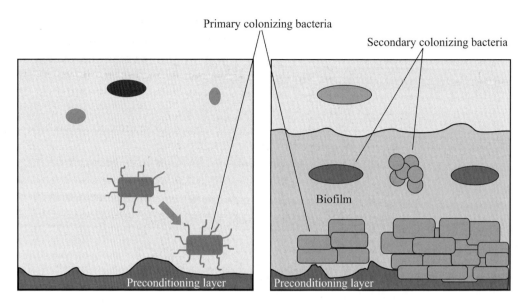

Fig. 9. Implant failure mechanism owing to bacterial infection.

substrate, which represents a negative free energy of adhesion ($-\Delta F_{adh}$), implying a favorable condition for the bacterium to adhere to the prosthetic surface.

FUTURE TRENDS IN PROSTHETIC DESIGN

The initial objective of modern biomaterial design was to produce materials that matched the properties of the replacement tissue with minimum toxic response. The past two decades have focused on developing materials that would slowly resorb and ultimately be replaced by regenerating tissues. The trend now is to synthesize polymers or other materials that may control the regeneration of tissue as well as mimic or enhance the healing process, ultimately producing a natural tissue. For example, superoxide dismutases are known to impart anti-inflammatory character. Udipi et al. *(96)* showed that incorporating superoxide dismutase-like substances on the surface of biomaterials significantly improves biocompatibility. For hemocompatiblity, modifying the surface of biomaterials to mimic the endothelium could render it nonthrombogenic. Phosphatidylcholine, a constituent of the lipid bilayer of the cell membrane, is recognized as one of the cell surface modifiers that render red blood cells and vessels nonthrombogenic. Thus, several studies have used phosphatidylcholine to coat biomaterial surfaces *(34,35)*. Resorbable polymeric scaffolds with growth factors for regeneration of natural functional tissues have been recently produced. Currently, prostheses are imagined as totally synthetic materials. In the future, however, prosthetics may be replaced with natural tissues obtained via tissue engineering. Thus, the future role of prosthetic devices may be to provide the initial scaffold, within which the natural tissue will grow to integrate and eventually absorb the scaffold.

REFERENCES

1. Bronzino JD. (ed.). The biomedical engineering handbook. Boca Raton, FL: CRC Press, 1995.
2. Gristina AG, Giridhar G, Gabriel BL, Naylor PT, Myrvik QN. Cell biology and molecular mechanisms in artificial device infections. Int J Artif Organs 1993;16:755–763.

along with their extracellular products, on the surface of a prosthesis. Development of a biofilm involves several steps: first is the deposition of proteins from blood or tissue fluids onto the surface of the biomaterial (the type and extent of protein deposition is a function of the surface characteristics of the biomaterial); second is the initial reversible attachment of the microbe to the conditioned surface, followed by permanent attachment; third is the growth and colonization of the bacterium, along with more deposition of extracellular polymers; finally, there is the formation of a functioning biofilm.

The second step is critical and several theoretical models have been developed to explain the complex interactions that occur among the microbe, the surrounding tissue fluid, and the conditioned prosthetic surface as the bacterium approaches. The two most popular models are the Derjaguin, Landau, Verwey, and Overbeck models, based on stability of colloidal dispersions *(92,93)*, and the surface free energy model, which utilizes the theory of thermodynamics *(94,95)*.

In the Derjaguin, Landau, Verwey, and Overbeck models, the total potential energy of interaction, (V_t), between charged colloidal surfaces is assumed to be the linear superimposition of the attractive London dispersion forces (V_a) and the electrostatic repulsion (V_e) forces and is a function of their separation. For a bacterium, modeled as a spherical colloid, the changes in potential energy as the bacterium approaches the surface (modeled as a plate) is represented as shown in the following equation *(92,93)*.

$$V_t = V_a + V_e$$

$$V_t = -AijR/6h + 4\pi\varepsilon\alpha^2\psi^2/R \exp(-kR)$$

where Aij is a coefficient characteristic of material i interacting with material j, R is the radius of the sphere, h is the distance of separation from the surfaces, ε is the permitivity of the fluid phase, α is the thickness of the electrical double layer, $R = (2\alpha+h)$, and k is the Debye-Huckle decay parameter, which is a function of the ionic strength and permitivity of the medium. The significance of the above equation is that the bacterium has to overcome at least two potential energy minima as it approaches the surface. Their magnitude is a function of the charge on the bacterium, the characteristics of the material, and the ionic strength of the fluid. The minimum of the weak London dispersion forces (or van der Waals interactions) usually occurs at large separation distances. As the bacterium advances closer to the surface, the balance between opposite electrostatic forces (i.e., attractive [hydrogen bondings, hydrophobic interactions] and repulsive forces [like electrostatic forces]) takes over. If and when the bacterium overcomes the electrostatic minimum, it adheres irreversibly to the conditioned surface (*see* Fig. 9).

The thermodynamic model is based on the concept that the measure of attraction between two solids across an interface is the reversible work of attraction W_{adh}. This quantity is given by the relationship of Duprè *(94,95)*.

$$W_{adh} = \gamma_{SF} + \gamma_{BF} + \gamma_{BS}$$

Where W_{adh} is the work required to separate two solids in a fluid environment, in our case the bacterium (B) from the surface (S) within the tissue fluid continuum (F). The interfacial energies involved in such a situation would include the bacterium–fluid, the conditioned prosthetic implant–surface fluid, and the bacterium–prosthetic interfaces. If W_{adh} is positive, then work has to be performed to separate the bacterium from the

vitronectin, C3b, and so on) present on the implants. This initiates their phenotypic transformation from monocytes to macrophages, a process characterized by increase in cell size, secretion of inflammatory mediators, and expression of membrane proteins (61–64,73,74). The products made by activated macrophages may mediate tissue injury and fibrosis. In the presence of prolonged inflammation, macrophages are maintained at the site by continued recruitment of monocytes from blood, sustained local proliferation *in situ*, and/or increased longevity and immobilization. The macrophage is considered the central figure (17) of chronic inflammation because it is capable of producing an impressive arsenal of mediators that can cause tissue destruction or fibrosis. Thus, it is not surprising that it has been implicated in a number of adverse effects, such as osteolytic changes around joint implants (75–77), stress cracking of pacemaker leads (66,78), degradation of biomaterial implants (79,80), and fibrosis surrounding breast prostheses and many other implant types (81).

One of the characteristic functions of the macrophage is debridement of devitalized tissues and foreign materials by phagocytosis. Degradable particles smaller than 60 μ can be phagocytized, if coated with opsonins (IgG or C3b), and digested. Large particles, or nondegradable ones, that cannot be processed, however, eventually cause local macrophages to fuse and form multinucleated FBGCs that surround and isolate the particle or prosthesis from the surrounding tissue. The fusion of macrophages is believed to be controlled by IL-4 and IL-13, by upregulation of mannose receptors (82), interactions among intercellular adhesion molecules (83,84), and transglutaminase activity (85). Foreign body multinuclear giant cells typically form 7 to 28 days after prosthesis placement through cell–cell fusion and produce IL-1α and TNF. After 2 months, the production of IL-1α and TNF decreases, and production of transforming growth factor (TGF)-β is very high, suggesting that multinuclear giant cells are an active source of inflammatory cytokines. The membranes of adherent macrophages fuse, resulting in multinucleated FBGCs that may contain hundreds of nuclei (86).

Kao et al. (65,87) have proposed a mathematical model describing the giant cell-size distribution, the density of adherent macrophages, and the kinetics of giant cell fusion. In this model, each macrophage is analogous to a monomer and the process of cell fusion is similar to the condensation polymerization process (88). In the cell-size distribution model, the number of FBGCs (Nx) with a cell area of x is defined as

$$Nx = p^{ax-3}(1-p)$$

where p is the ratio of the number of fused cells compared to the initial population, and 1/a is a constant relating the density of nuclei within the giant cell area. The rate constant of cell fusion (k) is given by

$$1/(1-p) = dtk + 1$$

where d is the density of adherent macrophages and t is the time. This model has been used successfully to analyze foreign body giant cell formation on polydimethylsiloxane, low-density polyethylene, polyetherurethanes, and poly(etherurethane urea) polymers (65).

Implant Failure Caused by Infection

Implant failure may also be attributed to bacterial infection. Most bacteria are found in multicellular biofilm communities anchored to the implant surface (Fig. 9). Biofilm (2,89–91) may be defined as a single- or multilayered aggregation of microorganisms,

of a sterile implant modulates this general homeostatic mechanism, usually culminating in the formation of FBGCs, the most striking histopathological feature seen on explanted prostheses. Similarly, the hallmark of an antibiotic-resistant, chronically infected implant is the presence of a biofilm. The following paragraphs highlight the molecular mechanisms that eventually culminate in giant cell or biofilm formation. In either case, the prosthesis may be destroyed, rendered dysfunctional, absorbed, walled off, or integrated with the normal tissue. The magnitude and duration of the inflammatory process has a direct impact on material biostability and biocompatibility *(61–66)*.

Acute Phase

Cells of patient or bacterial origin seldom attach irreversibly to the surface of implants; a layer of extracellular matrix is needed for permanent attachment. In the absence of any adsorbed proteins, the attachment of host cells is assumed to be reversible and transient. Thus, surface modification of the prosthesis is the first step in both biocompatibility and infection. Several changes occur at the interface where prosthetic materials come in contact with proteins. Although a large number of molecules can be adsorbed on implant surfaces, only certain ones have been identified as playing pivotal roles in tissue responses to the implants. These molecules mediate homeostasis and facilitate the migration of inflammatory cells toward them. There is overwhelming evidence that, in addition to tissue and plasma mediators, extracellular matrix molecules adsorbed on surfaces or in the vicinity of the implant also critically control the function of adjacent cells *(67,68)*. Examples of such molecules include glycoproteins (fibronectin, laminin, chondronectin, vitronectin, cytoactin), proteoglycans (dermatin, chondroitin, hyaluronic acid, and heparan sulfates), collagen, and fibrinogen. They interact with each other as well as with cell surface receptors. For example, fibronectin has binding sites for heparan sulfate, collagen, C3bi (complement), cytoactin, extracellular molecules, and hyaluronic acid. All of these are present on cell surfaces, are adsorbed on the surface of biomaterials, and may be involved in monocyte adhesion to biomaterials *(67,68)*. Although preadsorbed immunoglobulin (Ig)G plays no role in short-term (<24 hours) recruitment or adhesion of the macrophages in vivo, it enhances long-term macrophage adhesion in vitro *(69,70)*. The polymorphonuclear leukocytes adsorbed to the surface have a short lifetime, and usually disappear from the exudates. Both macrophage inflammatory protein 1α and monocyte chemoattractant protein 1 have been shown to participate in the chemotaxis of phagocytes toward biomaterial implants *(71,72)*.

Chronic Phase

Inflammation is considered chronic if it persists beyond a few weeks and is accompanied by elements of tissue destruction as well as repair. In contrast to acute inflammation, which is manifested by edema, vascular leakage, and neutrophilic infiltration, chronic inflammation is characterized by the presence of high local concentrations of mononuclear cells, such as lymphocytes, macrophages, and plasma cells. Histopathology of explanted prosthetic materials shows a fibrous capsule laden with macrophages and FBGCs. The precursor of the macrophage is the circulating monocyte. Monocytes migrate from the vascular system into tissues governed by the same factors that facilitate the migration of neutrophils during acute inflammation. Once at the target site these cells adhere via integrin–protein interaction to preadsorbed proteins (fibrinogen,

Table 3
Examples of Tissue Reactions to Prostheses

Prosthesis	Histolology	Infection/complication	References
Breast	T cells, foamy Mϕ, foreign body giant cells (FBGC), macrophage	Dense fibrovascular connective tissue	38,41
Cardiac valves and pacemakers	Endothelial cell ingrowth, macrophages, FBGC,	Calcification, disintegration	29–31
Hernia (ePTFE)	Fibroblasts, macrophages, rare FBGC, collagen deposition	Seroma, fistula, infection	52,53
Intraocular implants	FBGC	Posterior capsular opacification	24–27
Joint prosthesis	MNGC-FB, C3bi-receptor Mϕ,	Periprosthetic granulomatosis, loosening of the prosthetic stem, particulate debris	42–44,46, 47,51, 54, 55
Ossicular implant	Multinucleated FBGCS	Degradation of PE	56,57
Penile	Biofilm formation	Infection, partial extrusion, progressive neuropathy	58–60

sis and surrounding bone *(42)*. Methylmethacrylate is the material used most often as cement for prosthetic stabilization within bone. Polymethylmethacrylate, in cemented arthroplasty, although immunologically relatively inert, induces mononuclear-cell migration and is often associated with adverse lytic reactions. Organisms bound to it are more resistant to antibiotics than are those bound to metals *(43–45)*. In cementless cases, both polyethylene and titanium debris cause adverse reactions and may play a role in macrophage activation and the release of mediators of bone resorption *(46)*. Metallic debris is seen in the periprosthetic layer as early as 8 months postoperatively, usually within histiocytes or surrounded by FBGCs *(47)*. At sites of implant loosening, the G protein-activated fibroblasts localized in the periprosthetic granulomatous membrane are the source of macrophage chemoattractant factors and proinflammatory mediators *(48)*. In most biodegradable orthopedic fixation devices, inflammatory foreign body reaction is observed *(49,50)*. Because of the close proximity of most orthopedic implants to bone, macrophages in arthroplasty pseudomembranes are capable of forming osteoclasts, resulting in local bone resorption *(51)*. Table 3 summarizes the tissue reactions to prosthesis.

Molecular and Cellular Events in Tissue Reaction to Biomaterials

Surgical implantation of a prosthetic material inflicts a planned and controlled injury, which elicits a repair mechanism to maintain homeostasis. However, the introduction

ing degrees of nonspecific foreign body reaction were observed, with heparinized surfaces having the least number of adherent cells *(25–27)*. In retinal detachment surgery, a silicone buckle encircles the sclera. D'Hermies *(28)* observed explanted buckles to be extensively encapsulated with granulomatous foreign body giant cell reaction.

Cardiac and Vascular

The treatment for clinically significant heart valve disease often entails replacement with prosthetic valves. Some are totally synthetic, whereas others employ biological tissues. Histological examination of failed bioprosthetic heart valves indicates that failure is often due to invasion of microphages and accompanying infection or calcification *(29–31)*. In arterial grafts, four phases of the response to a prosthesis have been identified: thrombosis, inflammation, proliferation, and remodeling. The severity of tissue injury following deployment of intra-arterial stents correlates with the severity of inflammation and neointimal growth *(32,33)*. The major requirements for biocompatibility of vascular grafts are antithrombogenicity, compliance matching, and the formation of neointima without excessive thickening. For hemocompatiblity, modifying the surface of the prosthesis to mimic the endothelium would render it nonthrombogenic. Phosphatidylcholine, a constituent of the lipid bilayer of the cell membrane, is recognized as one of the cell surface modifiers that render red blood cells and vessels nonthrombogenic. Several studies have observed improved hemocompatibility after using phosphatidyl choline to coat biomaterial surfaces *(34–36)*.

Breast Implants

Millions of women have undergone implantation of breast prostheses for postmastectomy breast reconstruction or for the purpose of aesthetic augmentation. Although most patients with breast implants are asymptomatic, there exists a small cohort in whom the silicone breast implant has been associated with numerous medical problems requiring explanation. Histopathological analysis of these silicone breast implants has revealed exuberant capsular formation with fibrocollagen deposition, macrophages, granulomas, and foreign body giant cells (FBGCs) *(37)*. The periprosthetic capsule itself appears histologically similar to proliferative synovitis, exhibiting papillary villous synovial-like hyperplasia *(38,39)*. The level of tumor necrosis factor (TNF)-α expressed by the cells in the capsule is proportional to the number of macrophages *(40)*. CD44, an activation and intracellular adhesion marker, is frequently observed on these cell surfaces *(41)*.

Orthopedic

Surgery for musculoskeletal disorders ranges from spinal stabilization, fixation of simple fractures, and reconstruction of massive bone defects, to joint and ligament replacement or augmentation. In most of these cases, nonbiological engineering materials are used, ranging from metals, ceramics, and composites to degradable and nondegradable polymers. These materials and their debris interact with surrounding tissues and elicit a variety of host-adaptive responses, depending on whether they are placed in the intraosseous or extraosseous environment. In hip arthroplasty, fibroblasts, giant cells, and macrophages are the predominant cell types at the interface between prosthe-

from it into the surrounding inflamed tissue. Integrins mediate cell-extracellular matrix interactions, and selectins promote adhesion to carbohydrate residues. The chemotactic molecules include components of the complement system, particularly C5a, products of lipoxygenase pathway (LTB4), and cytokines (e.g., interleukin [IL]-8), which direct leukocytes to the site of inflammation. Platelets are also a source of vasoactive agents and may be activated by platelet-activating factors from neutrophils.

HISTOPATHOLOGY OF EXPLANTED PROSTHESES

Dental

The use of dental implants for the edentulous patient has become an important mode of treatment. Good integration of the implant-to-bone interface is a prerequisite for long-term stability. Failure rates in dental implants are generally low (<2%). The primary causes of failure are excessive mobility, peri-implantitis, or fractures *(17,18)*. In cases of early spontaneous mucosal perforation at the site of dental implantation, histology of explants reveals diffuse infiltration of chronic inflammatory cells *(19)*. For patients with tempomandibular joint dysfunction, silicone-containing prosthetic joints are often used. Histopathology of failed tempomandibular joint explants showed scattered fragments of silicone along with histiocytes and T lymphocytes in the vicinity *(20)*.

Ophthalmic Implants

More than 1 million cataract extractions are performed each year in the United States alone and, in almost all cases, the natural cataractous lens is replaced with an intraocular lens *(21)*. All implanted intraocular materials removed for various reasons show a mixture of cellular and proteinaceous deposits. The cells seen on the lens surface are granulocytes, macrophages, giant cells, and fibroblast-like cells. Their combined presence indicates that a foreign body reaction has occurred. These adherent cells are responsible for deposition and regulation of extracellular matrix *(22–24)*. In a prospective clinical trial evaluating the cellular reaction to various types of intraocular lens material, vary-

---

Fig. 8. *(opposite page)* Complement activation. The two pathways of complement activation are the classical and the alternative pathways. Alternative pathway: C3 is the major complement protein that circulates in the bloodstream. When it encounters a microbe, C3 cleaves into C3a and C3b. C3b attaches to the surface of the microbe. Another factor, factor B, breaks and forms a Bb fragment that forms a complex with C3b. This complex acts as a C3 convertase: it hydrolyses more and more C3 to produce C3b that attaches to the microbe. Complexes of C3bBbC3b are formed and these act as a C5 convertase that cleaves C5 into fragments that are important for later complement activation steps. Classical pathway: antibodies such as IgG bind to antigens on the microbial surface. C1 complement protein binds to the Fc region of the antibodies, and when it contacts two or more Fc regions, it becomes active enzymatically. As a result of this activity, C4b and C2a form a complex and bond with the antibody and the microbial surface. This acts as a C3 convertase, resulting in the production of C3b. The C3b forms a complex with C4b2a to function as a C5 convertase. So both pathways eventually result in the coating or opsonization of the microbe with C3b. The C5 convertase initiates the late steps of complement activation that leads to the formation of the Membrane Attack Complex, through which substances enter the cell, causing cell death. (Adapted from ref. *24a.*)

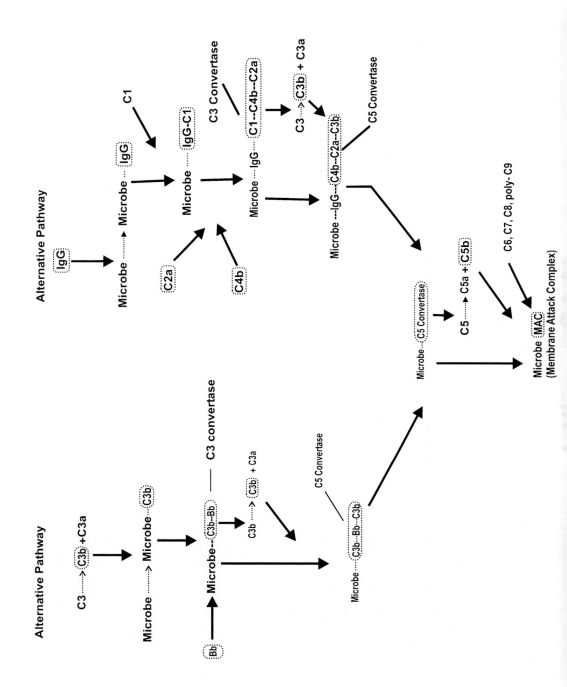

including the intrinsic and extrinsic coagulation cascades, the fibrinolytic pathway, the complement system, the kinin-generating systems, and platelets determine the rate and extent of clot formation and resolution. There are many ways in which coagulation factors and their byproducts play important roles in response to injury. For example, thrombin, in addition to its procoagulant properties, also acts as a growth factor and a cytokine that promotes monocyte, fibroblast, and endothelial cell influx into an area of recent injury, setting the stage for removal of cellular debris and for wound healing. Factor Xa, tissue factor, and fibrinogen fragments similarly have roles as inflammatory mediators and cell-growth regulators. In addition, contact factors may play a role in host defense mechanisms.

Complement System

The complement system is part of the immune system. It employs an enzyme-cascade process to discriminate between self and nonself. The complement system consists of 11 known component proteins that are found in greatest concentration in plasma. The complement cascade may be initiated by any one of several events, including activation of factor XII (thrombin), or the presence of necrotic tissue. In response to injury, a series of enzymatic reactions converts circulating C3, the cascade's most abundant component, to C3a and C3b. (Details of conversion mechanisms are shown in Fig. 8.) The fate of the surface-adsorbed C3b is critical in distinguishing self from nonself. Once C3b is formed on a self-cell, it is catabolized, preventing further amplification. In the case of nonself materials (such as prostheses) that lack intrinsic inhibitors, the deposited C3b acts as a binding site for other complement factors, resulting in increased deposition of C3b on the same surface, thus marking it as nonself. In general, components of the complement system increase vascular permeability, act as chemotaxins (attractants of leukocytes), and mediate the adhesion of polymorphonuclear leukocytes to complement-coated surfaces.

Cells and Signaling Molecules in Inflammation

Leukocytes are the primary cell types involved in inflammation. They are generally referred to as white blood cells and are usually involved in defense against infection or foreign bodies. Leukocytes include lymphocytes consisting of B and T cells, phagocytes consisting of mononuclear phagocytes, neutrophils, eosinophils, and auxilliary cells (basophils, mast cells, and platelets). Each produces and secretes only a particular type of cytokine or inflammatory mediator. For example, B cells produce antibodies, whereas mast cells produce inflammatory mediators. Under normal conditions, mast cells are the only leukocytes that reside outside of blood vessels; they lie outside of, but in close proximity to, a tissue's vessels. All other leukocytes are housed in the blood and are in constant transit between various tissues. The migration of these circulating leukocytes, particularly to the site of injury, is a complex process affecting the various cell types differently and is dependent on their state of activation and the local concentration of many signaling molecules. Cellular adhesion molecules, integrins, selectins, and chemotactic molecules are some examples of agents that orchestrate the movement and activity of cells during inflammation. The intercellular adhesion molecules are membrane-bound proteins. When expressed on vascular endothelium during injury, intercellular adhesion molecules help leukocytes adhere to the blood vessel and exit

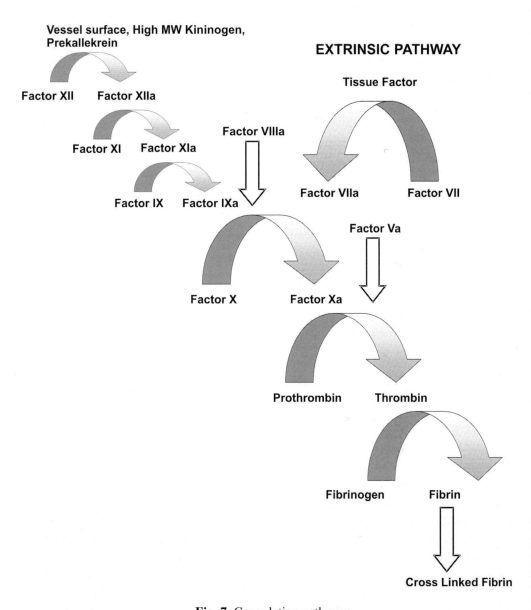

Fig. 7. Coagulation pathways.

Coagulation Pathway

The coagulation mechanism is conventionally divided into intrinsic and extrinsic pathways. Blood-borne proteins, along with factor XII, initiate the intrinsic pathway, whereas tissue lipoproteins and factor VII activate the extrinsic pathway. Both pathways independently produce factor X, which cleaves prothrombin to thrombin. Thrombin cleaves fibrinogen, forming fibrin, which is insoluble. This is the basis of the clot. The various steps of the coagulation pathway are shown in Fig. 7. Many factors,

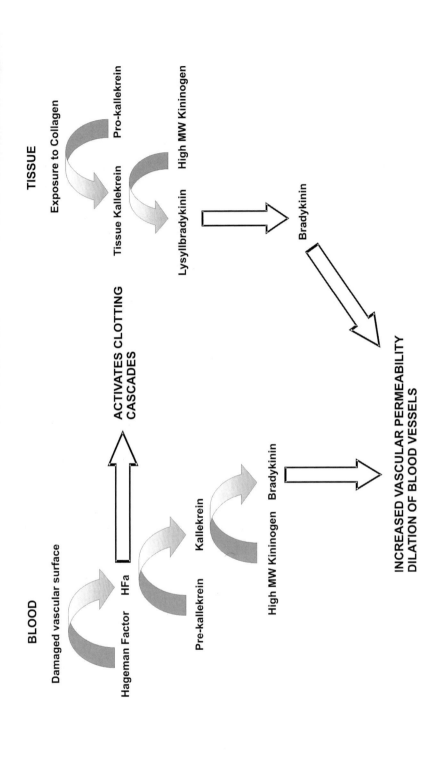

Fig. 6. Kinin system. High-molecular-weight kininogen and pre-kallekrein circulate in the plasma with the Hageman factor. Once they are exposed to extravascular tissue such as collagen, the Hageman factor is activated. The activated Hf triggers the coagulation cascade, and also helps convert pre-kallekrein to kallekrein. Kallekrein in turn helps for converting high-molecular-weight kininogen to bradykinin. Both Hfa and Kallekrein act as positive feedback factors, increasing the formation of Hfa. Bradykinin causes vasodilation and increased permeability of the blood vessels. (Adapted from http://nic.savba.sk/logos/books/scientific/node31.html http:\nic.sav.sk/logos/books/scientific/node3.html#Section0044600000000000000.)

in a volume change. Viscous forces in the fluid oppose this flow, and the stress relaxation time, t, for the poroelastic solid is given by:

$$t = d^2 / (\pi^2 D)$$

where t is time, d is the radius or thickness of the sample, and D is the collective diffusion rate.

MOLECULAR AND CELLULAR EVENTS OF HOMEOSTASIS

Classifications of injury to cells or tissues commonly employ such categories as physical, chemical, infectious, immunological, or age-related—terms that overlap but may be useful nonetheless. The inflammatory response is carefully choreographed by endogenous chemical mediators derived from tissue and plasma that direct and localize the response exclusively to the site of injury. Subsequent to an injury, the temporal sequence of the normal homeostatic phases usually progresses from acute inflammation, through formation of granulation tissue leading to fibrosis, to resolution or a state of chronic inflammation. In the acute phase, the injured tissue responds by controlling bleeding, identifying any foreign material, and recruiting inflammatory cells. The chronic phase is characterized by tissue scarring, along with hypertrophy, hyperplasia, metaplasia, dysplasia, or autophagy. If an implant is present, these processes cause the prosthetic device to be destroyed, rendered dysfunctional, absorbed, walled off, or integrated into the normal tissue.

Acute Processes

The initial response of injured tissue to bleeding is vasoconstriction, which is almost instantaneous, but also transient, and can only slow, rather than stop, the loss of blood. To ensure proper control of the coagulation process, the injured tissue recruits mediators from plasma and surrounding tissue. Examples of tissue-derived mediators are vasoactive amines (histamine, 5-hydroxytryptamine) and acidic lipids (prostaglandins). The plasma's interconnected mediator-producing systems include the kinin system, the complement system, and the clotting system. The kinin system dilates the vasculature to prepare the injured site for regeneration. The complement cascade facilitates the removal of bacteria and foreign bodies. The initiation and activation of the coagulation cascade results in the formation of a fibrin-based hemostatic clot to stop further bleeding. Roitt *(16)* and Cotran *(17)* provide a more thorough discussion.

Kinin-Generating Systems

The kinin system dilates blood vessels, increases vascular permeability, and is chemotactic for neutrophils and mononuclear phagocytes. Like the clotting and complement systems, the kinin system consists of a series of enzymatic steps that progressively amplify the number of molecules involved. The schema of the kinin system is shown in Fig. 6. It begins the activation of Hageman factor (factor XII) by contact with a host of substances, such as collagen, basement membrane, glass, and bacterial endotoxins. Once activated, factor XII triggers a cascade of events that culminate in the production of large amounts of bradykinin (kinin), which is primarily responsible for the dilation of blood vessels and their increased permeability. Activated Hageman factor also triggers the fibrinolytic system and the clotting cascades.

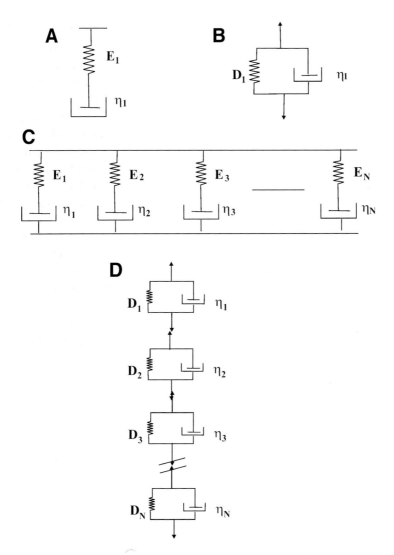

Fig. 5. Representation of mechanical analogs of visoelastic liquids and solids by Maxwell and Voigt models.

ered as functions of its composition and of the orientation of the constituent components. For example, the elastic moduli of tissues vary from <1 kPa (cell membranes) to >1 GPa (dense bones). This wide range of elastic moduli of tissue corresponds to those of polymers, metals, and ceramics used in prostheses. While viscoelastic models usually describe the mechanical properties of tissues quite well, it must be recognized that many tissues are actually poroelastic *(12–14)*. Cartilage, for example, may be modeled as a viscoelastic solid, but it is primarily poroelastic *(15)*. The main difference between a viscoelastic solid and a poroelastic one is that, in a poroelastic material, the stress relaxation time of a sample is proportional to the square of its characteristic linear dimension (e.g., the thickness of a disc-shaped sample or the radius of a spherical sample). This means that, when a poroelastic solid, such as a sponge, is subjected to compressive strain, the fluid within the pores flows out of the solid material that results

body, its strain will increase and continue to increase as a function of time. This behavior is called creep phenomenon (*see* Fig. 4C). The opposite of creep is stress relaxation, in which a body subjected to a step strain held constant experiences stress that decreases with time (*see* Fig. 4D). If a body is subjected to cyclic loading and unloading, the stress–strain plot of the loading cycle differs from that of the unloading cycle. This phenomenon is called hysteresis. In addition to creep (where stress is held constant) and stress relaxation (where strain is held constant) experiments, sinusoidal dynamic experiments are used to determine dynamic modulus values as a function of angular frequency, ω, rather than of time. The classic example of a viscoelastic material with a time-dependent property is Silly Putty. When Silly Putty is thrown rapidly against a wall, it bounces back like an elastic solid; however, when pulled slowly, it flows like a liquid.

Traditional depictions of viscoelastic behavior involve mechanical models comprised of springs (representing the elastic component) and dashpots (a cup filled with a Newtonian fluid of viscosity η, in which a piston is placed, representing the viscous component) in series or parallel (or in combination). The Maxwell and the Voigt models are two mechanical analogues of viscoelastic liquids and solids, respectively. The Maxwell model *(10,11)* is comprised of a spring and a dashpot in series (*see* Fig. 5A). In this model the stress is the same in both elements and the total strain equals the sum of the strains in each element. Hence, for the Maxwell model, the rate of strain is the sum of the strain rates of the spring plus the dashpot, as shown in the following equation:

$$\frac{d\varepsilon}{dt} = \frac{d\varepsilon_{spring}}{dt} + \frac{d\varepsilon_{dashpot}}{dt} = \frac{1}{dt}\frac{d\sigma}{dt} + \frac{\sigma}{\eta}$$

In the Voigt model *(10,11)*, the spring and dashpot are in parallel rather than in series (*see* Fig. 5B). Here, total stress is the sum of the stresses in each element and can be expressed as shown in the following equation:

$$\sigma(t) = \sigma(t)_{spring} + \sigma(t)_{dashpot} = \varepsilon(t)E + \eta\frac{d\varepsilon}{dt}$$

The Maxwell and Voigt models can be used to predict simple linear viscoelastic behavior. However, to describe the behavior of most viscoelastic fluids and solids, more generalized models are required. For fluids, the Maxwell-Weichart model *(10,11)* is a generalized model that is comprised of Maxwell elements connected in parallel (*see* Fig. 5C). In each of the individual elements, the strain is the same and the total stress is the summation of the individual stresses experienced by the individual elements. Similarly, for solids, the Voigt-Kelvin model *(10,11)* is comprised of Voigt elements connected in series (*see* Fig. 5D). Mathematically, the Maxwell-Weichart model is represented by this equation:

$$\frac{d\varepsilon}{dt} = \frac{1}{E_1}\frac{d\sigma_1}{dt} + \frac{\sigma_1}{\eta_1} = \frac{1}{E_2}\frac{d\sigma_2}{dt} + \frac{\sigma_2}{\eta_2} = \ldots\ldots = \frac{1}{E_N}\frac{d\sigma_N}{dt} + \frac{\sigma_N}{\eta_N}$$

The Voigt-Kelvin model is described by the following equation:

$$\sigma(t) = \sigma_1 + \sigma_2 + \ldots\ldots + \sigma_N$$

Biological tissues can be viewed as natural composites made of macromolecules, lipids, water, ions, and so on. The properties of a particular tissue can then be consid-

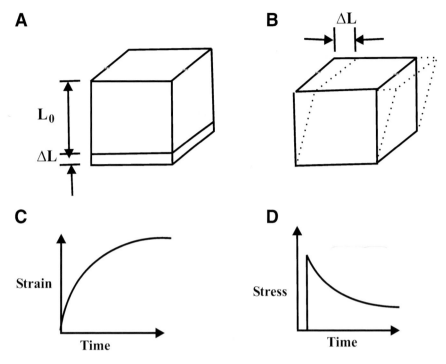

Fig. 4. Schematic representation of deformation of solids.

When a Hookean solid is subjected to simple shear stress, as denoted in Fig. 4B, the corresponding shear strain is termed γ where $\gamma = \tan \theta$. The shear modulus, or modulus of rigidity, G, and the shear compliance, J, are defined using the following equation:

$$G = \frac{\sigma_s}{\gamma} = \frac{1}{J}$$

The elastic moduli of polymers vary from 10^3 Pascals (1 kPa) to 10^6 Pascals (1 MPa) and are usually used to augment soft tissues. In contrast, most metals, metal alloys, and ceramics possess very high elastic moduli, ranging from 400 to 1500 MPa, which is considerably greater than the moduli of most carbon-based polymers.

In contrast to solids, a dominant characteristic of fluids is their viscous behavior. Here, stress is a function of strain rate. Stress and strain rate are related by the equation:

$$\sigma = \eta \frac{d\varepsilon}{dt}$$

where η is the viscosity of the fluid. Liquids that obey this equation are referred to as Newtonian fluids.

In reality, ideal behaviors, such as perfectly elastic deformation and perfectly viscous flow, are rarely seen, particularly in biological systems. Most biomaterials and tissues examined under normal laboratory conditions exhibit characteristics that usually fall between those of pure Hookean elastic solids and Newtonian viscous liquids. These intermediate materials are called viscoelastic and are said to possess time- or rate-dependent properties. If a constant force is suddenly applied to a viscoelastic solid

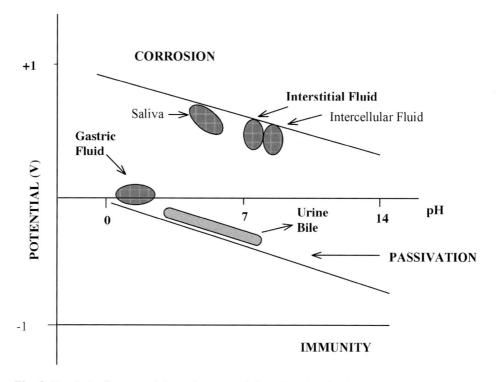

Fig. 3. Pourbaix diagram of the redox potential vs pH. The plot is demarcated into regions of corrosion, passivation, and immunity.

chanics, of the tissue being replaced or augmented. This section explains the behaviors of elastic solids and viscous fluids in preparation for understanding the behaviors of viscoelastic and poroelastic substances, because their properties are common both to prostheses and to the tissues they replace. For greater explanation than is provided in the following paragraphs, we recommend several textbooks (7–11).

Solid bodies deform when unidirectional tensile (or compressive) stress (termed σ where σ = force per unit area) is applied. This deformation is called strain and is represented by ε. This is a dimensional number ratio of the change in length to the original length ($\Delta L/L_0$, where L_0 is the initial length of the sample and ΔL is the change in length due to the application of the unidirectional tensile stress, see Fig. 4A). Whereas strain is dimensionless, a unit of stress is expressed in terms of newtons per square meter or dynes per square centimeter. For most engineering materials subjected to an infinitesimal strain in unidirectional stretching, the following equation is valid within a certain range of stresses:

$$E = \frac{\sigma}{\varepsilon} = \frac{1}{D}$$

The equation is called Hooke's law and materials obeying this law are termed Hookean solids. In the equation, E is a constant known as Young's modulus of elasticity (or the tensile modulus of the material). D is called the tensile compliance, which, as can be seen, is the reciprocal of the elastic modulus. The elastic moduli of most biomaterials may vary from 10^3 Pascals (1 kPa) to 10^9 Pascals (1 GPa).

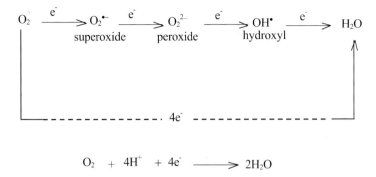

$$O_2 + 4H^+ + 4e^- \longrightarrow 2H_2O$$

Fig. 2. Pathways for the formation of reactive species from oxygen.

Table 2
Redox Potentials of Various
Metals Used as Prostheses

Reaction	ΔE° (volt)
Al ↔ Al^{+++}	−1.66
Ti ↔ Ti^{+++}	−1.63
Cr ↔ Cr^{++}	−0.56
Fe ↔ Fe$^+$	−0.44
Cu ↔ Cu^{++}	−0.34
H$_2$ ↔ 2H$^+$	0.00
Ag ↔ Ag$^+$	+0.80
Au ↔ Au$^+$	+1.68

by its electrochemical potential, the local pH, and the concentration of various ions in the medium. The interplay between electrochemical potential and pH is graphically summarized in Fig. 3 in a Pourbaix diagram *(5)*. The corrosion region is above an arbitrarily set equilibrium concentration of 10^{-6} M of the metal ion. At equilibrium concentrations below 10^{-6} M, metals corrode, but if the products of corrosion are inert, passivity occurs, preventing further corrosion.

Ceramics, unlike metal alloys, are refractory polycrystalline materials consisting of both metallic as well as nonmetallic atoms and are therefore represented structurally as ionic compounds. Their ionic bonds have very high dissociation energy and directionality, rendering them resistant to chemical and mechanical degradation and providing high mechanical strength and stiffness. Zirconia, porous aluminum oxides, and single-phase calcium-aluminates are examples of relatively bioinert bioceramics *(6)* that are not conducive to integration with tissues, whereas surface-reactive ceramics, such as porous hydroxyapatite and dense nonporous glass, promote the integration of prosthetic materials, forming strong bonds with adjacent cellular structures.

Mechanical Properties of Biomaterials and Tissues

Performance of any prosthesis in vivo requires not only biocompatibility at the tissue–prosthetic interface but also matching of the bulk properties, particularly biome-

consistency at a broad range of high temperatures, rather than melting into liquid form at a narrowly defined temperature. The average temperature at which this transition from solid to rubber occurs is referred to as the glass transition temperature (T_g).

Depending on the nature of the polymer, its surface properties (smoothness, etc.) may vary considerably from its bulk properties (hardness, brittleness, etc). This is particularly true for copolymers, in which one substance may migrate to the surface in preference to the other substance. Polymer surfaces may be dynamic, in which case they exhibit numerous sites of nonspecific, or low-energy binding, resulting in a very high total binding energy (3). The surface of a polymer can structurally adapt to adsorb proteins (4). This feature plays a very important role in surface biocompatibility, thwarting potential rejection of the prosthesis by the immune system. This is explained in greater detail in a later section of this chapter.

Metals, a second category, are typically crystalline inorganic materials with face-centered cubic structures. Their chemical bonds share highly mobile outer electrons with all atoms of the unit cell. Some metallic implants are pure, such as gold weights for eyelids, but most are alloys created by the inclusion of additional metals that are chosen to enhance specific physical or mechanical properties of the prosthesis. Most incite little antigenic or inflammatory response. As a result, they are used extensively in dentistry and orthopedics. The least reactive metals are titanium and tantalum. The alloys most commonly employed in medical devices are silver-tin-mercury-copper alloys that are used for dental amalgam, titanium alloys that are used in joint components, screws, and conductive leads, and chromium-cobalt alloys that are used in heart valves, joint components, and fracture fixation plates (1) (*see* Table 1).

The physiological environment within the human body (consisting of water, dissolved oxygen, chloride ions, hydroxides, amino acids, proteins, etc.) corrodes most metals. In this environment, particularly under alkaline conditions (pH 7.4 or higher), metals may hydrolyze, undergo redox reactions, or form organometallic complexes. The lowest free energy states in which metals can exist under physiological conditions are the free oxide state and soluble ligand complexes (amino acids and proteins). Metal complexes, by virtue of their mobile outer electrons, can reduce (donate electrons to) oxygen to produce reactive oxygen species, progressing from superoxide to hydrogen peroxide to hydroxyl radicals. Each intermediate step consumes one electron, resulting finally in the formation of water. Reducing agents, such as sugars, present in physiological tissue, may then reduce the oxidized metal complexes and thereby regenerate the original ligand complex—hence completing the redox cycle in which the metal complex acts as a catalyst. Corrosion resistance of a metallic implant is an important aspect of its biocompatibility because the four-electron reduction of oxygen, with its concurrent generation of reactive oxygen species (as shown in Fig. 2), may lead to severe local tissue injury and corrosion of the metal.

Thermodynamic considerations predict that the tendency of metals to corrode should correlate with their electrochemical potential (*see* Table 2). Experimentally, however, this order of reactivity is not usually observed because the reaction products of some metals, typically their metal oxides, form a thin surface film that inhibits further reaction, a property called passivity. Metals, such as silver and gold, that have redox potentials that are higher than that of hydrogen do not produce reactive oxygen species under physiological conditions. Whether a metal corrodes or exhibits passivity is determined

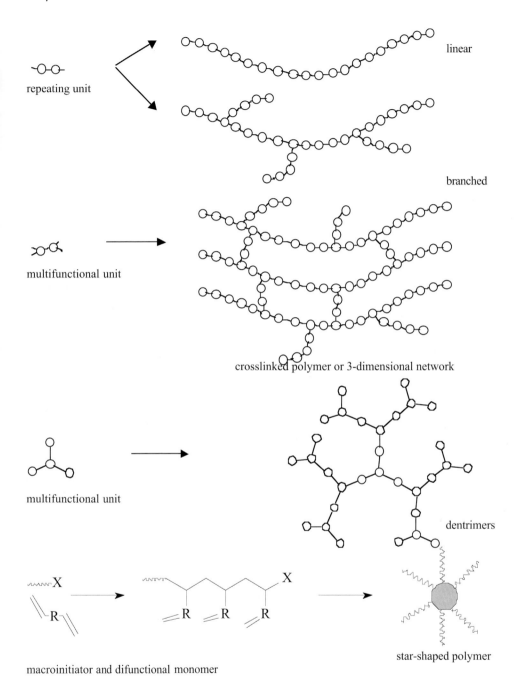

Fig. 1. Schematic representation of different type of polymer chain configurations.

Instead, they occur as semicrystalline or amorphous solids. Also, most polymers do not have a single unique molecular weight, exhibiting instead a distribution of molecular weights, which results in a characteristic distribution of physical, mechanical, and thermal properties. For instance, a solid synthetic polymer typically softens to a rubbery

Table 1
Examples of Types of Implants and Their Uses

Material	Applications
Polymethylmethacrylate	Orthopedics, ophthalmology, dentistry
Polyethylene	Nasal implants, maxillofacial, knee, and hip prostheses
Silicone	Mammary implants
Dacron polyester	Vascular grafts
Polytetrafluoroethylene	Abdominal hernias
Titanium alloys	Conductive leads, pacemaker cases
Cobalt-chromium alloys	Dental, heart valves, joint components, nails, screws
Stainless steel	Guide wires, vascular stents, fracture plates

degradable and nondegradable synthetic polymers, and composites. Typically, the end use dictates the type of prosthetic material chosen. Metals and ceramics are generally used for joint replacements, bone plates, and screws, whereas polymers and composites are usually used for soft tissue augmentation and heterologous tissues for cardiac valve replacement (*see* Table 1).

Biomaterials are now used in more disorders than ever before and often require interdisciplinary approaches to design and test them. Fifty years ago, implants were used primarily in geriatric patients with life-threatening conditions. Now they are used in healthy patients of all ages. Designing better devices, providing appropriate care for patients, and maintaining good function of the implants all require an understanding of the chemistry, physics, and mechanics of prosthesis and of the molecular and cellular events in tissue homeostasis. The first section of this chapter covers molecular, physical, and mechanical characteristics of various types of materials. The second deals with a variety of chemical reactions relevant to tissue homeostasis. The third highlights the histopathology of various implants and identifies key recurring features, whereas the final section incorporates all previously introduced concepts into a discussion of how biological tissue reacts to a prosthetic material.

MATERIALS USED IN PROSTHESES

Prosthetics can be totally manmade or only partially synthetic. Partially synthetic materials use chemically modified tissues, for example, porcine heart valves that have been treated to enhance biocompatibility. Polymers, ceramics, and metals are examples of totally synthetic materials.

Synthetic polymers are primarily formed by the linkage of monomers, giving rise to chains of chemical motifs. Linkage via addition reactions usually results in a nondegradable carbon backbone (such as polyethylene or polymethylmethacrylates), whereas linkage via condensation reactions typically creates a degradable heteroatomic backbone (e.g., polyurethane, polyesters, nylon). Both linkage techniques result in a variety of architectural forms, which may be linear, branched, multiarmed, star-shaped, or comb-like, and which dictate their mechanical properties (*see* Fig. 1). This wide variety of forms is exclusive to synthetic polymers that demonstrate other unique characteristics. For example, unlike small molecules, polymers are never totally crystalline.

8
Tissue Reaction to Prosthetic Materials

Nathan Ravi and Hyder Ali Aliyar

INTRODUCTION

Prostheses are manmade materials to replace or augment diseased or damaged body parts in a safe, reliable, economical, and physiologically acceptable manner. This chapter is restricted to materials that remain in intimate contact with blood, tissue, or body fluids for a prolonged time (weeks to years). Eyeglasses, hearing aids, wearable artificial limbs, and so on, although very important to rehabilitation, are not covered. Tissue reactions to disposable devices, such as contact lenses, or extracorporeal materials that are used briefly (dialysis equipment, etc.) are also outside the scope of this discussion. In addition, no particular distinction between blood vs tissue biocompatibility is made, as the molecular events in both are similar and assumed to be a part of the same physiological continuum. This chapter focuses on recent literature (1990 to present) that primarily deals with histopathological investigations of implants in human subjects. Because the interplay between tissues and prostheses involves various aspects of chemical, physical, and biological sciences, some basic concepts that will be useful to both the surgeon and the materials scientist are first introduced. These concepts are subsequently integrated to address the complex interplay between molecular and cellular reactions that occur when a foreign object is placed in the body.

Artificial materials have been used as tissue substitutes from the time of the ancient Greeks *(1)*. It was only around the 1860s, with the introduction of aseptic surgical technique, that the use of biomaterials became practical. At the beginning of the 20th century, metal alloys were the primary biomaterials used. The modern era of biomaterials may be traced back to the 1950s and 1960s, which saw significant advances in surgical technique and materials sciences. These innovations culminated in the introduction of new devices such as vascular prostheses, cardiac pacemakers, and acrylic bone cement, which vastly increased both the length and quality of life for patients. These early successes not only catalyzed the development of newer materials but also extended their applications.

In the past decade there have been significant increases in the number of implant procedures performed, the variety of materials used, and the types of tissues replaced or augmented. Currently, in the United States alone, more than 10 million such operations are performed every year, more than 3 million of which involve long-term implants *(2)*. Modern prostheses are now composed of metals, alloys, ceramics,

81. Rabinowitz J. A method for preserving confidentiality when linking computerized registries [letter]. Am J Public Health 1998;88:836.
82. Pear R. Bush accepts rules to guard privacy of medical records. The New York Times, 2001 March 13, p. A1.
83. Information Policy Committee IITF. Privacy and the National Information Infrastructure: principles for providing and using personal information. Available at: http://nsi.org/Library/Comm/niiprivp.htm. Accessed 2005.
84. Office for Civil Rights-HIPAA. Medical privacy-National Standards to Protect the Privacy of Personal Health Information. Available at: http://www.hhs.gov/ocr/hipaa. Accessed 2003.
85. The HIPAA Privacy Rule. Information for researchers. Available at: http://privacyruleandresearch.nih.gov. Accessed 2003.
86. Internal Market DG at the European Commission. Europa: The European Commission; Internal Market. Available at: http://europa.eu.int/comm/dg15/en/media/dataprot/index.htm. Accessed 2000.

55. Greenland S, Finkle WD. A case-control study of prosthetic implants and selected chronic diseases in Medicare claims data. Ann Epidemiol 1998;8:319–326.
56. Wheatley DJ, Crawford FA, Kay PH, et al. A ten-year study of the Ionescu-Shiley low-profile bioprosthetic heart valve. Eur J Cardiothorac Surg 1994;8:541–548.
57. Rabago G, Corbi P, Tedy G, et al. Five-year experience with the Medtronic Hall prosthesis in isolated aortic valve replacement. J Card Surg 1993;8:85–88.
58. Lafata JE, Koch GG, Ward RE. Synthesizing evidence from multiple studies. The role of meta-analysis in pharmacoeconomics. Med Care 1996;34(12 Suppl):DS136–DS145.
59. Temple R. Meta-analysis and epidemiologic studies in drug development and postmarketing surveillance. JAMA 1999;281:841–844.
60. Berlin JA, Colditz GA. The role of meta-analysis in the regulatory process for foods, drugs, and devices. JAMA 1999;281:830–834.
61. Da Costa A, Kirkorian G, Cucherat M, et al. Antibiotic prophylaxis for permanent pacemaker implantation: a meta-analysis. Circulation 1998;97:1796–1801.
62. Foote SB. Managing the medical arms race: public policy and medical device innovation. Berkley: University of California Press, 1992.
63. Institute of Medicine. Assessing medical technologies. 2. Washington, DC, National Academy Press, 1985.
64. Brook RH, Lohr KN. Efficacy, effectiveness, variations, and quality. Boundary-crossing research. Med Care 1985;23:710–722.
65. Ramsey SD, Luce BR. Technology assessment of medical devices. Am J Manag Care 1998;25:SP113–114.
66. Braslow NM, Shatin D, McCarthy DB, Newcomer LN. Role of technology assessment in health benefits coverage for medical devices. Am J Manag Care 1998;4 Spec No:SP139–SP150.
67. Goodman CS. Healthcare technology assessment: methods, framework, and role in policy making. Am J Manag Care 1998;4 Spec No:SP200–SP214.
68. Ramsey SD, Luce BR, Deyo R, Franklin G. The limited state of technology assessment for medical devices: facing the issues. Am J Manag Care 1998;4 Spec No:SP188–SP199.
69. Priester R. Using technology appropriately. Challenges for Minnesota's Health Technology Advisory Committee. Minn Med 1994;77:25–29.
70. Mendelson DN, Abramson RG, Rubin RJ. State involvement in medical technology assessment. Health Aff (Millwood) 1995;14:83–98.
71. ECRI. Tour Our World: ECRI; a nonprofit agency. Available at: http://www.healthcare.ecri.org/. Accessed 2000.
72. Medical Data International. Medical Data International: MDI Online. Available at: http://www.medicaldata.com. Accessed 2000.
73. Rettig RA. Health care in transition: technology assessment in the private sector. Santa Monica, CA: RAND, 1997.
74. Perry S, Thamer M. Medical innovation and the critical role of health technology assessment. JAMA 1999;282:1869–1872.
75. Mark DH, Glass RM. Impact of new technologies in medicine: a global theme issue [editorial]. JAMA 1999;282:1875.
76. Eisenberg JM. Ten lessons for evidence-based technology assessment. JAMA 1999;282:1865–1869.
77. National Institute for Clinical Excellence. National Institute for Clinical Excellence: NHS. Available at: www.nice.org.uk. Accessed 2000.
78. Bruning CO, III, Breslin DS, Morgentaler A, Staskin DR. Experience with the penile prosthesis and artificial urinary sphincter. J Long Term Eff Med Implants 1995;5:27–45.
79. Melton LJ, III. The threat to medical-records research. N Engl J Med 1997;337:1466–1470.
80. Pharmacoepidemiology. Available at: http://www.pharmacoepi.org/resources/cfr_parts.cfm. Accessed 2005.

30. Chernew M, Fendrick AM, Hirth RA. Managed care and medical technology: implications for cost growth. Health Aff (Millwood) 1997;16:196–206.
31. Steiner CA, Powe NR, Anderson GF, Das A. Technology coverage decisions by health care plans and considerations by medical directors. Med Care 1997;35:472–489.
32. Saksema S, Madan N, Lewis C. Implantable cardioverter-defibrillators are preferable to drugs as primary therapy in sustained ventricular tachyarrhythmias. Prog Cardiovasc Dis 1996;38:445–454.
33. Hawker GA, Coyte PC, Wright JG, Paul JE, Bombardier C. Accuracy of administrative data for assessing outcomes after knee replacement surgery. J Clin Epidemiol 1997;50: 265–273.
34. Steinberg EP, Whittle J, Anderson GF. Impact of claims data research on clinical practice. Int J Technol Assess Health Care 1990;6:282–287.
35. McDonald CJ, Overhage JM, Dexter P, Takesue BY, Dwyer DM. A framework for capturing clinical data sets from computerized sources. Ann Intern Med 1997;127(8 Pt 2): 675–682.
36. Waid MO. Brief summaries of Medicare and Medicaid: Title XVIII and Title XIX of the Social Security Act. Baltimore, Md; Health Care Financing Administration, Department of Health and Human Services, 1998.
37. Helbing C, Schieber G. Use of Medicare data in international comparisons. Health Policy 1990;15:45–66.
38. International Classification of Diseases, 9th Revision. Clinical Modification (ICD-9-CM). U.S. Department of Health and Human Services, ed. Washington, DC: Public Health Service, Health Care Financing Administration, 1991.
39. Medicare's National Level II Codes, 1998: HCPCS. Dover, DE: American Medical Association, 1997.
40. The USRDS and its products. Am J Kidney Dis 1999;34(2 Suppl 1):S20-S39.
41. Healthcare Cost and Utilization Project (HCUP-3), Nationwide Inpatient Sample. Agency for Health Care Policy and Research, ed. Release 3. Rockville, MD, 1994.
42. Cochran WG. Sampling techniques (3rd ed.). New York: Wiley, 1977.
43. Korn EL, Graubard BI. Analysis of health surveys. New York: Wiley, 1999.
44. Design and estimation for the National Health Interview Survey, 1995–2004. Hyattsville, MD: National Center for Health Statistics, Vital and Health Statistics, Series 2; 1997.
45. Brogan DJ. Software for sample survey data: misuse of standard packages. In: Armitage P, Colton T, eds. Encyclopedia of Biostatistics. New York: Wiley, 1998, pp. 4167–4174.
46. Shah BV, Barnwell BG, Bieler GS. SUDAAN User's Manual, Release 7.0. Research Triangle Park, NC: Research Triangle Institute, 1996.
47. Stata Statistical Software. College Station, TX: Stata Corporation, 1997.
48. Strom BL. Pharmacoepidemiology (3rd ed.). Sussex, UK: Wiley, 2000.
49. Woosley RL. Centers for education and research in therapeutics [see comments]. Clin Pharmacol Ther 1994;55:249–255. Overview: Centers for Education and Research on Therapeutics. Available at: http://www.certs.hhs.gov. Accessed 2005.
50. Brewer T, Colditz GA. Postmarketing surveillance and adverse drug reactions: current perspectives and future needs. JAMA 1999;281:824–829.
51. Blais R. Using administrative data bases for technology assessment in health care. Results of an international survey. Int J Technol Assess Health Care 1991;7:203–208.
52. Glikson M, Hyberger LK, Hitzke MK, Kincaid DK, Hayes DL. Clinical surveillance of a tined, bipolar, steroid-eluting, silicone-insulated ventricular pacing lead. Pacing Clin Electrophysiol 1999;22:765–768.
53. Chamberlain-Webber R, Barnes E, Papouchado M, Crick JP. Long-term survival of VDD pacing. Pacing Clin Electrophysiol 1998;21(11 Pt 2):2246–2248.
54. Greenland S, Finkle WD. A case-control study of prosthetic implants and selected chronic diseases [published erratum appears in 1997;7:367]. Ann Epidemiol 1996;6:530–540.

7. Garver D, Kaczmarek RG, Silverman BG, Gross TP, Hamilton PM. The epidemiology of prosthetic heart valves in the United States. Tex Heart Inst J 1995;22:86–91.
8. Moore RM, Jr., Hamburger S, Jeng LL, Hamilton PM. Orthopedic implant devices: prevalence and sociodemographic findings from the 1988 National Health Interview Survey. J Appl Biomater 1991;2:127–131.
9. Moore RM, Jr., Bright RA, Jeng LL, Sharkness CM, Hamburger SE, Hamilton PM. The prevalence of internal orthopedic fixation devices in children in the United States, 1988. Am J Public Health 1993;83:1028–1030.
10. Sharkness CM, Hamburger S, Kaczmarek RG, Hamilton PM, Bright RA, Moore RM, Jr. Racial differences in the prevalence of intraocular lens implants in the United States. Am J Ophthalmol 1992;114:667–674.
11. Sharkness CM, Hamburger S, Moore RM, Jr., Kaczmarek RG. Prevalence of artificial hips in the United States. J Long Term Eff Med Implants 1992;2:1–8.
12. Sharkness CM, Hamburger S, Moore RM, Jr., Kaczmarek RG. Prevalence of artificial hip implants and use of health services by recipients. Public Health Rep 1993;108:70–75.
13. Silverman BG, Gross TP, Kaczmarek RG, Hamilton P, Hamburger S. The epidemiology of pacemaker implantation in the United States. Public Health Rep 1995;110:42–46.
14. Silverman BG, Gross TP, Kaczmarek RG, Hamilton P, Hamburger S. Epidemiology of artificial knee implantation in the USA. The Knee 1995;2:95–102.
15. National Center for Health Statistics. Surveys and Data Collection Systems: National Health Interview Survey on Disability (NHIS-D). http://www.cdc.gov/nchs/about/major/nhis_dis/nhis_dis.htm. Accessed 2005.
16. National Center for Health Statistics. 1994 National Health Interview Survey on Disability, Phase I and Phase II. Available at: http://www.cdc.gov/nchs/about/major/nhis_dis/nhisddes.htm. Accessed 2000.
17. National Center for Health Statistics. Survey and data collection systems: National Health Interview Survey of Disability (NHIS-D). Available at: http://www.cdc.gov/nchs/about/major/nhis_dis/nhis_dis.htm. Accessed 2005.
18. National Center for Health Statistics. The mortality followback survey program. Available at: http://www.cdc.gov/nchs/about/major/nmfs/desc.htm. Accessed 2000.
19. National Center for Health Statistics. National Mortality Followback Survey. http://www.cdc.gov/nchs/about/major/nmfs/nmfs.htm. Accessed 2003.
20. National Center for Health Statistics. NHAMCS description. Available at: http://www.cdc.gov/nchs/about/major/ahcd/nhamcsds.htm. Accessed 2000.
21. National Center for Health Statistics. NAMCS description. Available at: http://www.cdc.gov/nchs/about/major/ahcd/namcsdes.htm. Accessed 2000.
22. National Center for Health Statistics. Survey instrument: national hospital ambulatory medical care and national ambulatory medical care surveys. Available at: http://www.cdc.gov/nchs/about/major/ahcd/surinst.htm. Accessed 2000.
23. Hospital Supply Index: Product analyses. Volume 1A. Plymouth Meeting, PA: IMS America, 1995.
24. Riordan P, Bickler G, Lyons C. Lessons of a hip failure. Registers of joint replacement operations should be set up [letter; comment]. BMJ 1998;316(7149):1985.
25. Miles J. National registry is also needed for neurological implants [letter; comment]. BMJ 1998;317(7173):1658–1659.
26. Product Performance Report: Bradycardia Products. Minneapolis, MN; Medtronic, Inc., 1998.
27. Rettig RA. The industrialization of clinical research. Health Aff (Millwood) 2000;19:129–146.
28. Eichenwald K, Kolata G. When physicians double as businessmen. The New York Times 1999; November 30, p. A1.
29. Ramsey SD, Pauly MV. Structural incentives and adoption of medical technologies in HMO and fee-for-service health insurance plans. Inquiry 1997;34:228–236.

use of such information for research purposes. Further information regarding HIPAA is available on the Internet *(83–85)*.

The issue of data privacy has also been a focus of the European Union, through the data protection directive, which has its own website *(86)*. This directive generally forbids processing of "special categories" of data without consent (including health) but does permit individual nations to establish exceptions with "suitable safeguards." Scientific research is noted as a basis for exception. Other countries, including Australia, are also developing more stringent requirements with respect to data privacy.

Summary

With populations of many countries throughout the world aging and increasing numbers of people over the age of 65, the need for information on medical devices and prosthetics will only accelerate. Those who place these devices and those who follow patients with implants long-term need information about the technical performance of devices, their effectiveness compared to alternative treatments, likely complications, costs, availability, and a host of other topics of interest. Many databases, as described in this chapter, exist to study the epidemiology of implanted devices. Others surely will be created in the near future. Sources of such information include Internet sites, noted throughout this chapter, as well as scientific and medical publications. Device epidemiology has a shorter history than pharmacoepidemiology. However, medical innovation and technological advances will create ever-greater demand for such knowledge on a worldwide basis. Some data will certainly be deemed to be proprietary information, which will limit access to it. Confidentiality issues will limit access to other data. But the proliferation of medical devices seems certain to continue and device epidemiology is here to stay. Global demand for prostheses to improve human health is growing. Technological developments, such as the electronic medical record, improved computer performance, and uniform coding practices insure that databases, both public and private, will become increasingly valuable tools for knowledge about prosthesis epidemiology for industry, government, medicine, and society at large.

ACKNOWLEDGMENT

The authors would like to acknowledge Carolyn Easton, RPh, MLIS, for her capable research contributions.

REFERENCES

1. Allen A. Medical Device Industry Fact Book (3rd ed.) Santa Monica, CA: Canon Communications, Inc., 1996.
2. Marwick C. Implant recommendations [news]. JAMA 2000;283:869.
3. Ahlbom A, Norell S. Introduction to modern epidemiology. Chestnut Hill, MA: Epidemiology Resources, 1984.
4. Moss AJ, Hamberger S, Moore RM, Jeng LL, Howie LJ. Use of selected medical device implants in the United States, 1988: Advance data from vital and health statistics. No. 191. 2000. Hyattsville, MD: National Center for Health Statistics.
5. Bright RA, Jeng LL, Moore RM. National survey of self-reported breast implants: 1988 estimates. J Long Term Eff Med Implants 1993;3:81–89.
6. Bright RA, Moore RM, Jr., Jeng LL, Sharkness CM, Hamburger SE, Hamilton PM. The prevalence of tympanostomy tubes in children in the United States, 1988. Am J Public Health 1993;83:1026–1028.

the Medicare Coverage Advisory Committee who will advise on coverage policy decisions. Among the six panels of experts is the Medical Devices and Prosthetics Panel. In 2003, the CMS created the Medical Technology Council to improve Medicare policies on coverage, coding, and payment for medical devices

Other countries that do have national technology assessment entities include Canada, Spain, Australia, Sweden, and France. In the United Kingdom, the National Institute for Clinical Excellence has conducted appraisals of prosthetic devices, including prostheses for primary hip replacement and coronary artery stents, in the treatment of ischemic heart disease *(77)*. A device to be reviewed in the future is the implantable cardioverter-defibrillator. Such appraisals assess clinical efficacy, cost effectiveness, and National Health Service implications.

FUTURE OF DATABASES FOR STUDYING THE EPIDEMIOLOGY OF IMPLANTED DEVICES

Increasing Technological Sophistication of Patient Data

As electronic data entry at the bedside becomes more common, fully electronic patient records are projected to become standard. Standardized formats will evolve deliberately and as a consequence of software development; interchangeable software from different vendors will be attractive to users. Data-processing capabilities also will increase as searching electronic medical records for key words becomes less expensive than hand-abstracting paper records and software algorithms are developed and commercialized. Simultaneously, developments in software analytical tools, such as neural nets and data mining, will further researchers' abilities to find and analyze health care information. Furthermore, as the current databases mature, longer follow-up of patients with prostheses will be possible. The *Journal of Long Term Effects of Medical Implants* is a source of long-term information about various types of implants. Articles in this journal usually concern materials and engineering aspects of device development, although some deal with complications of a specific device *(78)*.

Privacy Concerns

As electronic databases and research capabilities mature, concerns about patient privacy heighten *(79)*. One research-focused association (the International Society for Pharmacoepidemiology) has provided recommendations on data privacy in a public document on the Internet, cited on February 17, 2000 (http//www.pharmacoepi.org/resources/cfr_parts.cfm) *(80)*. Besides exercising the usual care of personal data with passwords and locked file cabinets, researchers are also developing more sophisticated methods of protecting patient identity *(81)*.

In the United States, concern about data privacy has been the subject of extensive media coverage as well as legislative efforts. In the Health Insurance Portability and Accountability Act of 1996 (HIPAA), Congress directed the Secretary of Health and Human Services to submit detailed recommendations on privacy standards with respect to individually identifiable health information. The federal government published a notice of proposed rulemaking in April 2000 that generated an extraordinary number of responses and the final rule in December 2000. The regulation went into effect in April 2001 *(82)* and was implemented on April 14, 2003. This legislation also addressed the

Analysis and Retrieval System of the US National Library of Medicine), the NIH Clinical Trials Registry, mandated by the FDA Modernization Act of 1997, and the National Guidelines Clearing House cosponsored by AHRQ, the American Medical Association (AMA), and the American Health Insurance Plans (AHIP). Ramsey et al. *(68)* also provide a table of select technology assessment groups that conduct effectiveness evaluations of medical devices as postmarketing activities. They rightly point out that device evaluations that are conducted may not be accessible by external parties because they may be conducted for internal use or are costly or are not readily cited. The Task Force on Technology Assessment of Medical Devices *(68)* has recommended guidelines for information exchange to assist coverage decisions and determine appropriate use for therapeutic medical devices, including open discussion between manufacturers, insurers, and providers. The Task Force has also recommended a Guideline for Technology Assessment of Therapeutic Medical Devices that includes assessment of effectiveness, cost, cost-effectiveness, and research integrity through publications.

Increasingly, states in the United States also have played a role in conducting technology assessment. The number of devices is rapidly increasing and the US federal government has been unable to meet the demand, so several states, including Minnesota, Oregon, and Washington, have conducted technology assessments of medical devices and worked collaboratively with the federal government and the private sector *(69)*. Under cost pressure, however, states are reevaluating their role *(70)*.

Third-party firms also are involved in technology assessment activities. Examples include independent technology assessments by ECRI *(71)*, subscriber services to assess market opportunities by Medical Data International *(72)*, and overall payer assessments for member organizations by Blue Cross/Blue Shield. Technology assessment sponsored by medical specialty societies includes the AMA Diagnostic and Therapeutic Technology Assessment Program and the Clinical Efficacy Assessment Project (by the American College of Physicians) *(68)*. Thus, technology assessment has increasingly become an important activity in the US private health care system. With no overall responsible federal agency, the evaluation of clinical practice for its clinical and cost-effectiveness has assumed greater market value *(73)*. Perry and Thamer provide an excellent summary of the shrinking role of the US government in providing technology assessment at the national level, with the consequent increasing growth in the US private sector *(74)*.

New medical technology, prominently including implants, continues to attract media attention. *JAMA* devoted its November 1999 issue to "Impact of New Technologies in Medicine" *(75)*. Dozens of other medical journals devoted space to this topic just before the year 2000 arrived. Greater international collaboration is fostered by the World Health Organization, the International Clinical Epidemiologic Network, the International Network of Agencies for Health Technology Assessment and others, attesting to the importance nations place on this subject. Linkages across the US federal government are occurring through the Quality Interagency Coordinating Task Force, including members from FDA, Centers for Disease Control, and numerous other federal agencies. Questions that need to be addressed include when technologies are most effective and for whom and how they should be implemented *(76)*.

Technology assessment at the US federal level is re-emerging within the context of Medicare, since CMS announced in September 1999 the complete list of members of

pharmacoepidemiology group at Vanderbilt University has linked Medicare, Medicaid, and state public health services data for Tennessee to study medications, vaccines, and health care economics; certain types of implant studies would also be possible *(48)*. Harvard University and the University of Pennsylvania have also used public data to create novel databases.

Meta-Analyses

There are a large number of publications regarding medical therapies and meta-analysis is a useful analytic tool to compare them. It permits the application of quantitative methods to summarize research findings. Lafata et al. have provided an extensive discussion of the use of this method *(58)* and Temple *(59)* has outlined its limitations. Both focused primarily on drugs, but they apply equally to prosthetic devices.

Berlin and Colditz *(60)* discuss use of meta-analysis in the regulatory process for devices as well as foods and drugs, noting that it allows us to understand the heterogeneity of treatment effects across studies and to identify sources of variability in effectiveness. For regulatory purposes meta-analysis is conducted to support claims, but this technique also permits generalization of findings across diverse patient populations using observational data. For example, Da Costa et al. reviewed many small trials of antibiotic prophylaxis for permanent pacemaker implantation and persuasively concluded, based on the large scope of the analysis, that such use reduced the incidence of complications owing to infection *(61)*.

Technology Assessments

With "the industrialization of clinical research" *(27)* and the "medical arms race" *(62)*, attention has now clearly focused on different mechanisms to conduct clinical trials in the arenas of devices, biotechnology, and drugs. The conduct of clinical trials frequently still involves academic medical centers but the evaluation of medical devices through "technology assessment" has broadened substantially.

Technology assessment may be defined specifically as "any process of examining and reporting properties of a medical technology used in health care, such as safety, efficacy, feasibility and indications for use, and cost-effectiveness, as well as social, economic, and ethical consequences, whether intended or unintended" *(63)*. We consider safety and efficacy as the province of clinical trials and focus instead on effectiveness or use in the general medical community *(64)*.

Health care companies, such as MCOs, are one of the constituencies involved in technology assessment. As private payers for medical technology, they are interested in the cost-effectiveness of new devices and their comparison with prior devices or alternate therapeutic modalities such as medications. A special issue of *The American Journal of Managed Care* was devoted to this topic *(65)*. Braslow et al. *(66)* provide a detailed example of the role of a health plan in technology assessment, from the perspective of a large national MCO. In general, health care delivery systems rely increasingly on evidence-based methods to decide whether to provide prosthetic implants to patients. Health plans use technology assessment to evaluate scientific evidence of effectiveness, in addition to cost considerations, when deciding about covered benefits for patients.

Goodman has compiled an extensive list of information resources for health care technology assessment *(67)*. These include MEDLARS databases (MEDical Literature

of service. As noted with respect to Medicare and Medicaid files, there are limitations imposed by differences in coverage policies across organizations and the generally low representation of specific device model numbers in the databases. Health plans, such as certain of those listed above, have participated as US FDA cooperative agreement sites to study pharmacoepidemiology since 1981. Another vehicle to study use of therapies are the Centers for Education and Research in Therapeutics (CERTs), first proposed by Woosley in 1994 and enabled by the 1997 FDA Modernization Act *(49)*. The CERTs are administered by AHRQ and were implemented in 1999 *(49)*.

Databases of administrative claims have the potential to contribute information on the effectiveness of implanted devices, particularly when supplemented by the use of medical records to obtain detailed clinical data. As Brewer and Colditz note, computerized databases and medical records, as well as clinical trials, should be used to evaluate adverse outcomes for devices as well as medications *(50)*. In the past, concern has been raised about the quality of the data and the ability to make comparisons across sites and times *(34)*. For example, in a survey of the use of claims data across 15 countries, only half of the databases provided linkages across time *(51)*. Outcome analyses were conducted using more than one-third of the databases in that survey. This number is undoubtedly higher today with the explosion of computerized databases and information system capabilities. A summary of use of claims databases noted the distinction between operational and analytic datasets *(35)*. A distinct advantage of private databases is that they often provide a view of the clinical experience with new devices in the general community, going beyond utilization information from clinical trials conducted in select settings.

Particularly in the cardiovascular area, there is a history of long-term studies following the introduction of a medical device into the market. As one example, Mayo Clinic and a clinic in Eau Claire, Wisconsin, followed a particular cardiac pacemaker lead implanted over 10 years and determined that it had a lower rate of complications than other leads *(52)*. Another study looked at long-term survival of patients with cardiac pacemakers using two alternative pacing methods *(53)*. An interesting series of postmarketing studies used an insurance claims database (Dun and Bradstreet Health Care Information) as well as Medicare data to analyze the relationship between prosthetic nonbreast implants and selected conditions. Possible associations between implants (silicone and metal) and specific neurological conditions were found *(54,55)*.

Numerous long-term studies of implants have been conducted using various data sets. Most heart valves have been followed for decades. A 10-year study of the Ionescu-Shiley valve, an international collaboration among researchers from the United States, United Kingdom, Scotland, and Canada, determined that implanted valves began to deteriorate after 7 years and recommended intensified patient monitoring after that time *(56)*. Similarly, clinical data for patients receiving the Medtronic Hall heart valve prosthesis were reviewed, documenting a low rate of thromboembolic events without structural failure *(57)*. These are only examples of the many published studies providing postmarketing experience with implanted prosthetic devices using private databases in the clinical setting.

Academic Databases of Public Data

A few academically based centers have obtained, linked, cleaned, and restructured public databases into more comprehensive databases useful for epidemiology. A

weights to researchers in the data files. In addition, because the unit sampled in the NHIS is a household rather than an individual, clustering of responses within households is expected. Clustering of responses is owing to the fact that responses from two people in the same household are more likely to be correlated than are responses from two people chosen at random.

To draw correct conclusions, statistical analyses of data from complex survey designs must take into account the sampling (and poststratification) weights and the possible clustering of responses. Use of the sampling weights in standard statistical software packages will allow correct estimation of means (e.g., mean age of implant recipients) and proportions (e.g., proportion of recipients who are female). Use of more sophisticated methods are required to correctly test hypotheses and compute estimates of variance (e.g., confidence intervals, p-values, and standard deviations) *(45)*. Software packages that have such capability are available, including SUDAAN *(46)* and Stata *(47)*.

Canada

The Canadian provinces administer government-sponsored health care coverage whereby electronic claims for covered care are collected and stored. The databases for Saskatchewan and Manitoba have been used regularly for epidemiological and health services research *(48)*. The main advantage of these data systems is the ability to follow patients across all types of covered care and over time. Disadvantages include the problem of patient migration in and out of the province and changes in coverage policy over time. Major surgery, such as prosthetic implantation for medical reasons, is likely to be noted on claims. However, as with any claims data, follow-up for conditions that are usually poorly indicated on claims would be problematic.

Databases Generated by Private Parties and Other Sources

Health Care Databases

Retrospective analyses using existing databases may be subcategorized as those based on claims databases and those based on independent efforts to aggregate data within an institution, as described below. Private databases that reflect experience with device implants may be based in for-profit or not-for-profit organizations. Databases from large managed care organizations (MCOs) are used effectively for the epidemiology of device implants. To date, much of the epidemiological work with these databases has focused on prescription medications. Several of these databases in the United States (Group Health Cooperative of Puget Sound, Harvard Pilgrim Health Care, Kaiser Permanente Medical Care Program, and UnitedHealth Group), the United Kingdom (The General Practitioners Research Database), and the Netherlands are described elsewhere *(48)*.

Large population-based MCO data, utilizing automated claims billing records, include both the number of implant patients and the number of enrollees for comparison. Given the size of the databases, uncommon adverse reactions to device implants may be detected. Further, it is possible to determine rates of reimplantations and subsequent health services utilization and cost. An additional strength of these databases, compared to administrative Medicare and Medicaid files described previously, is the ability to link claims longitudinally to follow care of patients over time and across sites

that is sponsored by the Agency for Healthcare Research and Quality (AHRQ, formerly Agency for Health Care Policy and Research) and has been performed every year since 1988. It is a stratified probability sample of more than 900 nonfederal hospitals in 17 participating states and is designed to approximate a 20% national sample. Administrative data are collected from all discharges for each sampled hospital. These patient-level data provide clinical and resource use information typically included in a discharge abstract. More than 7 million inpatient records are included in the 1997 database.

These data provide the opportunity to study utilization, rates of diagnoses and procedures, and rates of adverse events associated with implanted medical devices, to the extent that devices are reflected in procedure codes. More specific device information (brand, model, size, etc.) is not included. Because all ages and payers are represented in this database, results are generalizable to the US population treated at nonfederal hospitals. Specifically, rates of inhospital morbidity and mortality associated with device implant or explant procedures can be determined, controlling for numerous demographic, clinical, and hospital-related factors. An example of the importance of this information is the scenario in which a type of implanted device is known to have a high risk of failure, in which failure is associated with patient death, but in which the risk of removal of the device is unknown. Knowledge of the risk of explantation, stratified by patient age, sex, and co-morbid conditions, is essential to a determination of the risk–benefit ratio for patients with such an implanted device.

The NIS does not provide patient identifiers, and each discharge is a separate record that cannot be linked to other possible discharges for the same patient. Thus, longitudinal studies cannot be performed using this database.

IMPACT OR CONSEQUENCES OF COMPLEX SAMPLING

Many large-scale surveys make use of complex sampling methods to allow conclusions about a much larger population to be drawn from data on a (relatively) small sample. Although this is an important advantage, derivation of population estimates from data collected on a sample of persons requires the use of appropriate analytic techniques that account for the sampling methods used *(42,43)*. Applying standard methods of statistical analysis to data collected through complex sampling may lead to biased and erroneous results. Most standard statistical software packages, however, do not have the capability to appropriately account for the sampling techniques used in these types of surveys.

As an example of the use of complex sampling methods, the NCHS selected more than 40,000 households for inclusion in the 1997 NHIS *(44)*. These households were selected through a complex sample design involving stratification and multistage sampling. The selection method was designed so that the sampled population represented the entire noninstitutionalized population of the United States. In this method, each person in the population has a known nonzero probability of being sampled. The sampling probability for the NHIS varied across strata defined by age, sex, and race. This sampling probability is inversely proportional to the number of persons that the person sampled represents (the sampling "weight"). The weights assigned in the NHIS database also account for the proportion of persons in each stratum who were actually interviewed in the survey (poststratification weights). NCHS provides the assigned

researcher using coded data, whereas a researcher using clinical records could easily reach the opposite conclusion. For example, a vascular access device in a cancer patient may be removed as planned after the cancer has been cured or because it has been infected, but coded data may not distinguish between the indications. Important clinical covariates may not be identifiable through these codes, limiting the statistical adjustments or stratification that can be performed in analyses of these data. The use of standardized diagnostic and procedure codes also limits the detail of information available on any specific device. No manufacturer or model information is included in CMS codes, so researchers are confined to exploring questions concerning broad categories of devices (e.g., all mechanical aortic heart valves) rather than specific devices.

The population served by Medicare also limits the application of these data. Medicare files are not suitable for studies of device implants commonly used in persons under 65 years of age. As many implants such as artificial joints are used predominantly in older patients, this limitation does not preclude the use of these data for many interesting studies in this population.

CMS also provides additional opportunities for epidemiological research using publicly available data for special study populations. For instance, the United States Renal Data System (USRDS) provides administrative data on all patients (>900,000) in the United States with ESRD who were covered by Medicare from 1988 to the present. The diagnosis of ESRD makes all but a few patients eligible for Medicare, and USRDS currently includes approximately 93% of the ESRD population in the United States *(40)*. It is funded by the National Institute of Diabetes and Digestive and Kidney Diseases in conjunction with CMS. The USRDS Coordinating Center is operated under a contract with the Minneapolis Medical Research Foundation.

CMS has attempted to increase the number of researchers making use of its data by funding the Research Data Assistance Center (ResDAC). ResDAC is a consortium of universities, chaired by the University of Minnesota School of Public Health, which assists researchers in the academic, nonprofit and government sectors with the use of Medicare and Medicaid data. ResDAC offers periodic training courses on the use of administrative data, workshops at numerous national meetings, ad hoc assistance to researchers wishing to use Medicare or Medicaid databases, and training in using sample CMS databases.

Medicare files provide valuable information on other aspects of the use of implanted devices. The denominator of all persons participating in Medicare is available, and the number and rate of use of specific implants can be determined by age, sex, race, and other variables. Using other procedure codes, the number and rate of specific adverse events associated with implanted devices can be estimated, such as clinical outcomes associated with explantation of some devices. Even rare events can be studied, owing to the enormous size of the Medicare data files. These data have another unique advantage in that the costs associated with implant procedures, associated complications, and explant surgery can be determined for the Medicare system. This can serve as a surrogate for costs in other health care systems.

AGENCY FOR HEALTHCARE RESEARCH AND QUALITY
(HEALTHCARE COST AND UTILIZATION PROJECT NATIONWIDE INPATIENT SAMPLE)

Governmental bodies in the United States have made existing data available to researchers in other ways. The Nationwide Inpatient Sample (NIS) is the most widely recognized example *(41)*. The NIS is part of the Healthcare Cost and Utilization Project

purposes, patient, physician, and hospital identifiers also are important. Regrettably, certain information is lacking in many databases.

Government Databases

United States

CENTERS FOR MEDICARE AND MEDICAID SERVICES

Advances in electronic data storage and computing capability in the last two decades have allowed the collection and analysis of extremely large amounts of administrative data to become an important component of epidemiological and health services research. A notable example of this type of data source in the United States is the Medicare and Medicaid claims data files administered by the Centers for Medicare and Medicaid Services (CMS; formerly known as the Health Care Financing Administration or HCFA) of the Department of Health and Human Services.

Medicare is a national health insurance program for persons age 65 and older and certain other groups, such as those with end-stage renal disease (ESRD). Medicare is the single largest health insurance program in the United States, covering 33 million aged and another 5 million disabled individuals *(36)*. Approximately 500 million claims for payments are submitted to CMS each year *(37)*. In 1997, total disbursements for Medicare programs exceeded $213 billion. Medicare is composed of two insurance programs: Medicare Hospital Insurance (Part A), which covers inpatient hospital, skilled nursing facility, home health agency, and hospice services; and Medicare Supplementary Medical Insurance (Part B), which covers physician services, supplies, ambulatory surgical services, and other outpatient services. These two programs operate under different payment systems and provide different information for researchers.

Medicare files are available in three general forms: Research Identifiable Files (RIFs), Beneficiary Encrypted Files (BEFs), and Public Use Files (PUFs). RIFs contain person-specific data, including patient names and social security numbers. These data are protected by privacy provisions, are available to researchers only when necessary, and require a special request and strict adherence to CMS security measures. Because of the cost of processing the data and the privacy concerns, these files are used only for focused studies of relatively few patients. BEFs are created by systematically encrypting or deleting data elements that could identify patients. Claims for specific patients can be linked using an encrypted identification number. Because these records are based on claims for reimbursements and not on individuals, however, following patients over time to study the long-term effects of treatments, such as implants, can be difficult. PUFs are aggregated data that do not contain any patient-level information.

The administrative nature of the Medicare database limits the types of research questions that can be answered. Clinical diagnostic and procedure data are available in the form of diagnostic-related groups and International Classification of Disease (ICD-9) diagnosis and procedure codes *(38)* for inpatient hospital stays and Health Care Procedure Codes *(39)* for other claims. These codes have changed over time, making longitudinal studies of long-term effects of some implants difficult. Coding changes are especially problematic in the early stages of a newer technology because CMS may not have reimbursed providers for these devices (and thus codes may not have been assigned) until they became established in clinical practice. In addition, researchers must rely on the coding system to draw conclusions about possible device malfunctions. For instance, removal of an implant may be considered a device failure by a

GATHERING AND SYNTHESIS OF EXISTING PATIENT DATA AND STUDIES

Many sources of patient data provide mostly administrative details, collected for record-keeping or for processing claims. Others summarize existing clinical information obtained by reviews of discharge summaries or progress notes. These typically lack the ability to focus queries in a systematic manner. Because the data were not collected for purposes of research, they may lack important clinical covariates. They can, however, provide a much broader coverage of patients and contain sufficient data to allow meaningful analysis of rare events and conditions. This section first presents general issues in using existing databases for epidemiology, and then describes databases generated by government, databases and studies generated by private parties, meta-analyses, and technology assessments.

General Issues Related to Using Existing Databases for Epidemiology

Administrative data sources and data collected through reviews of existing clinical information have proven to be very useful in epidemiological research. Although they are not appropriate to address some specific questions, their large size, representative nature, and automated record-keeping systems make these sources ideal for investigation of issues that may not be addressed in more focused epidemiological surveys. These data are especially useful for determination of population-based rates of specific diagnoses, utilization of specific procedures in specific groups of patients, and costs.

Whether the data were originally developed for administrative or research purposes, there are generally nuances that are not apparent from the data documentation. Sometimes these nuances, such as coding or coverage policies or changes, are critical to correct interpretation of the data or analyses. An effective solution is to enlist as a study collaborator someone who is already intimately familiar with the database.

Making use of data that were abstracted from original records raises additional concerns regarding the validity of the data. Researchers must be careful to report the abstraction methodology, including the expertise of the personnel who abstracted the data, and any checks of variable ranges or validity, because the process of abstraction itself can introduce errors. For example, a study compared administrative data with data derived directly from hospital files *(33)* and noted undercoding of co-morbidities and inhospital complications in the abstracted data. Other studies document the pitfalls of secondary administrative data. A typical strategy to minimize such errors is to sample original documents, comparing data derived from them with the data derived from secondary sources.

Limitations of coding using claims databases include variations in coding, errors in coding, incompleteness in coding, limits in the specificity of available codes, and errors in clinical diagnosis *(34)*. In addition, new procedures may not have a code assigned. However, clinically relevant insights still can be drawn from such secondary data. McDonald et al. cite issues with comparing across sites owing to lack of standardization of coding *(35)*. Close collaboration across sites by researchers would facilitate the development of a comprehensive mapping system across private health care settings.

Device technology is rapidly evolving for all types of prostheses. Therefore, manufacturer and model information for each prosthesis is highly desirable. For tracking

and for neurological implants *(25)*. Certain registries may be viewed as a form of postmarketing surveillance because through long-term follow-up they may provide early warning systems for tracking device failures.

Device-Specific Studies

Manufacturer-Sponsored Studies

The device approval process in the United States and other countries necessitates that the manufacturer or sponsor conduct clinical trial(s) necessary to demonstrate safety and efficacy. Such research relies on the participation of physicians, often situated in academic centers, who are at the forefront of research and who conduct the trials necessary for device approval. These same physician-investigators develop a long-term interest in the outcome of the device since they continue to provide follow-up for implanted patients. The manufacturer may provide financial support to determine the short- or long-term safety and effectiveness of the implant after market approval under three possible mechanisms. In certain cases, as noted in Chapter 5, the FDA may require postmarketing studies as a condition of approval or, after approval, as a response to public health concern. An example of an effort that fulfills such a postmarket study requirement is the Medtronic, Inc. Product Performance Report that summarizes long-term experience with implanted pacemakers and leads, including survival analysis *(26)*. It was initiated before the requirement, which was made of multiple manufacturers, and had already been reporting more data than the FDA required. Third, even without such regulatory requirements, manufacturers may sponsor further research after government approval, both to determine long-term effects and to further elucidate the cost-effectiveness of the device.

Academic and Professional Society Studies

Physicians in academic medical centers typically are opinion leaders of their field and contribute to the medical and scientific literature. They often work with manufacturers to investigate new medical devices or new indications for existing devices. They may, in fact, suggest innovative designs for implantable technologies based on their clinical expertise and patient experience. Recently, however, the dominance of the academic investigator has begun to erode. Managed care entities are becoming more involved in clinical trials. Contract research organizations and other firms linked to traditional businesses and not affiliated with universities are increasingly used to outsource clinical trial activities *(27)*. The role of physicians, academic or not, as businessmen with vested interests in the outcomes of clinical trials was highlighted in a *New York Times* article *(28)*.

Academic research also has focused on financial coverage for expensive new technologies and how these innovations diffuse across the health system. Based on surveys, Ramsey and Pauly *(29)* and Chernew *(30)* concluded that managed care plans do not necessarily provide less coverage for new medical technologies than traditional indemnity insurers. Decisions on coverage are often based on medical acceptability, FDA approval, and cost-effectiveness *(31)*.

In addition, academic researchers may study comparisons across modalities, such as the effectiveness of a drug therapy versus a medical device. An example of such a comparison is use of anti-arrhythmic medications vs cardiac defibrillators *(32)*.

joints, orthopedic fixation devices, artificial heart valves, annuloplasty rings, pacemakers, lenses, silicone implants, infusion pumps, access ports, catheters, and shunts. There were also activity of daily living questions for specific prostheses *(19)*.

Other surveys periodically sample nonfederal hospital outpatient and emergency rooms (National Hospital Ambulatory Medical Care Survey [NHAMCS]) *(20)* and private offices (National Ambulatory Medical Care Survey [NAMCS]) *(21)* and ask for standard information about each patient visit during specified time periods. Data concerning the health care setting, health care practitioner, and patient demographics, complaints, diagnoses, procedures, medications, and payment method also are solicited. For the 1999 and 2000 surveys, prosthesis information was recorded in open-ended fields for diagnosis or nature of injury *(22)*.

The Consumer Product Safety Commission (CPSC) receives injury data from hospital emergency rooms that participate in its National Electronic Injury Surveillance System (NEISS). The FDA is part of the NEISS All Injury Program, under which any federal agency may obtain information about medical device injuries from CPSC. As this program continues, information about injuries related to implants detected in emergency rooms will accumulate.

Market Research Surveys

IMS Health Inc., is a market assessment firm that offers the Hospital Supply Index, which is derived from the paid invoices, to a national sample of US hospitals for all types of devices, including implants *(23)*. This is a good resource for estimating the number of implantations that occur in hospitals, with the caveat that both implants that are never used and varying storage life must be estimated and taken into account. In addition, implant procedures carried out in nonhospital settings are not captured.

Registries

Private registries are one source of prospective data on implanted devices and their long-term follow-up that have been developed in various regions of the world. Some register only the fact of the implant. Others obtain detailed clinical information as part of formal systematic efforts. These registries may be independent from government requirements and a number have followed patients for over a decade. Many countries maintain device registries, including the United States, the United Kingdom, Sweden, Norway, Spain, Italy, France, the Netherlands, Australia, and the Czech Republic. Examples of three US governmental registries are those maintained by the National Heart, Lung, and Blood Institute (cardiovascular devices), National Cancer Institute (silicone breast implants), and The Department of Veterans Affairs (implanted defibrillators).

The implanted devices followed in private registries include numerous cardiac devices (stent, pacemaker, defibrillator), hip prostheses, breast implants, penile prostheses, and cochlear implants. A number of professional societies maintain registries for medical devices implanted by their members, such as the American College of Cardiology, and the Australian Society of Cardiac and Thoracic Surgeons. Registries may be maintained in order to compare the efficacy of drugs and implants: antiarrhythmia drugs have been compared to implantable defibrillators in this way. In England, two specialty groups have called for national registries for hip implants *(24)*

- New data collection for a systematic study.
- Gathering and synthesis of existing patient data and studies.

Manufacturers and government regulators collect and maintain databases of adverse event reports. The FDA database is available to the public and can be used to signal possible problems with implants (*see* Chapter 5).

Databases that can be used for formal epidemiological research are described in the next two sections. The final section addresses the future of databases for device implant epidemiology.

SYSTEMATIC NEW DATA COLLECTION

New data collection is the classic basis of epidemiological studies. It generally is the result of discoveries in the basic sciences or clinical studies and often relies on interviews of patients, family members, or health care providers. New data collection is expensive but especially important when factors that are not noted in routine medical records, such as patient use of nonprescription treatments, must be accounted for by the study. We discuss three basic types: surveys, registries, and device-specific studies.

Surveys

National US Surveys

In the United States, the National Center for Health Statistics (NCHS) regularly conducts surveys of the general population and health care providers. The surveys are periodic and designed around themes. The NCHS attempts to follow leads in a timely way, balancing its desire to measure a parameter consistently over time with its tendency to improve focus by modifying the research strategy.

One survey focused on prostheses: The Medical Device Implant Supplement (MDIS) to the 1988 National Health Interview Survey (NHIS). The NHIS survey queries the noninstitutionalized population about general health, with supplements that vary yearly in number and theme. The Center for Devices and Radiological Health at FDA sponsored the MDIS *(4)*. As part of the 1988 NHIS, if the responder on behalf of all the household members answered positively to a screening question on whether any household members currently had an implant, the MDIS was administered to the household responder for the implantees. The questions asked whether the current implant was a replacement, about common complications, and about general reasons for the implant. MDIS answers can be linked to the answers to the core NHIS. Prostheses that were common enough to allow statistical descriptions were heart valves, pacemakers, hip joints, knee joints, orthopedic internal fixation devices, intraocular lenses, ear vent tubes, and breast implants *(5–14)*. In 1994 and 1995, the NHIS focused on disabilities *(15)*. In phase 1, administered over the 2-year period, disabled household members were identified. In phase 2, these individuals were given one of four detailed surveys for children, adults, elderly, and polio survivors *(16)*. The phase 2 surveys, excluding the one for polio survivors, collected information about implants *(17)*.

The 1993 National Mortality Followback Survey was based on a random sample of people who died in 1993 and was administered to a surviving respondent *(18)*. The questionnaire concerned the final year of life and requested information about implanted devices. Specific prostheses that were listed in the questions were artificial

7
Databases for Studying the Epidemiology of Implanted Medical Devices*

Deborah Shatin, Roselie A. Bright, and Brad Astor

INTRODUCTION

The worldwide market for implanted medical devices continues to accelerate from $41.7 billion in 1993 to $56.7 billion in 1995 *(1)*. Unlike epidemiological research for pharmaceuticals, which builds knowledge on clinical trials that are typically completed in several months, clinical trials for implanted medical devices require longer follow-up to determine safety and efficacy. Follow-up for the purpose of regulatory approval is usually complete in 1 or 2 years, but practitioners and patients are interested in the safety and efficacy of implants for the entire life span of patients. This was highlighted in the development of a National Institutes of Health (NIH) Technology Assessment Panel Consensus Statement on Implant Registries *(2)*. The panel noted that 20 to 25 million US patients have had a medical device implanted. Device epidemiology provides a scientific method to contribute to this area of inquiry.

Epidemiology, in general, is "the science of occurrence of diseases in human populations. Disease occurrence is measured and related to different characteristics of individuals or their environments" *(3)*. Although clinical trials are sometimes considered a subset of epidemiology, we restrict our discussion here to observational studies of patients who receive their implants as a result of routine care.

The epidemiology of implanted devices includes the following:

- Descriptions of patients with implants in terms of demographics, underlying disease, and concurrent disease.
- Assessment of efficacy and safety of the implant.
- Comparison of treatments.

There are three types of data typically used. The source of data utilized for a given study depends on the information being sought and the level of clinical detail necessary. A given study may use either one or a mix of types:

- Case series of adverse events (Food and Drug Administration [FDA] adverse event reports or manufacturer's collections of adverse event reports and product complaints).

---

*The views expressed here are those of the authors and are not the official position of the US Food and Drug Administration.

From: *The Bionic Human: Health Promotion for People With Implanted Prosthetic Devices*
Edited by: F. E. Johnson and K. S. Virgo © Humana Press Inc., Totowa, NJ

32. Kalleward M. Bjork-Shiley Convexo-Concave heart valves: the risk of outlet strut fracture and clinical implications. Utrecht: Drukkerij Elinkwijk B.V., 1997.
33. The bioengineer's obligations to patients. J Invest Surg 1992;5(3). (Originally presented at the Hunter Honors Colloquium in Bioengineering, Ethical Issues at the Interface Between Engineering and Medicine, Clemson, SC, 1990.)
34. IDR3 Implant Data: Record, Report, Review. Final Conference Report, Society for Biomaterials, 1996.
35. Malchau H, Herberts P, Anhfelt L, Johnell D. Prognosis of total hip replacement results from the National Register of Revised Failures 1978 to 1990 in Sweden—a ten-year follow-up of 92,675 THR (Booklet) (San Francisco, 1993), p. 9.
36. I have relied on expert testimony in "Are FDA and NIH Ignoring the Dangers of TMJ (Jaw) Implants?" Hearing before the Human Resources and Intergovernmental Relations Subcommittee of the Committee on Government Operations, House of Representatives, One Hundred Second Congress, Second Session, June 4, 1992.
37. MacLachlan A. The chemical industry: risk management in today's product liability environment. In: Hunziker JR, Jones TO, eds. Product liability and innovation: managing risk in an uncertain environment. Washington, DC: National Academy Press, 1994, pp. 47–53.
38. Citron P. Medical devices, component materials, and product liability. In: HunzikerJR, Jones TO, eds. Product liability and innovation: managing risk in an uncertain environment. Washington, DC: National Academy Press, 1994:54–61.
39. Fielder J. Ethical issues in managed care. Technology and Society 1996;15(3):28–31, 39.
40. Crouch, R. Letting the deaf be deaf: reconsidering the use of cochlear implants in prelingually deaf children. Hastings Center Report 1997;27(4),14–21
41. Osterager V, Salomon G, Jagd M. Age related hearing difficulties: II. psychological and sociological consequences of hearing problems—a controlled study. Audiology 1988;27: 179–192.
42. Ganz BJ, Tyler RS, Woodworth GG, Tye-Murray N, Fryauf-Bertschy H. Results of multichannel cochlear implants in congenital and acquired prelingual deafness in children: five year follow-up. Am J Otol 1994;15(Suppl. No. 2):1–7.
43. Miyamoto KI, Robbins AM, Todd S, Riley A, Kirk KI. Speech perception and speech production skills of children with multichannel cochlear implants. Acta Oto-Laryngologica 1996;116(2):240–243.
44. Preston T. The artificial heart. In: Dutton D, ed. Worse than the disease: pitfalls of medical progress. Cambridge: Cambridge University Press, 1988:91–126.
45. Fox R, Swazey J. Spare parts: organ replacement in American society. New York: Oxford University Press, 1992.
46. Claire S. A change of heart: a memoir. Boston, MA: Little Brown, 1997.
47. Youngner SJ, Fox RC, O'Connell LJ, eds. Organ transplantation: meanings and realities. Madison: University of Wisconsin Press, 1996, p. 49.
48. Wolfe GK. Instrumentalities of the body: the mechanization of human form in science fiction. In: Dunn TP, Erlich RD, eds. The mechanical god: machines in science fiction. Westport, CT: Greenwood Press, 1982, pp. 212–213.
49. Jones AH. The cyborg (r)evolution in science fiction. In: Dunn TP, Erlich RD, eds. The mechanical god: machines in science fiction. Westport, CT: Greenwood Press, 1982, pp. 203–204.
50. Kiernan V. The artificial heart is poised for resuscitation. The Chronicle of Higher Education, 1998;A20–A21.
51. Hogness J, van Antwerp M, eds. The artificial heart: prototypes, policies, and patients Washington, DC: National Academy Press, 1991, p. 111.
52. Fielder J. Ethical issues in clinical trials. Engineering in Medicine and Biology 1993;12(1).

3. Foote SB. Managing the medical arms race: innovation and public policy in the medical device industry. Berkeley: University of California Press, 1992.
4. Their S. Prioritizing biomedical technologies. In: Andrade JD, ed. Medical and biological engineering in the future of health care. Salt Lake City: University of Utah Press, 1994, p. 9.
5. Roberts MJ. Your money or your life: the health care crisis explained. New York: Doubleday, 1993, p. 39.
6. Lundberg G. Managed care. Arch Int Med 1995;155:2274.
7. Fielder J. Ethical issues in managed care. Technol Soc 1996;15 (3):28–31, 39.
8. Stoline AM, Weiner JP. The new medical marketplace: a physician's guide to the health care system in the 1990s. Baltimore, MD: Johns Hopkins University Press, 1993, pp. 88–91.
9. Williams DF, Roaf R. Implants in surgery. London: W.B. Saunders, 1973.
10. Heinmann CFL. Acceptable risk: politics, policy and risky technology. Ann Arbor: University of Michigan Press, 1997, p. 65.
11. Bordo S. Material girl: the effacements of postmodern culture. In: Bordo S, ed. Unbearable weight: feminism, western culture, and the body. Berkeley: University of California Press, 1993, pp. 245–277.
12. Gilman S. Making the body beautiful: a cultural history of aesthetic surgery. Princeton, NJ: Princeton University Press, 1999.
13. Yalom M. A history of the breast. New York: Knopf, 1997.
14. Haiken E. Venus envy: a history of cosmetic surgery. Baltimore, MD: Johns Hopkins University Press, 1997.
15. Angell M. Science on trial: the clash of medical evidence and the law in the breast implant case. New York, NY: W.W. Norton,1996.
15a. Beauchamp T, Childress J. Principles of biomedical ethics (4th ed.). New York, NY: Oxford University Press, 1996.
16. Zimmerman SM. Silicone survivors: women's experience with breast implants. Philadelphia, PA: Temple University Press, 1998, p. 39.
17. Foster K. Risk, testimony, and the burden of proof: science on trial. Minerva 1997;35: 73–81.
18. Daubert v. Merrell Dow Pharmaceuticals, No. 92-102, June 28, 1993, 113 S Ct 2768 1993.
19. Lefrak E, Starr A. Cardiac valve prostheses. New York: Appleton-Century-Crofts, 1979.
20. Bokros JC, Haubold AD, Akins RJ, et al. Haubold AD, Akins RJ. The durability of mechanical heart valve replacements: past experience and current trends. In: Bodnar E, Frater RWM, eds. Replacement cardiac valves. Pergamon Press, 1991, pp. 21–22.
21. Bokros JC. Carbon in biomedical devices. Carbon 1977;15:355–371.
22. Black J. Prosthetic materials. Encyclopedia of Applied Physics. New York: VCH Publishers, Inc., 1996:15:143.
23. Supplementary Minutes of the Circulatory Systems Devices Panel, February 2, 1979.
24. Bjork VO. The Bjork-Shiley tilting disc valve: past, present and future. Cardiac Surgery: State of the Art Reviews. 1987;1(2):188.
25. Letzing WG. Summary of safety and effectiveness for the Bjork-Shiley cardiac valve prosthesis modified with Convexo-Concave cccluder. Irvine, CA: Shiley Laboratories, Inc.
26. Circulatory Systems Device Panel. Food and Drug Administration, 1979, February 2:12.
27. Fielder J. Getting the bad news about your artificial heart valve. Hastings Center Report 23, March–April 1993, pp. 22–28.
28. Johnson RM, Nelson GE, Spears LD. [FDA] Report of the Shiley Task Group. December, 1990.
29. FDA and the Medical Device Industry, Hearing before the Subcommittee on Oversight and Investigations of the Committee on Energy and Commerce, House of Representatives, One Hundred First Congress, Second Session, February 26 1990;Serial No. 101-127.
30. FDA and the Medical Device Industry, pp. 410–411.
31. Farley D. Shiley saga leads to improved communication. FDA Consumer, January–February 1994, p. 16.

The fractures of hundreds of Bjork-Shiley artificial heart valves revealed that we had no system for registering and tracking patients with life-sustaining artificial heart valves. In fact, our fragmented health care system generally suffers from lack of basic data about devices and their performance. It also revealed the FDA's lack of enforcement authority when faced with a manufacturer determined to keep its product on the market. New questions also arose about notification and the role of patients, physicians, and the FDA.

PROFESSIONAL AND CONSUMER MODELS OF HEALTH CARE

Notification of patients suggested two models of how to proceed. The professional model emphasized the physician–patient relationship and communication from the FDA went to the physician and then to the patient. The consumer model emphasized the consumer–manufacturer relationship with the FDA determining when a manufacturer should directly advise consumers about problems with its product. The resulting policy moved from a professional model, which didn't work very well, toward a consumer model with some elements of the professional model (notification of physicians). These models will surely have a larger role to play in the health care system in the future.

The dominance of managed care has reinforced the consumer model, with Congress considering establishing a Patient's Bill of Rights. This is a form of consumer protection legislation, which acknowledges that the professional model was inadequate to protect patients' rights. A similar shift occurred with patient protection in clinical trials *(52)*. As much as it is needed, it is important to acknowledge what is lost. When patients are consumers and doctors are service providers, even though highly educated and well-paid ones, many of the values associated with the professional model will be lost. Medical care isn't just a service, like fixing appliances, and sick people aren't just consumers in the market for health care. Sick people are scared and need someone to trust. Doctors aren't just caregivers, they deal with our deepest anxieties about our bodies, illness, and death. The consumer approach tends to overlook this dimension of medicine and creates arrangements that do not adequately allow for it, such as sharply limiting time to listen to patients and commiserate with them.

Consumer-oriented health care also incorporates market mechanisms and assumptions. If health care is just another commodity, there is less concern about some consumers not being able to afford it. Profitability is attained by segmenting the market and picking those (healthy people) who are less likely to require services and avoiding those (the old, the unhealthy) whose conditions may lead to large expenditures. Although this is standard marketing wisdom, incorporating it into the health care system demeans both patient and physician.

These cases of implanted medical devices reveal a health care system burdened by its past and threatened by its future. It is also a system with many caring professionals trying to offer humane and effective health care despite a variety of obstacles. We owe them a saner, kinder place to help the sick people who seek them out.

REFERENCES

1. Pacey A. The culture of technology. Cambridge, MA: MIT Press, 1984.
2. Rothman D. Beginnings count: The technological imperative in American health care. New York: Oxford University Press, 1997.

not providing health insurance to 40 million of our mostly poorer and unluckier fellow citizens. Part of the concern about managed care is the dawning recognition among patients that rationing has at last arrived, even for those of us with jobs that provide good health benefits. The artificial heart will surely put us to the test. Will we adopt some sort of nonfinancial rationing that treats all patients equally? Or will we simply draw the circle tighter, avoiding rationing for a smaller segment of society and increasing the ranks of the medically underprivileged? History suggests the latter alternative, but that may not be politically possible this time around.

Conclusion

New technologies force us to reshape society in order to make a place for them. The artificial heart, if successful, will reshape the medical landscape, allowing more sick people to be helped and forcing us to either commit more resources for this expensive device or develop a rationing plan. Both options present enormous difficulties and opportunities for social conflict. The great unresolved issue of health insurance for all once again contributes to our difficulties with implanted devices.

SOCIETAL ISSUES IN IMPLANTED MEDICAL DEVICES

These cases illustrate some of the many social issues associated with implanted medical devices. Far from being just useful products that help sick people, implants are like social depth charges that sink into society, explode, and reshape it. The artificial heart has the potential to cause wrenching changes in our health care system. Its cost will force us to examine some of the basic principles and assumptions about health care. Politicians, advertisers, advocacy groups, and so on, will be back, as the major stakeholders seek to shape public opinion and policy.

If more proof were needed that implants have a cultural dimension, cochlear implants would provide it. The choice of whether to have a cochlear implant for a prelingually deaf child is a momentous one, placing the child either in the Deaf community with the need to learn sign language or, more precariously, in the hearing community. To choose is to strongly determine the identity of the child and the child's path in life.

Other implants have revealed anomalies and weaknesses in the social arrangements for health care. Patients who received the Vitek implant for their TMJ disorder find themselves in a medical limbo. Unable to sue the company that made the implants and often without medical benefits, they represent a hole in the system where neither law nor medicine can help them. Their plight led to the biomaterials crisis and passage of the Biomaterials Assurance Act of 1998, but it did not lead to assured medical care for their condition or compensation for their injuries. Problems with the Vitek implant reveal how easily implant patients can slip through the large cracks in our health care system.

The controversy over silicone-gel breast implants illuminates the uneasy relationship between medical science and law. To many observers the legal system failed to adequately incorporate science into the disposition of lawsuits claiming damage from breast implants. The issue was complicated by our regulatory history, which grandfathered pre-1976 devices, so they had no premarket review, and by our ambivalence about cosmetic surgery in general and breast implants in particular.

Costs

ABIOMED estimates today that its artificial heart will cost between $75,000 and $100,000, not including surgery and medical monitoring. An earlier article used a figure of $148,000, which does include surgery and 5 years of medical follow-up *(50)*. A less optimistic Institute of Medicine (IOM) report in 1991 estimated that 2.85 years of life with an artificial heart would cost $299,000 *(51)*. If there are unanticipated problems, estimates will have to be adjusted upward. In addition, would it have to be replaced after a period of time? A mechanical heart valve is expected to last the life of the patient, but a complete artificial heart is more complex and, presumably, more prone to fail.

How many patients would be candidates? There are about 700,000 deaths each year from heart disease and the IOM estimates that between 10,000 and 20,000 are potential candidates for an artificial heart. But this is only the primary patient group, those who are in greatest need. Assuming the success of the artificial heart, there is a secondary group of patients for which it would be beneficial. The size of this group is estimated to be as large as 200,000 *(51)*. Using a figure of $150,000 for cost and the lower estimate of primary candidates, the cost for each year would be $1.5 billion, a total that does not include any costs associated with unforeseen problems or for replacement. If we add more primary patients and some secondary ones, the costs rise steeply. When hemodialysis was introduced, the number of candidates originally estimated was far below the actual number of patients. This is not unusual; when a new technology is available, health care professionals often find additional applications for it to help patients who do not respond to standard treatments. Once a new drug is on the market, physicians often try off-label uses for their patients.

One reason to think costs will be high is that if people are dying and an available artificial heart will save them, it will be difficult to refuse to provide it. It is a perfect media story: a hard-working, likable patient, an active citizen in the community, lacks adequate heath insurance and will die unless an artificial heart is provided. Similar stories accompanied the first heart transplants as well as the first dialysis machines. In the case of transplants, the limited number of donors required a rationing process because there was no way to match the demand. The United Network of Organ Sharing has a complex set of rules and policies for distributing organs. But in the case of hemodialysis, we could create as many machines are there are patients who need them. This is the same logic that David Rothman pointed out in the provision of iron lungs for polio victims. It is a compelling logic: how can you turn your back on a dying person when he or she can be saved with a machine that is on the shelf and ready to use? Isn't that person's life worth $150,000? And this applies to the 40 million Americans that have no health insurance, too. It will be difficult to restrict this lifesaving technology only to the middle class with good health insurance, particularly because much of its development was funded by the government. The artificial heart could be another issue like hemodialysis, where a disproportionate amount of medical resources are reserved for a small part of the medical population.

Unless there is an unlimited budget for health care, money spent in one area is money not available elsewhere. Rothman believes that we have solved the problem of rationing so far by not rationing medical technology for the middle class and paying for it by

The artificial heart is not just another implanted device that will help patients. First, it replaces an organ that has deep cultural meaning. Clark's wife wondered if her husband would still love her after his heart was removed. Second, there is an extensive body of work in science fiction that explores the idea of human–machine hybrids. Television shows like *The Bionic Woman* and *The Six Million Dollar Man* explored this theme, but there are darker visions outside of television that explore the dehumanization of people who become machine-like. Third, because of its cost, the artificial heart has the potential to create another social and political crisis like the earlier introduction of hemodialysis.

The Heart and Culture

The heart has long been regarded as the seat of the emotions. It symbolizes courage (having heart), love (give your heart), anticipation (be still, my heart!), faithfulness (true heart), fortitude (lion's heart), robustness (hearty), uncaring (heartless), and so on. Thirty-five pages in *Bartlett's Familiar Quotations* are devoted to the heart. Of all our internal organs, only the heart is in constant motion, often reflecting our emotions in its pace. More than any other organ, it is us, and we call our physicians when it skips beats or has an unusual rhythm. Physicians assure us that it is just a pump, but that is not how we experience it. Replacing it with a machine will not be like getting an artificial hip or pacemaker or heart valve. It will cause a greater shift in our psychic universe.

Some heart transplant patients claim that they have taken on some of the qualities of the person who donated the organ. Sylvia Claire describes how she received a heart transplant and had dreams about the donor and changes in her tastes in food. Later she discovered the donor's identity and through conversations with family members found out that many of the things she had dreamed were characteristic of the donor, as were her changes in the kinds of food she preferred *(46)*. Whether one believes this story or not, it does present a subjective theme that is common in heart transplants, the idea that one has taken on personality traits of the donor, not just the organ. The impact of taking the organs of another person into your body is immense, sometimes described as a form of nonoral cannibalism in which the body parts of others are taken not for food but for power *(47)*. Implanting artificial hearts will surely create psychological responses in patients, particularly those associated with having one's identity changed by removing the warm, organic human heart and replacing it with hard, cold machinery.

Art, particularly science fiction, deals with human–machine relationships, projecting our deepest fears, hopes, and fantasies in imaginative stories. With their potential to conquer time and death, implanted devices, particularly the artificial heart, provide fertile ground for imaginative reflections on what life would be like when we become more like machines *(48)*. The literature on cyborgs (beings that combine human and machine characteristics) is filled with reflections on the impact of implants on patients. In some of these stories individuals who are repaired with prosthetic parts feel—and are regarded as—less than human. In others, mechanical replacement parts convey a degree of immortality, though often turning the person into a machine *(49)*. These metaphors for alienation and dehumanization will likely appear in some patients with implanted artificial hearts.

community emphasizes their disability and need to overcome it in order to be accepted in the hearing world.

Conclusion

What looked like a standard type of medical intervention—a cochlear implant—involves a more complicated question of cultural identity. It is far from obvious, given the state of development of these devices, that the best decision for the child is to turn to an implant. Apart from its technical limitations, the implant presents parents of prelingually deaf children with a cultural as well as a medical choice.

THE TOTAL ARTIFICIAL HEART

Background

In 1982, Barney Clark received the first implanted total artificial heart. Compressed air from an enormous console powered the heart through a pair of tubes that entered his abdomen. Clark lived for 112 days with his Jarvik-7 artificial heart, but during that time he had respiratory problems, convulsions, mental confusion (probably from strokes), and kidney failure. William Schroeder, the second recipient, lived longer—620 days—but not much better. *The New York Times* called the artificial heart the Dracula of medical technology. (For a history of the artificial heart *see* ref. *44*. For a more sociological account *see* ref. *45*.)

Much has changed since that time. Power supplies, microprocessors, and materials have all improved, and current models are much smaller than before. This is largely because of work on ventricular assist devices that do not replace the heart but supplement it. A typical left ventricular assist device is a pump that fits inside the abdomen. It does some of the work of the left ventricle and requires either a direct electrical or air connection or induction coils that transfer power directly through the skin without any openings. These devices have been used as a bridge to transplant, keeping the patient alive and healthy until a donor heart can be found. They are also used to help patients' hearts recover from injury or illness until they are strong enough to circulate blood without assistance. By increasing blood flow, they help patients regain energy and health lost to inadequate heart function. Because they do not replace the heart, ventricular assist devices pose fewer problems than the total artificial heart. Although these are generally not meant for permanent implantation, some are being developed as permanent implants.

Recently, ABIOMED began clinical trials of its AbioCor implantable replacement heart. It is powered by batteries that transmit electricity through the skin without the need for wires that pierce the surface and provide opportunity for infections. This allows patients to move about freely and even remove the external batteries for brief periods (e.g., for showers) while relying on a smaller, internal battery.

As of late 2003, ABIOMED had implanted 11 patients and planned 4 more by early 2004. The results are not as good as hoped but far better than the experiences of Barney Clark and his cohort. Of the 11 implants, 9 were successful, with an average duration of support of 5.6 months. One patient went home for 17 months before dying, whereas several others were able to make excursions out of the hospital. An expensive genie is emerging from its bottle at last.

acquisition of oral speech, cochlear implants allow them to *relearn* to hear through a process of rehabilitation. However, recognition of speech is not always possible without lip-reading and other visual cues. There are some success stories of patients who are able to converse on the telephone, but these are the exceptions.

The outlook for prelingually deaf persons is considerably less optimistic. Even after 5 years of extensive training in sound recognition and speech production, patients were able to correctly pronounce vowel sounds only 70% of the time *(42)*. In another study, a panel of three persons could understand only 40% of the words spoken by children with cochlear implants who had had more than 3 years of rehabilitation *(43)*. Acquisition of an adequate level of oral speech is a long, difficult process with no guarantee of success. At the present level of development, it is clear that, even with cochlear implants, many deaf children will not hear well.

No doubt both the performance of the device and the effectiveness of the rehabilitation process will improve over time, just as they have for other implanted devices. (For a recent discussion of the technical literature, see the May/June 1999 issue of IEEE *Engineering in Medicine and Biology* which contains several articles of cochlear implants and neural engineering.) Those who come in at the beginning of the learning curve get some benefit even though the device is imperfect. This is the standard approach to the use of implanted devices, but cochlear implants have an additional cultural dimension that raises difficult questions about its use.

Hearing parents of a deaf child naturally want their children to be able to hear so that they can participate in the oral language community of their parents and others. They want the child to be in their world, and they imagine their child's life to be similar to what their lives would be like if they suddenly lost their hearing. But the choice isn't between the hearing community and none. There is another option, the Deaf community, capitalized to indicate that it is a distinct cultural minority with its own language, history, and value system. Acquisition of American Sign Language not only places the child within a linguistic community, but it also serves as a bridge to understand written English and gain access to the resources of the hearing world. The film *Children of a Lesser God* powerfully tells the story of a deaf woman who refuses to be "rescued" from her "disability."

Members of the Deaf community do not regard themselves as disabled, just different. An eighth-grade boy put it this way: "I'm not disabled, just deaf" *(40)*. Similarly, members of the gay and lesbian community generally do not regard themselves as lacking heterosexuality, just as having a different sexual orientation. Like the deaf, they have their own community, history, and values. Both groups are asserting their identity as cultural minorities rather than members of the mainstream with disabilities.

Within this context, the decision to place a cochlear implant in a prelingually deaf child is not just a medical one. It is also a choice about which community the child will inhabit, the deaf or the hearing. Parents of these children can learn American Sign Language and raise the child in the Deaf community or they can, assuming eligibility, use a cochlear implant and raise the child in the hearing community. However, learning an oral language with an implant is difficult, even with good instructors, and there is the danger that the child will fail to become a functional member of the hearing community. Such children are in a kind of no-man's land between the hearing and Deaf communities, belonging to neither. In addition, education of the deaf for the hearing

A number of conferences and meetings were held to study and publicize the issue. Prominent biomaterials scientists worked with the Health Industry Manufacturers Association to lobby Congress for a legislative solution to the problem. The result was new legislation, the Biomaterials Access Assurance Act of 1998. The act protects biomaterials suppliers who are not manufacturers from lawsuits as long as the product meets the applicable contractual specifications and requirements. This allows suppliers to move for summary judgment on these lawsuits. Nothing was done about the plight of the patients with defective implants.

Did this solve the problem? Oddly enough, it isn't clear. Biomaterials are still available, but not necessarily because of the new law. The legislation hasn't been tested in court, so there is some doubt about its effectiveness. Other arrangements assuring the flow of biomaterials to manufacturers may be indemnification agreements between manufacturers and suppliers and creating smaller suppliers with few assets. A study is planned and we may know more in a few years.

Conclusion

The problems associated with the Proplast temporomandibular implant illustrates once again how the lack of a comprehensive system of health insurance contributes to problems with implanted devices. It is likely that there would be far fewer lawsuits against DuPont if the victims of the Vitek implant were assured of medical care for their condition. It is ironic that fear of government involvement in health care has put important components of health care in the hands of insurance companies who have traditionally succeeded by not insuring sick people. The result is a health care system with a built-in aversion to sick people *(39)*.

COCHLEAR IMPLANTS *(40)**

Background

The popular view of deafness is that it is a terrible disability, a life sentence of utter silence. Samuel Johnson called it "one of the most desperate of human calamities." Hearing parents who give birth to a deaf child feel cut off and disconnected from their child. It is no surprise that the gradual loss of hearing that often accompanies aging is regarded as a disability, so much so that it is commonplace to encounter people who dislike wearing their hearing aid because it is public evidence of their defect. Studies show that even where cost is not a consideration, the loss of hearing must be substantial before people will resort to a hearing aid *(41)*. Given this outlook, any implanted device that would provide some level of hearing would appear to be a godsend.

The cochlear implant is an electronic device consisting of two parts. The external part of the device picks up and transmits sounds to an implanted receiver connected to the auditory nerve. The receiver, in turn, stimulates the nerve so that sound is perceived. Not every deaf person is a candidate for a cochlear implant; one must meet NIH eligibility criteria. Even those who meet these criteria have only taken a step toward hearing. What is perceived with the implant are not the sounds that are familiar to people with unimpaired hearing. For persons who have lost their hearing after the

---

*I am greatly indebted to Robert Crouch's article for my treatment of this issue. *See* ref. *40*.

Legal Issues

One disturbing result of the Proplast problems is that the patients who have been injured by it have little legal recourse, once the assets of the company have been transferred and the president has left the country. No assets means no resources for legal settlements to pay for medical expenses and to compensate for pain and suffering. For patients without medical insurance the results are devastating. Once again, the absence of a comprehensive health insurance system in the United States leaves these patients without help for a debilitating condition.

There is another significant dimension of this issue. One of the strengths of the medical device industry is the number of relatively small firms that create innovative designs. A relatively small company can be more creative and move more quickly than a large one. Funded by venture capital, small companies develop new devices. Once the product is approved, these companies are often purchased by larger firms that have more resources for marketing and distribution. Although this makes the device industry extremely innovative, it also means that patients injured by a device made by one of these small companies may not be compensated for their injuries. This is a significant loophole in our social safety net for implanted medical devices.

The Biomaterials Crisis

When Vitek removed its assets and declared bankruptcy, lawyers representing plaintiffs injured by the Proplast implant looked for ways to help their clients. They sued Dupont, the manufacturer of the Teflon coating that failed. These suits had little merit, but there were hundreds of them, and DuPont had to respond to each one. Going to trial is always risky and unpredictable; an injured plaintiff and a sympathetic jury may bring a verdict at odds with the facts and common sense (recall Marcia Angell's criticisms of the breast implant litigation, discussed earlier). DuPont settled many of these cases for undisclosed, probably small, sums, but the legal costs were still significant.

This could be written off as just another anomaly in the US legal system except for its impact on the medical device industry. The core of the problem is financial: Dupont was spending much more money dealing with lawsuits associated with use of its raw materials in medical devices than the profits it made selling them. The obvious solution was to stop selling polyethylene for use in medical devices. Dupont sells an enormous amount of polyethylene and only a tiny fraction of it goes into medical devices. From a business standpoint this makes no sense. (For a discussion of this issue, see ref. *37; see also* ref. *38*).

DuPont's announcement that it would stop selling polyethylene and other materials for medical devices sent shock waves through the industry. Polyethylene is widely used in medical devices, particularly implants, and its loss would be devastating. Everything from hydrocephalus shunts to pacemaker leads would be affected. The materials DuPont and others supplied have been studied and tested for a long time and their characteristics and behavior in the body are fairly well known. Starting over with new materials would require additional research and testing that would set the industry back for years. Would patients be deprived of lifesaving devices because of these worthless lawsuits?

A similar failure occurred in the 1960s with Teflon used in hip prostheses. As with the Proplast implant, initial reports were favorable, but eventually the Teflon fragmented, causing severe foreign body reaction and bone lysis. Dr. Larry M. Wofford remarked that "The orthopedic experience could have predicted the long-term results described in the oral and maxillofacial surgery literature" *(36).*

Reports of failures of the Proplast implant with substantial foreign body reaction began to emerge in the mid-1980s and continued to increase. Vitek voluntarily removed the implants from the market in 1988 and declared bankruptcy in 1990. The president of the company, Dr. Charles Homsy, left the country and now resides in Switzerland, where he continues to defend his implant. The company's assets were transferred to several successor corporations and overseas corporations shortly before bankruptcy was declared.

Regulation of Devices

Regulation of medical devices has a history different from that of drugs. Drug manufacturers were required to submit premarket evidence of safety in 1938; proof of effectiveness was added in 1962. Devices did not require any premarket approval until 1976. One reason for this difference is that in 1938 the problems with devices were mostly about fraud. The Museum of Questionable Medical Devices in Minneapolis, MN, houses an amazing collection of devices, from phrenology machines to a contraption utilizing colored lights, invented by a man who described himself as a commander in the New York City Air Force. These devices, with the exception of those using radiation, rarely harmed people directly, although they often prevented them from getting legitimate medical attention. Compared to drugs, there were relatively few medical devices that were harmful, so premarket approval was not seen to be necessary until relatively recently.

When the 1976 legislation was passed requiring proof of safety and effectiveness, one of the issues was what to do with devices already on the market. Requiring premarket approval for all of them would be an enormous burden on the FDA, and the devices already had a record of performance. The solution was to "grandfather" these devices and allow them to be sold without premarket clearance, with the proviso that the FDA could reclassify them and call for a submission of evidence of safety and effectiveness. New devices that could be shown to be substantially equivalent to a grandfathered device could be approved without additional testing. This is what happened with the Vitek implant.

What counts as "substantial equivalence"? The FDA was criticized for approving the Proplast implant as substantially equivalent to Silastic despite the fact that Proplast is not the same as silicone and the originally approved material was for reconstructing facial contours, not for use in a joint. However, there was an article comparing the two, which was submitted to the FDA in the premarket submission. The more basic point is that regulatory standards evolve, and the degree of scrutiny that a premarket approval submission received at the time of the Congressional hearing (1992) was significantly higher than those in place in 1983, partly because of the problems with the Proplast implant. This is another example of "Tombstone Technology."

This is what is needed, not only here but on an international basis. Neither patients nor medical devices stay within national boundaries, and it would be foolish to ignore the wealth of data available from other countries.

Conclusion

The Shiley case raised fundamental questions about the roles of manufacturers, physicians, patients, and regulators in the provision of implanted medical devices. What arrangements should be established to deal with problems that develop after artificial heart valves have been implanted in patients? What are the rights of patients to be informed of new, unexpected risks of their implant? Should doctors always inform their patients or may the FDA intrude into the physician–patient relationship by ordering manufacturers to inform patients directly? Under what conditions should a responsible manufacturer withdraw a valve from the market for safety reasons? The Bjork-Shiley Convexo-Concave artificial heart valve forced us to ask and develop at least partial answers for some of these questions and substantially changed the regulatory arrangements and expectations associated with medical devices already on the market.

TEMPOROMANDIBULAR IMPLANTS AND THE BIOMATERIALS CRISIS

Background*

The temporomandibular joint (TMJ) in the jaw consists of three main parts: the mandible, the glenoid fossa of the temporal bone, and a disc that is located between them. TMJ disorder strikes between 500,000 and 1 million Americans every year. The symptoms are facial pain, often intense, dizziness, and limited jaw movement, which affects speech and eating. Symptoms can sometimes be controlled with medication, physical therapy, and other treatments. About 15–20% of patients with this condition choose surgical intervention.

One common surgical treatment is the replacement of the disc with an implant. One of these devices, sold under the name Proplast TMJ Interpositional Implant by Vitek, Inc., consisted of a Teflon film laminated to a porous composite material fabricated from polytetrafluoroethylene and aluminum oxide. It was approved by the FDA in 1983.

The TMJ is moved by powerful muscles needed for chewing, so there are relatively large mechanical forces on parts of the joint. The Teflon coating on the Vitek implant was unable to withstand these forces over time and began to wear excessively and delaminate. This process creates wear debris and subsequent foreign body reaction, inflammation, and bone resorption. In severe cases, the bones of the jaw and skull can disintegrate, allowing perforation into the brain and middle ear. Patients often become isolated because of constant pain, deformity, and dizziness. Standard methods of treatment are often ineffective for these patients, leaving them a grim future with little hope for relief. Some patients have committed suicide.

---

*I have relied on expert testimony cited in ref. *36.*

database that scientists would use to formulate research projects and policy makers would consult when making decisions about health care. Unfortunately, no such almanac exists. There is no centralized database to develop reliable answers to these questions. Although we can make estimates, we simply do not know exactly how many patients in our country have hip replacements or other types of implants. Without this denominator we cannot accurately estimate the percentage who have experienced certain kinds of failures, or the average time of service, or the relative performance of different types or materials, etc.

Because we lack this kind of basic information about medical devices generally, and not just hip replacements, our knowledge of the performance of devices is limited. Of course, there have been many limited studies that have provided useful information, and there is a wealth of clinical experience that practitioners can draw upon. But without basic census data it is impossible to generate the kind of statistically valid knowledge needed to make reliable, scientific judgments about device performance. This is also an ethical failure, for such knowledge is necessary for professions like surgery or biomedical engineering to meet their obligations to patients *(33)*. Not knowing the basic information necessary for scientifically valid performance studies is a critical omission in providing patients with adequate care.

Why Aren't the Data Available?

A series of conferences involving industry, regulators, and other interested parties revealed significant obstacles to developing a national database to gather this information *(34)*. Data of the kind needed are not available because there is no requirement to collect them. Everyone involved in the process—manufacturers, the FDA, surgeons, and hospitals—keeps records for their own purposes, but a program for collecting basic census data on medical devices is not one of them. Such a requirement would add to the administrative costs of everyone involved, but the greatest opposition is from manufacturers *(34)*. They do not see any comparable advantage to be gained and worry about potential losses, particularly use of this data in litigation. If we could reliably compare performance characteristics, some products would be below average, which makes them harder to sell. Not everyone reads *Consumer Reports*, but those who do go for the higher ranked products. No one doubts that managed care organizations would be very pleased to have data on comparative performance of medical products and that their purchasing decisions would follow them. But in the absence of data about failure and performance rates, each manufacturer can point to studies that emphasize the good qualities of the product and let the marketing department do the rest. Just as with the Shiley heart valve, in the absence of data, each stakeholder can put its own spin on the product's performance.

Sweden set up a National Registry for total hip replacements in 1979 and recently issued a report covering 92,675 instances from 1978 to 1990. That report states:

> The public has the right to be assured that artificial hip replacement procedures do not introduce unexpected hazards associated with poor design, incompatible materials, or deficient technique....We believe that the information contained in this study will enable us to improve on the indication for surgery, the implantation technique, and to recommend implants with a verified durability. *(35)*

Legislation

As a result of the weaknesses in dealing with defective devices, Congress passed new legislation, the Safe Medical Device Act of 1990. Life-sustaining devices like artificial heart valves are now required to be tracked so patients and physicians can be notified. Reporting requirements are also strengthened, and the FDA has authority to order postmarketing surveillance of medical devices. The FDA was also given greater authority to act quickly to remove devices from the market. Often it takes a catastrophe or some other dramatic failure before regulatory reform can take place, a process dubbed "Tombstone Technology."

Lack of Data

One of the problems in dealing with the Shiley valve was the lack of data. Even the most commonplace information, such as the names of the patients who had received the valve, was not available. Analysis and research were crippled by the difficulties in obtaining data about patients, their implants, and the failures. This was partly owing to the lack of regulatory requirements, such as a registry and tracking system for patients. But it was also a function of our fragmented health care system. In the Netherlands, where health care is provided by the government, researchers were able to identify every patient and conduct extensive studies of mortality, patient anxiety, rate of failures over time, associations between different welders and failures, and much more *(32)*.

Lack of information on such important matters means that the various stakeholders will fill the data vacuum with their own interpretations. Consumer groups, trade associations, the manufacturer and its lawyers and publicists, assorted plaintiff's lawyers, the FDA, the media, and interested members of Congress all put their spin on the problem, leaving the public confused and cynical. This is particularly disappointing because putting a medical product on the market is like a final clinical trial. Premarket testing is done on a limited number of patients and therefore cannot identify very low rates of failure or side effects. In addition, whole classes of patients cannot be represented in clinical trials for the same reason. Thus, many problems appear only after the product is on the market. In a clinical trial, participants are informed of any newly discovered risks that arise so they can decide if they want to continue to participate. Similarly, consumers should also be informed of any newly discovered risks, particularly with implants, so they can decide what to do. But no one can be informed about such risks unless there is an adequate postmarketing system to identify them and communicate information about them. The Shiley case moved us closer to an adequate postmarketing system, but there is still a long way to go.

Consider these questions:

- How many US patients have total hip replacements?
- How many of these have been performed on women under age 55?
- How many hip replacements use uncemented titanium alloy femoral components?
- Which biomaterials provide cheaper average quality of life in this application?

These are the kinds of questions you would expect to be answered in some kind of medical device almanac, a book where you could find up-to-date information about how many and what kind. It would be rather like the US census, with lots of statistical

These numbers certainly make the fracture rate seem very small. No mention is made of hidden fractures, nor the fact that the latter number is calculated on the basis of the number of valves distributed, not just implanted. Later communications, after the valve's final recall, presented information in a more understandable format of percentages of failures per year for different classes of valves.

Apart from the content of these letters, the Shiley case illustrated serious weaknesses regarding communication with interested parties. Shiley's letters were sent to surgeons, because they were the ones who ordered the valves, but patients rarely see their surgeons after leaving the hospital. No registry of patients or their cardiologists were kept. When you buy a car, the manufacturer is required to keep your name and address in a registry so you can be warned and instructed about what to do if there is a problem. It is ironic that no such system was in place for implanted life-sustaining medical devices when the Shiley valves began failing.

Notification is not just an academic issue in this case. Patients with heart valves need to know the symptoms of valve failure and to get the patient to a hospital that is capable of emergency open-heart surgery. Going to the wrong hospital causes delay and, often, death. In 1990, Shiley agreed to a program to locate patients with their valves and their physicians. Letters were sent to hospitals and heart specialists to obtain names of patients. Doctors were then sent information kits and letters to be sent to their patients advising them to set up a meeting to talk about, among other things, the symptoms of valve failure and what kind of hospital to go to. This arrangement was chosen because the FDA did not want to intrude into the doctor–patient relationship. Unfortunately, this did not work entirely as planned. Many patients were not notified, despite repeated communications to their physicians. Ultimately, patients were notified directly. An audit in 1992 revealed that some hospitals and heart specialists were not complying, and additional encouragement from the FDA was needed.

A mid-course evaluation of this program of notification found that 73% of patients were not told about the type of hospital they should go to, despite "Dear Doctor" letters that advised doctors to do so. Similarly, 40% of physician discussions with patients did not address symptoms of valve failure, which was also advised. New letters were sent to patients about these topics *(31)*. Further notifications were sent to both physicians and patients as new estimates of risk were developed. While patients were ultimately advised of what they needed to know, the process showed that notification via physicians alone was inadequate.

Once patients are notified, they are faced with the question of whether to have their valve replaced. The chances that a Shiley valve will fail are greatest for the large (>29 mm) valves, and dramatically higher for the 70° valve, which was not sold in the United States. Explantation carries a significant risk, particularly for older and sicker patients.

Litigation

Failures of the valve resulted in many lawsuits, including a class action suit. In 1992, Pfizer settled the class action by agreeing to pay $90 to $140 million for medical consultations for patients and $75 million for a research project at William Beaumont Hospital in Royal Oaks, Michigan, to identify patients with defective valves. Individual lawsuits have continued.

Regulation

As concern about the valve fractures at the FDA increased, regulators faced a difficult issue. Because all valves fractured in the same place, the valve clearly suffered from a design and/or manufacturing problem, which suggested that the valve should be withdrawn until the problem could be solved. On the other hand, the FDA initially believed that the valve was superior in preventing thromboembolism. Half of the patients with artificial heart valves die from their underlying heart disease and another 30% of deaths are unrelated to the artificial valve. The remaining 20% of deaths are caused by valve-related problems *(30)*. Thromboembolic phenomena are responsible for about half of valve-related deaths (10% of the total) and the rest are caused by anticoagulant-related hemorrhage, infection, leaks around the valve and mechanical failure. For the Convexo-Concave valve, aggregate mortality from the valve failures seemed to be offset by its reduced thrombogenicity. If the FDA forced the valve off the market, would the other valves available to surgeons be as good? It would make no sense to remove the Shiley valves if no better alternatives were available. In a world with very good data about these matters, one might be fairly confident in making such a decision. But in its absence, the situation was frightening.

The regulatory dilemma was also complicated by the fact that the valve failures were politically significant. Leaving the valve on the market put the FDA in the position of allowing surgeons to use an implant with a known pattern of failure which would be fatal to two-thirds of the patients whose valves fractured. Critics depicted the FDA as failing to act in the face of an obvious threat to the public health.

In this atmosphere, the FDA focused on the data regarding valve-related thromboembolic events. After much cogitation on this issue, the FDA finally concluded that the reduced thrombogenicity of the Shiley valve was not as great as previously believed, opening the way to its removal. But an involuntary recall was, at that time, a cumbersome process involving reconvening the Advisory Panel and, if necessary, going to court to show that the valve was misbranded or mislabeled.

Shiley, which had been purchased by Pfizer in 1980, finally voluntarily withdrew the valve in 1986 amid growing criticism and public concern. Ironically, the company's last fix may have worked; they report no mechanical failures in the last batch of "improved" valves.

Notifying Physicians and Patients

Shiley sent "Dear Doctor" letters to the surgeons who had purchased their valves, ostensibly to warn them of the failures. In its April 1982, letter Shiley first touted itself as "a responsible medical device manufacturer and the world's leading supplier of cardiac valve prostheses," then told readers about the worldwide distribution of its valves and their 50% reduction in thromboembolic episodes. After that came the first recall statistics: 0.000103 (1.03×10^{-4}) per implant month. Compared to this, the improved valves have a fracture rate of 0.0000106 (1.06×10^{-5}) per implant month, a result of "improvements in manufacturing and quality control inspection procedures." Despite this optimism, less than 1 month later Shiley sent out another letter again recalling its 29–33 mm valves.

During clinical trials one of the Convexo-Concave valves fractured, resulting in the death of the patient. The welded *outlet* strut had broken at the welds, allowing the disc to escape. When this happens, patients die in a short time unless the valve is quickly replaced. Even if emergency surgery is obtained fairly soon, the patient is likely to suffer some damage to the heart.

Shiley's new valve was very popular and production increased. But the shadow of the strut fracture in the clinical trial lengthened when another similar fracture was reported to Shiley in August of 1979, only a few months after the valve went on the market. Then another was reported in December 1979, and it looked as if a pattern might be developing. Although only a few fractures were reported, the number of reported fractures is obviously lower than the actual number of fractures because many will be undetected or hidden fractures. Without an autopsy or X-ray, it is difficult to distinguish valve failure from other forms of heart failure. Most patients with artificial heart valves die of their underlying heart disease, so it is easy to overlook mechanical valve failure.

Low rates of failure are difficult to reproduce in the laboratory. Accelerated testing of heart valves introduces new variables that complicate the analysis. Within these limitations, Shiley analyzed the failures in its Convexo-Concave valve, made some changes in the way the valve was manufactured, and began making valves with the new procedure. When a significant inventory had been produced, Shiley sent a "Dear Doctor" letter on February 5, 1980, to the surgeons who implanted Shiley valves asking them to return their unused valves for replacement valves made under the new procedures. By doing this, the corporation avoided having a product recall in which all its valves would be taken off the market. If the company had simply asked for the return of their valves with a promise to provide an improved valve soon, surgeons would have to turn to other valves for their patients and might not return to the fold when Shiley reintroduced its improved valve.

Manufacturers of medical devices often face conflicting goals. They are profit-making companies in a competitive market. To be successful they must persuade surgeons to buy their product. At the same time, they are producing products that will be used in the care of sick people. Physicians and patients both depend upon them for safe products and for information about performance and risks associated with them. Responsible manufacturers must take care that their marketing goals do not cause them to violate their ethical and legal obligations to physicians and patients. Many observers believed that Shiley was not sufficiently sensitive to its ethical obligations and leaned too far towards its marketing needs when dealing with problems in the Convexo-Concave valve. Critics have claimed that the "Dear Doctor" letters were misleading and did not adequately inform physicians about the danger of strut fracture. Nor was the FDA adequately informed of strut fractures *(27; see also* ref. *28;* Shiley's "Dear Doctor" letters and more are reprinted in ref. *29).*

Twice more in the next few years, Shiley developed new manufacturing and testing procedures and requested that old valves be returned for replacement with new ones. But the fractures continued and Shiley was eventually forced to recall all of its valves in 1986 as a result of growing public concern and the FDA's movement toward an involuntary withdrawal.

development of pyrolytic carbon, which is extremely hard, so that the hinges of the leaflets do not wear out. In an interesting example of technological cross-pollination, pyrolytic carbon was originally used to coat fuel pellets in nuclear reactors. It caught the attention of medical researchers and became an important biomaterial for artificial heart valves *(21)*. The most widely used bileaflet valve is the St. Jude valve, which was introduced in 1982, and has been implanted in more than 1 million patients.

The Biological Environment

Designing a successful heart valve poses substantial difficulties. To begin with, the biological environment for an artificial heart valve is "remarkably aggressive, combining high chemical activity, similar to that encountered in tropical marine applications, but enhanced by the presence of free radicals and active energy-using enzyme systems, with a highly variable spectrum of multidirectional mechanical stresses" *(22)*. The most important part of this biochemical activity is the tendency of blood to react to foreign bodies, such as artificial heart valves, by forming clots *(22)*. A thrombus on the valve may grow and ultimately prevent it from opening and closing properly, or emboli may form, causing a stroke or other disastrous consequences.

Artificial heart valves must be very robust, opening and closing about 40 million times a year. The mechanical heart valves must be highly resistant to wear and fatigue and must be able to function without maintenance for an indefinite period. It must be strongly resistant to corrosion and polymer degradation by blood components. Finally, the materials used must minimize the host response to them. Because the most significant host response is the formation of blood clots, materials have been developed to reduce thrombogenicity. Pyrolytic carbon, in addition to being extremely hard, is also less thrombogenic than other materials. Shiley and other tilting disc valve manufacturers made their discs out of pyrolytic carbon when it became available in the 1970s.

It is no wonder that many heart valve designs failed to meet these demanding criteria. Designing an artificial heart valve presents a formidable challenge to bioengineers, one that they have met with a number of successful designs. Initially the most successful of the tilting disc valves was the one designed by Shiley in the late 1960s. In 1974, he developed the spherical disc valve, consisting of a ring with two welded wire struts that hold the disc in place and allow it to swing open and closed in response to blood flow. Of the 200,000 spherical valves sold, 11 experienced fracture of the inlet strut owing to fatigue, so Shiley developed a new version of the spherical disc valve, the Convexo-Concave valve *(23)*.

Valve Fractures

This valve was an improved version of the earlier valve, developed with the assistance of Dr. Viking Bjork, a noted heart surgeon. The disc opened wider (60°) and had other features that enhanced blood flow through the valve *(24; see also* ref. *25)*. The inlet strut, which is subject to greater force, was made an integral part of the ring, while the smaller, outlet strut was still welded to it. The valve was approved for sale by the FDA in 1979. In a prophetic comment, Dr. David Wieting noted that "It is rare when you change a design to improve the overall performance of it on the desirable characteristics, that you don't have something undesirable coming along with it" *(26)*.

decision, it will be strongly and publicly attacked by those who don't agree with its decision.

Conclusion

The controversy over silicone-gel breast implants reveals the complex array of institutions, policies, and history that surrounds implanted medical devices. The actions of the various players have to be understood in the light of the institutional framework in which they had to act. Significant alterations of the framework are difficult because it is an ecological system in which one aspect can't be changed without affecting many others, some in unanticipated and unpleasant ways.

THE BJORK-SHILEY CONVEXO-CONCAVE ARTIFICIAL HEART VALVE

Background

Heart valve disease is a serious condition in which a valve does not open sufficiently (stenosis) or fails to close adequately so that some blood leaks back through the closed valve (insufficiency). Either way, the heart's pumping function is impaired. Reduced cardiac output has serious medical consequences. In severe cases it ends in death. Currently in the United States, more than 100,000 people have their diseased heart valves replaced each year. (This total includes both mechanical and bioprosthetic heart valves made from the heart valves of pigs.)

The development of artificial heart valves was crucially dependent on the introduction of the heart-lung machine in the 1950s. Once an artificial means for circulating and oxygenating blood had been perfected, open-heart surgery could be performed to repair or replace diseased heart valves.

Most initial heart valve designs attempted to imitate the three flaps of a natural heart valve *(19)*. Elaborate attempts were made to define the characteristics of the materials needed and their functional characteristics, but bioengineers were unable to overcome problems of wear and flexibility in imitating natural heart valves. This period in heart valve design was analogous to the first attempts to fly by imitating the wings of birds. In the First National Conference on Prosthetic Heart Valves in 1960, only 2 of 29 chapters in the published proceedings were devoted to rigid mechanical valves.

Just as the Wright brothers succeeded by *not* trying to fly like a bird, Dr. Charles Hufnagel succeeded by designing a rigid mechanical valve modeled after an old-fashioned ball-and-cage bottle stopper *(19)*. He implanted an artificial heart valve of this type in 1952, and by the second National Conference on Prosthetic Heart Valves in 1968 only rigid valves were discussed *(20)*. Although the valve did not adequately replace the form and function of a natural valve, Hufnagel's design ultimately led to a very successful ball-and-cage valve, the Starr-Edwards, which is still in use today.

The second generation of successful mechanical heart valves utilized a tilting disc (the occluder) within a metal ring. In these valves, the disc is held in place with struts that allow it to open and close. Because it is held by the struts but not attached to them, the disc can rotate as well, evenly distributing wear. Don Shiley, a pioneer in heart valve development, designed a tilting disc heart valve in the late 1960s, which was widely used and led to the development of the Bjork-Shiley Convexo-Concave tilting disc valve. A third generation of heart valves is based on a bileaflet design in which hinged leaflets open and close. This revolutionary design was made possible by the

However much one deplores the culture of beauty and sexual attraction that emphasizes large breasts *(13; see also* ref. *14)*, shouldn't individual women have the right to decide this issue for themselves? Critics suggest that the FDA should have left breast implants on the market while safety studies were taking place because there was no epidemiological evidence that, in the 30 years silicone-gel breast implants had been on the market, the incidence of autoimmune diseases had significantly increased. These critics argued that patients could be informed about what is known and not known and make their own decisions *(15; see also* ref. *15a)*. Although this proposal has merit, it is hard to see how the FDA could have left the implants on the market in the face of a widespread outcry and uncertainty about their safety. In a democracy, government agencies are understandably sensitive to public opinion, partly because Congress has the same sensitivity. Congress also has the power to publicly criticize agencies in hearings and can impose legislative and budgetary constraints.

Law and Medicine

Kessler's decision in 1992 to ban silicone-gel breast implants unleashed a flood of additional litigation. But given the difficulty of determining a statistically significant link between breast implants and autoimmune diseases, how did juries decide that breast implants caused the injuries alleged by the plaintiffs? For many in the medical community, this revealed a deeply flawed legal system that is weighted against science and allows unscrupulous lawyers to prey on physicians and device manufacturers. Dr. Marcia Angell, a former editor of the *New England Journal of Medicine (see* ref. *15)*, thinks that juries just disregarded science, a manifestation of a general anti-science attitude in our culture. Furthermore, scientific testimony in the courtroom is provided by expert witnesses who may be just hired guns. Why not have judges appoint neutral scientific experts to testify and take it out of the adversarial system? This is an attractive idea, but in the absence of conclusive scientific evidence, scientists must make judgments, and they will often disagree. Not all scientists agree with Angell about the significance of the large studies that found no link between breast implants and autoimmune diseases *(16)*. The Mayo Clinic study examined 749 implant recipients, far fewer than needed to provide statistically reliable data about rare autoimmune diseases.

Even if it were possible to obtain neutral scientific experts, juries are free to ignore evidence that scientists would regard as dispositive (the O. J. Simpson trial is a good example). Short of revolutionary changes in the role of juries, it is hard to see how science can have the kind of influence in the courtroom that Angell would like (for a critique of Angell's book, *see* ref. *17)*. Judges have appointed expert panels in the breast implant litigation, but this augments rather than replaces the role of expert witnesses now in place. In addition, courts have made it more difficult for unqualified individuals to be used as expert witnesses *(18)*. Although these are improvements, the basic system is still intact.

The Return of Silicone-Gel Breast Implants?

An FDA Advisory Panel recently approved, by a divided vote, a silicone-gel breast implant. Although the FDA usually follows the recommendations of its panels, this is a controversial and politically charged implant with a potent history. Regardless of its

elastomer. The overall effect of the released documents was to depict Dow Corning as deceptive and uncaring. In a controversial decision, Dr. David Kessler, then the head of the FDA, banned silicone-gel implants until their safety could be established. However, he allowed women who needed implants for reconstructive surgery to obtain them if they consented to enroll in a study. Women seeking implants for cosmetic reasons could obtain saline-filled models (which pose no danger if they rupture but do not mimic breast tissue as well).

Doing the Right Thing When Things Go Wrong

What to do when things go wrong is a recurrent problem with implanted devices. If a drug is recalled, the patient simply stops taking it, but an implanted device presents more difficult choices. Explantation entails surgical risks, discomfort, and possibly financial loss. In addition, some patients may have perfectly safe implants removed solely because of the fear generated by publicity. Regulators must take into account both the question of safety and the impact on the public and the medical community.

By its nature, the FDA is risk-averse; better to err in removing a safe device than to fail to remove an unsafe one. The medical and political consequences of failing to remove an unsafe device are far more severe than prematurely recalling a safe one. Patients who lose benefits from the recalled device are largely statistical, while harmed patients file lawsuits, go on talk shows, and testify in Congress *(10)*.

Risks and Benefits

One of the controversies associated with Kessler's decision was his risk–benefit analysis. He reasoned that because breast implants for cosmetic reasons had no medical benefit, very little risk would be acceptable. Because safety is defined in terms of benefits outweighing risks, the absence of benefits means that no risks are acceptable. Some see this as a paternalistic attitude that demeans women. Because millions of women have had silicone-gel breast implants, it is obvious that increasing the size of their breasts is very important to many women. Are they dupes of a cultural norm that is so powerful that women will undergo surgery to meet it? Or are they empowered women who now have the opportunity to take control of their bodies and shape them as they please *(11)*?

People have an ambivalent attitude toward cosmetic surgery. Initially, it was confined to serious cases of deformity and injury, particularly those suffered in war. It was a natural extension of this idea to correct undesirable physical features that prevented individuals from entering certain social groups. Noses that were too Jewish or Irish could be altered to allow their owners to "pass" into social groups that would otherwise exclude them *(12)*. In Brazil, breast-reduction surgery is used to distinguish its recipients from the lower, more "black" classes with larger breasts.

There is a dark side to cosmetic surgery. It is not unusual for patients to think that, once their noses are fixed or their breasts are enlarged, their social problems will be solved. Alas, these physical changes do not always overcome the disadvantages of undesirable personality traits, which often are the result of compensating for physical deficits. If you believe that people will like you once your nose is fixed or breasts are larger, it can be crushing to find out that not much has changed afterward.

used in transformers. They, like many scientists and physicians, believed that silicone was inert and incapable of causing significant health problems.

American plastic surgeons adopted silicone injection from the Japanese in the 1960s and 1970s, although they did not have to rob transformers to get it. Soon, reports of serious problems emerged, the result of the body's reaction to the presence of silicone and the migration of silicone to other parts of the body. Cases appeared in the medical literature describing women with ulcerated breasts, breast deformities, and health problems caused by bits of silicone drifting into their abdomens, arms, back, and lungs. Some deaths were associated with silicone pulmonary emboli. Nevada and California made silicone injection illegal, and the FDA followed suit. Today, there are reports in the press that silicone injection is still being used, especially for the "filling in" of wrinkles. Former First Lady Nancy Reagan is reported to have received this treatment.

Scientists at Dow Corning developed a silicone gel by combining a rubbery polymer and a liquid one. This gel was put into a plastic container, called an elastomer, to create a silicone-gel breast implant. An estimated 2 million American women have received silicone-gel breast implants. Four out of five do so for cosmetic reasons; the remainder are primarily women who have lost breasts to cancer.

There are problems associated with this implant. A capsule of scar tissue forms around the implant and may contract into a hard, painful mass. Low-molecular-weight silicone "bleeds" through the elastomer; and rupture is common. What is not contained in the capsule may migrate through the body via the lymph system. Not much was known about the medical consequences of silicone being released into the body, and manufacturers undertook few long-term studies until relatively recently. Gel implants also interfere with mammograms, a significant issue given the relatively high incidence of breast cancer in women.

Because breast implants were on the market before the 1976 Medical Device Act, which requires FDA approval before a device can be marketed, systematic testing that is now required for medical devices was not carried out for early breast implants. Some animal studies were done, but they provided nothing like the kind of evidence that would now be required for FDA approval. An additional element of the controversy was the publication of Dow Corning memos indicating that they had no long-term data about the safety of their product.

Serious medical and political controversy developed when some women claimed that the implants had caused serious autoimmune and connective tissue diseases. Lawyers sued the manufacturers and won some large awards. Media and Congressional attention followed. Autoimmune diseases associated with breast implants are very rare, so that only very large studies would show any statistically significant correlation. During the development of the controversy it was uncertain whether silicone-gel implants caused these diseases. Nonetheless, many Americans were persuaded that silicone-gel prostheses are dangerous.

Public interest groups lobbied the FDA to remove them from the market, whereas plastic surgeons and others lobbied to keep them available. The FDA required manufacturers to submit evidence of safety and efficacy and determined that there was insufficient evidence for their safety. The release of a number of Dow Corning internal memos raised the stakes. One memo instructed sales people to wash the implant before showing it to doctors to remove the greasy coating caused by gel bleeding through the

insurance. If the payment for a total hip replacement is reduced, hospitals naturally look for ways to save on their costs for this procedure. One method is to reduce the number of different brands and models of hips available for replacement. Buying larger numbers of fewer different kinds of hips allows hospitals to save on equipment and other costs associated with different brands of hips as well as exert more economic leverage on vendors. Like group purchasing agreements, this saves money, but it also reduces the choices available to surgeons and, consequently, their patients. Some patients may receive somewhat lesser quality implants as a result of this arrangement, which also reduces physician autonomy.

IMPLANTED TECHNOLOGY

Surgically placed prosthetic devices represent fairly recent developments. Although bone fixation devices were widely used in the late 19th century, almost all of the devices described in this book were developed in the last half of the 20th century *(9)*. The body is a hostile environment that relentlessly attacks foreign bodies within it. Materials had to be found that would both function in the body and not cause uncontrolled harmful reactions. Devices had to be designed that would assist or replace bodily functions without causing harm. Today, we are reaping the benefits of decades of research, clinical experience with new materials and miniaturization of implanted medical devices. Medtronic's first pacemaker was a large, bulky device that had to be pushed around on a cart; today's versions are about the size of a half-dollar. These and other medical devices have done an enormous amount of good, but not without social costs. In this chapter, I focus on the social changes and problems associated with implanted medical devices by examining a number of cases that illustrate and illuminate them. Because not all readers will be familiar with the history and background of these cases, I begin each case with a brief overview. These cases are complex and controversial; consequently, any brief account must omit many events and qualifications. My summary is designed to highlight the social issues associated with these implants, not to present a complete account. Readers wishing to critically evaluate the actions of the parties involved should consult the substantial literature on these cases. It is useful to start with one of the most controversial devices, the breast implant.

BREAST IMPLANTS

Background

In the 1930s, scientists at Corning Glass were looking for something to use as mortar between the glass blocks it manufactured. They were especially interested in silicone, a polymer that withstood temperature extremes and could take on a variety of forms. Some polymers are runny liquids, others are elastic (like rubberbands) or solid. Unfortunately, silicone is not a good mortar for glass bricks. It remained a scientific curiosity and was marketed in the 1950s as a toy, the familiar Silly Putty.

During World War II, silicone was extensively used as an insulator in electrical transformers. After the war, the history of silicone took a bizarre twist. Japanese cosmetologists, knowing that American servicemen preferred women with larger breasts than were typical in Japanese women, experimented with breast-enlarging substances for Japanese prostitutes. They tried injecting paraffin, goat's milk, and the silicone

of their negotiated products. Some physicians complain that this practice results in the purchase of low-end implants, with little option for choosing a more effective, and more expensive, device. This is a form of rationing, but the major rationing efforts come from managed care organizations.

MANAGED CARE

In 1980, most insured Americans obtained medical care in a system dominated by fee-for-service. Doctors determined what treatments patients needed, prescribed drugs or medical devices, and sent patients to the hospital or to specialists. Payers were generally passive, simply providing reimbursement for these services. One writer compared this to having "restaurant insurance." If insurance companies paid for meals at restaurants this way, restaurant owners and employees would have a strong incentive to serve ever more exotic and expensive foods to better serve their customers and, not inconsequentially, increase their income. Diners would not complain because their needs were being met and someone else was paying the bill *(5)*. This kind of system provides no incentives for anyone to hold down costs. Not surprisingly, health care costs have risen rapidly until they have become politically unacceptable.

In the United States, the primary response to uncontrolled costs has been managed care. Although health care has always been managed, what is new are the financial and structural arrangements that have been developed to reduce costs. The variety and complexity of managed care plans is great, and their characterization here is greatly simplified. As Victor Cohn of *The Washington Post* remarked, "If you've seen one managed care plan, you've seen one managed care plan" (cited in ref. 6). However, managed care organizations share many common features that generate a characteristic set of ethical problems *(7)*.

In such systems, payers exercise control over medical costs by eliminating or modifying fee-for-service with forms of payment by capitation. Under this arrangement, the managed care organization pays doctors a fixed yearly amount to take care of a patient. Typically, some of the doctor's income (10–20%) is withheld and the amount returned at the end of the year is based on provider efficiency, the amount of funds remaining in a risk pool, productivity, percentage of profits of the organization, or some combination of these incentives *(8)*. Two other widely used cost-reducing measures are utilization review, in which hospitalization or expensive diagnostic tests must be approved by the organization, and requiring authorization for specialist care by the primary physician, who acts as a "gatekeeper."

Unlike fee-for-service, which encourages doctors to provide whatever care might have some benefit for the patient, capitation sets up financial incentives for doctors to reduce care. Physicians who take on more patients and provide less care for each will make more money. Although few physicians treat patients simply as a unit of income production, economic incentives do have an effect on behavior. By changing financial incentives, using utilization review, and establishing gatekeepers, managed care organizations have greatly slowed the growth of health care costs and physician incomes in recent years.

Another aspect of managed care that directly affects implants is the reduction of payments to hospitals and doctors. These organizations have the power to control reimbursement rates because hospitals and doctors need access to patients who have health

cine, technology, and the role of private initiatives. As with hemodialysis, the question was what to do when a lifesaving technology was available, but in short supply. Without an iron lung, thousands of children would die, so we either had to set up a rationing arrangement or provide one for every child who needed it. Heroic efforts by the March of Dimes successfully raised enough money to avoid rationing, using the argument that no one should die for lack of access to the machine. Although this was an admirable effort that saved the lives of many children, it also conveyed two important lessons: rationing technology is unnecessary and unacceptable, and private initiatives can solve such problems. This and other historical decisions concerning the role of private organizations in health care have made it more difficult to develop a health care system that adequately serves all of our citizens. Among the industrial nations, only the United States and South Africa lack a national health insurance system. More than 40 million Americans have no health insurance and no assured access to health care. According to Rothman, we chose to avoid rationing health care for the middle class by not providing the same level of care to everyone else. As is seen here, this has a significant effect on patients, physicians, and manufacturers of implanted medical devices.

COSTS

Health care costs relentlessly continue to rise and now account for about 14% of our gross national product. In 1965, it was less than half of what it is today. Medical technology accounts for a significant portion of this increase, accounting for as much as 40% of our health care expenses *(3)*. Social policy both fosters and restricts innovation and distribution of medical devices. The National Institutes of Health (NIH) funds most device research, whereas FDA regulations and cost-saving arrangements impose restrictions. Demand for innovative lifesaving devices seems insatiable, and this provides powerful incentives in a market economy for ever more costly technology. The result is a supply-driven "medical arms race" in which companies are constantly developing new products that the government and managed care organizations are trying to control. The reimbursement policies of Medicare, Centers for Medicare and Medicaid Services, and managed care organizations are central to this effort and will critically affect the growth and diffusion of medical technology, including implants. Beneath this proliferation of policies of control are the important questions of how we want to live and die, who gets what, fair distribution of medical resources, what happens when things go wrong, and how much we are willing to pay for the kind of health care we want *(3)*.

The way health care costs are counted tends to highlight initial costs and hide savings. An implantable cardioverter defibrillator has a high initial cost, but large potential savings in emergency care, hospital visits, long-term care, and time returned to work, to say nothing about the years of life not lost to fatal arrhythmias. Patients whose sudden cardiac death is prevented can continue to be productive and avoid medical costs associated with their illness *(4)*. However, these are savings based on statistical projections from data. In contrast, someone has to write an actual check to pay for the defibrillator and its associated costs.

Many hospitals use the services of a group purchasing organization to reduce their costs. These organizations represent many hospitals and can negotiate lower prices through bulk purchasing. However, the contracts tend to require exclusive (95%) use

6
Societal Issues

John H. Fielder

Medical devices that are implanted in patients are also implanted in a complex array of social institutions, cultural values, and history. By the time a surgeon places a device, it already has a long history of scientific research, investment, engineering design, testing, Food and Drug Adminitration (FDA) regulation, marketing, and, often, litigation. An implant is thus a social as well as a technological artifact, which bears the marks of its society and may reshape society as well.

TECHNOLOGY AND SOCIETY

New technologies require that we make a place for them in our social institutions, values, and culture. (A good treatment of this issue may be found in ref. *1*.) Transformative technologies, such as anesthesia for surgery, hemodialysis, and the total artificial heart require substantial social changes in order to do this. Anesthesia changed surgery from a ghastly ordeal, performed at breakneck speed, to an unpleasant but bearable experience that permitted an enormous increase in medical interventions. These, in turn, required changes in hospitals, medical schools, research programs, and professional societies. Hemodialysis forced us to confront the ethical problem of rationing a scarce, lifesaving technology. The decision to have the government provide treatment of end-stage renal disease (ESRD) to anyone who needs it is still causing reverberations today, as roughly 6.4% of the Medicare budget goes to pay for the less than 0.1% of its beneficiaries with ESRD. The total artificial heart is again under development after its initial debacle in the 1980s. Hopefully, it will help many patients with heart disease, but it will also raise the same issues as dialysis, only at a much higher level of expense, and within a health care system striving to hold down relentless increases in costs. Policymakers would do well to remember the financial consequences of providing hemodialysis to everyone when the estimated 10,000 to 20,000 primary candidates and up to 200,000 secondary ones clamor for a lifesaving artificial heart.

TECHNOLOGY AND HISTORY

Americans have a fondness for technology and particularly for medical technology. David Rothman *(2)* tells the story of the iron lung, used to treat polio, as an example of a lesson learned that determined the direction of American social thinking about medi-

From: *The Bionic Human: Health Promotion for People With Implanted Prosthetic Devices*
Edited by: F. E. Johnson and K. S. Virgo © Humana Press Inc., Totowa, NJ

47. Title 21 of the U.S. Code of Federal Regulations Part 821.30(b).
48. Food and Drug Administration. Guidance on Medical Device Tracking. Rockville, MD: U.S. Food and Drug Administration, Office of Compliance, 1998. Available at: http://www.fda.gov/cdrh/pdp/420.html.
49. Kaczmarek RG, Beaulieu MD, Kessler LG. Medical device tracking: results of a case study of the implantable cardioverter defibrillator. Am J Cardiol 2000;85:588–592.
50. Section 519(f)(1) of the U.S. Food, Drug, and Cosmetic Act and Title 21 of the U.S. Code of Federal Regulations Part 806.
51. Title 21 of the U.S. Code of Federal Regulations Part 7.
52. Title 21 of the U.S. Code of Federal Regulations Part 806.2(j).
53. Food and Drug Administration. Guide to Inspections of Quality Systems. Rockville, MD: U.S. Food and Drug Administration, Office of Compliance, August 1999.
54. Food and Drug Administration. Regulatory Procedures Manual. Rockville, MD: U.S. Food and Drug Administration, August 1997.
55. Title 21 of the U.S. Code of Federal Regulations Part 821.
56. Global Harmonization Task Force. Available at: http://www.ghtf.org.
57. Title 21 of the U.S. Code, Section 360(j)(b).

20. Section 520 of the U.S. Food, Drug, and Cosmetic Act.
21. Food and Drug Administration. Guidance for Industry: Contents of a Product Development Protocol. Rockville, MD: U.S. Food and Drug Administration, Office of Device Evaluation, 1998. Available at: http://www.fda.gov/cdrh/pdp/420.html
22. Title 21 of the U.S. Code of Federal Regulations Parts 809 and 812.
23. Food and Drug Administration. Information Sheets: Guidance for Institutional Review Boards and Clinical Investigators. Rockville, MD: U.S. Food and Drug Administration, Office of Health Affairs, 1998.
24. O'Neil RT. Assessment of safety. In: Peace KE, ed. Biopharmaceutical Statistics for Drug Development. New York: Marcel Dekker, 1988:543.
25. Kessler DA. Introducing MEDWatch: A new approach to reporting medication and device adverse effects and product problems. JAMA 1993;269:2765–2768.
26. Title 21 of the U.S. Code of Federal Regulations Part 803.
27. ECRI. Medical device problem reporting for the betterment of healthcare. Health Devices 1998;27(8):277–292.
28. Cardiology. FDA alerts device manufacturers. Cardiology 1997;26:7.
29. Dickinson's FDA Review. FDA detective work on implants gets grip on vacuum hazard. Dickinson's FDA Review 1997, September 22.
30. Food and Drug Administration. Vacuum loss in resonating components. FDA Dear Manufacturer Letter. Rockville, MD: U.S. Food and Drug Administration, 1997.
31. Title 21 of the U.S. Code of Federal Regulations Part 814.82(a)(9).
32. Food and Drug Administration. Designing a medical device surveillance network. FDA Report to Congress. Rockville, MD: U.S. Food and Drug Administration, September 1999.
33. The National Institutes of Health Technology Assessment Conference on Improving Medical Implant Performance through Retrieval Information: Challenges and Opportunities, Jan 10–12, 2000, Natcher Conference Center, NIH, Bethesda, MD.
34. Title 21 of the U.S. Code of Federal Regulations Part 814(a)(2).
35. Food and Drug Administration. Guidance on Criteria and Approaches to Postmarket Surveillance. Rockville, MD: U.S. Food and Drug Administration, Office of Surveillance and Biometrics; 1998. Available at: http://www.fda.gov/cdrh/pdp/420.html.
36. Hester TR Jr., Ford NF, Gale PJ, et al. Measurement of 2,4-toluenediamine in urine and serum samples from women with Meme or Replicon breast implants. Plast Reconstr Surg 1997;100:1291–1298.
37. DoLuu HM, Hutter JC, Bushar HF. A physiologically based pharmacokinetic model 2,4-toluenediamine leached from polyurethane foam-covered breast implants. Environ Health Perspectives 1998;106:393–400.
38. Food and Drug Administration. TDA and polyurethane breast implants. Food and Drug Administration Talk Paper. Rockville, MD: U.S. Food and Drug Administration, June 1995.
39. Friedman GD. Primer of epidemiology. New York: McGraw-Hill, 1980.
40. Faich GA. Adverse drug-reaction monitoring. NEJM 1986;314:1589–1592.
41. Lilienfeld AM, Lilienfeld DE. Foundations of epidemiology. New York: Oxford University Press, 1980.
42. Marinac-Dabic D, Kennard ED, Torrence ME, et al. Palmaz-Schatz stenting in women: acute and long-term outcomes. APHA 126th Annual Meeting abstract. Washington, DC, November 1998.
43. Silverman BG, Brown SL, Bright RA, et al. Reported complications of silicone gel breast implants: an epidemiologic review. Ann Intern Med 1996;124:744–756.
44. Brown SL, Langone JJ, Brinton LA. Silicone breast implants and autoimmune disease. J Am Med Women Assoc 1998;53:21–24.
45. Section 518(e) of the U.S. Food, Drug, and Cosmetic Act.
46. Title 21 of the U.S. Code of Federal Regulations Part 821, as amended.

For practical purposes, a reproducible, manufactured device is precluded from meeting the statutory custom device exemption. A manufacturer may change certain features to a legally available device by modifying the materials, shape, or sizes so as to meet a physician's unsolicited request. The modification of attributes to a legally marketed device with a premarket notification constitutes "customizing."

If a device does not qualify for the custom device exemption, but was in fact used under that errant belief, the device is in violation of FDA requirements and subject to legal remedies. The agency could also ask the manufacturer to consider the submission of an IDE so data from a well-controlled study could be used to support a premarket submission.

REFERENCES

1. Food and Drug Administration. Frequently asked questions: FDA (general). Rockville, MD: U.S. Food and Drug Administration, March 1999. Available at: http://www.fda.gov.
2. Food and Drug Administration. Index: FDA's mission. Rockville, MD: U.S. Food and Drug Administration; March 1996, October 1998. Available at: http://www.fda.gov.
3. Gallivan M. The 1997 global medical technology update: the challenges facing U.S. industry and policy makers. Washington, DC: Health Industry Manufacturers Association, 1997;37–51.
4. Forces reshaping the performance and contribution of the U.S. medical device industry. New York, NY: Wilkerson Group, 1995;3–31.
5. Section 201[h] of the U.S. Food, Drug, and Cosmetic Act, Title 21 U.S. Code §321.
6. Food and Drug Administration. Managing the risks from medical product use: creating a risk management framework. Report to the FDA Commissioner from the Task Force on Risk Management. Rockville, MD: U.S. Food and Drug Administration, May 1999.
7. Hearings before the Subcommittee on Health, of the Committee on Labor and Public Welfare, United States Senate, 93rd Congress, Medical Device Amendments, September 13 and 17, 1973.
8. Section 502 of the U.S. Food, Drug, and Cosmetic Act.
9. Title 21 of the U.S. Code of Federal Regulations Part 807.92(a)(3).
10. Section 520(f) of the U.S. Food, Drug, and Cosmetic Act.
11. Food and Drug Administration. Guidance on IDE Policies and Procedures. Rockville, MD: U.S. Food and Drug Administration, Office of Device Evaluation, 1998. Available at: http://www.fda.gov/cdrh/ode/idepolicy.html.
12. Food and Drug Administration. Guidance for Spinal System 510(k)s. Rockville, MD: U.S. Food and Drug Administration, Office of Device Evaluation; 1999. Available at: http://www.fda.gov/cdrh/ode/idepolicy.html.
13. Food and Drug Administration. Guidance for Resorbable Adhesion Barrier Devices for Use in Abdominal and/or Pelvic Surgery. Rockville, MD: U.S. Food and Drug Administration, Office of Device Evaluation; 1999. Available at: http://www.fda.gov/cdrh/ode/idepolicy.html.
14. Section 513(a)(1)(A) of the U.S. Food, Drug, and Cosmetic Act.
15. Section 513(a)(1)(B) of the U.S. Food, Drug, and Cosmetic Act.
16. Section 513(a)(1)(C) of the U.S. Food, Drug, and Cosmetic Act.
17. Section 513 of the U.S. Food, Drug, and Cosmetic Act.
18. Title 21 of the U.S. Code of Federal Regulations Part 860.7.
19. Food and Drug Administration. Draft Replacement Heart Valve Guidance. Rockville, MD: U.S. Food and Drug Administration, Office of Device Evaluation, 1994.

and risks), technology development, and a product's regulatory history. Each assessment is to provide clinicians, consumers, and industry with product information leading to optimal use of medical devices and enhanced public health. The vehicles for information dissemination may be workshops, teleconferences, and peer-reviewed journal articles, among others.

The neurological shunts STAMP is a case in point. A 1-day international conference entitled *Cerebrospinal Shunt Technology: Challenges and Emerging Directions* was convened among relevant stakeholders including, among others, neurosurgeons, patient advocates, government clinicians and scientists, and industry representatives. Presentations were made and a variety of issues were explored regarding different approaches to improving patient outcome. Issues identified for further work included development of labeling and/or patient alert cards that provide patients with specific shunt information (e.g., type of shunt implanted and size of ventricles); improvement in test methods to simulate clinical performance of these devices, possibly in conjunction with development of animal models; creation of improved test methods to evaluate anti-infective coatings and assess immunological effects (e.g., from silicone); and development of outcomes databases to aid the formulation of practice guidelines for shunt selection and antibiotic use.

Custom Device Exemption

A custom device exemption provided by the Act permits a manufacturer to make and distribute a particular device and remain exempt from the FDA's premarket approval or IDE requirements *(57)*. A custom device is a new, unmarketed device with unique specifications with little or no prior use. Consequently, investigational clinical studies are not feasible.

For a device to qualify for a custom device exemption, the FDA applies five statutory criteria to the facts of the situation. Each statutory criterion must be met to qualify for the exemption. The criteria include the following:

1. Necessarily deviates from a performance standard or PMA requirements. For example, a clinical study is not feasible given the rare patient population using the same or similar device in a 1-year period.
2. Not generally available in finished form for purchase or for dispensing upon prescription. For example, the manufacturer lacks existing design specifications and manufacturing procedures to reproduce the device. Devices available under an IDE do not qualify.
3. Not offered through labeling or advertising. For example, the device is not solicited through any electronic, hardcopy literature or promotional material, or anecdotal testimonials that constitute labeling or advertising. An unsolicited request by a physician who specifies unique design inputs, in the absence of a commercially available alternative would probably meet this statutory criterion.
4. Not generally used by other physicians, dentists or specially qualified individuals. For example, the device must be a new concept, not a device already available that is simply "customized" for a patient, like a contact lens.
5. To meet a specific patient form, or the special needs of a practitioner. For example, the device is formed to the patient's physiological needs, such as with a prosthetic device; or the device is made to meet the unique needs of a specially qualified practitioner with a professional practice.

postmarket), Guidance Documents, and New Device Information, and under "Special Interest Items" one can find Breast Implants, Device Alerts, and International Issues, among others. For those who wish to directly interact with CDRH, the topical heading "Interacting with CDRH" lists a variety of avenues from "Contacts in CDRH on Specific Issues" to "Meetings, Presentations, and Videoconferencing" to "Report Problems with Medical Devices," among others. In addition, this home page keeps visitors up-to-date on breaking issues and events and provides opportunity for feedback. An alphabetized topic index is also available for direct access to specific topics.

Before much of this information was available via the Internet (e.g., access to AE reports), it could only be obtained by submitting a written Freedom of Information (FOI) request. The latter still applies, however, for information not normally prepared for public distribution and thus not found on the Internet. Despite this, certain records may be withheld in whole or part if they fall within one of nine FOI Act exemptions including trade secret and confidential commercial or financial information or individual information that would constitute clearly unwarranted invasion of privacy. Information about submitting and paying for FOI requests is available under the "Interacting with CDRH" pull-down menu.

Global Harmonization

A Global Harmonization Task Force (GHTF) was established in 1992 to respond to the increasing need for international harmonization (explained below) in the regulation of medical devices *(56)*. The GHTF is a voluntary international consortium of public health officials, responsible for administering national medical device regulatory systems, and representatives from regulated industry. The GHTF acts as a vehicle for convergence in regulatory practices related to ensuring the safety and effectiveness and quality of medical devices and promoting technological innovation as well as facilitating international trade. This is principally accomplished through publication and dissemination of harmonized guidance documents on basic regulatory practices.

To accomplish its mission, the GHTF conducts much of its work through four study groups whose focus is on countries with developed regulatory systems. Study group 1 is responsible for comparing operational medical device regulatory systems around the world and is also responsible for developing a standardized format for premarket submissions and harmonized product labeling requirements. Study group 2 is charged with reviewing current adverse event reporting, postmarket surveillance and other forms of vigilance for medical devices and performing an analysis of different requirements with a view to harmonizing data collection and reporting systems. Study group 3 is responsible for examining existing quality system requirements and identifying areas suitable for harmonization. Study group 4 is charged with examining quality system auditing practices and developing guidance documents laying down harmonized principles for the medical device auditing process.

Systematic Technology Assessment of Medical Products

The nascent Systematic Technology Assessment of Medical Products (STAMP) program seeks to assess the current status of the agency's knowledge of marketed devices that present particularly challenging medical and public health issues. As part of the evaluation, each STAMP project addresses such issues as clinical applications (reflected in on- and off-label use, patient selection and population exposure, benefits

stantial harm to the public health. Typically, however, an individual AE report will not lead to remedial action; rather, remedial action is generally based on an AE case series, pattern, or trend.

Manufacturers must implement procedures to collect and analyze information about their medical devices to identify potential product and quality problems that may need preventive action or that require recall consideration *(53)*. A manufacturer's assessment may be based on analysis of information provided through AE reporting, quality control records, or internal quality audit reports. During an on-site inspection of a manufacturer, an FDA field investigator will determine whether the firm analyzes data concerning a nonconforming product (i.e., one that does not meet specifications). The FDA investigator will also review the firm's statistical control techniques and other data relevant to failure investigations including nonstatistical methods such as *ad hoc* committee reviews. Where a firm's failure investigation identifies a root cause, the agency will assess the adequacy of the firm's corrective action. Corrections and removals are considered a form of corrective action.

Recalls usually are conducted on a voluntary basis in order to protect the public health or correct a violation of the act. In 1990, Section 519(f) of the Act was amended by the SMDA, which authorized FDA to issue reporting and record-keeping requirements concerning the recall activities. Congress established this statutory requirement because it believed that device manufacturers and importers would conduct recalls without notifying the FDA in a timely fashion, if at all.

When a firm fails to correct or remove dangerous devices from the market promptly, and the FDA finds that there is a reasonable probability that a device would cause serious adverse health consequences or death, the FDA will issue an order, under Section 518 of the Act, to require the appropriate person to (a) immediately stop distribution of the device, (b) immediately notify health professionals and device user facilities of the order, or (b) instruct them to stop using the device *(54)*. When a device is the subject of a mandatory recall order and the device has been subject to medical device tracking requirements, the agency expects the responsible firm to use the relevant tracking information required by the tracking regulation *(55)*. If necessary, the agency can order the firm to produce product distribution records, obtained under the tracking authority, and use that information to ensure that a mandatory recall is effective and prompt.

Firms are expected to conduct effectiveness checks to make sure the users or consignees have received notice of the correction or removal and that they have taken the appropriate action. FDA field staff will audit the firm's effectiveness checks. When a correction or removal involves a device that poses substantial risk of serious injury or death, the agency will contact each user or consignee to ensure the appropriate action has been taken.

SPECIAL TOPICS

Access to Information

A wealth of information about all aspects of the regulation of medical devices is readily available on the Internet *(1)*. Through a series of pull-down menus, visitors to the CDRH's site can access information on a variety of specific topics via broad topical headings. For instance, under "Popular Items" one can find listings for CDRH Databases (both pre- and

the manufacturer, among other items, the name, address, and telephone number of the patient to whom it distributed the device, as well as the prescribing physician and physician who regularly follows the patient *(47)*.

The agency has issued tracking orders to manufacturers of the following implantable devices: abdominal aortic aneurysm stent grafts; dura mater; implantable infusion pumps; implanted diaphragmatic/phrenic nerve stimulators; AICDs; mechanical replacement heart valves; implantable pacemaker pulse generators and electrodes; mandibular condyle prostheses; and temporomandibular joint prostheses. Manufacturers have used their tracking system to conduct a voluntary recall. The FDA, however, has not ordered a manufacturer to produce tracking information in conjunction with an FDA-ordered recall. FDA guidance on medical device tracking is available on the Internet *(48)*.

A recently published case study of medical device tracking highlights its important public health function *(49)*. Ventritex, a manufacturer of implantable cardioverter defibrillators, used its tracking database to alert patients and their physicians of an immediate need to reprogram model V-110 and V-112 implantable cardioverter defibrillators. This action was taken in response to reports received by the firm of a fatal tachycardia. Subsequent analysis of the device indicated the potential for a failure mode that could affect device performance in a variety of ways. Of the approximately 5600 patients implanted with the device and alive at the time of the alert, 98.7% were successfully located and their devices reprogrammed within the first 60 days of the notification. Ultimately, more than 99.8% of devices in patients were reprogrammed. This case study demonstrated that most tracked device recipients could be located and receive medical intervention.

Recall Authority/Field Inspections

Recall activities involving medical devices are termed *corrections and removals* by the FDA. A "correction" covers a number of activities, including the repair, modification, adjustment, relabeling, destruction, inspection, or patient monitoring of a device, even without physical removal from its point of use. A "removal" also covers a number of activities, including the physical removal of a device from its point of use to some other location for repair, modification, adjustment, relabeling, destruction or inspection.

Manufacturers and importers must report to the FDA information required by the "corrections and removals" regulation when action is undertaken (a) to reduce a risk to health posed by the device or (b) to remedy a violation of the Act caused by a device which may present a risk to health *(50)*. No report to the FDA is required if the correction or removal does not present a risk to health; however, a firm may voluntarily report a correction or removal as part of the FDA's voluntary recall policy *(51)*.

The definition of "risk to health" plays an important role in whether a correction or removal must be reported to the agency. The phrase "risk to health" means: (a) a reasonable probability that use of, or exposure to, the product will cause serious adverse health consequences or death or (b) that use of, or exposure to, the product may cause temporary or medically reversible adverse health consequences, or an outcome where the probability of serious adverse health consequences is remote *(52)*.

Reporting requirements of corrections and removals may appear to overlap AE reporting requirements. This is particularly germane for AE reports that involve death or serious injury that require remedial action to prevent an unreasonable risk of sub-

Postmarket Enforcement

Enforcement activities for the FDA's premarket and postmarket medical device programs focus on developing and implementing legal and administrative remedies to effect compliance with applicable federal statutory and regulatory requirements. Regulatory and enforcement policies address specific issues, such as device labeling, manufacturing practices, AE reporting, recalls, advertising and promotion, and performance standards for medical and nonmedical electronic products that emit radiation. Enforcement actions usually are directed at a manufacturer or importer, but may also include a sponsor of a PMA, an IRB or, on rare occasions, a particular individual.

Postmarket enforcement activities, however, are not the agency's preferred venue for effecting compliance. The agency speaks with the device industry through formal and grassroots meetings to explain policies, to resolve issues in dispute, to review industry complaints, and to provide training and guidance to new and existing firms. Enforcement activities also require the agency to remain up-to-date on industry practices, trade developments, scientific innovations, the scientific literature, and the activities of professional associations. As the need arises, the agency amends regulatory requirements to keep pace with changes in trade practices, technology and professional practices. The FDA's medical device tracking program (*see* later) is an example of a postmarket program that is periodically amended due to technological advancements and changes in professional practices.

Tracking

The FDA's medical device tracking program serves as a regulatory safety net to ensure that manufacturers of certain devices establish a tracking system. The system will enable them, on their own initiative, to promptly locate and remove devices in commercial distribution and provide patient/physician notification (regarding the safety concerns that may require special clinical management) if that is the recall strategy. Tracking also augments the FDA's recall authority to order a mandatory recall *(45)*.

By law, the agency may require tracking for a class II or class III device (a) the failure of which would be reasonably likely to have serious adverse health consequences; or (b) which is intended to be implanted in the human body for more than 1 year; or (c) which is life-sustaining or life-supporting and used outside a user facility. The agency also uses additional nonbinding factors to consider when requiring tracking. Those factors are similar to those identified in the MDR regulation that comprise serious injuries (i.e., likelihood of sudden catastrophic failure; likelihood of significant adverse clinical outcome; and need for prompt professional intervention). The agency may add or remove devices from the list of tracked devices as a result of its review of premarket applications, postmarket surveillance (including reports of AEs), recalls, or other information coming to its attention.

Manufacturers that receive tracking orders must implement tracking procedures and collect information required by the tracking regulation *(46)*. Permanently implanted devices and life-sustaining or life-supporting devices that are intended for use by a single patient over the life of the device must be tracked to the patient using the device. Manufacturers are required to audit their tracking system, which requires effective communication through the chain of distribution. Manufacturers need to ensure that distributors and hospitals comply with their information reporting obligations. Final distributors of a tracked device, which includes doctors and hospitals, must report to

epidemiology program serves a vital postmarket function and serves to inform CDRH and FDA device policy, address relevant scientific questions, assess effectiveness of regulatory approaches, provide risk assessments, influence patient management, develop new postmarket data resources, and provide important public health information (e.g., through peer-reviewed publications).

To accomplish its mission, the epidemiology program has made use of a variety of databases. In addition to database access and analysis, and in response to device issues, the program has developed and conducted surveys, made use of and expanded existing device registries, helped design and analyze mandated studies, and assessed the literature.

As true of any tool, there are advantages and disadvantages of epidemiological (observational) studies. The advantages are many. These studies (whether cohort, case–control, cross-sectional, or otherwise) fill the knowledge gap between individual case reports and randomized, controlled clinical trials. In contrast to the latter, observational studies tend to be less expensive, more timely (especially if retrospective), and are better equipped to detect and/or assess rare events (and provide meaningful frequency measures) *(39,40)*. In addition, subjects captured in postmarket observational studies tend to include patients more representative of "real-world" device experience in contrast to more narrowly defined clinical trial subjects. Thus, those outside certain age ranges or with more complex disease or co-morbid conditions might be included in the former, but not the latter. In addition, if valid data are available, other risk factors (as confounders or effect modifiers) may be more effectively explored in postmarket studies given the greater breadth of patient population.

The downside to the observational approach, in particular for devices, is that existing databases may not capture relevant device and/or patient data (e.g., lack of model and/or brand device identification) or, if they do, may not be large enough to appropriately assess rare events. In addition, observational studies are subject to a variety of biases (e.g., selection and recall bias) and may succumb to difficulties in patient follow-up (e.g., in prospective cohort studies) or obtaining valid data (e.g., in retrospective approaches) *(41)*. Finally, on a practical note, studies may be too resource intensive or cannot be done within certain time constraints.

The role of epidemiology is exemplified by the following two cases. Based on both AEs and reports in the literature of possible gender-related effects, the program undertook a collaborative effort, with principal investigators of the NIH-funded New Approaches to Coronary Intervention Registry, to expand the registry to include a much larger proportion of women and to analyze data on short- and long-term outcomes related to use of the Palmaz-Schatz stent *(42)*. The results of the study were reassuring for women, showing that the stent was effective and revealing a low overall rate of major cardiac events during hospitalization. The results were equally encouraging in long-term follow-up. The epidemiology program was also very much involved in assessing the overall risk of connective tissue and autoimmune disease related to silicone gel-filled breast implants *(43,44)*. This important public health issue had been addressed through multiple published observational studies, many of which suffered from some of the weaknesses mentioned earlier. CDRH epidemiologists performed a systematic review of that literature to date, and determined that, if there were a risk, the risk was small.

to cover 510k products as well. Unless there are unusual circumstances, the section 522 authority is typically reserved for the latter.

Prior to issuing an order, the FDA will discuss the public health concern with the firm. The concern may arise from questions about a product's long-term safety, about performance of a device in general use or involving a change in user setting (e.g., professional to home use), or notable AEs. Upon receiving an order, the firm has up to 30 days in which to submit its study plan and, by statute, studies are limited to 3-year patient follow-up (or longer if agreed to by the firm). (The FDA soon plans to issue a regulation clearly specifying the requirements for a study plan, conduct, and follow-up.)

In October 1998, the FDA issued guidance on criteria used in considering order issuance as well as possible study approaches. Briefly, the criteria include: the public health issue must be important; other postmarket mechanisms cannot effectively address the issue; the study must be practical (i.e., feasible, timely, not cost-prohibitive); and the issue is of high priority. Details are posted on the Internet *(35)*. The possible study approaches vary widely (designed to capture the most practical, least burdensome approach to produce a scientifically sound answer) and include detailed review of complaint history and the literature; nonclinical testing of the device; telephone or mail follow-up of a patient sample; use of registries; observational studies; and, rarely, randomized controlled trials.

Generally speaking, these mandated postmarket studies (both PMA conditions of approval and section 522) require the participation of both firms and the clinical community. Problems, however, may arise in the conduct of these studies if, for instance, it is difficult to recruit physician investigators or accrue patients or if industry lacks incentive. These issues particularly resonate with rapidly evolving technologies, where rapid device evolution may make studies of prior models obsolete by the time they are completed.

Although there may be difficulties in study conduct, an example of a section 522 discretionary study reveals the authority's public health importance and its risk assessment role. In 1991, FDA scientists demonstrated that it was possible for polyurethane to break down under laboratory conditions to form 2,4-toluenediamine (TDA). TDA had been shown to be an animal carcinogen. Prior to this it was thought that breakdown could only occur at very high temperatures and pH extremes. The firm that manufactured polyurethane foam-coated breast implants ceased sales in 1991 and agreed to a clinical study under section 522. The study involved comparing TDA levels in urine and serum samples from women with and without the implants. Although minute amounts of TDA were found in the majority of women with the implants, the increase in cancer risk was determined to be vanishingly small (1 in 1 million) *(36,37)*. The FDA issued a public health correspondence (FDA Talk Paper) on the results and their reassuring implications *(38)*.

Applied Epidemiological Research

Postmarket surveillance and risk assessment would not be complete without epidemiology, a discipline that provides the means and methods to elucidate potential AEs and device use in a population context. Through employing methods of observational (as opposed to experimental) study, epidemiologists help refine AE signals, characterize subgroups at risk, test hypotheses, and evaluate device performance and use. The

mentioned earlier, clinical trials are usually short compared to the expected lifetime of a device. Under these conditions, only the worst design defects will be uncovered. In addition, as noted previously, there are limits to what reported AEs can tell us about device failure, including those involving implants.

One way in which the data needed to assess long-term reliability can be obtained is through implant retrieval and device analysis. Historically, there have been many impediments to the implementation of a national implant retrieval and analysis program. The main deterrents have been cost, ethical concerns, and legal issues. Many of these issues were discussed at a recent National Institutes of Health (NIH)Technology Assessment Conference *(33)*. It was the consensus of the expert panel at this conference that many of the core issues were intractable at this time and that the best that could be done is to inform the public through educational programs of what can be expected of medical devices under normal conditions of use. This is not to say that implant retrieval and analysis does not occur. In fact, manufacturers are required to perform failure analyses. Depending upon the nature of the problem, these analyses may lead to a redesign of the product. In addition to manufacturers, there are certain disciplines, such as cardiovascular and orthopedic surgery, in which product safety and utility are everyday concerns. Professional groups in these disciplines may undertake such analyses. However, these programs usually reside at large teaching university hospitals and are driven by the academic interests of the faculty. The coordination of the data among these academic institutes usually occurs through the publishing of journal articles and at national meetings and inconsistencies among investigators can go unresolved for several years. The FDA must balance the data received from the open literature, AE reporting systems, failure mode analyses performed by manufacturers, and epidemiological studies (*see* later) in order to make an assessment of the safety and effectiveness of a new medical device design.

Mandated Postmarket Studies

Another "tool" that the FDA uses to achieve its surveillance and risk-assessment goals is the mandated postmarket study, conducted under either PMA conditions of approval or FDAMA (section 522) authorities. As noted in the section on premarket review, a sponsor may be required to perform a postapproval study as a condition of approval for a PMA (34). The study questions may relate to longer-term performance of an implant, or focus on specific safety issues that may have been identified during review of the product and for which additional postmarket information is felt to be needed. Results from these studies may be included as revisions to the product's labeling (including patient- and clinician-related material).

In addition to the PMA authority, the agency may, under section 522 of FDAMA, impose postmarket study requirements on certain devices. The latter provision, originally mandated in 1990 under SMDA, allows the agency, under its discretion and for good reason, to order a manufacturer of a class II or class III device to conduct a postmarket study if the device (a) is intended to be implanted in the human body for more than 1 year; (b) is life-sustaining or life-supporting (and used outside a device user facility); or (c) failure would reasonably be likely to have serious adverse health consequences. Although this discretionary authority overlaps the PMA postapproval authority for some products (e.g., PMA implants), it effectively extends FDA authority

or uniqueness or publicity and litigation; events are generally underreported and this, in combination with lack of denominator data, precludes determination of event incidence or prevalence; and causality cannot be inferred from any individual report. (With regard to causality assessment, device retrieval and analysis data are often inadequate or lacking. Although there are no current reporting authorities that systematically require such data, FDA Quality System regulations do require investigation of any failure of a device to meet its performance specifications.) The system also has many strengths including: provision of nationwide safety surveillance from a variety of sources, thus providing insight into AEs related to "real-world" use; relatively low cost considering the scope of surveillance; and uniformity of data collected in terms of a standardized form with prespecified data elements. This system is one of only a few means to detect rare AEs, is accessible, and the information is open to the public.

Supplementing this reporting system are PMA conditions of approval. As noted previously, all products with approved PMAs have conditions of approval, one of those being submission of information on AEs outside the MDR regulatory requirements *(31)*. Examples of this include labeled AEs occurring with unexpected severity or frequency. This requirement helps the agency cast a wider "safety net" in its surveillance of AEs.

Prior to discussing postmarket studies, brief mention should be made of MedSuN *(32)*. This network of user facilities is currently in its conceptual/pilot phase. Its principal objective is to increase the utility of user facility reporting by recruiting a cadre of well-trained and motivated facilities. It is envisioned that, in addition to enhancing the detection of emerging device problems, the network would act as a two-way communication channel between the FDA and the clinical community and serve as a setting for applied clinical research on device issues. To succeed, the effort must train staff in the recognition and reporting of AEs, assure confidentiality to reporters, minimize burden of participation, and provide timely feedback.

Implant Retrieval, Failure Analysis, and Device Reliability

All engineered systems have a finite lifetime, whether they are electronic, aerospace, or medical device systems. When reliability engineers look at systems failures they note that failure rates have what have been described as "bathtub"-shaped curves. That is there is usually a higher rate of failures within a short period of use followed by a long and relatively low rate of failure followed by a rise in the rate of failure as the system reaches its end of life. The early failures are usually associated with a defect that occurs during the manufacture or assembly of a device or system. The builder of the system has direct control of this phase of the life cycle of a product and, by implementing the proper measurement and control systems, can greatly reduce the likelihood of these "infant failures." The long relatively stable period in the middle section of the life cycle of a product is the useful life of the product. The onset of an increase in the rate of failure as the product or system reaches its end of life is influenced by many factors including the type and harshness of the environment in which the system was used.

One cannot accurately predict product lifetime because a good model system to fully test the performance of a device does not exist. Animal models are used in the initial screening of a device, but the physiological make-up of animals does not fully mimic the biological or physical stresses that a human will place on a device or system. As

Since its inception in 1973, the FDA's database of voluntary and mandatory reports of device AEs has received slightly more than 1 million reports and currently averages approximately 150,000 per year, with mandatory reports accounting for about 98% of the total. As noted previously, the reports are submitted on standardized forms that capture information on device specifics (e.g., brand name, model number), event description, pertinent dates (e.g., event date), and patient characteristics. The FDA has devised methods for report triage to enhance signal detection of previously unforeseen or not-well-characterized adverse events; thus, only about 24% of reports received require individual review. Well-characterized and understood AEs are either automatically computer-screened or submitted as periodic tabular summaries by the manufacturer (for detection of trends). A clinical staff person individually reviews the reports from a variety of perspectives including the potential for device failure (e.g., poor design, manufacturing defect), use error (e.g., device mis-assembly, incorrect clinical use, misreading instructions), packaging error, support system failure, adverse environmental factors, underlying patient disease or co-morbid conditions, idiosyncratic patient reactions (e.g., allergy), maintenance error, and adverse device interaction (e.g., electromagnetic interference) *(27)*.

Several actions, aside from routine requests for follow-up information, may be taken and include directed inspections of manufacturers (which may ultimately lead to label changes or product recall and rarely product seizure or injunction), internal expert meetings (which may lead to public notifications such as safety alerts and public health advisories or additional postmarket study, pending deliberations by FDA experts), the use of additional postmarket controls (e.g., tracking requirements), and alerting regulatory authorities outside the United States. Public health notifications can be issued by the firm ("Dear Doctor" or "Dear Patient" letters) or by the FDA depending on the specific device issues. Other uses of the AE data are widespread and include input into classification and monitoring of recalls; product classifications and 510k exemptions; standards efforts; premarket review (by providing human factors insights and information on product experience in the general population); educating the clinical community through newsletters, literature articles (peer-reviewed and professional and trade journals), and teleconferences; and as a general information resource for health care providers and the general public.

A recent example of reports of AEs typifies the system in action. The agency received reports alerting us to events (including deaths) related to malfunction of crystal resonating components of a model of automatic implantable cardioverter defibrillators (AICDs). In follow-up with the firm for failure analyses, it was learned they planned a voluntary recall. To elucidate the failure mechanism, agency engineers (in concert with the company) discovered that vacuum loss in the crystal housing, secondary to substandard hermeticity, led to crystal oscillation malfunction resulting in degradation of timing functions and electrical signal synchronization, a failure mode that could be catastrophic *(28,29)*. Subsequent to this discovery, and because this failure mode could pertain to other implantable medical devices, a public notification ("Dear Manufacturer" letter) was issued by the agency *(30)*.

As is typical of passive surveillance systems, the FDA's system has notable weaknesses including the following: data may be incomplete or inaccurate and are typically not independently verified; data may reflect reporting biases driven by event severity

Postmarket Surveillance and Risk Assessment

The goals of postmarket surveillance and risk assessment are identification of previously unknown or not well-characterized adverse events (AEs)/product problems ("signals"), identification and characterization of subgroups at risk, collection, and evaluation of information on issues not directly addressed in premarket submissions (e.g., long-term effectiveness), and development of a public health context to interpret these data. The postmarket "tools" to achieve these goals are: (a) AE/product problem reporting (through the Medical Device Reporting system, MEDWatch, PMA conditions of approval, and the pilot Medical Product Surveillance Network [MedSuN]); (b) implant retrieval and failure analysis; (c) mandated postmarket studies (including conditions of approval and section 522 studies); and (d) applied epidemiological research.

Adverse Event/Product Problem Reporting

The FDA monitors postmarket device-related AEs, through both voluntary and mandatory reporting, to detect "signals" of potential public health safety issues. Voluntary reporting to the FDA began in 1973 and presently continues under MEDWatch (25), a program created in 1993 to encourage voluntary reporting by all interested parties (but principally among health care professionals), as a critical professional and public health responsibility.

It was not until 1984 that the FDA implemented mandatory reporting per the MDR regulation. This regulation required device manufacturers and importers to report device-related deaths, serious injuries, and malfunctions to the FDA. Additional legislative initiatives in the 1990s resulted in significant changes to mandatory reporting. Under the SMDA 1990, universal reporting of adverse events by user facilities (hospitals, nursing homes, ambulatory surgical facilities, and outpatient diagnostic and treatment facilities) and distributors was enacted. Under FDAMA (17), and in response to experience with distributor and user facility reporting, Congress mandated that distributor reporting be repealed and that universal user facility reporting be limited to a "subset of user facilities that constitutes a representative profile of user reports." The conceptual framework for these sentinel sites, collectively referred to as the MedSuN, is discussed later.

To better understand reporting of AEs under the current MDR regulations governing mandatory reporting (26), requirements should be noted and terms defined. Manufacturers and importers are currently required to submit reports of device-related deaths, serious injuries, and malfunctions. User facilities are required to report deaths to the FDA and deaths and serious injuries to the manufacturer. Serious injuries are defined as life-threatening events, events that result in permanent impairment of a body function or permanent damage to a body structure, and events that require medical or surgical intervention to preclude permanent impairment or damage. Malfunctions are defined as the failure of a device to meet its performance specifications or otherwise perform as intended. The term *device-related* means that the event was or may have been attributable to a medical device, or that a device was or may have been a factor in an event, including those occurring as a result of device failure, malfunction, improper or inadequate design, poor manufacture, inadequate labeling, or use error. Guidance is issued to reporting entities as needed to more clearly define the reporting of specific events (e.g., implant failures).

BIMO inspections conclude with an exit interview to discuss the findings and resolve any misunderstandings. Where departures from regulatory requirements are observed, the FDA will identify those items in writing at the conclusion of an inspection. IRBs and clinical investigators should make an effort to let the FDA know that the inspectional findings will be or have been corrected.

POSTMARKET OVERSIGHT

Overview

As noted previously, the FDA requires that medical devices be reasonably safe and effective prior to market entrance. "Reasonably safe," however, is not synonymous with "risk-free." The agency, therefore, approves a device when it deems that the product's benefits outweigh its risks for the intended population and use. Thus, when device marketing begins, there is reasonable assurance that the product will be useful while not posing unacceptable risks to patients.

For the majority of marketed products, however, little or no clinical data are required. Even when clinical trial information is provided, these data have inherent limitations. Device clinical trials are typically conducted with limited numbers of patients (<1000); therefore, detection of rare adverse events or product problems may be unlikely. For instance, if an adverse event occurs in 1 in 10,000 exposed patients, there is only a 9.5% probability of observing that event at least once in a trial of 1000 patients *(24)*. Trials are typically of short duration (<3 years); thus, long-term effects may go undetected, whether of long-latency or late-term (e.g., related to durability). Inclusion criteria for many trials tend to describe a more homogeneous patient population than the population for which the device is indicated and may be restricted in age (no children, elderly), gender (no pregnant women), co-morbid conditions, and disease complexity. Once a product goes on the market, however, less stringent diagnostic and other criteria are typically applied, reflecting either nonoptimal product choice or off-label use, the latter a hallmark of the evolving practice of medicine. Finally, the investigators in premarket clinical trials tend to be those physicians at the "cutting edge" of product development and who are most familiar with the device characteristics and applications. Once in the marketplace, the devices may be used by a wide array of physicians and other clinicians of varying skill levels, training, and experience.

Because no device is free of adverse events and product problems, and because premarket clinical data are limited, postmarket oversight is needed as a "safety net" to assure the continued safety and effectiveness of marketed products. Postmarket oversight refers to both postmarket surveillance and risk assessment as well as postmarket enforcement. The former refers to the systematic process of adverse event/product problem reporting, monitoring, and evaluation as well as the further, more formal, assessments of identified potential patient risks. The latter refers to investigations of a device firm's compliance with statutory and regulatory requirements. Characteristics of both these processes (surveillance and enforcement) are to (a) disseminate information regarding newly emerging device problems to appropriate stakeholders (particularly health care professionals and the public); (b) incorporate the information into the device approval process; and (c) provide findings to the device industry to aid in product corrections and improvements. Both processes are integral to product development and evolution. The remainder of this section focuses on the programs constituting postmarket oversight, beginning with postmarket surveillance.

apply for a study of an approved device for a new indication. The IDE allows a manufacturer to ship and use unapproved medical devices for clinical investigation involving human subjects. All clinical investigations of medical devices require IRB approval and patient informed consent. In general, clinical investigations with medical devices are classified as having significant risk or nonsignificant risk; all studies involving implants are considered to be significant risk studies. All significant risk studies require advance FDA approval

An unapproved medical device may normally only be used on a patient in the context of an approved clinical study. However, there are important access mechanisms to enable a health care provider to use an unapproved device to save the life of a patient, prevent irreversible morbidity, or treat a serious disease or condition for which no alternative exists. Prior FDA approval is needed unless the use of the device fits the criteria for emergency use, which provides, in part, that the treatment must be needed to prevent irreversible morbidity with insufficient time for prior FDA approval.

Bioresearch Monitoring

The purpose of FDA's bioresearch monitoring (BIMO) program is twofold: to ensure protection of the rights and welfare of human subjects involved in clinical research, and to assure the validity of clinical data submitted to the FDA in support of a marketing application. The BIMO program uses prearranged on-site inspections of IRBs and clinical investigators to assess conformance with relevant legal requirements that establish the procedures, responsibilities and human subject safeguards involving the clinical study of investigational medical devices not yet available for commercial distribution *(23)*.

An IRB inspection helps to ensure that safeguards are in place to protect the rights and welfare of human subjects involved in clinical research. BIMO inspections focus on an IRB's organization, membership, procedures, and informed consent requirements. The inspection activity typically involves the review of records that show whether the IRB approved and periodically reviewed on-going clinical research conducted under the IRB's auspices. The agency pays particular attention to the adequacy of the corresponding informed consent procedures and the substantive text of the informed consent document. The FDA may review an IRB's records concerning a number of studies or those of a particular clinical investigator to determine whether the IRB and the clinical investigator are meeting their regulatory responsibilities. Of particular importance is how the IRB and clinical investigator manage, document, and report adverse reactions and unexpected events. Failure to report untoward adverse events raises serious concerns about the adequacy of the IRB's operations and corresponding safeguards intended to protect the safety of human subjects.

The inspection of a clinical investigator's activity addresses the quality and integrity of the data used to support a sponsor's premarket submission to the Office of Device Evaluation. Clinical investigator inspections may be study-oriented or investigator-oriented. Study-oriented inspections include review of basic information relevant to the conduct of a study (e.g., delegation of authority, records, study monitoring, and patient information). The FDA uses this information to audit the data submitted to the agency in support of a premarket submission. An investigator-oriented inspection may be needed when the agency receives serious adverse reports about a clinical investigator from the sponsor or from the subjects of a particular study.

Determining the appropriate clinical study duration to evaluate safety and effectiveness of a long-term implant is important. Often, the benefit of an implant cannot be fully assessed in a short-term period. For example, the benefit of a total joint arthroplasty will not become apparent until the patient has undergone appropriate rehabilitative care. In unmasked studies, in which results may be confounded by a placebo effect, longer study duration may lessen this effect. A common concern regarding study duration is that the benefit may accrue earlier than the time at which failures may occur. This type of concern may be addressed by postapproval study requirements. The postmarket surveillance system is also an important element of additional longer term postapproval information about safety and/or effectiveness problems.

The challenge of determining when to use a surrogate endpoint in a trial is not unique to device trials. Many surgical trials are not larger than a few hundred patients. Because of this and because of the longer interval until patient benefit is seen for some types of intervention, surrogate measures are an attractive option. However, the utility of the surrogate measure as a predictor of the outcome of interest needs to be validated.

Even with the most detailed protocol, interpretation of data can be confounded by variations in surgical technique or postoperative care. For example, the alignment of a joint arthroplasty device can affect both its functioning and long-term survival. How soft tissues are handled can affect adhesion barrier formation and affect the outcome of a surgical procedure in which an adhesion barrier is used.

Many device trials are designed as "equivalence" trials. This means that the study objective is to demonstrate equivalence in outcomes among patients with the device under study and those with the control treatment. Although a detailed discussion of the statistical model of this type of trial is beyond the scope of this chapter, there is one important point to be made here. If a device (or treatment) is to be shown to be equivalent to another device or treatment, the design of the study should include a definition of what is often called the delta, or the realm of clinical indifference. This delta refers to the amount by which the patient outcome with the study device (or treatment) could actually be worse than patient outcome with the control device (or treatment) and still be considered equivalent from a clinical perspective. This delta will not be the same for all studies. Determining this value is often the subject of considerable discussion between the FDA and sponsors and is a question on which advisory panel input is sometimes sought.

The objectives of many of these studies are not necessarily stand-alone determinations of safety and effectiveness but are rather means to answer more focused questions. For example, if a device is approved for open surgical placement, and the sponsor wants to modify the label to include laparoscopic use, it may be appropriate that the study focus on short-term safety outcomes associated with this method of device placement. A modification to the device itself may call for a more focused study than the original study on which device approval was obtained. A modification of the device to improve handling may require only a study to look at device handling, if it is not felt that other safety and effectiveness parameters would be affected by the proposed change.

Investigational Device Exemption

All clinical studies with unapproved devices performed in the United States must be conducted in accordance with the IDE regulations *(22)*. These regulations would also

entry. Working in concert with industry, the FDA developed guidance for industry and review staff on the PDP process *(21)*. Since that time, the FDA has had a number of inquiries and discussions with sponsors about this application process. PDPs for which the plan has been approved, but the data collection has not yet been completed, are confidential. There are several products that have gained market entry through successful completion of this pathway. These include a rate-responsive pacemaker and a penile prosthesis.

EVALUATION OF AUTOMATIC CLASS III DESIGNATION

Devices that cannot be found substantially equivalent to a predicate device because such a device does not exist, are automatically placed into class III by statute. FDAMA provided the FDA with a new mechanism to reclassify these products to class I or II if general or general and special controls can provide reasonable assurance of safety and effectiveness.

Examples of products that have gained market entry by this mechanism include nitric oxide delivery systems and a cranial orthosis.

Clinical Trials in Device Approval

Clinical data are often required in support of a marketing application when preclinical testing alone may not answer all the questions about device performance.

As noted previously, not all data reviewed in support of medical device approval are from randomized trials. Clinical trials, are, however, frequently carried out to investigate safety and effectiveness of medical devices. This section describes some particular challenges of medical device trials, the Investigational Device Exemption (IDE) regulations, and the FDA's program to ensure data integrity in clinical trials.

Challenges in Medical Device Trials

Medical device trials present particular challenges that may not apply in pharmaceutical trials. Masking may not be possible in many device trials. Determining appropriate study duration for a long-term implant can be a challenge to the FDA and sponsors alike. Variations in surgical technique and postoperative care can confound results. Many device trials are not designed as superiority trials but equivalence trials aimed at demonstrating equivalent patient outcomes. For this type of trial, arriving at a suitable definition of equivalence is not simple. Some trials are designed to look at one or several focused aspects of device performance. All of these aspects of device trials present their own challenges, and each of these topics is discussed briefly here.

For most trials assessing the performance of an implant, it is not possible to mask the implanting physician. For a comparative trial of one implant to another, it may be possible to mask the patient provided the postoperative care is the same for both devices. Patients are not masked when the trial is a comparison of treatment with an implant to a noninvasive treatment. A crossover trial of an implant designed to introduce energy (such as electrical stimulation) may be difficult to mask as an effective stimulus may be perceptible to the patient. The difficulty in masking the treating physician or the patient may also apply to a third-party assessor. For example, a radiologist performing an assessment of lucencies around a hip prosthesis will not be masked even in a comparative trial if the radiological appearance of the study hip prosthesis is not identical to the control device.

goal is achieved through the HDE regulation. A device that meets certain criteria is exempt from the effectiveness requirement that would ordinarily apply for marketing that type of product.

The criteria for a device to be granted an exemption are as follow:

1. "the device is designed to treat or diagnose a disease or condition that affects fewer than 4,000 individuals in the United Sates…"
2. "there is no comparable device…available to treat or diagnose such disease or condition…" and
3. "the device will not expose patients to an unreasonable or significant risk of illness or injury and the probable benefit to health from the use of the device outweighs the risk of injury or illness from its use, taking into account the probable risks and benefits of currently available devices or alternative forms of treatment" (20).

Thus, the criteria for approval of products by the HDE are a demonstration that the device is for a small population with limited alternatives, and there is a reasonable assurance of safety and probable benefit.

Since the HDE regulation was published, the FDA has approved a number of these devices. A recent approval of an application for an endovascularly placed device for closure of patent foramen ovale illustrates this regulatory route. Cryptogenic stroke in a patient with patent foramen ovale can represent a dilemma for the treating physician. Anticoagulation is a common medical treatment, but a patient may have a stroke despite this treatment. A patient may be offered open surgical repair of his or her patent foramen ovale. For the patient for whom closure of this defect is indicated, a less-invasive means of closure may be of benefit. Because of the small numbers of patients, however, demonstration of the benefits of any treatment, medical or surgical, may be a challenge. The HDE for this treatment was based on clinical data, including a small series of patients in which relatively short-term endpoints, including safety of the procedure for placement of the device, were assessed.

Other approved HDE devices include fetal bladder stents, stimulation devices for patients with neurogenic bowel or bladder, and artificial grafts for certain patients undergoing coronary artery bypass grafting.

Except in an emergency, devices approved under this exemption may only be used at a facility with an Institutional Review Board (IRB) that approves the use of the device in their facility and will provide appropriate oversight.

PRODUCT DEVELOPMENT PROTOCOL

As an alternative mechanism for gaining access to the market, Congress included the PDP in the 1976 MDA. The PMA mechanism calls for sponsors to submit data in support of an application after that data has been developed. In contrast, the PDP mechanism allows for sponsors to interact early with the FDA on a plan for data (preclinical and clinical) to support a marketing application. In this process, the FDA and the manufacturer agree both on the preclinical and clinical testing to be performed and on the acceptance criteria. The requirements for the sponsor to go to market with that particular application are that the testing be completed and that the test results meet the agreed on success criteria.

This provision of the law was not successfully used for the first 20 years. However, in 1997, the CDRH re-evaluated this potential mechanism for devices to gain market

qualified by training and experience to evaluate the effectiveness of the device, from which investigations it can fairly and responsibly be concluded by qualified experts that the device will have the effect it purports or is represented to have under the conditions of use prescribed, recommended, or suggested in the labeling of the device" or based on other "valid scientific evidence" *(17)*. The FDA has published regulations defining "valid scientific evidence." This hierarchy of evidence is described in the Code of Federal Regulations *(18)* as "evidence from well-controlled investigations, partially controlled studies, studies and objective trials without matched controls, well-documented case histories conducted by qualified experts, and reports of significant human experience with a marketed device."

The FDA has used all of these forms of valid scientific evidence in making decisions on PMA products. For some novel types of products, such as resorbable adhesion barriers for use in abdominal and pelvic surgery, all products that have been presented at meetings of the General and Plastic Surgery Advisory Panel to date have included data from a randomized controlled pivotal clinical study. In contrast, prosthetic heart valves are products for which there is substantial cumulative experience. The FDA *Draft Replacement Heart Valve Guidance* suggests a prospective single-arm clinical study in which results with the sponsor's valve can be compared to target safety goals, termed "objective performance criteria" for certain critical safety parameters *(19)*. The FDA called for PMAs for constrained hip prostheses in 1996. These products had a long history of use and were pre-amendments devices (thus the call for PMAs). The clinical data for the applications that were submitted contained either literature articles or clinical case series describing the long-term clinical performance of these devices. The applications also reviewed the types and frequency of reported adverse events. They were approved on the basis of this and other information, allowing these devices to remain on the market.

On occasion, the FDA has approved a PMA based on a demonstration of reasonable assurance of safety and effectiveness by the sponsor, but has requested additional postmarket studies as a condition of approval. These studies can be nonclinical or clinical, and are typically directed at answering a focused question. One such example is the condition of approval for spinal fusion cages. These products are hollow metal cylinders or cages designed to be placed in the intervertebral space and packed with bone. These devices are intended to stabilize the spine and promote fusion in the target population. The devices have been approved on the basis of 2-year studies. Sponsors have been asked to provide longer term follow-up on a cohort of patients, and conduct device retrieval analysis, to gain a better long-term understanding of the product performance.

For a novel type of PMA product, the FDA will seek advisory panel input on the safety and effectiveness of the device. This input is obtained at an open public meeting.

HUMANITARIAN DEVICE EXEMPTION APPLICATIONS

Development of definitive safety and effectiveness information for a class III product intended for a small patient population can be a challenge. The limited number of patients eligible for treatment with a device may make it difficult to perform a full clinical study. The 1990 SMDA directed the FDA to provide a mechanism to encourage the discovery and use of devices that would benefit these small populations. This

particular, the proposed device must have the same intended use as the predicate device, and also either the same technological characteristics, or if the characteristics are different, the device must be as safe and effective as the predicate device and not raise different questions of safety and effectiveness. Thus, a manufacturer needs to provide a description of the device proposed for market, the predicate device, and a comparison of the technological differences between the products in a marketing application. For technological differences, a discussion and possibly bench testing and/or clinical data regarding the effect of these differences on device performance may be needed.

For example, for a new device with the same intended use as a predicate, if the new device and the predicate device have the same material specifications, the biocompatibility of the materials in the new device would probably not need to be independently demonstrated. If the new device used different materials, biocompatibility testing on the new device might need to be performed.

Although the majority of these products are marketed without the necessity of a premarketing clinical study, a clinical study may be needed. For example, cerebrospinal fluid shunts are regulated as class II products and require a 510k application. If a manufacturer were to propose a benefit in terms of reducing complications, by adding a new type of coating or manufacturing the shunt out of a new material, a clinical study would be needed to demonstrate that the device could in fact perform as intended. Because of the limitations of the ability of preclinical testing alone to provide the expected performance of the device, a new type of shunt valve design might also require a clinical study.

Class III Products

There are a number of regulatory paths a sponsor may follow to market a class III product that is subject to a PMA or Product Development Protocol (PDP). If the sponsor has completed data collection on the product, this information would be provided in a PMA. A manufacturer that has not completed testing but wants approval of the plan for development of the data may submit a PDP. The Humanitarian Device Exemption (HDE) path is an alternative for devices that are indicated for limited patient populations. Each of these market paths is described in this section. For all three of these types of marketing applications, a summary of the preclinical and clinical testing that form the basis of the regulatory decision is available on the FDA website (http://www.fda.gov/cdrh/programs.html). These summaries are well worth attention from any clinician caring for a patient with an implant that has been approved through one of these marketing paths.

In addition to these three marketing paths, a sponsor may submit a petition for reclassification. One of these mechanisms for reclassification, the *de novo* reclassification, is also discussed here.

PREMARKET APPLICATION

A manufacturer of a class III product subject to a PMA needs to demonstrate reasonable assurance of safety and effectiveness in order to market the product. The safety and effectiveness of a device are determined with respect to the target population and conditions of use.

The effectiveness of the device is to be determined based on "well-controlled investigations, including one or more clinical investigations where appropriate, by experts

cient information to establish special controls" *(15)*. Examples of class II devices include absorbable polyglycolide/L-lactide sutures, intramedullary fixation rods, and aneurysm clips.

CLASS III DEVICES

For some types of devices, there may be insufficient information to determine that the application of general controls alone or general and special controls are sufficient to provide reasonable assurance of safety and effectiveness. If such a device is also life-sustaining, life-supporting, or used for a substantial importance in preventing impairment of human health, or presents a potential unreasonable risk of illness or injury, then that device is placed into class III *(16)*. Examples of class III devices include spinal fusion cages, silicone gel-filled breast implants, and heart valves.

Over time, additional information may become available about a generic type of device that may enable it to be regulated in a lower risk category. There are a number of regulatory mechanisms for reclassification; one of these is discussed later in this chapter.

The Medical Device Advisory Committee

The Medical Device Advisory Committee consists of 16 advisory panels, based on clinical device applications. The members of the committee are individuals recognized in the academic or clinical communities for their scientific or technical expertise. This Advisory Committee was originally formed to advise the FDA on classification of medical devices. Over time, the FDA has classified approximately 1700 device types. Most of these generic device types are classified into class II.

The FDA relies on the advisory panels for advice on safety and effectiveness of medical products, advice on clinical study design, and other matters regarding medical devices. The recommendations from advisory panels play an important role in assisting the FDA to realize its goal of sound regulatory decisions based on good science.

Marketing Applications

Manufacturers are required to obtain premarket clearance or approval unless the device is specifically exempt. The marketing application will vary depending on the device classification and regulatory history. What follows is a description of these and the types of devices eligible. An example is provided for each application type. As is clear from this section, the information that is needed about a specific device, and how it is evaluated, relates to the marketing application that needs to be submitted.

510k Premarket Notification

For class II products, for the few class I products that are not exempt from the need for a marketing application, and for class III pre-amendments products for which the FDA has not called for safety and effectiveness data, a 510k marketing application is the appropriate mechanism for market entry.

This type of application, named after Section 510k of the Act, requires a sponsor to provide evidence that the device proposed for market is "substantially equivalent" to a predicate device. (In general, class I and class II devices can serve as predicate devices. In some cases, class III devices can also serve as predicate devices if they have been on the market since before 1976.) For a device to be substantially equivalent to a previously marketed device means that the product is as safe and effective as its ancestor. In

vided on the contents of an application for a specific device type, or group of devices, as is provided, for example, in the *Guidance for Spinal System 510ks (12)*. Guidances are available on clinical study design for devices to treat certain indications. One example, the *Guidance for Resorbable Adhesion Barrier Devices for Use in Abdominal and/or Pelvic Surgery* provides guidance to sponsors in the development of preclinical and clinical data to support a marketing application for this type of device *(13)*. The FDA has procedures in place for development, issuance, and use of guidance documents. These procedures are, in part, designed to ensure public participation in the development of these documents and to ensure that they are made available to the public.

With the passage of FDAMA, special controls were amended to include both national and international consensus standards (e.g., test methods). Unlike the previously cited performance standards, these standards are developed through accredited standards development organizations, such as the American Society of Testing and Materials or the Association for Advancement of Medical Instrumentation, with the full participation of the government, industry, and academia. FDAMA gave the agency the ability to recognize all or part of a published standard. As of January 2000, the FDA has recognized nearly 500 standards. Most of these standards pertain to test methods that can be used to evaluate a device or material specifications that give the type and quality of the materials used in the manufacture of the devices. A manufacturer may choose to declare conformity to one of the FDA-recognized standards in a new device application and the FDA is legally bound to accept that declaration. It then becomes incumbent on the manufacturer to maintain records that demonstrate conformity to the standard. By declaring conformity to a recognized standard, a manufacturer may be able to eliminate the need to submit some of the detailed information to the agency during the application for approval to market a device. The FDA recognizes that the device manufacturers are operating in a global economy and that the use of standards is a means of achieving global harmonization in the development of safe and effective medical devices. It is for these reasons that the FDA participates in standards development and actively encourages industry to do the same.

Classification of Medical Devices

Medical devices are classified into three classes based on the ability of the controls just noted to provide reasonable assurance of safety and effectiveness. As a reminder, the term *device* incorporates both the physical product and its intended use. Thus, a scalpel that was labeled to "cut tissue" could be placed in one regulatory class, whereas the same scalpel that was labeled to "cure cancer" would possibly be classified differently. The definition of each of the three regulatory classes is provided next.

CLASS I DEVICES

Class I devices are devices for which general controls alone are sufficient to provide reasonable assurance of safety and effectiveness *(14)*. Examples of class I devices include manual surgical instruments for general use, hot or cold disposable packs, and limb orthoses.

CLASS II DEVICES

Class II devices are products for which "general controls alone are insufficient to provide reasonable assurance of safety and effectiveness, but for which there is suffi-

general and special controls, are discussed. Finally, the role of an advisory panel in device classification, and in device regulation, is highlighted.

General and Special Controls

GENERAL CONTROLS

General controls are the provisions of the 1976 MDA that provide the FDA with the basic regulatory tools to ensure device safety and effectiveness. These include provisions relating to labeling, registration and listing, premarket notification, good manufacturing practices (now called quality systems regulation), and records and reports. Each of these are briefly discussed here:

1. Labeling. The label for any prescription device, which includes implantables, needs to contain information conveying the "intended uses of the device and relevant warnings, precautions, side effects, and contraindications" *(8)*. Note that, in approving a medical device, the agency takes into consideration its intended use.
2. Registration and listing. All medical device manufacturers are required, before device manufacturing begins, to register their facility with the FDA and list each generic type of medical device manufactured at that facility.
3. Premarket notification. Unless specifically exempt, any manufacturer or specification developer intending to market a medical device must submit an application at least 90 days before beginning commercial distribution. The agency then determines if the device is substantially equivalent to a predicate device. The concept of a predicate device, substantial equivalence, and how these are evaluated will be explained in the next section. A predicate device is officially defined as: a device that was legally marketed prior to May 28, 1976, or a device which has been reclassified from class III to class II or I, or a device that has been found to be substantially equivalent through the 510(k) premarket notification process *(9)*.
4. Manufacturing practices. Manufacturing requirements "govern the methods used in, and the facilities and controls used for, the design, manufacture, packaging, labeling, storage, installation, and servicing of all finished devices intended for human use" *(10)*. The corresponding comprehensive quality systems regulation is designed to help ensure that finished devices will be safe and effective.
5. Records and reports. Under Section 519 of the Act, the FDA is authorized to require of manufacturers the maintenance of records and the submission of reports and such information as is necessary to ensure, among other things, a device's safety and effectiveness.

SPECIAL CONTROLS

For some devices, general controls alone may not be adequate to ensure safety and effectiveness. For a subset of these devices, there may be sufficient information to establish special controls (*see* later) in addition to general controls to provide this assurance. Although this concept originally meant that performance standards could be established, the term *special controls* has broadened to include such controls as patient registries, postmarket surveillance, guidances, and standards.

FDA guidance documents play an important role in the premarket review process, frequently serving as special controls. They are available on the Internet at http://www.fda.gov/cdrh/guidance.html. Although guidance documents are nonbinding on industry and the FDA, they assist industry in preparing regulatory submissions and also assist FDA staff in the review process. They cover a range of topics. A guidance document may be in the area of interpretation of regulatory requirements. An example of this is the *Guidance on IDE Policies and Procedures (11)*. Guidances are also pro-

previously) and required premarket submissions for medical devices introduced into interstate commerce after 1976. It also allowed FDA to require manufacturers, importers, and distributors to report device-related product defects and adverse reactions.

Additional regulations and changes to the Act were made during the 1980s and 1990s. The Medical Device Reporting (MDR) regulation of 1984, based on the provisions in the 1976 MDA, finally established requirements for device manufacturers and importers for reporting device-related adverse events. In 1990, the Safe Medical Devices Act (SMDA) was passed, which placed even greater emphasis on postmarket oversight, as did the 1992 amendments.

One of the most significant pieces of legislation to affect device regulation is the Food and Drug Administration Modernization Act (FDAMA) of 1997. This legislation affects regulation of medical devices during all phases of its life cycle. For example, some provisions in the law increase physician and patient access to devices in the investigational stage, exempt some device types from the need for a premarket submission, and provide some additional premarket review tools (e.g., the evaluation of automatic class III designation that is discussed later). Additionally, there were significant changes in postmarket oversight, including study and tracking requirements, some of which are noted later in the section that deals with postmarket issues. It is also important to note that, in addition to the specific sections modifying regulation of a product throughout its life cycle, FDAMA also provided formal mechanisms for collaboration with industry in product development and review, and encouraged other types of activities such as global harmonization.

In keeping with the theme of a product's life cycle, this chapter details the agency's premarket product evaluation process (including preclinical and investigational studies), its postmarket surveillance and risk-assessment functions, its pre- and postmarket enforcement activities, and addresses special topics that bridge pre- and postmarket domains (e.g., global harmonization or information on the Internet).

PREMARKET REVIEW

Background

As noted in the introduction, marketing of medical devices before 1976 did not require a premarket submission. This changed in 1976 with the passage of the MDA. At that time, devices on the market prior to 1976 (pre-amendments devices, e.g., silicone gel-filled breast implants) were "grandfathered." The provisions outlined in the MDA called for the FDA to classify all medical devices into one of three classes based on risk and assign an appropriate level of regulatory control to ensure safety and effectiveness. Class III is the highest risk category, requiring the most stringent regulatory control. The FDA was directed to call for safety and effectiveness data in the form of a Premarket Approval Application (PMA) for any grandfathered devices placed in class III. This law also provided for some transitional devices to be regulated as class III devices requiring approval prior to commercial distribution.

Regulatory Controls, Device Classification, and the Medical Devices Advisory Committee

This section describes the current general and special regulatory controls for medical devices. The definition of the three classes of medical devices, and the concept of

intended purposes through chemical action within or on the body of man or other animals and which is not dependent on being metabolized for the achievement of any of its primary intended purposes" *(5)*.

The agency's mandate is carried out through both premarket product evaluation and postmarket oversight that continues over the lifetime of the product, from early design to widespread use and, ultimately, to obsolescence. At major junctures of a product's life cycle, the FDA must weigh the product's benefits and risks. Central to this risk-management function is the FDA's decision for marketing, one that must ensure that beneficial medical products are available (and labeled with adequate information on their benefits and risks) while protecting the public from unsafe products or false claims *(6)*. Once marketed, a product's continued safety and effectiveness must be ensured not only by oversight on the part of industry and the FDA but, most importantly, by health care providers' and patients' appropriate product selection and use based on the product's labeling. Comprehensive regulation of food and drugs in the United States began with the passage of the Pure Foods and Drug Act of 1906. Medical devices were not included in this initial regulation. (In fact, the FDA was part of the Bureau of Chemistry in the Department of Agriculture at that time.) More comprehensive legislation providing for regulation of food and drugs, including additional enforcement authority, was passed in 1938 as the Act. Because of increasing public and congressional concern regarding the use of fraudulent medical devices, the Act provided for regulation of medical devices (which were defined as "instruments, apparatus and contrivances...intended [1] for use in the diagnosis, cure, mitigation, treatment, or prevention of disease...or [2] to affect the structure or any function of the body..."). However, as described here, this initial regulation was minimal.

Medical device manufacturers did not need to obtain authorization from the FDA in advance of commercial distribution. The Act prohibited shipment in interstate commerce of adulterated or misbranded devices and provided the FDA with enforcement authority to remove unsafe or fraudulent devices from the market. The burden of proof for these actions was on the agency to prove that the device was unsafe or ineffective, not on the manufacturer to prove the contrary. Consequently, regulation of medical devices was on a product-by-product basis.

Over the next several decades, the application of the term *drug* came to include some items, now termed *transitional devices*, that appeared also to fit the definition of a medical device. These products included absorbable sutures, contact lenses, and bone cement. Regulation of these products as drug entities gave the FDA the ability to require a premarket application, which, as noted earlier, was not required for devices at that time. This would change in 1976 with the passage of the Medical Device Amendments (MDA).

As medical devices became more sophisticated, there was the potential that some devices would produce more benefit and some devices more harm. The Dalkon Shield (an intrauterine device), with its reported link to deaths, miscarriages, and infertility, was a widely publicized example of the latter *(7)*. Legislative concern, previously centered on prohibiting fraudulent medical devices from the market, began to turn to safety issues.

Increased statutory authority for the regulation of medical devices was provided to the FDA in the 1976 MDA. This amendment broadened the definition of devices (noted

5
A View From the US Food and Drug Administration*

Thomas P. Gross, Celia M. Witten, Casper Uldriks, and William F. Regnault

INTRODUCTION

The US Food and Drug Administration (FDA) is first and foremost a public health and consumer protection agency. It regulates products, including implantable prosthetic devices, worth more than $1 trillion per year and accounting for one-fourth of all dollars spent annually by US consumers *(1)*. The agency is responsible, through enforcement of the federal Food, Drug, and Cosmetic Act (the "Act") and several related public health laws, for ensuring that (a) foods are safe, wholesome, sanitary, and properly labeled; human and veterinary drugs are safe and effective; there is reasonable assurance of the safety and effectiveness of devices intended for human use; cosmetics are safe and properly labeled; and public health and safety are protected from electronic product radiation; (b) regulated products are honestly, accurately, and informatively represented; and (c) these products are in compliance with the law and FDA regulations; noncompliance is identified and corrected; and any unsafe or unlawful products are removed from the marketplace *(2)*.

The Center for Devices and Radiological Health (CDRH) is that part of the agency that helps ensure that medical devices are safe and effective and helps reduce unnecessary exposure to radiation from medical, occupational, and consumer products. The industry regulated by CDRH accounted for $129.5 billion in global business in 1996 and consisted of 3000 product lines and 84,000 individual products *(3,4)*. Medical devices are officially defined as "an instrument, apparatus, implement, machine, contrivance, implant, in vitro reagent, or other similar or related article, including a component part, or accessory which is (a) recognized in the official National Formulary, or the US Pharmacopoeia, or any supplement to them; (b) intended for use in the diagnosis of disease or other conditions, or in the cure, mitigation, treatment, or prevention of disease, in man or other animals; or (c) intended to affect the structure or any function of the body of man or other animals, and which does not achieve any of its primary

---

*Disclaimer: The opinions or assertions presented herein are the private views of the authors and are not to be construed as conveying either an official endorsement or criticism by the US Department of Health and Human Services, the Public Health Service, or the US Food and Drug Administration.

From: *The Bionic Human: Health Promotion for People With Implanted Prosthetic Devices*
Edited by: F. E. Johnson and K. S. Virgo © Humana Press Inc., Totowa, NJ

25. Mann CC. Can meta-analysis make policy? Science 1994;266:960–962.
26. Bruinvels DJ, Stiggelbout AM, Kievit J, Van Houwelingen HC, Habbema JDF, van de Velde CJH. Follow-up of patients with colorectal cancer: a meta-analysis. Ann Surg 1994;219:174–182.
27. Woolf SH. Practice guidelines: a new reality in medicine I. Recent developments. Arch Int Med 1990;150:1811–1818.
28. Woolf SH. Practice guidelines: a new reality in medicine II. Methods of developing guidelines. Arch Int Med 1992;152:946–952.
29. Woolf SH. Practice guidelines: a new reality in medicine III. Impact on patient care. Arch Int Med 1993;153:2646–2655.
30. Parmley WW. Clinical practice guidelines. Does the cookbook have enough recipes? J Am Med Assn 1994;272:1374–1375.
31. Grilli R, Apolone G, Marsoni S, Nicolucci A, Zola P, Liberati A. The impact of patient management guidelines on the care of breast, colorectal, and ovarian cancer patients in Italy. Med Care 1991;29:50–63.
32. Virgo KS, Naunheim KS, Coplin MA, Johnson FE. Lung cancer patient follow-up: motivation of thoracic surgeons. Chest 1998;114:1519–1534.
33. Darouiche RO. Treatment of infections associated with surgical implants. New Engl J Med 2004;350:1422–1429.
34. Reefhuis J, Honein MA, Whitney CG, et al. Risk of bacterial meningitis in children with cochlear implants. N Engl J Med 2003;349:435–445.
35. Anonymous. Surgical implants and other foreign bodies. International Agency for Research on Cancer (IARC) monographs on the evaluation of carcinogenic risks to humans. Volume 74. World Health Organization publication, 1999.
36. Eddy DM. Health system reform. Will controlling costs require rationing services? J Am Med Assn 1994;272:324–328.
37. Nowak R. Genetic testing set for take-off. Science 1994;265:464–467.
38. National Advisory Council for Human Genome Research. Statement on use of DNA testing for presymptomatic identification of cancer risk. J Am Med Assn 1994;271:785.
39. Lowden JA. Genetic testing. Science 1994;265:1509–1510.
40. Eddy DM. Principles for making difficult decisions in difficult times. J Am Med Assn 1994;271:1792–1798.
41. Phelps CE, Parente ST. Priority setting in medical technology and medical practice assessment. Med Care 1990;28:703–723.

REFERENCES

1. Pagel W. William Harvey's Biological Ideas. New York: Hafner; 1967. (Cited in Schultz SG. William Harvey and the circulation of the blood: the birth of a scientific revolution and modern physiology. News Physiol Sci 2002;17:175–180.)
2. Randal J. Randomized controlled trials mark a golden anniversary. J Nat Cancer Inst 1999;91:10–12.
3. Ault A. Clinical research. Climbing a medical Everest. Science 2003;300:2024–2025.
4. Field MJ, Lohr K, eds. Guidelines for Clinical Practice: From Development to Use. Washington, DC: National Academy Press, 1992.
5. Chassin MR, Galvin RW. The urgent need to improve health care quality: Institute of Medicine National Roundtable on Health Care Quality. JAMA 1998;280:1000–1005.
6. Lembcke PA. Measuring the quality of medical care through vital statistics based on hospital service areas. Am J Public Health 1952;42:276–286.
7. Brook RH, Ware JE, Rogers WH, et al. Does free care improve adults' health? New Engl J Med 1983;309:1426–1434.
8. Earle CC, Chapman RH, Baker CS, et al. Systematic overview of cost–utility assessments in oncology. J Clin Oncol 2000;18:3302–3317.
9. Kmietowicz Z. Companies offer surgeons incentives to use their products. Brit Med J 2004;328:1091.
10. Grimshaw JM, Russell IT. Effect of clinical guidelines on medical practice: a systematic review of rigorous evaluations. Lancet 1993;342:1317–1322.
11. Maisel WH, Sweeney MO, Stevenson WG, Ellison KE, Epstein LM. Recalls and safety alerts involving pacemakers and implantable cardioverter-defibrillator generators. JAMA 2001;286:793–799.
12. O'Meara JJ, McNutt RA, Evans AT, Moore SW, Downs SM. A decision analysis of streptokinase plus heparin as compared with heparin alone for deep-vein thrombosis. N Engl J Med 1994;330:1864–1869.
13. Kassirer JP. Incorporating patients' preferences into medical decisions. N Engl J Med 1994;330:1895–1896.
14. Jetter A. Warning! The medical "miracles" that may be hazardous to your health. Good Housekeeping Magazine, Hearst Communications, Inc., New York, NY, 2004:138, et seq.
15. Jeffrey NA. The bionic boomer. Wall Street Journal, August 22, 2003, p. W1.
16. May M. Bionics from biochips. HMS Beagle–The BioMedNet Magazine, Issue 75, posted March 31, 2000, on the Internet, at URL: http://www.biomednet.com/hmsbeagle/75/reviews/insitu.
17. Geary J. The body electric. An anatomy of the new bionic senses. New Brunswick, NJ: Rutgers University Press, 2002.
18. Gion M, Cappelli G, Mione R, et al. Variability of tumor markers in the follow-up of patients radically resected for breast cancer. Tumor Biol 1993;14:325–333.
19. GIVIO Investigators. Impact of follow-up testing on survival and health-related quality of life in breast cancer patients. J Am Med Assn 1994;271:1587–1592.
20. Chen AY, Escarce JJ. Quantifying income-related inequality in healthcare delivery in the United States. Med Care 2004;42:38–47.
21. Powell TM, Thompsen JP, Virgo KS, et al. Geographic variation in patient surveillance after radical prostatectomy. Ann Surg Oncol 2000;7:339–345.
22. Romano PS, Roos LL, Luft HS, Jollis JG, Doliszny K, and the Ischemic Heart Disease Patient Outcomes Research Team. A comparison of administrative versus clinical data: coronary artery bypass surgery as an example. J Clin Epidemiol 1994:47:249–260.
23. Anderson C. Measuring what works in health care. Science 1994;263:1080–1082.
24. Janowsky EC, Kupper LL, Hulka BS. Meta-analyses of the relation between silicone breast implants and the risk of connective tissue disease. New Engl J Med 2000;342:781–790.

3. Because financial resources are limited, it is necessary to set priorities.
4. A consequence of priority setting is that it will not be possible to cover from shared resources every treatment that might have some benefit.
5. The objective of health care is to maximize the health of the population served, subject to the available resources.
6. The priority a treatment should receive should not depend on whether the particular individuals who would receive treatment are our personal patients.
7. Determining the priority of a treatment will require estimating the magnitudes of its benefits, harms, and costs.
8. To the greatest extent possible, estimates of benefits, harms, and costs should be based on empirical evidence. A corollary is that when empirical evidence contradicts subjective judgments, empirical evidence should take priority.
9. Before it should be promoted for use, a treatment should satisfy three criteria. There should be convincing evidence that, compared with no treatment, the treatment is effective in improving health outcomes. Compared with no treatment, its beneficial effects on health outcomes should outweigh any harmful effects on health outcomes. Compared with the next best alternative treatment, the treatment should represent a good use of resources in the sense that it satisfies principle 5.
10. When making judgments about benefits, harms, and costs, to the greatest extent possible, the judgments should reflect the preferences of the individuals who will actually receive the treatments.
11. When determining whether a treatment satisfies the criteria of principle number 9, the burden of proof should be on those who want to promote the use of the treatment.

SUMMARY

Medical care is better than ever, but its very success breeds new problems, and post-treatment surveillance is no exception. Rising health care costs affect the economy at large, resulting in pressure to contain them. New ways to analyze the benefits of care result in changes in the delivery of care. Better understanding of pathological processes promises to improve surveillance and introduce new and fascinating questions. Physicians will require a close relationship with their patients and a clear appreciation of societal desires in order to reap the benefits of these trends for their patients, while also avoiding the inevitable hazards.

Clinical trials of alternative strategies for care of people with medical implants are not currently regarded as being high-priority investigations by pertinent funding agencies. However, until funding is provided for such trials, practice will continue to rest on relatively low-quality evidence. Further efforts to generate high-quality evidence are warranted. Certainly, the current practice is quite variable and very expensive. Organizations concerned with health promotion should become active advocates for the funding of clinical trials for these individuals. The more successful our initial therapy for these patients becomes, the more people will enter follow-up programs. The opportunities to improve patient management after primary therapy are numerous. Trials will be able to address big questions because so few have been done so far, and improvements in the quality of patient care will surely result. They are important both to society and to individual patients. Previous research has estimated that the economic returns on such investments exceed the expenditures by one to two orders of magnitude *(41)*. The generation of high-quality evidence to rationalize practice is an important priority for the medical community, although trials are expensive to mount and take a long time to complete.

mon and influential in medical practice. It is possible, for example, that such devices could be used to remove silicone from a ruptured breast implant, deliver antibiotics in the vicinity of an infection, or repair the fractured strut of a Björk-Shiley prosthetic heart valve. The mechanical nature of prosthetic devices renders these possibilities quite attractive. Genetic testing also holds promise as a way to detect diseases in apparently healthy people, which could render expensive traditional surveillance schemes obsolete. Such testing may allow us to discriminate which patients deserve particularly vigorous surveillance from those who do not. However, gene-based screening must take into account the large number of possible variations in each pertinent allele and the relative risk of disease associated with each. Because gene-based diagnostic testing carries with it the aura of infallibility, a positive test result can have shattering consequences. Patients may become uninsurable, undergo prophylactic excision of the organ at risk, refrain from childbearing, or suffer grave psychological stress. Thus, it will be crucial not to interpret a harmless genetic polymorphism as a dangerous mutation *(37)*.

The National Center for Human Genome Research at the National Institutes of Health has addressed the important issue of genetic testing and has issued a statement defining several fundamental questions that should be answered before genetic testing for common disease-risk genes is offered to high-risk individuals or the general public as accepted medical practice *(38)*. These questions include the following:

- How many mutations of each gene conferring risk exist?
- What is the clinical risk associated with each?
- What is the frequency of false-positive and false-negative test results?
- How can technical quality of genetic testing be ensured?
- How are clinicians and patients to use the test results—in other words, what interventions can be carried out to prevent or treat disease in affected populations?
- How can patients be educated about the implications of DNA testing in order to ensure that consent for testing is informed?
- How is genetic counseling to be given?
- How can discrimination against those found to harbor disease-risk genes be avoided?

The insurance industry, for obvious reasons, is very interested in this topic, although insurers do not carry out such tests themselves *(39)*. The political process is working to modify the outcomes flowing from genetic testing *(37)*. The National Center for Human Genome Research has begun to gather information and establish methods to carefully introduce the powerful tools of genetic testing into clinical practice *(38)*.

Rationing of care given to asymptomatic patients after primary therapy will likely be easier for the general public to accept, at least at first glance, than rationing of care intended for symptomatic, ill individuals. Justification of any diagnostic or therapeutic intervention should ideally be considered by reference to widely agreed on assumptions. This will restrict the tendency for pressure groups to petition providers of health care for special treatment and will introduce rationality rather than political tactics into the decision-making process. Assumptions and criteria that might best govern a future care-rationing process are as follow *(40)*:

1. The financial resources available to provide health care to a population are limited.
2. Because financial resources are limited, when deciding about the appropriate use of treatments, it is both valid and important to consider the financial costs of the treatments.

modern multimodality treatment should alter surveillance practices is almost completely unknown. This is caused, in part, by our near-total ignorance of the actual practice patterns of clinicians managing patients with prostheses of any sort. In this area, where data are lacking and the variables are so numerous, a rational policy can probably not be devised at the current time. Modern techniques of decision analysis *(12,13)* involving sophisticated computer modeling are likely to prove valuable, however. Computer modeling has already demonstrated its value in predicting such complex phenomena as the behavior of national economies, chemical reactions, and the weather. As is often the case, help may also be arriving from a more profound understanding of biological processes in general. As information regarding the human genome becomes available, we should be able to predict the development of new disorders in patients who have undergone treatment for a particular problem.

There is little empirical evidence suggesting that follow-up strategies should vary by the type of primary treatment. In part, this reflects the difficulty of mounting comparative clinical trials, and in part it represents a lack of interest in postoperative surveillance strategy as a research area worthy of investigation.

Managed care may affect the practice of patient follow-up. In managed care settings, more and more guidelines are being promulgated by diverse regulatory bodies, ranging from insurance companies to health maintenance organizations to the federal government, which restrict physicians' autonomy. This has probably affected the pattern of care in countries where socialized medicine has been adopted with fixed budgets and seemingly limitless demands for medical attention. Health care executives have been forced to search for evidence that a given policy pays off in terms of patient benefit before writing it into patient care guidelines. Although this has not been well documented, strategies appear to vary appreciably among health care settings (such as Department of Veterans Affairs Medical Center hospitals, private hospitals, etc.). Eddy *(36)*, an influential analyst of clinical decision-making theory, proposes that cost control in medicine will lead to rationing of services. The costs of health care, measured as a fraction of the gross domestic product of any country, cannot rise indefinitely. The rewards to those who are able to rein in costs are so large, and the adverse consequences to an economy of continued failure to do so are so severe, that the imposition of effective cost-control measures can be taken as inevitable. The greater the delay in instituting such measures, the deeper the cuts will eventually have to be. It will not be possible to control costs by increased managerial efficiency and elimination of wasteful medical practices alone, indicating that rationing is unavoidable. Whether quality of care can be maintained irrespective of rationing is open to question *(36)*.

Scientific advances and improved techniques may help maintain or improve medical care quality. The trend toward less-invasive surgery will also probably have implications for follow-up. Sizeable trials, not always well controlled, are currently being carried out to assess this issue and a consensus is evolving that the same surgical objectives should be demanded of endoscopic procedures as standard surgical procedures. The emergence of limited surgery as therapy for various disorders seems likely to continue and even accelerate. There is little evidence that follow-up for patients treated with limited surgery should be different from that offered patients treated with more extensive conventional surgery. In the future, micro-machines will probably become com-

major co-morbid conditions. To some extent, this clashes with ethical and medico-legal concerns because withholding of otherwise-indicated surveillance testing for elderly, demented, or otherwise impaired patients may be considered discriminatory and constitute grounds for a lawsuit. Nonetheless, it seems to be common practice among physicians.

DURATION OF SURVEILLANCE

The intensity of postoperative surveillance typically diminishes with time, as documented throughout this book. The appropriate duration of follow-up is almost impossible to calculate. For most implants, the threat of infection is a permanent concern and this alone should drive follow-up decisions (33). This is treated in more detail in Chapter 9. Other concerns, such as the detection of late effects of prosthesis placement, may be important as well. The development of connective tissue diseases that has been attributed to silicone breast prostheses is undoubtedly an example, although current evidence does not support a causal relationship (24). A causal relationship does clearly exist between cochlear implants and meningitis, however, and preventive measures such as vaccination are warranted (34). Certainly, the progression of the disease for which the implant has been placed is another consideration. A patient with a prosthetic hip may soon require a prosthetic knee owing to progression of arthritis, for example. All of this is affected by patient age and the projected life span. Most patients who have undergone prosthetic implantation are old and suffer from chronic illnesses. Detection of malfunction of prostheses can easily be incorporated into standard health maintenance efforts.

The length of follow-up bears a strong relationship to the type of patient and type of prosthesis being followed. Patients with arterial grafts are at risk for progression of atherosclerosis for the rest of their lives. Others, such as those with intramedullary rods for a long bone fracture, do not have this sort of risk.

SECONDARY BENEFITS OF FOLLOW-UP

Depression, guilt, fear of death, and a sense of vulnerability are common patient emotions after the rigors of initial therapy are past. A traditional role for the physician providing posttreatment surveillance is to provide a nonthreatening, nonjudgmental outlet for these feelings. Patients are known to value such an outlet. With time, if the implant works well, the depression and feelings of loneliness and isolation tend to diminish. In addition, follow-up should assess the sequelae of treatment. This could include incisional hernia, implant extrusion, and the like. The possibility of late effects of the implant (e.g., cancer) warrants surveillance in some instances (35). In addition to the prosthesis-specific aspects of surveillance, physician input is important is assessing the full return of the patient to good health, assuring follow through with rehabilitation plans, and regular assessment of posttreatment concerns such as wound healing.

PRIMARY PREVENTION OF DISEASE

A relative newcomer on the list of indications for posttreatment care of the patient with an implant is primary disease prevention. This discipline is in its infancy, although physicians have long advised their patients to avoid certain practices. How

codify medical care can easily leave out approaches that are in an individual patient's best interests. Guidelines may be used as statements of the standard of care in medicolegal cases, tending to compel doctors to avoid valuable management options not spelled out in the cookbook. Parmley *(30)* also points out that all guidelines are relatively conservative documents that cannot reflect late-breaking evidence of benefit or harm resulting from treatment decisions and are only useful if used with wisdom and flexibility. Guidelines seem likely to be used in caring for bionic patients because they have complex problems. However, stringent enforcement offers the prospect of harm to patients, paradoxically increased costs, and unfair legal or administrative judgments against physicians who deviate from them *(29)*. The limited literature to date suggests that clinical guidelines do improve clinical practice *(10)*, but the improvement is often disappointingly small *(31)*. Increasingly, physician concerns are being superseded by the financial aspects of modern medical care in which cost–benefit analyses are paramount and authoritative opinions from wise and experienced physicians may count less than the formula prescribed by impersonal committees constructing practice parameters.

ETHICAL AND LEGAL ISSUES

Physicians, particularly in the United States but to an increasing extent in other countries, must be concerned with medico-legal issues. Everyone knows of patients who seem impossible to satisfy and from whom lawsuits may arise. Clinicians never have perfect results of therapy in all patients, whether this is measured in terms of cosmesis, function, wound pain, or in other ways. One way to avoid medico-legal problems is to obtain appropriate consultation when patients have suboptimal treatment outcomes, with the hope that responsibility for the poor outcome will be diffused among many individuals and with the expectation that the consultants will underwrite the correctness of the treatment plan undertaken. In the United States, fear of being sued has assumed legendary proportions, not easily comprehended by practitioners in other countries where lawsuits are much less common. Undoubtedly, defensive medicine does play a role in the decision making of individual doctors caring for their patients, and this clearly extends to bionic patients. Medico-legal concerns are also great in the US business system. Silicone-filled breast implants, intrauterine contraceptive devices, and others come readily to mind. Doctors, hospitals, and commercial firms have been driven to bankruptcy by lawsuits. Consequently, businesses entering into the field of implanted material must take these concerns seriously and acquire adequate financial reserves and corporate insurance.

The effect of fear of litigation on the intensity of postoperative surveillance *(32)* is not often measured, but a considerable amount of current diagnostic testing is undoubtedly defensive maneuvering by physicians eager to avoid legal entanglements with their patients. Although they are unpopular topics, often considered by physicians only subconsciously and without much discussion with patient or family, the implications of patient age, projected life span, and co-morbid conditions must also be incorporated into any rational analysis of surveillance strategy. When surveillance is rather extensive and sometimes invasive, risky, and uncomfortable, such as after arterial bypass surgery, the intensity is likely to be diminished if the patient is suffering from other

There is little doubt that testing can reveal problems before they are clinically apparent, and many can be managed successfully. Well-designed, randomized clinical trials are very persuasive and may alter the way physicians practice, but the large practical problems in carrying them out have led to a rapid expansion in meta-analyses and outcomes research as low-cost alternatives (22). Outcomes research typically involves sifting information stored in large computer-based data sets to discern how a medical intervention received by patients with a given condition is linked with various outcomes. The goal is to find out what intervention works best for which patients. Dominant forces behind this kind of retrospective data analysis have been industry and government officials who want to encourage physicians to use the most cost-effective treatments. The perception that physicians, bureaucrats, and accountants often cannot determine which medical maneuvers work best (23) has led to the idea that searching the medical records maintained by government agencies, insurance companies, and hospitals could generate a next-best idea of how well medical interventions pay off, avoiding costly, cumbersome, prospective clinical trials. Use of administrative data to conduct medical research clearly has large advantages in terms of the number of evaluable patients, availability of information about cost, representativeness of the population studied, and substantial freedom from biases affecting randomized trials (22). Equally clearly, research using these techniques cannot easily discern and analyze the reasons a particular physician chooses a particular strategy for a particular patient. Even the harshest critics of this research technique concede that it is often the only way to obtain information on uncommon conditions, uncommon treatments, or special populations of patients. Advocates insist that such retrospective research has a long record of producing results—far longer than randomized controlled trials—and is clearly better than nothing, which is what would be available if prospective trials were relied on to address every issue of interest (23). The statistical tool of meta-analysis, which has been available for nearly a century, has experienced explosive growth recently. This technique allows data from many small studies, each with little statistical power and often with conflicting results, to be combined and reanalyzed. It promises to resolve nettlesome questions at the interface of medicine and social policy (24–26).

Another common method of affecting medical decisions is the generation of practice guidelines, which are used to limit inappropriate or harmful medical practice, reduce geographical variation in medical practice patterns, and eliminate waste. This is a powerful and broad-based trend in modern medicine (27). The main motivation underlying this process is undoubtedly financial, but guidelines are also vehicles of physician education. Many methods of developing guidelines exist (28) and evaluating the success of guidelines can utilize many different yardsticks (29). There is a real concern that the promulgation of guidelines could harm patient care (29). Enforcement schemes, which inflict punishment on doctors who do not follow such guidelines, are of particular concern because guidelines cannot define optimal care for all patients under all circumstances. When guidelines become more detailed, comprehensibility is often sacrificed to specificity. As generalization decreases, cumbersomeness increases. If "cookbook" medicine replaces sound medical judgment—a fear often expressed—it is inconceivable that any cookbook of clinical guidelines will have enough recipes to encompass the complexities of human life and human disease (30). The impulse to

the more complicated issues such as defining sensitivity, specificity, predictive power, and the like for clinicians to confirm. This can be quite difficult, as biological and analytical factors both contribute to variability even among single patients.

The critical difference between final test results is degree of change needed to be considered clinically meaningful. This can be expressed as $K(X^2 + Y^2)^{1/2}$, where X is the coefficient of test variation owing to analytical variability, Y is the coefficient of test variation due to biological variability, and K is a constant that is chosen on the basis of the probability level desired (18). Calculation of analytic variability is commonly done rigorously but biological variability, which is the degree of fluctuation of a test result in a single patient measured at various times, is often not evaluated. It is likely that biological variability will be more carefully documented in the future.

As physicians devise schemes to care for their bionic patients, they often fail to consider the unpleasantness of undergoing testing. For certain patients, the clinic visit itself generates anxiety. Some patients fear venipuncture, and others dread the claustrophobic feelings encountered during CT scanning. More invasive tests often involve pain, which can be a real barrier. Patients' adherence to recommended surveillance schemes is often poor, and this is probably related to these factors. On the other hand, some patients suffer such anxiety that they request, even demand, frequent surveillance. Coping with these difficult-to-quantify considerations is a matter of concern to physicians, particularly in the current climate of shrinking physician autonomy. When quality of life has been studied in well-designed trials of more intensive vs less intensive follow-up for cancer patients (19), the frequency of testing seems to have little impact on quality of life, but patients do want regular follow-up by their physicians; the same is probably true of patients with prostheses. This preference may simply reflect the trust patients have in modern medicine, particularly because little accurate information exists on the risks, benefits, and costs of various follow-up strategies. Certainly, patients cannot be expected to have detailed knowledge of this area. Patients typically derive comfort from posttreatment follow-up and have come to expect it. Detrimental effects of frequent testing can also come into play because patients may fear the discovery of an adverse event every time testing is performed or a clinic visit occurs. A minimalist approach to surveillance for complications may not be achievable in wealthy societies because patients and physicians employ diagnostic tests to evaluate even minor symptoms or physical signs, even though physicians are well aware of test limitations, such as low sensitivity, low specificity, and high cost. Controversies about cost and benefits continue to appear in the scientific literature. It is not surprising that well-off patients, appropriately concerned with their own welfare, expect (and receive) intensive follow-up (20).

The impact of managed care on patient follow-up strategies has proven to be relatively difficult to study but there is some evidence on this point. In the area of cancer patient follow-up, the degree of penetration of managed care organizations into a community may have little impact on the intensity of follow-up (21) and it seems reasonable to expect that the same will hold true when substantial investigations of postimplantation surveillance for patients with prosthetic implants are carried out. After all, attention to preventive measures and early intervention to maintain health is one of the putative central strengths of managed care.

comes-based decision making makes it increasingly important to assess each patient's preferences and is a new reason for a close and trusting patient–physician relationship.

One would think that a major factor shaping clinical strategy after prosthesis implantation should be patient input. In a sense, patient and physician concerns are difficult to separate because the physician is morally bound to design and carry out a management plan for the patient that is in the patient's best interest. Nonetheless, physicians are typically only dimly aware of many patient and societal concerns. For example, physicians often ignore the logistical difficulties patients experience in complying with the requested posttreatment surveillance tests. However, some patients are naÔve about medical topics and rely on the physician's judgment. Other patients are highly motivated, intelligent, and aware of the natural history of the disorder for which the prosthetic device was implanted and risks and benefits of therapy. It is fairly common these days to find that the patient arrives in the clinic with extensive information obtained from Internet sources. There are plenty of articles on this topic in the popular press (14). Such patients may be quite interested in planning their own posttreatment strategy. The subject of how patient's wishes and fears should and do affect clinical management has been studied relatively little.

Society has good reason to be concerned about this as well. Medical care costs have gone through the roof in recent decades for many reasons. Patients are living longer. Patient demand for prosthetic devices is escalating (15). Complicated procedures, such as prosthetic device implantation, are carried out regularly and patients can be salvaged from the complications of these procedures. The array of devices in clinical use is also likely to expand (16). Electronic chips will probably figure prominently in next-generation prostheses. Research on how to connect such chips to the brain, eye, and limb is now being done at universities and in industrial laboratories (16,17). The expense of care, which can be considerable, is outlined in Chapter 3. This is particularly true for lesions for which modern medicine has some effective therapy, such as an implant infection. Treatment costs are of particular concern for the uninsured individual. Currently, in Western industrial societies, the cost of such care is often largely assumed by entities other than the patient, creating a situation for these entities referred to as moral hazard. As a result, patients are willing to undergo more extensive and expensive postoperative testing and care than they might if costs were directly assessed to them. Doctors and patients also have concerns about the inconvenience and periodic emotional upheaval associated with frequent surveillance testing. Thus, patient choices can sway a physician in the direction of more intensive or less intensive surveillance.

Physicians are concerned about test validity. For a test to be valid, it must have high reproducibility (precision) and accuracy (concordance with a known standard). Precision evaluation requires testing of multiple batches of identical samples on the same machine and evaluating machine-to-machine variation using multiple analytic machines. Estimation of test linearity requires measurement of the analyte of interest at varying concentrations. Run-to-run precision determination requires testing of known standards at several time points. Often, the estimates of the analyte require comparison with reference methods to ensure comparability because the test methodology may involve a different sort of chemical reaction. These multiple validation procedures are usually carried out by the instrument manufacturer before introduction into the market, leaving

this topic is scarce. The government has a role to play here as it monitors safety hazards and may mandate device recalls and safety alerts *(11)*.

At this point it is worthwhile to distinguish screening from early detection. Screening tests usually provide only clues that may possibly indicate the presence of a disorder in an apparently well population but do not usually result in a firm diagnosis. An elevated erythrocyte sedimentation rate, in a patient with a hip prosthesis, for example, may not be to the result of a prosthetic infection. Early detection requires a definitive test but not necessarily an invasive one. X-ray tests often provide the definitive diagnosis of prosthetic failure. Examples include computed tomography (CT) evidence of silicone leak from breast prostheses, angiographic evidence of thrombosis of vascular grafts, and the like. Ideal screening and early detection tests should be simple, minimally invasive, quick to administer, accurate, reproducible, inexpensive, quantitative, acceptable to patients, and directed at a disorder with a distinct preclinical phase. The disorder should be one for which an effective intervention exists. Major patient concerns are comfort and convenience. Physicians are more concerned with test specificity, sensitivity, and safety. Society is concerned with costs and outcomes.

In addition, modern methods of decision analysis *(12)* reveal that patients and physicians view possible outcomes of medical management differently. It is becoming clearer that cost–benefit or cost–utility analysis, although complicated to carry out, may be too simplistic and that a more global view of health care outcomes analysis may be preferable. Such analyses should simultaneously take into account the overall health of an entire population (rather than an individual patient), the clinical outcome for each specific patient, the degree of patient satisfaction with possible or actual outcomes, and costs to patients and society. This is a tall order and likely to be warranted for only a few problems because of the difficulty in carrying it out. The current tendency in medicine to seek standardization, consensus building, and codification of practice promises to run afoul of differences among patients, especially when decisions involve major differences among possible strategies and corresponding differences among possible outcomes. Physicians should ideally honor patient preferences *(13)*, but there are obvious pitfalls of unconventional patient choices. Resistance must be expected from those who pay the bills when patient preferences involve excessive costs. In the future, many more analyses of patient preference can be anticipated, using tools from game theory such as the standard gamble, analog scales, and time trade-off methods *(12,13)*. Where possible, patients should actively participate in medical decision making, and this is usually possible once the anxiety of the surgery has died down and emotions are no longer at flood tide.

Kassirer *(13)* has identified a number of attributes of "utility-sensitive" decisions (those strongly influenced by how patients value various possible outcomes). Such decisions arise when the possible choices involve large differences in possible outcomes, possible risks of intervention, trade-offs between near- and long-term events, small differences in outcomes despite large differences among interventions, and variations among patients in assigning values to specific possible outcomes or processes. Because considerations based on cost and effectiveness are increasing in prominence, it is becoming more important to identify what the patient values in life so that he or she is not shortchanged by impersonal, intrusive rules. This shift toward cost- and out-

medical advances. We believe that the value of this textbook will extend beyond physicians to such patients.

Many of the wealthier nations of the world are now able to provide access to modern medicine to all or most of their citizens, with concomitant increases in the duration and quality of life. A by-product of the increase in the average life span has been a large increase in the total number of people who are candidates for one sort of implant or another, as the incidence and prevalence of various disorders that can be treated with an implant usually increases with age. Costs of care also have risen dramatically, and therein lies one of the conundrums of modern clinical practice: how to use costly resources wisely. Determination of the value of a particular clinical strategy is one of the ways society can rationally decide how to use resources.

The term *value* is often used in an informal sense, but it has acquired a somewhat different, more precise meaning in the realm of quantitative outcomes research and economic analysis. It is indisputable that individual patients derive value—in the conventional sense of the term—by detection and treatment of implant malfunction, but it is not at all clear whether populations of patients derive value from a high-intensity surveillance strategy rather than a low-intensity one. Determining an optimal strategy for the population of people with implanted prosthetic devices has proved to be difficult. What amount of dollars is society willing to expend per year of additional life attributable to follow-up? In wealthy nations that number is probably about $50,000 to $100,000 per quality-adjusted life year *(8)*.

In Western industrial societies there are additional variables. Individuals in university hospitals, for example, may be investigated somewhat more intensely because they have access to new prostheses, the complications of which have not been fully worked out; therefore, follow-up is driven by investigational protocols and tends to be more structured and intense. In health maintenance organizations, Department of Veterans Affairs Medical Center, or military hospitals, care patterns may be influenced by mandates from higher central authorities. In private practices, however, patient expectations, financial realities, and local customs are typically strong motivators. One of the more troubling motivators is physician self-interest. Many hospitals in the United Kingdom, for example, choose hip prostheses they would not otherwise purchase because of financial incentives from manufacturers. Many consultant surgeons, similarly, have accepted incentives from manufacturers of these prostheses *(9)*.

How physicians reach clinical decisions of any sort is difficult to decipher and decisions about the care of patients with prosthetic implants are no exception. Guidelines formulated by groups of experts, such as the American Heart Association, have probably been influential *(10)* but the extent of their influence cannot be measured easily. The controlled clinical trial is accepted as the most persuasive type of evidence, but such trials are expensive, time-consuming, and difficult to carry out. Many trials deal with the surgical procedure itself, but few deal with postoperative care. The clinician typically devises a strategy for each patient based on many factors, conscious as well as unconscious. There is clear evidence that early detection of certain device complications, such as heart valve failure, has a large impact on survival, as discussed elsewhere in this book. For other devices, detection of problems in asymptomatic patients during routine care may not affect survival duration dramatically, and empirical evidence on

large, costly problem *(5)*. More than 50 years ago, Lembcke stated that "the best measure of quality is not how well or how frequently a medical service is given, but how closely the result approaches the fundamental objectives of prolonging life, relieving distress, and preventing disability" *(6)*. This clear, simple definition is widely accepted today. Subsequent evidence, based on a randomized trial, has shown that the health of a population is not linearly related to health care expenditures *(7)*. The optimal level of care would maximize patient welfare at the least cost. Patients are harmed if a clinical strategy is too intensive because they are unnecessarily exposed to hazards such as radiation, they undergo uncomfortable tests, and they incur treatment and opportunity costs, among other risks. If high-intensity surveillance testing provides no improvement in duration or quality of life, society is also harmed by the waste of resources. If the strategy is not intensive enough, patients who develop device malfunction and are potential candidates for effective salvage treatment may die or suffer disability needlessly. Society is also harmed in this case because the costs to social programs (such as Medicare and Social Security) of treating device failures are typically quite high, ill-wage earners are unable to work, their quality of life decreases, their children may grow up without a parent, and other difficult-to-quantify ripple effects take place. For most patients in poor nations and for those without access to care in wealthy ones, of course, these calculations have little relevance, adding an additional layer of complexity to the issue. What is to be done for those with few resources?

The design of trials to gather evidence is time-consuming, expensive, intellectually demanding and methodologically exacting. Patients often do not want to be regarded as "guinea pigs." For these and other reasons, there is a lack of evidence on which doctors can base their decisions. Yet decisions must be made. Choosing to forego health maintenance in patients with prosthetic implants represents a decision that will have consequences as surely as a decision to actively promote health in these patients. The magnitude of the consequences is currently unclear.

For those who have never had surgery, it is difficult to describe the change in body image that results from the implantation of a prosthetic device. Patients request surgery in hopes of feeling better, living longer, having an improved quality of life, and so on. They tolerate the pain, costs, and inevitable risks and have strong motivation to preserve the gains made by the implantation procedure, so patients are typically strongly motivated to comply with tests and treatments described as health promotion measures. If only their doctors knew what tests to request! The sparsity of high-quality evidence regarding health maintenance after prosthesis implantation affects patients' expectations as well. Think of how difficult it must be for a patient with a Björk-Shiley valve who knows of the propensity of the valve to suddenly and catastrophically fail, as discussed elsewhere in this book.

Delivery of medical care is also a function of patient expectations and cultural values. What is appropriate in one civilization may not be the expectation in another. However, as globalization of medical information and homogenization of medical care practices continues, one can expect that these differences may narrow. Nonetheless, there are stark gaps in the distribution of income and medical sophistication among the world's nations and these account for huge disparities in medical care delivery. However, in every society there is a middle class and an upper class with access to modern

4
Factors That Should Shape Clinical Strategies

Frank E. Johnson

Since the end of the Middle Ages and the demise of the Galenic model of human biology and illness, a reductionist, mechanistic, evidence-based paradigm has prevailed. During the several centuries since this transformation in our worldview, there has been a complete revision in our concepts of how to conceptualize health and manage human disorders. Galileo's pronouncement, "Measure all that is measurable, and make those things measurable which have hitherto not been measured" has found applications unimaginable by Galileo himself *(1)*. The contents of this book are an extension of this concept, as the authors have quantified their current recommendations about medical care for a well-defined group of patients. We expect that these recommendations will serve as the starting points for the generation of evidence concerning their actual value.

The optimal regimen for health promotion in patients with implanted prosthetic devices of all sorts is unknown, unfortunately. The practices espoused in this book by well-credentialed experts are sometimes contradictory and have not been translated into uniform care. In an attempt to rationalize care, the concept of the randomized trial was introduced in the mid-20th century. Credit for this is generally given to Austin Bradford Hill, a British epidemiologist *(2)*. The results of over one million randomized trials have been published since then *(3)*, but very few have dealt with the care of patients with implanted devices of any sort. This poses an important problem because there are many individuals at risk, the costs of managing them are high, and the price of failure to detect device malfunction can be very great indeed—whether measured in monetary units, quality-of-life units, or the familiar terms of mortality and morbidity rates.

It should not be surprising that the topic of primary care and health promotion for patients with implanted prosthetic devices suffers from a lack of evidence on which to base decisions. A recent analysis by the Institute of Medicine (IOM) of the US National Academies of Sciences estimated that quality of scientific evidence supporting health services in general is strong for only approximately 4% of patient care, modest for approximately 45%, and weak or absent for the remainder *(4)*. The IOM recognized that the level of agreement among doctors might be quite high even if the level of evidence-based support is low. In the case of people with implants, the actual practice of the clinicians who care for them does not even meet the criterion of general consensus. Why is it important to rationalize follow-up strategies? To minimize overuse, underuse, and misuse of medical resources, which has been identified by the IOM as a

From: *The Bionic Human: Health Promotion for People With Implanted Prosthetic Devices*
Edited by: F. E. Johnson and K. S. Virgo © Humana Press Inc., Totowa, NJ

REFERENCES

1. Marwick C. Implant recommendations. JAMA 2000;283:869.
2. InterStudy Competitive Edge. Managed Care Industry Report Fall 2004: St. Paul, MN.
3. Bero L, Rennie D. The Cochrane Collaboration: preparing, maintaining, and disseminating systematic reviews of the effects of health care. JAMA 1995;274:1935–1938.
4. Robinson A. Research, practice, and the Cochrane Collaboration. Can Med Assn J 1995;152:883–889.
5. Chalmers I, Haynes B. Reporting, updating, and correcting systematic reviews of the effects of health care. British Med J 1994;309:862–865.
6. Health Care Financing Administration. Part B Medicare Annual Data (BMAD) file, 2003.
7. U.S. Bureau of Labor Statistics. Consumer Price Index—All Urban Consumers, U.S. city average, medical care component, multiple years.

devices has achieved considerable consensus (e.g., hernia, cardiac valve, osseo-integrated dental; Chapters 12, 19, and 22, respectively).

Cost analyses of surveillance after implantation of prosthetic devices have rarely been conducted. This is not to say that studies of the lifetime costs and benefits of implantation have not been conducted. There is a wealth of studies, especially for cochlear implants and vascular prostheses. The difficulty arises in attempting to tease out surveillance costs for asymptomatic patients. Generally, the initial costs of implantation and costs associated with reimplantation and treating complications are combined with follow-up costs. The studies are clearly well done, but do not permit the analyst to compare the costs of varying intensities of follow-up after prosthetic device implantation and the associated benefits.

This chapter begins to address this gap in the literature. Because this book summarizes the recommendations of experts in both the United States and Europe for the follow-up of 14 prosthetic devices after implantation, it has provided an excellent opportunity to use one methodology to estimate costs and thereby facilitate comparisons across strategies. Although many assumptions were made in calculating cost estimates, the data represent a conservative starting point for understanding how expensive follow-up for patients implanted with prosthetic devices can be and how minor modifications in strategy can result in major modifications in cost. As explained in detail in Chapter 2, such analyses should ideally incorporate not only the direct costs of follow-up as calculated in this chapter, but also the indirect costs such as time lost from work, transportation costs, child- and adult-care costs, and so on. This is more readily done in a prospective fashion. Quality-of-life data can then also be incorporated. Quality of life may be either directly or indirectly impacted by the intensity of the follow-up regimen. Although there is a rich literature exploring quality of life after implantation with prosthetic devices, the impact of follow-up intensity after implantation on quality of life has not been examined.

One limitation of the current analysis is that actual patients were not followed prospectively to estimate costs. A large prospective study would allow one to collect data on the costs of many factors assumed away in many analyses, such as diagnosis and treatment costs for patients with symptoms or positive test results.

A second limitation is that this chapter mainly examines follow-up conducted by the clinician implanting the prosthetic device. Some patients are followed by their primary care physician beginning rather soon after implantation. Variation in follow-up practices among primary care physicians is probably much greater, unless practicing in a managed care system or national health care system. Thus, the cost ramifications of current follow-up practice for this patient population may be even greater.

The cost differentials among surveillance strategies identified in this chapter will be increasingly difficult to sustain in the current competitive medical practice environment in which cost containment is such a dominant force. Even in those instances where the variation in costs is moderate, the number of patients implanted with these devices is large and, therefore, the total costs associated with each annual cohort of newly implanted patients are staggering. Future research in the form of clinical trials is clearly needed to compare intensities of follow-up and determine if higher costs are substantiated by improved quality of life and longer survival.

sites was 498,617-fold (SD = 1,815,400), 13,595 fold (SD = 49,007) when vascular access devices were treated as an outlier, and 2.4-fold (SD 2.5) when both vascular access devices and intravascular filters and stents were treated as outliers.

Not surprisingly, those strategies requiring 60 or more office visits over the 5-year period were at the high end of the cost distribution. "No follow-up" strategies and those regimens that consist mainly of seven or fewer office visits over the 5-year surveillance period and a single form of diagnostic testing, usually X-ray, were the least expensive. Across all prosthetic device types, the surveillance strategy with the greatest frequency of either visits or any single test was a vascular access device strategy that required 520 office visits after implantation of a tunneled silastic catheter with an external hub (Table 11). Other strategies with high frequencies of office visits or any single test include a strategy for follow-up after implantation of hemodialysis grafts with 60 office visits and 60 venous pressure measurements (Table 8) and a strategy for follow-up after implantation of ports with attached silastic catheters consisting of 60 office visits (Table 11; primary and European counterpoint using Frankfurt data). The next most intensive strategy, also for follow-up after implantation of ports with attached silastic catheters, consisted of 30 office visits (Table 11; European counterpoint using Heidelberg data). The most intensive 5-year strategy across the greatest number of tests was the infrainguinal vein bypass graft surveillance strategy recommended by Vallabhaneni et al. (Table 8, European counterpoint) with 12 office visits, 10 serum blood urea nitrogen (BUN) levels, 10 serum cholesterol levels, 10 serum creatinine levels, 10 serum electrolyte levels, 10 complete blood counts (CBCs), and four duplex scans, for a total of 66 visits and tests, although total charges for the 5-year follow-up period were only $2151. Another somewhat high-intensity strategy is the Gleva strategy for follow-up after implantation of cardiac pacemakers (Table 13; US counterpoint) with five office visits, 1 chest X-ray, 30 transtelephonic monitoring sessions, and 10 pacemaker interrogations for total charges of $2084.

Discounting the 2004 charge figures for each strategy at 3 and 5% to account for the time value of money changes none of the relationships discussed above. For example, the most and least expensive strategies remain the same. Only the dollar values and the difference between any two dollar values are reduced.

COMMENT

Few attempts have been made to measure the relevant costs associated with many medical practices and follow-up after implantation of medical devices is no exception. As this analysis indicates, charges for follow-up can vary widely. Although variation in the range of charges for follow-up is largely dependent on the type of implant, variation as high as 6,803,914-fold or 176,701-fold within device categories cannot easily be justified in this era of cost-containment and health care reform. Even if surveillance strategies consisting of no follow-up are considered outliers and eliminated from analysis, charge differentials of 7.7-fold are similarly difficult to justify. The problem is an especially interesting one because little is known about how outcomes vary when components of the follow-up strategy are altered, and optimal follow-up testing intervals have not been defined by well-designed trials. After implantation of most prosthetic devices, no one surveillance strategy has been established as more efficacious than any other in terms of survival and quality of life, yet follow-up for many types of prosthetic

plus two counterpoints) was an inclusion criterion for the analysis, several specific surgeries demonstrated a narrow range of total Medicare-allowed charges across recommendations. These surgeries included internal fixation of skeletal trauma (Table 14; $140) and infrainguinal vein bypass grafting (Table 8; $231). Breast augmentation (Table 1; $633) and transjugular intrahepatic portosystemic shunt placement (Tables 8 and 10; $911) were the next highest in total charges.

Few authors varied their follow-up recommendations by diagnosis. Examples included inguinal/umbilical hernia vs incisional/large inguinal or umbilical hernia, normal vs abnormal detrusor function in patients with artificial urethral sphincters, and cholesteatoma or otosclerosis vs middle ear and mastoid disease for patients with ossicular devices. There were no instances where the authors of both the primary chapter and the two corresponding counterpoints provided recommendations based on the diagnosis criterion.

Only Lorenz et al. (Table 11, vascular access European counterpoint) provided follow-up strategies by geographic area. These follow-up strategies were for patients with implant ports and attached silastic catheters that are not actively used. No other author used geographic criteria to demonstrate variation in follow-up.

Focusing once again on the broad prosthetic device categories only, the range of total Medicare-allowed charges for follow-up after implantation of prosthetic devices was widest for vascular access devices (Table 11) where charges varied by $68,039 between Whitman's strategy of no follow-up unless symptoms arise (US counterpoint) to Compton's strategy for follow-up after implantation of tunneled silastic catheters with an external hub (primary chapter). The ranges of charges for vascular prostheses (Table 8; $7697) was also high but was nowhere near the range for vascular access devices. Specifically the high cost of surveillance for patients who have undergone hemodialysis grafts, compared with the relatively low cost for follow-up of patients after aortic graft placement, accounted for the large variation in charges for vascular prostheses. Excluding hemodialysis grafts, follow-up charges for vascular prostheses varied by approximately $1561. The average range of charges across all 14 broad prosthetic device categories was $68,039 (SD = $7760). When Compton's strategy for follow-up after implantation of tunneled silastic catheters with external hubs (Chapter 21) was classified as an outlier and excluded from analysis, the average range of charges was $8696 (SD = $1775).

The magnitude of the difference in total Medicare-allowed charges by broad prosthetic device categories was then examined. As expected, the largest charge differential, 6,803,914-fold, was for vascular access devices (Table 11). For intravascular filters and stents (Table 10), the magnitude of the charge differential was also high (176,701-fold), although the range of charges was only $1767. Both prosthetic device categories with high charge differentials included a strategy of no follow-up unless symptoms arise. The next closest charge differentials were 7.7-fold for vascular prostheses (Table 8), 5.9-fold for ossicular devices (Table 7), 4.7-fold for artificial urethral sphincters (Table 4), and 3-fold for penile prostheses (Table 3). Forty-three percent of prosthetic device categories had low charge differentials (less than twofold). These device categories included cerebrospinal fluid shunts, cochlear implants, cardiac valves, osseointegrated dental implants, cardiac pacemakers, and joint prostheses and internal fixation devices (Tables 5, 6, 9, 12–14). The average charge differential across all 14

Table 12
Osseointegrated Dental Implants (Chapter 22)
Total 5-Year Follow-Up Charges[a]

	Medicare-allowed charges	Discounted at 3%	Discounted at 5%
Primary (Table 1, Chapter 22)	$1083.66	$1016.16	$975.48
US counterpoint (Table 1, Chapter 22)	$1268.19	$1185.18	$1135.26
European counterpoint (Table 1, Chapter 22)	$1083.66	$1016.16	$975.48
Charge differential	0.2		

[a]Using 2003 charge data inflated to 2004 figures using the medical care component of the Consumer Price Index (7).

Table 13
Cardiac Pacemakers (Chapter 23)
Total 5-Year Follow-Up Charges[a]

	Medicare-allowed charges	Discounted at 3%	Discounted at 5%
Primary (Table 3, Chapter 23)	$1806.66	$1656.59	$1567.31
US counterpoint (Table 1, Chapter 23)	$2084.03	$1908.25	$1803.62
European counterpoint (Table 2, Chapter 23)	$915.91	$839.13	$793.62
Charge differential	1.3		

[a]Using 2003 charge data inflated to 2004 figures using the medical care component of the Consumer Price Index (7).

Table 14
Joint Prostheses and Internal Fixation Devices (Chapter 24)
Total 5-Year Follow-Up Charges[a]

	Medicare-allowed charges	Discounted at 3%	Discounted at 5%
Primary (Tables 1, 5, Chapter 24)			
After total joint arthroplasty	$1121.26	$1050.10	$1007.26
After internal fixation of skeletal trauma	$560.63	$544.30	$533.93
US counterpoint (Tables 1, 2, Chapter 24)			
After total joint arthroplasty	$1111.94	$1041.06	$998.39
After internal fixation of skeletal trauma	$420.47	$408.22	$400.45
European counterpoint (Tables 1–4, Chapter 24)			
After total hip arthroplasty	$977.86	$915.76	$875.16
After total knee arthroplasty	$976.48	$909.33	$869.04
After revision hip arthroplasty	$699.24	$663.15	$641.07
After revision knee arthroplasty	$692.35	$656.65	$634.81
After internal fixation of skeletal trauma	$429.78	$417.00	$408.90
Charge differential	1.7		

[a]Using 2003 charge data inflated to 2004 figures using the medical care component of the Consumer Price Index (7).

Table 10
Intravascular Filters and Stents (Chapter 20)
Total 5-Year Follow-Up Charges[a]

	Medicare-allowed charges	Discounted at 3%	Discounted at 5%
Primary (Tables 1, 2, Chapter 20)[b]			
Patients with vena caval filters	$1147.67	$1051.20	$993.76
Patients with venous stents	$1767.02	$1653.63	$1585.38
US counterpoint (Table 1, Chapter 20)[c]	—	—	—
European counterpoint (Table 2, Chapter 20)[b]			
After transjugular intrahepatic portosystemic shunt placement	$1647.85	$1509.33	$1426.87
Charge differential 176,701			

[a]Using 2003 charge data inflated to 2004 figures using the medical care component of the Consumer Price Index (7).
[b]No recommendations provided for devices other than those presented here.
[c]No specific follow-up recommended unless symptoms arise.

Table 11
Vascular Access Devices (Chapter 21)
Total 5-Year Follow-Up Charges[a]

	Medicare-allowed charges	Discounted at 3%	Discounted at 5%
Primary (Tables 3, 4, Chapter 21)			
Patients with tunneled silastic catheters with external hubs	$68,039.15	$62,319.88	$58,914.78
Patients with implanted ports with attached silastic catheters	$7850.67	$7190.76	$6797.86
US counterpoint (Table 1, Chapter 21)[b]	—	—	—
European counterpoint (Tables 2–4, Chapter 21)[c]			
Patients with implanted ports with attached silastic catheters not actively used at facilities in:			
Frankfurt, Germany	$7850.67	$7190.76	$6797.86
Heidelberg, Germany	$3925.34	$3595.38	$3398.93
Zurich, Switzerland	$2616.89	$2396.92	$2265.95
Charge differential 6,803,914			

[a]Using 2003 charge data inflated to 2004 figures using the medical care component of the Consumer Price Index (7).
[b]No specific follow-up recommended unless symptoms arise.
[c]No recommendations provided for devices other than those presented here.

Table 8 (Continued)
Vascular Prostheses (Chapter 18)
Total 5-Year Follow-Up Charges[a]

	Medicare-allowed charges	Discounted at 3%	5%
After aortic graft placement (open repair) for aneurysmal or occlusive disease	$998.30	$924.37	$880.27
After aortic endografts	$1759.60	$1641.72	$1570.72
After infrainguinal vein bypass grafts	$1919.13	$1798.64	$1725.87
After infrainguinal prosthetic bypass grafts	$1177.60	$1103.66	$1059.01
After transjugular intrahepatic portosystemic shunt placement	$2558.84	$2367.18	$2252.57
US counterpoint (Tables 1–10, Chapter 18)			
After carotid endarterectomy or carotid angioplasty/stenting	$1508.32	$1404.96	$1342.93
After hemodialysis grafts	$8695.68	$7964.74	$7529.55
After renal revascularization (bypass or angioplasty/stent)	$1436.99	$1335.10	$1274.04
After mesenteric revascularization	$1444.86	$1343.72	$1283.07
After iliac PTA/stenting	$1623.50	$1511.62	$1444.31
After aortic graft placement (open repair) for aneurysmal or occlusive disease	$998.30	$924.37	$880.27
After aortic endografts	$1686.79	$1573.09	$1504.68
After infrainguinal vein bypass grafts	$1919.13	$1798.64	$1725.87
After infrainguinal prosthetic bypass grafts	$1177.60	$1103.66	$1059.01
After transjugular intrahepatic portosystemic shunt placement	$2558.84	$2367.18	$2252.57
European counterpoint (Table 3, Chapter 18)[b]			
After infrainguinal vein bypass grafts	$2150.58	$2002.28	$1913.31
Charge differential 7.7			

[a]Using 2003 charge data inflated to 2004 figures using the medical care component of the Consumer Price Index *(7)*.
[b]No recommendation provided for procedures other than infrainguinal vein bypass grafts.
PTA, percutaneous transluminal angioplasty.

Table 9
Heart Valves (Chapter 19)
Total 5-Year Follow-Up Charges[a]

	Medicare-allowed charges	Discounted at 3%	5%
Primary (Table 2, Chapter 19)	$1182.26	$1097.43	$1046.62
US counterpoint (Table 1, Chapter 19)	$1182.26	$1097.43	$1046.62
European counterpoint (Table 20, Chapter 19)	$949.89	$886.29	$848.08
Charge differential 0.2			

[a]Using 2003 charge data inflated to 2004 figures using the medical care component of the Consumer Price Index *(7)*.

Table 6
Cochlear Implants (Chapter 16)
Total 5-Year Follow-Up Charges[a]

	Medicare-allowed charges	Discounted at 3%	Discounted at 5%
Primary (Table 4, Chapter 16)	$2102.66	$1971.53	$1892.50
US counterpoint (Table 2, Chapter 16)			
Patients followed with cochlear view X-ray	$3380.85	$3225.00	$3129.53
Patients followed with ear CT	$3417.81	$3260.89	$3164.73
European counterpoint (Table 1, Chapter 16)	$2597.72	$2411.04	$2299.13
Charge differential	0.6		

[a]Using 2003 charge data inflated to 2004 figures using the medical care component of the Consumer Price Index *(7)*. CT, computed tomography.

Table 7
Ossicular Implants (Chapter 17)
Total 5-Year Follow-Up Charges[a]

	Medicare-allowed charges	Discounted at 3%	Discounted at 5%
Primary (Table 2, Chapter 17)	$1253.30	$1159.42	$1103.29
US counterpoint (Tables 1, 2, Chapter 17)			
Patients with cholesteatoma/otosclerosis	$732.26	$675.00	$640.81
Patients with other middle ear and mastoid disease	$208.88	$202.80	$198.94
European counterpoint (Table 1, Chapter 17)	$1436.96	$1337.73	$1278.20
Charge differential	5.9		

[a]Using 2003 charge data inflated to 2004 figures using the medical care component of the Consumer Price Index *(7)*.

Table 8
Vascular Prostheses (Chapter 18)
Total 5-Year Follow-Up Charges[a]

	Medicare-allowed charges	Discounted at 3%	Discounted at 5%
Primary (Tables 2–7, 9–11, 14, Chapter 18)			
After carotid endarterectomy or carotid angioplasty/stenting	$1295.08	$1197.93	$1139.85
After hemodialysis grafts	$8695.68	$7964.74	$7529.55
After renal revascularization (bypass or angioplasty/stent)	$1519.39	$1415.10	$1352.51
After mesenteric revascularization	$1527.26	$1423.72	$1361.54
After iliac PTA/stenting	$1705.90	$1591.61	$1522.78

(continued)

Table 3
Penile Prostheses (Chapter 13)
Total 5-Year Follow-Up Charges[a]

	Medicare-allowed charges	Discounted at 3%	Discounted at 5%
Primary (Table 1, Chapter 13)	$915.91	$853.30	$815.72
US counterpoint (Table 1, Chapter 13)	$1046.76	$980.33	$940.33
European counterpoint (Table 4, Chapter 13)	$261.69	$254.07	$249.23
Charge differential 3.0			

[a]Using 2003 charge data inflated to 2004 figures using the medical care component of the Consumer Price Index (7).

Table 4
Artificial Urethral Sphincters (Chapter 14)
Total 5-Year Follow-Up Charges[a]

	Medicare-allowed charges	Discounted at 3%	Discounted at 5%
Primary (Tables 1, 2, Chapter 14)			
AUS and normal detrusor function	$654.22	$599.23	$566.49
AUS and abnormal detrusor function followed with IVP	$3673.42	$3364.64	$3180.80
AUS and abnormal detrusor function followed with renal ultrasound	$3739.87	$3425.50	$3238.34
US counterpoint (Table 1, Chapter 14)	$1065.13	$997.16	$956.24
European counterpoint (Table 4, Chapter 14)	$1052.63	$982.83	$940.87
Charge differential 4.7			

[a]Using 2003 charge data inflated to 2004 figures using the medical care component of the Consumer Price Index (7). AUS, artificial urethral sphincters; IVP, intravenous pyelography.

Table 5
Cerebrospinal Fluid Shunts (Chapter 15)
Total 5-Year Follow-Up Charges[a]

	Medicare-allowed charges	Discounted at 3%	Discounted at 5%
Primary (Table 1, Chapter 15)	$1308.48	$1216.46	$1161.29
US counterpoint (Table 1, Chapter 15)	$1584.90	$1478.16	$1414.06
European counterpoint (Table 1, Chapter 15)	$1439.32	$1343.49	$1285.91
Charge differential 0.2			

[a]Using 2003 charge data inflated to 2004 figures using the medical care component of the Consumer Price Index (7).

As a demonstration of the similarity across strategies focusing initially on the broad prosthetic device categories only, the range of total Medicare-allowed charges was narrowest for hernia implants (Table 2; $131), osseointegrated dental implants (Table 12; $185), heart valves (Table 9; $232), and cerebrospinal fluid shunts (Table 5; $276). The next highest in total charges were joint prostheses and internal fixation devices (Table 14; $701) and penile prostheses (Table 3; $785). When type of surgery was factored in and receipt of follow-up recommendations from all three authors (primary

Table 1
Breast Prostheses (Chapter 11)
Total 5-Year Follow-Up Charges[a]

	Medicare-allowed charges	Discounted at 3%	Discounted at 5%
Primary (Tables 4, 5, Chapter 11)[b]			
After breast augmentation	$1098.66	$1020.69	$973.96
After breast reconstruction	$1622.04	$1492.88	$1415.83
US counterpoint (Tables 1, 2, Chapter 11)[b]			
After breast augmentation	$706.13	$671.82	$650.77
After breast reconstruction	$836.97	$791.57	$763.79
European counterpoint (Table 1, Chapter 11)			
After breast augmentation[c]	$465.63	$449.06	$438.72
After breast reconstruction[d]	—	—	—
Charge differential 2.5			

[a]Using 2003 charge data inflated to 2004 figures using the medical care component of the Consumer Price Index (7).
[b]Recommendations for mammography only pertain to women 40 years of age and older.
[c]Recommendations for mammography only pertain to women 50 years of age and older.
[d]No recommendation provided.

Table 2
Hernia Prostheses (Chapter 12)
Total 5-Year Follow-Up Charges[a]

	Medicare-allowed charges	Discounted at 3%	Discounted at 5%
Primary (Table 1, Chapter 12)	$261.69	$254.07	$249.23
US counterpoint (Tables 2, 3, Chapter 12)			
After inguinal or umbilical hernia repair	$130.84	$127.03	$124.61
After incisional hernia repair or large inguinal or umbilical hernia repair	$261.69	$254.07	$249.23
European counterpoint (Table 1, Chapter 12)	$130.84	$127.03	$124.61
Charge differential 1.0			

[a]Using 2003 charge data inflated to 2004 figures using the medical care component of the Consumer Price Index (7).

Calculating total charges of follow-up for patients after implantation of venous stents was somewhat problematic because the stent placement site (superior vena cava, subclavian vein, femoral artery) varies among patients and charges vary by site for X-ray. Because none of the site-specific charges were applicable to all patients, an average charge was calculated across the three most likely sites (chest, abdomen, and leg).

Also problematic was the calculation of total follow-up charges after total joint replacement and after internal fixation of skeletal fractures. The reason was basically the same as for venous stents. The site of joint replacement or fracture fixation varies among patients and charges vary by site for X-ray. Though knee and hip are the predominant sites for joint replacement, shoulder, wrist, elbow, and ankle joint replacements are also conducted and need to be factored in. Thus, for joint replacement, an average X-ray charge was calculated across all of the six sites. The same average X-ray charge was used for fracture fixation.

Many assumptions were necessary in order to achieve consistency across the follow-up strategies recommended by the many authors in this text. In the case of liver function tests (LFTs), a full panel of tests was assumed unless the author specified only certain tests. In the case of audiometry in the follow-up of patients after implantation with cochlear or ossicular devices, it was assumed that a comprehensive audiometry threshold evaluation was performed.

In addition, a best-case scenario was assumed in calculating charges. For example, it was assumed patients were healthy and that additional workup based on either symptoms or positive test results was not required. Furthermore, it was assumed that each patient survived for 5 years after implantation because the purpose of the analysis was to measure the cost of the initial 5 years of follow-up. It is true that some patients may not have lived for 5 years posttreatment, but this assumption can be easily relaxed in the sensitivity analysis. Indirect costs, such as time lost from work, transportation charges, and child- or adult-care charges, were not factored into this analysis. Similarly, treatment costs for explantation, reimplantation, device repair, and any other conditions detected during surveillance, such as infection, were ignored, although they may impose massive additional expenses for individual patients. Under these assumptions, the resulting cost estimates should be considered very conservative and constitute baseline estimates of follow-up costs. All costs assumed away in this analysis would be considered add-ons to the estimates presented. Total Medicare-allowed charges were compared across surveillance strategies within each site.

RESULTS

Tables 1 through 14 display nationwide average Medicare-allowed charges for 2004 associated with the strategies recommended in this text for follow-up after implantation of 14 different prosthetic devices for the 5-year period posttreatment. For each follow-up strategy, total charges discounted at both 3 and 5% are also presented. For no broad prosthetic device category was there total agreement across all recommended follow-up strategies, although consensus was nearly achieved for many categories with two of three authors in agreement. Analyses at the prosthetic device category level do not control for surgery type. Even for prosthetic device categories for which no two authors agreed on a common surveillance strategy, the recommended strategies were not totally dissimilar, as an examination of charge ranges reveals.

METHODS

Nationwide average charges for 2003 associated with the 5-year follow-up strategies recommended in this text for patients implanted with prosthetic devices were computed at the individual patient level. These recommendations have been categorized into 14 broad prosthetic device categories. The charges associated with alternative strategies recommended in the counterpoints to each chapter were also computed. Charge data were obtained from the 2003 Part B Medicare Annual Data (BMAD) file (6). Average allowed charges nationwide for physician services were extracted from the BMAD file, which contains Medicare Part B data categorized by Current Procedural Terminology (CPT) code and place of service (such as inpatient hospital, outpatient hospital, office, etc.). Charges were then inflated to 2004 levels using the medical care component of the Consumer Price Index (CPI) (7). As the percentage change for the 12-month-period 2003 to 2004 was not available at the time this chapter was written, the percentage change for the first 6 months of 2004 (4.4%) was used as a proxy. Total charges were then discounted by 3 and 5% to account for the time value of money.

Not all tests needed for the current analysis were present in the 2003 BMAD data as some tests were not performed frequently enough in 2003 to provide valid charge data. Previous years of BMAD data were referenced to obtain charges figures for the two affected tests: voiding pressures and venous pressures. Relevant years of the medical care component of the CPI were used to inflate test charges to 2004 levels.

Place of service was selected based on frequency of test use. For example, for laboratory tests, independent laboratory was the obvious choice because test utilization was generally higher for independent laboratory than for any other place of service. Office was the place of service for tests such as cochlear implant programming, venous pressures, transtelephonic monitoring, and pacemaker interrogation. Outpatient hospital was the place of service for most radiologic and nuclear medicine tests with the exception of simple chest X-rays and electrocardiograms for which office was considered the place of service.

No additional charges were assessed for examinations considered to be part of the history and physical examination, such as blood pressure checks, ear microscopy, assessment of shunt function, ankle-brachial index, and dental prophylaxis, though charges for the office visit itself were calculated. Charges for follow-up modalities unrelated to surveillance after implantation of the prosthetic device, such as general health screening and smoking cessation consultation, were also excluded. If alternative modalities were recommended as equal substitutes for one another, total charges for the strategy were recalculated using each modality. For example, in Chapter 14, intravenous pyelogram and renal ultrasound were considered equal substitutes for the follow-up of patients after implantation of urethral sphincters. Likewise, if certain diagnostic tests were indicated for only certain subsets of the population, total charges were calculated for each subset individually. Subsets were generally defined by type of surgery or diagnosis. If no follow-up was recommended unless the patient was symptomatic, resulting in zero charges, a value of 1 cent was substituted to permit calculation of the magnitude of the charge difference among a range of follow-up strategies. This was done because division by zero is not possible.

3
Costs of Follow-Up After Implantation of Prosthetic Devices

Katherine S. Virgo

Follow-up after implantation of prosthetic devices is a much neglected research topic. There are few widely accepted practice guidelines, weak attempts at measuring costs, and the potential exists for unrestrained costs. It is estimated that 20 to 25 million patients have some type of implanted medical device *(1)*. Of these patients, the percentage that will develop infections or require device repair or reimplantation varies widely by type of prosthetic device.

Because of the lack of objective data, physicians often adopt the surveillance regimen specific to their residency training program without questioning the impact on patient survival or quality of life. Although this practice is not surprising, it is difficult to justify, particularly in an environment increasingly dominated by managed care. With more than 68.8 million people currently enrolled in health maintenance organizations in the United States and another 109 million enrolled in preferred provider organizations in the United States *(2)*, cost-effectiveness has become the yardstick for determining whether these patients will be allowed access to high-cost or highly utilized health care services. Increasing emphasis on the practice of evidence-based medicine *(3–5)* and the implementation of medical practice guidelines will serve to further restrict the current tendency of physicians to follow patients intensively.

With disagreement over the appropriate frequency of follow-up and no solid data on which to base surveillance strategies, it is not surprising that few attempts have been made to ascertain the costs associated with follow-up after implantation of prosthetic devices. The costs of patient workup prior to implantation of the device, costs associated with the implantation itself, and costs of reoperation are frequently analyzed, given their sheer magnitude, whereas the costs of surveillance after implantation are rarely even mentioned and are generally treated as nonsignificant. In an attempt to fill this void, this chapter provides per-patient cost data for every 5-year follow-up strategy discussed in this book across all 14 prosthetic device types. Although many assumptions are made in calculating these estimates of costs, the data represent a conservative starting point for understanding how expensive follow-up can be after prosthetic device implantation and how minor modifications in strategy can result in major modifications in cost.

From: *The Bionic Human: Health Promotion for People With Implanted Prosthetic Devices*
Edited by: F. E. Johnson and K. S. Virgo © Humana Press Inc., Totowa, NJ

47. Patrick DL, Starks HE, Cain KC, Uhlmann, RF, Pearlman RA. Measuring preferences for health states worse than death. Med Decis Making 1994;14:9–18.
48. Torrance GW. Social preferences for health states: an empirical evaluation of three measurement techniques. Socio Econ Plan Sci 1976;10:129–136.
49. von Neumann J, Morgenstern O. Theory of games and economic behavior. New York: John Wiley, 1953.
50. Sonnenberg FA. U-maker 1.0 [computer program]. Microcomputer utility assessment program. New Brunswick, NJ, 1993.
51. Gafni A. The standard gamble method: what is being measured and how it is interpreted. Health Serv Res 1994;29:207–224.
52. Torrance GW, Thomas WH, Sackett DL. A utility maximization model for evaluation of health care programmes. Health Serv Res 1972;7:118–133.
53. Weinstein MC. Principles of cost-effective resource allocation in health care organizations. Int J Technol Assess Health Care 1990;6:93–103.
54. Johannesson M, Pliskin JS, Weinstein MC. A note on QALYs, time trade-off, and discounting. Med Decis Making 1994;14:188–193.
55. Carr-Hill R. Assumptions of the QALY procedure. Soc Sci Med 1989;29:469–477.
56. Carr-Hill R. Allocating resources to health care: is the QALY a technical solution to a political problem? Int J Health Serv 1991;21:351–363.
57. Loomes G, McKenzie L. The use of QALYs in health care decision making. Soc Sci Med 1989;28:299–308.
58. Wagstaff A. QALYs and the equity–efficiency trade-off. J Health Econ 1991;10:21–41.
59. Johannesson M, Pliskin JS, Weinstein MC. Are healthy-years equivalents an improvement over quality-adjusted life years. Med Decis Making 1993;13:281–286.
60. Robinson R. Cost-effectiveness analysis. BMJ 1993;307(c):793–795.

18. Hanley JA, McNeil BJ. The meaning and use of the area under a receiver operating characteristic (ROC) curve. Radiology 1982;143:29–36.
19. Hanley JA, McNeil BJ. A method of comparing the areas under receiver operating characteristic curves derived from the same cases. Radiology 1983;148:839–843.
20. Metz CE, Kronman HB. Statistical significance tests for binormal ROC curves. J Math Psych 1980;22:218–243.
21. Moise A, Clement B, Ducimetiere P, Bourassa MG. Comparison of receiver operating curves derived from the same population: a bootstrapping approach. Comput Biomed Res 1985;18:125–131.
22. DeLong ER, DeLong DM, Clarke-Pearson DL. Comparing the areas under two or more correlated receiver operating characteristic curves: a nonparametric approach. Biometrics 1985; 44:837–845.
23. McClish DK. Comparing the areas under more than two independent ROC curves. Med Decis Making 1987;7:149–155.
24. Sox HC, Blatt MA, Higgins MC, Marton KI. Medical decision making. Boston, MA: Butterworths, 1988.
25. Pasanen PA, Eskelinen M, Partanen K, Pikkarainen P, Penttila I, Alhava E. Receiver operating characteristic (ROC) curve analysis of the tumour markers CEA, CA 50, and CA 242 in pancreatic cancer: results from a prospective study. Br J Cancer 1993;67:852–855.
26. Pauker SG, Kassirer JP. The threshold approach to clinical decision making. N Engl J Med 1980;302:1109–1117.
27. Nease RF, Owens DK, Sox HC. Threshold analysis using diagnostic tests with multiple results. Med Decis Making 1989;9:91–103.
28. Glasziou P. Threshold analysis via the Bayes' nomogram. Med Decis Making 1991;11:61–62.
29. Warner KE, Luce BR. Cost–benefit and cost-effectiveness analysis in health care. Ann Arbor, MI: Health Administration Press, 1982.
30. Drummond MF, Stoddart GL, Torrance GW. Methods for the economic evaluation of health care programmes. Oxford: Oxford Medical Publications, 1987.
31. Weinstein MC. Foundations of cost-effectiveness analysis for health and medical practices. N Engl J Med 1977;296:716–721.
32. Dasgupta AK, Pearce DW. Cost–benefit analysis: theory and practice. London: Macmillan, 1972.
33. Mishan EJ. Cost–benefit analysis. London: George Allen and Unwin, 1975.
34. Sugden R, Williams AH. The principles of practical cost–benefit analysis. Oxford: Oxford University Press, 1979.
35. Eastaugh SR. Medical economics and health finance. Dover, MA: Auburn House, 1981.
36. Robinson R. Cost-utility analysis. BMJ 1993;307:859–862.
37. Strom BL. Pharmacoepidemiology. New York: Churchill Livingstone, 1989.
38. Rapoport J, Robertson RL, Stuart B. Understanding health economics. Rockville, MD:, 1982.
39. Krahn M, Gafni A. Discounting in the economic evaluation of health care interventions. Med Care 1993;31:403–418.
40. Muller A, Reutzel TJ. Willingness to pay for reduction in fatality risk: an exploratory survey. Am J Public Health 1984;74:808–812.
41. Zeckhauser R. Procedures for valuing lives. Public Policy 1975;23:419–464.
42. Jacobs P. The economics of health and medical care. Rockville, MD: Aspen, 1987.
43. Robinson R. Cost–benefit analysis. BMJ 1993;307(b):924–926.
44. Mehrez A, Gafni A. Quality adjusted life years and healthy year equivalents. Med Decis Making 1989;9:142–149.
45. Mehrez A, Gafni A. Healthy-years equivalents versus quality-adjusted life years. Med Decis Making 1993;13:287–292.
46. Sackett DL, Torrance GW. The utility of different health states as perceived by the general public. J Chron Dis 1978;31:697–704.

SUMMARY

This chapter provides a review of the tools needed to assess and compare the performance of diagnostic tests, to determine thresholds for diagnostic testing and treating, and to calculate the total costs of follow-up after implantation of prosthetic devices. These tools allow the clinician to gather data, thus permitting more informed decision making regarding the composition of the chosen strategy. The concepts presented here lay the groundwork for the next chapter, which applies cost-evaluation methodology to the management of patients with implanted prosthetic devices, assigning dollar values to the follow-up strategies suggested in subsequent chapters. This chapter also lays the groundwork for Chapter 4, which discusses clinical, legal, economic, and ethical issues that impact how decisions should be made regarding the composition of follow-up strategies.

REFERENCES

1. Mausner JS, Kramer S. Epidemiology—an introductory text (2nd ed.). Philadelphia, PA: W.B. Saunders, 1985.
2. Fisher LD, van Belle G. Biostatistics: a methodology for the health sciences. New York: John Wiley, 1993.
3. Benish WA. Graphic and tabular expressions of Bayes' theorem. Med Decis Making 1987;7:104–106.
4. Fagan TJ. Nomogram for Bayes' formula. N Engl J Med 1975;293:257.
5. Matchar DB, Simel DL, Geweke JF, Feussner JR. A Bayesian method for evaluating medical test operating characteristics when some patients' conditions fail to be diagnosed by the reference standard. Med Decis Making 1990;10:102–111.
6. Irwig L, Glasziou PP, Berry G, Chock C, Mock P, Simpson JM. Efficient study designs to assess the accuracy of screening tests. Am J Epidemiol 1994;140:759–769.
7. Fletcher RH, Fletcher SW, Wagner EH. Clinical epidemiology: the essentials (2nd ed.). Baltimore, MD: Williams & Wilkins, 1988.
8. Vecchio TJ. Predictive value of a single diagnostic test in unselected populations. N Engl J Med 1966;274:1171–1173.
9. Friedman GD. Primer of epidemiology (2nd ed.). New York: McGraw-Hill Book, 1980.
10. Woolson RF. Statistical methods for the analysis of biomedical data. New York: John Wiley, 1987.
11. Knottnerus JA, Leffers P. The influence of referral patterns on the characteristics of diagnostic tests. J Clin Epidemiol 1992;45:1143–1154.
12. Wilson JM, Jungner F. Principles and practice of screening for disease. Geneva: World Health Organization, Public Health Papers; 1968: No. 34.
13. Feinstein AR. Clinical epidemiology: the architecture of clinical research. Philadelphia, PA: W.B. Saunders, 1985.
14. Metz CE. Basic principles of ROC analysis. Semin Nucl Med 1978;8:283–298.
14a. Sox HC, Blatt MA, Higgins HC, Marton KI. Medical decision making. Boston, MA: Butterworths, 1988.
15. Swets JA. Signal detection and recognition by human observers. New York: John Wiley, 1964.
16. Centor RM. Signal detectability: the use of ROC curves and their analysis. Med Decis Making 1991;11:102–106.
17. van der Schouw YT, Straatman H, Verbeek AL. ROC curves and the areas under them for dichotomized tests: empirical findings for logistically and normally distributed diagnostic test results. Med Decis Making 1994;14:374–381.

health is initially set at 100% and the probability of death is set at 0%. After the individual makes a choice, the probabilities are changed to 100% probability of death and 0% probability of perfect health, the pie graph is changed accordingly, and the question is posed again. This process continues until the individual is indifferent between living in the chronic health state for life and the gamble. This indifference point is the individual's utility for the health state.

In the time trade-off method, the patient is given a choice between living in a given health state for a given period followed by death vs being healthy for a shorter period of time followed by death *(52)*. After the individual makes a choice, the times are varied and the process repeats continuously until an indifference point is reached. That indifference point is the individual's utility for the health state.

Converting Utilities to QALYs

Regardless of the method used to calculate utilities, the final step in a utility analysis is to convert these utilities into QALYs and interpret the results. Assume a group's utility for living with a pacemaker averages 0.45, and a new follow-up strategy has been demonstrated to result in a 1.5-year increase in survival over the status quo. The group's QALYs would be 0.68 (0.45 × 1.5). The analyst would then calculate the cost of the follow-up strategy per QALY, discount costs and benefits to net present value if costs and benefits accrue over a period longer than 1 year, and use these data to determine if the new strategy was worth the investment *(31,53,54)*.

The use of QALYs is not without criticism, however. It has been suggested that QALYs discriminate against the elderly, equity issues are disregarded, and the resulting quality-of-life scores are biased *(44,55–59)*.

Sensitivity Analysis

No matter which method of cost analysis is chosen, certain assumptions need to be made in relation to causation. These assumptions must be carefully delineated. In addition, the analyst should determine how sensitive the results of the cost analysis are to the assumptions made. For example, if there is known imprecision in any of the estimates used, both conservative as well as liberal alternative estimates should be constructed and the sensitivity of the results to the varying estimates should be tested.

There are three major forms of sensitivity analysis: simple sensitivity analysis, extreme scenarios, and probabilistic sensitivity analysis. Simple sensitivity analysis involves varying one or more of the assumptions on which the economic evaluation is based to determine the effect on the results. The extreme-scenarios approach consists of analyzing the extremes of the distribution of costs and effectiveness and determining whether the results hold up under the most optimistic and pessimistic assumptions. Probabilistic sensitivity analysis assigns ranges and distributions to variables, using computer programs to select values at random from each range and measure the effects. This approach can handle a large number of variables and basically generates confidence intervals for each option *(60)*. Irrespective of the type of sensitivity analysis conducted, the goal is to measure whether large variations in the assumptions result in significant variations in the results of the cost evaluation. If significant variations are not the result, more confidence can be placed in the study's results. If significant variations are the result, an attempt should be made to either reduce uncertainty or improve the accuracy of crucial variables *(30)*.

QALYs saved relates to the quality of life of the patient whose life was saved. To have one's life saved and be in a wheelchair should be valued much differently than to have one's life saved and be healthy. The results of utility analyses are usually expressed as cost per QALY gained.

There are several circumstances in which utility analysis has particular applicability. These circumstances are first, if quality of life is an important outcome; second, if both morbidity and mortality are affected by the intervention and the preference is for a single outcome combining both effects; and, third, if multiple alternative programs are being compared with a wide variety of outcome measures, the use of utility analysis would simplify the evaluation by converting all outcomes to one unit of measure *(30)*.

Utility Values for Health States

The most time-consuming task in a utility analysis is determining utility values for health states. Utility is broadly defined in economics as the value an individual assigns to a given option. In health care it is generally defined as the level of well-being experienced in a given health state. Although one could estimate or possibly obtain these values from the literature, the best way to determine utility values for health states is to measure them directly *(46)*. There are different schools of thought on what populations should be used to measure utilities. One approach is to identify a population with the condition of interest and measure the population's utility for the condition. The analyst needs to keep in mind that patients have a tendency to exaggerate the disutility of their condition. The second approach is to identify a population without the condition, provide a scenario of what the life of a patient with the condition is like, and measure the population's utility for the condition. The methodological difficulty with the second approach is determining how much detail to provide, what media to use to describe the condition, and how to describe the condition without biasing the result. It is suggested that the level of detail be kept to a minimum and that a balanced presentation of the condition be provided, showing both positive and negative implications of the condition.

Utilities are generally measured on a scale from 0 to 1, with 1 representing healthy and 0 representing dead. Health states often viewed as worse than death, such as dementia and coma, are assigned negative values *(47)*. The three methods currently in use for measuring utility values are the rating scale, standard gamble, and time trade-off *(48)*. The rating scale method is normally depicted as a line on a page segmented into gradations by multiples of 10. A single chronic health state or multiple chronic health states and a single age of onset of illness are described to the individual whose utility for the various health states is being measured. A state of perfect health and a state of death are also described to the individual as points of reference. The individual is asked to select from among the various health states the most preferred and least preferred, which become the ends of the scale. The individual is then asked to locate the chronic health states relative to each other on the scale.

The standard gamble method is generally displayed as two circles, one representing a chronic health state and the other representing the gamble as a pie graph *(49–51)*. The individual is given a choice between two alternatives. The first alternative is a definite probability of living in a particular chronic health state for life. The second alternative, or the gamble, depicts the individual returning to normal health and living for an additional number of years or dying immediately. For the gamble, the probability of perfect

The willingness-to-pay approach is based on how much one is willing to pay to avoid sacrificing lives. The two methods of valuing lives under the willingness-to-pay approach are the questionnaire method and the risk premium method. The questionnaire method is self explanatory in that it entails surveying individuals to determine their willingness to pay. The problem with the questionnaire method is that respondents have little incentive to answer truthfully. However, it is easy to obtain data in the format required for analysis because the analyst has control over design of the instrument.

In contrast to the questionnaire method in which individuals are surveyed regarding their willingness to pay, the risk premium method entails observing actual behavior. For example, if people work in riskier jobs, do they really get paid more and what does that say about how they value life? Often people assume high-risk jobs because they have few job opportunities elsewhere. Other examples of high-risk behavior include smoking, drinking alcohol, and eating hazardous foods. Although activities that definitely increase the risk of death are generally intolerable to individuals, activities that may or may not increase the risk of death are apparently not *(42)*.

An alternative method for valuing life is the human capital approach, which is productivity-based and ignores the costs associated with the pain and suffering avoided by averting illness and prolonging life. The term *human capital* refers to the fact that individuals, like capital equipment, can be expected to yield productive activity over their lifetimes that can be valued at their wage rate *(43)*. There are three methods of valuing human life under the human capital approach: discounted future earnings, discounted consumption, and discounted net production. The discounted future earnings method involves discounting to present value all earnings that would be realized as a result of the prolongation of life or the avoidance of a disabling illness. The advantages of the discounted earnings method are that it is reasonably objective and easy to compute. The discounted consumption method calculates a person's value of life by estimating a person's lifetime consumption of goods and services and discounting it back, resulting in a conservative estimate of the value of life. The discounted net production method, a method often used in malpractice suits, combines discounted consumption with discounted earnings. The problem with this method is that the result may be a negative number, because the present value of an individual's future consumption may be more than the present value of an individual's future earnings. Of all the human capital approach methodologies, the discounted net production method clearly results in the lowest estimate of the value of life.

Utility Analysis

Utility Analysis vs CEA

Utility analysis or cost–utility analysis is very similar to CEA and is often treated as a special type of CEA. Relevant benefits include final outcomes, such as years of life saved or days of disability averted. The difference is that in utility analysis these benefits must be converted into QALYs or, as some have suggested, healthy-years equivalents *(44,45)*. Therefore, some benefits that would be included in a CEA, such as cases found or patients correctly treated, cannot be considered in a utility analysis, because they cannot be converted into QALYs. The difference between years of life saved and

1. Define the problem and the objective(s) to be attained.
2. Identify alternative solutions.
3. Identify the costs of solving the problem under each alternative and all relevant benefits.
4. Assign monetary values to the costs and benefits.
5. Discount future streams of costs and benefits to net present value if costs and benefits accrue over a period longer than 1 year.
6. Compare total present values.
7. Interpret the results *(38)*.

The first three steps are identical to those in CEA. However, the assignment of monetary values to all costs and benefits represents a major difference between CBA and CEA. Discussed in more detail in a separate section, CBA requires that a dollar value be assigned to life years saved. This issue has been quite controversial over the years.

The next step is to discount all future streams of costs and benefits to their net present value if costs and benefits accrue over a period longer than 1 year. Discounting is particularly important if one of the alternatives being compared has future costs and benefits and the other does not. The concept of discounting derives from the fact that time makes a difference. A dollar received today is worth more than a dollar to be received next year. This premise is true because a dollar received today could be invested and earn interest so that, by next year, it would be worth $1.05, assuming a 5% interest rate. On the other hand, a dollar to be received next year is only worth $0.95 today, assuming the same rate of interest. The equation for calculating the present value is:

$$PV = FV / (1+r)^t$$

where PV = present value, FV = costs or benefits to be incurred in the future, r = discount rate, and t = the number of years into the future when the costs or benefits are expected to be incurred. The selection of the appropriate discount rate should be a function of the rate of inflation, the perspective of the analysis, and the political process as a means of reflecting social values *(39)*.

The last two steps in the analysis are to compare present values and interpret the results. A benefit–cost ratio can be calculated as the present value of total benefits divided by the present value of total costs. In comparing two interventions, the intervention with the highest benefit–cost ratio would be considered as returning greater benefit per dollar of cost. The difference between the present value of total benefits and the present value of total costs (the net benefit) is another measure commonly used to compare interventions.

Valuation of Life

A controversial issue that often comes up in CBA is how to estimate the value of human life in dollar terms *(40,41)*. This is an extremely difficult task. There are a number of suggested methodologies in the literature, with no single method considered the most correct. The two major approaches for valuing life are the willingness-to-pay approach and the human capital approach. Factors to consider in valuing life include income potential, age, quality of life, number of dependents, productivity, personal preference (religion), and personal habits.

utilization. Indirect benefits are earnings not lost owing to avoidance of premature death or disability. Intangible benefits include pain, discomfort, and grief averted not only by the patient, but by family and friends as well. Depending on the perspective from which the analysis is performed, an item may be a cost in one analysis but a benefit (i.e., a cost averted) in the next.

The next three sections describe in greater depth each of the three types of economic evaluation. CEA is presented first.

Cost-Effectiveness Analysis

To understand CEA, one first needs to understand the term *cost-effective*. For an intervention to be cost-effective, it must be worth the money required to conduct it. Cost-effective is not always synonymous with the terms inexpensive or technically efficient, although the term is often used in this fashion. Cost-effectiveness is based on the concept of opportunity cost. The real cost of an intervention or treatment is the value of the alternative uses of the same resources *(29)*.

The goal in CEA is to determine which alternative intervention or treatment yields the greatest benefits for the lowest cost. There is no requirement that costs and benefits be measured in the same units. Some benefits are measured in nonmonetary units, such as years of life saved or disability days avoided. Indirect economic benefits are generally ignored. Unlike CBA, CEA does not allow a comparison of interventions or treatments with different outcome measures. In addition, it does not generate sufficient results to determine what dollar value per year of life saved is an acceptable level of investment.

The steps in CEA are as follows:

1. Define the problem and the objective(s) to be attained.
2. Identify alternative solutions.
3. Identify the costs of solving the problem under each alternative and all relevant benefits.
4. Compare the alternatives on the basis of prespecified criteria and select the best alternative.

Although CEA is considered a simpler approach than CBA because benefits do not need to be expressed in monetary terms, this does not mean that CEA is without its share of methodological problems. The first difficulty arises if there is more than one benefit and different units of measure apply to each benefit. The benefits are not additive and, therefore, must be analyzed separately. The next problem is how to interpret the results if the separate analyses produce contrary results. A third difficulty in CEA arises when costs and benefits accrue over a period longer than 1 year. Both costs and benefits would need to be discounted to present value. This can easily be achieved for costs, as explained in the CBA section. The problem arises when discounting benefits, because these are not measured in monetary terms.

Cost–Benefit Analysis

Cost–Benefit Analysis vs Cost-Effectiveness Analysis

CBA was previously considered a superior analysis to CEA because of its simplicity in valuing all costs and benefits in dollars. CBA is now considered inferior to CEA by some researchers because it ignores the noneconomic aspects of a program or intervention.

The steps in CBA are as follows:

the analyst to begin calculating costs and benefits. Complexity is not synonymous with accuracy, though models often become quite complex rather quickly.

Warner and Luce *(29)* mention six issues that must be considered in the development of the production model. First, economies of scale may exist that cause fewer and fewer inputs to be required as sample size increases to produce the same level of output per person. For example, an intervention that cost $25,000 for 1000 patients may cost only $35,000 for 5000 patients. Second, if technological change is occurring or is expected, this must be built into the model. If, while projecting future costs of follow-up after implantation of prosthetic devices, using currently available diagnostic tests, preliminary results are published of a new follow-up methodology that may replace one or more of the existing tests, the effect of substituting this test must be factored into the analysis. Third, market characteristics may affect the inputs required to produce a given output, causing the required inputs to vary by such factors as geographic location. For example, geographic differences in rates of pay for health care personnel or variation in the supply of personnel may cause an intervention to be more expensive in one city than in another. Fourth, different populations may respond differently to the same intervention, specifically in terms of compliance. More costly follow-up mechanisms may then be required to achieve the desired effect. Fifth, efficiency cannot always be assumed. The fact that a task has always been done one way does not mean that it is the most efficient. Sixth, some inputs are unique to a particular facility. If attempting to model an intervention at a new facility after an existing one at a different facility, one must examine carefully all inputs to ensure that these inputs are available or can be made available at the new facility.

Once the production function is specified, costs and benefits can be calculated. Costs can be direct, indirect, or intangible. Direct costs are defined as variable costs plus fixed (overhead) costs. In the health sector, the terms direct medical costs and direct nonmedical costs are often used. Direct medical costs are the costs directly related to the provision of care and usually involve monetary transactions, such as physicians' fees, nurses' salaries, drug purchases, equipment purchases, and independent laboratory processing fees. Direct nonmedical costs are costs incurred in the process of seeking care, such as the patient's costs of transportation to the hospital or clinic, parking costs, hotel costs if the patient cannot return home each evening because of distance, the costs of special equipment to modify one's home to accommodate a disabled family member, and child-care costs *(37)*.

Indirect costs are defined as the costs of foregone opportunities. These include the costs of morbidity and mortality. The indirect costs of morbidity are typically measured as time lost from work and the resulting wages foregone or production losses. In addition, morbidity would include the costs associated with an increased risk of complications. Similarly, the indirect costs of mortality can also be measured as time lost from work because premature death causes permanent removal from the workforce. Intangible costs are defined as the psychological costs of illness such as pain, suffering, and grief. These are the most difficult costs to measure.

Benefits can also be divided into the three categories of direct, indirect, or intangible. Benefits are often phrased as savings in costs. Direct benefits are tangible savings in health resource utilization, such as decreased length of stay or diagnostic test

monetary units. CEA is usually the method of choice unless there is no single quantifiable unit by which alternatives can be compared, in which case CBA is the method of choice.

CBA requires the valuation of all outcomes in economic terms, including lives or years of life and morbidity *(32–34)*. This analysis, which is often viewed as a subset of CEA, assumes a goal of economic efficiency. Economic efficiency is defined as providing each unit of output at minimum possible cost *(35)*. In CBA, total costs minus total benefits equals net benefit.

Utility analysis or cost–utility analysis is very similar to CEA and is often treated as a special type of CEA. The main difference between the two is that benefits must be converted into quality-adjusted life years (QALYs) for utility analysis. QALYs is a measure of years of life gained from a procedure or intervention that is then weighted to reflect the quality of life in that year *(36)*.

Subjectivity is an important issue in any type of economic evaluation. Different analysts performing basically the same analysis can easily reach very different conclusions. This can be confusing to the novice. However, the variation in conclusions is tied directly to differences in the assumptions made in the design of the evaluation. Different conclusions do not imply that one analysis is correct and the other incorrect. They just imply that different assumptions were made.

Selection of Perspective

A first step in any economic evaluation is the selection of the perspective from which the analysis will be performed. Such analyses are generally performed from a societal perspective, but other, narrower perspectives may often apply, such as the provider's perspective, the payor's perspective, or the patient's perspective. Which perspective is selected guides the identification of costs and benefits. For example, an insurance company will evaluate a new health prevention from a payor's perspective with respect to the change in total future costs, whereas a societal perspective would assign some inherent value to illness prevented.

Specification of the Problem

The second step in any economic evaluation is specification of the problem, objective(s), and alternatives. This would seem to be rather obvious at first glance. However, if the problem is not well delineated, the range of alternatives selected to address the problem may be too narrow, ignoring important alternatives. For example, if, in the case of patients undergoing implantation of prosthetic devices, the problem is defined as the suffering undergone by current patients, only treatment alternatives will be considered. However, if the problem is defined as affecting both current and future patients, options such as delaying or preventing the onset of device malfunction or infection will also be considered in the analysis.

Production Function

The third step in economic evaluation is describing the production function. In economics, the production function is the relationship between the output of a good or service and the inputs required to produce it. The goal of this step is to specify the resources that would be utilized under each of the alternatives, the way in which the resources would be combined, and the expected result. Completion of this step allows

$$p_1 = \frac{p^* \times \text{FPR} - p^* \times (U[\text{Test}]/(U[D-T-]-U[D-T+]))}{p^* \times \text{FPR} + (1-p^*) \times \text{TPR}}$$

$$p_2 = \frac{p^* \times \text{TNR} + p^* \times (U[\text{Test}]/(U[D-T-]-U[D-T+]))}{p^* \times \text{TNR} + (1-p^*) \times \text{FNR}}$$

where FPR = the false positive rate or 1 – specificity, FNR = the false-negative rate or 1 –sensitivity, TNR = the true-negative rate or specificity, TPR = the true-positive rate or sensitivity, and U [Test] represents the net utility of the diagnostic test as determined by the patient's assessment of the test regarding such factors as cost, potential side effects, the unpleasantness of the test itself, and any reassurance having the test performed provides to the patient (24). (The remaining variables have already been defined.)

Below p_1, treatment is never preferred because the information to be gained from additional testing would not increase the probability of disease sufficiently to cross the treatment threshold. Above p_2 treatment is always preferred because the information gained from additional testing would not decrease the probability of disease sufficiently to cross the treatment threshold. Only for the range of disease probabilities between p_1 and p_2 could an abnormal test result have enough of an influence on disease probability to cross the treatment threshold and change patient management.

The same analysis developed in the discussion of treatment thresholds and expanded in the discussion of testing thresholds can be expanded still further to permit choosing among two or more diagnostic tests or selecting combinations of diagnostic tests. For a more in-depth discussion of these topics, refer to Sox et al. (24). For an application of threshold analysis to cases where a single diagnostic test provides information about more than one event, see Nease et al. (27). For an adaptation of Bayes' nomogram to threshold analysis, see Glasziou (28).

ECONOMIC PRINCIPLES OF FOLLOW-UP EVALUATION

Overview of Cost Analysis

Clinicians, health administrators, and decision makers in general are constantly faced with questions regarding how to appropriate limited resources to cover what seem to be an ever-growing number of health needs. For what illnesses should every patient be automatically screened? How much follow-up is sufficient after primary treatment of a condition? If personnel dollars are short, where should cuts be made and what trade-offs should decision makers be willing to make? The need to clarify the decision-making process and promote efficiency are the reasons economic evaluation (also known as efficiency evaluation) methodologies were developed. The three most widely used methods for assessing the relative merit of alternative courses of action are cost-effectiveness analysis (CEA), cost–benefit analysis (CBA), and utility analysis (29,30). This section of the chapter provides an overview of these methods, followed by a discussion of concepts common to all three. Each of these methods is then discussed in greater detail in subsequent sections, restricting the discussion to differences across methods.

Briefly, CEA places priorities on alternative expenditures without requiring that the dollar value of life and health be assessed (31). Some benefits are measured in non-

Chapter 2 / Surveillance Test Performance and Cost

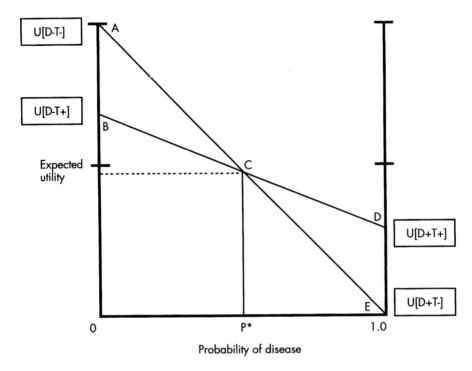

Fig. 3. Treatment threshold or the point of intersection between providing treatment and withholding treatment, whether disease is present or absent. (From ref. *14a.*)

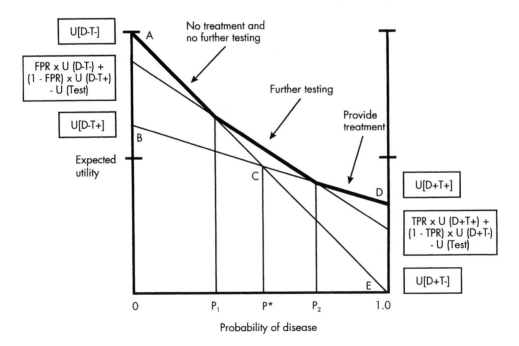

Fig. 4. Depiction of the no-treatment–test threshold, p_1, and the treatment–test threshold, p_2. (From ref. *14a.*)

margin for error is wide, the willingness to make a diagnosis will also be high. However, if the probability of disease is low and the margin for error is slim, the need for more information will be high and the willingness to make a diagnosis will be low.

A term commonly used to describe the dividing line between a decision to treat or not to treat is the treatment threshold. The treatment threshold, p^*, is the probability of disease at which the clinician is indifferent between treating and withholding treatment *(24)*. If the probability of disease for a given patient is above the treatment threshold, treatment will be selected because the acquisition of more information will not change the diagnosis. If the probability of disease is below the treatment threshold, treatment will be withheld and more testing may be ordered (or no action may be taken).

The treatment threshold can be depicted graphically with the probability of disease on the horizontal axis and the expected utility of the treatment on the vertical axis (Fig. 3). Utility is defined here as the value or the level of well-being an individual assigns to a given option. The treatment threshold is calculated by solving for p^* in the following equation:

$$\frac{p^*}{1-p^*} = \frac{U[D-T-]-U[D-T+]}{U[D+T+]-U[D+T-]}$$

where U = utility, $D-$ = absence of disease, $D+$ = presence of disease, $T-$ = withholding treatment, $T+$ = providing treatment, $U[D-T-]$ = the utility of withholding treatment in the absence of disease, $U[D-T+]$ = the utility of providing treatment in the absence of disease, $U[D+T+]$ = the utility of providing treatment in the presence of disease, and $U[D+T-]$ = the utility of withholding treatment in the presence of disease. The line defined by the points A, C, and E represents the utility of withholding treatment irrespective of whether disease is present or absent. The line defined by the points B, C, and D represents the utility of providing treatment irrespective of whether disease is present or absent. The point of intersection between these two lines is the treatment threshold. At this point, the utilities of the two choices are equal.

The above equation can also be rephrased in terms of costs and benefits, still solving for the treatment threshold, p^*. The difference in utility between treating and not treating patients without disease can be considered a cost, C, because no benefit derives from treating these patients. Similarly, the difference between treating and not treating patients with disease can be considered a benefit, B. The previous equation would then be simplified to $p^* = C / (C + B)$ *(24)*.

THRESHOLDS FOR TESTING

Up to this point, only the threshold between treating and withholding treatment has been discussed. There are two other thresholds: the no-treatment–test threshold and the treatment–test threshold *(26)*. The no-treatment–test threshold, p_1, is the probability of disease at which there is indifference between no treatment and further diagnostic testing. The treatment–test threshold, p_2, is the probability of disease at which there is indifference between treatment and further diagnostic testing. A third line can be plotted on Fig. 3 to depict these testing thresholds (Fig. 4). The testing thresholds are calculated as follows:

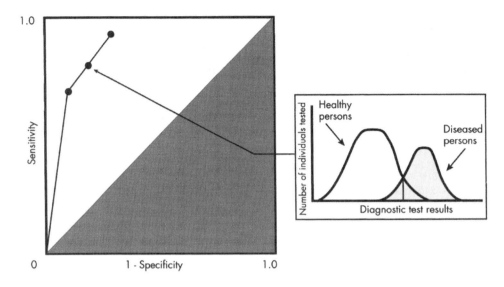

Fig. 2. Receiver operating characteristics curve depicting the trade-off between sensitivity and 1 – specificity. (From ref. *14a*.)

sible cutoffs between normal and abnormal. Derived from signal detection theory, ROC curves plot sensitivity on the vertical axis and 1 – specificity on the horizontal axis (Fig. 2) *(15)*. For all points along the 45-degree line, sensitivity equals 1 – specificity. Points on this line have no impact on the probability of disease. The probability of disease increases for points above the 45-degree line and decreases for points below the line. The points on the curve are calculated as sensitivity / (1 – specificity). Each point on the curve represents a different selected cutoff point between normal and abnormal. The perfect curve would extend straight from the origin to the upper left-hand corner and then over to the upper right-hand corner, maximizing the area under the curve (AUC) *(16)*. The AUC is considered an index of diagnostic performance *(17)*. If two tests are being compared statistically, the test with the greater AUC is considered the better test *(18–23)*. A perfect diagnostic test has an AUC of 1.0. Some authors consider the AUC concept, and ROC curve analysis in general, not very useful because prevalence is not incorporated *(2)*.

ROC curve analysis can also be used to identify the appropriate cutoff point between normal and abnormal *(24,25)*. Clinicians generally use the "upper limit of normal" provided by the laboratory. Sox et al. *(24)* suggest that the ROC curve method is better but its use is severely limited by the time needed to perform the analysis.

THRESHOLDS FOR TREATMENT

Once a decision has been made regarding whether a test is normal or abnormal, the next issue to be dealt with is whether sufficient testing has been completed to make a diagnosis. If no further testing is required, treatment can begin. The goal here is to determine at what point the acquisition of additional information would have no effect on the diagnosis. Major determining factors in this decision are the probability of disease and the penalty for being wrong. If the probability of disease is high and the

and very low test results than on borderline results when attempting to determine the odds that a disease is really present. In comparison to sensitivity and specificity measures, another advantage of likelihood ratios is that diagnostic test performance is quantified as one measure rather than two. A disadvantage of likelihood ratios is that the conversion from probability to odds and back again can be difficult.

REQUIREMENTS FOR ESTABLISHING A SCREENING PROGRAM

There are several major issues identified by Wilson and Jungner (12) that should serve as a prerequisite for the establishment of a screening program. Among these are that the health problem must be important, the disease should have either a latent stage or an early treatable stage, a diagnostic test acceptable to the population should be available, the natural history of the condition should be sufficiently understood, treatment should be available for identified cases, there should be clarity regarding which cases can be curatively treated, and screening should be cost effective.

RECEIVER OPERATING CHARACTERISTIC CURVE

Once the need for a screening program is established and the appropriate diagnostic test is selected, the next step is to clarify how test results will be interpreted. Complicating the situation is that factors such as age, sex, race, and nutrition can all impact laboratory test results. For example, what is normal for a 70-year-old male may not be normal for a 25-year-old female. Although what is considered normal can vary by patient, the distribution of clinical measurements for an individual is generally represented by a normally distributed (bell-shaped) curve (13). The dispersion of values around the mean in a normal distribution is due to random variation alone.

In addition to variation across subjects in terms of what is normal and abnormal, the cutoff between normal and abnormal for a given diagnostic test can be varied given the goals of the particular screening intervention. If the goal is to correctly identify, for example, 95% of all cases of disease, the range of values constituting an abnormal test result can be expanded until this goal is reached. Unfortunately, doing so causes the number of FP to increase, thus decreasing specificity, because sensitivity and specificity are inversely related. Similarly, if the goal is to correctly identify 95% of all cases without disease, the range of values constituting a normal test result can be expanded until this goal is reached. However, increasing specificity is achieved at the expense of decreasing sensitivity.

When diagnostic test results by patient are depicted graphically for both healthy and diseased individuals (Fig.1), there is usually a range of values that is clearly normal and another range of values that is clearly abnormal. However, there is also a range of values that could easily represent either normal or abnormal results, as depicted by the overlapping bell-shaped curves. Figure 1 depicts how the selected cutoff between normal and abnormal determines the sensitivity and specificity of a test.

Receiver operating characteristic (ROC) curves are used to depict the trade-off between TPR, or sensitivity, and FPR, or 1 – specificity (14). Unlike the limited information provided by a single estimate of sensitivity and specificity for one possible cutoff point between normal and abnormal, ROC curves are more useful, because they depict the complete range of all possible TPR/FPR trade-offs corresponding to all pos-

test. Variation related to the interpretation of the results can be classified into two types. Intraobserver variation is variation caused by one person interpreting the results differently on different occasions. Interobserver variation is variation across different persons interpreting the results *(9)*. Such variation can be substantially reduced through training seminars and the use of independent observations on a subsample of cases.

Yield

The third important characteristic of screening tests is yield, which refers to the number of cases with previously undiagnosed disease that are detected and treated as a result of the screen. Yield is affected by the sensitivity of the diagnostic test, the prevalence of unrecognized disease, whether the screening is multiphasic (multiple diagnostic tests were administered), screening frequency, and the number of positive screens who actually receive treatment *(1)*. The effect of sensitivity on yield is that, if few TPs are identified, the other factors become immaterial, because yield will be low. If the prevalence of unrecognized disease is low, because of such factors as high medical care availability or a recent screen of the population, the yield will be low.

The ability to identify risk factors for the disease and narrow down the number of individuals who must be screened will increase yield. Another way to increase yield is through multiphasic screening in which a variety of tests are used to screen for multiple conditions during one visit.

Frequency of Screening

On the issue of frequency of screening, the literature is not very clear in many instances. Frequency should be dictated by the natural history of the disease, the incidence of disease, and risk factors. Whether a patient with identified disease will consent to treatment is determined by whether the patient views there to be a serious threat to health, whether the patient feels vulnerable, and whether the patient decides that seeking treatment will be beneficial *(1)*.

LIKELIHOOD RATIOS

Another way to measure the performance of a diagnostic test, which has not yet been discussed, is through the use of likelihood ratios. To understand likelihood ratios, which are a type of odds ratio, the difference between probability and odds must be clear. Probability ranges from zero to 1 and measures the likelihood that a particular outcome will occur. A value close to zero indicates little chance of occurrence; a value close to 1 indicates a large chance of occurrence. If an experiment is conducted n times and the event of interest occurs m times, the probability of that event occurring is calculated as m/n *(10)*. Sensitivity and specificity are both measures of the probabilities of specific events occurring.

Odds are ratios of two probabilities and are calculated as the probability of an event / (1 – the probability of an event) *(7)*. One can also work backward and calculate probability from odds using the following equation: odds / (1 + odds). Likelihood ratios measure how much more likely it is that a diagnosis will be made in the presence of disease as in the absence of disease and can be defined for any number of test results over the entire range of possible values. For positive and negative test results, the respective likelihood ratios are sensitivity / (1 – specificity) and (1 – sensitivity) / specificity *(11)*. Use of likelihood ratios has the advantage of placing more weight on very high

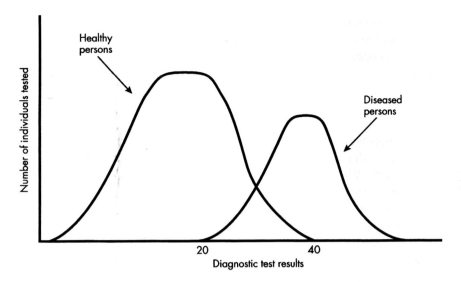

Fig. 1. Distribution of diagnostic test results.

redrawn so that 40 and above is considered abnormal and all other values are considered normal, the screening intervention will have low sensitivity and high specificity because all patients with diagnostic test results in the 39 and below range will be treated as TN.

Other factors that influence measurement of the validity of a test are severity of disease and the presence of co-morbid conditions. With some diagnostic tests, such as the serological test for syphilis, the probability of FN is very high in the early or very late stages of disease *(1)*. The presence of co-morbid conditions and drugs taken for these conditions can also greatly influence diagnostic test results.

The ability of a diagnostic test to correctly discriminate between the presence or absence of disease is also dependent on the prevalence of disease, in addition to a test's specificity and sensitivity. The greater the prevalence of disease, the greater the predictive value (PV) of a positive test, which is the probability that a positive test result is accurately predicting disease. As prevalence approaches zero, the PV of a positive test approaches zero. The PV of a positive test is calculated as $a / (a + b)$ or TP / (TP + FP). According to Bayes' theorem of conditional probabilities, the PV of a positive test can also be calculated as (sensitivity × prevalence) / [(sensitivity × prevalence) + (1 − specificity) × (1 − prevalence)] *(7)*. The PV of a negative test is $d / (c + d)$ or TN / (FN + TN). According to Vecchio *(8)*, the prevalence of a disease must be at least 15 to 20% to reach an acceptable PV (70–80%).

Reliability

The second important characteristic of screening tests is reliability or precision. Reliability measures whether the same test administered more than once to the same person will produce the same results repetitively. The two types of variation that can occur are variation in the method itself and variation related to the person(s) interpreting the results. Variation in method can be the result of mechanical fluctuations (fluctuations in the testing apparatus) or fluctuations in the substance being measured by the diagnostic

Table 1
Derivation of Sensitivity and Specificity

Screening test result	Disease category	
	Disease present	Disease absent
Positive	a (TP)	b (FP)
Negative	c (FN)	d (TN)

Sensitivity = a / (a + c) = TP / (TP + FN)
Specificity = d / (b + d) = TN / (FP + TN)
TP, true-positive; FN, false-negative; TN, true-negative; FP, false-positive.

Table 1 depicts the derivation of sensitivity and specificity *(2)*. Patients who are correctly predicted by the diagnostic test of interest to have disease are referred to as true-positives (TP). Similarly, those patients who are correctly predicted to be disease free are referred to as true-negatives (TN). Those patients falsely predicted to have disease are false-positives (FP). Those patients falsely predicted to be disease free are false-negatives (FN). Once the sensitivity and specificity of a diagnostic test are known, clinicians can use these estimates to revise original estimates of the probability of disease made prior to the ordering of a diagnostic test (pretest probability). According to a principle known as Bayes' theorem, posttest probability can be calculated as:

$$P_r = \frac{(P_i)(\text{sensitivity})}{(P_i)(\text{sensitivity}) + (1-P_i)(100\% - \text{specificity})}$$

where P_r is the posttest probability and P_i is the pretest probability. Tabular and graphic expressions of Bayes' theorem are also available, such as Bayes' nomogram and Benish's tables, which permit one to look up the posttest probability once the pretest probability, sensitivity, and specificity are known *(3,4)*.

Sensitivity and specificity are derived by comparing the results from the test in question (the index test) with those of a definitive test (a gold standard test). Irrespective of the results of the screen (positive or negative), in most cases, every person screened must be tested using the gold standard to establish or rule out disease *(5,6)*. The optimal test would be 100% specific and 100% sensitive. Unfortunately, this will not be observed in practice because sensitivity and specificity are usually inversely related. In other words, sensitivity can be improved, but only at the expense of specificity, and specificity can be improved, but only at the expense of sensitivity.

To understand this, consider that range of diagnostic test results that can be considered either normal or abnormal, as depicted by the overlapping bell-shaped curves in Fig. 1. If the range of overlapping values is 20 to 40 and the line distinguishing normal from abnormal test results is drawn so that 20 and above is considered abnormal, the screening intervention will have high sensitivity because all patients with diagnostic test results in the 20 and above range will be treated as TP. However, the intervention will have low specificity because many of the results treated as positive will turn out to be FP. Alternatively, if the line distinguishing normal from abnormal test results is

EPIDEMIOLOGICAL PRINCIPLES FOR EVALUATING DIAGNOSTIC TEST PERFORMANCE IN SCREENING FOR DEVICE MALFUNCTION OR INFECTION

Overview

Epidemiology is the study of the frequency and determinants of disease and injury in human populations *(1)*. Whereas clinical medicine focuses on the delivery of medical care to patients, epidemiology analyzes why different populations have differing incidences and prevalences of disease. Incidence refers to the probability that individuals without disease will develop disease over a given period of time and is calculated as the number of new cases of disease divided by the population at risk. Prevalence refers to the number of people in a population who already have the disease and is calculated as the number of existing cases of disease divided by the total population. Clinical epidemiology focuses on the application of epidemiological principles to the practice of clinical medicine. This section uses basic principles of epidemiology to assess diagnostic test performance in screening for device malfunction or infection.

Screening

Asymptomatic patients rarely seek care unless participating in a regular surveillance program. When symptoms do appear, patients often delay seeking care for an extended period, during which time the condition worsens. It is generally believed that early detection through the use of screening tests improves the probability of repairing device malfunction or treating infection and reduces the probabilities of both death and disability.

Screening is the use of tests or examinations to distinguish asymptomatic individuals with a high probability of disease from asymptomatic individuals with a low probability of disease. Some screening programs are designed to identify individuals who might not have disease now but who have a high probability of developing it in the future. Screening tests are usually quick, minimally invasive, and inexpensive. Usually, screening or surveillance is performed among populations who have not previously been diagnosed with the disease under evaluation. However, the term *surveillance* can also be used to refer to the follow-up of patients after implantation of prosthetic devices to detect device malfunction or infection. Although the term *screening* is used throughout this section, the same concepts apply to both screening and surveillance.

DIAGNOSTIC TEST CHARACTERISTICS

Validity

Important characteristics of screening tests are validity, reliability, and yield. Validity refers to the ability of a test to distinguish between those who have disease and those who do not. The two measures of validity are sensitivity and specificity. Sensitivity, often referred to as the true positive rate (TPR), measures the ability of the test to correctly identify those who actually have disease. Sensitivity is calculated as the percentage of all patients with disease who screen positive for disease. Specificity, also referred to as the true negative rate (TNR), measures the ability of the test to correctly identify those who do not have disease. Specificity is calculated as the percentage of all patients without disease who screen negative for disease *(1)*.

2
Assessment of Surveillance Test Performance and Cost

Katherine S. Virgo

Surveillance test performance and costs are important considerations for a book summarizing the state of the art in patient management after implantation of prosthetic devices for several reasons. First is the growing concern about more efficient use of limited resources, made ever more apparent by the ongoing health care reform debate. Second is the expanding role of "gatekeepers" wielding increasing control over the total cost of health care. Third is the push for clinicians to develop guidelines, driven by the idea that reimbursement could subsequently be tied to adherence to these guidelines.

Given such issues and the increasing cost of high-tech prosthetic devices, one would expect to see a plethora of articles assessing various follow-up strategies for efficacy, efficiency, and cost-effectiveness, but few exist. One reason for the shortage of articles may be that the appropriate patient-management strategy after implantation is not well delineated for many devices. Much of the existing literature on patient follow-up after implantation of prosthetic devices consists of articles that suggest strategies based on either inadequate sample size or data from a single institution. Very few proposed patient-management strategies are based on the results of large retrospective analyses of secondary data sets or prospective, randomized clinical trials.

Another reason for the shortage of articles may be that few clinicians sufficiently understand epidemiological and cost-analysis methodologies. Therefore, this chapter provides a review of the tools needed to weigh alternative follow-up strategies against one another. The epidemiology section describes how to determine whether screening for disease is appropriate in a given population, assess the performance of individual diagnostic tests, compare performance across diagnostic tests, and determine whether further diagnostic testing is required. The economics section specifies how to calculate and compare the costs and benefits of individual diagnostic tests or entire follow-up strategies.

From: *The Bionic Human: Health Promotion for People With Implanted Prosthetic Devices*
Edited by: F. E. Johnson and K. S. Virgo © Humana Press Inc., Totowa, NJ

2094 Gaither Road
Rockville, MD 20850
Fax: 301-594-4792

REFERENCES

1. Wangensteen OH, Wangensteen SD. The rise of surgery. Minneapolis: University of Minnesota Press, 1978.
2. Rosenberg SA. Principles of cancer management: surgical oncology. In: DeVita VT, Hellman S, Rosenberg SA, eds. Cancer—principles and practice of oncology (5th ed.). Philadelphia PA: Lippincott-Raven, 1997.
3. US Federal Food, Drug, and Cosmetic Act. 61 Fed. Reg. 44,396, et seq. (1996) Federal Food, Drug and Cosmetic Act (FFDCA) 21 U.S.C. 301.
4. US General Accounting Office. Medical devices—early warning of problems is hampered by severe underreporting. GAO Publication PEMD-87 1: US Government Printing Office, Washington, DC, 1986.
5. Anonymous. Improving patient care by reporting problems with medical devices. MedWatch Continuing Education Articles 1997;9:1 (US FDA, Rockville, MD).
6. Maki DG, Stolz SM, Wheeler S, Mermel LA. Prevention of central venous catheter-related bloodstream infection by use of an antibiotic-impregnated catheter. A randomized, controlled trial. Ann Int Med 1997;127:257–266.
7. Angell M. Science on trial. New York: W.W. Norton, 1996.
8. Nelson N. Institute of Medicine finds no link between breast implants and disease. J Nat Cancer Inst 1999;91:1191.
9. Rosenbaum JT. Lessons from litigation over silicone breast implants: a call for activism by scientists. Science 1997;276:1524–1525.
10. Gabriel SE, Woods JE, O'Fallon WM, Beard CM, Kurland LT, Melton LJ. Complications leading to surgery after breast implantation. N Engl J Med 1997;336:677–682.
11. Grady D. Cosmetic breast enlargements are making a comeback. The New York Times, July 21, 1998, p. B13.
12. Dajani AS, Taubert KA, Wilson W, et al. Prevention of bacterial endocarditis. Recommendations of the American Heart Association. JAMA 1997;277:1794–1801.
13. Bruck SD. Journal of Long-Term Effects of Medical Implants, Begell House Inc.
14. Vacanti C, Mikos A, eds. Tissue engineering. Larchmont, NY: Mary Ann Liebert.
15. Anonymous. 2000 Medical device register. Montvale, NJ: Medical Economics, 2000.
16. Sugarman B, Young EJ, eds. Infections associated with prosthetic devices. Boca Raton, FL: CRC Press, 1984.
17. Wise DL, Trantalo DJ, Lewandrowski K-U, Gresser JD, Cattaneo MV, Yaszemski MJ, eds. Biomaterials engineering and devices: human applications. Vol. I: Fundamentals, vascular, and carrier. Vol. 2: Orthopedic, dental and bone graft applications. Totowa, NJ: Humana Press, 2000.
18. Johnson FE, Virgo KS. Cancer patient follow-up. St. Louis, MO: Mosby, 1997.
19. McCarthy MJ, Olojugba DH, Loftus JM, Naylor AR, Bell PRF, London NJM. Lower limb surveillance following autologous vein bypass should be life long. Brit J Surg 1998;85:1369–1372.
20. Patel ST, Kuntz KM, Kent KC. Is routine duplex ultrasound surveillance after carotid endarterectomy cost-effective? Surgery 1998;124:343–352.
21. Naunheim KS, Virgo KS, Coplin MA, Johnson FE. Clinical surveillance testing after lung cancer operations. Ann Thorac Surg 1995;60:1612–1616.
22. Oh J, Colberg JW, Ornstein DK, et al. Current follow-up strategies after radical prostatectomy: a survey of American Urological Association urologists. J Urology 1999;161:520–523.

APPENDIX *(Continued)*

FDA Regional Offices

The Public Affairs Specialists of the FDA can provide information about local resources as well as the FDA's role in regulating implants. To listen to the most recent summary of FDA information, contact FDA's Office of Consumer Affairs Information Line at 1-888-INFOFDA, 24 hours a day. You may also speak with a Consumer Affairs Specialist by calling this number between the hours of 10 A.M.–4 P.M. eastern time, Monday through Friday.

Current Bibliographies in Medicine

The National Library of Medicine (NLM) is pleased to offer publications in its *Current Bibliographies in Medicine* (CBM) series free of charge through the World Wide Web. Bibliographies in the CBM series are produced by staff of NLM's Reference Section in collaboration with subject specialists from the National Institutes of Health and elsewhere. Each bibliography is prepared by searching a variety of online databases and covers a separate topic of current interest. The result is a subject-categorized list of citations to the recent literature, primarily journal articles and books. CBMs may be retrieved from the Library's website at http://www.nlm.nih.gov/pubs/resources.html
Examples of publications that may be of interest include:
92-6 *Silicone Implants*, citations from the literature from January 1989 through August 1992.
94-8 *Silicone Implants*, 825 citations from August 1992 through August 1994.
96-5 *Breast Cancer Screening* in Women Ages 40–49, 334 citations from January 1985 through November 1996.
96-1 *Cervical Cancer*, 926 citations from January 1993 through March 1996.
Questions on the CBM series may be answered by calling 1-888-FINDNLM or by sending an E-mail to ref@nlm nih.gov.

FDA Freedom of Information

If you need more technical information, please submit a written request to the FDA Freedom of Information staff.
The Freedom of Information Act allows anyone to request FDA records.
Your letter should include your name, address, and telephone number, and a statement of the desired records, identified as specifically as possible. A request for specific information that is releasable to the public can be processed much more quickly than a request for all information on a particular subject. There are fees for searching and reviewing the information plus a charge of 10 cents per page. You are billed after your request for information has been filled. Agency documents regarding silicone-gel and saline-filled breast implants are available from FDA and may be obtained by submitting a written request to the following address:

Food and Drug Administration
Freedom of Information Staff (HFI-35)
5600 Fishers Lane
Rockville, MD 20857
Fax: 301-443-1726
Voice Mail Message: 301-827-6500

If you have submitted a request and you have questions relating to its status your Freedom of Information request, please write to:

Freedom of Information Staff (HFZ-82)
Center for Devices and Radiological Health
Food and Drug Administration

(continued)

APPENDIX *(Continued)*

Plaintiffs' Steering Committee
MDL-926
410 PNC Bank Tower
4th and Vine Streets
Cincinnati, OH 45202
Phone: 205-252-6784
E-mail: schilds800@aol.com

Public Citizen Health Research Group
1600 20 Street, N.W.
Washington, D.C. 20009
Scleroderma Foundation
89 Newbury Street
Danvers, MA 01923
Phone: 978-750-4499
Fax: 978-750-9902
URL: http://www.scleroderma.com

Federal Government

Institute of Medicine
National Academy of Sciences
    Committee on the Safety of Silicone
    Breast Implants
2101 Constitution Avenue, N.W.
Washington, D.C. 20418
URL: http://www2.nas.edu/hpdp/22f6/htm

National Cancer Institute
Office of Cancer Communications
Building 31, Room 10A-24
9000 Rockville Pike
Bethesda, MD 20892
Phone: 1-800-4-CANCER (422-6237)

FDA, Center for Devices
    and Radiological Health
Office of Health and Industry Programs
HFZ-210
1350 Piccard Drive
Rockville, MD 20850
FDA, Center for Devices
    and Radiological Health
Division of Mammography Quality
    and Radiation Program
1350 Piccard Drive, HFZ-240
Rockville, MD 20850

Silicone Scene
1050 Cinnamon Lane
Corona, CA 91720

Y-ME National Breast Cancer Organization
212 West Van Buren
Chicago, IL 60607-3908
Phone: 1-800-221-2141

US Food and Drug Administration
Office of Consumer Affairs
5600 Fishers Lane, Rm. 16-59
Rockville, MD 20857
Phone: 1-800-532-4440

US Food and Drug Administration
Office of Women's Health
5600 Fishers Lane, Rm. 14-62
Rockville, MD 20857

Medicare Hotline
Phone: 1-800-638-6823

MDL926 Breast Implant Litigation (for attorneys)
URL: http://www.fjc.gov/breimlit/mdl926.htm

(continued)

APPENDIX (Continued)

CHEMOcare
231 North Avenue West
Westfield, NJ 07090-1420
Phone: 1-800-552-4366

Command Trust Network
256 South Linden Drive
Beverly Hills, CA 90212
East Coast Connection
7000 Boulevard East 21A
Guttenberg, NJ 07093
Phone: 502-897-2774
Fax: 502-893-6200

CANDO (Chemically Associated
   Neurological Disorders)
P.O. Box 682633
Houston, TX 77268-2633
Phone: 281-444-0662
Fax: 281-444-5468

National Alliance of Breast Cancer
   Organizations
9 East 37th Street, 10th Floor
New York, NY 10016
Phone: 212-889-0606

National Breast Implant Task Force
P.O. Box 210503
West Palm Beach, FL 33414
Phone: 561-791-2625
Fax: 561-791-4419

National Chronic Fatigue Syndrome
   and Myalgia Association
3521 Broadway
Kansas City, MO 64114
Phone: 816-931-4777

National Women's Health Network
514 Tenth Street, N.W.
Suite 400
Washington, D.C. 20004
Phone: 202-347-1140

Implant Survivors
2425 Parental Home Road
Jacksonville, FL 32216

Kentucky Women's Health Network
P.O. Box 5471
Louisville, KY 40255-0471

La Leche League
1400 North Meachum Road
Schaumburg, IL 60173-4840
Phone: 708-519-7730
Fax: 708-519-0035

Lupus Foundation of America
1717 Massachusetts Avenue, N.W.
Washington, D.C. 20026
Phone: 1-800-558-0121

Support in Silicone Incorporation
P.O. Box 641255
Kenner, LA 70064

Toxic Discovery Network, Inc.
1906 Grand Lane
Columbia, MO 65203
Phone: 573-445-0861
Fax: 573-445-8539

United Scleroderma Foundation
P.O. Box 399
Watsonville, CA 95077-0399
Phone: 1-800-722-HOPE

United Silicone Survivors of the World
Houston Chapter
12615 Misty Valley
Houston, TX 77066
Phone: 281-448-9760 or 281-350-5634
Fax: 281-448-4330

WASP-Wisconsin
Chetek, WI 54728
Phone: 715-924-3691

(continued)

APPENDIX *(Continued)*

Physician and Nursing Groups

American Academy of Cosmetic Surgery
401 North Michigan Avenue
Chicago, IL 60611-4267
Phone: 312-527-6713
URL: http://www.facs.org

American Medical Association
515 North State Street
Chicago, IL 60610
Phone: 312-464-4370
Fax: 312-464-5896
Fax: 202-554-2262

American Society of Plastic and
   Reconstructive Surgeons
444 E. Algonquin Road
Arlington Heights, IL 60005
Phone: 1-800-635-0635

Consumer and Patient Information

Arthritis Foundation
P.O. Box 7669
Atlanta, GA 30357-0069
Phone: 404-872-7200, Ext. 6350
Phone: 1-800-283-7800
URL: http://www.arthritis.org

American Cancer Society
   Reach to Recovery
Phone: 1-800-ACS-2345
Phone: 1-800-525-3777

Boston Women's Health Book Collective
P.O. Box 192
West Somerville, MA 02144
Phone: 617-625-0271
Phone: 512-837-5254

Children Afflicted by Toxic Substances
   (CATS)
60 Oser Avenue
Suite 1
Hauppauge, NY 11788
Phone: 619-270-0680
E-mail: ilena@san.rr.com

American College of Surgeons
633 N. Saint Clair Street
Chicago, IL 60611-7211
Phone: 312-202-5000

American Nurses Association
600 Maryland Avenue, S.W.
Suite 100W
Washington, D.C. 20024-2571
Phone: 202-554-4444, Ext. 270

American Society of Plastic
   and Reconstructive Surgical Nurses
East Holly Avenue, Box 56
Pitman, NJ 08071-0056
Phone: 609-256-2340

National Medical Association
1012 10th Street, N.W.
Washington, D.C. 20001
Phone: 202-347-1895
Fax: 202-842-3292

American Cancer Society
1599 Clifton Road N.E.
Atlanta, GA 30329
Phone: 404-320-3333

AMC Cancer Research Center
1600 Pierce Street
Denver, CO 80214

Central Texas Silicone Implant
   Support, Inc.
1900 A Gracy Farms Lane
Austin, TX 78578

Humantics Foundation for Women
Breast Implants:
   Recovery and Discovery
1380 Garnet #444
San Diego, CA 92109

Implant Information Foundation
P.O. Box 2907
Laguna Hills, CA 92653

(continued)

efit of follow-up *(21)*. Many physicians believe that follow-up minimizes the risk of medical malpractice suits *(22)*.

The follow-up encounter with a patient allows for risk counseling. In some instances, where the prosthesis has been implanted for a congenital condition, the risk counseling may extend to the family members. For example, many breast prostheses are placed after surgery for breast cancer. Relatives of breast cancer patients are themselves at higher than average risk for breast cancer. Cochlear implants are placed for deafness. This is often a hereditary condition and genetic counseling can be offered during follow-up visits. Patients with vascular prostheses may be counseled regarding lipid lowering, smoking cessation and weight reduction. Other examples will undoubtedly come readily to mind.

Finally, this book will touch on optimal prosthetic design to minimize risks of complications and prosthesis malfunction. This is a societal concern which draws in politics and regulatory laws. It affects businesses and their quality control processes. Thus, this topic has remarkably wide implications and the editors believe that this book will fill an important niche in the medical literature.

APPENDIX

Breast Implant Resource Groups

Current Manufacturers of Silicone Gel-Filled Implants

McGhan Medical Corp.
700 Ward Drive
Santa Barbara, CA 93111-2936
Phone: 1-800-862-4426

Mentor Corp.
5425 Hollister Avenue
Santa Barbara, CA 93111
Phone: 1-800-525-0245

Former Manufacturers

Baxter Healthcare Corp.
1385 Centennial Drive
Deerfield, IL 60015
Phone: 1-800-323-4533

Bioplasty, Inc.
1385 Centennial Drive
St. Paul, MN 55113
Phone: 1-800-328-9105

Dow Corning Corporation
P.O. Box 994
Midland, MI 48686-0994
Phone: 1-800-442-5442

Medical Engineering Corp.
(A Bristol-Myers Squibb Company)
2317 Eaton Lane
Racine, WI 53404
Phone: 414-632-3717

Porex Technologies
500 Bohannon Road
Fairburn, GA 30213
Phone: 1-800-241-0195

Surgitek (replaced by Medical Engineering Corp.)
3037 Mt. Pleasant Street
Racine, WI 53404

Current Manufacturers of Saline-Filled Implants

Hutchinson International, Inc.
7949 Jefferson Highway
Baton Rouge, LA
Phone: 504-927-6800

McGhan Medical, Inc.
700 Ward Drive
Santa Barbara, CA 93111
Phone: 1-800-862-4426

Mentor Corp.
5425 Hollister Avenue
Santa Barbara, CA 93111
Phone: 1-800-525-0245

Poly Implants Prostheses/USA
75 NE 39th Street
Miami, FL 33137
Phone: 305-573-6700

(continued)

The editors have searched the medical literature but found no textbooks on the topic. Suggestions regarding care of patients with implanted medical devices are sprinkled throughout the medical literature. There is a relatively new journal on a related topic *(13)* and another has recently been announced *(14)*. The Medical Devices Register is a multivolume resource listing every supplier licensed to sell these items in the United States *(15)*. There are more than 12,000 companies and 65,000 products covered. One volume deals with non-US devices and suppliers. Other books may be of value also *(7,16,17)*.

The topic of postoperative surveillance of asymptomatic patients is receiving increased attention. Textbooks devoted to the topic of cancer patient follow-up, for example, are available *(18)* and the goals of follow-up have been clarified. They include detection of prosthetic device failure at a time when intervention can be successful, for example, infection of a vascular prosthesis or mechanical failure of a heart valve. Another goal of follow-up is detection of progression of the disorder for which the prosthesis was placed. For example, the patient with a vascular prosthesis typically has advanced atherosclerosis. What is the role of surveillance for other clinically important manifestations of this systemic disease that, after all, is not cured by the vascular prosthesis? McCarthy et al. reported a prospective study of patient surveillance using color duplex ultrasound scan plus office visit on a regular schedule for patients who had received infrainguinal arterial bypass surgery *(19)*. The patient population of 326 patients was shown to have a hazard of new graft stenosis or arterial stenosis of 10% each year. The authors concluded that lifelong surveillance is warranted, citing other similar trials reaching similar conclusions. Others have examined the cost-effectiveness of routine surveillance with imaging tests after vascular surgery and concluded that this practice is not warranted *(20)*.

Another important, but often overlooked, goal of follow-up is the detection of other conditions of medical significance. For example, routine surveillance of patients with intraocular lens implants may detect glaucoma at an early and treatable stage. Glaucoma may be unrelated to the intraocular lens implant but is a common medical condition in older patients who are the usual recipients of intraocular lens implants. Patients with cardiac pacemakers may benefit from a general cardiac evaluation when they have pacemaker function assessment.

Another important goal of follow-up is to provide patient education. Individuals with artificial joints, for example, although they may be able to lead a relatively normal life, should avoid certain activities likely to lead to mechanical failure or prosthesis dislocation. Individuals with cardiac pacemakers should be wary of entering areas with a strong magnetic field. In addition, follow-up can provide important evidence regarding adverse outcomes and areas for technical improvement in devices. Thus, the creation of registries of patients to detect complications is an important and growing undertaking. An example of this is the detection of catastrophic failure in Björk-Shiley Convexo-Concave cardiac valves. This is a rare but life-threatening event. Quantitation of the dimensions of risk has been greatly aided by registry data. Patient follow-up is also useful for rehabilitation, psychological support, and maintenance of rapport with the patient. Maintenance of rapport with referring physicians can be an important ben-

try may travel to another country to have an implant, suffer a complication of the implant on returning home, but may seek legal redress in the country where the manufacturer of the device is located, or perhaps in a country where the rewards of a settlement are likely to be greatest. International law has had to deal with these issues. Legal proceedings, particularly class action lawsuits, can be lengthy and expensive. This book does not deal with such legal issues in great depth as they are dealt with elsewhere. In many cases, individuals who suffer bad outcomes from their implants have recourse to support groups as well. These are often organized informally but many have become large enough and are well enough funded to become nonprofit organizations. The authors of each chapter have been asked to list support groups, governmental agencies, nonprofit organizations, and so on, that can provide assistance to those who want more information. Many groups are specific for a particular type of prosthesis and/or disorder, but not well known, so we include details of importance to readers such as street address, telephone number, fax number, E-mail address, Internet address, and the like. Major organizations providing support and information have been compiled by the FDA. Some are listed in the appendix to this chapter. This compilation is regularly updated. This document is also available on the Internet at the following URL: http://www.fda.gov/cdrh/breastimplants/indexbid.html. State departments of health may abstract portions of the FDA database and republish it. The Missouri Department of Health recently released a 65-page booklet concerning breast implants, for example, derived largely from information collected by the FDA.

What is not so clear is how to manage these patients. In some areas of medicine, practice is based on high-quality evidence and is well accepted despite its complexity. Perhaps the best example concerns the prevention of bacterial endocarditis in patients with prosthetic heart valves. The American Heart Association has devoted a large amount of effort to this topic, including liaison with the American Dental Association, the Infectious Diseases Society of America, the American Academy of Pediatrics, and the American Society for Gastrointestinal Endoscopy *(12)*. Funding for research dealing with this topic has been similarly impressive and has been provided by the National Institutes of Health, pharmaceutical corporations, and manufacturers of valves, sutures, and other devices used during the implantation. Providing care for patients with other types of prostheses is often haphazard, embodying intuition, extrapolation from other medical conditions or other devices, and tradition. In most cases, well-designed clinical trials are lacking.

The reasons for this disparity are not clear. Perhaps the more devastating the complication, the more research is stimulated. Infection of a heart valve is typically more dramatic and more often lethal than infection of prosthetic mesh used in a hernia repair, for example, and this presumably has much to do with the disparity in the quality of evidence guiding practice. Perhaps fear of lawsuits and unfavorable publicity motivates other research. This book is designed to provide information on how to care for patients with implanted prosthetic devices that are working well. It is intended to be a resource providing an in-depth analysis of practices at prestigious medical centers throughout the world, including a justification of the practices employed. We discuss care of patients who are asymptomatic and who have recovered from the implantation procedure. The focus is not on patients who have overtly malfunctioning prostheses.

eliminating a risk factor is another way to avoid infections. A patient with ulcerative colitis facing implantation of a mechanical heart valve may be offered colectomy to avoid a source of chronic bacteremia, for example, as this could lessen the chance of developing a later valve infection.

For other alleged complications, such as connective tissue disease in patients with breast implants, causality has been difficult to prove *(7,8)*. Some of the problems putatively attributable to implanted medical appliances are probably related to other causes. The authors of this book discuss this in other chapters.

In all medical interventions, particularly invasive ones, bad outcomes can occur. The explosive increase in the use of prosthetic devices is evidence that many patients view the benefits as outweighing the risks. Nonetheless, the consequences of prosthetic failure can be disastrous. This has created opportunities for lawyers who pursue financial compensation for their clients. The prospect of enormous settlements has created a niche for some lawyers who may pursue a legal claim for years or even decades. Although such lawsuits are often well publicized, we often forget details of their magnitude and ramifications. The story of silicone breast implant litigation provides an excellent case study.

Silicone is a term given to a family of synthetic polymers of silicon, oxygen, and carbon. The details of the polymer chosen specify whether it will be liquid, solid, or gel at body temperature, and various sorts are used in items as diverse as artificial joints, antacids, cardiac valves, and testicular prostheses. Breast implants typically employ a flexible silicone envelope containing saline. Other silicone polymers or natural oils have been used to fill the envelope also. Breast augmentation using these implants began in 1962 *(9)* and about 2 million women have received them in the United States alone *(8)*. Presumably, several million more women have had implants throughout the world. Complications prompting additional surgery are common. In a well-publicized study of 749 women from Olmsted County, Minnesota, Gabriel et al. reported that 208 underwent 450 additional procedures related to their breast implants during a follow-up period of 25.8 years *(10)*. Lawsuits have been filed for local complications (poor cosmetic result, infection, pain, etc.) and systemic complications (autoimmune disease, cancer, etc.). An award in an individual lawsuit of $7 million was recorded in 1991 *(7)*. Media attention to this issue prompted the commissioner of the FDA to impose a moratorium that restricted breast implant surgery dramatically. The number of implants in the United States was estimated at 120,000 to 150,000 per year in 1990. This dropped to about 33,000 per year by 1992, the year following the moratorium, but rebounded to 122,000 per year by 1997 *(11)*. A tidal wave of lawsuits inundated surgeons, implant manufacturers, hospitals, and various other entities. Rosenbaum estimated that $40 to $60 billion was at stake in this process and provided details of the surprising ramifications of this specific problem *(9)*.

There are related implications for business, of course. Classes of implants such as intrauterine contraceptive devices are no longer available in the United States because of these financial issues. Complications allegedly arising from intrauterine contraceptive devices and silicone-filled breast implants have given rise to lawsuits that have even driven large companies to bankruptcy. Some are listed later in this chapter. This, in turn, reverberates throughout the world economy. For example, a citizen in one coun-

call to yet another bureaucrat are typically viewed as very low-priority undertakings. We all realize that problems with implants can be quite serious, even fatal, and initiatives to improve reporting have received considerable attention recently. Clearly, actively monitoring for and reporting such events will continue to be emphasized by industry, government agencies, public health groups, patient advocacy groups, and professional organizations until good quality data is regularly and easily available to guide quality improvement work. Reporting problems to the manufacturer and relevant government agency (such as the FDA) helps maximize the safety and effectiveness of implants *(5)*.

The progress of surgery of all sorts has been facilitated by the introduction of antibiotics, intravenous fluids, sophisticated monitoring tools, and the like. Progress in implants has been uniquely dependent on the development of specialized medical plastics, metal alloys, batteries, and so on. There are exceptions, of course, such as simple metal internal fixation devices for fractured bones, but the dramatic proliferation of implants in the last several decades has been made possible by the development of sophisticated materials that could be tolerated by human tissue. These developments, not surprisingly, caught the public eye. In America, a television show called *The Bionic Man* became popular. The hero in this series could perform superhuman feats using implanted devices. The editors have appropriated this neologism for the title of the book. The term *bionic* means "aided by the implantation of prosthetic devices." Of course, high technology is implicit. In any case, the word has entered the English language and we believe that readers will understand our meaning.

Not surprisingly, shortly after the implantation of various devices, problems arose that were unique to the devices. Undoubtedly, many were the result of inadequate materials used in experimental prototypes and were not publicized as they were never introduced as commercial products. Prosthetic–tissue interactions are discussed later in this book (Chapter 8). Those products in contact with the bloodstream were plagued by thrombogenicity. This complication has proved fairly simple to study. Our understanding of its nature has allowed us to take measures to avoid it, typically chronic anticoagulation. Engineering a device to decrease its thrombogenicity is another. Many other devices that were otherwise successful developed disastrous infections. This is still a frequent problem and is addressed throughout this text. There are several ways to avoid it. Administering prophylactic antibiotics during periods of high risk is one. Engineering devices to resist infection by incorporating antibiotics into the plastic used is another. Use of central venous catheters impregnated with chlorhexidine-silver sulfadiazine, for example, has been shown in a well-controlled trial to reduce the risk of catheter-related infection *(6)*. Because there are about 5 million such catheters used in the United States each year, and because catheter-related bloodstream infections occur with 3 to 7% of catheters, the number of patients who have this event exceeds 250,000 per year. Factoring in the reported 10 to 25% fatality rate associated with catheter-associated bacteremia, the societal implications are large. Maki et al. carried out a cost–benefit analysis and concluded that routine use of antibiotic-impregnated catheters need probably permit overall cost savings *(6)*. Similar considerations should apply with other devices in which the consequences of infection are even greater, such as total joint replacements, heart valves, and the like. Modifying a patient's lifestyle or

and implantation of prosthetic devices has been another. Here this textbook must draw a line. We do not deal with transplanted tissues or organs, as this is dealt with elsewhere. This textbook deals solely with implanted manufactured devices. We employ the US Federal Food, Drug and Cosmetic Act *(3)* definition of a medical device as

> an instrument, apparatus, implement, machine, contrivance, implant, in vitro reagent, or other similar or related article, . . . which is . . . intended for use in the diagnosis of disease or other conditions, or in the cure, mitigation, treatment, or prevention of disease . . . , or intended to affect the structure or any function of the body . . . , and which does not achieve any of its principal intended purposes through chemical action within or on the body . . . and which is not dependent upon being metabolized for the achievement of any of its principal intended purposes.

There are about 1800 types of medical devices defined by this language, ranging from home diagnostic kits, bandages, lipstick, and X-ray machines to infusion pumps, heart valves, and breast implants. These obviously vary enormously in complexity, cost, and risk to patient health. We deal only with material surgically inserted into the body, thus eliminating from consideration most of the devices covered in the Food and Drug Administration (FDA) definition. We do not treat certain types of prosthetic items such as conventional artificial limbs and most partially implanted devices such as tracheostomy tubes. We do, however, deal with other partially implanted devices such as Hickman catheters. Certain wholly internal implants, such as intrauterine contraceptive devices, are not treated at all and others, such as diaphragmatic pacemakers, are ignored because they are so uncommon. These boundaries are arbitrary, but the editors believe that the devices treated in this book are the most important ones and rely on input from readers to guide future editions. Doctors and nurses are the main users of some of these devices (e.g., vascular access devices for administration of drugs) but most of the prostheses we consider in this book are intended to replace a defective body part (mechanical heart valve, prosthetic joint, etc.) or enhance some body function (cochlear implants, breast implants, and the like). The number of prosthetic devices implanted per year is surprisingly difficult to discern. The information is often treated as a trade secret by the corporations producing the devices and is typically not widely publicized. In other instances, several companies scattered throughout the world make similar items, often of varying quality and technical specifications. The authors of each chapter have been asked to estimate both the total number of patients now alive with a particular type of implant and the number of devices implanted per year throughout the world. Clearly, many millions of patients who have various surgically implanted devices are alive today. Equally clearly, the devices are steadily improving in quality and variety.

The US General Accounting Office reported that most problems with medical devices are found by doctors and nurses *(4)* but this topic has received little rigorous attention in the medical literature. How many of the problems picked up by doctors and nurses occur in asymptomatic patients and how many are found as a result of an investigation triggered by a patient complaint? Certainly, one goal of routine postimplantation management of bionic patients should be the identification and timely reporting of adverse events possibly related to the implant. Unfortunately, doctors' lives are all too often crisis-driven and hectic. Filling out yet another form and making yet another phone

1
Overview

Frank E. Johnson

Throughout the evolution of medical practice, treatments for various disorders have flowed from cultural expectations and current understanding of disease causation. It must be supposed that ill individuals have wished for replacement of their malfunctioning organs since the dawn of consciousness in our species but no progress on this front was possible until the scientific foundations of medicine had been laid. The Galenic theory of disease causation emphasized an imbalance of four bodily humors and dominated Western medical thinking for 1000 years, from about 150 ad until the Middle Ages. This model sought to restore health by adjusting humors rather than treating individual body parts. It was eventually overturned on the basis of evidence from many sources. The Renaissance period saw a flowering of medical knowledge based on an empirical, reductionist philosophy of science and methods of investigation emphasizing direct observation, hypothesis generation, and model-building. These are essentially the same ones used at present.

This paid huge dividends. Direct investigation of human anatomy by means of dissections was introduced by Vesalius in the 16th century. The notion that microscopic organisms exist and that all living tissue is composed of cells was demonstrated by Leeuwenhoek in the 17th century. William Harvey was the first to understand the nature of blood circulation in the 17th century. All later proved crucial in the development of modern medical concepts and surgical techniques.

Surgery has been performed since the beginning of recorded history, as deduced from physical evidence and ancient writings. It was simple, crude, hazardous, very painful and often ineffective until relatively recently. In large part, knowledge of how the body works was sufficient in the 19th century to permit a dramatic advancement in surgery (1). Effective anesthesia was introduced in the mid nineteenth century and led to a surge in the number of operations (2). Rosenberg indicates that 385 operations were performed at the Massachusetts General Hospital in the decade before the introduction of ether anesthesia (about 1835 to 1845). There were 200,000 operations performed at the same hospital in the decade 1890–1900 and they were much more daring. Surgery was transformed from a last-ditch strategy with grave risks and immense suffering to a major tool in the conquest of disease, a relatively safe one with reasonable pain control.

Now the stage was set for the idea of replacement of body parts. The very idea was revolutionary. Organ transplantation has been one very successful offshoot of this idea

From: *The Bionic Human: Health Promotion for People With Implanted Prosthetic Devices*
Edited by: F. E. Johnson and K. S. Virgo © Humana Press Inc., Totowa, NJ

Tony Mundy, MD • *Institute of Urology, London, United Kingdom*
J. Gail Neely, MD • *Department of Otolaryngology, Head and Neck Surgery, Washington University School of Medicine, St. Louis, MO*
Thomas J. Otto, MD • *Department of Orthopaedic Surgery, Saint Louis University School of Medicine, St. Louis, MO*
Christian E. Paletta, MD • *Division of Plastic and Reconstructive Surgery, Saint Louis University School of Medicine, St. Louis, MO*
Mary C. Proctor, MD • *Department of Surgery, University of Michigan Hospital, Ann Arbor, MI*
John H. Raaf, MD, DPhil • *Department of Surgery, Case Western Reserve University, Cleveland, OH*
Nathan Ravi, MD, PhD • *Department of Veterans Affairs Medical Center, St. Louis, MO*
David J. Rea, MD • *Division of Urology, University of North Carolina at Chapel Hill, Chapel Hill, NC*
William F. Regnault, PhD • *Center for Devices and Radiological Health, US Food and Drug Administration, Rockville, MD*
Edward Rustamzadeh, MD • *Department of Neurosurgery, University of Minnesota School of Medicine, Minneapolis, MN*
Forrest S. Roth, MD • *Division of Plastic Surgery, Baylor College of Medicine, Houston, TX*
Issam Saliba, MD • *Department of Otolaryngology, Centre Hospitalier Universitaire de Toulouse, Toulouse, France*
Luis A. Sanchez, MD • *Departments of Surgery and Interventional Radiology, Washington University School of Medicine, St. Louis, MO*
Spyros Sgouros, MD • *Department of Neurosurgery, Birmingham Children's Hospital, Birmingham, United Kingdom*
Deborah Shatin, PhD • *Center for Health Care Policy and Evaluation, UnitedHealth Group, Minneapolis, MN*
Ramona E. Simionescu, MD • *Department of Internal Medicine, Saint Louis University School of Medicine, St. Louis, MO*
Ewout W. Steyerberg, PhD • *Department of Public Health, Erasmus MC–University Medical Center Rotterdam, Rotterdam, The Netherlands*
Anthony J. Summerwill, BDS • *Department of Restorative Dentistry, Liverpool University Dental Hospital, Liverpool, United Kingdom*
Thomas H. Tung, MD • *Division of Plastic and Reconstructive Surgery, Washington University School of Medicine, St. Louis, MO*
Casper Uldriks, JD • *Center for Devices and Radiological Health, US Food and Drug Administration, Rockville, MD*
S. Rao Vallabhaneni • *Royal Liverpool University Hospital, Liverpool, United Kingdom*
Suzie N. Venn, MBBS, MS, FRCS • *Institute of Urology, London, United Kingdom*
Katherine S. Virgo, PhD • *Department of Surgery, Saint Louis University School of Medicine and Department of Veterans Affairs Medical Center, St. Louis, MO*
Eric D. Whitman, MD • *Department of Surgery, Washington University School of Medicine, St. Louis, MO*
Celia M. Witten, MD, PhD • *Center for Devices and Radiological Health, US Food and Drug Administration, Rockville, MD*

Contributors

ANNELLE V. HODGES, PhD • *Department of Otolaryngology, University of Miami School of Medicine, Miami, FL*
DAVID M. HOVSEPIAN, MD • *Department of Radiology, Washington University School of Medicine, St. Louis, MO*
FRANK E. JOHNSON, MD • *Department of Surgery, Saint Louis University School of Medicine and Department of Veterans Affairs Medical Center, St. Louis, MO*
KARTHIKESHWAR KASIRAJAN, MD • *Division of Vascular Surgery, Emory University School of Medicine, Atlanta, GA*
JAMES A. KEENEY, MD • *Department of Orthopedic Surgery, Washington University School of Medicine, St. Louis, MO*
DONALD J. KENNEDY, MD • *Department of Internal Medicine, Saint Louis University School of Medicine, St. Louis, MO*
JOHN N. KENT, DDS • *Department of Oral and Maxillofacial Surgery, Louisiana State University School of Dentistry, New Orleans, LA*
JOHN E. KING, MD • *Department of Otolaryngology, University of Miami School of Medicine, Miami, FL*
SREENIVAS KOKA, DDS, MS, PhD • *Department of Dental Specialties, Mayo Clinic, Rochester, MN*
ARTHUR J. LABOVITZ, MD • *Department of Cardiology, Saint Louis University School of Medicine, St. Louis, MO*
TERRY C. LAIRMORE, MD • *Division of Surgical Oncology, Scott and White Hospital, Texas A & M University System Health Sciences Center, College of Medicine, Temple, TX*
CORNELIUS H. LAM, MD • *Department of Neurosurgery, University of Minnesota Medical School, Minneapolis, MN*
MARK LANGSFELD, MD • *Division of Vascular Surgery, University of New Mexico School of Medicine, Albuquerque, New Mexico*
JOHN P. LAVELLE, MD • *Division of Urology, University of North Carolina at Chapel Hill, Chapel Hill, NC*
COLES E. L'HOMMEDIEU, MD • *Department of Orthopaedic Surgery, Saint Louis University School of Medicine, St. Louis, MO*
MATTHIAS LORENZ, MD • *Department of General and Vascular Surgery, Johann-Wolfgang Goethe University, Frankfurt, Germany*
MICHAEL P. MANNING, MBChB, M CH ORTH, FRCS • *Whiston Hospital NHS Trust, Prescot, Liverpool, United Kingdom*
JOHN MAREK, MD • *Division of Vascular Surgery, University of New Mexico School of Medicine, Albuquerque, NM*
BRIAN MATTESON, MD • *Division of Vascular Surgery, University of New Mexico School of Medicine, Albuquerque, NM*
STEVEN R. MOBLEY, MD • *Department of Otolaryngology, University of Miami School of Medicine, Miami, FL*
MARC R. MOON, MD • *Division of Cardiothoracic Surgery, Washington University School of Medicine, St. Louis, MO*
JOHN J. MULCAHY, MD, PhD • *Department of Urology, Indiana Cancer Pavilion, Indianapolis, IN*

DEREK T. CONNELLY, BSC, MD • *The Cardiothoracic Centre, Liverpool NHS Trust, Liverpool, United Kingdom*

CATHARINE M. DARCY, MBChB, FRCSEd, FRCSG • *Whiston Hospital NHS Trust, Prescot, Liverpool, United Kingdom*

ARA W. DARZI, MD, FRCSI, KBE • *Department of Surgery, Imperial College, St. Mary's Hospital, London, United Kingdom*

JOSHUA L. DOWLING, MD • *Department of Neurological Surgery, Washington University School of Medicine, St. Louis, MO*

STEVEN E. ECKERT, DDS, MS • *Department of Dental Specialties, Mayo Clinic, Rochester, MN*

HESSAM ELFIKI, MD • *Department of Otolaryngology, University of Miami School of Medicine, Miami, FL*

AMR EL-SHAFEI, MD • *Division of Cardiology, Department of Veterans Affairs Medical Center, St. Louis, MO*

ADRIEN A. ESHRAGHI, MD • *Department of Otolaryngology, University of Miami School of Medicine, Miami, FL*

CHRISTINE M. EVANS • MD, FRCS • *Retired Consultant Urologist, Glan Clwyd Hospital, Rhyl, United Kingdom*

JOHN H. FIELDER, PhD • *Philosophy Department, Villanova University, Villanova, PA*

BERNARD FRAYSSE, MD • *Department of Otolaryngology, Centre Hospitalier Universitaire de Toulouse, Toulouse, France*

GOETZ GEYER, MD • *Department of Otolaryngology, Head and Neck Surgery, Municipal Hospital, Solingen, Germany*

G. E. GHALI, DDS, MD • *Department of Oral and Maxillofacial Surgery, Louisiana State University Health Sciences Center, Shreveport, LA*

SOUMEN GHOSH, MB • *Whiston Hospital NHS Trust, Prescot, Liverpool, United Kingdom*

MARYE J. GLEVA, MD • *Cardiovascular Division, Department of Medicine, Washington University School of Medicine, St. Louis, MO*

KRISTINE J. GULESERIAN, MD • *Division of Cardiothoracic Surgery, Washington University School of Medicine, St. Louis, MO*

DEREK A. GOULD, MBChB, DMRD, FRCP, FRCR • *Department of Radiology, Royal Liverpool NHS Trust, Liverpool, United Kingdom*

DAVID J. GRAY, MD • *Department of Surgery, Saint Louis University School of Medicine, St. Louis, MO*

LAZAR J. GREENFIELD, MD • *Department of Surgery, University of Michigan School of Medicine, Ann Arbor, MI*

THOMAS P. GROSS, MD • *Center for Devices and Radiological Health, US Food and Drug Administration, Rockville, MD*

CARSTEN N. GUTT, MD • *Department of General, Visceral, and Trauma Surgery, Ruprecht-Karls University, Heidelberg, Germany*

CONSTANTINOS HAJIVASSILIOU, BSC, MBChB, MD, FRCSEd, FRCS • *University Department of Paediatric Surgery, Royal Hospital for Sick Children, Glasgow, United Kingdom*

KEVIN HANCOCK, FRCS • *Whiston Hospital NHS Trust, Prescot, Liverpool, United Kingdom*

A. E. HEALEY, BSC, MBChB • *Department of Radiology, Royal Liverpool NHS Trust, Liverpool, United Kingdom*

STEFAN HEINRICH, MD • *Visceral and Transplant Surgery, University Hospital of Zurich, Zurich, Switzerland*

Contributors

GIANNI D. ANGELINI, MD, FRCS • *Department of Cardiac Surgery, Bristol University, Bristol, United Kingdom*

HYDER ALI ALIYAR, PhD • *Department of Ophthalmology and Visual Sciences, Washington University School of Medicine, St. Louis, MO*

M. DAVID ARYA, MD • *Department of Cardiology, Saint Louis University School of Medicine, St. Louis, MO*

BRAD ASTOR, PhD, MPH • *Department of Epidemiology, Bloomberg School of Pubic Health, Johns Hopkins University, Baltimore, MD*

RICCARDO A. AUDISIO, MD, FRCS • *Department of Surgery, University of Liverpool, Liverpool, United Kingdom*

THOMAS J. BALKANY, MD • *Department of Otolaryngology, Miller School of Medicine, University of Miami, Miami, FL*

ARTHUR E. BAUE, MD • *Former Chairman of Surgery, Yale University School of Medicine, Former Vice-President, Saint Louis University School of Medicine, St. Louis, MO*

R. IVAN BERETVAS, MD • *Department of Surgery, Saint Louis University School of Medicine, St. Louis, MO*

PREBEN BJERREGAARD, MD • *Department of Cardiology, Saint Louis University School of Medicine, St. Louis, MO*

STEVEN B. BRANDES, MD • *Department of Urological Surgery, Washington University School of Medicine, St. Louis, MO*

JOHN A. BRENNAN, MD • *Royal Liverpool University Hospital, Liverpool, United Kingdom*

ROSELIE A. BRIGHT, ScD • *Center for Devices and Radiological Health, US Food and Drug Administration, Rockville, MD*

L. MICHAEL BRUNT, MD • *Siteman Cancer Center, Washington University School of Medicine, St. Louis, MO*

ALAN J. BRYAN, DM, FRCS • *Department of Cardiac Surgery, Bristol University, Bristol, United Kingdom*

CULLEY C. CARSON, III, MD • *Division of Urology, University of North Carolina at Chapel Hill School of Medicine, Chapel Hill, NC*

JOHN I. CAWOOD, FDSRC • *Grosvenor Nuffield Hospital, Chester, United Kingdom*

AVRIL A. P. CHANG, MBBS, FRACS • *Department of Surgical Oncology and Technology, Imperial College, London, United Kingdom*

JOHN C. CLOHISY, MD • *Department of Orthopedic Surgery, Washington University School of Medicine, St. Louis, MO*

CHRISTOPHER N. COMPTON, MD • *Department of Surgery, Case Western Reserve University School of Medicine, Cleveland, OH*

23 Cardiac Pacemakers
    Preben Bjerregaard and Amr El-Shafei 633

    US Counterpoint
    Marye J. Gleva 645

    European Counterpoint
    Derek T. Connelly 649

24 Joint Prostheses and Internal Fixation Devices
    Thomas J. Otto and Coles E. L'Hommedieu 655

    US Counterpoint
    James A. Keeney and John C. Clohisy 679

    European Counterpoint
    Michael P. Manning 686

    Index 693

Contents

17 Ossicular Implants
    *Adrien A. Eshraghi, Hessam Elfiki, Steven R. Mobley,
    and Thomas J. Balkany* 413

    US Counterpoint
    J. Gail Neely 424

    European Counterpoint
    Goetz Geyer 429

18 Vascular Prostheses
    *Karthikeshwar Kasirajan, Brian Matteson, John Marek,
    and Mark Langsfeld* 437

    US Counterpoint
    Luis A. Sanchez 473

    European Counterpoint
    S. Rao Vallabhaneni and John A. Brennan 480

19 Cardiac Valves
    M. David Arya and Arthur J. Labovitz 489

    US Counterpoint
    Kristine J. Guleserian and Marc R. Moon 523

    European Counterpoint
    Alan J. Bryan and Gianni D. Angelini 528

20 Intravascular Filters and Stents
    Lazar J. Greenfield and Mary C. Proctor 533

    US Counterpoint
    David M. Hovsepian 542

    European Counterpoint
    A. E. Healey and Derek A. Gould 547

21 Vascular Access Devices
    Christopher N. Compton and John H. Raaf 561

    US Counterpoint
    Eric D. Whitman 588

    European Counterpoint
    Matthias Lorenz, Carsten N. Gutt, and Stefan Heinrich 593

22 Osseointegrated Dental Implants
    Steven E. Eckert and Sreenivas Koka 603

    US Counterpoint
    G. E. Ghali and John N. Kent 619

    European Counterpoint
    Anthony J. Summerwill and John I. Cawood 624

11 Breast Prostheses: *Current Recommendations for Care of Patients After Implantation of Breast Prostheses*
   Forrest S. Roth, David J. Gray, and Christian E. Paletta 231

   US Counterpoint
   Thomas H. Tung 255

   European Counterpoint
   Catherine M. Darcy and Kevin Hancock 259

12 Prostheses for Hernia Repair
   R. Ivan Beretvas 271

   US Counterpoint
   L. Michael Brunt 277

   European Counterpoint
   Avril A. P. Chang and Ara W. Darzi 284

13 Penile Prostheses
   John J. Mulcahy 289

   US Counterpoint
   Steven B. Brandes 300

   European Counterpoint
   Christine M. Evans 302

14 Artificial Urethral Sphincters
   David J. Rea, John P. Lavelle, and Culley C. Carson, III 313

   US Counterpoint
   Steven B. Brandes 322

   European Counterpoint
   Suzie N. Venn, Constantinos Hajivassiliou, and Tony Mundy 324

15 Cerebrospinal Fluid Shunts
   Edward Rustamzadeh and Cornelius H. Lam 333

   US Counterpoint
   Joshua L. Dowling 359

   European Counterpoint
   Spyros Sgouros 363

16 Cochlear Implants
   Adrien A. Eshraghi, John E. King, Annelle V. Hodges, and Thomas J. Balkany 379

   US Counterpoint
   J. Gail Neely 404

   European Counterpoint
   Issam Saliba and Bernard Fraysse 408

Contents

Foreword .... *v*
Preface .... *vii*
Contributors .... *xiii*

1 Overview
   Frank E. Johnson 1

2 Assessment of Surveillance Test Performance and Cost
   Katherine S. Virgo 13

3 Cost of Follow-Up After Implantation of Prosthetic
   Devices
   Katherine S. Virgo 33

4 Factors That Should Shape Clinical Strategies
   Frank E. Johnson 47

5 A View From the US Food and Drug Administration
   Thomas P. Gross, Celia M. Witten, Casper Uldriks,
   and William F. Regnault 61

6 Societal Issues
   John H. Fielder 89

7 Databases for Studying the Epidemiology
   of Implanted Medical Devices
   Deborah Shatin, Roselie A. Bright, and Brad Astor 115

8 Tissue Reaction to Prosthetic Materials
   Nathan Ravi and Hyder Ali Aliyar 133

9 Prevention of Infection in Prosthetic Devices
   Ramona E. Simionescu and Donald J. Kennedy 159

   European Counterpoint
   Soumen Ghosh and Riccardo A. Audisio 186

10 Computer Modeling
   Ewout W. Steyerberg 205

Preface

There are currently few well-accepted standards for patient management after implantation of prosthetic devices. Few research studies have documented patient benefit from postimplantation care, as measured by extended patient survival time, device function, or improved quality of life. *The Bionic Human: Health Promotion for People With Implanted Prosthetic Devices* is intended to provide clinicians with guidance regarding the follow-up of these patients by compiling the strategies recommended by acknowledged experts. Current strategies of practicing clinicians and the factors influencing variability in current practice are reviewed in the chapters dealing with particular prosthetic devices. Both currently available and promising new diagnostic tests are evaluated. Some of the consequences of device failure are explained. Concluding each of these chapters are counterpoints from distinguished authors at major centers in the United States and Europe. The followup recommendations provided by experts at each institution are summarized in standardized format, which should encourage comparative analysis and stimulate discussion. We hope this will lay the foundation for controlled clinical trials and the eventual establishment of evidence-based guidelines on which consensus can be based.

One difficulty we encountered in putting together this book was in the selection of devices that could feasibly be included in this first edition. Common ones, such as intraocular lenses, and rare ones, such as diaphragm-pacing systems, although excluded from the current edition, are candidates for inclusion in the next edition.

The image on the book cover is derived from an original work of art by Cameron Slayden. It graced the cover of the February 8, 2002, issue of Science, the official journal of the American Association for the Advancement of *Science* (AAAS). The publisher, editors, and authors are grateful to the AAAS for allowing us to use this image.

We take great personal satisfaction in the completion of this project, as it is one so long overdue. We extend thanks to Dr. Arthur Baue for his input and guidance. We also express our gratitude to Ms. Judy Feldworth, who served as editorial assistant on the project. She accomplished the mammoth task of copyediting all manuscripts, ensuring consistency in style and format throughout with intelligence, skill, and good humor. The book could not have been completed without her hard work and attention to detail. We also thank Mrs. Michele Graser, who helped with the required correspondence and other organizational matters. Her computer skills were useful every step of the way. Her calm, competent personality helped keep the contributors from losing their tempers and the editors from losing their minds.

Frank E. Johnson, MD
Katherine S. Virgo, PhD

Foreword

The Bionic Human is evidence for the incredible scientific achievements of recent years. Implantation of prosthetic devices was once only science fiction, but is now a reality. The book describes what can help a patient stay alive and enjoy life after having an implant. Devices, implants, and replacement parts presently not only sustain and support life, but also make it livable. Some of these new parts, such as heart valves, are critical for life. Others, such as vascular prostheses for obstructions caused by arteriosclerosis, keep parts of the body viable. Others, such as penile prostheses, improve the quality of life. Similarly, a breast implant after a mastectomy for cancer is much more than cosmetic for a woman; it aids in restoring her sense of womanhood.

The Bionic Human is written for patients and their families who have, or are contemplating, such replacement parts. It will help them sort out what is important for them—what to know, what to be careful about, and where to go for help and further information. The book will also be a valuable resource for doctors who are not involved in performing such procedures, but must advise their patients about these possibilities. Should the patient have it done? Is it urgent? Is it elective? Is it necessary? Do the potential complications and risks outweigh the benefits? A particularly valuable chapter is that written by members of the Center for Devices and Radiological Health of the FDA, the group that protects us from devices that are unsafe or ineffective. They describe their procedures, reviews, and follow-up in clear terms. Each chapter describes the device and its pros and cons. In addition, there are counterpoints from North America and Europe. This format, featuring perspectives from several centers, is unique. It provides a balance, as experiences may vary and different authors may have different perspectives. Help is provided about where to go for further information, including patient support groups, commercial associations, websites, and the like. The editors are to be complimented for bringing together authorities in these fields to share their experiences.

The Bionic Human emphasizes the care of patients who have recovered from implant surgery and are without symptoms. For many devices, surveillance by physicians, including X-rays, blood studies, and so on, is necessary. For some implants, such as a hip replacement, no further attention is needed other than metal detector alerts at security posts in airports. Each chapter suggests where to go for help with prosthesis malfunction or with symptoms that develop later. *The Bionic Human* is a great story of achievement and a valuable resource.

Arthur E. Baue, MD
Former Chairman of Surgery, Yale University School of Medicine
Former Vice-President, Saint Louis University Health Sciences Center

© 2006 Humana Press Inc.
999 Riverview Drive, Suite 208
Totowa, New Jersey 07512

All rights reserved.

www.humanapress.com

All rights reserved. No part of this book may be reproduced, stored in a retrieval system, or transmitted in any form or by any means, electronic, mechanical, photocopying, microfilming, recording, or otherwise without written permission from the Publisher.

The content and opinions expressed in this book are the sole work of the authors and editors, who have warranted due diligence in the creation and issuance of their work. The publisher, editors, and authors are not responsible for errors or omissions or for any consequences arising from the information or opinions presented in this book and make no warranty, express or implied, with respect to its contents.

Due diligence has been taken by the publishers, editors, and authors of this book to assure the accuracy of the information published and to describe generally accepted practices. The contributors herein have carefully checked to ensure that the drug selections and dosages set forth in this text are accurate and in accord with the standards accepted at the time of publication. Notwithstanding, since new research, changes in government regulations, and knowledge from clinical experience relating to drug therapy and drug reactions constantly occur, the reader is advised to check the product information provided by the manufacturer of each drug for any change in dosages or for additional warnings and contraindications. This is of utmost importance when the recommended drug herein is a new or infrequently used drug. It is the responsibility of the treating physician to determine dosages and treatment strategies for individual patients. Further, it is the responsibility of the health care provider to ascertain the Food and Drug Administration status of each drug or device used in their clinical practice. The publishers, editors, and authors are not responsible for errors or omissions or for any consequences from the application of the information presented in this book and make no warranty, express or implied, with respect to the contents in this publication.

This publication is printed on acid-free paper. ∞

ANSI Z39.48-1984 (American Standards Institute) Permanence of Paper for Printed Library Materials.

Production editor: Robin B. Weisberg and Melissa Caravella

Cover design by Patricia F. Cleary

Cover Illustration: Reprinted with permission from *Science* 02-08-2002 cover image © 2002 AAAS.

For additional copies, pricing for bulk purchases, and/or information about other Humana titles, contact Humana at the above address or at any of the following numbers: Tel.: 973-256-1699; Fax: 973-256-8341; E-mail: orders@humanapr.com; or visit our Website: www.humanapress.com

Photocopy Authorization Policy:
Authorization to photocopy items for internal or personal use, or the internal or personal use of specific clients, is granted by Humana Press Inc., provided that the base fee of US $30.00 per copy is paid directly to the Copyright Clearance Center at 222 Rosewood Drive, Danvers, MA 01923. For those organizations that have been granted a photocopy license from the CCC, a separate system of payment has been arranged and is acceptable to Humana Press Inc. The fee code for users of the Transactional Reporting Service is: [0-89603-959-5/06 $30.00].

Printed in the United States of America. 10 9 8 7 6 5 4 3 2 1

eISBN: 1-59259-975-3

Library of Congress Cataloging-in-Publication Data

The bionic human : health promotion for people with implanted prosthetic devices / edited by Frank E. Johnson, Katherine S. Virgo ; associate editors, Terry Lairmore, Riccardo Audisio.
       p. cm.
  Includes bibliographical references and index.
  ISBN 0-89603-959-5 (alk. paper)
  1. Prosthesis--Physiological aspects. 2. Implants, Artificial--Physiological aspects. 3. Surgery--Patients--Health. 4. Health education. I. Johnson, Frank E., 1943 II. Virgo, Katherine S.
  RD130.B57 2005
  617.9'5--dc22
                                                                                            2005006247

The Bionic Human

*Health Promotion for People
With Implanted Prosthetic Devices*

Edited by

Frank E. Johnson, MD

Katherine S. Virgo, PhD

*Saint Louis University School of Medicine
and Department of Veterans Affairs Medical Center
St. Louis, MO*

Associate Editors

Terry C. Lairmore, MD

*Scott and White Hospital
Texas A & M University System Health Sciences Center
College of Medicine, Temple, TX*

Riccardo A. Audisio, MD, FRCS

University of Liverpool, United Kingdom

Foreword by

Arthur E. Baue, MD

*Former Chairman of Surgery, Yale University School of Medicine
Former Vice-President, Saint Louis University Health Sciences Center*

HUMANA PRESS ✳ TOTOWA, NEW JERSEY

The Bionic Human

US Counterpoint to Chapter 11

Thomas H. Tung

INTRODUCTION

Chapter 11 provides a nice overview of the issues concerning the use of breast prostheses and the long-term management of patients following breast augmentation and reconstruction with implants. The history of breast implants is very interesting and the developments leading to the current generation of implants are well summarized.

At the Barnes-Jewish Hospital at Washington University, an average of 67 breast augmentation and 61 breast reconstructive procedures using tissue expanders or implants were performed each year for the last 4 years. The most common type used at our institution for both breast augmentation and reconstruction is the smooth, round saline implant. As the first and classic shape, it has stood the test of time and provides a predictable and pleasing outcome. Placement is technically easy and concerns about displacement and rotation postoperatively do not exist as they do for anatomic implants *(1–3)*.

For a given volume, it has a wider base diameter than an anatomic implant and provides more medial fullness, producing more "cleavage," which many patients desire. Concerns of collapse and folding of the upper pole leading to compromise of shell integrity have not been clearly substantiated by an increased rate of leak or failure in this type of implant *(4)*. However, because anatomic implants are designed to produce a more natural shaped breast, many women prefer and request these implants. For a given volume, they provide more projection, especially in the lower pole, and the intuitive appeal is obvious. Because of their shape, they are necessarily textured, and more attention has to be given to precise symmetrical placement. Similarly, any significant postoperative displacement or rotation will generally require surgical correction *(5–9)*.

More recently, high- and/or low-profile designs of both the round and anatomic implants are now available to provide more or less projection respectively for a given volume, with corresponding changes in base diameter *(10,11)*.

Our preferences for route and position of implant placement are overwhelmingly an inframammary incision and submuscular placement. The inframammary incision provides direct exposure without the need for endoscopic instruments or experience, and in most patients is inconspicuous. Because there is direct visualization and exposure, there is better control of implant position and hemostasis. Submuscular implant place-

From: *The Bionic Human: Health Promotion for People With Implanted Prosthetic Devices*
Edited by: F. E. Johnson and K. S. Virgo © Humana Press Inc., Totowa, NJ

ment is preferred, especially for saline implants, to minimize rippling and edge visibility and palpability. Advantages also include a decreased potential for capsular contracture, and less interference with mammography.

Immediate breast reconstruction with implants following mastectomy conventionally requires placement of a tissue expander in a subpectoral position followed by gradual tissue expansion over several months. Implant exchange with placement of the permanent implant is then performed, and any procedure on the contralateral breast for symmetry is commonly done at the same time. Prostheses designed to function as both tissue expander and permanent implant can be used to minimize the need for additional surgery (12). The use of these implants at our institution remains limited because tissue expansion is still required, contralateral breast procedures for symmetry are usually still needed, and modification of the implant pocket and position is frequently desired. Although a good cosmetic outcome can be obtained in some patients, especially those with small breasts, the outcome is often suboptimal. This is owing to, in large part, the contraction of the skin and soft tissue envelope that occurs before the expander can be fully inflated. The technique of immediate implant reconstruction with a permanent prosthesis was described in the 1980s but interest has waned. Refinements in the technique using a fully inflated saline or gel prosthesis have recently appeared in the literature as a means of preventing contracture, fully utilizing the soft tissue envelope and maintaining breast shape (13–15). The use of a postoperatively adjustable implant allows fine-tuning of the desired volume and size. Lower pole fullness and contour appear to be better with this approach, and the need for prolonged tissue expansion and implant exchange is eliminated. This technique is cautiously being used at our institution to a limited extent. The potential for poor healing, skin flap necrosis, and dehiscence leading to implant exposure remain significant with immediate placement of a fully expanded implant (16). The vascularity of the mastectomy skin flaps can be unpredictable, and tension provided by the full implant will further compromise the blood supply. Indications for this approach may include patients with small breast size undergoing skin-sparing mastectomy.

Follow-up at our institution is usually for two to three years, depending on the procedure, and then as needed. For asymptomatic and uncomplicated patients following breast augmentation, follow-up consists of three office visits within the first year after surgery and one visit during the second year (Table 1). Patients with implants for breast reconstruction following mastectomy are seen once more during the third postoperative year (Table 2). All patients should begin or remain on the same schedule of routine mammography as recommended for their age and risk group (17–19).

Regular surveillance for cancer recurrence and new primary cancer is recommended and maintained by the oncological surgeon. Additional evaluation by the plastic surgeon in the long term is mainly to evaluate potential complications such as a change in implant shape or size, capsular contracture, or infection.

The most common long-term complications are implant failure and capsular contracture. As stated in Chapter 11, magnetic resonance imaging (MRI) is the imaging modality of choice for the evaluation of implant leak or rupture because of its high sensitivity and good specificity (20–22). At our institution MRI remains the preferred mode of evaluation, followed by ultrasound. The diagnosis of capsular contracture is made clinically using the Baker classification. Our practice for the treatment of sys-

Table 1
Routine Follow-Up After Breast Augmentation With an Implant for Asymptomatic Patients at Barnes-Jewish Hospital, St. Louis, MO

	Postoperative year					
	1	2	3	4	5	10
Office visit	3	1	0	0	0	0
Mammogram[a]	1	1	1	1	1	1

[a]The number of times per year mammography is recommended only pertains to women 40 years of age. Additional office visits and tests are indicated only if signs or symptoms warrant them.

Table 2
Routine Follow-Up With Plastic Surgeon After Breast Reconstruction With an Implant for Asymptomatic Patients at Barnes-Jewish Hospital, St. Louis, MO

	Postoperative year					
	1	2	3	4	5	10
Office visit	3	1	1	0	0	0
Mammogram[a]	1	1	1	1	1	1

[a]The number of times per year mammography is recommended only pertains to women 40 years of age. Additional tests and office visits are indicated only if signs or symptoms warrant them.

temic infections in patients with breast prostheses is consistent with the recommendations given in Chapter 11. There is no clear benefit from the use of different antibiotics, higher dosages, or a prolonged course of therapy to prevent seeding of the implant.

The role of breast prostheses in cosmetic and reconstructive breast surgery is well established and continues to provide significant benefits to the patient. There are many unique considerations and potential sequelae, which will continue to change as newer generations of implants and designs are developed. Chapter 11 provides a thorough overview of the statistics and considerations that every patient who is considering or who has breast implants should know. The importance of the fully informed and educated patient in the preoperative decision-making process, as well as early and long-term postoperative management and surveillance, cannot be overemphasized. A comprehensive list of support groups, societies, and manufacturers is provided by the authors and should prove to be invaluable to all patients.

REFERENCES

1. Baecke JL. Word of caution about teardrop breast implants. Cosmetic Surg Times 2000;3:3.
2. Hobar PC, Gutowski K. Experience with anatomic breast implants. Clin Plast Surg 2001;28:553–558.
3. Baecke JL. Warning about anatomical breast implants. Plast Reconstr Surg 2000;106:740.
4. Tebbets JB. The other side of the story—reply (letter). Plast Reconstr Surg 1994;101:875.

5. Baecke JL. Teardrop implants: enough to make a grown man cry (reply). Plast Reconstr Surg 2001;107:1914.
6. Adams WP. Breast deformity caused by anatomical or teardrop implant rotation. Plast Reconstr Surg 2003;111:2110–2111.
7. Baecke JL. Breast deformity caused by anatomical or teardrop implant rotation. Plast Reconstr Surg 2002;109:2555.
8. Tebbets JB. Warning about warning about anatomical breast implants. Plast Reconstr Surg 2001;107:1912.
9. Baecke JL. Teardrop implants advisory. Ann Plast Surg 2000;44.
10. Tebbets JB. Breast augmentation with full-height anatomic saline implants: the pros and cons. Clin Plast Surg 2001;28:567–577.
11. Spear SL. Breast augmentation with reduced-height anatomic implants: the pros and cons. Clin Plast Surg 2001;28:561–565.
12. Gui GP, Tan SM, Faliakou EC, Choy C, A'Hern R, Ward A. Immediate breast reconstruction using biodimensional anatomical permanent expander implants: a prospective analysis of outcome and patient satisfaction. Plast Reconstr Surg 2003;111:125–138.
13. Hudson DA, Skoll PJ. Complete one-stage, immediate breast reconstruction with prosthetic material in patients with large or ptotic breasts. Plast Reconstr Surg 2002;110:487–493.
14. Spear SL, Spittler CJ. Breast reconstruction with implants and expanders. Plast Reconstr Surg 2001;107:177–187.
15. Malata CM, McIntosh SA, Purushotham AD. Immediate breast reconstruction after mastectomy for cancer. Br J Surg 2000;87:1455–1472.
16. Vandeweyer E, Deraemaecker R, Nogaret JM, Hertens D. Immediate breast reconstruction with implants and adjuvant chemotherapy: a good option? Acta Chir Belgica 2003;103:98–101.
17. Shons AR. Breast cancer and augmentation mammaplasty: the preoperative consultation. Plast Reconstr Surg 2002;109:383–385.
18. Miglioretti DL, Rutter CM, Geller BM, et al. Effect of breast augmentation on the accuracy of mammography and cancer characteristics. JAMA 2004;291:442–450.
19. Smalley SM. Breast implants and breast cancer screening. J Midwifery Women's Health 2003;48:329–337.
20. Belli P, Romani M, Magistrelli A, Masetti R, Pastore G, Costantini M. Diagnostic imaging of breast implants: role of MRI. Rays 2002;27:259–277.
21. Topping A, George C, Wilson G. Appropriateness of MRI scanning in the detection of ruptured implants used for breast reconstruction. Br J Plast Surg 2003;56:186–189.
22. Kneeshaw PJ, Turnbull LW, Drew PJ. Current applications and future direction of MR mammography. Br J Cancer 2003;88:4–10.

European Counterpoint to Chapter 11

Catharine M. Darcy and Kevin Hancock

USE OF BREAST IMPLANTS IN THE UNITED KINGDOM

In the United Kingdom, in 2001, the National Breast Implant Registry registered 13,402 women who underwent breast implant surgery involving 23,910 breast implants. Registration forms are available in all units where breast implants are carried out. They are filled in for insertion, replacement, and removal procedures. After the form is completed it is forwarded to the Registry. The process is voluntary and there is thought to be only 70% compliance with the Registry. Thus, many procedures are not registered. Of the 13,402 women registered in 2001, 11,096 underwent primary procedures and 2306 underwent replacement procedures. The most common reason for primary surgery was cosmetic augmentation (8569 or 77%), followed by reconstruction after mastectomy for cancer (1765 or 16%), congenital or developmental abnormalities (601 or 5%), reconstruction after mastectomy for benign disease or prophylactic mastectomy (63 or 0.6%), and other (96 or 0.9%).

The most common type of implant used was the silicone-gel prosthesis. This accounted for 21,125 (88%) of the implants used; saline-filled implants accounted for 1265 (5%). Expander/implants, defined as those that contain a fillable chamber into which saline may be injected via an external port (that may itself be detached, leaving a nontemporary prosthesis), accounted for 1420 (6%). One hundred eighty-five patients had explantation of a breast prosthesis (removal without replacement). Since the start of the Implant Registry in 1993, the yearly report shows an increase in the number of registrations. There are now more than 50,000 women in the United Kingdom who have breast implants.

While the silicone safety controversy was taking place in the United States, in the United Kingdom an Independent Expert Advisory Group was established to review the information on connective tissue disease. The group published its findings in 1992 and found no evidence for an increase in connective tissue disease in women with breast implants. Surgeons in the United Kingdom were thus able to continue using silicone breast implants with no restrictions. The group also recommended that a voluntary register be set up and the National Breast Implant Registry, funded by the Department of Health, was established in 1993.

From: *The Bionic Human: Health Promotion for People With Implanted Prosthetic Devices*
Edited by: F. E. Johnson and K. S. Virgo © Humana Press Inc., Totowa, NJ

The Independent Review Group (IRG) was set up by the chief medical officer of the National Health Service in 1998 to review the health issues associated with silicone-gel breast implants. IRG group members were selected for their independent views and included professors of rheumatology, immunology, medicine, pathology, rheumatic disease, and epidemiology; a consultant plastic surgeon; and a senior law lecturer. They obtained an immense amount of written and oral information from a wide range of relevant sources and made a full report with a series of conclusions and recommendations. The report is available from the address given at the end of the chapter. The full report with the evidence that was considered is available on the website (*see* Appendix).

The IRG concluded that silicone-gel breast implants are not associated with any greater health risk than other surgical implants. Although it is recognized that there are a number of local complications, such as capsular contracture and gel bleed, the incidence of ill health in women with silicone-gel breast implants is no greater than in the general population. In particular, there is no evidence of an association with an abnormal immune response or connective tissue diseases. Children of implanted women are not at increased risk of connective tissue disease. Areas that require further research include the incidence of rupture and the investigation of conditions such as low-grade infection, which may account for some of the nonspecific illnesses noted in some implanted women. The group feels that there is currently no justification for routine regular breast investigation to detect rupture. The group recommended that this subject should be kept under review and the decision revisited in the light of possible new information and technical advances relating to imaging techniques used in the detection of rupture.

In the 1990s, implants with new filler materials were introduced to the UK market. Trilucent™ (soya bean oil-filled implants manufactured by Lipomatrix in Switzerland) gained a "CE" marking (French for *Conformité Européene*) in 1995. This mark indicates that the manufacturer has satisfied all assessment procedures specified by law for its product to be sold on the European market. More than 9000 of these implants were inserted into almost 5000 women in the United Kingdom. In March 1999, they were voluntarily withdrawn from the market while the Medical Device Agency in the United Kingdom gathered information relating to the long-term toxicity of these implants. In June 2000, the Medical Devices Agency recommended that women with Trilucent implants have them removed after further testing revealed that the breakdown products of the soya bean oil may be genotoxic. Women with these implants were contacted and advised to have their Trilucent implants removed and appropriate replacements implanted. In a reported series of patients who had their implants removed, most chose to have a replacement with an alternative implant, whereas a small number chose to have explantation only *(1)*.

PIP hydrogel® breast prostheses, made in France, were implanted in approximately 4000 women in the United Kingdom from 1994 to 2000. These were filled with a hydroxypropyl cellulose hydrogel (polysaccharide) gel. The Medical Devices Agency reviewed the manufacturer's biological safety assessment in 2000 and found it was inadequate. There were particular concerns over the uncertainty of the metabolic fate of the filling material. After the agency issued an alert in 2000, the manufacturer voluntarily withdrew them from the market and is performing further tests to obtain adequate information to address the concerns. The agency also issued an alert in

Chapter 11 / European Counterpoint   261

December 2000 to stop the use of Novagold™ breast implants manufactured by Novamedical/Somatech Medical Limited. These are filled with polyvinylpyrrolidone hydrogel and guar gum gel. Approximately 250 women in the United Kingdom had them implanted between 1996 and 2000 but they are no longer being marketed in the United Kingdom.

The indications for breast expanders, implants, and expander/implants in the United Kingdom are to correct congenital absence or deformity of one or both breasts, to regain breast shape or size after having children, to correct breast asymmetry (congenital or acquired), to enlarge the breasts in women who are dissatisfied with the shape and size of their breasts, and to reconstruct the breast after mastectomy.

We consider first augmentation and later discuss breast reconstruction after mastectomy.

BREAST AUGMENTATION

In the United Kingdom, a common incision for breast augmentation implant insertion is the inframammary incision. This gives a good exposure, is technically straightforward, and allows placement of the implant in the submammary or subpectoral plane. The resulting scar may be more conspicuous than with periareolar, axillary, or transumbilical incisions. The prosthesis most often used in the United Kingdom is the silicone implant with a textured surface. This usually gives a good result when placed in the submammary position but in very thin women the upper edge of the implant may be visible and there may be visible wrinkling of the implant. This unfavorable result is reduced with anatomically shaped cohesive gel implants and/or a subpectoral placement.

Patients are given full counseling regarding the type of implant to be used, the incision and the position of the implant, the expected outcome, the risks, and the complications. Surgeons are advised to provide free written material, which can be obtained from the Internet from a site prepared by the Department of Health as a booklet entitled *Breast Implants*.

There are published guidelines for follow-up for women who have had implants inserted for cosmetic augmentation *(2)*. This advice comes from the IRG, which recommended that women should be followed up for a minimum of 1 year with the option of longer follow-up at the woman's request (Table 1).

The surgeon typically sees the patient several times in the first year, depending on the physician's own protocol, and assesses the patient for complications and adequacy of cosmetic appearance. The surgeon generally advises the patient to return for review if there is any change in the shape of the breast or the development of pain or hardening of the breast. A typical schedule for office visits is at 1 day, 2 weeks, 6 weeks, 6 months, and 1 year postoperatively for breast augmentation.

Patients are generally advised to lead a healthy lifestyle and to allow several weeks for healing, avoiding strenuous exercise and driving during this period. Most surgeons advise these patients to wear a brassiere night and day for 4 to 6 weeks to support the implants in position. The aim is to avoid implant migration and to reduce the incidence of adverse capsular contracture, Baker grades II–IV. Breast massage postoperatively to reduce capsular contracture has many proponents and several techniques have been described.

Table 1
Surveillance Schedule After Breast Augmentation With an Implant at Mersey Regional Plastic Surgery and Burn Centre

	Postoperative year					
	1	2	3	4	5	10
Office visit	3	0	0	0	0	0
Mammograms[a]	1	0	0	1	0	0

[a]After age 50, once every 3 years as part of the National Breast Screening program. Other tests are obtained only when clinical findings warrant them. Patients who have breast reconstruction with an implant have a more frequent schedule of office visits and mammograms jointly managed by the plastic surgeon and the general surgeon with breast surgery as their specialized area.

Some, but not all, surgeons advise women that they will require antibiotic prophylaxis if they undergo dental or surgical procedures in the future. This is to prevent bacteria settling on the implant during a transient bacteremia. Women are advised to carry out regular breast self-examination. If a woman has an infection elsewhere in the body, for example a urinary tract infection or a chest infection, no alteration in treatment is required merely because she has breast implants.

Women with breast implants are advised to have screening mammograms and always to inform the radiographer that they have breast implants in place. The radiographer will then use a displacement technique to avoid compression of the implants and to obtain maximal visualization of the breast tissue *(3,4)*. In the United Kingdom, women are called by the National Breast Cancer Screening Programme at the age of 50 and then once every 3 years until the age of 64. Women over this age may continue on the screening program at their own request. Breast implants obscure a proportion of breast tissue during imaging. However, a prospective analysis of more than 3000 women with breast cancer showed that women with breast implants did not have a more advanced cancer at the time of diagnosis or a worse outcome than those without implants *(4)*. Women who have had surgery for breast cancer undergo mammography more frequently at time intervals determined by their breast surgeon.

Reconstruction After Mastectomy

The breast reconstruction may be immediate (at the same time as the mastectomy) or delayed (after healing of the mastectomy wound). It may be performed using an implant alone or an implant in conjunction with a latissimus dorsi musculocutaneous flap. If an implant is used alone, it is usually preceded by an expander. This is placed beneath the pectoralis muscle, the rectus abdominus, and the serratus anterior muscle. Complete muscle coverage of the expander protects it from extrusion if the skin flaps become nonviable. After tissue expansion is completed, then, at a second operation, the tissue expander is removed and replaced by a permanent implant. Alternatively, a one-stage operation can be performed by using an expander/implant that can also act as a permanent implant.

Reconstruction using a latissimus dorsi musculocutaneous flap in conjunction with an implant can be performed as a one-stage operation using an expander/implant or a two-stage operation using a tissue expander and later replacement with a permanent implant. The implant lies beneath the latissimus dorsi muscle and is protected from exposure if the skin flaps do not survive. This technique is beneficial because it brings extra skin into the breast area. It is often used when postoperative radiotherapy has been given to a mastectomy site. Occasionally, tissue expansion is not needed and a permanent implant can be inserted at the same time the myocutaneous flap is performed. This is a major operation with postoperative risks such as bleeding and gives the woman an additional scar on the back. In a recent report of the results using a one-stage biodimensional anatomical expander/implant system for immediate breast reconstruction, there was good postoperative symmetry, high levels of patient and surgeon satisfaction and low complication rates (5). There were 129 breast reconstructions, 68 reconstructions involved submuscular expander/implant placement alone and 61 reconstructions involved an expander/implant in conjunction with a latissimus dorsi flap. The infection rate was 6.2%; hematomas occurred in 1.6% and implant loss occurred in 3.9%.

Responsibility of postoperative care following reconstructive surgery for breast cancer is shared between the breast surgeon/oncologist and plastic surgeon.

Mammography After Mastectomy

Although it appears unnecessary to perform mammography on a breast that has been reconstructed after a total mastectomy, it can be valuable. Four hundred-fifty women who had breast reconstruction were advised to undergo routine surveillance mammography of their reconstructed breasts; three occult local recurrences were picked up by routine surveillance mammography. These three patients had reconstruction with subpectoral implants and, thus, there was tissue overlying the implants that could be incorporated into the mammogram (4,6).

The complication rate after breast implant surgery is low, but unfavorable cosmetic results such as upper pole rippling, asymmetry and unsatisfactory position of implant can occur. Infection, capsular contracture, and implant rupture or extrusion may have serious consequences and can be difficult to manage. Others, such as seroma, superficial epigastric vein thrombosis (Mondor's syndrome), and altered nipple sensation, are usually simple to manage.

INFECTION AND BREAST IMPLANTS

The reported incidence of infection around breast implants ranges from less than 1% to 7% (5,7). The most common organism is *Staphylococcus aureus* followed by *Staphylococcus epidermidis*. Less common infecting organisms are *Streptococcus* and *Pseudomonas*. There are reports of infections with *Mycobacteria, Pasteurella,* and fungi (8,9). Life-threatening toxic shock syndrome has occurred after implantation (10–13).

Infection around a breast implant usually presents within 6 weeks of insertion with local swelling, pain, heat, and redness. There may be discharge from the wound. The woman may have systemic signs of infection such as fever and general malaise.

Low-grade infection around an implant may be a cause of malaise and generalized symptoms in some women. As discussed by other authors of this book, the source of the infection may come from contamination at the time of implant insertion. Organisms on the patient's own skin may adhere to the implant when it is inserted. Bacteria released from the milk ducts as the pocket is developed may contaminate the implant. Later infection may gain access through the port injection site when injection of an expander is performed.

There is evidence for the value of antibiotic prophylaxis and normally Flucloxacillin or Cefuroxine is given at induction of anesthesia *(14)*. It is important to educate the patient on breast hygiene before surgery and treat any local skin infection before surgery. Prevention of hematoma helps reduce infection rates as does eradication of loose fat globules by washout of the cavity. A betadine washout of the pocket, before insertion of the implant, may possibly reduce the risk of infection. Implants placed in a subpectoral position have a lower rate of infection than those placed in the submammary position. The use of nipple shields has been shown to reduce the number of bacteria present on the nipple *(15)*.

Using a "no-touch" technique of handling the implant reduces the risk of infection. Mladick used this technique for placement of submuscular saline implants and reduced the infection rate to zero for 2863 implants *(16)*.

To reduce contamination of the implant, it is kept in the manufacturer's packet until just before use and only the surgeon handles the implant. Fresh betadine is poured over the implant before insertion. All sharp instruments and needles are removed from the patient at the time of implant insertion to avoid accidental puncture or damage to the implant. Some surgeons choose to instill an antibiotic solution into the breast cavity. Surgical drains are used according to the preference of the surgeon and the amount of bleeding into the implant cavity at the time of surgery. Because the source of infection may be the patient's own skin or the gloves of the surgeon, the scrub nurse or the assistant, some surgeons take the plastic bag supplied with the implant and place the implant within it and position the other end of the plastic bag in the cavity for the breast implant. The implant is then pushed into the cavity by manipulating the bag and does not come into contact with the surgeon's gloves, the assistant's gloves, or the patient's skin.

If there is a clear indication of infection around the implant, the patient is admitted for intravenous antibiotics and an ultrasound scan may be performed. A blood test for white blood cell count and C-reactive protein may be done. The infection may settle with intravenous antibiotics followed by a course of oral antibiotics. When the infection fails to settle, the patient has surgical removal of the implant. A few months after the infection has settled, the patient can have insertion of a new implant. There have been some cases of infection treated by antibiotics and drainage and the implants have remained in place but the final results have been firm breasts *(17)*.

CAPSULAR CONTRACTURE

There are many papers analyzing the cause of capsular contracture and its prevention but few are randomized controlled trials. The body forms scar tissue in response to foreign material; when it occurs around a breast prosthesis, it is called a capsule. It may remain soft but when it contracts and becomes dense the implant feels firm. If contraction continues, the implant becomes hard and visible. This capsular contracture is graded by Baker *(18)* and has been more recently modified by Spear *(19)*.

The etiology of capsular contracture is unknown but probably multifactorial. There has been association with hematoma, periprosthetic infection and mechanical irritation. The use of vitamin E, periprosthetic steroids, and tissue expansion to reduce capsular contracture is controversial. Preventive measures that have been proposed include steroids, antibiotics, prevention of hematoma and infection, immobilization of the implant, massage of the breast after surgery, the use of a textured silicone shell and double-lumen prosthesis, submuscular placement, a wide surgical pocket, and use of low-bleed implants and avoidance of using gloves with talc. Burkhardt and Eades studied the effect of povidone-iodine irrigation on capsular contracture in a prospective controlled, blinded 4-year trial on 60 patients. They found that betadine washout and textured implants both independently reduced capsular contracture *(20)*. Tarpila and others carried out a randomized, double-blind trial investigating the capsular contracture rate with textured vs smooth-walled, saline-filled implants after breast augmentation. The study was carried out in 21 women; Baker's grade III capsular contracture occurred in 33% of the breasts at the end of the study. There was no apparent difference between the two groups *(21)*.

It has been shown that textured silicone implants produce a much lower rate of adverse capsular contracture than smooth silicone implants at short- and long-term follow-up *(22–25)*. Hakelius compared the capsular contracture rate in smooth and textured silicone implants at 1-year follow-up. The assessment was made using patient opinion, the Baker classification and tonometry. The textured silicone implants had a lower contracture than the smooth silicone implants *(22)*. Collis reported the capsular contracture rate for silicone implants at 10 year follow-up. The textured silicone implants had an 11% incidence of capsular contracture and the smooth silicone implants had a 65% incidence of capsular contracture. Fifty-three patients were entered into this double-blind randomized study *(23–25)*.

Fagrell and others investigated the capsular contracture rate around saline-filled implants. Twenty healthy women were enrolled into the study and the follow-up period was 7.5 years. All breast implants were placed in a subglandular position. The implants used were Siltex textured surface saline-filled prostheses placed in one breast. A smooth-surfaced shell saline-filled prosthesis was placed in the other breast. This study showed no significant difference of contracture rate with smooth vs fine-textured implants for implants filled with saline. The majority of patients preferred smooth saline-filled implants *(26)*.

Pollock examined 98 patients and, in a retrospective study of textured vs smooth silicone implants, found that the capsular contracture rate was 21% for the smooth silicone implants and 4% for the textured silicone implants. The follow-up period was less than 2 years *(27)*.

In a further effort to decrease capsular contracture rate, it is important to reduce to the absolute minimum the number of bacteria contaminating the implant. Clear statistical evidence to prove that infection or bacterial contamination causes capsular contracture is lacking. Camirand reported more than 100 women who had insertion of saline implants in a subpectoral position for augmentation. They all followed a postoperative regime of daily breast massage. The capsular contracture rate was zero *(27,28)*. However, the assessment was not based on well-established methods of establishing or classifying capsular contracture.

Treatment of capsular contracture depends on the patient's symptoms and a full discussion between the patient and her surgeon. There are options that include removal of the implant with capsulotomy and replacement of the same or a new implant, removal of the capsule-capsulectomy and replacement of the implant, or removal of the implant with no replacement. There is a significant risk of capsular contracture recurring.

Accidental puncture of the implant may occur at the time of insertion of the implant or postoperatively if a needle is inserted into the breast for aspiration of a seroma. As mentioned elsewhere in this chapter, the failure of an implant may also be due to a tear in the breast implant shell or a failure at joining points of the silicone shell. If an intact thick fibrous capsule surrounds a ruptured implant, the silicone is contained within the capsule and this intracapsular breast implant rupture may cause no clinical symptoms and no change in breast shape. If the capsule is not intact and the implant filler is a liquid form of silicone, then silicone leaks out into the breast tissues. This is known as an extracapsular rupture and the silicone may spread from the breast tissues to the axilla or further away into other tissues. The silicone may incite a vigorous inflammatory reaction and lead to the formation of tender swellings that may mimic breast cancer. It can be difficult to remove the silicone at this stage because it has extensively infiltrated the tissues and it may be impossible to remove it all. Thus, an extracapsular rupture leads to a change in breast shape and may cause pain and lumps in the breast. The newer form of silicone filler material is cohesive gel silicone and it has a semisolid consistency. This may be an advantage if a rupture of the shell occurs because the cohesive gel maintains its shape and does not spread like a liquid through the breast tissue. A ruptured implant is treated by surgical removal and, if the patient desires, replacement with a new implant.

Magnetic resonance imaging (MRI) diagnoses intracapsular rupture more accurately than ultrasound scan. Beekman and others showed in their prospective study of 35 single-lumen silicone-gel implants that MRI has a sensitivity of 88% and specificity of 100%, with an accuracy of 94% for rupture of the implant *(27,29)*. By comparison, they found the ultrasound scan had a sensitivity of 44%, specificity of 87%, and accuracy of 66%. They regard MRI as the "gold standard" in the evaluation of rupture of breast implants filled with silicone gel. When MRI is not readily available, ultrasound is an acceptable alternative *(27,30)*. Reynolds and others compared the value of mammography, ultrasound, and MRI in the detection of silicone-gel breast implant rupture and found none was clearly superior *(31)*. Rohrich and others retrospectively reviewed 180 women who underwent explantation of 357 silicone gel-filled implants from 1991 to 1995. The age of the implants ranged from 6 months to 24 years. They found the frequency of implant rupture significantly increased with implant age; the average implant age at rupture was 13.4 years *(32)*. Robinson studied a series of 300 patients who had explantation due to the controversy about the safety of silicone. The implant age was between 1 and 25 years. Overall, only 50% of women could expect to have their prostheses intact by 12 years and they recommended that implants should be electively replaced at 8 years *(33)*. Analysis of mammography and MRI as diagnostic tools for differentiating intact implants, implant leakage, and implant rupture showed that mammography was less reliable in diagnosing implant leak or rupture and the sensitivity of MRI was 72% with a specificity of 82% *(32)*.

The life span of silicone gel-filled breast prostheses is limited. Beekman studied 182 patients who had their breast prostheses replaced, repositioned, or removed one to three times between 1988 and 1995. Common indications for surgery were capsular contraction, dislocation, pain, paresthesia, or suspected rupture and in this selected group of patients approximately 50% of the breast prostheses with an implant age of 7 to 10 years showed gel bleed or rupture. When the authors created a survival curve from their data, they found that 50% of breast implants might be expected to bleed or be ruptured 15 years after insertion. Rupture of implants was observed more frequently than gel bleed (34).

A woman with a ruptured implant usually presents with a breast that has suddenly changed shape, become tender, and has hard knots. The breasts tend to be uneven with symptoms of tenderness, tingling, swelling, numbness, burning, and the appearance of a lump or lumps in the breast or axilla. The first investigation may be MRI of the breast or ultrasound scan. This may reveal a tear in the capsule and a leak of silicone. Ultrasound scan is often available more quickly and is not as expensive as an MRI scan. When an implant shell has failed, the implant is removed.

APPENDIX: SOURCES OF ADDITIONAL INFORMATION

Department of Health
P.O. Box 777
London SE1 6XH
Web site: http://www.doh.gov.uk/bimplants.htm
E-mail: doh@prolog.uk.com

Report of the Independent Review Group
Silicone Gel Breast Implants available from:
Silicone Gel Breast Implants IRG
9th Floor, Hannibal House
Elephant and Castle
London SE1 6TQ
Telephone: 020 7972 8077
Web site: http://www.silicone-review.gov.uk

National Breast Implant Registry
Salisbury District General Hospital
Salisbury SP2 8BJ
Telephone: 01722 425059
Fax: 01722 429343

Consumer Interest Groups:
Action Against Silicone Gel, UK
Can be contacted through:
Patients' Association
Helpline: 0208 423 8999

Breast Cancer Care
Kiln House
210 New Kings Road
London SW6 4NZ
Telephone: 0207 384 2984

Breast Implant Information Society (BIIS)
P.O. Box 1084
Mitcham
Surrey CR4 4ZU
Telephone: 07041 471225

(continued)

APPENDIX: SOURCES OF ADDITIONAL INFORMATION (Continued)

CancerBACUP
3 Bath Place, Rivington St.
London EC2A 3DR
Telephone: 020 7920 7231

Patients' Association
P.O. Box 935
Harrow
Middlesex HA1 3YJ
Helpline: 0208 423 8999

Silicone Support UK
9 Casimir Road
Clapton
London E5 9NU
Telephone: 0208 806 6923

Professional Interest Groups:

British Association of Aesthetic Plastic Surgeons (BAAPS)
Royal College of Surgeons
36 Lincoln's Inn Fields
London WC2A 3PN

British Association of Cosmetic Surgeons (BACS)
17 Harley Street
London W1N 1DA

British Association of Plastic Surgeons (BAPS)
Royal College of Surgeons
36 Lincoln's Inn Fields
London WC2A 3PN

British Association of Surgical Oncology
Breast Care Group
Royal College of Surgeons
36 Lincoln's Inn Fields
London WC2A 3PN

General Medical Council
178 Great Portland Street
London WC1N 6JE

RCN Breast Care
Nurse Forum
Royal College of Nursing
Cavendish Square
London W1M OAB

REFERENCES

1. Baecke JL. Word of caution about teardrop breast implants. Cosmetic Surg Times 2000;3:3.
2. Hobar PC, Gutowski K. Experience with anatomic breast implants. Clin Plast Surg 2001; 28:553–558.
3. Baecke JL. Warning about anatomical breast implants. Plast Reconstr Surg 2000;106:740.
4. Tebbets JB. The other side of the story—reply (letter). Plast Reconstr Surg 1994;101:875.

5. Baecke JL. Teardrop implants: enough to make a grown man cry (reply). Plast Reconstr Surg 2001;107:1914.
6. Adams WP. Breast deformity caused by anatomical or teardrop implant rotation [comment]. Plast Reconstr Surg 2003;111:2110–2111.
7. Baecke JL. Breast deformity caused by anatomical or teardrop implant rotation. Plast Reconstr Surg 2002;109:2555.
8. Tebbets JB. Warning about warning about anatomical breast implants. Plast Reconstr Surg 2001;107:1912.
9. Baecke JL. Teardrop implants advisory. Ann Plast Surg 2000;44:678–679.
10. Tebbets JB. Breast augmentation with full-height anatomic saline implants: the pros and cons. Clin Plast Surg 2001;28:567–577.
11. Spear SL. Breast augmentation with reduced-height anatomic implants: the pros and cons. Clin Plast Surg 2001;28:561–565.
12. Gui GP, Tan SM, Faliakou EC, Choy C, A'Hern R, Ward A. Immediate breast reconstruction using biodimensional anatomical permanent expander implants: a prospective analysis of outcome and patient satisfaction. Plast Reconstr Surg 2003;111:125–138.
13. Hudson DA, Skoll PJ. Complete one-stage, immediate breast reconstruction with prosthetic material in patients with large or ptotic breasts. Plast Reconstr Surg 2002;110:487–493.
14. Spear SL, Spittler CJ. Breast reconstruction with implants and expanders (comment). Plast Reconstr Surg 2001;107:177–187.
15. Malata CM, McIntosh SA, Purushotham AD. Immediate breast reconstruction after mastectomy for cancer. Br J Surg 2000;87:1455–1472.
16. Vandeweyer E, Deraemaecker R, Nogaret JM, Hertens D. Immediate breast reconstruction with implants and adjuvant chemotherapy: a good option? Acta Chir Belgica 2003;103:98–101.
17. Shons AR. Breast cancer and augmentation mammaplasty: the preoperative consultation. Plast Reconstr Surg 2002;109:383–385.
18. Miglioretti DL, Rutter CM, Geller BM, Cutter G, Barlow WE, Rosenberg R, Weaver DL, Taplin SH, Ballard-Barbash R, Carney PA, Yankaskas BC, Kerlikowske K. Effect of breast augmentation on the accuracy of mammography and cancer characteristics. JAMA 2004;291:442–450.
19. Smalley SM. Breast implants and breast cancer screening. J Midwifery Women's Health 2003;48:329–337.
20. Belli P, Romani M, Magistrelli A, Masetti R, Pastore G, Costantini M. Diagnostic imaging of breast implants: role of MRI. Rays 2002;27:259–277.
21. Topping A, George C, Wilson G. Appropriateness of MRI scanning in the detection of ruptured implants used for breast reconstruction. Br J Plast Surg 2003;56:186–189.
22. Kneeshaw PJ, Turnbull LW, Drew PJ. Current applications and future direction of MR mammography. Br J Cancer 2003;88:4–10.

12
Prostheses for Hernia Repair

R. Ivan Beretvas

INTRODUCTION

The use of mesh implants to repair or reinforce abdominal wall defects has become increasingly common in the past half century. There are about 375,000 inguinal hernia defects and 45,000 incisional hernias repaired with mesh in the United States per year. Two million procedures are performed annually in the United States requiring an abdominal incision; 2 to 11% of these patients later develop an incisional hernia eligible for repair with mesh *(1)*. The reason for the increased use of mesh is because primary (suture, without mesh) repair of abdominal wall hernias has a significant recurrence rate that can be as high as 52% *(2)*. Mesh repair has been shown to reduce recurrence rates in a well-controlled trial *(3)*. Mesh is used to decrease the tension placed on tissues and sutures, which, in turn, is thought to decrease the incidence of recurrence.

Initial reports of mesh repair of hernias described the use of silver filigree, tantalum gauze, and steel mesh *(4)*. In 1894, Phelps approximated the abdominal wall layers over silver coils and relied on the foreign body reaction and tissue fibrosis induced by the coils for the reinforcement of the repair. Soon afterward, in Germany, Witzel and Goepel used silver filigree, which was probably the first utilization of a prosthetic mesh to reinforce the abdominal wall. The disadvantages associated with these types of mesh include lack of flexibility, breakage, migration of the mesh, and erosion into the lumen of the bowel *(5,6)*. By the mid-20th century synthetic meshes had been developed, including polypropylene (Fig. 1), polyester, and expanded polytetrafluoroethylene *(7)*. These allowed more flexibility, better incorporation into tissues, and less tissue reaction. The ideal mesh should resist infection, become well incorporated into host tissue, while avoiding the formation of intraperitoneal adhesions, remain flexible, have a high tensile strength and have low reactivity with the host tissues *(4)*. It should also be commercially available, easy for surgeons to handle, inexpensive and have no potential carcinogenicity or other toxicities. Currently available products function far better than silver filigree, accounting for their wide acceptance. Studies have compared the various types of mesh *(1)*. No mesh is ideal. Composite meshes have been developed that purportedly decrease the risk of intraabdominal adhesions *(4,8)* but no long-term studies have been reported and the separate disadvantages of the components the

From: *The Bionic Human: Health Promotion for People With Implanted Prosthetic Devices*
Edited by: F. E. Johnson and K. S. Virgo © Humana Press Inc., Totowa, NJ

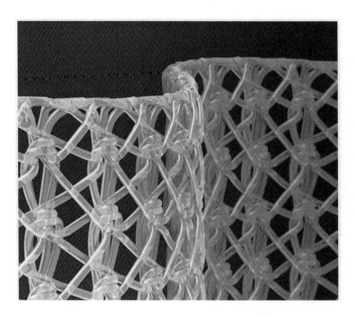

Fig. 1. Polypropylene mesh (4× magnification).

Fig. 2. Sepramesh composite of polyprophylene and sodium hyaluronate (GenzymeCorp.) (4× magnification).

composite mesh may be present. Sepramesh (Genzyme Corp.) (Fig. 2) is a composite of polypropylene on one side and sodium hyaluronate and carboxymethylcellulose on the other. The polypropylene side allows tissue ingrowth, whereas the other side forms a bioresorbable barrier to decrease adhesions. Composix mesh (Davol, Inc.) (Fig. 3) consists of polypropylene with expanded polytetrafluoroethylene bonded on one side. The expanded polytetrafluoroethylene forms the adhesion barrier. Dualmesh (W.L.

Fig. 3. Composix mesh composite of polypropylene and expanded polytetrafluoroethylene (Davol, Inc.) (4× magnification).

Fig. 4. Dualmesh two-sided expanded polytetrafluoroethylene (W.L. Gore and Associates) (4× magnification).

Gore and Associates) (Fig. 4), is a two-sided expanded polytetrafluoroethylene mesh with a smooth surface to prevent adhesions and a ridged surface to allow tissue ingrowth.

There is much discussion in the literature concerning the relative benefits and complications associated with intraperitoneal vs extraperitoneal placement of mesh *(1,3,9–12)*. Intraperitoneal placement can lead to formation of adhesions between the viscera and mesh. This may lead to erosion into bowel *(13)* or difficulty in reentering the abdomen if additional abdominal explorations are needed. Many authors recommend placing omentum between the mesh and the abdominal contents. Intra-abdominal placement is virtually mandatory when repairing hernias laparoscopically. It is very difficult to repair hernias, particularly postoperative ventral ones, laparoscopically using sutures. Laparoscopic placement of mesh is relatively easy, especially with the advent of staples or small corkscrew-like devices used for tacking the mesh to the abdominal wall. Patients undergoing laparoscopic hernia repair tend to have less pain postoperatively compared with those who undergo the open procedure, possibly owing to the decreased size of the incisions or because of the use of a tension-free repair or a combination of both *(2)*. This will probably lead to an increased rate of laparoscopic ventral hernia repair in the future.

There does not appear to be any literature defining the optimal follow-up for these patients. Our recommendations are similar to those used for patients whose hernias have been repaired without mesh. However, even these are individualized according to the individual circumstances. A survey of surgeons in St. Louis revealed no difference in follow-up scheduling for those patients whose hernias were repaired with mesh compared to those without mesh. No additional testing was performed postoperatively. The majority of (8 of 11) surgeons would not treat a patient with mesh differently from one without mesh in the event of a subsequent unrelated intra-abdominal infectious process (e.g., appendicitis). The remaining three surgeons would alter the type and duration of antibiotic coverage. There is little data in the literature supporting either position. This issue is dealt with in Chapter 9 also.

PROSTHETIC FAILURE AND INFECTION

Prosthetic failure may be detected relatively easily by the patient as a recurrence of the hernia but, in rare instances, patients may present with signs of intestinal obstruction secondary to a recurrence previously unknown to the patient. Recurrence is generally owing to disruption of the attachment of the mesh to the native tissues as opposed to breaking of the mesh itself. Detection of failure by the clinician may be through physical examination alone. Occasionally, recurrence of hernias, infection, or other complications are incidental findings on ultrasound, computed tomography, or magnetic resonance imaging. Sometimes patient symptoms may prompt imaging tests as well.

Prosthetic infection may present with relatively nonspecific symptoms including fever or pain around the site of the implant, or even general malaise. Signs of mesh infection include erythema at the incision site and drainage of fluid (either serous or purulent) from the incision. It is not rare for these signs to go unreported until a scheduled office visit. This provides a rationale for scheduled visits especially in high-risk patients. When infections are more chronic, discharging sinuses may develop *(14)*. There is no consensus about the treatment of mesh infections. Some authors advocate immediate removal of the mesh. Others treat with antibiotics with or without superficial debridement, while leaving the mesh in place. Some articles suggest the use of an absorbable mesh when the operative field is contaminated *(4)* but this generally leads to a recurrence of the hernia.

Table 1
**Patient Surveillance After Implantation
of Mesh for Hernia Repair at Saint Louis University**

	Postoperative year					
	1	2	3	4	5	10
Office Visit[a]	2[b]	0	0	0	0	0

[a]Number of times each year the modality is recommended
[b]At about 2 and 6 weeks postoperatively. No additional X-rays, scans, blood tests, or other tests are routinely obtained unless clinical findings warrant them.

PREVENTIVE MEASURES

At the time of implantation of the mesh, perioperative antibiotics are commonly used to decrease the risk of infection. They should be administered within 2 hours prior to surgery start in order for tissue levels to be adequate *(15,16)*. Some surgeons advocate irrigation of the mesh with antibiotic solution, but this appears to be based on anecdotal experience rather than on randomized trials. This is dealt with more fully in Chapter 9.

It is obviously essential that prostheses are manufactured in a sterile fashion and packaged appropriately to remain sterile. Mesh implantation can be the source of significant infection, particularly as the mesh may be protected from the host defense mechanisms, with the formation of a relatively impervious polysaccharide slime surrounding the mesh *(15)*.

New meshes currently being used include composite meshes described above, with one surface that is less adherent to tissues than the other. The less adherent surface is the side that would be placed intraabdominally if necessary. These meshes have also been modified to have an incorporated antibiotic substance that may include silver carbonate and chlorhexidine diacetate. Another new mesh is made of porcine intestinal submucosa *(17)*, which may be better incorporated into tissues with less adhesion formation. However there are no long-term studies available. Multi-institutional prospective trials would certainly help identify the ideal mesh and may also assist in the determination of best pre- and postoperative management of these patients.

Patients are generally advised to lose weight prior to hernia repair in addition to stopping smoking as this tends to improve the chances of an uncomplicated recovery. They are advised to avoid heavy lifting (more than 10 pounds) for at least 6 weeks postoperatively and then gradually to return to usual activity. There are no information hotlines, support groups, and the like, available.

REFERENCES

1. Leber GE, Garb JL, Alexander AI, Reed WP. Long term complications with prosthetic repair of incisional hernias. Arch Surg 1998;133:378–382.
2. Heniford BT, Park A, Ramshaw BJ, Voeller G. Laparoscopic ventral and incisional hernia repair in 407 patients. J Am Coll Surg 2000;190:645–650.

3. Luijendijk RW, Hop WCJ, van den Tol M, et al. A comparison of suture repair with mesh repair for incisional hernia. NEJM 2000;343:392–398.
4. DeBord JR. The historical development of prosthetics in hernia surgery. Surg Clin North Am 1998;78:973–1006.
5. Majeski J. Migration of wire mesh into the intestinal lumen causing an intestinal obstruction 30 years after repair of a ventral hernia. South Med J 1998;91:496–498.
6. Dieter RA. Mesh plug migration into scrotum: a new complication of hernia repair. International Surgery 1999;84:57–59.
7. Usher FC, Gannon JP. Marlex mesh, a new plastic mesh for replacing tissue defects. Arch Surg 1959;78:131–137.
8. Amid PK, Shulman AG, Lichtenstein IL, Sostrin S, Young J, Hakakha M. Experimental evaluation of a new composite mesh with the selective property of incorporation into the abdominal wall without adhering to the intestines. J Biomed Mat Research 1994;28: 373–375.
9. Arnaud JP, Tuech J-J, Pessaux P, Hadchity Y. Surgical treatment of postoperative incisional hernias by intraperitoneal insertion of dacron mesh and an aponeurotic graft: a report on 250 cases. Arch Surg 1999;134:1260–1262, with correspondence: Arch Surg 2000;135:238.
10. Schlechter B, Marks J, Shillingstad B, Ponsky JL. Intraabdominal mesh prosthesis in a canine model. Surg Endosc 1994;8:127–129.
11. Delany HM. Intraperitoneal mesh—a word of caution. Surg Endosc 1994;8:287–288.
12. Gurski RR, Schirmer CC, Wagner J, et al. The influence of reperitonization on the induction of formation of intraperitoneal adhesions by a polypropylene mesh prosthesis. Int Surg 1998;83:67–68.
13. Miller K, Junger W. Ileocutaneous fistula formation following laparoscopic polypropylene mesh hernia repair. Surg Endosc 1997;11:772–773.
14. Taylor SG, O'Dwyer PJ. Chronic groin sepsis following tension free inguinal hernioplasty. Br J Surg 1999;86:562–565.
15. Deysine M. Pathophysiology, prevention, and management of prosthetic infections in hernia surgery. Surg Clin North Am 1998;78:1105–1115.
16. Classen DC, Scott Evans R, Pestonik SL. The timing of prophylactic administration of antibiotics and the risk of surgical wound infections. N Engl J Med 1992;326:281–286.
17. Clarke KM, Lantz GC, Salisbury SK, Badylak SF, Hiles MC, Voytik SL. Intestine submucosa and polypropylene mesh for abdominal wall repair in dogs. J Surg Res 1996;60: 107–114.

US Counterpoint to Chapter 12

L. Michael Brunt

The use of mesh prostheses for repair of abdominal wall hernias has become commonplace over the last 15 years. Chapter 12 highlights this fact and provides many useful tips and guidelines for the management of patients who either have mesh implants or who are being considered for prosthetic repair. Several reasons exist for the now widespread use of mesh in hernia repairs of all types. These include (a) reduced hernia recurrence rates associated with a tension-free mesh repair technique, (b) more rapid return to unrestricted physical activity after tension-free mesh repair, (c) the wide variety and quality of mesh products now available, and (d) the overall safety and low complication rates associated with mesh prostheses. In this counterpoint, I review and emphasize many of the salient aspects regarding mesh hernia repairs, as already addressed in Chapter 12, and provide additional perspective based on my experience and clinical practice and review of the available literature.

Repair of abdominal wall hernias is the most common category of operations performed by general surgeons today. In 1996, it was estimated that more than 1 million people in the United States underwent repair of abdominal wall defects *(1)*. There were more than 750,000 inguinal/femoral hernias, 166,000 umbilical hernias, 97,000 incisional hernias, and about 76,000 epigastric hernias and hernias at other sites repaired. In 1985, the use of mesh for repair of groin hernias was uncommon and most surgeons trained in that era or before were schooled in the primary repair methods of Bassini, McVay, and Shouldice. In contrast, by 1996, just 11 years later, it was estimated that only 15% of inguinal/femoral hernia repairs were performed as primary tissue repairs, whereas 85% of these repairs employed one of the various tension-free mesh techniques *(1)*. As a result, there are now large numbers of patients who harbor implanted mesh prosthetic devices. It is necessary to consider some site-specific issues of these various hernias in order to address results, outcomes, risks, and recommendations for follow-up.

INGUINAL HERNIA REPAIR

The use of mesh in the repair of groin hernias should be considered separately from mesh applications in other abdominal wall sites for two reasons. First, mesh in the groin is not in direct contact with the abdominal viscera, and, second, a number of

Table 1
Mesh Approaches and Repair Techniques for Groin Hernia

Repair technique	Mesh product
Lichtenstein repair	Polypropylene or PTFE patch
Patch-and-plug repair	Various mesh plugs +/− onlay patch
Bilayer repair	Prolene hernia system mesh (Ethicon, Inc.)
Kugel repair	Bard Kugel patch
Laparoscopic repair	Polypropylene, Visilex, Bard 3-D mesh

PTFE, polytetrafluoroethylene.

groin-specific mesh prostheses are available. The various mesh approaches and types of mesh repair for groin hernia are listed in Table 1. Of the open anterior approaches to inguinal hernia repair, three basic mesh repairs have been utilized: the Lichtenstein repair, the patch-and-plug technique, and the bilayer mesh repair. All three techniques use polypropylene mesh in different configurations. The Lichtenstein group was the first to popularize the use of mesh for groin hernia repair *(2)* and this is the most commonly performed inguinal hernia repair in the United States today *(1)*. The patch-and-plug repair involves placement of a mesh plug through the hernia defect, either at the internal ring or direct space defect where it is anchored by sutures, and then an onlay mesh is placed over this to further reinforce the inguinal floor. The bilayer patch consists of two pieces of polypropylene mesh connected by a cylindrical mesh plug *(3)*. One leaflet of the mesh is placed through the defect to serve as a posterior buttress and the other leaflet is placed anteriorly such that this approach to some extent combines the anterior and posterior repairs. Regardless of the type of mesh, it should be inserted with enough laxity to account for the approximately 20% mesh contracture that may occur *(4)*. Excellent results have been reported with these various mesh prosthetic repairs from dedicated hernia centers *(2,3,5,6)*. Recurrence rates have been extremely low, from 0 to 0.2%. Whereas mesh techniques for inguinal hernia have been shown to have superior outcomes to primary tissue repairs in terms of recurrence rates *(7)*, data comparing the different types of mesh repairs by proponents of these techniques are limited. In one study that compared the Lichtenstein to the patch-and-plug repair, the patch-and-plug technique was associated with somewhat less pain and a slightly shorter operative time *(8)*. However, there were no differences in pain medication, return to activity, or days of work missed between groups.

The posterior repairs consists of laparoscopic inguinal hernia repair, the Kugel repair, and the Stoppa repair. Laparoscopic inguinal hernia repair can be performed either transabdominally or using a total extraperitoneal approach. The Kugel patch consists of a preformed polypropylene mesh with a stiff ring around the edges that is placed in the preperitoneal space using an open incision *(9)*. These techniques are based on the open preperitoneal repair devised by Stoppa in which a large piece of mesh is used to cover the entire myopectineal orifice of the inguinal floor *(10)*. This latter technique is now mainly reserved for patients with multiple recurrent hernias who are not candidates for a laparoscopic approach. Recurrence rates have been low for the laparoscopic repair *(11,12)*, although this procedure is somewhat more difficult to learn than the open anterior mesh techniques. Preliminary results suggest good outcomes with the Kugel repair *(9)*. It would appear that placement of mesh in the preperitoneal space over the inguinal floor is safe and mesh infections at this site have been rare. Long-term follow-up studies for the laparoscopic and Kugel repairs are lacking, however.

A detailed analysis of the results and relative advantages and disadvantages of the various groin mesh repair techniques is beyond the scope of this review. However, it would appear that recurrence rates with mesh repairs are generally much lower than those reported with primary tissue repairs both historically and in comparative trials *(7)*. Additional concerns about the potential for infection of mesh placed in the groin have been unfounded, provided appropriate precautions are taken. These precautions should include administration of antibiotics intravenously prior to the skin incision and minimizing contact of the mesh with the skin during the procedure. Unrelated infections that develop at distant sites should be treated aggressively with appropriate antibiotics to avoid seeding of the mesh. If the mesh has become well incorporated into the native tissue, the risk of its becoming secondarily infected from a remote site is low. There are no data to support more aggressive treatment of such infections in patients with implanted mesh than in patients without implanted mesh, however.

Other complications that may be mesh-related include pain resulting from nerve entrapment by the mesh or from scarring around the mesh *(13)*. Long-term follow-up studies regarding chronic groin pain have been reported with primary tissue repairs *(14)* but are lacking in the era of mesh inguinal hernia repair. The use of mesh in the anterior groin does not appear to result in impairment of testicular perfusion or sexual function *(15)*. Following any of the tension-free groin mesh hernia repairs described above, patients may resume normal activities as soon as comfort allows and may return to manual type labor in 7 to 14 days. In Amid's series *(16)*, 75% of desk workers and 50% of manual laborers had returned to work by 1 week following repair. Only 10% of manual laborers required more than 2 weeks to return to work.

INCISIONAL HERNIA REPAIR

The second major abdominal wall hernia category in which mesh is commonly used is that of incisional hernia. Recent data suggest that the incidence of incisional hernia after a midline laparotomy is 20% or higher *(17,18)*. Because obesity is a major risk factor for the development of incisional hernia *(17)*, the epidemic of obesity in the United States has the potential to lead to a secondary epidemic of incisional hernia. Recommendations for patients to reduce weight and stop smoking preoperatively are commendable but are not often complied with.

The high recurrence rates associated with primary repair of incisional hernias have led to increasing usage of mesh in these patients. In Washington State, the use of synthetic mesh in incisional hernia repairs increased from 34.2% in 1987 to 65.5% in 1999 *(19)*. The array of mesh products and their general characteristics have been described in this chapter. Recently, a new polyester mesh coated with a resorbable hydrophilic film (Parietex Composite, Sophradim Inc., France) to minimize tissue attachment has become available as an additional option for these patients *(20)*. The closure of defects in contaminated fields or in patients who must undergo removal of infected mesh poses a special challenge. Absorbable meshes (Vicryl, Dexon) have been used to temporize in these patients but their use invariably leads to hernia formation. A new generation of nonsynthetic products that has promise for management of this difficult group of patients has recently been introduced. Surgisis (Cook Surgical, Bloomington, IN) is a porcine small intestine submucosa product that promotes normal tissue ingrowth and has been used successfully in contaminated settings *(21)*. Alloderm (Lifecell Corp., Branchburg, NJ) is a decellularized human skin preparation that acts as a matrix for revascularization, cellular ingrowth, and incorporation into surrounding tissue. Both are available for clinical use and will need to be studied carefully in order to determine the indications for usage and outcomes for incisional hernia repair.

In addition to the different mesh choices, the technique for incisional hernia repair may be open or laparoscopic and may vary by the level of the abdominal wall at which the mesh is placed (intraperitoneal, retrorectus, anterior onlay). No consensus exists on the optimal approach for ventral hernia repair, but, in general, one should avoid placement of polypropylene mesh in direct contact with the bowel because of the increased tissue incorporation properties of polypropylene and the potential for adhesions and even fistulization to the bowel. Simple suturing of the mesh to the edges of the fascial defect is also not recommended because of the high failure rate *(22)*. Onlay techniques where the mesh is placed over the fascia and anchored superficially to it after the fascia has been closed are also not recommended and may predispose to mesh infection *(23)*. The Rives-Stoppa approach in which a large piece of mesh is implanted in a retrorectus location has been used with good results *(24,25)* and is the method on which laparoscopic ventral hernia repair is based. This technique also limits contact of the mesh with the subcutaneous space where it may be more vulnerable to secondary infection.

Seroma is the most common complication of ventral hernia repair (both open and laparoscopic) and has been reported in 5 to 30% of cases *(26)*. The placement of closed suction drains in the subcutaneous dead space left by the hernia after open incisional herniorrhaphy may reduce the incidence of seroma formation. Infection has been reported in approximately 5% of open incisional hernia repairs and appears to be even less common after the laparoscopic approach *(27)*. Wound or seroma infections may necessitate removal of the mesh if it has been placed as an onlay patch in continuity with the subcutaneous space. Mesh in the abdominal midline may make subsequent laparotomy for unrelated reasons more difficult because of underlying adhesions to the mesh. In those circumstances, entry into the peritoneal cavity should be carried out above or below the level of mesh placement and at closure the mesh should be reapproximated with a nonabsorbable suture because healing of the two edges of the mesh will not occur.

Evidence-based guidelines do not exist for follow-up activity and lifting recommendations after incisional hernia repair. However, at a minimum, patients should be seen in the clinic within 2 to 3 weeks of surgery to assess for early wound complications and again at 2 to 3 months to evaluate the integrity of the repair (Tables 2 and 3). Further follow-up is as clinically indicated. Because of the higher recurrence rate after incisional hernia repair, my practice has been to allow patients to return to normal activities as soon as comfort allows but to avoid heavy lifting more than 20 pounds or vigorous sporting activities for approximately 3 months postoperatively.

Table 2
Patient Surveillance After Implantation of Mesh for Inguinal Hernia or Umbilical Hernia Repair at Washington University

	Postoperative year					
	1	2	3	4	5	10
Office visit[a]	1[b]	0	0	0	0	0

[a]Number of times each year modality is recommended.
[b]One visit at 2 to 3 weeks postoperatively.

Table 3
Patient Surveillance After Incisional or Large Inguinal or Umbilical Hernia Repair

	Postoperative year					
	1	2	3	4	5	10
Office visit	2[a]	0	0	0	0	0

[a]Visits at 2 and 12 weeks postoperatively.

OTHER ABDOMINAL WALL HERNIAS

Most umbilical and epigastric hernias are small defects that can be repaired primarily with good results. Mesh should be reserved for defects larger than 2 cm or patients with attenuated fascia at the margins of the defect. In one randomized trial, mesh repair of umbilical hernias was associated with a lower recurrence rate (1%) compared to primary suture repair (11% recurrence rate) *(28)*. However, patients in this study were not stratified according to hernia size.

Parastomal hernias are a common consequence of intestinal stoma formation and pose a challenging problem for repair because of the nature of the stoma and potential for infectious and other complications if mesh is used. However, because recurrence rates after primary repair of parastomal hernias of up to 50% have been reported *(29)*, some groups have used mesh to repair these defects with acceptable results. Steele and associates *(30)* reported mesh-related complications in 36% of 58 patients who underwent parastomal hernia repair with polypropylene, although none of the patients required mesh removal. Hernia recurrence developed in 26% of their patients. Whether newer mesh products or techniques for repair such as a laparoscopic approach will improve outcomes in these patients is unknown.

Finally, the role of mesh repair in large hiatal hernias merits discussion. Repair of large hiatal hernias has been associated with an incidence of recurrent hiatal abnormalities in 22–33% of cases in carefully documented series *(31–33)*. However, surgeons have been reluctant to use mesh at the esophageal hiatus because of the potential for erosion into the esophagus or stomach. Despite some small series reporting successes *(34)*, polypropylene should generally be avoided at the hiatus because of its erosive potential. A recent prospective randomized trial that compared a polytetrafluoroethylene (Gore-Tex) patch to simple suture repair of the crural defect in large hiatal hernias showed a recurrence rate of 22% in the primary repair group compared to no recurrences in the Gore-Tex group *(35)*. No infectious or erosive complications were observed over a mean follow-up period of 3.3 years. Further studies are needed to verify these results and to develop more durable methods of hiatal closure in this difficult group of patients.

SUMMARY

Synthetic mesh is now widely employed in the management of patients with various abdominal wall defects. A new and improved array of mesh products and configurations to facilitate repair has been developed over the last decade and repair techniques continue to evolve as well. Tension-free mesh repairs of abdominal hernias have a lower recurrence rate, compared to primary tissue repairs, and available evidence suggests that mesh results in acceptable complication rates and few long-term problems. Extended follow-up and outcome studies are still needed to determine the optimal mesh products and techniques for repair.

REFERENCES

1. Rutkow IM. Epidemiologic, economic and sociologic aspects of hernia surgery in the United States in the 1990s. Surg Clin N Amer 1998;78:941–951.
2. Amid PK, Lichtenstein IL. Critical scrutiny of the open "tension-free" hernioplasty. Am J Surg 1993;165:369–371.
3. Gilbert AI, Graham MF, Voigt WJ. A bilayer patch device for inguinal hernia. Hernia 1999;3:161–166.
4. Amid PK. How to avoid recurrence in Lichtenstein tension-free hernioplasty. Am J Surg 2002;184:259–260.
5. Rutkow IM, Robbins AW. "Tension-free" inguinal herniorrhaphy: a preliminary report on the "mesh-plug" technique. Surgery 1993;114:3–8.
6. Millikan KW, Cummings B, Doolas A. A prospective study of mesh-plug hernioplasty. Am Surg 2001;67:285–289.
7. Friis E, Lindahl F. The tension-free hernioplasty in a randomized trial. Am J Surg 1996;172:315–319.
8. Kingsnorth AN, Porter CS, Bennett DH, Walker AJ, Hyland ME, Sodergren S. Lichtenstein patch or Perfix plug-and-patch in inguinal hernia: A prospective double-blind randomized controlled trial of short term outcome. Surgery 2000;127:276–283.
9. Kugel RD. Minimally invasive, nonlaparoscopic, preperitoneal, and sutureless inguinal herniorrhaphy. Am J Surg 1999;178:298–302.
10. Stoppa RE. The use of Dacron in the repair of hernias in the groin. Surg Clin N Amer 1984;64:269–286.
11. Ramshaw B, Shuler FW, Jones HB, et al. Laparoscopic inguinal hernia repair: Lessons learned after 1224 consecutive cases. Surg Endosc 2001;15:50–54.
12. Schultz C, Baca I, Gotzen V. Laparoscopic inguinal hernia repair. Surg Endosc 2001;15:582–584.
13. Amid PK. A 1-stage surgical treatment of postherniorrhaphy neuropathic pain. Arch Surg 2002;137:100–104.
14. Cunningham J, Temple WJ, Mitchell P, Nixon JA, Preshaw RM, Hagen NA. Cooperative hernia study. Pain in the post-repair patient. Ann Surg 1996;224:598–602.

15. Zieren J, Beyersdorff D, Beier KM, Muller JM. Sexual function and testicular perfusion after inguinal hernia repair with mesh. Am J Surg 2001;181:204–206.
16. Amid PK, Lichtenstein IL. Technique facilitating improved recovery following hernia repair. Contemp Surg 1996;49:62–66.
17. Sugerman HJ, Kellum JM, Reines HD, DeMaria EJ, Newsome HH, Lowry JW. Greater risk of incisional hernia with morbidly obese than steroid-dependent patients and low recurrence with prefascial polypropylene mesh. Am J Surg 1996;171:80–84.
18. Winslow ER, Fleshman JW, Birnbaum EH, Brunt L. Wound complications of laparoscopic vs open colectomy. Surg Endosc 2002;16:1420–1425.
19. Flum DR, Horvath K, Keoepsell T. Have outcomes of incisional hernia repair improved with time? A population-based analysis. Ann Surg 2003;237:129–135.
20. Balique JG, Alexandre JH, Arnaud JP, et al. Intraperitoneal treatment of incisional and umbilical hernias: Intermediate results of a multicenter prospective clinical trial using an innovative composite mesh. Hernia 2000;4 (S):S10–S16.
21. Gonzalez J. Prosthetic biomaterial for hernia repair shows promise. General Surgery News September 2002:42.
22. Wantz GE, Chevrel JP, Flament JB, Kingsnorth AN, Schumpelick V, Verhaege P. Symposium: Incisional hernia: The problem and the cure. J Am Coll Surg 1999;188:429–447.
23. Santora TA, Roslyn JJ. Incisional hernia. Surg Clin N Amer 1993;73:557–570.
24. Stoppa RE. Treatment of complicated groin and incisional hernias. World J Surg 1989;13:545–554.
25. Temudom T, Siadati M, Sarr MG. Repair of complex giant or recurrent ventral hernias by using tension-free intraparietal prosthetic mesh (Stoppa technique): Lessons learned from our initial experience. Surgery 1996;120:738–744.
26. Morris-Stiff GJ, Hughes LE. The outcomes of nonabsorbable mesh within the abdominal cavity: Literature review and clinical experience. J Am Coll Surg 1998;186:352–367.
27. Goodney PP, Birkmeyer CM, Birkmeyer JD. Short-term outcomes of laparoscopic and open ventral hernia repair: A meta-analysis. Arch Surg 2002;137:1161–1165.
28. Arroyo A, Garcia P, Perez F, Andreu J, Candela F, Calpena R. Randomized clinical trial comparing suture and mesh repair of umbilical hernia in adults. Br J Surg 2001;88: 1321–1323.
29. Sugarbaker PH. Peritoneal approach to prosthetic repair of parastomy hernias. Ann Surg 1985;201:344–346.
30. Steele SR, Lee PWR, Martin MJ, Mullenix PS, Sullivan ES. Is parastomal hernia repair with polypropylene mesh safe? Am J Surg 2003;185:436–440.
31. Hashemi M, Peters JH, DeMeester TR, et al. Laparoscopic repair of large type III hiatal hernia: Objective follow-up reveals high recurrence rate. J Am Coll Surg 2000;190:553–561.
32. Mattar SG, Bowers SP, Galloway KD, Hunter JG, Smith CD. Long-term outcome of laparoscopic repair of paraesophageal hernia. Surg Endosc 2002;16:745–749.
33. Diaz S, Brunt L, Klingensmith ME, Frisella PM, Soper NJ. Laparoscopic paraesophageal hernia repair, a challenging operation: Medium-term outcome of 116 patients. J Gastrointest Surg 2003;7:59–67.
34. Carlson MA, Condon RE, Ludwig KA, Schulte WJ. Management of intrathoracic stomach with polypropylene mesh prosthesis reinforced transabdominal hiatus hernia repair. J Am Coll Surg 1998;187:227–230.
35. Frantzides CT, Madan AK, Carlson MA, Stavropoulos GP. A prospective, randomized trial of laparoscopic polytetrafluoroethylene (PTFE) patch repair vs simple cruroplasty for large hiatal hernia. Arch Surg 2002;137:649–652.

European Counterpoint to Chapter 12

Avril A. P. Chang and Ara W. Darzi

PROSTHESES FOR HERNIA REPAIR

Hernia repair in Europe and the United Kingdom is similar to that in the United States. By far the most common hernia repaired with a mesh prosthesis is the inguinal hernia. Other hernias that are repaired with mesh are, in decreasing order, incisional hernias, large umbilical or paraumbilical hernias, femoral hernias, occasionally large epigastric hernias, and less common hernias like large paraesophageal and lumbar hernias. Each year, 70,000 to 80,000 inguinal hernia operations are performed in the United Kingdom (1). Of these, the majority (80–90%) (2,3) have prosthetic mesh inserted to reinforce the repair. Incisional hernias have also increasingly been repaired with mesh rather than sutures, especially large and recurrent hernias (4). Mesh repairs have overtaken simple suture repairs because they yield better results. For suture repairs, there is a quoted recurrence rate of at least 10%, in some series as high as 40% (5). Inguinal hernia repair with mesh has been shown to reduce recurrence rates to as low as 0.1% (6). The history of the evolution of mesh repair follows that of the American experience. The most common mesh used nowadays is made of nonabsorbable polypropylene (Prolene™) or polyester (Mersilene™).

There are currently no published guidelines in the United Kingdom for follow-up of patients who have had mesh repairs. Most surgeons do not differentiate between patients who have had a mesh repair and those who have not. The standard practice of 10 surgeons surveyed in two London hospitals is to have only one review or, increasingly common with some of them, even no formal review postoperatively for healthy patients with uncomplicated hernia operations. These patients are usually given advice about the symptoms of prosthetic mesh infection and are asked to present to their general practitioner for a wound check. Generally, the patient is discharged from further review if there are no complications at that time. There are no routine blood tests, scans or imaging tests performed as follow-up for the asymptomatic patient postoperatively (Table 1).

Early recurrence of a hernia is usually due to a technical problem with the repair rather than the prosthetic implant. Our experience is similar to the American one in that failure is usually detected by the patient as a hernia recurrence. The mesh itself is not usually disrupted, but herniation often occurs around the periphery of the mesh. This

From: *The Bionic Human: Health Promotion for People With Implanted Prosthetic Devices*
Edited by: F. E. Johnson and K. S. Virgo © Humana Press Inc., Totowa, NJ

Table 1
Patient Surveillance After Implantation of Mesh for Hernia Repair at Two Teaching Hospitals in London

	Postoperative year					
	1	2	3	4	5	10
Office visit[a]	1[b]	0	0	0	0	0

[a] No routine X-rays, blood tests, or other investigations are performed for the follow-up of asymptomatic patients.
[b] Usually at 2 to 4 weeks there is one routine outpatient visit.

may be due to the patient's tissues being disrupted or poor placement of the mesh (7,8). Detection of recurrences is usually based on clinical examination but can be confirmed or primarily diagnosed by ultrasonography, computed tomography, or magnetic resonance imaging scanning. Other complications include wound infection, development of seromas, and neuralgia. However, a large systematic review showed that there is no significant difference in the rates of these complications compared with patients who did not have mesh in their repairs (9).

Prosthetic infection can occur. A single perioperative dose of antibiotics like Cefuroxime or Augmentin is thought to reduce the rate of wound infection, although this has not been the subject of a randomized trial. There is no guideline regarding irrigation of the mesh with antibiotics or antiseptic, and individual surgeon preferences determine whether they soak or irrigate the mesh with any solutions.

IMPORTANCE OF PROSTHESIS DESIGN AND QUALITY OF MANUFACTURE

I agree with my American colleague that it is of vital importance that meshes be sterile, nontoxic, nonallergenic, and nonimmunogenic. The meshes we currently use for inguinal hernia repairs fulfill these requirements, although we have no long-term information about possible immunogenic consequences of leaving a foreign body in over a long period of time (>10 years). To date, there have been no reports of adverse effects long term, apart from physical migration of the mesh. Although soft-tissue tumors have been induced experimentally in animals, there have been no reports of any malignant tumours developing from mesh implants (10).

Each mesh is packaged with stickers with the manufacturer's information. These are placed into the patient record and also the theater register. This provides a means by which patients can be traced should any problems with a particular material or manufacturing line be discovered in the future.

Implantable mesh comes in many shapes and sizes and manufacturers have been innovative in coming up with different designs for different requirements. Cone-shaped devices and gently curved shapes to follow the contours of the abdominal wall are already readily available but are costlier than normal flat meshes. The search for better materials continues and composites of nonabsorbable and absorbable meshes have promise in decreasing the potential complications of adhesion formation and lack of flexibility while still maintaining strength (11).

PREVENTIVE MEASURES

Preventive measures to avoid hernia formation vary according to the types of hernia involved. Many are congenital or difficult to prevent. For example, indirect inguinal hernia is caused by an inherent weakness in the abdominal wall around the exit of the spermatic cord as it travels toward the scrotum. This means men are much more susceptible to the problem, but there is little that can be done to prevent a hernia from forming. Hernia is exacerbated by increased abdominal pressure, for example lifting heavy weights, chronic coughing, or straining (e.g., at micturition or defecation). Sometimes an underlying problem is treatable (e.g., treating an underlying chest problem to prevent chronic cough, treating a prostatic problem to relieve a urinary outlet obstruction, and putting a patient on a high fiber diet to prevent constipation). People known to lift heavy weights can also be advised to wear a "hernia belt," which is a wide belt placed across the lower abdomen. It can bear the increased abdominal pressure when the person does heavy lifting and prevents excessive force being exerted across the deep inguinal ring. To avoid complications from hernia repair, patients are advised to lose weight, stop smoking, and control their medical problems (e.g., diabetes) prior to having their operation.

Current multi-institutional trials have concentrated on the difference in outcomes between mesh repair using laparoscopic vs open technique (rather than whether or not to use a mesh). Large trials would be useful to answer questions of whether newer generation meshes work better than current ones. For the present, however, it is generally accepted that the use of mesh in hernia repairs markedly reduces the incidence of hernia recurrence, and meshes are increasingly replacing primary suture in many types of hernia repair.

REFERENCES

1. Anonymous. Table B 15. Hospital Inpatient Activity: Finished consultant episodes, the most frequently performed operative procedures 1994 to 1998. In: Health and Personal Social Services Statistics, Statistics Division, Department of Health Publication 2003. http://www.doh.gov.uk/HPSSS/TBL_B15.HTM.
2. Metzger J, Lutz N, Laidlaw I. Guidelines for inguinal hernia repair in everyday practice. Ann R Coll Surg Engl 2001;83:209–214.
3. Hair A, Duffy K, McLean J, et al. Groin hernia repair in Scotland. Br J Surg 2000;87: 1722–1726.
4. Korenkov M, Sauerland S, Paul A, Neugebauer EA. Incisional hernia repair in Germany at the crossroads: a comparison of two hospital surveys in 1995 and 2001. Zentralbl Chir 2002; 127:700–704.
5. The Danish hernia database-the first year. Ugeskr Laeger 2000;162:1552–1555.
6. Amid PK, Shulman AG, Lichtenstein IL. Open "tension-free" repair of inguinal hernias: the Lichtenstein technique. Eur J Surg 1996;162:447–453.
7. Junge K, Klinge U, Prescher A, Giboni P, Niewiera M, Schumpelick V. Elasticity of the anterior abdominal wall and impact for reparation of incisional hernias using mesh implants. Hernia 2001;5:113–118.
8. Amid PK, Shulman AG, Lichtenstein IL. Open "tension-free" repair of inguinal hernias: the Lichtenstein technique. Surg Today 1995;25:619–625.
9. Scott NW, McCormack K, Graham P, Go PM, Ross SJ, Grant AM. Open mesh versus non-mesh for repair of femoral and inguinal hernia. Cochrane Database Syst Rev 2002;4: CD002197.

10. Michael Ghadimi B, Langer C, Becker H. The carcinogenic potential of biomaterials in hernia surgery. Chirurg 2002;73(8):833–837.
11. Klinge U, Klosterhalfen B, Conze J, et al. Modified mesh for hernia repair that is adapted to the physiology of the abdominal wall. Eur J Surg 1998;164:951–960.

13
Penile Prostheses

John J. Mulcahy

INTRODUCTION

Erectile dysfunction is a significant quality-of-life issue that affects both genders. One in four men have difficulty achieving an erection by age 55. The incidence of this problem is 50% in men by age 65, and the majority of men over age 70 cannot achieve an erection suitable for intercourse *(1)*. The penis has two elastic chambers that are central to erection. They are covered by an elastic membrane, the tunica albunginea, and contain cul-de-sacs surrounded by smooth muscle. Various stimuli influence this smooth muscle to relax. As this occurs, blood flows into the cul-de-sacs, expanding the tunica albuginea to its limits of stretch and compressing venules that run underneath it, thus preventing blood from leaking out of the chambers. This gives increased size and rigidity to the penis. Other stimuli cause this smooth muscle to contract, relaxing the erection and maintaining the penis flaccid.

In youth, erections are an accepted part of life and a man usually does not think of being unable to have one. As a man ages, the proper environment becomes more important. When he is rested, his erections are better than if he is tired. If he is distracted by pain, worried about money, or otherwise preoccupied, his erections may be poor. A partner's influence is also important in the aging male. If the partner remains attractive and positive about sexual activity and is helpful in stimulating the erection, these will be achieved more easily than if the partner becomes unattractive or disinterested. With aging, hardening of the arteries and narrowing of blood vessels progresses throughout the body. Diseases such as hypertension, hyperlipidemia, and diabetes mellitus and habits such as cigaret smoking accelerate this. This results in less blood flow to the penis and erections become softer and shorter in duration. Elasticity in all parts of the body, including the penis, decreases with age. The tunica albuginea stretches less readily, resulting in a shorter and narrower erection. Certain medications interfere with erections. In the past, antihypertensives were notorious in this respect. Today, 20% or more of patients taking selective serotonin reuptake inhibitors, such as Prozac for depression, notice aspects of sexual dysfunction. Neurological diseases, such as multiple sclerosis, spinal cord injury, temporal lobe epilepsy, and Parkinson's disease disrupt neural input to the penis. This affects blood flow and thus erection. Pelvic surgery, which interrupts nerves or blood supply, is a common cause of impotence. Penile diseases, such as priapism or Peyronie's disease, which result in scarring of erectile

From: *The Bionic Human: Health Promotion for People With Implanted Prosthetic Devices*
Edited by: F. E. Johnson and K. S. Virgo © Humana Press Inc., Totowa, NJ

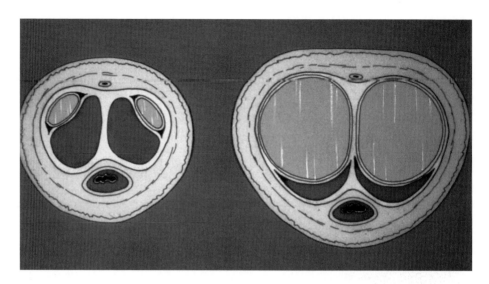

Fig. 1. Cross section of penis: cylinders placed inside erectile bodies (left, deflated; right, inflated).

bodies, may reduce the elasticity of the tunica albuginea or replace cavernosal smooth muscle with fibrotic tissue, again causing the same end result.

Most erectile difficulties, especially in aging patients, are organic in nature but emotional or environmental influences can impede ability to achieve an erection as well. If these are long standing, and if repeated failure at intercourse occurs, the performance anxiety often exacerbates the original emotional problem. The more he fails, the more a man worries and this anxiety contributes to his poor performance.

Erectile dysfunction has frustrated men and their partners through the ages. However, in former centuries, life expectancy was relatively short and age-related sexual problems were therefore infrequent. Recall that the average life span in 1900 was just 49 years. In 1933, when Social Security was introduced in the United States, only 2% of the population was older than age 65. The figure is 12% today and expected to be 20% by the year 2020. As life expectancy has increased and couples look forward to the golden years, the times might not be as golden as they would wish if this disability exists.

In the mid-20th century, it was thought by Kinsey and others that most problems in this area were psychological (2). Behavioral sex therapy and various forms of psychotherapy were tried but were largely ineffective. The first effective treatment of erectile dysfunction, the penile prosthesis, was introduced in the early 1970s (3,4). In placing the penile implant, the spongy smooth muscle, sinusoids, and blood vessels of the two erectile bodies are pushed to the sides by dilating instruments to make room for cylindrical supports which are placed in these chambers (Fig. 1). There are two basic types of supports, semirigid rods and inflatable cylinders. The rod-type devices consist of a wire core surrounded by silicone or a series of articulating polyethylene segments held together by a cable attached to a spring (Fig. 2). Either variety of rod is positionable by bending the penis to the desired position. There are two vendors of rod prostheses in the United States. American Medical Systems markets the 650M and Dura II; Mentor

Chapter 13 / Penile Prostheses

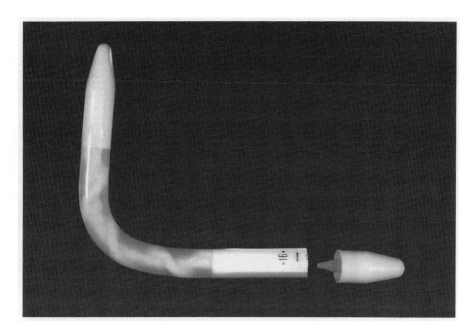

Fig. 2. Malleable rod penile implant.

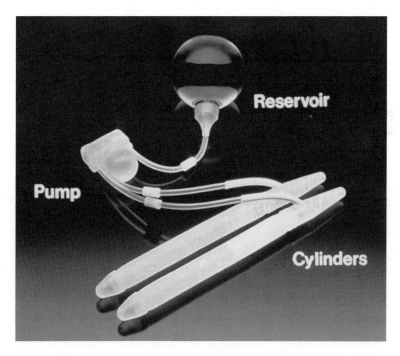

Fig. 3. Three-piece inflatable penile implant.

Corporation sells the Genesis device. The inflatable type of implant consist of inner cylindrical tubes, which are placed in each erectile body and a device to fill them with saline for penile firmness, then remove the saline for penile flaccidity (Fig. 3). The current marketplace offers two three-piece prostheses and one two-piece prosthesis. The three-piece type is comprised of two inflatable cylinders, one in each corporal body, a pump located in the dependent portion of the scrotum, and a reservoir positioned behind the rectus abdominis muscles in front of the urinary bladder. The pump controls the transfer of fluid from the reservoir into the penis for inflation (rigidity) and out of the penis back to the reservoir for deflation (flaccidity). Very good penile rigidity can be achieved because the reservoir can have as large a capacity as needed. The two-piece prosthesis consists of a reservoir-pump combination in the scrotum and two cylinders. A proximal portion of each cylinder also contains a reservoir cavity. One pumps fluid into the cylinder to achieve rigidity. Increasing pressure in the cylinder for 10 seconds opens the valve between the cylinder and proximal reservoir cavity, allowing part of the cylinder's fluid to pass out of the chamber. Some fluid remains in the cylinder, however, and the cylinder becomes easily bendable but not totally flaccid in the deflated state. American Medical Systems sells two varieties of the three-piece prosthesis, the 700 Ultrex (which expands in both girth and length) and the 700 CX (which expands in girth only). Mentor Corp. sells the Titan prosthesis, which expands in girth only. Both vendors also supply a narrower prosthesis (the 700 CXR of American Medical Systems and the Titan Narrow Base of Mentor Corp.), which is more suitable for the narrow or scarred penis. There is only one two-piece prosthesis on the market, the Ambicor of American Medical Systems. The inflatable implants provide a more natural erection than the bendable rod types and are more frequently chosen. Patients with limited manual dexterity are encouraged to choose the semirigid rod type because it is easier to maneuver.

The penile implant provides a firm penis suitable for use for intercourse but does not improve sex drive or sensitivity in the penis. Sensitivity is usually unchanged following surgery but a few patients may note diminished sexual feeling after surgery. Nerve endings change with aging and this can affect sexual feeling even without surgery, of course, though feelings, such as touch and pain, typically remain intact. The size of the erect penis is frequently shorter after penile prosthesis placement as a sheath of scar forms around the cylinders. This is part of normal healing. The scar is not stretchable and may actually contract in time resulting in a shorter phallus. Even cylinders that expand distally (AMS 700 Ultrex) cannot stretch the scar and hence shortening results. With repeated injury to the penis more scar forms and the erection becomes shorter. Aging itself, even without surgery or injury, results in a degree of penile shortening as the abundant elastic tissue in the youthful tunica albuginea is replaced with less expansile tissue such as collagen.

The penile implant has the greatest satisfaction rate of all treatments of erectile dysfunction. However, it is the most invasive, most expensive, and least frequently chosen option. Improved understanding of physiology and pharmacology has led to several effective treatment options that are less invasive than permanent implants. Patients tend to select the least invasive effective treatments of their problem. The recent introduction of Viagra in 1998, and subsequently Levitra and Cialis, broadened the market many-fold *(5)*. The vast majority of patients who elect to treat their erectile dysfunction select this option and it works effectively in about two-thirds of them. When Viagra is ineffective, other medications, such as intraurethral prostaglandin E-1 (MUSE) *(6)* or intracorporal prostaglandin E-1 (Caverject, Edex) *(7)*, may be tried. In addition, intracorporal injections of papaverine and Phentolamine *(8)*, either alone or in combination with prostaglandin E-1 (Trimix), are useful in certain circumstances. Vacuum devices are also available. They draw blood into the erectile bodies by negative pressure created in a large plastic cylinder placed over the penis *(9)*. Blood is then trapped in the engorged erectile bodies by quickly slipping a heavy rubber band from the base of the cylinder onto the shaft of the penis. This provides a workably rigid but hinged erection. These devices do not have widespread appeal and may be unsuitable for the patient with a corpulent body habitus. Patients offered these alternatives often find the prospect of a needle in the penis abhorrent or a hinged erection obviously mechanical and repulsive. Since the penile implant gives a natural appearance with a predictable, reliable, and very satisfactory result it has the highest satisfaction rate. Eighty to 90% of patients are pleased with the results. The main reason for disappointment is the shorter penile size following the surgery.

When a patient contacts his physician regarding erectile dysfunction, the doctor begins with a comprehensive history of the development of the problem. The partner is encouraged to participate in the discussion, especially when treatment options are mentioned. However, fewer than half of the partners actually attend these sessions. The initial visit includes inquiry regarding disease processes, habits, medications, previous injuries, or surgery that may contribute to the problem. Testing to monitor the presence or absence of nighttime erections *(11)* or penile blood flow *(12)* was more extensively performed in the past. Recently a more empiric approach has been taken with the availability of effective oral medication. The first approach to therapy has been to try Viagra if the patient has no contraindication, such as severe heart disease or medications containing nitrates. If Viagra is ineffective, a test dose of intracorporal prostaglandin E-1 is given. If a positive response occurs, the patient is offered this option or the intraurethral prostaglandin E-1 delivery system (MUSE). A vacuum erection device may also be offered, depending on body habitus and motivation. A penile implant is offered if other less invasive modalities fail or the patient is not interested in trying them. A few patients are motivated to opt for a penile prosthesis by a friend's positive experience with such a device.

After a decision is made to proceed with placement of a penile implant, financial aspects are discussed and prior approval obtained from the patient's insurance company. Preliminary tests, such as complete blood count, electrocardiograph, and urinalysis, may be obtained in the days shortly before the surgery. The patient arrives at the surgical suite the morning of the procedure. He is advised to clean the genital area with a strong soap in the days before surgery. Shaving and a thorough antiseptic prep of the surgical area are performed in the operating room after the induction of anesthesia. Most penile implant operations take between 45 minutes and 2 hours. The patient may leave the surgical suite after recovering from the anesthetic, or the following morning, when the effects of anesthesia have worn off and the possibility of immediate problems, such as bleeding or difficulty urinating, is remote. Some physicians leave a urinary catheter in place overnight for patient convenience although postoperative urinary retention is unusual.

Depending on the type of prosthesis and the decision of the surgeon, two types of incisions are usually used: below the penis in the scrotum or above the penis in the area in front of the pubic bone. Occasionally an incision similar to a circumcision (subcoronal) is used for semirigid rod placement and rarely two incisions may be needed in complex cases. If excessive scar tissue is present in the erectile tissues or in the elastic tunica covering the erectile bodies, innovative techniques may be necessary to create cavities for the cylinders or correct curvature in the resulting erection once the cylinders are placed. Frequent irrigations of the wound are performed during the procedure to help reduce the incidence of infection. Some surgeons use systemic and topical prophylactic antibiotics but others do not.

After surgery, pain and swelling are to be expected. The pain is located in the scrotum in the area of the pump and in the shaft of the penis where the cylinders are located. The pain and swelling usually subside promptly and by 6 weeks the patient can usually use the pump without significant discomfort. The pain is usually easily controlled with oral analgesics during the postoperative recovery time. At 6 weeks, the patient is taught how to work his new device and can commence using it for intercourse. There may be some discomfort initially with use but by 6 months most patients hardly know the device is there when they are not using it. The patients are instructed on the positioning of the device and encouraged to cycle the inflatable models frequently to gain familiarity with their operation. The patient is usually seen 2 months later to discuss the prosthesis to determine if it is meeting his expectations and to answer questions.

Penile implants, the first effective treatment of erectile dysfunction, have been on the market for more than 30 years. Inflatable and semirigid models were introduced in 1972 and sales grew exponentially. By the late 1980s, about 25,000 implants were placed worldwide each year. Eighty percent of these were in the United States, where insurance frequently covers the cost of the surgery. In other countries patients typically pay cash. The figure of about 25,000 prostheses yearly was maintained until 1998, the year of the introduction of Viagra. In that year, the implantation of prostheses plummeted to around 12,000 units. Patients who have failed trials of Viagra have now come forward for treatment and the incidence of prosthesis implantation is once again on the rise. Currently, about 18,000 are placed annually in the United States. The vast majority (90%) are the inflatable types. Although penile prostheses have the highest satisfaction rate, they only comprise about 0.5% of the treatments of erectile dysfunction.

The frequency of follow-up visits after placement of a penile prosthesis varies among individual surgeons. Many surgeons see the patient back the week following surgery for suture removal and inspection of the wound. If absorbable sutures are used, the patient is often seen at about 6 weeks, especially if he lives a considerable distance away. Cycling the prosthesis and using it for intercourse usually commence at this juncture and the patient is seen at a follow-up visit 1 to 2 months later to document the effectiveness of the device and to afford time to answer questions. Thereafter, the patient is advised to return yearly for evaluation (Table 1). At these yearly office visits, device function is documented and patient satisfaction is noted. The penis is inspected for signs of tenderness and thinning of skin over prosthetic parts and fixation of prosthetic parts to the overlying skin. These are signs of wearing of the tissue or the possible development of infection. In several series the reliability of both inflatable and malleable devices has been very good. It is rare for modern devices to fail but certainly breakage can occur with excessive use or bending. The inflatable devices have a mechanical repair rate of about 15% at 5 years and about 30% at 10 years *(13)*. As mechanical products go, these repair rates are very good. The harder one pumps his

Table 1
Patient Surveillance After Implantation of a Penile
Prosthesis at Indiana University Medical Center

	Postoperative year				
	1	2	3	4	5	10
Clinic visit	3[a]	1	1	1	1	1

[a]At 6 weeks, 3 months, and 12 months postoperatively.
No X-rays, blood tests, etc., are done unless symptoms or signs warrant them.

device, and the more frequently and more aggressively he uses it, the more rapidly it wears. Some patients, in an attempt to increase the size of the penis, may be prone to pump to excessive firmness. It is possible with the three-piece inflatable devices to inflate to pressures of 28 pounds per square inch. This is similar to the pressure in an automobile tire. If the patient consistently pumps to high pressures and uses excessive force applied to the end of the penis, prosthesis wear is rapid. Failure of the device can occur in two ways and these can be subtle. Parts will wear with resulting loss of fluid from the system. This fluid is isotonic saline and is readily absorbed by the body if it does leak out of the prosthesis. When this occurs the device simply will not inflate. Also, the device may migrate from the body cavity in which it is placed. This may occur distally, with the cylinder tips extruding outside the erectile body under the foreskin or into the urethra. From trauma to the penis a curvature to the erection may develop, which makes intercourse difficult. Rod prostheses are prone to perforate proximally such that they can sometimes be felt in the buttocks. Pump migration may occur after the initial placement of the device but this is uncommon after the first month because it is fixed in place by scar tissue. Reservoir herniation through the abdominal wall or erosion into the bowel or bladder has been reported but is extremely rare. The common denominator for all of these unfavorable excessive occurrences is mechanical tension or pressure.

All the currently available penile implants in the United States have a limited lifetime warranty on the parts. Hence, if a mechanical problem develops, surgery to repair the problem is necessary and, at the same time, all parts of the prosthesis are generally changed. If the repair occurs within a few months of the original surgery, and there is a minor problem that is easily correctable, then the remaining parts may be left in place. This is especially recommended for penile prosthesis cylinders since a change of cylinders may result in shortening of the resulting erection, as mentioned earlier. Various repair techniques using natural tissue have been devised to take care of the problems of erosion and perforation of parts of the device that may occur *(14)*.

Prosthesis infection is a serious complication. The incidence of infection associated with the devices is low (lifetime risk 1–3%) *(15)*. Caution with preoperative cleansing of the skin, antiseptic preparation of the skin prior to surgery, use of copious irrigations during the procedure, use of antibiotic-coated implants, and possibly also the use of prophylactic antibiotics has combined to keep the infection rate in this low range. When an infection does occur, systemic antibiotics are ineffective *(16)*. In the past this necessitated removal, a prolonged period of healing, and possible reinsertion at a later date. The most common organism associated with implant infection is *staphylococcus epidermidis* (coagulase-negative *staphylococcus) (16)*. This organism produces a slime or biofilm that surrounds the prosthesis and that can provide a hiding place for bacteria. The slime also reduces phagocytosis. The relatively avascular capsule that forms around the prosthesis tends not to permit good penetration of antibacterials into the infected tissues. To completely eradicate the infection the entire device, plus any foreign material or permanent sutures, should be removed. The wound is then copiously irrigated with a series of antiseptic solutions. A new penile implant is placed at the same sitting. This new approach, termed a "rescue" or "salvage" procedure, successfully eradicates the infection in 80–90% of patients *(16)*. The advantage of salvage is that it usually maintains the length of the penis and allows easier insertion of the cylinders. The traditional approach of removing the prosthesis and returning for prosthesis replacement 6 months after eradication of the infection and healing of the wound results in considerable penile shortening (about 2 inches less than the original length of the erection after the original penile implant placement). In addition, it is much more difficult to place prosthesis cylinders in scarred corporal bodies than when the cylinder cavity is patent during the salvage procedure. Most failures of salvage have occurred with infections caused by aggressive organisms, such as methicillin-resistant *staphylococcus aureus* and *Pseudomonas aeruginosa*, and have occurred very shortly after placement of the penile prosthesis. They are typically associated with extensive cellulitis. Salvage can be successful when pus is drained and the cellulitis is eradicated by systemic antibiotics for a few days before and after the salvage procedure. In these circumstances the systemic antibiotics sterilize the tissue, the salvage procedure sterilizes the cavity, and an optimal result is usually achieved. Yearly follow-up visits can be helpful in detecting symptoms of infection, such as tenderness over prosthesis parts or fixation of the pump to the scrotal wall, before the infection progresses to the point where urethral erosion occurs. In this circumstance a salvage procedure is impossible.

There have been isolated reports of seeding of infection to a penile prosthesis from distant sites such as following dental procedures. Although the incidence of infection spreading is extremely low, some urologists recommend prophylactic antibiotics during dental manipulation or colon examination when a penile implant is in place, as well as when an active infection such as an abscess or pneumonia is present. We use ampicillin and gentamicin in most circumstances.

If the patient gains considerable weight after placement of the penile prosthesis, the penis will apparently become shorter. For every 35 pounds of weight a man gains, 1 inch of penile length is lost, as the growing panniculus encroaches on the shaft of the penis. This may not interfere with the function of the device but the penis apparently shrinks because of this phenomenon and intercourse may become difficult especially if the partner is heavyset as well. Weight loss can be a goal of follow-up, with the expectation of good results. Erectile dysfunction, although a common accompaniment of aging, may also be a symptom of significant systemic disease. This often warrants investigation during follow-up visits. If indicated, testing for various contributing disease entities, such as diabetes mellitus, hormonal dysfunction, vascular disease, or neurological disease, should be undertaken. The patient's primary care physician usually performs this. At follow-up visits to the urologist no routine testing, such as X-rays or blood work, is necessary. Proper functioning of prosthetic parts and physical examination of the device documents that the prosthesis is working well. If there is a question of extrusion of parts, a magnetic resonance imaging scan is the most accurate test for determining the location and proper sizing of cylinders, reservoir, and pump. This, however, is a costly test and in most instances physical examination will be able to detect any problems. Although the penile prosthesis chambers are close to the urinary tract, penile implants do not usually interfere with urination unless there is erosion into the urethra. Slowing of the stream, spraying on urination, small amounts of blood at the meatus, and tenderness of the glans penis are symptoms of this complication. They are occasionally so subtle that the patient does not complain of them.

When younger men, such as medical students, learn about erectile dysfunction, they frequently ask what they can do to prevent or forestall it. Our advice is to avoid risk factors for vascular disease, use medications judiciously, avoid smoking, drink alcohol sparingly or avoid it altogether, eat a prudent diet, exercise regularly, and remain slender. In addition, he should have intercourse regularly with a partner who does things to excite him and stimulate him sexually.

Recently the Food and Drug Administration requested American Medical Systems and Mentor Corp. to conduct prospective clinical trials to document the reliability and safety of their inflatable penile implants and patient satisfaction with them. Problems with silicone devices, such as breast implants, prompted this request although penile prostheses had already been approved for many years. Both vendors complied with the request and conducted these trials with satisfactory outcomes. The mechanical reliability of the implants and the high degree of patient satisfaction was confirmed in these trials. There are a number of sources of information about erectile dysfunction or penile prostheses (*see* Appendices A and B). Physicians interested in erectile dysfunction frequently conduct support group meetings and provide both information about this problem and an opportunity to hear the stories of patients who have experienced it firsthand and undergone treatment.

In summary, erectile dysfunction is a very common problem, affecting over 25% of the male population in the United States. Treatments range from simple oral medication to a penile prosthesis. Although a penile implant is the most invasive and costly option, it reliably provides an erection. Prudent follow-up can minimize the failure rate and permit early recognition of complications.

APPENDIX A: ORGANIZATIONS WITH ERECTILE DYSFUNCTION INFORMATION

American Foundation for Urologic Disease
1128 N. Charles Street
Baltimore, MD 21201-5559
phone (410) 468-1806 (800) 242-2383
Web site: www.impotence.org

(continued)

APPENDIX A: ORGANIZATIONS
WITH ERECTILE DYSFUNCTION INFORMATION (Continued)

Impotence Anonymous
(800) 669-1603

SIECUS—The Sexuality Information and Education Council of the U.S.
130 West 42nd Street, Suite 350
New York, NY 10036
Phone: (212) 819-9770

American Diabetes Association, Inc.
1660 Duke Street
Alexandria, VA 22314
Phone: (800) 232-3472

Impotence World Association
Phone: (800) 669-1603

PROSTATE CANCER SURVIVORS GROUPS:

Man to Man
American Cancer Society
(800) 227-2345

US Too International, Inc.
(800) 808-7866

APPENDIX B: VENDORS OF PENILE IMPLANTS IN THE UNITED STATES

American Medical Systems
10700 Bren Road West
Minnetonka, MN 55343
1-800-328-3881

Mentor Corp.
201 Mentor Drive
Santa Barbara, CA 93111
1-800-525-0245

REFERENCES

1. Feldman HA, Goldstein I, Hatzichristou, D, et al. Impotence and its medical and psychosocial correlates: results of the Massachusetts Male Aging Study. J Urol 1994;151:54–61.
2. Kinsey AC, Pomeroy WB, Martine CE. Sexual behavior in the human male. Philadelphia, PA: W.B. Saunders Co., 1948.
3. Scott FB, Bradley WE, Timm GW. Management of erectile impotence: use of implantable inflatable prosthesis. Urology 1973;2:80–82.
4. Small MP, Carrion HM, Gordan JA. Small-Carrion Penile Prosthesis: new implant for management of impotence. Urology 1975;5:479–486.
5. Goldstein I, Lue TF, Padma-Nathan H, et al. Oral sildenafil citrate in the treatment of erectile dysfunction. New Eng J Med 1998;338:1397–1404.

6. Padma-Nathan H, Hellstrom WJG, Kaiser FE, et al. Treatment of men with erectile dysfunction with transurethral alprostadil. New Eng J Med 1997;336:1–7.
7. Stackl W, Hasun R, Marberger M. Intracavernous injection of prostaglandin E1 in impotent men. J Urol 1988;140:66–68.
8. Zorgniotti AW, Lefleur RS. Auto-injection of the corpus cavernosum with vasoactive drug combination for vasculogenic impotence. J Urol 1985;133:39–41.
9. Witherington R. Vacuum constriction device for management of erectile dysfunction. J Urol 1989;141:320–322.
10. Fallon B, Ghanem H. Sexual performance and satisfaction with penile prostheses in impotence of various etiologies. Int J Impotence Res 1990;2:35–42.
11. Levine LA, Carroll RA. Nocturnal penile tumescence and rigidity in men without complaints of erectile dysfunction using a new quantitative analysis software. J Urol 1996;152:1103–1107.
12. Broderick GA, Arger PA. Duplex Doppler ultrasonography: noninvasive assessment of penile anatomy and function. Semin Roentgenol 1993;28:43–56.
13. Carson CC, Mulcahy JJ, Govier FE. Efficacy, safety, and patient satisfaction outcomes of the AMS 700 CX inflatable penile prosthesis: results of a long-term multicenter study. J Urol 2000;164:376–380.
14. Mulcahy JJ. Distal corporoplasty for lateral extrusion of penile prosthesis cylinders. J Urol 1999;161:193–195.
15. Carson CC. Infections in genitourinary prostheses. Urol Clin North Am 1989;16:139–147.
16. Mulcahy, JJ. Long-term experience with salvage of infected penile implants. Urol 2000;163:481–482.

US Counterpoint to Chapter 13

Steven B. Brandes

INTRODUCTION

Patients who elect to get a penile implant today are sildenafil nonresponders, those who are physically unable to use, do not desire to use, or are nonresponders to intraurethral alprostadil, injection therapy, and/or a vacuum erection device.

For tetraplegics or patients with poor manual dexterity, we typically place a malleable implant. The two-piece implant (i.e., Ambicor) generally gives a poor cosmetic and functional result and so we see little reason to place such implants. The presumed advantage of the two-piece implant is that, without a reservoir, one does not have to be troubled with placing a reservoir in a scarred prevesical space. Practically, this space is rarely so scarred after a radical prostatectomy that a reservoir cannot be placed. In the rare cases where the prevesical space is obliterated, we place the reservoir beneath the rectus muscle. After a cystectomy or pelvic exenteration, without the bladder in place, the potential for bowel injury is higher for prevesical reservoir placement. In such instances we routinely place the reservoir subrectus.

Superior cylinder inflation, axial rigidity, deflation flaccidity, and patient satisfaction are achieved with three-piece implants. In the preoperative consult, the patient is instructed in implant deflation and given an instructional video and materials. In preparation for scheduled penile implant surgery, patients require a thorough medical evaluation since most suffer from multiple co-morbid conditions, such as diabetes, hypertension, peripheral vascular disease, and coronary artery disease. In general, the penile implant infection rate for diabetics is 3%, for those with spinal cord injury is 9%, for diabetics undergoing revision surgery is 18%, and for nondiabetic patients receiving revision procedures is 8% *(1)*. In contemporary series, the risk for implant infection does not correlate with the level of glycosylated hemoglobin *(1)*.

Another important aspect of penile implant surgery is preoperative teaching. To avoid dissatisfaction, we disclose that erections in patients with prostheses are shorter than natural erections. Also, there is no glans engorgement. The cylinders commonly autoinflate during the first 12 weeks. It is worth emphasizing that libido will not improve and that achieving orgasm is typically more difficult.

We have the patient wash his genitals and penis with a washcloth and antibacterial soap at least once daily for 2 weeks prior to surgery. Cephazolin and gentamicin are

From: *The Bionic Human: Health Promotion for People With Implanted Prosthetic Devices*
Edited by: F. E. Johnson and K. S. Virgo © Humana Press Inc., Totowa, NJ

Table 1
Patient Surveillance After Penile Prosthesis Implantation at Washington University

	Postoperative year					
	1	2	3	4	5	10
Office visit	4 [a]	1	1	1	1	1

[a] Follow-up visits at 2 weeks, 6 weeks, 3 months, and 12 months after surgery.

given intravenously 1 hour before surgery. We previously gave intravenous vancomycin but institutional delays in antibiotic availability and long infusion times often resulted in skin incision before adequate antibiotic infusion. Because most of our patients are overweight, we have found the penoscrotal approach to be technically easier and quicker. We usually close the corporotomies with preplaced interrupted sutures (to minimize the risk of injury to cylinders) and leave the phallus deflated. We routinely place a closed-suction scrotal drain for 23 hours. A small (14 F) Foley catheter is also placed for 23 hours. The patient is discharged with a 2-week supply of an oral first-generation cephalosporin (i.e., Keflex®).

All patients are routinely seen 2 weeks after surgery to inspect the wounds, to confirm that the cylinders and the pump are properly placed, and to ensure the implant is not autoinflating. At this time, we reinstruct the patient in the technique of implant deflation. We do not have the patient cycle the implant until 6 weeks after surgery. At 6 weeks, we reassess again for proper cylinder and pump location and assess ease of use at inflation and deflation. We obtain follow-up in all patients at 2 weeks, 6 weeks, 3 months, and then yearly after surgery (as long as the patient is asymptomatic and able to have intercourse. *See* Table 1.)

Implant infections are typically indolent and caused by a *Staphylococcal* species. A red, inflamed, and obviously infected penis in the initial postoperative period is rare. Symptoms are typically mild and do not appear for 3 to 6 months. The hallmarks of infection are persistent penile pain, scrotal skin fixation to the implant, sinus tracts to the skin, and persistent swelling of the penis or scrotum. Blood at the meatus after 3 months postoperative is also a sign of infection. Prolonged antibiotic use may be effective in treating a few subclinical infections. When implant infection is obvious, however, the prosthesis should be removed. The salvage technique detailed by Mulcahy is effective and successful at avoiding the corporal fibrosis that typically occurs with implant removal. When the implant is grossly infected, we typically use a Simpulse® to irrigate the wounds and corpora, and exchange the infected three-piece for a malleable implant (instead of another three-piece). Three to 6 months later, in a staged fashion, we exchange the malleable for a three-piece implant. We feel this staged salvage technique maximizes success and minimizes reinfection.

With proper follow-up *(1)*, three-piece inflatable implants have high rates of performance durability and patient satisfaction, as well as a low rate of mechanical failure.

REFERENCE

1. Wilson SK, Delk JR II. Inflatable penile implant infection: Predisposing factors and treatment suggestions. J Urol 1994;153:659–661.

European Counterpoint to Chapter 13

Christine M. Evans

INTRODUCTION

Chapter 13 provides a detailed account of the development and management of erectile dysfunction, including treatments and the need for penile prostheses. Owing to the greater number of patients presenting with erectile dysfunction in the United States, American surgeons have more experience in placing penile prostheses. The advent of oral treatment in 1998 reduced the need for surgery, but the figures are again rising. The percentage of patients is still small; however, the success rate is uniformly very good. This is the most successful form of treatment for erectile dysfunction, the satisfaction rate being about 90% for patients and slightly less for partners.

In the 1950s, synthetic material was implanted into the penis, but erosion and draining sinuses meant that these implants were unsuccessful as was the technique described by Pearman *(1)* of placing silastic prostheses between Buck's fascia and the tunica albuginea. Thereafter, the implants were placed intracorporeally. The first recognizably modern silastic prosthesis to be developed was the Small-Carrion *(2)*. It was low cost and trimmable, but was poorly concealed and not very rigid. It is no longer available. The subsequent "Flexirod" prosthesis (Surgitek Implants, Bristol-Myers Squibb) *(3)* was more concealable. This was also a silicone rod and hinged but was unstable and too flexible. To allow the prosthesis to change shape, newer models were made with a silicone exterior and a metal wire inside, first described by Jonas in 1983 *(4)*, then marketed as the AMS 600 (American Medical Systems, Minneapolis, MN), and later the Mentor (Mentor Urology, Santa Barbara, CA). In 1987, the Duraphase and Omniphase devices (Dacomed Corp., Minneapolis, MN) were developed, both comprising a collection of segments held together by a cable in the center. Although patient acceptability was high, at moderate cost and with few mechanical problems, all these prostheses had problems of concealment and variable rigidity.

The multipart inflatable devices developed by Scott (AMS) in 1974 *(5)* have improved over the years, with kink-free tubing, better connectors, reinforced cylinders to prevent aneurysms and an ability to elongate. Since 1983, Mentor has also marketed inflatable prostheses. These have been revised and now combine the pump and reservoir into one unit. Models with self-contained cylinders began with the Flexiflate (Surgitec, no longer trading), Hydroflex (AMS), and later the Dynaflex (AMS), which produce erection by pumping the distal end, the fluid being transferred from a reservoir or other chamber to a nondistensible inner chamber, causing rigidity. These models have the advantage of being easy to implant and have better concealability than the semirigid prostheses, although not as good as the multipart prostheses.

SAFETY AND OUTCOME

There are many published series for all varieties of prosthesis *(6–11)*. In a 2-year follow-up of patients using Mentor Alpha-1 *(6)*, there was a prosthesis malfunction rate of 2.5% requiring revision; explantation was required in an additional 4.4%. There were no aneurysms of the cylinder. The satisfaction rate of the patients (both in confidence and for device rigidity) was 80%, and this was also the satisfaction rate in their partners. In another series of 150 patients followed up for a mean of 19 months *(7)*, 145 had no complications, 2 had infections, and 3 had cylinder aneurysms. The insertion of a prosthesis in patients with Peyronie's disease is more difficult. In a series of 33 patients using AMS 700CX, additional plaque surgery was necessary in 40% at the time of surgery. There was one case of glandular ischaemia and a 12% wound infection rate. However, 70% of the patients subsequently had straight penises, with 79% of patients and 75% of their partners satisfied at a mean follow-up time of 17 months *(8)*.

INDICATIONS FOR IMPLANTING PENILE PROSTHESES

Since the advent of penile prostheses in the mid-1970s, both the indications for use and the prostheses themselves have changed (Table 1) *(12,13)*. Failure or lack of acceptance of oral drug treatments, intracorporeal pharmacotherapy, external appliances such as vacuum devices, and improved topical preparations to treat impotence have made the insertion of a prosthesis an operation for which there are definite indications. Most are inserted in patients with an organic or physical cause of impotence (e.g., radical prostatectomy, total cystectomy, abdominoperineal excision of rectum). Some patients have a combination of aging, an organic cause and a psychological condition, and never fully respond to other treatment modalities; they eventually resort to an implant. Some patients have psychological problems alone and, if all other treatments have been exhausted, then an implant can be justified. The benefit of using a prosthesis in men impotent with Peyronie's disease is that the deformity is easily corrected at the same time by excision/incision of the plaque. The surgery in patients with Peyronie's disease, those with complications after priapism and those with reinsertion after prior insertion/failure/explantation, however, is often difficult and needs experienced hands. The choice of prosthesis (Table 2) depends on the preference of the surgeon. They are all well made and tested, and both AMS and Mentor provide excellent in-theater assistance by trained representatives. Facilities for learning how to insert prostheses are also available.

Cost aside, the main considerations are the wishes of the patient. The semirigid/malleable prostheses and the inflatable cylinders protrude and therefore are unsuitable for the younger man with young or teenage children, for those participating in swimming or sporting events, and for naturists. The cost of the inflatable prostheses may appear prohibitive if the patient is treated under the National Health Service (NHS) and some NHS Trusts will not countenance them. However, they can often be persuaded to

Table 1
Indications for Penile Prosthesis

Organic impotence
— Patients unwilling to consider, failing to respond to, or unable to continue with intra cavernosal drugs and external devices
— Postinjection penile fibrosis
— Peyronie's disease with impotence
— Priapism: surgery within 1 to 5 days recommended in medically untreatable priapism
— Postphalloplasty (6)
— Neuropathic bladder requiring condom-continence systems (7)
— Psychological impotence after all treatments and counseling exhausted

Table 2
Types and Costs (Exclusive of VAT) of Prostheses Available in 2003

Type	Cost –(£ Sterling)
Semirigid/malleable	
AMS 650	866
Acuform	743
Inflatable cylinders	
Inflatable two-piece	
AMS Ambicor	2779
Inflatable three-piece	
AMS 700 CX	3258
AMS 700 Ultrex	3863
AMS 700 Ultrex plus	4225
Mentor Alpha-1	3316

Note: AMS Dynaflex, the Duraphase, and Omniphase are no longer available in the United Kingdom.

pay for them in special circumstances, especially for the young diabetic patient with impotence and a potentially long lifetime of sexual activity. It is worth stressing that marriages may fail in such young men because of their erectile dysfunction. There are no particular extra problems in immunosuppressed or anticoagulated patients, and any type of prosthesis can be inserted. The inflatable prostheses are better cosmetically, but the surgery is a little more extensive. The satisfaction rate is similar for the different prostheses. Reinforced inflatable cylinders should be used in patients with Peyronie's disease or fibrosis. If it is difficult to establish space in the corpora; sometimes only a semirigid prostheses can be fitted. It is advisable to warn the patient beforehand that he may have to accept what can be fitted. The expanding cylinders of the AMS Ultrex have the advantage of easier insertion for long cylinders, because there is up to 2 cm of elongation at 6 months, which occupies any space and stretches the corpora.

The insertion of prostheses in Europe is not as widespread as in the United States (Table 3). Only 1% of all patients requesting management of impotence/erectile dys-

Table 3
Penile Implants Performed in Various European Countries in 2001

	Inflatable	Malleable
Benelux (Belgium, Holland, and Luxembourg)	46	8
France	350	50
Germany	420	35
United Kingdom	150	95
Italy	70	60
Spain	300	90
Switzerland	20	5
Czech Republic	50	0

Note: This assessment is based on market knowledge from product providers.

function are offered this treatment, partly because it is considered the final treatment. However, in the NHS, the delay in getting to surgery, especially in younger patients who fail all treatment, is excessive and it might be opportune to speed up the consultation process before serious marital friction occurs.

Both the patient and his partner should be involved in the preoperative assessment and discussion of their expectations *(14)*. It is important to inform them that the prosthesis does not involve the glans but only makes the two corpora cavernosa rigid, and that the erection is not as good as in the potent man but should be very adequate for vaginal penetration. The penis will be equivalent to the stretched length of the penis, but will not otherwise extend and therefore will not lengthen as much as the patient's original youthful erections.

The patient should be made aware of the risk of infection, which might require prosthesis removal. Similarly, they should be told that, if the inflatable prosthesis develops a mechanical problem, it can be corrected or the part replaced. They should appreciate that, with replacement for either infection or malfunction, the operation becomes more difficult, with a higher incidence of infection. The most important aspect to stress is that the patient's corporal tissue is being compressed to accommodate the prosthesis and therefore the patient should use his erectile tissue for as long as possible. The partner also needs to know that the prosthesis can feel cold and that the erect penis will not be as long as it was, but should be satisfactory. Ejaculation will not be affected if it was present prior to surgery; in the neuropathic patient it is usually absent. Additional erection of the glans occurs in many patients. The prostheses will fit into the glans but not into the tip; therefore, additional help may be needed by the partner for penetration. Manual dexterity needs to be checked prior to implanting an inflatable three-piece device. A semirigid prosthesis can be tucked upward so it becomes less obvious but, on the whole, semirigid prostheses are not recommended in the younger man with younger families or those who participate in sporting activities (e.g., swimming), as they are too obvious.

Patients are considered for penile prosthesis when all other methods of treatment have been exhausted. Oral treatments, intracorporeal pharmacotherapy, intraurethral therapy, and vacuum devices should be offered, tried, and found wanting first. In the

United Kingdom, only about 1% of patients with erectile dysfunction are eventually considered for an implant. Certain conditions, however, predispose to selection. Insulin-dependent diabetics whose vascular condition progresses and who therefore try all treatments may initially respond but later fail. Patients with previous pelvic surgery respond poorly (less than 50% success rate) to medical treatments. Patients with extensive Peyronie's disease often have problems with filling of the corpora and need a prosthesis in conjunction with correction of the deformity. The degree of distal filling can be assessed by color Doppler ultrasonography with intracorporeal pharmacotherapy. Extensive fibrosis related to the sequelae of priapism usually makes a penile prosthesis the only treatment available. Priapism that has not been treated for 48 hours causes extensive intracorporeal thrombosis (which can be confirmed by color Doppler ultrasonography); early implantation of a penile prosthesis is advised before fibrosis sets in *(15)*.

Preoperative preparation involves measurement of urinary flow, lower bowel clearance, and antibiotic cover routinely. The choice is the surgeon's, but cover against anaerobes and both Gram-negative bacilli and Gram-positive cocci are needed (e.g., rectal metronidazole, intravenous cephalosporin or gentamicin, and ampicillin). No infection occurred in a series of patients using vancomycin and gentamycin for 48 hours *(16)*. Pubic shaving in theater is now recommended, and a catheter is not usually needed as it is only an added source of infection.

A subcoronal incision is used for semirigid prostheses; circumcision is often needed unless the foreskin is very lax. A penoscrotal incision is used for inflatable cylinders, two-part inflatable prostheses, and for difficult fibrotic corpora and reoperations. An infrapubic incision (a single incision) is easier for multipart inflatables. The operation may be performed under regional or general anaesthesia. Local anaesthesia has found little favor in the United Kingdom *(17)*. After the initial skin incision, the tunica albuginea is opened between stays, each corpus cavernosus is dilated with a Hegar dilator, both proximally and distally, and penile length is measured with a measuring tool. It is important not to attempt to implant a prosthesis that is too long, and rear tip extenders can be used to shorten or lengthen the prostheses as needed. The tip of the prostheses should fit snugly into the glans. The corpora are then closed.

When inflatable prostheses are used, the reservoir can be placed extraperitoneally though a separate groin incision or by careful blunt dissection through the external ring, or by opening the linea alba in the lower midline. The reservoir is filled with 50 to 100 mL of saline, the pump is placed in the scrotum, and the parts connected. At the end of the operation the whole prosthesis is checked for function and satisfactory fit and left deflated.

Problems can arise during surgery. Cross-dilatation is often encountered, especially in difficult fibrotic corpora, as the proximal corpora, before they separate, are joined in the midline. If the cylinders are difficult to fit, then this condition should be suspected. It can be assessed by inserting two Hegar dilators, one on each side. Very rarely will a second incision be needed to rectify this problem. Tunical disruption is not rare. Distal disruption is only important with semirigid prostheses due to the possibility of a prosthesis eroding through the glans. Proximal disruption is more important as the semirigid prosthesIs can migrate posteriorly when used. The breach should be closed through a separate incision. The hole in the tunica (usually at its attachment to the bone) is closed, incorporating either the tip of the prosthesis or the rear tip-extenders into the suture line. Ensure that the patient does not use the prosthesis for at least 6 to 8 weeks to allow extra healing. The use of a short Dacron sleeve over the proximal end of the prosthesis *(18)* inserted from the original incision has also been advocated, although primary closure is probably safer. Disruption of the urethra usually occurs near the meatus when the dilatation has been difficult or the operator heavy-handed. It is advisable to abandon the procedure if this occurs and reoperate a few weeks later when the breach has healed. Failure to recognize this problem leads to infection and removal of the prosthesis. If there is any doubt, the patient should undergo urethroscopy during the procedure. If the hole in the urethra is small and accessible near the incision site and reoperation is not advisable, closure of the urethral defect could be considered.

Poor fit can be caused either by inadequate or difficult dilatation. Most corpora are of similar size, but a difference of 1 cm is acceptable and is not noticed by the patient. However, if the tips of the prostheses are offset, this is noticed and unsightly. If cylinders do not enter the glans, a flaccid glans or "droop" results. These problems may be correctable by additional dilatation, taking care not to disrupt the tunica. If not, the patient may have to tolerate a flaccid glans. Difficulty in closing the corpora usually occurs in slim or fibrotic penises. Slim prostheses (9 mm diameter) are available for such patients. In patients with extensively fibrotic corpora, it may be necessary to close the incision with synthetic material, e.g., Goretex *(19)*. If the penis is curved by a Peyronie's plaque or fibrosis related to intracorporeal injections, this should be corrected. Puncture of the cylinders is usually caused by carelessness. If there is any suspicion of this, one should replace the component. Postoperatively, pain relief is very important. Disrupting the corporal tissues and inserting a prosthesis, especially a semi-rigid one, is very uncomfortable. This can continue for a few weeks and does not necessarily herald infection. Adequate broad spectrum antibiotics (usually oral) for a week after surgery are advisable, especially in diabetics and immunosuppressed patients, the choice being the surgeon's. Patients with voiding difficulties (e.g., those with a neuropathic bladder) initially have a catheter inserted. Those using clean intermittent self-catheterization can continue to do so after 48 hours. Older patients with unexpected voiding problems are more difficult to manage and an indwelling catheter is safer than lower tract surgery in the first month after implantation. The patient will be sufficiently uncomfortable not to want to use the prosthesis for 4 weeks; this delay is needed before the pump mechanism in the scrotum is usable. Thereafter, they should be taught to pump regularly two to three times a week to expand the tunica.

Infection is the most serious postoperative complication. It invariably leads to reoperation, loss of the implant and difficult further implant surgery *(20)*. In various series, the infection rate varies from 2 to 16% per implant lifetime *(21–23)* It is much higher (8–18%) in reoperations *(24)* and is increased in spinally injured and immunosuppressed patients on steroids, although not in diabetics *(25)*. Most infections are thought to occur through contamination at surgery. The commonest infecting organism is *Staphylococcus epidermidis,* followed by Gram-negative bacilli, *S. aureus,* and anaerobic organisms. Other authors indicate that *S. aureus* is the most commonly cultured organism *(24)*. Prevention of infection by prophylactic antibiotics is important, but it is also important to eradicate infections. Examples include urinary tract infections in the neuropathic patient and local skin sepsis and balanitis in diabetics. Prophylactic antibiotics may be chosen by the surgeon but aminoglycosides will be effective against Gram-negative bacilli, cephalosporins against *S.aureus* and *S.epidermidis,* and metronidazole against the anaerobic organisms usually found in the perineal region.

Infection may be suspected if the penis or the scrotal region are painful or if there is erosion of a cylinder or pump. The presence of a purulent discharge leaves three options *(21).* The first is removal of the infected part. Certainly, if one semirigid cylinder is infected, removal is justified, taking a culture from the prosthesis and using adequate and appropriate antibiotics. The other cylinder has a reasonable chance of surviving in position and can be usable alone for intercourse. The second is removal of the infected implant, irrigation with antibiotic solution, and insertion of a new device. This has been successful in about 85% of cases *(26).* The third option is to remove all parts of the prosthesis, insert drains and close the wound *(27),* with the intent to insert a second prosthesis after six months. The disadvantage of this is that the cavity and corpora will fibrose and contract, making it much more difficult to insert another prosthesis.

Erosion/migration is more common with rigid prostheses and is associated with the need for an indwelling catheter or intermittent catheterization. It also occurs in patients who have a semirigid prosthesis or reduced penile sensation (paraplegics), those who have had radiotherapy, urethral strictures or are on steroids *(27).* It also occurs when too long a prosthesis (semirigid) is inserted or if there has been an unsuspected urethral breach. The erosion of the cylinder may or may not be caused by infection but, if not, it will certainly become rapidly infected. When erosion is suspected, urethroscopy should be performed or, when it can be seen at the meatus, the prosthesis must be removed, leaving the other cylinder in place to maintain penile length. At least a 4-month delay is needed before a second cylinder is reinserted. To prevent erosion it is advisable to avoid semirigid prostheses in patients with predisposing factors; inflatable cylinders are preferred.

Erosion of the cylinder usually presents as an emergency. If proximal erosion or migration occurs and the cylinder is not exposed, then, through a perineal incision, the proximal crus can either be resutured or rebuilt using a Dacron cup sutured to the crus and tunica albuginea *(27).* Erosion of a viscus by the reservoir is uncommon. To avoid this, a balloon should be placed away from the bladder or bowel. If it occurs, the site of the reservoir should be changed and the defect in the viscus closed.

A flaccid glans at the completion of the operation is caused by insufficient dilatation of the distal corpora and the insertion of too short a prosthesis. This is generally addressed by removal and implantation with a longer prosthesis. If the cylinders lengthen (AMS Ultrex), some of the floppiness may be corrected over the following months. Alternatively, the glans can be fixed *(18).* Tucking sutures are placed through the subglandular tunica on the dorsal aspect of the penis, avoiding the nerve and blood supply to the glans, and then the sutures are brought through the glans. The use of intraurethral pellets of prostaglandin E1 (medicated urethral system of erection [MUSE]) may well improve glandular erection in this case *(28).* Glandular ischemia/ penile necrosis is a disastrous complication. It is very rare *(29,30)* and typically occurs shortly after surgery. Predisposing factors are severe diabetes, indwelling catheters, and the use of compression bandages. Perineal pain can be a major clinical problem. It may reflect a smoldering periprosthetic infection or too long a prosthesis *(29).* Persistent penile curvature in patients with Peyronie's disease after surgery indicates that the release of the fibrotic tissue was inadequate *(29)*. Straightening procedures, incision of the plaque or ellipse of the tunica should only be undertaken if the angulation precludes sexual intercourse and then only after an adequate period has elapsed to determine whether the condition can correct itself spontaneously.

Since 1981, the incidence of mechanical problems with the inflatable prosthesis has decreased markedly to 5% within the last 5 years *(27)* and has since remained infrequent. If the prosthesis leaks, it usually stops working or rigidity is insufficient *(31)*. The leaks tend to occur at the connections or where the cylinder tubing enters the cylinders. The former problem is much less common since the advent of improved connectors (quick connects, AMS). On exploration those components not affected or leaking can be detected using an electrical resistance meter *(27)* and retained or replaced accordingly. The whole system must be well irrigated to eradicate any debris. If the prosthesis has not been functioning for a couple of weeks, or if the whole prosthesis has been implanted for a few years, the whole device should be replaced.

Autoinflation occurs at early and late follow-up. This can be prevented at the time of surgery by checking that there is adequate space for the reservoir and that, after filling the reservoir and before disconnecting the fluid-filled syringe, there is no back filling of the syringe. The cylinders should be left completely deflated for four weeks to allow a capsule to surround the reservoir. If autoinflation occurs, the capsule which has formed around the partially filled reservoir may be broken by frequent pumping and release. An explanation of what has happened may be enough to reassure the patient; replacement is unlikely to be necessary.

Cylinder aneurysm is a bulbous dilatation of the cylinder. It now rarely occurs in the present generation of reinforced prostheses. Cylinders can rupture through unrecognized damage on closure of the tunica. This is detected when the prosthesis is inflated at 4 weeks. Late rupture can occur and is treated by replacement of the device.

The author currently inserts 20 penile prostheses per year in North Wales. Half are new implants and half are reimplants. Of the 10 reimplants in a recent year, 4 were for mechanical breakdowns and 6 were reimplants following removal of prostheses in other centers. These were removed for malfunction or infection. All prostheses except one were inflatable multipart varieties.

Early complications are detected in hospital. Most complications are detected at 1-month follow-up. Infection usually occurs within the first month and presents as pain or overt purulent discharge. Autoinflation can present within the first few days/months. Floppy glans is often evident after the first month when the prosthesis is inflated. Mechanical failure is uncommon with modern equipment. The pump fails to refill or fills up with air if there is a leak or disconnection. If noted early, a defective single part can be replaced, or the tubing reconnected after refilling the prosthesis. If the failure occurs after some months all parts should be replaced as these prostheses are now under full warranty. Semirigid prostheses do not have problems with failure, but they are more prone to erosion. This is almost always the result of sepsis. It is worth just removing the single eroded cylinder unless overt sepsis is obvious in the second cylinder. Replacement of malfunctioning/infected or eroded prosthesis, using an inflatable prosthesis, even with penile fibrosis, is successful in 95% at 1 year *(32)*. Additional information is available from several sources (*see* the Appendix).

Table 4
Patient Surveillance After a Penile Prosthesis at Glan Clwyd Hospital, North Wales, UK

	Postoperative year					
	1	2	3	4	5	10
Clinic visit	2[a]	0	0	0	0	0

[a] At 1 month, the patient is taught how to operate the device. At 3 months, the visit is to check that it is working properly. Further visits are arranged as necessary. The patient can inform the hospital of problems thereafter by telephone or by letter. No other follow-up tests are performed routinely.

PROSTHETIC DESIGN/QUALITY OF MANUFACTURE

The design of the leading prostheses AMS and Mentor have changed little in the last 10 years, but all explanted prostheses are returned to the makers where they are evaluated. Two improvements in design have recently been reported. One is a lockout valve to prevent autoinflation *(33)* and the other is an antibiotic-impregnated device to prevent infection *(34)*.

APPENDIX: SOURCES OF ADDITIONAL IINFORMATION

AMS UK	9 Ironbridge House Windmill Place 2-4 Windmill Lane Hanwell Middlesex UB2 4NJ UK Tel. 02086069955
AMS EUROPE	AMS EurAope Straatweg-66H 3621BR Breukelen Amsterdam The Netherlands E-mail: amseuropebv@ams.com
MENTOR UK	The Woolpack Church Street Wantage OX12 8BL UK Tel. 01235768758
Impotence Association	P.O. Box 10296 London SW17 9WH UK

REFERENCES

1. Pearman RO. Treatment of organic impotence by implantation of penile prosthesis. J Urol 1967;97:716.
2. Small MP. Small-Carrion penile prosthesis: a report on 160 cases and review of the literature. 1978. J Urol 2002;167:1191–1194.
3. Finney RP, Sharpe JR, Sadlowski RW. Finney hinged penile implant: experience with 100 cases. J Urol 1980;124:205–207.
4. Jonas U. Silicone-silver penis prosthesis (Jonas-Eska), long-term experiences. A critical assessment. Urologe A 1991;30:277–281.
5. Diagnostic and therapeutic technology assessment. Penile implants for erectile impotence. JAMA 1988;260:997–1000.
6. Goldstein I, Newman L, Baum N, et al. Safety and efficacy outcome of Mentor alpha-1 inflatable prosthesis implantation for impotence treatment. J Urol 1997;157:833–839.
7. Garber B-B. Inflatable penile prosthesis. Results of 150 cases. Br J Urol 1996;78:933–935.
8. Montorsi F, Guazzoni G, Barbieri L, et al. AMS 700 CX inflatable penile implants for Peyronie's disease. Functional results, morbidity and patient partner satisfaction. Int J Impot Res 1996;8:81–85.
9. Montague D-K, Angermeier KW, Lakin MM, Ingleright BJ. AMS 3-piece inflatable penile prosthesis implantation: comparison of CX and Ultrex cylinders. Urology 1996; 156: 1633–1635.
10. Merrill DC. Clinical experience with Mentor inflatable penile prosthesis in 301 patients. J Urol 1988;140:1424–1427.
11. Wilson SK, Cleves M, Delk JR. Long term results with Hydroflex and Dynaflex penile prosthesis: device survival comparison to multi component inflatables. J Urol 1996;155: 162–163.
12. Jordan GH. Alter GJ, Gilbert DA, Horton GE, Devine CJ Jr. Penile prosthetic surgery implantation in total phalloplasty. J Urol 1994;152:410–414.
13. Gross A-J, Sauerwein DH, Kutzenberger J, Ringert RH. Penile prosthesis in paraplegic men. J Urol 1996;78:262–264.
14. Evans C. The use of penile prostheses in the treatment of impotence. Br J Urol 1998;81: 591–598.
15. Rees R, Goorney S, Peters J, Ralph D. The management of low-flow priapism with the immediate insertion of a penile prosthesis. Br J Urol 2002;90:893.
16. Malfezzini M, Capone, M, Ciampalini S, Stephanis D, Sinonato A, Carnignani G. Antibiotic prophylaxis in prosthetic penile surgery. Critical assessment of results in 75 consecutive patients. Int J Impot Res 1996;8:87–89.
17. Randrup E, Wilson S, Mobley D, Suarez G, Mekras G, Baum N. Clinical experience with Mentor Alpha 1 inflatable penile prosthesis; report on 333 cases. Urology 1993;42:305–308.
18. Fishman IJ. Corporal reconstruction procedures for complicated penile implants. Urol Clin North Am 1991;16:73–90.
19. George VK, Shah GS, Mills R, Dhabuwala CB. The management of extensive penile fibrosis a new technique of minimal scar tissue excision. Br J Urol 1996;77:282–284.
20. Thomalla JV, Thompson ST, Rowland RG, Mulcahy JJ. Infectious complications of penile prosthetic implants. J Urol 1987;138:65–67.
21. Carson CC. Management of penile prosthesis infection. Prob Urol 1993;7:368–380.
22. Kessler R. Complications of inflatable penile prostheses. Urology 1981;81:470–472.
23. Radomski SB, Herschorn S. Risk Factors associated with penile prosthesis infection. J Urol 1992;147:383–335.
24. Wilson SK, Delk JR. Inflatable penile implant injection rate predisposing factors and treatment suggestions. J Urol 1995;153:659–661.

25. Montague DK, Angermeier KW, Lakin MM. Penile prosthesis infections. Int J Impot Res 2001;13:326–328.
26. Furlow WL, Goldwasser B. Salvage of eroded inflatable penile prosthesis—a new concept. J Urol 1987;138:312–314.
27. Mulcahy JJ. The management of complications of penile implants. Prob Urol 1991;5:608–627.
28. Chew KK, Stuckey BC. Use of transurethral alprostodil (Muse) for glans tumescence in a patient with penile prosthesis. Int J Imp Res 2000;12:195–196.
29. Lopez TM, Martins FE, Dias JC, Vickers MA. Complications of penile prosthetic surgery. EBU Update Ser 1995;4:58–63.
30. Maclean DS, Masih BK. Gangrene of the penis as a complication of penile prosthesis. J Urol 1985;133:862–863.
31. Parulkar BG, Lamb B, Vickers MA. The detection of leakage in urinary devices. J Urol 1995;153:358A.
32. Wang R, Kay B, Cancillero VA, et al. Complex penile implantation. J Urol 2001;165: abstract 1050.
33. Hakim LS, Nehra A. Further experience with the Mentor Lock-out valve in preventing auto inflation post inflatable penile prosthesis. J Urol 2001;165:abstract 1043.
34. Brock G, Bochinski D, Mahoney CB. Inhibizone treatment: The first antibiotic treatment impregnated into the tissue–contacting surface of an inflatable prosthesis. J Urol 2001; 165:abstract 1047.

14
Artificial Urethral Sphincters

David J. Rea, John P. Lavelle, and Culley C. Carson, III

INTRODUCTION

Since the introduction of the first artificial urethral sphincter (AUS) in 1973, approximately 50,000 AUS devices have been implanted. The development of new biomaterials, surgical techniques, and new prosthetic sphincter design has correspondingly evolved over this period such that the previously seen problems of tissue inflammation, erosion, infection, and poor device design are fortunately now less common. The currently available AUS available from American Medical Systems (AMS) is the AMS Sphincter 800™ (Fig. 1). It is estimated that approximately 5000 AUS devices are implanted worldwide each year.

In 1972, the presence of a large number of patients who had failed to respond to conventional continence treatments led to the development of the AMS model 721, the first commercially available AUS (1). Collaboration between the medical device industry and the National Aeronautics and Space Administration allowed incorporation of high reliability aerospace components into these devices and hastened development of the modern AUS (2). The AS 721 was in use for approximately 3 years. Difficulties with this device arose from the large number of components and overly complex design that relied on multiple spring-loaded valves to function normally. Because of problems with mechanical reliability and the prevalence of urethral erosions, the device underwent multiple design modifications. These improved its versatility and increased the overall numbers of patients achieving continence. A pressure-regulating balloon reservoir was incorporated to overcome the problem of urethral erosion caused by overinflation of the cuff. Other modifications have reduced the number of malfunctions associated with spring-valve technology. The simplified designs provide for easier implantation, more reliable function, and flexibility in the selection of the site for cuff placement. In addition, the concept of delayed cuff activation was introduced. This permits the device to be primed but deactivated for several weeks to allow for tissue healing. It is then activated without the need for a second surgical priming procedure (3).

The AS 800 is implanted around the urethra using the chosen size of cuff and a reservoir with appropriate plateau pressure that can vary from 51 to 80 cm H_2O. In

From: *The Bionic Human: Health Promotion for People With Implanted Prosthetic Devices*
Edited by: F. E. Johnson and K. S. Virgo © Humana Press Inc., Totowa, NJ

Male Patient

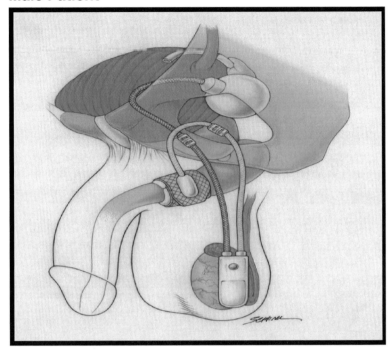

Female Patient

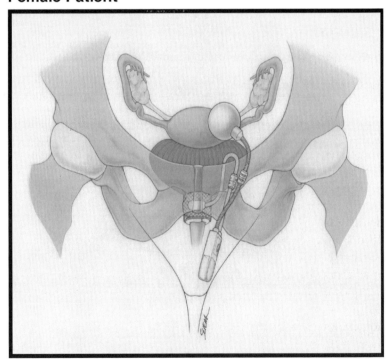

Fig. 1. The AMS Sphincter 800™ urinary prosthesis. (Courtesy of American Medical Systems, Inc. Minnetonka, MN. Illustrations by Michael Schenk. www.visitAMS.com.)

Table 1
Asymptomatic Patients With Artificial Urethral Sphincters and Normal Detrusor Function: Patient Surveillance After Implantation of an Artificial Urethral Sphincter at UNC Chapel Hill

	Postoperative year					
	1	2	3	4	5	10
Office visit[a]	1	1	1	1	1	1

[a] The number in each cell indicates the number of times each evaluation is recommended for a particular year.

1987, the cuff design was changed to provide a more even distribution of pressure. Concurrent with this design change, the process used to manufacture the synthetic material used for the fluid reservoir and pump was improved to make these components more resistant to fracture and subsequent mechanical failure. These changes resulted in almost a 50% reduction in the rates of mechanical failure (4). The current model is suitable for male and female patients and is implanted fully primed with radiopaque fluid in its deactivated state (cuff empty and deactivation button engaged). An interval of 6 weeks is allowed for tissue healing before activating the device for the first time. The device is activated by giving the pump a quick forceful squeeze. The cuff closes in 3 to 5 minutes. Pumping of the bulb moves fluid from the cuff into the reservoir and allows the patient to void. The cuff closes automatically once again after 3 to 5 minutes when micturition is complete.

PUBLISHED GUIDELINES FOR FOLLOW-UP

Few papers comment on follow-up of patients who have functioning devices and who are free from complications. Most have concentrated on the complications arising from AUS placement. In order to better understand how to follow these patients, it is helpful to divide patients into two broad groups: patients with pure sphincteric insufficiency (including men with incontinence following transurethral resection of the prostate [TURP] or radical prostatectomy, women with postpartum incontinence and those with congenital anomalies of the lower urinary tract, or traumatic urethral disruption) and patients with sphincteric insufficiency with abnormal detrusor function (neurogenic bladder dysfunction).

The AMS operating room manual (5) advises that the patient should contact the implanting surgeon at least once each year to evaluate the function of the device (Table 1). We have found this to be satisfactory for patients with pure sphincteric incompetence.

Patients who have sphincteric insufficiency with abnormal detrusor function require more careful long-term follow-up (Table 2) because of the potential for renal failure in patients who develop bladder hypertonicity after AUS placement. This is particularly true for children with myelodysplasia without bladder augmentation who have an AUS. At the University of North Carolina School of Medicine, we recommend urine culture every 3 months, annual excretory urography or renal ultrasound, and annual postvoid residual urine measurement. In addition, biochemical assessment of renal function

Table 2
Asymptomatic Patients With Artificial Urethral Sphincters and Abnormal Detrusor Function

	Postoperative year					
	1	2	3	4	5	10
Office visits to implanting urologist[a]	4	4	4	4	4	4
IVP or renal ultrasound	1	1	1	1	1	1
Postvoid residual	1	1	1	1	1	1
Serum BUN and creatinine clearance	1	1	1	1	1	1
Urine culture and sensitivity	4	4	4	4	4	4

[a]The number in each cell indicates the number of times each evaluation is recommended for a particular year. IVP, intravenous pyelogram; BUN, blood urea nitrogen.

should be performed every 1 to 2 years *(6,7)*. Female patients who have received an AUS and who become pregnant should be advised to receive treatment according to normal obstetric practice *(8,9)*.

DETECTION OF PROSTHETIC FAILURE

The most common presentation of AUS failure is recurrence of urinary incontinence, which is occasionally detected at scheduled routine visits. It is important to first ensure that the device has not inadvertently been placed in the deactivated state. Incontinence may also occur because of mechanical failure when there is loss of hydraulic fluid owing to connector malfunction or device component failure. Presenting symptoms suggesting erosion of the cuff into the urethra include bloody urethral discharge, pain and swelling in the perineum or scrotum, acute incontinence, and symptoms of urinary tract infection. Most urologists would elect to perform radiographs with the cuff inflated and deflated in the clinic to examine for system leakage or tubing kinks. If no contrast is seen on the plain radiograph, then leakage of hydraulic fluid has occurred and surgery to explore the device is warranted. Most leaks occur at the cuff. However, it is wiser to explore the reservoir and control assembly first for leaks as they are easier to isolate. If necessary, electrical continuity testing may be employed to locate the site of the leak *(10)*. If the inflate and deflate radiograph is normal, cystourethroscopy should be performed to assess for cuff erosion. Urodynamics may be performed to assess for detrusor hyperreflexia.

DETECTION OF PROGRESSION OF THE DISORDER FOR WHICH THE PROSTHESIS WAS PLACED

Implantation of an AUS does not in any way inhibit assessment of the lower urinary tract for conditions for which the AUS may have been placed, including postprostatectomy incontinence, previous unsuccessful anti-incontinence surgery in the female patient, congenital anomalies of the lower urinary tract, trauma-induced urethral rupture, or neurogenic bladder dysfunction. All physicians whose practice includes these patients should be aware of the presence of the device and know how to deactivate it, should the need arise.

DETECTION OF OTHER CONDITIONS OF MEDICAL SIGNIFICANCE

The presence of a functioning implanted AUS does not preclude detection of other conditions of medical significance. Clinical assessment of the external genitalia is not impaired or impeded by the presence of the device. Endoscopic assessment of the lower urinary tract is the same as for the nonimplanted patient once the device has been deactivated and appropriate antibiotic prophylaxis administered. Other than the contraindication to magnetic resonance imaging, no specific precautions for imaging of the pelvis or genitourinary system for other medical conditions are required. As these patients are likely to be seen on an annual basis, other conditions of medical significance may be detected at an earlier stage of development than would normally be expected. This is a function of the follow-up that these patients undergo rather than any benefit accruing from placement of an AUS.

WHAT IS KNOWN ABOUT ACTUAL STRATEGIES OF PRACTICING CLINICIANS?

Little is known about the actual strategies of individual practicing physicians who follow patients after AUS placement when they are asymptomatic and have recovered from the implantation procedure. The vast majority of the published literature on AUS follow-up concentrates on the management of complications, but follow-up of asymptomatic patients can and should emphasize measures to avoid problems. Several authors have recommended that patients who are dry at night in a recumbent position deactivate their sphincter prior to falling asleep in order to reduce the risk of ischemia to the underlying urethra or bladder neck *(11–13)*. It is the responsibility of each implanting clinician to ensure that patients are educated and aware of several clinical scenarios in order to maintain health and prevent device-related complications. Patients need to be aware that sphincter cuff deflation is needed prior to bladder catheterization in order to prevent unintentional damage from medical staff who do not normally look after such patients. Many patients with functioning AUS devices wear Medic-Alert bracelets to alert medical staff in the event that they are involved in an accident in order to prevent iatrogenic urethral injury in a trauma management setting. Patients also need to receive appropriate antibiotic prophylaxis when undergoing any invasive procedure to reduce the risk of hematogenous seeding of the prosthetic device *(14)*.

IMPORTANCE OF PROSTHESIS DESIGN AND QUALITY OF MANUFACTURE

The last 23 years of AUS manufacture have seen many changes and improvements in prosthesis design. The group of artificial sphincter devices patented and produced by AMS have remained the only clinically and commercially viable implantable products. They have been redesigned seven times (AMS 721, AMS 761, AMS 742 [A,B,C], AMS 792, and AMS Sphincter 800—the current version). Simplification of its hydraulic circuit, reduction in the number and complexity of components, and incorporation of the pressure-regulating balloon reservoir to limit the pressure applied to the urethra have improved its mechanical reliability and reduced the probability of device erosion.

VARIATION IN FOLLOW-UP BY PROSTHETIC TYPE

The vast majority of devices implanted in the last 18 years has been of the AMS 800 type. In our experience, no specific follow-up other than that discussed above is required for asymptomatic patients. Many patients who had early AUS models have had surgery to revise or replace them owing to device failure, erosion, or infection. However, if an AUS device that predates the AMS 800 model is functioning well, the same follow-up scheme as for the current model is adequate.

Role of Office Visits, Imaging Tests, Markers, Inspection, and Nonspecific Tests

Patients who have normal detrusor function with a functioning AUS and are asymptomatic require no specific imaging or laboratory investigations. Any subsequent urinary symptoms in these patients should be investigated as for patients without a device, with two exceptions. Appropriate antibiotic prophylaxis should be given for interventional procedures and the device cuff should be deactivated if cystoscopy or catheterization is required. Patients who have abnormal detrusor function require careful long-term follow-up. Lifelong surveillance of the urinary tract is mandatory because of the potential for renal failure in patients who develop bladder hypertonicity after AUS placement, as mentioned earlier.

What Primary Prevention Measures Should Be Carried Out?

The primary prevention measures required to reduce the numbers of patients receiving AUS are multiple, widespread, and cover areas of medicine as diverse as urological operative technique, obstetric care, public health, and education.

For adult male patients, reported postoperative incontinence rates for radical perineal or retropubic prostatectomy range from 0.5 to 60% (15–17). For TURP, the rate is 1.7% for postoperative stress incontinence and 0.4% for significant or total incontinence (18). These figures vary because of differing definitions of what constitutes continence and wide differences in the methods of evaluation of continence. With the massive increase in the number of cases of prostate cancer being detected, and the decrease in the mean age of this population presenting for treatment, urologists are striving to improve outcomes for these patients in terms of both duration and quality of life. The advent of nerve-sparing radical prostatectomy and the increasing subspecialization within urology offers the prospect of a decrease in the numbers of patients rendered incontinent from these procedures. The risk of incontinence arising from transurethral resection of the prostate is already low but might be improved by better patient selection and training in endoscopic surgery with strict attention to operative detail.

Reducing the number of women requiring an AUS depends on minimizing the incidence of stress incontinence. Trauma, parturition injury, radiation, pelvic surgery, and sacral cord lesions may cause stress incontinence in women. These conditions may be exacerbated by obesity, multiple failed anti-incontinence procedures, and urogenital atrophy. Targeting preventable risks such as technical errors during pelvic surgery and parturition is a rational strategy.

Primary prevention measures aimed at preventing congenital neurological deficits such as spina bifida have recently focused on periconceptual folic acid ingestion. In 1999, the American Academy of Pediatrics endorsed the US Public Health Service recommendation that all women capable of becoming pregnant consume 400 µg of folic acid daily to prevent neural tube defects such as spina bifida *(19)*. Studies have demonstrated that periconceptional folic acid supplementation can prevent at least 50% of neural tube defects. At present, fewer than one in three women of childbearing age consume the recommended amount of folic acid *(20)*. The primary preventative measure most likely to be successful in this instance is patient education.

No specific health maintenance counseling is recommended after AUS placement other than advising patients that they will require appropriate antibiotic prophylaxis for interventional procedures and making them aware that the device cuff should be deactivated if cystoscopy or catheterization is required. Patients must be educated to pass on this vital information to their health care providers as some providers will not be familiar with AUS devices and may inadvertently cause damage to the patient or to the sphincter for this reason. In general terms, patients with AUS devices should try not to gain excess weight because, in the event of a device malfunction or failure, subsequent revision surgery will be more complicated and may have an increased risk of infection. Patients who have an AUS device placed, and whose employment requires them to do heavy manual work, should be warned that some degree of incontinence can result from heavy lifting when intra-abdominal and consequent intravesical pressure exceeds that generated by the AUS.

No specific support groups exist for patients who have had an AUS implanted. There are, however, several support groups for people with continence-related problems. These are listed in the Appendix.

ARE MULTICENTER PROSPECTIVE TRIALS NEEDED?

Few practicing urologists or patients would argue that AUS devices are problem-free. However, multicenter prospective clinical trials do not seem likely for several reasons. First, there are only a modest number of patients. Second, there are no concepts fundamentally different from the current one ready for evaluation in clinical trials. At present, surgical implantation of an AUS for moderate to severe urinary incontinence is the standard of care against which other treatments are compared. However, if new methods to treat moderate to severe urinary incontinence were to become available, clinical trials might be warranted. Further clinical research is required to find ways to prevent loss of renal function in patients with unrecognized increased bladder outflow resistance from the AUS device. Patients who have neurological deficits such as spina bifida and who have not had bladder augmentation are particularly vulnerable.

Future advances in the development of cuff design should concentrate on avoiding the "triple cushion effect," which would improve the efficacy and durability of the device. This would mean fewer reoperations for cuff leakage. Leakage now occurs most commonly at the site of folds that form on inflation of the cuff. Advances in device design should concentrate on developing a means to alter the hydraulic pressure within the system without the need for reoperation. A new device design concept has been described that may accomplish this *(21)*. It uses a mechanical compressive coil encompassed in a polytetrafluoroethylene sheath to establish a compressive force, thereby occluding the urethra and restoring continence. If properly developed, this could eliminate up to 50% of the reoperations seen with the current AMS 800 device.

APPENDIX: SUPPORT GROUPS FOR PEOPLE WITH CONTINENCE-RELATED PROBLEMS

National Association for Continence (NAFC)
P.O. Box 8310
Spartanburg, SC 29305-8310
Telephone: 864-579-7900
Toll Free: 1-800-BLADDER
Fax: 864-579-7902
E-mail: memberservices@nafc.org
Web site: www.nafc.org

Simon Foundation, Inc.
Box 835
Wilmette, IL 60091
Telephone: 708-864-3913
Toll Free: 1-800-23-SIMON
Fax: 847-864-3913
E-mail: simoninfo@simonfoundation.org
Web site: www.simonfoundation.org

Continence Restored, Inc.
407 Strawberry Hill Ave
Stamford, CT 06902
Telephone: 212-879-3131
Fax: None
E-mail: Under development
Web site: Under development

REFERENCES

1. Scott FB, Bradley WE, Timm GW. Treatment of urinary incontinence by implantable prosthetic sphincter. Urology 1973;1:252–259.
2. Rouse DJ, Brown JN, Jr, Whitten RP. Methodology for NASA technology transfer in medicine. Med Instrum 1981;15:234–236.
3. Furlow WL. Implantation of a new semiautomatic artificial genitourinary sphincter: experience with primary activation and deactivation in 47 patients. J Urol 1981;126:741–744.
4. Elliott DS, Barrett DM. Mayo Clinic long-term analysis of the functional durability of the AMS 800 artificial urinary sphincter: a review of 323 cases. J Urol 1998;159:1206–1208.
5. American Medical Systems. AMS Sphincter 800 Urinary Prosthesis Operating Room Manual. American Medical Systems Publication, Order No. 22000017c, Dec 1995.
6. Gonzalez R, Merino FG, Vaughn M. Long-term results of the artificial urinary sphincter in male patients with neurogenic bladder. J Urol 1995;154:769–770.
7. Levesque PE, Bauer SB, Atala A, Zurakowski D, Colodny A, Peters C, et al. Ten-year experience with the artificial urinary sphincter in children. J Urol 1996;156:625–628.
8. Fishman IJ, Scott FB. Pregnancy in patients with the artificial urinary sphincter. J Urol 1993;150(2 Pt 1):340–341.
9. Creagh TA, McInerney PD, Thomas PJ, Mundy AR. Pregnancy after lower urinary tract reconstruction in women. J Urol 1995;154:1323–1324.
10. Webster GD, Sihelnik SA. Troubleshooting the malfunctioning Scott artificial urinary sphincter. J Urol 1984;131:269–272.
11. Kowalczyk JJ, Spicer DL, Mulcahy JJ. Erosion rate of the double cuff AMS 800 artificial urinary sphincter: long-term follow-up. J Urol 1996;156:1300–1301.

12. Goldwasser B, Furlow WL, Barrett DM. The model AS 800 artificial urinary sphincter: Mayo Clinic experience. J Urol 1987;137:668–671.
13. Abbassian A. A new operation for insertion of the artificial urinary sphincter. J Urol 1988; 140:512–513.
14. Carson CC, Robertson CN. Late hematogenous infection of penile prosthesis. J Urol 1988; 39:50–53.
15. Fowler FJ, Jr, Barry MJ, Lu-Yao, G. Effect of radical prostatectomy for prostate cancer on patient quality of life: results from a Medicare survey. Urology 1995;45:1007–1012.
16. Steiner MS, Morton MA, Walsh PC. Impact of anatomical radical prostatectomy on urinary incontinence. Urology 1996;48:769–775.
17. Davidson PJT, van den Ouden D, Schroder FH. Radical prostatectomy: prospective assessment of mortality and morbidity. Eur Urol 1996;29:168–173.
18. Mebust WK. A review of transurethral resection of the prostate complications and the AUA national cooperation study. AUA Update Series 1989; Vol. VIII: Lesson 24:186–191.
19. American Academy of Pediatrics. Committee on Genetics. Folic acid for the prevention of neural tube defects. Pediatrics 1999;104:325–327.
20. Centers for Disease Control and Prevention. Knowledge and use of folic acid by women of childbearing age—United States, 1995 and 1998. JAMA 1999;281:1883–1884.
21. Elliott DS, Timm GW, Barrett DM. An implantable mechanical urinary sphincter: a new nonhydraulic design concept. Urology 1998;52:1151–1154.

US Counterpoint to Chapter 14

Steven B. Brandes

INTRODUCTION

Proper patient selection is critical to a successful outcome for an artificial urinary sphincter (AUS). The first key issue we determine is if the urinary incontinence is the result of intrinsic sphincter deficiency. We perform a urodynamics study, as well as leak-point pressure studies to distinguish a bladder etiology from a urinary sphincter etiology. A pressure flow study and flexible cystoscopy are done to evaluate the bladder outlet for obstruction or urethral stricture disease. The presence of overactive or neurogenic bladder by urodynamics demands initial treatment with anticholinergic therapy before any consideration of AUS. Patients who receive an AUS should have good manual dexterity and enough intellectual capacity to properly work the device. This should be evaluated preoperatively.

We also use intravenous prophylactic antibiotics prior to surgery and immediately postoperatively for 24 hours. We leave a 12 Fr Foley catheter overnight and perform a voiding trial afterward. A small percentage of patients are not able to void after implantation; this is owing to urethral swelling. In such patients, we usually place a temporary suprapubic tube until the swelling resolves. Some have advocated temporary intermittent catheterization; however, such catherization risks cuff injury and erosion. We send our AUS patients home on a 2-week course of oral antibiotics (i.e., Keflex®). We also typically leave the prosthesis deactivated for at least 6 weeks after implantation. By waiting 6 weeks, discomfort in the scrotum is allowed to resolve and revascularization and healing at the cuff site allowed to mature. To avoid iatrogenic trauma to the cuff, which seems to happen all too often in the cardiac catheterization suite, we routinely ask our AUS patients to wear a Medic-Alert bracelet. We tell each of our patients that they should have "no urethral instrumentation without AUS deactivation."

We typically counsel our patients about the risk of urine leakage caused by cuff compression when sitting on a hard surface (i.e., hard chair). We also discourage bicycle riding and horse-back riding, which again can compress and potentially injure the cuff. In order to prevent delayed urethral atrophy, we instruct our patients to deactivate the sphincter (cuff open) overnight, if they do not suffer from enuresis. If they do leak at night (when recumbent), we usually have the patients use an alarm clock so they can deflate the cuff every 4 hours.

From: *The Bionic Human: Health Promotion for People With Implanted Prosthetic Devices*
Edited by: F. E. Johnson and K. S. Virgo © Humana Press Inc., Totowa, NJ

Table 1
**Patient Surveillance After Placement
of Artificial Urinary Sphincters at Washington University**

	Postoperative year					
	1	2	3	4	5	10
Office visit	4[a]	1	1	1	1	1
Urinalysis	1	1	1	1	1	1
AUA symptom score	2	1	1	1	1	1

[a] Follow-up visits at 2 weeks, 6 weeks, 3 months, and 12 months after surgery.

Complications after surgery are typically discovered on routine follow-up office visits and by the use of pelvic radiographs. Because we routinely fill the AUS with a radio-opaque and saline fluid mix (i.e., Cysto-conray II®), troubleshooting is facilitated by observing for loss of contrast, indicating a leak. Electrical testing has also been reported to be useful for localizing a leak; we have little experience and success with this technique. When a leak or malfunction is determined in the first 2 years after implantation, we replace the faulty part; after 2 years, we typically replace the entire prosthesis. Post-AUS urinary leakage can also be caused by tissue atrophy, which occurs in roughly 10% of cases, and usually causes problems 1 to 5 years after implantation (1). We make the diagnosis of tissue atrophy by excluding other causes. Evaluation includes cystoscopy and urodynamics. Infection and incontinence can also be the result of cuff erosion. This is typically owing to iatrogenic injury to the corpus spongiosum adventitia at the time of implantation. If we find gross pus at exploration, we remove all the sphincter components. For cuff erosion with no infection, we remove the cuff only.

Our typical follow-up regimen after AUS implantation is to check the perineal and abdominal incisions at 2 weeks (Table 1). We make sure the pump is nicely descended in the scrotum. At 6 weeks, we activate the sphincter and instruct the patient in detail in the care and management of the AUS. Overall, the patient satisfaction rate exceeds 90% with AUS (1).

REFERENCE

1. Venn SN, Greenwell TJ, Mundy AR. The long term outcome of artificial urinary sphincters. J Urol 2000;164:702–707.

European Counterpoint to Chapter 14

Suzie N. Venn, Constantinos Hajivassiliou, and Tony Mundy

There are few conditions more socially devastating than urinary incontinence. The incidence of incontinence is underreported and underestimated. It represents an enormous social and psychological problem. Many methods of treatment have been tried, including penile clamps *(1,2)*, external collecting devices, indwelling catheters, surgery to increase the resistance of the bladder neck and urethra, electrical stimulation *(3,4)*, and supravesical urinary diversion *(5,6)*, but none are ideal *(6)*. For those patients whose incontinence has failed to respond to conventional means, the best option is usually implantation of an artificial urinary sphincter (AUS).

The only commercially available AUS is the American Medical Systems (AMS) device, the AMS 800®. The original design was released in 1974 *(7)*. Its use was assessed in eight mongrel dogs for up to 6 months *(8)*. Satisfactory continence does not appear to have been achieved even in the dogs. Nevertheless, this device was implanted in humans and was in use for approximately 3 years. It was difficult to implant because of the large number of components, which necessitated extensive tunneling and dissection. A pressure-regulating balloon reservoir was incorporated in a subsequent modification of the device (AMS 761) to overcome the problem of high pressure applied to the urethra. The complexity of the design and the critical central role of a collection of spring-loaded valves of different characteristics led to many complications. It was therefore redesigned in stages, culminating in the current model (AMS 800) that was introduced into clinical practice in 1983.

This device consists of three components—a cuff, a pump, and a pressure-regulating balloon. The cuff consists of an outer, firm monofilament knitted polypropylene (Dacron) backing and a pliable inner silicone cuff shell that is in contact with the tissues. It is available in a variety of sizes and is implanted around the bladder neck in either sex or bulbar urethra in males. The pump is placed in one of the labia majora in females and the scrotum in males. Pumping of the bulb moves fluid into the reservoir and empties the cuff, thus allowing the urethra to open. The cuff closes automatically once again after several minutes, when micturition is complete. Cuff refilling can be deactivated by pressing the deactivation button on the pump before the cuff refills. The balloons are available with plateau pressures of 51–60, 61–70, or 71–80 cm H_2O and are placed in the extraperitoneal space. The unit is implanted fully primed with isotonic radio-opaque fluid in its deactivated state (cuff empty and deactivation button engaged). After approximately 3 to 4 weeks to allow for tissue healing, the device is activated for the first time by sharply squeezing the bulb. The AS 800 underwent some design modifications in 1987, but has otherwise been in its present form since then.

The AUS is generally used in complex incontinence where other treatments have failed. The common indications for an AUS are postprostatectomy incontinence, neuropathy, stress incontinence, and congenital disorders such as exstrophy-epispadias, cloaca, and urinary undiversion. It is estimated that more than 20,000 devices have been implanted worldwide. A recent publication *(9)* analyzed all published results, encompassing approximately 10% of the total implanted worldwide. The overall rate of full continence was 73%, with an 88% rate of improved continence.

The most common use for the AUS worldwide is postprostatectomy incontinence. The results in this group are excellent, with a medium term (3-year) "social continence" rate of 96% *(10)*. No other treatment for this condition approaches these results. The risk of removal for infection and erosion in this group is also low, as the AUS can be inserted around the bulbar urethra. The Mayo Clinic *(11)*, where the majority of devices (70%) are inserted for postprostatectomy incontinence, reported that 88% were still functioning at a mean of 6.5 years later.

The AUS has revolutionized the quality of life of patients with neuropathic bladders. The most common cause of neuropathy leading to the need for AUS is spina bifida. It is important to ensure that these patients have an adequate capacity and a stable bladder prior to AUS insertion. Otherwise, deterioration of the upper tracts is likely to occur *(12)*. Continence is usually achieved in this group with a combination of augmentation cystoplasty and an AUS *(13)*. It has been shown that there is no increased risk of infection of the AUS with simultaneous cystoplasty *(14)*. Patients in this group must be able to reliably perform clean intermittent self-catheterization, as they are at risk for all the complications of poor bladder emptying, including perforation of the bladder, if they fail to do so *(15)*.

The AUS is considered by some authors to be indicated in women with stress incontinence caused by intrinsic sphincter deficiency (type III). Early results were published in 1985, with Donovan *(16)* reporting success in two-thirds of patients and Light and Scott *(17)* reporting a 92% success rate. Scott *(18)* reported an overall success rate in all females of 84%. Results have been variable, with continence rates up to 92% in some centers at 2.5 years *(19)* but as low as 60% in others *(20)*. There are few publications on the use of AUS in stress incontinence, raising doubts as to the real results in this group of patients.

Recently, 10-year results in 100 patients have been reported from one of our units *(21)*. The indication for insertion of a sphincter was neuropathic bladder dysfunction in 59, postprostatectomy incontinence in 23, and stress incontinence and other indications in the remaining 18 patients. The overall continence rate at 10 years was 84%. The continence rate was 92% in male patients with bulbar urethral cuffs, 84% in males with bladder neck cuffs, and 73% in all female patients. The prosthesis survival rate at 10 years, if it did not have to be removed for infection or erosion, was 66% (Fig. 1). The use of the AUS in current practice in the above groups of patients is now the

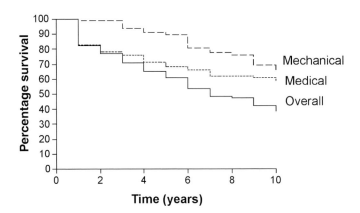

Fig. 1. Kaplan–Meier survival curves for all artifical urinary sphincter devices. Overall: survival of the original device with adequate performance. Medical: survival of the original device with adequate performance, excluding mechanical failure. Mechanical: survival of the original device with adequate performance, excluding failures owing to infection or erosion into the urethra. (Data from ref. Venn et al. *J Urol* 2000;164:702–7–60)

standard of care, as it has been shown to be an effective treatment for severe incontinence with reasonable long-term outcome.

Revisions of the AUS are of three main types: early revisions, early explantations, and late replacements. Early revisions are usually for mechanical problems. These present immediately, with the device never working or not providing satisfactory continence. The usual treatment is implantation of a balloon with a higher pressure range. Other early revisions are for minor technical problems such as bubbles in the system or kinking of the tubing. Failure of the device itself is very rare. Early explantations are for infection or erosion. Infection is either introduced with the device at the time of implantation or owing to injury of the urethra or vagina during insertion that may have gone unrecognized. Infection can also be introduced because of erosion of the urethra under the cuff. Infection and erosion generally occur in the first 2 years, but late erosion has been reported up to 10 years after implantation. Replacement is usually required when a device suddenly stops working effectively, having been effective for some years.

The first and last of these complications are easily corrected by replacing the defective component (early) or the whole device (late). Infection or erosion requires removal of the entire device, a period of recovery and a further implant some months later, if possible and desirable.

In the previously mentioned review *(9)*, the complications reported in the world literature and in the Food and Drug Administration (FDA) database were analyzed. The FDA database had 4130 entries for 3508 patients. Most revisions took place during the first year postimplantation. The world literature included reports on 2606 patients (10% of the total implanted), with revisions reported in 32%. Mechanical failure occurred in approximately 14% and erosion and infection in 16%. Most of these reports evaluate very short term follow-up and therefore probably underestimate the risk of late replacement for mechanical failure and late erosion.

Table 1
Overall (N = 100) Outcome of Original Artifical Urethral Sphincters

	No.	Median duration of AUS in years (range)
Remain *in situ*	36	11 (2–15)
Replaced for late mechanical failure	27	7 (3–12)
Removed for infection or erosion	37	see Table 2

Note: The outcome of 100 patients who had artifical urethral sphincters placed for mechanical failure more than 10 years ago showing the number still *in situ*, the number replaced having failed after working satisfactorily for a number of years, and the number removed owing to infection or erosion.
Data from ref. *21*.

Table 2
Subsequent Outcome of Patients Whose Artifical Urethral Sphincters (AUS) Were Removed for Infection or Erosion, Stratified by Reason AUS Was Placed

Reason AUS was placed	AUS successfully replaced	Dry without AUS	Diverted	Wet	Other
Neuropathy	8	8	2	5	
Postprostatectomy	2			2	
Stress incontinence	1	1		2	
Hysterectomy and radiotherapy			3		
Post-cystectomy					1 died from metastases
Renal tubular acidosis	1				
Cloaca			1		

Data from ref. *21*.

Early revisions for mechanical failure were required in a quarter of the patients reported by our unit *(21)*. This was mainly done to increase the pressure in the balloon when the cuff was inserted around the bulbar urethra–usually to replace a 51–60 cm H_2O balloon with a 61–70 cm H_2O balloon. Over 10 years, 27 patients had their AUS replaced for late mechanical failure. The median time was 7 years (range 3–12) after primary AUS placement. All of these patients are currently dry after one AUS replacement (24 patients) or two (3 patients). The AUS was removed for infection or erosion in 37 patients, of which 23 were within 2 years of placement. The remaining 14 were replaced up to 10 years after insertion (Table 1). Of these patients, the AUS was subsequently successfully replaced in 12, and a further 9 remained socially continent after removal of their prosthesis (Table 2).

In this series, the site of the cuff and the sex of the patient were important in outcome. Placing the cuff around the bulbar urethra reduces the risk of infection and ero-

Table 3
5- and 10-Year Survival of the Artifical Urethral Sphincter (AUS) in All Patients, Female Patients, and Male Patients With Bladder Neck and Bulbar Cuffs

		5 Year (%)	10 Year (%)
All patients	Overall	60	39
	Medical	68	59
	Mechanical	90	66
Female bladder neck AUS	Overall	40	18
	Medical	48	39
	Mechanical	85	50
Male bladder neck AUS	Overall	68	42
	Medical	74	63
	Mechanical	92	68
Male bulbar cuff	Overall	71	52
	Medical	79	72
	Mechanical	90	70

Overall: survival of the original device. Medical: survival of device excluding prosthesis failure. Mechanical: survival of the prosthesis excluding those removed for infection or erosion.
Data from ref. *21*.

sion of the device (Table 3). Risk factors for explantation were female sex and previous surgery or other trauma at the bladder neck. All three female patients who had an AUS inserted for incontinence after a hysterectomy and radiotherapy had their device removed later. Although it is reported to be possible to insert an AUS after radiotherapy, the risk of failure requiring revision is higher *(22)*. It is possible to implant an AUS after radiotherapy to the bladder or prostate in men and get satisfactory results if the cuff is placed around the bulbar urethra, presumably because that area has not been irradiated. However, the risk of erosion in female patients after pelvic irradiation is high.

FOLLOW-UP

The follow-up of patients with AUS depends in part on their underlying condition. Patients with neuropathic bladder dysfunction, for example, need follow-up appropriate for their condition. The purpose of follow-up of a patient with an AUS is to detect device failure, evidence of erosion and infection around the device, and to exclude secondary consequences of the presence of the AUS. These include upper tract dilatation, excessive postvoid residual urine volume and stone formation in the bladder.

During the first year, after successful activation of the device, close follow-up is required to exclude infection and erosion and secondary consequences of the AUS (Table 4). Initially, a plain X-ray is obtained as a baseline. An upper urinary tract ultrasound is also obtained to ensure that the presence of the AUS has not caused dilatation or excessive postvoid residual urine in the bladder. After the first year, follow-up is generally confined to a yearly outpatient visit. Patients are examined for swelling and tenderness around the pump and to ensure it is functioning normally. A yearly urine culture is done to exclude infection. Further investigation is only undertaken if the

Table 4
**Patient Surveillance After Implantation
of Artifical Urethral Sphincters at the Institute of Urology, London**

	Postoperative year					
Year after implantation	1	2	3	4	5	10
Office visit with urine culture	3	1	1	1	1	1
Plain X-ray and ultrasound	1	0	0	0	0	0

Note: The numbers in this table indicate the number of times each modality is requested during the indicated postoperative year. No other tests are requested unless indicated by clinical symptoms or signs.

patient reports incontinence. A typical evaluation for incontinence includes videourodynamics to assess the function of the device and cystoscopy to look for urethral erosion. Although follow-up is important to detect subtle problems, patients usually present quickly if the device fails with recurrent incontinence. The need for other investigations depends on the underlying condition for which the AUS was inserted. For example, patients with neuropathic bladder dysfunction require regular surveillance of their upper urinary tracts.

If a patient with an AUS develops an infection at a distant site—pneumonia, for example—we treat the patient just as we would if no AUS were present. We do, however, give a 1-day course of broad-spectrum antibiotics, including gentamicin, to patients undergoing instrumentation of the lower urinary tract. Antibiotics are started before the procedure and may continue for a dose or two afterward.

FUTURE RESEARCH

Although the long-term outcome of this device is generally satisfactory, there are a number of improvements to the device that would be of value. The pressure in the balloon cannot be controlled accurately and this may result in an unnecessarily high pressure. This, in turn, leads to atrophy of the underlying tissue and may result in eventual erosion of the urethra. The balloons are available with plateau pressures of 51–60, 61–70 or 71–80 cm H_2O. If the balloon is changed to the one with the next pressure range, the new pressure could be anything between 1 and 19 cm H_2O higher. This critical parameter should be set to a tighter tolerance (e.g., 5 cm H_2O). Much of our knowledge on pressure tolerance of tissue beneath the AMS 800 sphincter comes from experiments on rabbits by Engelmann and his group *(23,24)*. Pressure-related complications were seen mainly in high-pressure groups. Further research into the actual pressure under the cuff in humans is needed so that pressures known to be safe for the human urethra can be established.

Early infection or erosion may be reduced by impregnating the device with antibiotics during manufacture (a method used in the manufacture of other devices, e.g., ventriculoperitoneal shunt components). The life expectancy of the AUS is limited because, in most cases, the system eventually leaks. Leaks usually result from a perforation within a crease in the cuff. Manufacture of the cuff using a curved rather than a flat template (to produce a truly circular cuff that, on inflation, would uniformly constrict the lumen) would reduce this. The current system responds poorly to rises in intra-abdominal pressure. This results in stress incontinence or higher than necessary pressure exerted continuously on the urethra. A new design to allow for this would be of value. The results in female patients are not as good as in the male. Research into a new cuff that does not need to be passed around the posterior aspect of the female urethra might reduce the excess risk of removal for infection and erosion. Finally, accurate documentation of outcome is required so that results can be monitored closely. This is one of the reasons these patients should only be cared for in a specialized unit.

In conclusion, the AUS has a valuable role in the management of complex incontinence. It is durable and the long-term outcome is excellent. However, there are significant complications and a high revision rate necessitating close follow-up of these patients. Their care should be in specialist units, where complications can be identified and treated and the experience gained can be used to improve the appropriate selection of patients and management of complications. Future research is needed to document safe and effective cuff pressures on the urethra and to develop new designs to compensate for stress and allow alteration of cuff pressure.

Additional information resources are available (*see* the Appendix).

APPENDIX: ADDITIONAL RESOURCES FOR INFORMATION ABOUT ARTIFICIAL URETHRAL SPHINCTERS

http://www.incontinenceurinary.com

National Institute of Diabetes and Digestive and Kidney Disease
National Kidney and Urologic Diseases Information Clearinghouse
Attn: BCW
3 Information Way
Bethesda, MD 20892-3580
Fax (301) 907-8906
http://www.niddk.nih.gov/health/urolog/uibcw/index.htm

National Association for Continence
P.O. Box 8310
Spartanburg, SC 29305-8310
Tel (864) 579-7900
Fax (864) 579-7902
http://www.nafc.org/site2/index.html

Continenceworldwide (Web site for the International Continence Society)
Addresses for the national associations are on this Web site:
http://www.continenceworldwide.org

REFERENCES

1. Foley EBF. An artificial sphincter: a new device and operation for control of enuresis and urinary incontinence. J Urol 1947;58:250–295.
2. Foley EBF. Artificial Sphincter and Method. United States Patent 1948;2455859.
3. Brindley GS. Treatment of urinary and faecal incontinence by surgically implanted devices. Ciba Found Symp 1990;151:267–282.
4. Brindley GS, Polkey CE, Rushton DN. Sacral anterior root stimulators for bladder control in paraplegia. Paraplegia 1982;20:365–381.

5. Light JK, Hawila M, Scott FB. Treatment of urinary incontinence in children: the artificial sphincter versus other methods. J Urol 1983;130:518–521.
6. Cass AS, Luxenberg M, Johnson CF, Gleich P. Management of the neurogenic bladder in 413 children. J Urol 1984;132:521–525.
7. Buuck RE. Incontinence system and methods of implanting same. US Patent 1975; 4222377:1–8.
8. Timm GW, Bradley WE, Scott FB. Experimental evaluation of an implantable externally controllable urinary sphincter. Invest Urol 1974;11:326–330.
9. Hajivassiliou CA. A review of the complications and results of implantation of the AMS artificial urinary sphincter. Eur Urol 1999;35(1):36–44.
10. Singh G, Thomas DG. Artificial urinary sphincter for postprostatectomy incontinence. Br J Urol 1996;77(2):248–51.
11. Elliott DS, Barrett DM. Mayo Clinic long-term analysis of the functional durability of the AMS 800 artificial urinary sphincter: a review of 323 cases. J Urol 1998;159(4):1206–1208.
12. Murray KH, Nurse DE, Mundy AR. Detrusor behaviour following implantation of the Brantley Scott artificial urinary sphincter for neuropathic incontinence. Br J Urol 1988; 61(2):122–128.
13. Stephenson TP, Mundy AR. Treatment of the neuropathic bladder by enterocystoplasty and selective sphincterotomy or sphincter ablation and replacement. Br J Urol 1985;57(1): 27–31.
14. Singh G, Thomas DG. Artificial urinary sphincter in patients with neurogenic bladder dysfunction [see comments]. Br J Urol 1996;77(2):252–255.
15. Couillard DR, Vapnek JM, Rentzepis MJ, Stone AR. Fatal perforation of augmentation cystoplasty in an adult. Urology 1993;42(5):585–588.
16. Donovan MG, Barrett DM, Furlow WL. Use of the artificial urinary sphincter in the management of severe incontinence in females. Surg Gynecol Obstet 1985;161(1):17–20.
17. Light JK, Scott FB. Management of urinary incontinence in women with the artificial urinary sphincter. J Urol 1985;134(3):476–478.
18. Scott FB. The use of the artificial sphincter in the treatment of urinary incontinence in the female patient. Urol Clin North Am 1985;12(2):305–315.
19. Webster GD, Perez LM, Khoury JM, Timmons SL. Management of type III stress urinary incontinence using artificial urinary sphincter. Urology 1992;39(6):499–503.
20. Duncan HJ, Nurse DE, Mundy AR. Role of the artificial urinary sphincter in the treatment of stress incontinence in women. Br J Urol 1992;69(2):141–143.
21. Venn SN, Greenwell TJ, Mundy AR. The long term outcome of artificial urinary sphincters. J Urol 2000; 164: 702–707.
22. Perez LM, Webster GD. Successful outcome of artificial urinary sphincters in men with post-prostatectomy urinary incontinence despite adverse implantation features. J Urol 1992;148(4):1166–1170.
23. Engelmann UH, Felderman TP, Scott FB. The use of the AMS-AS800 artificial sphincter for continent urinary diversion. I. Investigations, including pressure-flow studies, using rabbit intestinal loops. J Urol 1985;134(1):183–186.
24. Engelmann UH, Felderman TP, Scott FB. Evaluation of AMS 800 artificial sphincter for continent urinary diversion using intestinal loops. Urology 1985;25(6):620–621.

15
Cerebrospinal Fluid Shunts

Edward Rustamzadeh and Cornelius H. Lam

HISTORY

Hydrocephalus is a congenital or acquired condition in which cerebrospinal fluid (CSF) accumulates in the ventricles and the subarachnoid space around the brain (Fig. 1). It can lead to an increase in intracranial pressure. It has existed since primitive man roamed the earth. Man's understanding of anatomy and physiology has often consisted mainly of myth and superstition. The earliest recording of the awareness of CSF can be found in the Edwin Smith Papyrus (2200 BC) *(1,2)* that describes the "spillage of clear fluid from the inner brain" in ancient Egypt. Hippocrates (460–377 BC) is credited with attributing the enlargement of the head to a buildup of fluid. His treatment consisted of drilling a hole in the skull and puncturing the meninges to allow the fluid to escape *(3)*. Erasistratus of Alexandria (280 BC) was the first author to describe the ventricular system based on human dissection. He believed that the seat of the soul lay within the fourth ventricle *(4)*. The great Roman physician Galen (129–210 AD) described CSF as a waste product of the "animal spirit" that was discharged from the nose as the "pituita" *(5)*.

For the next 1300 years, little progress was made in elucidating the anatomy of the brain. This was in part owing to the era of the Dark Ages and the edict of 1163, *Ecclesia abhorret a sanguine* ("the Church abhors the shedding of blood") *(6)*. However, as the Renaissance began, analytical assessment of the human body began in earnest. In 1543, Andreas Vesalius published *De Humani Corporis Fabrica,* in which he accurately described what we now refer to as noncommunicating hydrocephalus:

> The water had not collected between the skull and its outer surrounding membrane, or the skin (where the doctors' books teach that water is deposited in other cases), but in the cavity of the brain itself, and actually in the right and left ventricles of the brain. The cavity and breadth of these had so increased—and the brain itself was so distended—that they contained about nine pounds of water *(7)*.

In 1768, Whytt further divided hydrocephalus into internal and external types. He postulated that the cause of hydrocephalus lay in an imbalance between the production of interstitial fluid by the cerebral arteries and its absorption by the veins *(8)*.

In 1876, Key and Retzius were the first to show the intricate connection of the ventricular system by injecting dye into the lateral ventricles of autopsies *(9)*. Building on

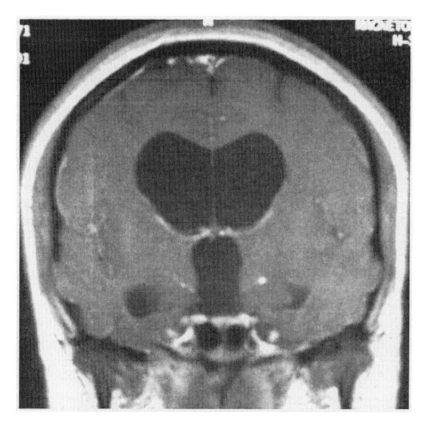

Fig. 1. T1-weighted post-contrast coronal magnetic resonance imaging showing enlarged lateral and third ventricles.

this knowledge, Dandy and Blackfan reproduced a hydrocephalic condition in dogs by obstructing either the foramen of Monro or the Sylvian aqueduct *(10)*.

Until the early part of the 20th century, the treatment of hydrocephalus ranged from primitive measures, such as bleeding, purging, injection of astringents, head wrapping, and application of potions to the head to more sophisticated treatments, with insertion of ventricular setons, cannulas, and lumbar punctures *(11)*. Most of these procedures resulted in disastrous outcomes, mainly due to infection. In order to decrease the risk of infection, attention was focused on surgical interventions leading to internal drainage of the CSF. In 1917, Sharpe reported passage of a linen thread from the subarachnoid space to the ventricle *(12)*. Although 13 of his 41 patients died soon after the operation, 28 of the patients were reported to have good or excellent outcomes. With this newly found success, surgeons were encouraged in the development of various new procedures, including diversion into the cisterna magna, the sagittal sinus, peritoneum, jugular vein, ureter, gallbladder, pleural space, paranasal air sinuses, thoracic duct, and right atrium of the heart *(13,14)*. Owing to inferior components, many of these attempts did not result in long-term solutions to the problem. Until the middle of the 20th century the materials for shunt catheters were not ideal. Some were too rigid. Others were not biologically inert. A few were toxic. Tubes of silver, rubber, calf vein, nylon, and autogenous skin all were proposed and in most cases actually used in patients *(3,15–18)*. Furthermore, early shunts were valveless and relied on a favorable pressure gradient for CSF flow.

Nulsen introduced the first valvular shunt in the early 1950s (19). In 1947, Ingraham (20) published his experimental work on using a synthetic polyethylene catheter, which subsequently became the standard material in shunting procedures. Initially, this catheter was successfully used in diverting CSF by means of a lumbar–ureterostomy bypass (21). However, because of the complexity of this procedure, that entailed removing a kidney, it was associated with serious complications. Thus, this procedure was abandoned in favor of a ventriculoperitoneal diversion (22). However, the use of polyethylene tubing in the peritoneum was also associated with a high incidence of complications, specifically kinking, viscus perforation, fracturing, and peritonitis. This led to abandonment of the ventriculoperitoneal shunt in favor of a ventriculoatrial shunt (23). This, too, was associated with serious cardiac complications such as cor pulmonale, pulmonary hypertension, and cardiac perforation. With the introduction of silicone, the ventriculoperitoneal shunt once again regained favor. Silicone is an extremely flexible material initially believed to be inert. However, experience with silicone tubing has shown that it can lead to a delayed hypersensitivity reaction (24,25). By the late 1970s, silicone shunt catheters were being used increasingly in ventriculoperitoneal shunt procedures, especially after reports of high morbidity and mortality associated with ventriculoatrial shunts. In the late 1980s, a programmable shunt was introduced that allowed the surgeon to change the opening pressure of the valve not only at the time of surgery, but also after the procedure (26). Unfortunately, the valve setting can change if the patient is near a strong magnetic field, such as is used for magnetic resonance imaging (MRI).

EPIDEMIOLOGY

Hydrocephalus can occur at any age. It may present during the prenatal period or much later in adulthood. It is difficult to determine accurately the incidence of congenital hydrocephalus because the statistical data in numerous studies is confounded by patient selection. The estimated prevalence of hydrocephalus in all ages is between 1 and 1.5%. The incidence of congenital hydrocephalus ranges from 0.5 to 1 per 1000 births (27–31). The database of the International Society for Pediatric Neurosurgery indicates that 73% of the first time shunt placements are in patients 6 months of age or younger (32). Hydrocephalus detected in the fetus generally carries a poor prognosis. The fetal death rate associated with hydrocephalus has been estimated at 25% (30,33). The neonatal death rate associated with hydrocephalus has been estimated to be 17–41% (30,33).

A 1988 national health survey revealed that there were more than 127,000 patients with CSF shunts in the United States (34). About 45,000 to 50,000 shunts are placed each year in the United States (35). Data obtained from a survey of US hospital discharge diagnoses revealed that about 69,000 patients are admitted for hydrocephalus annually. Of these, about 33,000 are newly diagnosed with hydrocephalus and 42% of the remainder is admitted for shunt revision. Based on the 1988 US National Health Survey, the yearly cost of placing shunts, excluding hospital stay, was estimated at $9.4 million. The total cost of treating hydrocephalus in 1988 was 0.02% of the total US health care expenditure (34).

The etiologies of hydrocephalus are numerous. Intraventricular bleeding is a common cause of acquired hydrocephalus, with 20–50% of infants with intraventricular hemorrhage requiring a shunt *(36)*. Up to 45% of patients with a ruptured aneurysm eventually need permanent shunting *(37)*. Congenital conditions, such as aqueductal stenosis, myelomeningocele, and Dandy-Walker malformation, are other common causes. Of patients with myelomeningocele, 5 to 10% require a shunt at the time of birth and the majority (65–85%) require a shunt at some point during childhood *(38)*. One of the sequelae of bacterial meningitis is hydrocephalus. Approximately 8% of patients with bacterial meningitis later require a shunt *(39)*. Twenty percent of children undergoing resection of a posterior fossa tumor require a shunt postoperatively for abnormal CSF dynamics *(39)*. It is estimated that 10–15% of older patients who are diagnosed with early onset dementia actually have a condition known as normal pressure hydrocephalus. This syndrome features a triad of clinical findings: dementia, gait difficulty, and urinary incontinence. Although computed tomography (CT) and MRI images of the brain are consistent with hydrocephalus, intracranial pressure monitoring reveals normal pressure. CSF shunting can improve the clinical symptoms in a select group of patients with normal pressure hydrocephalus.

The treatment of hydrocephalus has come a long way. As late as 1962, infants with hydrocephalus had a 64% mortality rate and only 20% survived into adulthood *(40)*. The current mortality rate for infants with congenital hydrocephalus is 3–10% *(41,42)*. Sixty percent of children with shunts attend normal school and an additional 20% require special classes. Only 10% of children require significant assistance throughout their lives. A popular new procedure in treating patients with hydrocephalus caused by aqueductal stenosis is third ventriculostomy. This procedure involves placing a scope into the third ventricle, then perforating the anterior wall to allow CSF to drain into the cistern. Thus, the patient is shunt-free.

Hydrocephalus can present as part of a syndrome (e.g., Crouzon, Apert, Hurler, Fanconi, or trisomy 9, 13, or 18) or as a nonsyndromic form (e.g., hydranencephaly, Arnold-Chiari, or myelomeningocele) *(43)*. The risk of congenital hydrocephalus in future offspring depends on the etiology. The empiric risk without an obvious cause is between 0 and 4% *(44)*. However, depending on the cause, it can be as high as 50% *(43)*.

PATHOPHYSIOLOGY

Understanding the physiology of CSF formation, circulation, and absorption is important in diagnosing and managing hydrocephalus. It is a clear, colorless fluid that contains protein, glucose, and electrolytes. It has a specific gravity slightly higher than water and a pH of 7.35, similar to serum. Approximately 75% of CSF is formed by the choroid plexus in the lateral, third, and fourth ventricles *(45)*. The remaining 25% is formed by the capillary endothelium within the brain, nerve root sleeves, and the ependymal lining of ventricles. In adults, it is produced at a rate of 0.3–0.35 mL per minute, or approximately 450 mL per day. At any one time, there is approximately 150 mL of CSF circulating in the brain, spine, and ventricles. In contrast, an infant produces 25 mL of CSF per day and has approximately 50 mL of CSF bathing the central nervous system at any time *(43)*. By 1 year of age the differences in the rate of CSF production and absorption compared to an adult become negligible *(46)*. Interestingly, the rate of CSF production is relatively resistant to changes in the intracranial pressure, except at the extremely high range when the rate of production decreases dramatically *(47)*. CSF formation is a two-step process. The first involves an active process linked to sodium transport and bicarbonate metabolism. Drugs such as acetazolamide and furosemide that affect the active transport mechanism reduce the rate of CSF formation by 60% *(48,49)*. There is a correlation between cerebral blood flow and CSF production. Nilsson et al., using MRI, noted 3.5 times more CSF production in the morning than at night, reflecting more cerebral blood flow in the morning and less at night *(50)*. Elderly people produce less CSF than young adults, which also correlates with their lower cerebral blood flow *(51)*. Vasoconstrictors decrease CSF formation, again resulting from decreased blood flow *(52)*. The only known pathological entity that can increase the production of CSF is a rare tumor known as choroid plexus papilloma *(53)*. Whether this tumor is sufficient to induce hydrocephalus is still controversial *(54)*. Surgical excision of choroid plexus as a primary therapy for hydrocephalus has been disappointing, with a 15% mortality rate and only a temporary decrease in the rate of CSF formation *(55,56)*.

The intracerebral circulation of CSF is from the lateral ventricles through the foramen of Monro into the third ventricle, then into the fourth ventricle by way of the Sylvian aqueduct. CSF leaves the fourth ventricle by means of the foramen of Lushka and Magendie and flows over the brain surface. It occupies the basal cisterns and the subarachnoid space and surrounds the spinal cord and proximal nerve roots within the subarachnoid space. Obstruction of CSF flow at any of the above sites can cause hydrocephalus. Unilateral occlusion of CSF flow across the foramen of Monro is quite rare, but can occur as a result of congenital atresia or stenosis of the foramen, secondary aqueductal gliosis owing to an intrauterine infection, or germinal matrix hemorrhage *(57)*. Obstruction of the foramen can also result from intraventricular masses such as colloid cysts, giant cell subependymal astrocytomas, hypothalamic tumors, or craniopharyngiomas *(35)*.

Genetically induced aqueductal stenosis is seen primarily in male infants as a result of X-linked recessive syndromes such as Mental retardation, Aphasia, Shuffling gait, Adducted thumbs *(58)* or Hydrocephalus with Stenosis of the Aqueduct of Sylvius *(59)*. Approximately 5% of cases of congenital hydrocephalus can be accounted for by these genetic disorders *(43)*. Missense coding of the gene responsible for production of L1, a cell adhesion molecule, is the most frequent known cause of congenital aqueductal stenosis. It also results in nonformation of the pyramidal and corticospinal tracts *(60,61)*. Structurally, congenital aqueductal stenosis can present as one of four types: (a) forking of the aqueduct with noncommunication between the two ends, (b) periaqueductal stenosis owing to subependymal proliferation, (c) septated aqueduct, and (d) histologically normal aqueduct with anatomical narrowing *(62)*. The second most common reason for aqueductal obstruction is a tumor such as tectal glioma or pineal tumor. Patients with these tumors clinically present with an upward gaze and convergence palsy known as Parinaud's syndrome.

Congenital entrapment of the fourth ventricle is the cause of hydrocephalus in 2.4% of patients and is associated with a Dandy-Walker malformation, resulting in atresia of the foramen of Magendie and Luschka *(39)*. Meningitis can also cause obstruction of the outlet of the fourth ventricle. The most common cause of obstruction is due to posterior fossa tumors, which are the most common brain tumors in children. Medulloblastomas, or juvenile pilocytic astrocytomas, can compress the fourth ventricle. Choroid plexus carcinoma, dermoid, and subependymoma can grow within the fourth ventricle *(63)*. With tumor resection, most children can expect to regain normal CSF flow through the fourth ventricle, whereas 20% require permanent CSF diversion *(39)*.

Obstruction of the cisterns surrounding the brainstem can also lead to hydrocephalus because the CSF flow from the ventricles to the subarachnoid space and finally to the arachnoid villi for absorption into the venous system is impeded. This form of obstruction, known as communicating hydrocephalus, can be seen following infections such as tuberculosis or syphilitic meningitis and also is presumed to be the underlying pathology in normal pressure hydrocephalus in adults *(64)*.

Anomalies in the absorption of CSF can also lead to hydrocephalus. CSF is absorbed by the arachnoid villi that extend into the dural venous sinuses. This phenomenon is related to the pressure differential between the arachnoid space that contains the CSF and the venous pressure within the sinuses. The sagittal sinus pressure averages about 5 mmHg in the recumbent position and drops to –10 mmHg when standing *(65)*. This acts as an internal valve to control the rate of absorption of CSF. In the erect position, when the venous pressure falls below atmospheric pressure, the jugular veins collapse, overriding the internal valve mechanism *(66)*. In fact, the relationship between the rate of CSF formation and absorption becomes linear after the opening pressure of 5 mmHg is reached, and the two come into equilibrium at 10 mmHg *(47)*. Therefore, any etiology that results in an increase in intracranial venous pressure can change the equilibrium between CSF absorption and formation, leading to hydrocephalus. Examples of this are dural venous thrombosis, impingement of the jugular foramen, obstruction or absence of arachnoid villi, and vein of Galen malformations *(67)*. Venous congestion can increase the cerebral blood volume and decrease the compressibility of the brain, thus increasing the brain turgor *(68)*. One form of hydrocephalus that is believed to be due to this phenomenon is pseudotumor cerebri *(69)*. Although alternative sites of CSF absorption such as paracervical lymphatics, endolymphatics, cranial and spinal nerve dural sleeves have been postulated, none of these are sufficient to prevent hydrocephalus *(70)*.

CSF SHUNT HYDRODYNAMICS

To understand how shunts are designed, one has to assess the hydrodynamics of CSF. There are four major components that determine CSF hydrodynamics. The first is pressure. Pressure is simply force per unit area. Shunts are designed to address the range of pressure gradients that exist across the ventricular wall from a supine to a standing body position. The intraventricular pressure (IVP) is referenced to the atmospheric pressure, which is set at zero. In clinical practice this is recorded in mmHg or cm H_2O (1 mm Hg = 1.365 cm H_2O). A simple way to look at the pressure differential is to use an analogy of a column of fluid to describe hydrostatic pressure (HP) *(71)*. The same volume of fluid will exert a higher pressure when raised to a new height; therefore, HP is equivalent to the height of the column of fluid or, in other words, the length of the catheter in a CSF shunt. In the supine position, the intraperitoneal pressure (IP) has been reported to range from slightly above atmospheric to subatmospheric *(72,73)*. Thus, in the supine position, assuming IP and HP are zero, IVP is equivalent to the opening pressure of the valve (OPV). In the erect position, the IP increases by 8–12 mmHg *(74,75)*; however, the HP increases by 20–30 mmHg. Using the formula IVP = OPV – HP + IP, one discovers that a potential problem known as siphoning can occur, leading to shunt complications that will be discussed later. Although this formula is helpful in empirically determining the hydrodynamics of CSF flow with a shunt, it does not take into account the possibility of occlusion of the distal catheter opening and the fact that intra-abdominal pressure varies with respiration and digestion.

Flow (Q) is the change in volume over time. Flow can be either turbulent or laminar, depending on the system's inherent Reynolds number, which is directly proportional to the length of the tube, velocity, and density of the fluid, and inversely proportional to the viscosity of the fluid *(76)*. Flow can be related to pressure and resistance in terms of Q = ΔP/R, where ΔP is the change in pressure and R is the resistance *(77)*. Assuming that the inherent resistance of the tubing is constant, then the flow of CSF in a shunt is equivalent to IVP = OPV – HP + IP. In the erect position, this could be as high as 20 mL per hour. This leads to a concept known as siphoning in which the flow of CSF is based on differences in the vertical length of the catheter and not on the opening pressure of the shunt valve. Flow continues and intraventricular pressure decreases until the compliance of the brain (Δ volume/Δ pressure) has been reached, at which time the negative ventricular pressure approximates the height of the catheter *(77–79)*. This negative IVP continues until the volume is replenished.

Resistance (R) can be thought of as a force in the opposite direction of flow. It is dependent on the length (L) of the tubing, viscosity (v) of the fluid, density (D) of the fluid, gravity (g), and the radius (r) of the tubing. It can be expressed as R = $8Lv/Dgr^4$, also known as Poiseuille's law *(80)*. The importance of this law is that resistance is affected greatly by small changes in the radius. In designing a flow-control shunt valve, manufacturers have used this principle by using a moveable cone-shaped pin to reduce the radius of a valvular ring to control the flow through the shunt.

Finally, the property of viscosity (v) must be taken into consideration in designing a shunt. Viscosity is the resistance of the fluid to shear force (i.e., how the fluid resists motion). Viscosity can be defined as v = (F/A)/(s/L), where F is the perpendicular force, A is the area, s is the velocity, and L is the length of the tubing. Viscosity is inversely related to temperature. Thus, in process of designing a shunt, the flow characteristics at the mean human core temperature of 37°C must be considered *(71)*.

Drake and Sainte-Rose *(71)* described a mathematical model that is used in designing shunts, which incorporate all of the above CSF hydrodynamic parameters. As mentioned earlier, the change in volume (ΔV) of CSF within the ventricles is a function of CSF formation (f), flow (Q), and absorption (a). With respect to time (Δt) this can be expressed as ΔV/Δt = f-a-d. The absorption (a) of CSF is directly proportional to the differential pressure between the intraventricular space (p) and the sagittal sinus (p_v). This can be stated as a = $(p - p_v)/R$. The change in the IVP as a function of volume can be described by the formula $\Delta p/\Delta V = Bp(e^{k|\Delta V|})$, where Bp is the baseline IVP, k is a constant, and $|\Delta V|$ is the absolute change in volume. The flow rate of CSF is zero when the IVP is less than the OPV, and equal to (p – OPV)/R when it is greater than the

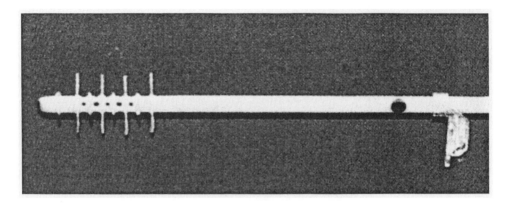

Fig. 2. Ventricular catheter with flanged tips.

opening pressure. To describe the change in IVP over time we use the following formula: $\Delta p/\Delta t = (\Delta p/\Delta V)(\Delta V/\Delta t)$ or $\Delta p/\Delta t = (Bp\ e^{k|DV|})(f-a-s)$.

SHUNT COMPONENTS AND DESIGN

Shunt devices consist of four basic parts: a ventricular or lumbar catheter, a reservoir, a valve, and a distal catheter. In some instances, an antisiphon component can be added to the system or can be incorporated by the manufacturer into the valve. The manufacture of shunts for sale in the United States must follow standards of good manufacturing practices, which are a set of rules enforced by the Food and Drug Administration. Shunt components are tested under sterile conditions. A variety of materials such as silicone, titanium, synthetic ruby, barium sulfate, stainless steel, and various plastics are used in the construction of shunts *(71)*. The devices are then subjected to three sets of tests. The first is the biocompatibility of the system. This is composed of two phases: (a) acute, which determines the cytotoxicity, genotoxicity, allergenicity, and hemocompatibility; and (b) chronic, which concerns late toxicity and carcinogenicity *(81)*. The mechanical properties of the shunt system are then addressed. The shunt system undergoes a rigorous set of tests to assess the physical properties of the device *(82,83)*. Only the systems that pass the initial tests are subjected to clinical testing.

Ventricular catheters are primarily silicone-based. The inner diameter varies between 1 and 1.6 mm and the outer diameter ranges from 2.1 to 3.2 mm *(71)*. The catheter comes in either a straight or an angled configuration. Holes are usually in the distal 1–1.5 cm of the catheter. In a special J-shaped catheter, the holes are placed in the inner groove of the catheter. Another type of ventricular catheter has flanged tips perpendicular to the holes to prevent proximal obstruction by ingrowth of the choroid plexus (Fig. 2). Unfortunately, the experience with this particular catheter has been disappointing *(84)*. Whereas the straight catheter can be cut to any length intraoperatively, the angled catheters have a set length.

There are basically two types of valves: pressure-regulating and flow-regulating. The variations on this theme are large. The basic mechanism in a differential pressure valve involves a predetermined pressure setting, which is called the opening pressure. Standard opening pressures are set at 5, 10, and 15 cm H_2O. When the pressure gradi-

Codman®
Holter® Valve, Elliptical

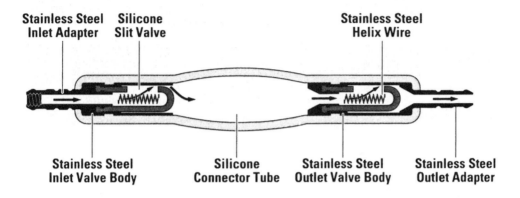

Fig. 3. Codman Holter Valve.

ent exceeds this setting, the valve opens. There is little control over the rate of CSF flow and, as mentioned earlier, when the patient is in the erect position, the gradient can far exceed this limit. There are four common designs in a differential pressure valve. The first is a slit valve, which consists of a slit made either in the distal catheter tube or within a proximal valve. The length of the slit and the properties of the material determine its opening pressure. Examples of this include the Codman Holter Valve, Phoenix Holter-Hausner valve, and Radionics standard shunt (Fig. 3).

The second type is known as a duckbill or miter valve and consists of two concave leaflets of silicone, either at the distal end or incorporated in the proximal valve. Once again, when a pressure gradient exists across the valve, the leaflets are pushed apart, allowing CSF to drain. An example of this is the Mueller Heyer Schulte In-Line valve. Both the slit valve and the miter valve have a drawback in that the opening pressure might not be the same as the closing pressure. This is primarily owing to two reasons. The first is that silicone is not a purely elastic material and can exhibit some deformity when pressure is applied; therefore, the magnitude of the opening and closing forces can be different. Furthermore, the actual design of the valves allows for adherence between the edges, which, in turn, could also explain the difference between the opening and closing pressures.

A third type employs a diaphragm that responds to a pressure gradient. The diaphragm can be attached to a piston that pushes the diaphragm to open or close the chamber, thus preventing CSF flow (e.g., the PS Medical Flow-Control valve, and the Radionics Contour Flex valve [Fig. 4]). A simple membrane that overlies the inlet or outlet connectors can also function as a diaphragm that occludes or allows CSF flow based on the differential pressure applied across the membrane (e.g., Codman Accu-Flo valve and the Mueller Heyer Schulte Pudenz Flushing valve [Fig. 5]).

Finally, a ball-and-spring valve design can function in a similar manner. When a pressure differential exists across the inlet and outlet connectors, the ball is forced to

Medtronic PS Medical
CSF-Flow Control Valve, Contoured

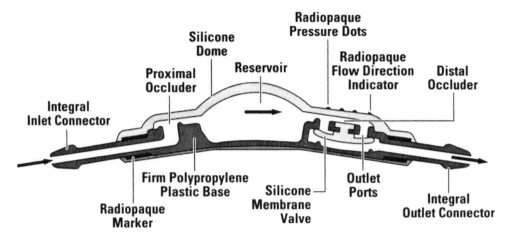

Fig. 4. Medtronic PS Medical Valve.

Heyer-Schulte®
Low Profile Valve-LPV®

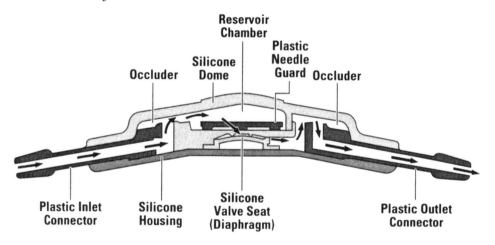

Fig. 5. Heyer Schulte Valve.

compress the spring and allow CSF to flow. Opening pressures can be set by either increasing the resistance of the spring coil to compress or by adding additional balls, which, in turn, requires a larger force to compress the spring coil. Currently, there exists a standard prefixed opening pressure setting, such as the Cordis Hakim valve (Fig. 6), Cordis Horizontal-Vertical Lumboperitoneal valve, and newer valves, such as the Codman Medos Programmable valve (Fig. 7) and the Sophysa Adjustable valve

Cordis®
Hakim Valve System

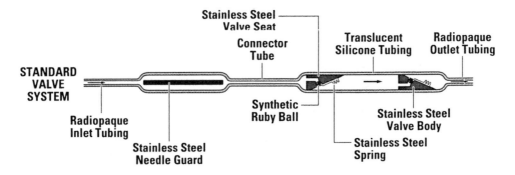

Fig. 6. Cordis Hakim Valve.

Codman® -Medos®
Programmable Hakim™ Valve System

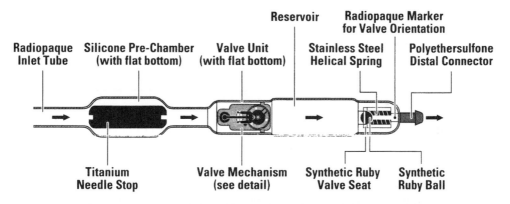

Fig. 7. Codman Medos Programmable Valve.

(Fig. 8), which employ a magnetic rotor that alters the position of the fulcrum in relation to the spring. This accomplishes the same result mentioned earlier. The advantage of this system is that the shunt can be programmed postoperatively to new settings without requiring a second procedure to change the valve. The disadvantage is the need to obtain skull X-rays after every MRI to ensure that there has been no change in the setting.

Flow-regulating shunts prevent the siphoning that occurs with the differential pressure valve. The concept involves a conical occluder that moves into a circular opening. Because of the conical shape, the flow rate can be kept relatively constant at low and medium differential pressures; however, at high-pressure gradients, flow can cease. Examples of this are the PS Medical Delta valve and the Orbis Sigma valve (Fig. 9)

Sophysa®
Sophy® Programmable Pressure Valve Model SU8

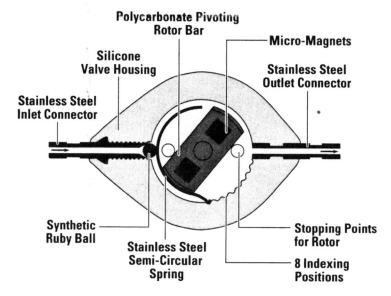

Fig. 8. Sophy Programmable Valve.

(85). As its name implies, the Orbis Sigma valve has a sigmoidal response curve on the pressure-flow diagram. At the lower level, it acts as a standard differential pressure valve. Between the first and second inflection points, flow regulation is essentially pressure-independent. When the pressure increases to a critical level, resistance to flow diminishes and a steep increase in CSF flow occurs. Regulation of flow can also be implemented with the addition of an antisiphon device to an existing differential pressure valve. Antisiphon devices use a membrane that is depressed if the pressure in the distal catheter is negative. An example of this is the Heyer Schulte antisiphon device. Tokoro and Chiba *(86)* discovered that, if the antisiphon device is placed less than 10 cm from the valve, it overcompensates and decreases CSF flow excessively, resulting in inadequate drainage. If the antisiphon device is placed further than 10 cm from the valve, excessive CSF flow occurs. Thus, appropriate positioning of the antisiphon device is crucial.

Most distal catheter tubings are silicone-based. The most commonly used type has a cylindrical opening distally. Other configurations include a slit opening and a sleeve opening. These have fallen from favor because of high rates of shunt malfunction *(84)*.

OPERATIVE PROCEDURES

The decision to shunt a patient is made on clinical symptoms and radiographic correlation. In infants, head circumference measurements are used as an indirect measure of the severity of the underlying hydrocephalus. Young et al. *(87)* studied the relationship between the thickness of the cerebral mantle and outcome in terms of intelligence

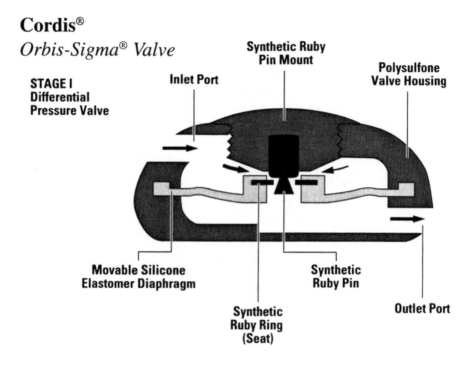

Fig. 9. Cordis Orbis Sigma Valve.

in shunted hydrocephalic infants. They discovered that, if the infant is shunted early and the cerebral mantle is at least 3 cm by 6 months of age, the likelihood of normal intelligence is fairly good. Concurrently, if hydrocephalus is diagnosed *in utero* by ultrasound and the cerebral mantle is less than 1 cm, the prognosis for independent function is very poor *(88)*. After the decision to shunt the patient has been made, the next questions are which type of shunting procedure to perform and which type of valve to use. A multicenter trial *(89)* compared three types of shunts: a standard differential pressure valve, a Delta valve, and an Orbis Sigma valve. The results indicated that there was no statistical difference between these valves in terms of 1-year shunt failure rates. Because no particular type of valve seems to be clearly superior to another, the decision about which valve to use should be tailored to the individual patient. For example, a medium-pressure valve might be better suited than a low-pressure/high-pressure valve for an infant or adult with large ventricles because the risk of subdural hemorrhage and slit ventricles is higher with the latter. Similarly, a high-pressure valve is usually better suited than a medium- or low-pressure valve for a patient with hydrocephalus owing to pseudotumor cerebri.

The most common type of operative shunt procedure is the ventriculoperitoneal bypass (Fig. 10). The patient is placed in a supine position under general anesthesia. The head is usually turned to the left for a right frontal or occipital horn entry site. Studies to determine which site is optimum have reported conflicting results *(90,91)*. The important point in decreasing proximal catheter occlusion is not the site of entry but the amount of CSF surrounding the catheter (i.e., the atrium of ventricle and the body of

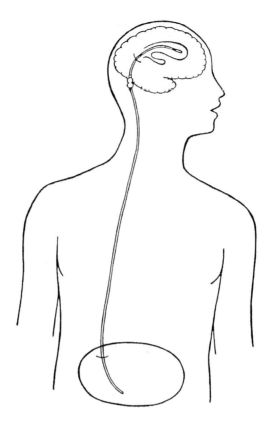

Fig. 10. Depiction of a ventriculoperitoneal shunt system.

ventricle) *(84)*. The entire right half of the body, down to the abdomen, is prepped and draped under sterile conditions. An incision is made in the scalp and the periosteum is cleared for a burr hole. The dura is opened only enough to allow the catheter to pass into the brain. A subgaleal pocket is made for the valve. The ventricular catheter is passed through the brain with a stylet. When CSF return is noted, the stylet is removed. If a straight ventricular catheter is used, a Rickham reservoir is connected to it to allow a 90° angle to connect with the valve. The valve is placed in the subgaleal pocket and a subcutaneous tunnel is formed with a passer connecting the head with the abdomen. The distal shunt tubing is brought out through the abdominal incision and passed intraperitoneally.

If a patient has had previous abdominal operations or peritonitis, the risk of passing a catheter into the abdomen becomes higher, and the surgeon might opt to place a ventriculoatrial shunt (Fig. 11). The procedure is similar except that a neck incision is made in order to pass the distal catheter into the right atrium via the internal jugular vein. Intraoperative X-ray or fluoroscopy is used to determine the location of the distal catheter. The sixth thoracic vertebrae is the anatomical landmark used to judge the location of the junction of the superior vena cava and the right atrium. Borgbjerg et al. *(92)* compared the revision rate and the durability of ventriculoatrial shunts to ventriculoperitoneal shunts. They noted a statistically significantly higher rate of revision of ventriculoatrial shunts (51%) compared to ventriculoperitoneal shunts (38.5%).

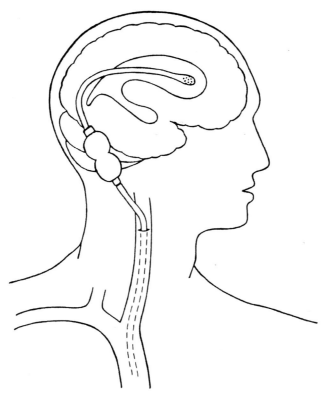

Fig. 11. Depiction of a ventriculoatrial shunt system.

However, the durability rates of the two varieties were not statistically significantly different. When stratified by age groups, the rates of revision were not statistically significantly different in adults. Although both procedures can result in complications, the severity of the complications is higher for ventriculoatrial shunts with a reported mortality rate of 3% *(93)*. These complications include cor pulmonale, endocarditis, cardiac tamponade, and shunt nephritis. The last condition involves renal failure owing to embolization of immune complexes to the glomeruli *(94)*. Urine analysis reveals proteinuria and hematuria, and the patient may or may not have oliguria.

Lumboperitoneal shunts have been used for more than 100 years (Fig. 12) *(100)*. This procedure involves passing a shunt catheter from the lumbar subarachnoid space to the peritoneal cavity. This can be done by two means. The first involves placing the patient in a lateral decubitus position. A small incision is made at the level of the second lumbar spinous process and carried down to the lamina where a small hemilaminectomy is made. The underlying dura is incised and a T-tube catheter is placed into the subarachnoid space. The distal catheter is then tunneled subcutaneously to a small abdominal incision and passed into the peritoneal cavity *(95)*. The second method also positions the patient in a similar manner but uses a Touhy needle to percutaneously access the subarachnoid space and then pass a straight catheter into position. The distal portion of the catheter is then passed into the peritoneal cavity, as described above *(96)*. A comparison of malfunction-free survival curve between the two procedures shows a statistically significant difference at 5 years with more than 60% of T-tube

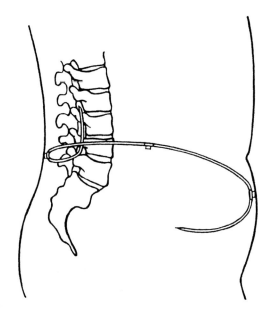

Fig. 12. Depiction of a lumboperitoneal shunt system.

lumboperitoneal shunts functioning vs 10% of percutaneously placed lumboperitoneal shunts *(97)*. This procedure can be used to treat communicating hydrocephalus but is especially helpful in managing pseudotumor cerebri, where the ventricles can be slit-like, making cannulation via ventricular catheter difficult or impossible.

Lumboperitoneal shunts have the same complications as other shunts (obstruction, overdrainage, shunt-tubing fracture, and infection), plus a few that are unique to them (arachnoiditis, shunt-tube migration, and tonsillar herniation). Although the rate of arachnoiditis has gone down to 3% with the use of silicone-based catheters *(97)*, this complication can cause chronic back pain and radicular pain. In children, lumboperitoneal shunts tend to migrate out of the subarachnoid space over time as the children grow. This is partially related to the type of procedure used in implanting it. Chumas et al. *(97)* reported a mean time for migration out of the thecal sac of 40 days for percutaneously placed catheters compared to 6.62 years in the T-tube type. Furthermore, a possible fatal complication of this type of shunt is tonsillar herniation owing to differential pressure between the foramen magnum and the lumbar subarachnoid space. Up to 70% of patients with these shunts have some degree of tonsillar herniation *(98)*; 5% are symptomatic and require further cranial surgery.

FOLLOW-UP CARE

Follow-up care begins as soon as the shunt has been placed (Table 1). It is routine to obtain shunt series X-rays (Fig. 13) and a CT scan the following day to document that the shunt is intact and the location of the ventricular tip. In an infant, the anterior fontanelle is examined to see if it is depressed and soft, indicating that the shunt is working properly. Patients are discharged within 48 hours of surgery. Family members and the patient are instructed to watch for fever, drainage or erythema at the incision sites, headaches, lethargy, nausea and vomiting, and especially an increase in the head cir-

Table 1
Recommended Schedule of Follow-Up Care After Placement of Cerebrospinal Fluid Shunt

	Postoperative year					
	1	2	3	4	5	10
Office visit[a]	3[b]	1	1	1	1	1
Head computed tomography[c]	2[d]	1	1	1	1	1

[a] Office visits include assessment of shunt function.
[b] At 1 week postoperatively, 3 months, and 1 year.
[c] Other tests and treatments are recommended only if clinical evidence warrant them.
[d] At 3 months and 1 year.

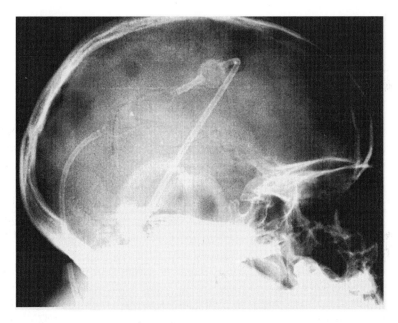

Fig. 13. Postoperative head X-ray showing shunt valve.

cumference above a rate of 1 cm per week. Another issue about which parents are frequently concerned is the requirement of prophylactic antibiotics before any dental procedure. The literature is not clear about this (99,100). Because many patients with prosthetic valve devices who are about to undergo procedures involving risk of bacteremia (see Chapter 9) are given prophylactic antibiotics, we recommend that patients with a ventriculoatrial shunt also receive prophylactic antibiotics before such procedures.

The next follow-up is within 1 to 2 weeks and involves suture removal and assessment of the wound. Studies regarding the optimum time of follow-up CT scan have indicated two reasonable time points (89,101). The first is 3 months after surgery, when maximum change in the ventricle size is expected (Figs. 14 and 15). The next is 1 year after surgery, at which time further decrease in ventricular size is often observed. There are two exceptions to the above protocol. The first is in the infant who has a cerebral

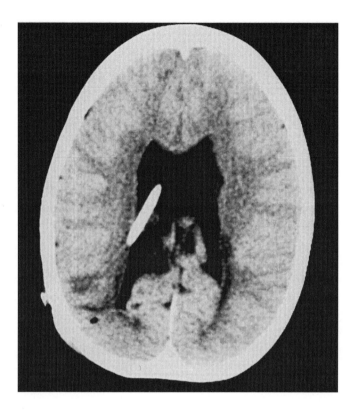

Fig. 14. Computed tomography of the head showing newly placed occipital ventricular catheter.

mantle thickness of 1–3.5 cm at the time of shunting. Young et al. *(87)* recommend another follow-up CT scan at 5 months to reassess mantle thickness. This allows the surgeon to judge whether or not the shunt valve needs to be changed to a lower pressure setting. With an Orbis Sigma valve, ventricular size often does not change rapidly *(34)*. After this period, except for spina bifida patients (who require yearly scans), there is no need to obtain annual scans without a clinical indication.

Annual follow-ups are recommended to keep accurate records of patient performance and to detect shunt malfunction as early as possible. Studies analyzing the well-being of shunted patients receiving this follow-up care by different providers have provided conflicting results. One study *(102)* showed no difference in results if the patient had follow-up care with a provider other than his or her neurosurgeon, whereas another study indicated increased risk to the life and well-being of the patient if a different health care provider was used for follow-up care *(103)*. The key lies in the comfort level of the health care provider in dealing with shunts and their complications. A close relationship between the health care provider and the primary neurosurgeon is important, especially if care is provided by a physician other than the primary neurosurgeon.

It is not uncommon for families to have concern regarding their loved one's limitations in participating in sports, outdoor activities, type of clothing attire, and flying in a plane. Although it is true that someone with a shunt does have to take extra precaution when participating in certain sports, such as hockey, football, or wrestling, there are professional players who have shunts and still participate in such activities. Wearing

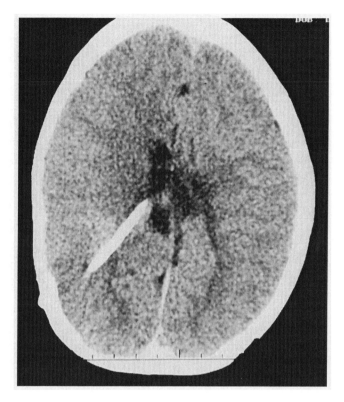

Fig. 15. Computed tomography of the same patient 3 months after surgery showing markeddecrease in the ventricular size.

tight clothing is unlikely to pose a potential problem because the amount of force required to obstruct a shunt tube is quite high. There are no limitations on flying on commercial planes because the cabin is pressurized.

DIAGNOSIS AND TREATMENT OF SHUNT COMPLICATIONS

Even with all the advances in shunt design, implementation of new synthetic material, and meticulous attention to detail intraoperatively, shunt complications are so common that neurosurgeons have come to expect them (Fig. 13). The shunt survival rate is about 60% at 1 year, 50% at 2 years, and 20–30% at 10 years *(35,103,104)*. There is an inherent risk of seizures after placement of a shunt. The risk in the first year is approximately 5.5%, decreasing to 1% per year by the third year *(105)*.

Shunt malfunction can be placed in two categories: mechanical and infectious. Mechanical malfunction accounts for 81–85% of shunt malfunctions *(101,102)*. Proximal ventricular catheter obstruction is the commonest form of obstruction, accounting for 56–63% of mechanical complications *(35,84)*. The second most common cause of mechanical failure is distal obstruction of the peritoneal catheter. These two entities tend to present in different time periods, with proximal obstruction usually occurring early and distal obstruction later. Other causes of mechanical malfunction include fracture, migration, disconnection, and overdrainage. The last complication can result in slit ventricles, subdural hematomas, and orthostatic hypotension.

Much has been written about the assessment of possible shunt malfunction. The most common approach is to obtain a head CT or ultrasound and a shunt series. The CT or ultrasound is compared to a previous scan to assess the size of the ventricles. A shunt series is a set of radiographs obtained to evaluate for any fractures, migration, or disconnection. A shunt can be interrogated by tapping the reservoir with a 25-gage needle and connecting a manometer to assess opening pressure, closing pressure, and ease of CSF drawback. Another test for assessing shunt function involves pressing down on the shunt reservoir while occluding the proximal portion. If the distal tubing is working properly, the reservoir should collapse immediately. Releasing the proximal occlusion should result in inflation of the reservoir dome, if the proximal portion is not occluded. The reported sensitivity of this test is 18–20% with a negative predictive value of 65–81% *(106)*. Piatt *(107)* assessed the utility of changes in the characteristics of pumping the shunt as a predictor of malfunction. In his study, the sensitivity of this test was 50% with a specificity of 64% with a positive predictive value of only 12%. In 1966, Di Choro and Grove introduced shuntography for assessing shunt patency *(108)*. This involves injecting an isotope into the shunt reservoir and scanning the patient to assess the patency of the system. Reports indicate a false negative rate of 14–40% *(109,110)*. The high false-negative rate was owing to partial obstruction of the shunt system. Inadequate volume injected and/or force of the injection can all lead to a false-negative outcome. Technical details differ among shunt types. A less commonly used technique to assess shunt patency involves use of two thermistors *(111)*. The gold standard is to take the patient to the operating room to interrogate and, if necessary, revise the shunt.

Shunt infections can present in many ways. Symptoms include headache, nausea, vomiting, nuchal rigidity, and diffuse abdominal pain. Of shunt malfunctions, 3 to 15% are the result of shunt infection *(112)*. The majority of shunt infections (70–85%) occur in the first 6 months after surgery. The next peak occurs at 1 year postsurgery, and thereafter the rate is approximately 1% each year *(113)*. The mortality of shunt infection is estimated to be more than 30% *(112)*. The lifetime incidence of shunt infection is 5–10% *(115)*. Shapiro et al. reported an infection rate of 4% over a 3-year period and analyzed the types of organisms cultured from the shunt *(114)*. Forty-five percent of the organisms were *Staphylococcus epidermidis*, 20% were *Staphylococcus aureus*, and 20% were Gram-negative bacilli, with the remaining 15% equally divided between *Corynebacterium*, α-*Streptococcus*, and *Micrococcus*. In this study, only 20% of the organisms isolated were identical to those isolated from the patient's skin. In diagnosing shunt infections, routine blood work and cultures, both aerobic and anaerobic, of blood, urine, and CSF from the shunt are sent. It is important that both aerobic and anaerobic cultures be sent. One particular species, *Propionibacterium acnes*, needs to be specifically cultured in thioglycollate broth. This species, although accounting for less than 2% of shunt infections, needs to be sought out, especially if initial cultures are negative, because it can take up to 14 days of incubation before this organism is isolated *(116)*. Kulkarni et al., in a prospective study, analyzed the relationship between numerous variables and shunt infection and found three variables that correlated strongly with shunt infection: presence of postoperative CSF leak, patient age, and the number of times the shunt system was exposed to a breached surgical glove *(110)*.

Currently, there is no standard protocol for the treatment of a shunt infection. The American Society of Pediatric Neurosurgeons recently published the results of a survey of its members concerning the management of CSF shunt infections. The survey revealed that treatment differed based on the organism isolated. If the organism were *Staphylococcus epidermidis*, 60% would remove the shunt and place an external ventricular device until the infection had cleared. If the organism were *Staphylococcus aureus*, 64% would do the same, and if the organism were a Gram-negative rod, 68% would do similarly. The remainder would treat the infection with externalization of the distal portion of the shunt and antibiotics. The duration of antibiotic treatment also varied from 5 to 24 days. The combination of antibiotics was also variable. The number of days of consecutive sterile cultures required before reshunting the patient ranged from 2 to 37 days *(117)*. Recent shunt devices have antibiotics impregnated into the silicone tubing; whether or not this will lower the infection rate is still to be determined. There is no evidence that a patient with a CSF shunt who develops an infection elsewhere warrants higher doses of antibiotics, a more prolonged course of antibiotics, or different antibiotics than a patient without a shunt in order to avoid shunt infection.

SUPPORT ORGANIZATIONS

With the advent of the Internet, the amount of information available to patients has exploded. Today's patients can investigate their illness through the World Wide Web and be more informed consumers of their health care. It is important to point out that not all the information found on the web is correct or regulated. Most states have a hydrocephalus support group. Patients or family members can find a local chapter by asking their healthcare provider or by searching the internet. We have listed two national hydrocephalus websites that can help patients and family members with answers to questions regarding hydrocephalus (*see* the Appendix).

APPENDIX:
RECOMMENDED HYDROCEPHALUS SUPPORT ORGANIZATIONS

Hydrocephalus Association
870 Market Street, Suite 955
San Francisco, California 94102
Telephone: 415-732-7040
Fax: 415-732-7044
E-mail: mailto:hydroassoc@aol.com
Web site: http://www.hydrocephalus.org

The Hydrocephalus Association provides support, education, and advocacy for families and professionals. It has local chapters and a national directory of hydrocephalus support groups. It publishes a quarterly newsletter.

National Hydrocephalus Foundation
12313 Centralia Road
Lakewood, California 90715
Telephone: 562-402-3532
E-mail: mailto:hydrobrat@aol.com
Web site: http://www.nhfonline.org

REFERENCES

1. Breasted JH. The Edwin Smith Surgical Papyrus (published in facsimile and hieroglyphic transliteration with translation and commentary). Chicago, IL: University of Chicago Press, 1930, pp. 164–174.
2. Elsberg CA. The Edwin Smith surgical papyrus and the diagnosis and treatment of injuries to the skull and spine 5000 years ago. Ann Med Hist 1931;3:271–279.
3. Davidoff LE. Treatment of hydrocephalus. Arch Surg 1929;18:1737–1762.
4. Skinner HA. The origin of medical terms (2nd ed.). New York: Hafner, 1970.
5. Fisher RG. Surgery of the congenital anomalies. In: Walker EA, ed. A history of neurological surgery. Baltimore, MD: Williams and Wilkins, 1951, pp. 334–347.
6. Amundsen DW. Medieval canon law on medical and surgical practice by the clergy. Bull Hist Med 1978;52:22–44.
7. Vesalius A. De Humani Corporis Fabrica Libri Septem. Basel: Ex Officina Ioannis Oporini, 1543.
8. Whytt R. Observations on the most frequent species of hydrocephalus internus. In: The Works of Robert Whytt, MD, (3rd ed.). Philadelphia, PA: B and T Kite, M. C. Hopkins, 1809.
9. Key A, Retzius G. Studien in der Anatomie des Nervensystems und des Bindgewebes. Stockholm, Sweden: Samson and Wallin, 1876.
10. Dandy WE, Blackfan KD. Internal hydrocephalus: An experimental, clinical, and pathological study. Am J Dis Child 1914;8:406–482.
11. Drake JM, Saint-Rose C. History of cerebrospinal fluid shunts, In: Drake JM, Saint-Rose C. The shunt book. Cambridge, MA: Blackwell Science, 1995, pp. 1–12.
12. Sharpe W. The operative treatment of hydrocephalus: A preliminary report of forty one patients. Am J Med Sci 1917;153:563–571.
13. Torkildsen A. A new palliative operation in cases of inoperable occlusion of the sylvian aqueduct. Acta Chir Scand 1939;82:117–124.
14. Matson DD. A new operation for the treatment of communicating hydrocephalus. J Neurosurg 1949;6:238–247.
15. Ferguson AH. Intraperitoneal diversion of the cerebrospinal fluid in cases of hydrocephalus. NY Med J 1898;67:902.
16. Haynes IS. Congenital internal hydrocephalus: Its treatment by the drainage of the cisterna magna into the cranial sinuses. Ann Surg 1913;57:449.
17. Heile B. Zur chirurgischen Behandlung des Hydrocephalus internus durch Ableitung der Cerebrospinalflussigkeit nach der Bauchhohle und nach der Pleurakuppe. Archiv fur Klinische Chirurgic 1914;105:501–516.
18. Chaptal J, Gros C, Jean R, Campos C, Vlahovitch B. Resultas obtenus dans le traitment chirurgical de l'hydrocephalie progressive en l'enfant. Pediatrie 1955;10:415–420.
19. Nulsen FE, Spitz EB. Treatment of hydrocephalus by direct shunt from ventricle to jugular vein. Surg Forum 1952;2:399–403.
20. Ingraham FD, Alexander E, Matson DD. Polyethylene, a new synthetic plastic for use in surgery; experimental application in neurosurgery. J Neurosurg 1947;42:679–682.
21. Matson DD. A new operation for the treatment of communicating hydrocephalus-report of a case secondary to generalized meningitis. J Neurosurg 1949;6:238–247.
22. Cone WV, Lewis RD, Jackson IJ. Shunting of cerebrospinal fluid into the peritoneal cavity. Presented at a meeting of American College of Physicians, Montreal, Canada, 1949.
23. Davidson RI. Peritoneal bypass in the treatment of hydrocephalus: historical review and abdominal complications. J Neurol Neurosurg Psychiatry 1976;39:640–646.
24. Traynelis VC, Powell RG, Koss W, Schochet SS, Kaufman HH. Cerebrospinal fluid eosinophilia and sterile shunt malfunction. Neurosurg 1988;23:645–649.
25. Gower DJ, Lewis JC, Kelly DL. Sterile shunt malfunction: A scanning electron microscope perspective. J Neurosurg 1984;61:1079–1084.

26. Yamashita N, Kamiya K, Yamada K. Experience with a programmable valve shunt system. J Neurosurg 1999;91:26–31.
27. Lemire RJ. Neural tube defects. JAMA 1988;259:558–562.
28. Fernell E, Hagberg B, Hagberg G, Wendt L Von. Epidemiology of infantile hydrocephalus in Sweden. 1. Birth Prevalence and general data. Acta Paediatr Scand 1986;75: 975–981.
29. McAllister PJ. Hydrocephalus: an update on incidence and cost. Hydrocephalus 1996;1:4–5.
30. Stein SC, Feldman JG, Stewart A, et al. The epidemiology of congenital hydrocephalus: a study in Brooklyn, NY. 1958–1976. Child Brain 1981;8:253.
31. Stoll C, Alembik Y, Dott B, Roth MP. An epidemiologic study of environmental and genetic factors in congenital hydrocephalus. Eur J Epidemiol 1992;8:797–803.
32. Di Rocco C, Marchese E, Valerdi F. A survey of the first complication of newly implanted CSF shunt devices for the treatment of non-tumoral hydrocephalus: cooperative survey of the 1991–1992 Education Committee on the ISPN. Child'sNerv Syst 1994;10(5):321–327.
33. Hozgreve W, Feil R, Louwen F, Miny P. Prenatal diagnosis and management of fetal hydrocephaly and lissencephaly. Child's Nerv Syst 1993;9:408–412.
34. Bondurant C Jimenez D. Epidemiology of cerebrospinal fluid shunting. Pediatr Neurosurg 1995;23:254–258.
35. Rekate HL. Treatment of hydrocephalus. In: Albright AL, Pollack IF, Adelson PD, eds. Principles and practice of pediatric neurosurgery. New York: Thieme Publishing, 1999, pp. 47–73.
36. Fishman MA, Dutton RY, Okumura S. Progressive ventriculomegaly following minor intracranial hemorrhage in premature infants. Dev Med Child Neurol 1984;26:725–731.
37. Auer LM, Mokry M. Disturbed cerebrospinal fluid circulation after subarachnoid hemorrhage and acute aneurysm surgery. Neurosurgery 1990;26:804–809.
38. Stein SC, Schut L. Hydrocephalus in myelomeningocele. Childs Brain 1989;5:413–419.
39. Amacher AL, Wellington J. Infantile hydrocephalus: Long-term results of surgical therapy. Childs Brain 1984;11:217–229.
40. Laurence K, Coates S. The natural history of hydrocephalus: detailed analysis of 187 unoperated cases. Arch Dis Child 192;37:345–362.
41. Watkins L, Hayward R, Andar U, et al. The diagnosis of blocked cerebrospinal fluid shunts: a prospective study of referral to a pediatric neurosurgical unit. Childs Nerv Syst 1994;10:87–90.
42. Lumenta CB, Skotarczak U. Long-term follow-up in 233 patients with congenital hydrocephalus. Childs Nerv Syst 1995;11:173–175.
43. Schrander-Stumpel C, Fryns JP. Congenital hydrocephalus: nosology and guidelines for clinical approach and genetic counseling. Eur J Pediatr 1998;157:355–362.
44. Burton BK. Recurrence risks for congenital hydrocephalus. Clin Genet 1979;16:47–53.
45. Behrman RE, Kliegman RM, Arvin AM. Nelson's textbook of pediatrics (15th ed.). Philadelphia, PA: W.B. Saunders, 1996, pp. 1683–1685.
46. Sato O, Bering EA. Extraventricular formation of cerebrospinal fluid. Brain Nerv 1967;19: 883–835
47. Cutler RWP, Page LK, Galicich J, et al. Formation and absorption of cerebrospinal fluid in man. Brain 1968;91:707–720.
48. Lorenzo AV, Hornig G, Zavala LM, et al. Furosemide lowers intracranial pressure by inhibiting CSF production. Z Kinderchir 1986;41(Suppl 1):10–12.
49. Faraci FM, Mayhan WG, Heistad DD. Vascular effects of acetazolamide on the choroid plexus. J Pharmacol Exp Therap 1990;254:23–27.
50. Nilsson C, Stahlberg F, Thomsen C, et al. Circadian variation in human cerebrospinal fluid production measured by magnetic resonance imaging. Am J Physiol 1992;262:R20—R24.
51. May C, Kaye J, Atack J, et al. Cerebrospinal fluid production is reduced in healthy aging. Neurology 1990;40:500–503.

52. Faraci FM, Mayhan WG, Heistad DD. Effect of vasopressin on production of cerebrospinal fluid: possible role of vasopressin (V1)-receptors. Am J Physiol 1990;258:R94—R98.
53. Rekate HL, Erwood S, Brodkey JA, et al. Formation of ventriculomegaly in choroid plexus papilloma. Pediatr Neurosci 1986;12:196–201.
54. Eisenberg H, McComb J, Lorenzo A. Cerebrospinal fluid overproduction and hydrocephalus associated with choroid plexus papilloma. J Neurosurg 1974;40:381–385.
55. Milhorat T. Failure of choroid plexectomy as treatment for hydrocephalus. Surg Gynecol Obstet 1974;139:505–508.
56. Scarff JE. Treatment of hydrocephalus: an historical and critical review of methods and results. J Neurol Neurosurg Psychiatr 1964;16:1–26.
57. Hill A, Rozdilsky B. Congenital hydrocephalus secondary to intra-uterine germinal matrix/intraventricular hemorrhage. Dev Med Child Neuro 1984;26:509–527.
58. Bianchine JW, Lewis RC Jr. The MASA syndrome: a new heritable mental retardation syndrome. Clin Genet 1974;5:298–306.
59. Bickers DA, Adams RD. Hereditary stenosis of the aqueduct of Sylvius as a cause of congenital hydrocephalus. Brain 1949;72:246–262.
60. Hlavin ML, Lemmon V. Molecular structure and functional testing of human L1. Genomics 1991;11:416–423.
61. Kenwrick S, Jouet M, Donnai D. X-linked hydrocephalus and MASA syndrome. J Med Genet 1996;33:59–65.
62. Nag TK, Falconer MA. Non-tumoral stenosis of the aqueduct in adults. Brit Med J 1966;2:1168–1170.
63. Morrison G, Sobel DF, Kelley WM, et al. Intraventricular mass lesions. Radiology 1984;153:435–442.
64. Ojemann RG. Normal pressure hydrocephalus. Clin Neurosurg 1971;18:337–370.
65. Olivero WC, Rekate HL, Chizeck HJ, et al. Relationship between intracranial and sagittal sinus pressure in normal and hydrocephalic dogs. Pediatr Neurosci 1988;14:196–201.
66. Hakim CA. The physics of physicopathology of the hydraulic complex of the central nervous system (the mechanics of hydrocephalus and normal pressure hydrocephalus), thesis. Cambridge: Massachusetts Institute of Technology, 1985.
67. Zerah M, Garcia-Monaco R, Rodesch G, et al. Hydrodynamics in vein of Galen malformation. Childs Nerv Syst 1992;8:111.
68. Rekate HL. Brain turgor (KB): intrinsic property of the brain to resist distortion. Pediatr Neurosurg 1992;18:257–262.
69. Karahalios DO, Rekate HL, Khayata MH, et al. Elevated intracranial venous pressure as a universal mechanism in pseudotumor cerebri of varying etiologies. Neurology 1996;46:198–202.
70. McComb JG. Recent research into the nature of CSF formation and absorption. J Neurosurg 1983;59:369–383.
71. Drake JM, Sainte-Rose C. The shunt book. Cambridge, MA: Blackwell Science, 1995.
72. Harman PK, Kron IL, McLachlan HD, et al. Elevated intra-abdominal pressure and renal function. Ann Surg 1982;196:594–597.
73. Emerson H. Intra-abdominal pressures. Arch Int Med 1911;7:754–784.
74. Drye JC. Intraperitoneal pressure in the human. Surg Gynecol Obstet 1948;87:472–475.
75. Iberti TJ, Lieber CE, Benjamin E. Determination of intra-abdominal pressure using a transurethral bladder catheter: clinical validation of the technique. Anesthesiology 1989;70:47–50.
76. Hobbie RK. Intermediate physics for medicine and biology (3rd ed.). New York: Springer-Verlag, 1997.
77. Chapman PH, Cosman ER, Arnold MA. The relationship between ventricular fluid pressure and body position in normal subjects and subjects with shunts: a telemetric study. Neurosurg 1990;26:181–189.

78. Magnaes B. Body position and cerebrospinal fluid pressure. Part 2: Clinical studies on orthostatic pressure and the hydrostatic indifferent point. J Neurosurg 1976;44:698–705.
79. Foltz EL, Blanks JP. Symptomatic low intracranial pressure in shunted hydrocephalus. J Neurosurg 1988;68:401–408.
80. da Silva MC, Drake JM. Effect of subcutaneous impantation of antisiphon devices on CSF function. Pediatr Neurosurg 1990–91;16:197–202.
81. International Standards Organization. ISO/DIS 10993-1.
82. Agranoff J, ed. Modern plastics encyclopedia (Vol. 57:10A). New York: McGraw-Hill, 1980.
83. Baumeister T. Mark's standard handbook for mechanical engineers (8th ed.). New York: McGraw-Hill, 1978.
84. Sainte-Rose C, Piatt JH, Renier D, et al. Mechanical complications in shunts. Pediatr Neurosurg 1991;17(1):2–9.
85. Sainte-Rose C, Hooven MD, Hirsch JF. A new approach to the treatment of hydrocephalus. J Neurosurg 1987;66:213–226.
86. Tokoro K, Chiba Y. Optimum position for an antisiphon device in a cerebrospinal fluid shunt system. Neurosurg 1991;29(4):519–525.
87. Young HF, Nulsen FE, Weiss MH, Thomas P. The relationship of intelligence and cerebral mantle in treated infantile hydrocephalus (IQ potential in hydrocephalic children). Pediatr 1973;52(1):38–60.
88. Rosseau GL, McCullough DC, Joseph AL. Current prognosis in fetal ventriculomegaly. J Neurosurg 1992;77:551–555.
89. Drake J, Kestle J, Milner R, et al. Randomized trial of cerebrospinal fluid shunt valve design in pediatric hydrocephalus. Neurosurg 1998;43:294–303.
90. Albright AL, Haines SJ, Taylor FH. Function of parietal and frontal shunts in childhood hydrocephalus. J Neurosurg 1988;69(6):883–886.
91. Bierbrauer KS, Storrs BB, McLone DG, Tomita T, Dauser RA. A prospective, randomized study of shunt function and infections as a function of shunt placement. Pediatr Neurosurg 1990;16(6):287–291.
92. Borgbjerg BM, Gjerris F, Albeck MJ, Hauerberg J, Borgesen SV. A comparison between ventriculo-peritoneal and ventriculo-atrial cerebrospinal fluid shunts in relation to rate of revision and durability. Acta Neurochir 1998;140:459–465.
93. Lundar T, Langmoen IA, Hovind KH. Fatal cardiopulmonary complications in children treated with ventriculoatrial shunts. Child's Nerv Syst 1991;7:215–217.
94. Wald SL, McLaurin RL. Shunt-associated glomerulonephritis. Neurosurg 1978;3:146–150.
95. Eisenberg HM, Davidson RI, Shillito J. Lumboperitoneal shunts: review of 34 cases. J Neurosurg 1971;35:427–431.
96. Spetzler RF, Wilson CB, Schulte R. Simplified percutaneous lumboperitoneal shunting. Surg Neurol 1977;7:25–29.
97. Chumas PD, Kulkarni AV, Drake JM, Hoffman HJ, Humphreys RP, Rutka JT. Lumboperitoneal shunting: A retrospective study in the pediatric population. Neurosurg 1993;32(3):376–383.
98. Chumas PD, Armstrong DC, Drake JM, et al. Tonsillar herniation—the rule rather than the exception after lumboperitoneal shunting in the pediatric population. J Neurosurg 1993;78(4):568–573.
99. Norden CW. Antibiotic prophylaxis in orthopedic surgery. Rev Infect Dis 1991;13(Suppl 10):S842–S846.
100. Dajani AS, Bisno AL, Chung KJ, et al. Prevention of bacterial endocarditis: recommendations by the American Heart Association. JAMA 1990;264:2919–2922.
101. Steinbok P, Boyd M, Flodmark CO, Cochrane DD. Radiographic imaging requirements following ventriculoperitoneal shunt procedures. Pediatr Neurosurg 1995;22:141–146.
102. Sgouros S, Malluci C, Walsh AR, et al. Long-term complications of hydrocephalus. Pediatr Neurosurg 1995;23:127–132.

103. Kestle J, Drake J, Milner R, Sainte-Rose C, et al. Long-term follow-up data from the shunt design trial. Pediatr Neurosurg 2000;33:230–236.
104. Dan NG, Wade MJ. The incidence of epilepsy after ventricular shunting procedures. J Neurosurg 1986;65:19–21.
105. Piatt JH. Physical examination of patients with cerebrospinal fluid shunts. Is there useful information in pumping the shunt? Pediatrics 1992;89:470–473.
106. Piatt JH. Pumping the shunt revisited: A longitudinal study. Pediatr Neurosurg 1996;25:73–77.
107. Di Choro G, Grove AS. Evaluation of surgical and spontaneous cerebrospinal fluid shunts by isotope scanning. J Neurosurg 1966;24:743–748.
108. Vernet O, Farmer JP, Lambert R, Montes JL. Radionuclide shutogram: adjunct to manage hydrocephalic patients. J Nuc Med 1996;37(3):406–410.
109. French BN, Swanson M. Radionuclide-imaging shuntography for the evaluation of shunt patency. Surg Neurol 1981;16(3):173–182.
110. Chiba Y, Ishiwata Y, Suzuki N, Muramoto M, Kunimi Y. Thermosensitive determination of obstructed sites in ventriculoperitoneal shunts. J Neurosurg 1985;62:363–366.
111. Kulkarni AV, Drake JM, Lamberti-Pasculli M. Cerebrospinal fluid infection: a prospective study of risk factors. J Neurosurg 2001;94:195–201.
112. Kaufman B. Management of complications of shunting. In: McLone DG, ed. Pediatric neurosurgery: surgery of the developing nervous system. Philadelphia, PA: WB Saunders; 2001, pp. 529–547.
113. Walters BC, Hoffman HJ, hendrick EB, et al. Cerebrospinal fluid shunt infection. Influences on initial management and subsequent outcome. J Neurosurg 1984;60:1014–1021.
114. Blount JP, Campbell JA, Haines SJ. Complications in ventricular cerebrospinal fluid shunting. Neurosurg Clin North Am 1993:633–656.
115. Shapiro S, Boaz J, Kleiman M, Kalsbeck J, Mealey J. Origin of organisms infecting ventricular shunts. Neurosurg 1988;22:868–872.
116. Everett ED, Eickhoff TC, Simon RH. Cerebrospinal fluid shunt infections with anaerobic diphtheroids (Propionibacterium species). J Neurosurg 1976;44:580–584.
117. Whitehead WE, Kestle JR. The treatment of cerebrospinal fluid shunt infections. Results from a practice survey of the American Society of Pediatric Neurosurgeons. Pediatr Neurosurg 2001;35(4):205–210.

US Counterpoint to Chapter 15

Joshua L. Dowling

INTRODUCTION

In Chapter 15, Rustamzadeh and Lam provided a thorough discussion of cerebrospinal fluid (CSF) shunting and have proposed rational recommendations for long-term care and follow-up of shunted patients. Many, if not most, patients do not conform well to a standardized schedule of care. Table 1 shows our current practice for routine care. Several factors may modify the management of individual patients. CSF diversion is indicated for a variety of diagnoses, and numerous surgical techniques and hardware choices exist. In addition, patients often become more complex with each shunt revision. These variables preclude entirely uniform recommendations for long-term care and follow-up. Beyond routine care, a number of situations require particular attention. Suspected shunt malfunction or infection prompts the majority of unscheduled office or emergency room visits and have been thoroughly reviewed. Pregnancy in shunted patients warrants further discussion. Although much of the care of shunted patients involves addressing issues that arise unexpectedly, routine follow-up plays an important role in minimizing morbidity in these patients.

The specific diagnosis for which a CSF shunt is implanted is a major determinant in formulating a care plan for shunted patients. Shunts are placed for communicating and obstructive hydrocephalus, both of which can have a number of causes. Pseudotumor cerebri and normal pressure hydrocephalus are other common indications. In most cases, the follow-up is tailored to the diagnosis. For example, in a patient with obstructive hydrocephalus related to a tumor, management is usually dictated by the tumor more than the shunt. Likewise, a shunted spina bifida patient may have other neurosurgical issues such as tethered cord or Chiari II malformation. Patients with obstructive hydrocephalus may be more prone to rapid deterioration and therefore may benefit from closer monitoring. Pseudotumor cerebri patients usually require some visual field testing not indicated in other shunted patients. Thus, the pathophysiology of the patient's disease is more important to management than the shunt itself.

There are several types of CSF shunts and, for each of these types, a number of different hardware components are available. Ventriculoperitoneal, ventriculoatrial, and lumboperitoneal are the most frequently implanted. Less common are ventriculopleural, thoracopleural, and syringosubarachnoid, and syringopleural shunts. Numerous valves are available including fixed pressure and programmable valves. Other

Table 1
Patient Surveillance After Implantation of a Cerebrospinal Fluid Shunt at Washington University School of Medicine[a]

	Postoperative year					
	1	2	3	4	5	10
Office visit	4	1	1	1	1	1
Shunt series	2	1	1	1	1	1
Head computed tomography	2	1	1	1	1	1

[a]The number in each cell denotes the number of times per year the modality is recommended for a patient without clinical evidence of a shunt problem.

features such as on–off switches and antisiphon devices are available. All of these variables should be considered in the care of a shunted patient because certain concerns are specific to the shunt type. Ventriculoatrial shunting, for instance, can be complicated by shunt nephritis *(1)* and may require evaluation of renal function, especially when infection is suspected. Lumboperitoneal shunts, which are commonly used for pseudotumor cerebri, but can also be used for communicating hydrocephalus, carry the risk of acquired Chiari malformation *(2)*. Magnetic resonance imaging might therefore be indicated at some point to assess for tonsillar descent. A programmable valve allows extraoperative manipulation of the valve pressure *(3–5)*. Patients in whom adjustments are being made toward a desired clinical effect usually require more frequent monitoring. Because the term "cerebrospinal fluid shunt" refers to multiple device types, uniform recommendations for follow-up are not practical.

The treatment of infections involving shunts has been thoroughly reviewed in the primary chapter. As pointed out earlier in Chapter 15, there is no evidence to support altering the antibiotic treatment of a shunted patient with an infection elsewhere in the body simply because a shunt is present. Clear guidelines for the use of secondary prophylaxis, such as for dental procedures, are also lacking. Nevertheless, ventriculoperitoneal and ventriculoatrial shunts may differ in the risk of infection from transient bacteremia. After healing, the ventriculoperitoneal shunt is essentially excluded from the circulation, whereas a ventriculoatrial shunt is intravascular. Therefore, extrapolating recommendations for other intravascular prosthetic devices, although not rigidly scientific, may be prudent. In the case of ventriculoatrial shunts, the application of recommendations of the American Heart Association for prosthetic heart valves is reasonable *(6)*.

When women with CSF shunts become pregnant, a number of issues can arise. Some published reports suggest a high incidence of complications in shunted patients *(7,8)*, although others emphasize that pregnancy and delivery proceed normally in most cases *(9–11)*. Women with CSF shunts can become symptomatic from increased intracranial pressure during pregnancy, especially in the third trimester. Most symptoms resolve postpartum *(8)* if the shunt itself is intact. Because the incidence of shunt revision in the peripartum period appears to be increased *(7,8)*, shunt evaluation, including a shunt series, is indicated after delivery.

Asymptomatic patients often present to the neurosurgeon for the first time with a shunt already implanted. This occurs frequently when the patient moves to a new area or changes insurance carrier. Also, patients followed by a pediatric neurosurgeon may eventually transfer care to an adult neurosurgeon. Establishing care with a new neurosurgeon should ideally be done while the patient is asymptomatic as elective evaluation at that point can facilitate emergent workup if the patient later becomes symptomatic. Crucial information includes the reason for shunting, the type of shunt and hardware implanted, prior revisions for malfunction or infection, and the clinical pattern of previous malfunction. Baseline shunt series and head computed tomography (CT) should also be obtained. The plain X-rays confirm the type of shunt and hardware as well as screen for an asymptomatic break in the tubing. The head CT is necessary to document the ventricular size for future comparison during suspected malfunctions. As with all visits, patient education should be reinforced.

Although CSF shunting can be one of the simplest technical procedures performed by a neurosurgeon, shunted patients are often among the most complex to manage. The numerous variables and high rate of complications seen in this group render a uniform schedule of care impractical. As patients become older and their clinical course more complicated, their management becomes progressively more individualized. Nevertheless, routine follow-up is essential to minimize the morbidity associated with shunt complications (12). Education of patients and their families is as important as screening for shunt problems during these visits. Instruction is especially needed as the patient transitions into adulthood and becomes responsible for his or her own care. Also, new family members, mainly spouses, need to understand the signs and symptoms of shunt malfunction. Because delay in treatment can have serious consequences, it is usually advisable to review a specific plan of action if a shunt problem is suspected. Although there may be situations in which a provider other than the neurosurgeon can perform routine follow-up care (13), other health care providers will have varying levels of experience in dealing with shunts and their complications. From the neurosurgical perspective, routinely delegating follow-up to others carries risk to the patient.

The Appendix lists resources for learning more about hydrocephalus.

APPENDIX:
RESOURCES FOR LEARNING MORE ABOUT HYDROCEPHALUS

Hydrocephalus Association
870 Market Street, Suite 955
San Francisco, CA 94102
Phone: (415) 776-4713

National Hydrocephalus Foundation
400 N. Michigan Ave., Suite 1102
Chicago, IL 60611-4102
Phone: (815) 467-6548

Spina Bifida Association of America
4590 MacArthur Boulevard, NW
Suite 250
Washington, D.C. 20007-4226
(202) 944-3285
(800) 621-3141

REFERENCES

1. Wald SL, McLaurin RL. Shunt-associated glomerulonephritis. Neurosurgery 1978;3:146–150.
2. Payner TD, Prenger E, Berger TS, Crone KR. Acquired Chiari malformations: incidence, diagnosis, and management. Neurosurgery 1994;34:429–434.
3. Zemack G, Romner B. Adjustable valves in normal-pressure hydrocephalus: a retrospective study of 218 patients. Neurosurg 2002;51:1392–1400.
4. Zemack G, Bellner J, Siesjo P, et al. Clinical experience with the use of a shunt with an adjustable valve in children with hydrocephalus. J Neurosurg 2003;98:471–476.
5. Yamashita N, Kamiya K, Yamada K. Experience with a programmable valve shunt system. J Neurosurg 1999;91:26–31.
6. Dajani AS, Taubert KA, Wilson W, et al. Prevention of bacterial endocarditis. Recommendations by the American Heart Association. JAMA 1997;22:1794–1801.
7. Liakos AM, Bradley NK, Magram G, Muszynski C. Hydrocephalus and the reproductive health of women: the medical implications of maternal shunt dependency in 70 women and 138 pregnancies. Neurol Res 2000;22:69–88.
8. Wisoff JH, Kratzert KJ, Handwerker SM, et al. Pregnancy in patients with cerebrospinal fluid shunts: report of a series and review of the literature. Neurosurgery 1991;29:827–831.
9. Cusimano MD, Meffe FM, Gentili F, Sermer M. Management of pregnant women with cerebrospinal fluid shunts. Pediatr Neurosurg 1991–1992;17:10–13.
10. Yu JN. Pregnancy and extracranial shunts: case report and review of the literature. J Fam Pract 1994;38:622–626.
11. Samuels P, Driscoll DA, Landon MB, et al. Cerebrospinal fluid shunts in pregnancy. Report of two cases and review of the literature. Am J Perinatol 1988;5:22–25.
12. Vinchon M, Fichten A, Delestret I, Dhellemmes P. Shunt revision for asymptomatic failure: surgical and clinical results. Neurosurg 2003;52:347–356.
13. Sgouros S, Malluci C, Walsh AR, et al. Long-term complications of hydrocephalus. Pediatr Neurosurg 1995;23:127–132.

European Counterpoint to Chapter 15

Spyros Sgouros

EPIDEMIOLOGY

Ventriculoperitoneal shunts have revolutionized the outcome of hydrocephalus at all ages and have improved quality of life. Before shunts were widely available, hydrocephalus carried an overall 5-year survival rate of 20% and a very poor intellectual outcome for those who survived *(1,2)*. Wide introduction of shunts after the 1950s improved the 5-year survival rate to 80% *(1,3–7)*. The intellectual outcome among survivors improved dramatically as well; more than 70% of treated hydrocephalic children now complete school training *(3,5,7,8)*. But, at the same time, introduction of shunts created new socioeconomic circumstances as they are prone to a significant number of complications with substantial impact on individuals and health systems alike.

The incidence of infantile hydrocephalus in Europe is estimated at 1 to 3 cases per 1000 live births, with regional variations. Availability of statistical data varies in different regions, with a relative lack of information from eastern European countries. The peak ages of presentation in early life are the first few weeks of life, 4 to 8 years, and early adulthood. The latter two peaks represent delayed presentations of infantile hydrocephalus. The incidence of hydrocephalus secondary to a previous neurological insult is estimated at 10% of the patients who survive following subarachnoid haemorrhage and 4% of those who survive head injury *(9)*. The peak ages of presentation for this group vary according to the original pathology but tend to be in the second to fifth decade of life. The prevalence of "normal pressure hydrocephalus" is estimated at 1 in 100,000 live people, but it could be even higher. It accounts for up to 6% of cases of dementia *(10)*. Most of the patients diagnosed with hydrocephalus at any age require operative treatment, which is likely to be in the form of a shunt, although endoscopic third ventriculostomy has recently regained popularity for some forms of hydrocephalus. A small proportion of patients will prove to have compensated hydrocephalus, which may not require treatment.

An estimated 750,000 people suffer from hydrocephalus worldwide and 160,000 ventriculoperitoneal shunts are implanted each year worldwide. These figures are derived largely from sales of shunts from the main commercial companies. There are no robust population-based statistical data worldwide, and it is conceivable that the prevalence of this condition is much higher as ready access to diagnosis and treat-

From: *The Bionic Human: Health Promotion for People With Implanted Prosthetic Devices*
Edited by: F. E. Johnson and K. S. Virgo © Humana Press Inc., Totowa, NJ

ment is not available in certain parts of the world. A more palpable measure of the number of implanted shunts can be obtained by observing the level of activity in a relatively "closed" population base. The two neurosurgical units of Birmingham, UK (pediatric and adult) serve a population of 1.2 million children and 2.5 million adults, with little interchange from outside the area. It is a metropolitan area with a mixed economy. During 1999, the two units implanted 71 shunts in children (32 new and 39 revisions) and 108 shunts in adults (77 new, 31 revisions). Diagnoses in the pediatric new-shunt group included aqueduct stenosis (25%), spina bifida (20%), congenital malformations (10%), posthemorrhagic (20%), postmeningitic (14%), and various other causes (9%). Diagnoses in the adult new-shunt group included tumor-related (24%), postsubarachnoid hemorrhage (13%), "normal pressure hydrocephalus" (9%), and various other causes. Another 28 patients (15 children and 13 adults) had endoscopic neurosurgical treatment for hydrocephalus owing to aqueduct stenosis without requiring a permanent shunt.

Hydrocephalus has different clinical profiles and underlying pathophysiological mechanisms in children and adults. Although some of the shunt problems are common to both groups, the overall prognosis of the condition tends to be different.

Infantile hydrocephalus can be associated with congenital anomalies such as aqueduct stenosis, spina bifida, and Chiari II malformation, and less common conditions such as Dandy-Walker syndrome, and encephalocele or acquired conditions such as intraventricular hemorrhage, meningitis, or brain tumors. Of these, the most frequent congenital cause is aqueduct stenosis, which results from atresia (partial or complete) or compression of the aqueduct. Despite the increasing use of endoscopic third ventriculostomy, a large proportion of children with aqueduct stenosis worldwide are still treated with shunts. The outcome of hydrocephalus due to aqueduct stenosis is in general good provided that follow-up is diligent *(4,5,7,8,11,12,14)*. At least half of the affected children are expected to complete normal schooling and achieve social independence *(7)*. The social and educational outcome of the surviving spina bifida patients is also good, similar to that of any other group of surviving hydrocephalics who reach adult years, making allowance for the physical disabilities that result from the associated paraplegia *(7)*. Other rare congenital syndromes have variable prognoses, depending on the extent of the deformity or malformation of the brain. The most common acquired conditions associated with infantile hydrocephalus are perinatal intraventricular hemorrhage and meningitis. The outcome varies with the extent of the underlying brain damage sustained during the acute event. In general, posthemorrhagic and postmeningitic etiologies carry the most adverse prognoses of all forms of infantile hydrocephalus, with only 10% of the children achieving independent status later on in life *(5,7)*. They are also associated with a high incidence of epilepsy *(5,7,15)*. When hydrocephalus is associated with intracranial tumours adjacent to the cerebrospinal fluid (CSF) pathways (ventricular system, aqueduct/pineal region, posterior fossa), outcome is dictated by the natural history of the tumour itself and hydrocephalus usually does not dominate the clinical picture *(16)*. For all hydrocephalic patients, shunt infection has a detrimental effect on intelligence and development *(5,7,8,17–19)*.

Hydrocephalus can develop at any age secondary to another primary neurological insult such as head injury, subarachnoid haemorrhage, or intracranial infections owing to obstruction at the level of the arachnoid granulations from either blood (head injury, subarachnoid hemorrhage, craniotomy) or increased protein in the CSF (intracranial infections) *(9,20–24)*. Although hydrocephalus can appear at any time after the initial event, early onset usually implies increased severity of the underlying disease and hence forecasts a worse outcome. In general, outcome in such clinical settings is determined by the extent of neurological damage sustained during the acute event, and treatment of hydrocephalus optimises any chance for recovery.

In adults, the term "normal pressure hydrocephalus" refers to the clinical syndrome consisting of gait disturbance, mental deterioration, and urinary incontinence associated with enlargement of the ventricular system. It is distinct from other forms of hydrocephalus seen in adult patients, secondary to hemorrhage or trauma, for example *(9,10,20,22,25–27)*. Controversy still surrounds the diagnostic methods that should be employed for patient selection as the pathophysiology of the syndrome has not been completely described *(9,10)*. It is often considered in the differential diagnosis of cerebral atrophy.

Most patients diagnosed with normal pressure hydrocephalus are treated with shunts. Of the three prominent clinical features mentioned already, gait disturbance is the one most likely to improve after shunting and may be considered a good prognostic feature when it is the main presenting symptom *(9,10,25-27)*. Dementia is a less prominent feature of symptomatic adult hydrocephalus. When dementia dominates the clinical picture, another underlying condition such as Alzheimer's disease should be suspected. Urinary incontinence is usually a late sign and responds variably to treatment. The outcome of normal pressure hydrocephalus improved after the introduction of ventricular shunts. A significant proportion of these patients were originally considered to suffer from primary dementia, but shunting improved their mental and physical status. There has been a renewed interest in this condition because the advent of variable pressure shunts allows the neurosurgeon to titrate the effect of shunting against the improvement of symptoms.

EVOLUTION OF CLINICAL PRACTICE

The development of the CSF shunt in the 1950s revolutionized the treatment of hydrocephalus and dramatically improved outcome *(1,3–7)*. Until the late 1970s and early 1980s, ventriculoatrial shunting was the common practice. Gradually, appreciation of long-term problems of this type of shunts, such as endocarditis, cor pulmonale, and diffuse glomerulonephritis from septic emboli (shunt nephritis) *(4,28–33)*, led to a switch to ventriculoperitoneal shunting. Now very few surgeons perform ventriculoatrial shunts as a first choice. Prior to computed tomography (CT) scanning, diagnosis of hydrocephalus and follow-up were difficult, as it was necessary to employ invasive diagnostic methods such as air encephalography or ventricular cisternography. Advent of CT scan in the late 1970s and early 1980s promoted shunt surgery, as diagnosis and follow-up became easily available. In the early 1990s, the advent of magnetic resonance scanning offered improved visualization of the anatomy of the ventricular system, which led to a revival of endoscopic treatment. This was closely followed by technological improvements of the equipment used to perform the operations. Although the enthusiasts predicted that shunts would become obsolete, the current practice has evolved to a situation in which approximately one-third of the patients receive endoscopic treatment and the rest get CSF shunts. In a small proportion of patients with communicating hydrocephalus and in most patients with benign intracranial hypertension in whom medical treatment has failed, lumboperitoneal shunt is preferred, especially when it appears from the CT scan that it may be difficult to directly cannulate the ventricles due to their small size *(34,35)*. This practice is particularly applicable to adult patients and is used less frequently in children as it has been shown that lumboperitoneal shunts can cause secondary hindbrain hernia in children *(36)*.

Since the 1950s, the main improvements have been in the materials that constitute the shunt parts. Most shunts are silicone-based, although a few are either totally made of metal, usually titanium, or have metal parts in them. As a result of improved materials, late mechanical complications such as fracture or migration of shunt catheters *(37–40)* as a result of material disintegration have decreased in frequency. There are still unresolved problems that relate to the in vivo interaction of the shunt with its environment. Obstruction of catheters and valves by cellular debris *(41)* results in functional deterioration of valves with time *(42)*. Progress in the design of shunt valves has been modest during the last five decades, and several studies have shown only small differences in shunt survival between different types of shunts *(7,13,14,43–45)*. Appreciation of the "slit-ventricle" syndrome *(6,46–51)* led to the concept of siphoning and the subsequent development of antisiphon devices and flow-control valves in the early 1980s *(52–57)*. In the last decade, the concept of an adjustable valve gained popularity, allowing change of the opening pressure of the valve percutaneously to optimize symptoms *(11,14,22,26,27,58–60)*.

FOLLOW-UP MANAGEMENT

Most surgeons remove the sutures a week after surgery, and some even use absorbable sutures that do not need removal, especially in young children. The average inpatient stay following uncomplicated shunt insertion or revision is three to five days. Most surgeons prefer to obtain a CT in the first few days after surgery that shows the position of the catheter and gives an indication of shunt function from the reduction of the ventricular size *(61–63)*. In the first week after surgery, the typical reduction in ventricular volume is about 20–30% *(45)*, which is easily appreciated by the trained observer. Some flow-control valves allow much slower reduction of ventricular size than differential pressure valves, and this should be kept in mind when surgeons and patients are trying to interpret early scans. In the unlikely event that the catheter is wrongly positioned, the surgeon may have to perform revision surgery at the same hospitalisation. If the catheter is in part inside the ventricles, most surgeons elect to leave it in place and observe the evolution of the ventricular size with frequent clinical examinations and repeat CT scans. Apart from a CT scan, there is no need for any other tests postoperatively. Assuming that the patient was not anemic preoperatively and that there were no untoward intraoperative complications, there is usually very little blood loss during shunt insertion or revision, and there is no need for postoperative blood tests.

Postoperative follow-up of shunted patients has attracted some controversy in the past, especially with respect to how often there is a need for CT scanning. Table 1 shows our plan of follow-up. Most surgeons would like to review the patient 4 to 6 weeks following shunt surgery. By that time, the wounds should have healed well and the patient should have returned to his or her usual activities. Most surgeons advise

Table 1
Patient Follow-Up After Implantation of Ventricular Shunt at the Birmingham Children's Hospital

	Postoperative year					
	1	2	3	4	5	10
Office visit[a]	4[b]	1	1	1	1	1
Computed tomography scan	2[c]	1	1	1	1	1

[a]Office visit includes assessment of visual function with fundoscopy.
[b]At 6 weeks postoperatively, 3 months, 6 months, and 1 year.
[c]At 1 week or no later than 3 months postoperatively and at 12 months.
If the patient has an adjustable valve, he or she may need a skull radiograph every time the valve setting is changed.

paediatric patients to return to school at 4 to 6 weeks following surgery, usually part time at first and gradually increasing to full time. Excessively energetic activities such as contact sports are usually allowed 3 months after surgery. Outpatient reviews always include fundoscopic ophthalmic examination for all patients, both those who had papilledema preoperatively and those who have never had it, as one of the features of impending shunt obstruction is the development of papilledema. This can lead to optic atrophy and blindness if unchecked. In most occasions, the outpatient review serves as an opportunity for the patient to voice any subtle symptoms that may be creeping in and for the neurosurgeon to remind the patient of the possible symptoms of shunt failure. In practice, acute shunt obstructions rarely coincide with outpatient visits! But the gradual development of shunt overdrainage is one of the problems that can be monitored with regular visits. In the absence of any symptoms implying shunt infection, there is no need for any blood tests during outpatient follow-up. One CT scan should be obtained in the first 3 months after surgery if it has not been obtained in the immediate postoperative period *(61–63)*. After this time, another CT scan should be obtained at 12 months and subsequently every 12 months, or when clinical symptoms indicate any possible complications. Several studies have demonstrated that the ventricular size continues to decrease in the first months after implantation and stabilises at 6 to 12 months postoperatively *(64,65)*. It is unnecessary to perform repeat scans during this period in the absence of any clinical symptoms, because the scans do not alter the clinical management *(63)*, and the patient receives unnecessary irradiation. Some adjustable valves require a lateral skull X-ray to confirm the valve setting. This should be performed every time the setting is changed in an outpatient visit.

Although ventricular shunting has saved many lives, it has presented clinicians with a new set of problems associated with the various forms of shunt failure. The majority are related to the almost inevitable blockage of some part of the shunt system and a significant part of technological innovation is still directed toward lowering the incidence of blockage. Almost 40% of shunts fail at 1 year after implantation, more than 60% fail by 5 years, and 80% fail by 10 years *(12–14,26,43,44,65–67)*. This very high failure rate provides a clear rationale for regular follow-up. Early failure during the first year after implantation is usually owing to catheter obstruction at the ventricular level and less often due to valve or distal catheter obstruction *(65–67)*. As the shunt drains the ventricles, the ventricular wall collapses around the ventricular catheter as early as 3 months after implantation in differential pressure valves *(45)*. This leads to growth of ependyma or choroid plexus cells through the catheter holes and subsequent obstruction *(41)*. It presents with acute development of symptoms of raised intracranial pressure within 24 to 48 hours *(68)*. In a small proportion of patients, intermittent shunt obstruction occurs over a period of weeks or months prior, and they complain of episodic headaches *(46)*. In patients with normal pressure hydrocephalus, obstruction usually presents differently, with gradual symptomatic deterioration to the preshunting levels. Overall, shunt failure caused by obstruction is more common among children than adults. On average, a child who has had shunt implantation early in life requires two to three operations for revision of the shunt in the first 15 years of life *(7)*. For every shunt revision, there is typically another occasion when the child is admitted with symptoms suggestive of shunt obstruction, which settle without requiring surgery *(69)*. Suspicion of shunt obstruction prompts CT scanning, which in most cases, shows ventricular dilatation when the shunt is blocked. It is always helpful to examine a previous CT, obtained when the patient is clinically well, for comparison. Depending on the level of organization of health services, in some countries complete records are kept in the local hospitals, whereas elsewhere the practice of offering the patient copies of the scans circumvents other organizational difficulties.

Another early complication is development a few weeks after implantation of subdural haematoma, unilateral or bilateral, caused by shunt overdrainage. This is commonly seen in patients with differential pressure valves *(6,13,26,27,43,57,60,70)*. Children with large ventricles before shunting and elderly patients with normal pressure hydrocephalus and a significant degree of brain atrophy are particularly susceptible. It manifests with progressive signs of raised intracranial pressure and the CT scan is diagnostic. It is treated usually with a change of valve to a flow-regulating or an adjustable pressure one or introduction in the valve system of an antisiphon device. Early complications include shunt infection, which is discussed later.

Late complications include obstruction of the ventricular catheter, valve or distal tube owing to deposition of debris, and mechanical complications such as fracture or migration *(43,44)*. Fracture occurs because the shunt tubing receives continuous mechanical stress in the most mobile parts of the head, neck, and chest and gradually disintegrates. It is especially common when a child who was shunted in infancy goes through a growth spurt and increases in height *(37–40)*. Fracture of the shunt catheters is easily visualized in plain radiographs, as most catheters are radio-opaque and it is for this reason that radiographs of the entire system should be obtained if the shunt has been in place for a long time (years). Other uncommon late complications include abdominal pseudocysts and erosion of the skin overlying the shunt in areas of longstanding pressure.

In the upright position, most shunts overdrain CSF from the ventricular system, resulting in the disabling "slit-ventricle" syndrome, one of the most important late shunt complications *(6,46–53)*. Prior to the advent of antisiphon devices, the main method to treat slit-ventricle syndrome was subtemporal craniectomy *(46–51)*. This is necessarily resective and rather mutilating surgery, aiming to remove part of the skull and thus to improve the compliance of the brain. As it is not designed to counteract the cause of overdrainage, its effect tends to diminish with time as new bone grows over the craniectomy site and symptoms recur. Utilization of antisiphon devices has dramatically reduced the need for subtemporal decompression and this procedure is rarely performed now in departments with a modern approach to hydrocephalus *(54–57)*. The development of some programmable shunt devices, the opening pressure of which can be altered percutaneously, is another attempt to counteract overdrainage symptoms and "tailor" the shunt to the individual needs *(11,14,22,26,27,58–60)*. The development of slit-ventricle syndrome is seen almost exclusively in children. It usually takes place over several weeks or months, although it can appear as early as 3 months after implantation *(13)*. The child reports intermittent periodic symptoms of overdrainage such as headaches in the upright position or chronic headaches associated with visual disturbance and declining school performance and behavior. It is treated with a change of valve to a flow-control or an adjustable pressure or introduction in the valve system of an antisiphon device *(51)*.

Most shunt types have similar complication rate. This is presumably owing to the fact that, despite differences in design, all shunts are subjected to the same deposition of debris in their critical parts. It has been shown that only a minority of shunts perform to specification in the long term, indicating that in vivo interaction alters performance *(42)*. The multitude of commercially available shunts, all of which claim technological superiority but, in practice, suffer from similar problems, reflects the fact that no design is clearly superior. The latest trend of most manufacturers is toward adjustable valves, but it is not clear whether these valve types are superior. Their usefulness may be particularly applicable to normal pressure hydrocephalus patients where the option to alter the pressure setting of the valve may allow the surgeon to minimize the occurrence of chronic subdural hematomas by setting the original pressure at a higher level *(26,27)*. In any case, there is the perception that newer shunt designs have improved the quality of life of patients by reducing symptoms such as chronic headaches. These are notoriously difficult to quantify and even more difficult to treat and, for that reason, neurosurgeons tend to ignore them even though they have a great impact on patients' lives.

Several attempts have been made in the past to devise a noninvasive test that could reliably identify a functioning shunt. Various types of "shuntograms" have been tried, including injection of contrast or radionuclide, use of temperature sensors, standard and color Doppler ultrasound, and even magnetic resonance scanning *(71–76)*. Unfortunately, none have stood the test of time. The main problems have been reliability and reproducibility, as well as bedside availability out of hours, when the patient presents with acute symptoms. As mentioned already, current common clinical practice for the investigation of suspected shunt failure consists mainly of a CT scan that, in most hospitals, can be obtained easily out of hours when the clinical need arises.

DETECTION AND MANAGEMENT OF SHUNT INFECTION

Shunt infection is seen in 5 to 10% of patients and commonly presents in the first 3 months from implantation but can appear also much later, even years after the original surgery. Shunt infection is most common in babies under the age of 6 months at implantation and in immunocompromised patients *(17,18,77–80)*. Most efforts to minimize shunt infection have concentrated on refining surgical technique, administering prophylactic antibiotics, and educating neurosurgeons to perform shunt surgery first in their operating lists, with as few people as possible in the operating room and with as correct technique as possible *(81–83)*. Recent efforts have also focused on producing antibiotic-impregnated shunt material, mainly ventricular catheters, which are meant to discourage bacterial adhesion on the material immediately after surgery *(84,85)*.

Shunt infection usually causes symptoms of shunt obstruction such as headaches and vomiting, symptoms of infection such as malaise, and signs of infection such as pyrexia. Often these are not fulminant and at times the diagnosis is not appreciated. Once the clinical suspicion has risen, blood tests should be carried out, including full blood count obtained to assess for leucocytosis, erythrocyte sedimentation rate, and C-reactive protein (CRP). Even in the presence of infected shunt, although the white cell count is usually normal or only mildly elevated, erythrocyte sedimentation rate and CRP may be only mildly elevated. The ventricular system is isolated from the rest of the body in immunological terms and a generalized response to infection is not always present. It is not uncommon to have elevated CRP in the CSF because in infection there is a breakdown of the blood–brain barrier. The only reliable method of detection of shunt infection is CSF sampling from the shunt itself. Such a procedure should be carried out ideally by neurosurgeons familiar with different types of shunts, who can avoid puncturing the wrong part of the shunt system, which would damage it. Most modern shunt systems incorporate a reservoir of some type especially devoted to such an emergency. Not uncommonly, a lateral skull radiograph proves helpful in determining the type of shunt present in patients where previous records are not available. Diagnosis of an infected shunt usually requires removal of the shunt, insertion of an external ventricular drain that secures CSF drainage and decompression, treatment with appropriate systemic and intraventricular antibiotics for 1 to 2 weeks, and shunt reimplantation in a fresh site. In the presence of a functioning shunt, systemic and intraventricular antibiotic administration through the shunt may avoid initial shunt removal. When sterilization of CSF has been achieved, a new shunt is inserted in a new site, and the previous one is removed *(86–90)*. Not uncommonly shunt infection can present as an abdominal emergency, with abdominal pain and peritonitis, due to an infected CSF "pseudocyst" *(91–93)*. Shunt removal, and sometimes ultrasound-guided or open cyst drainage, associated with antibiotic treatment, usually settles the problem, allowing new shunt implantation in 1 to 2 weeks. In rare patients the peritoneum fails to absorb CSF even in the absence of infection. This presents with abdominal dilatation and eventual shunt failure and is postulated to be due to delayed hypersensitivity to the silicone material of the tube. Conversion from ventriculoperitoneal to ventriculoatrial shunt usually settles the problem.

If a patient with a shunt develops infection elsewhere in the body, such as pneumonia, there is a risk of developing metastatic shunt infection as a result of bacteremia. Theoretically, bacteria can gain access to the CSF and colonize the shunt, although this is usually very rare. To prevent such an occurrence, it is advisable to administer a wide spectrum antibiotic such as cefalosporin for 1 week, in addition to the antibiotic specific to the systemic infection.

Over the years, researchers have performed in vitro and in vivo studies of shunt function for various types of shunts and manufacturers have tested their own products in order to achieve commercial certification. However, until the early 1990s, there was no independent facility that could systematically test a large variety of shunt devices with a high degree of accuracy and reproducibility. There are now two independent shunt-testing laboratories. In Cambridge, UK, the U.K. Shunt Evaluation Laboratory was created and is housed at Addenbrooks Hospital with financial support from the Medical Devices Agency, with the initial aim to test on behalf of the agency a variety of shunt implants in order to assist with certification procedures *(53,94)*. The original scope of the laboratory has been expanded and it provides currently in vitro computerized testing for a large variety of shunt implants, intracranial pressure sensors, and other similar commercial products. In parallel, in Heidelberg, Germany, a similar computerized shunt-testing laboratory was created at Kupfe Klinik, which has independently tested and reported on almost all commercially available shunts, even some obscure designs. The Heidelberg laboratory specializes particularly in long-term in vitro testing and has provided technical evidence of shunt performance after 1 year of simulation function, something that the manufacturers are not requested to provide *(52,95)*. Both laboratories have provided evidence that almost all commercial designs have at some point deviated from original specifications, a fact that has generated some anxiety among commercial vendors. There is now a wealth of evidence that, after a prolonged period of function, almost all shunts deviate significantly from their original specifications, and this is thought to be due to material "memory" and "fatigue" *(52,53,94,95)*. The contribution of both facilities to the understanding of shunt function has been tremendous, although everybody acknowledges that simulated in vitro performance may not be directly comparable to in vivo performance.

PRIMARY AND SECONDARY PREVENTION

There is currently no means of preventing development of hydrocephalus. Its pathogenesis is not at all clear, and the term includes different pathophysiological mechanisms. The earliest that hydrocephalus can be diagnosed is the last few weeks of intrauterine life, when ultrasound examination can show a dilated ventricular system. Magnetic resonance scan can offer considerable anatomical detail. The traditional management of prenatally diagnosed hydrocephalus is to simply stand by and observe the progress, hoping that a normal labor will take place. A large head circumference often dictates Cesarean section. Close postnatal observation by a neurosurgeon usually results in optimal treatment of hydrocephalus. Attempts have been made to drain the ventricular system *in utero* by endoscopically inserting a pigtail catheter, puncturing the soft skull and gaining access to the lateral ventricle. This treatment method has not gained popularity and is rarely performed. On the other hand, *in utero* closure of an open spina bifida has been performed in a small number of centres in the United States in about 200 patients so far and is reported to decrease the likelihood that the patient will need a ventricular shunt later in life *(96)*. Ethical considerations and practical difficulties have restricted the wide establishment of this treatment to date. It is conceivable that, in the future, endoscopic technology and materials will improve enough to allow successful endoscopic *in utero* treatment of hydrocephalus, although there will always be the inclination from the clinician's point of view to allow babies to be born, and institute treatment later, when needed, after close observation. So far, only a handful of specialized centres worldwide have accumulated the necessary diagnostic expertise to allow correct decisions every time. Even fewer have the necessary equipment and surgical skills for *in utero* surgery.

MULTI-INSTITUTION PROSPECTIVE TRIALS

An inherent problem in the development of CSF shunts is the fact that any benefit from a new design requires at least 5 years of follow up to demonstrate its effectiveness, as most of the currently unsolved problems are related to the long-term presence of the shunt, and its inherent inability to simulate the normal flow of CSF. After decades of research and development, the main issues relating to patient survival have been addressed, but most shunt designs have similar clinical performance in reported series *(3,11,13,22,26,27,45,58,60,64,67)*. Because of this, the choice of shunts by neurosurgeons is mostly based on personal beliefs and preconceived perceptions. There are few large randomized controlled studies but three have been completed recently, two of which were instigated by the same driving force.

The first attempted to compare clinical outcomes between two large groups of different shunt designs: adjustable vs nonadjustable valves *(14)*. It did not find any difference between the two different types of valves. This could be of interest to funding bodies as adjustable valves cost almost twice as much as fixed pressure ones.

The Shunt Design Trial attempted to assess the differences in outcomes as measured by shunt complications, at 1 and 4 years, among three different types of shunts, an ordinary differential pressure valve, a differential valve with an antisiphon device and a flow-control valve *(43,44)*. The study failed to demonstrate any statistically significant difference between the three different shunt types in both early and late follow-up.

The Endoscopic Shunt Insertion Trial attempted to address the issue of ventricular catheter placement, a less controversial issue but equally important from the economic point of view *(97)*. The traditional technique of catheter placement is "blind" insertion using natural anatomical landmarks. In a significant proportion of instances, this results in incorrect placement, requiring further surgery to correct it. Manufacturers have developed a fine fiberoptic endoscope that can be inserted through the lumen of a ventricular catheter and allow the surgeon to see the ventricular cavity directly and thus decide on the adequacy of catheter placement. This, in theory, should reduce the rate of wrong catheter placement, save patients from further operations to reposition the catheter, and decrease the overall cost of treatment. The results of the trial showed that the endoscopic placement is not superior to the ordinary "blind" placement, reminding us that an idea may be good but its implementation in clinical practice may be fraught with problems *(97)*.

All three trials attracted some criticism on methodological issues and highlighted the difficulties of conducting randomized controlled trials of surgical device involving several surgeons over many centers. Despite these criticisms, all trials focused the attention of neurosurgeons on the need for high-quality data to guide their choice of shunt.

Although the organization of shunt trials has been difficult, the establishment and maintenance of shunt registries can also provide some valuable data. Such a national registry has been kept for several years at the Department of Neurosurgery in Addenbrookes Hospital, Cambridge, UK *(98)*. It was funded by the Department of Health of the United Kingdom and the Medical Devices Agency and started collecting data in 1994 as a pilot and in 1995 definitively. It takes advantage of the high degree of regulation of neurosurgical practice in the United Kingdom, as in practice the overwhelming majority of shunt surgery takes place in closely monitored state hospitals. It recruits information from more than 60 departments. In practice, this captures data from most practicing neurosurgeons. After each operation, the surgeon fills out a paper "shunt form," which is sent by surface mail to the U.K. Shunt Registry, in exchange for a symbolic monetary reimbursment of $1.81! Although it is difficult to ascertain what percentage of operations are recorded and to what degree of accuracy, nevertheless the collected information has provided useful data about trends of current clinical practice. The U.K. Shunt Registry provides regular feedback with progress reports to all the neurosurgeons who submit data and presents these data regularly in national and international meetings.

Across Europe, there are very few associations that the patients can turn for help with other nonsurgical problems related to hydrocephalus (*see* the Appendix for contact details for some of them). Through their communications material they provide very useful information on the impact of hydrocephalus on daily living and function as a port of call for any help that patients may need with social and practical issues. The quality of the information and guidance they provide is ensured, as they work in close collaboration with medical experts in the field.

APPENDIX: SUPPORT GROUPS IN EUROPE

Society for Research into Hydrocephalus and Spina Bifida (SRHSB)
Current Secretary: Dr Terry Cubbitt, The Health Centre, Anstley Road, Alton, Hampshire, GU34 2QX, England
Tel: -44-1420-542542
Fax: -44-1420-549466
Web site: http://www.srhsb.org

Association for Spina Bifida and Hydrocephalus (ASBAH)
42 Park Road, Peterborough, PE1 2QU,
England
Tel: -44-1733-555988
Fax: -44-1733-555985
Web site: http://www.asbah.org

Arbeitsgemeinschaft Spina bifida und Hydrocephalus e.V. (ASbH e.V)
Bundesverband, Münsterstr 13, D-44145, Dortmund, Germany
Tel: -49-231-861050-0
Fax: -49-231-861050-50
Web site: http://www.asbh.de

REFERENCES

1. Foltz EL, Shurtleff DB. Five years comparative study of hydrocephalus in children with and without operation (113 cases). J Neurosurg 1963;20:1064–1079.
2. Laurence KM, Coates S. The natural history of hydrocephalus. Detailed analysis of 182 unoperated cases. Arch Dis Child 1962;37:345–362.
3. Hayden PW, Shurtleff DB, Stuntz TJ. A longitudinal study of shunt function in 360 patients with hydrocephalus. Develop Med Child Neurol 1983;25:334–337.
4. Keucher TR, Mealey J. Long-term results after ventriculoatrial and ventriculoperitoneal shunting for infantile hydrocephalus. J Neurosurg 1979;50:179–186.

5. Kokkonen J, Serlo W, Saukkonen A-L, Juolasmaa A. Long-term prognosis for children with shunted hydrocephalus. Childs Nerv Syst 1994;10:384–387.
6. Pudenz RH, Foltz EL. Hydrocephalus: overdrainage by ventricular shunts. A review and recommendations. Surg Neurol 1991;35:200–212.
7. Sgouros, Malluci CL, Walsh AR, Hockley AD. Long-term complications of hydrocephalus. Pediatric Neurosurg 1995;23:127–132.
8. Dennis M, Fitz CR, Netley CT, et al. The intelligence of hydrocephalic children. Arch Neurol 1981;38:607–615.
9. Pickard JD. Adult communicating hydrocephalus. Br J Hosp Med 1982;27:35–44.
10. Vanneste JAL, Augustijn P, Dirven C, Tan WF, Goedhart ZD. Shunting normal pressure hydrocephalus: do the benefits outweigh the risks? A multicenter study and literature review. Neurology 1992;42:54–59.
11. Benesch C, Friese M, Aschoff A. Four-year follow-up study of 146 patients with programmable Medos Hakim valve shunt system. Childs Nerv Syst 1994;10:475.
12. Hirsch JF. Surgery of hydrocephalus: past, present and future. Acta Neurochir (Wien) 1992;116:155–160.
13. Jain H, Sgouros S, Walsh AR, Hockley AD. The treatment of infantile hydrocephalus: "differential pressure" or "flow-control" valves? A pilot study. Childs Nerv Syst 2000;16: 242–246.
14. Pollack IF, Albright AL, Adelson PD. A randomized, controlled study of a programmable shunt valve versus a conventional valve for patients with hydrocephalus. Hakim-Medos Investigator Group. Neurosurgery 1999;45:1399–1408.
15. Dan NG, Wade MJ. The incidence of epilepsy after ventricular shunting procedures. J Neurosurg 1986;65:19–21.
16. Dias MS, Albright AL. Management of hydrocephalus complicating childhood posterior fossa tumors. Pediatr Neurosci 1989;15:283–289.
17. Ammirati M, Raimondi AJ. Cerebrospinal fluid shunt infections in children. A study on the relationship between the etiology of hydrocephalus, age at the time of shunt placement, and infection rate. Childs Nerv Syst 1987;3:106–109.
18. McLone DG, Czyzewski D, Raimondi A, Sommers R. Central nervous system infections as a limiting factor in the intelligence of children with meningomyelocele. Pediatrics 1982;70:338–342.
19. Walters BC, Hoffman HJ, Hendrick EB, Humphreys RP. Cerebrospinal fluid infections. J Neurosurg 1984;60:1014–1021.
20. Benzel EC, Pelletier AL, Levy PG. Communicating hydrocephalus in adults: Prediction of outcome after ventricular shunting procedures. Neurosurgery 1990;26:655–660.
21. Milhorat TH. Acute hydrocephalus after subarachnoid hemorrhage. Neurosurgery 1987;20: 15–19.
22. O'Reilly G, Williams B. The Sophy valve and the El-Shafei shunt system for adult hydrocephalus. J Neurol Neurosurg Psychiatry 1995;59:621–624.
23. Van Gijn J, Hijdra A, Wijdicks EFM, Vermeulen M, Van Crevel H. Acute hydrocephalus after aneurysmal subarachnoid hemorrhage. J Neurosurg 1985;63:355–362.
24. Yasargil MG, Yonekawa Y, Zumstein B, Stahl HJ. Hydrocephalus following spontaneous subarachnoid hemorrhage. Clinical features and treatment. J Neurosurg 1973;39:474–479.
25. Black PM. Idiopathic normal pressure hydrocephalus. J Neurosurg 1980;52:371–377.
26. Zemack G, Romner B. Seven years of clinical experience with the programmable Codman Hakim valve: a retrospective study of 583 patients. J Neurosurg 2000;92:941–948.
27. Zemack G, Romner B. Adjustable valves in normal-pressure hydrocephalus: a retrospective study of 218 patients. Neurosurgery 2002;51:1392–1400; discussion, 1400–1402.
28. Forrest DM, Cooper DGW. Complications of ventriculoatrial shunt. A review of 455 cases. J Neurosurg 1968;29:506–512.

29. Lundar T, Langmoen IA, Hovind KH. Fatal cardiopulmonal complications in children with ventriculoatrial shunts. Childs Nerv Syst 1991;7:215–217.
30. Noonan JA, Ehmke DE. Complications of ventriculovenous shunts for control of hydrocephalus. New Engl J Med 1978;269:70–74.
31. Piatt JH, Hoffman HJ. Cor pulmonale: a lethal complication of ventriculoatrial CSF diversion. Childs Nerv Syst 1989;5:29–31.
32. Stickler GB, Slim MM, Burke EC, Holley KE, Miller RH, Segar WE. Diffuse glomerulonephritis associated with infected ventriculoatrial shunt. New Engl J Med 1968;279:1077–1082.
33. Zamora I, Lurbe A, Alvarez-Garijo A, Mendizabal S, Siman J. Shunt nephritis. A report on five children. Childs Brain 1984;11:183–187.
34. Aoki N. Lumboperitoneal shunt: Clinical applications, complications, and comparison with ventriculoperitoneal shunt. Neurosurg 1990;26:998–1004.
35. Selman WR, Spetzler RF, Wilson CB, Grollmus JW. Percutaneous lumboperitoneal shunt: review of 130 cases. Neurosurgery 1980;6:225–227.
36. Chumas PD, Armstrong DC, Drake JM, et al. Tonsillar herniation: the rule rather than the exception after lumboperitoneal shunting in the pediatric population. J Neurosurg 1993;78:568–573.
37. Boch A-L, Hermelin E, Sainte-Rose C, Sgouros S. Mechanical dysfunction of ventriculoperitoneal shunt due to calcification of the silicone rubber catheter. J Neurosurg 1998;88:975–982.
38. Cuka GM, Hellbusch LC. Fractures of the periton real catheter of cerebrospinal fluid shunts. Pediatr Neurosurg 1995;22:101–103.
39. Echizenya K, Satoh M, Murai H, Ueno H, Abe H, Komai T. Mineralization and biodegradation of CSF shunting systems. J Neurosurg 67:584–591, 1987.
40. Shimotake K, Kondo A, Aojama I, Nin K, Tashiro Y, Nishioka T. Calcification of ventriculoperitoneal shunt tube. Surg Neurol 1988;30:156–158.
41. Collins P, Hockley AD, Woollam DHM. Surface ultrastructure of tissues occluding ventricular catheters. J Neurosurg 1978;48:609–613.
42. Brydon HL, Bayston R, Hayward R, Harkness W. Removed shunt valves: reasons for failure and implications for valve design. Br J Neurosurg 1996;10:245–251.
43. Drake JM, Kestle JRW, Milner R, et al. Randomised trial of cerebrospinal fluid shunt valve design in paediatric hydrocephalus Neurosurgery 1998;43:294–305.
44. Kestle J, Drake J, Milner R, et al. Long-term follow-up data from the Shunt Design Trial. Pediatr Neurosurg 2000;33:230–236.
45. Xenos C, Sgouros S, Natarajan K, Walsh AR, Hockley AD. Influence of shunt type on ventricular volume changes in children with hydrocephalus. J Neurosurg 2003;98:277–283.
46. Epstein F, Marlin AE, Wald A. Chronic headache in the shunt-dependent adolescent with nearly normal ventricular volume: diagnosis and treatment. Neurosurgery 1978;3:351–355.
47. Holness RO, Hoffmann HJ, Henrik EB. Subtemporal decompression for the slit-ventricle-syndrome after shunting in hydrocephalic children. Childs Brain 1979;5:137–144.
48. Kiekens R, Mortier W, Pothmann R, Bock WJ, Seibert H. The slit-ventricle syndrome after shunting in hydrocephalic children. Neuropediatrics 1982;13:190–194.
49. Oi S, Matsumoto S. Infantile hydrocephalus and the slit ventricle syndrome in early infancy. Childs Nerv Syst 1987;3:145–150.
50. Serlo W, Saukkonen AL, Heikkinen E, von Wendt L. The incidence and management of the slit ventricle syndrome. Acta Neurochir (Wien) 1989;99:113–116.
51. Walsh JW, James HE. Subtemporal craniectomy and elevation of shunt valve opening pressure in the management of small ventricle-induced cerebrospinal fluid shunt dysfunction. Neurosurgery 1982;10:698–713.
52. Aschoff A, Benesch C, Kremer P, Fruh K, Klank A, Kunze S. Overdrainage and shunt-technology. A critical comparison of programmable, hydrostatic and variable-resistance valves and flow-reducing devices. Childs Nerv Syst 1995;11:193–202.

53. Czosnyka Z, Czosnyka M, Richards HK, Pickard JD. Posture-related overdrainage: comparison of the performance of 10 hydrocephalus shunts in vitro. Neurosurgery 1998;42: 327–333.
54. Gruber R, Jenny P, Herzog B. Experiences with the antisiphon device (ASD) in shunt therapy of pediatric hydrocephalus. J Neurosurg 1984;61.156–162.
55. Portnoy HD, Schulte RR, Fox JL. Antisiphon and reversible occlusion valves for shunting in hydrocephalus and preventing post-shunt subdural hematomas. J Neurosurg 1973;38: 729–738.
56. Sainte-Rose C, Hooven MD, Hirsch JF. A new approach in the treatment of hydrocephalus. J Neurosurg 1987;66:213–226.
57. Tokoro K, Chiba Y, Abe H, Tanaka N, Yamataki A, Kanno H. Importance of anti-siphon devices in the treatment of pediatric hydrocephalus. Childs Nerv Syst 1994;10:236–238.
58. Lumenta CB, Roosen N, Dietrich U. Clinical experience with a pressure-adjustable valve SOPHY in the management of hydrocephalus. Childs Nerv Syst 1990;6:1–6.
59. Will BE, Müller-Korbsch U, Buchholz R. Experience with the programmable Sophy SU-8 valve. Childs Nerv Syst 1994;10:476.
60. Zemack G, Bellner J, Siesjo P, Stromblad LG, Romner B. Clinical experience with the use of a shunt with an adjustable valve in children with hydrocephalus. J Neurosurg 2003;98: 471–476.
61. Schellinger D, McCullough DC, Peterson RT. Computed tomography in the hydrocephalic patient after shunting. Radiology 1980;137:693–704.
62. Schönmayr R, Zierski J, Agnoli AL. CT-follow-up of hydrocephalus in children. Adv Neurosurg 1980;8:164–166.
63. Steinbok P, Boyd M, Flodmark CO, Cochrane DD. Radiographic imaging requirements following ventriculoperitoneal shunt procedures. Pediatr Neurosurg 1995;22:141–146.
64. Tuli S, O'Hayon B, Drake J, Clarke M, Kestle J. Change in ventricular size and effect of ventricular catheter placement in pediatric patients with shunted hydrocephalus. Neurosurgery 1999;45:1329–1333.
65. Di Rocco C, Marchese E, Velardi F. A survey of the first complication of newly implanted CSF shunt devices for the treatment of nontumoral hydrocephalus. Childs Nerv Syst 1994; 10:321–327.
66. Sainte-Rose C. Shunt obstruction: A preventable complication? Pediatr Neurosurg 1993; 19:156–164.
67. Sainte-Rose C, Piatt JH, Renier D, Pierre-Kahn A, Hirsch J-F, Hoffman HJ, Humphreys RP, Hendrick EB. Mechanical complications of shunts. Pediatr Neurosurg 1991–1992; 17:2–9.
68. Cinalli G, Sainte-Rose C, Simon I, Lot G, Sgouros S. Sylvian aqueduct syndrome and global rostral midbrain dysfunction associated with shunt malfunction. J Neurosurg 1999; 90:227–236.
69. Watkins L, Hayward R, Andar U, Harkness W. The diagnosis of blocked cerebrospinal fluid shunts: a prospective study of referral to a paediatric neurosurgical unit. Childs Nerv Syst 1994;10:87–90.
70. Moussa AH, Sharma SK. Subdural haematoma and the malfunctioning shunt. J Neurol Neurosurg Psychiatry 1978;41: 759–761.
71. Deway RC, Kosnik EJ, Sayers MP. A simple test of shunt function. The shuntogram. J Neurosurg 1976;44:121–126.
72. Drake JM, Martin AJ, Henkleman RM. Determination of cerebrospinal fluid shunt obstruction with magnetic resonance phase imaging. J Neurosurg 1991;75:535–540.
73. French BN, Swanson M. Radionuclide-imaging shuntography for the evaluation of shunt patency. Surg Neurol 1980;16:173–182.
74. Go KG, Melchior HJ, Lakke JPWF. A thermosensive device for the evaluation of the patency of ventriculoatrial shunts in hydrocephalus. Acta Neurochir (Wien) 1968;19: 209–216.

75. Pople IK. Doppler flow velocities in children with controlled hydrocephalus: reference values for the diagnosis of blocked cerebrospinal fluid shunts. Childs Nerv Syst 1992;8: 124–125.
76. Sgouros S, John P, Walsh AR, Hockley AD. The value of Colour Doppler Imaging in assessing flow through ventricular shunts. Childs Nerv Syst 1996;12:454–459.
77. James HE, Walsh JW, Wilson HD, Connor JD, Bean JR, Tibbs PA. Prospective randomized study of therapy in cerebrospinal fluid shunt infection. Neurosurgery 1980;7: 459–463.
78. Pople IK, Bayston R, Hayward RD. Infection of cerebrospinal fluid shunts in infancy: a study of etiological factors. J Neurosurg 1992;77:29–36.
79. Renier D, Lacombe J, Pierre-Kahn A, Sainte-Rose C, Hirsch J-F. Factors causing acute shunt infection. Computer analysis of 1174 operations. J Neurosurg 1984;61:1072–1078.
80. Spanu G, Karrusos G, Adinolfi D, Bonfati N. An analysis of shunt infections in adults. A clinical experience of twelve years. Acta Neurochir (Wien) 1986;80:79–82.
81. Choux M, Genitori L, Lang D, Lena G. Shunt implantation: reducing the incidence of shunt infection. J Neurosurg 1992;77:875–880.
82. Djinddjan M, Fevrier MJ, Otterbein G, Soussy JC. Oxacillin prophylaxis in cerebrospinal fluid shunt procedures: results of a randomized open study in 60 hydrocephalic patients. Surg Neurol 1986;25:178–180.
83. Schmidt K, Gjerris F, Osgaard O, et al. Antibiotic prophylaxis in cerebrospinal fluid shunting: a prospective randomized trial in 152 hydrocephalus patients. Neurosurgery 1985; 17:1–5.
84. Bayston R, Grove N, Siegel J, Lawellin D, Barsham S. Prevention of hydrocephalus catheter colonisation in vitro by impregnation with antimicrobials. J Neurol Neurosurg Psychiatry 1989;52:605–609.
85. Hampl J, Jansen B, Schierholz J, Aschoff A. In vitro and in vivo efficacy of rifampicin-loaded silicone catheter for the prevention of CSF shunt infections. Acta Neurochir (Wien) 1995;133:147–152.
86. Gombert ME, Landesman SH, Corrado ML, Stein SC, Melvin ET, Cummings M. Vancomycin and rifampicin therapy for staphylococcus epidermidis meningitis associated with CSF-shunts. J Neurosurg 1981;55:633–636.
87. Mates S, Glaser J, Shapiro K. Treatment of cerebrospinal fluid shunt infections with medical therapy alone. Neurosurg 1982;11:781–783.
88. McCracken GH, Mize SG, Threlkeld N. Intraventricular gentamycin therapy in gram-negative bacillary meningitis of infancy. Report of the Second Neonatal Meningitis Cooperative Study Group. Lancet 1980;1(8172):787–791.
89. Osborn JS, Sharp S, Hanson J, MacGee E, Brewer JH. Staphylococcus epidermidis ventriculitis treated with vancomycin and rifampicin. Neurosurgery 1986;19:824–827.
90. Swyne R, Rampling A, Newsom SW. Intraventricular vancomycin for treatment of shunt-associated ventriculitis. J Antimicrob Chemother 1987;19:249–253.
91. Bryant MS, Bremer AM, Tepas JJ, Mollitt DL, Nquyen TQ, Talbert JL. Abdominal complications of ventriculoperitoneal shunts. Am Surg 1988;54:50–55.
92. Gutierrez FA, Raimondi AJ. Peritoneal cysts: a complication of ventriculoperitoneal shunts. Surgery 1976;79:188–192.
93. Rekate HL, Yonas H, White RJ, Nulsen FE. The acute abdomen in patients with ventriculoperitoneal shunt. Surg Neurol 1979;11:442–445.
94. Czosnyka Z, Czosnyka M, Richards HK, Pickard JD. Laboratory testing of hydrocephalus shunts—conclusion of the U.K. Shunt evaluation programme. Acta Neurochir (Wien) 2002; 144:525–538.
95. Oikonomou J, Aschoff A, Hashemi B, Kunze S. New valves-new dangers? 22 valves (38 probes) designed in the nineties in ultralong-term tests (365 days). Eur J Pediatr Surg 1999; 9(Suppl 1):23–26.

96. Bruner JP, Tulipan N, Paschall RL, et al. Fetal surgery for myelomeningocele and the incidence of shunt-dependent hydrocephalus. JAMA 1999;282:1819–1825.
97. Kestle JR, Drake JM, Cochrane DD, et al. Endoscopic Shunt Insertion Trial participants. Lack of benefit of endoscopic ventriculoperitoneal shunt insertion: a multicenter randomized trial. J Neurosurg 2003,98.284–90.
98. O'Kane MC, Richards H, Winfield P, Pickard JD. The United Kingdom Shunt Registry. Eur J Pediatr Surg 1997;7(Suppl 1):56.

16
Cochlear Implants

Adrien A. Eshraghi, John E. King, Annelle V. Hodges, and Thomas J. Balkany

Cochlear implants are electronic devices introduced surgically into the inner ear that directly stimulate the auditory nerve in response to sound. They can benefit severely, profoundly, or totally hearing-impaired patients who derive little or no benefit from hearing aids. The implant consists of an external component and a surgically implanted internal component. The external portion includes a microphone, microprocessor-based speech processor, and radio-frequency transmitting coil (Fig. 1). The implanted portion houses a radio-frequency receiver coil, microprocessor-based stimulator, and multichannel electrode array (Fig. 2).

Devices currently being used around the world include those manufactured by Advanced Bionics Corporation (United States), Cochlear Corp. (Australia), Medical Electronics Corp. (Austria), and Digisonic (France). These implants do not restore normal hearing but enable their recipients to function at a level similar to less hearing-impaired patients who are successful hearing aid users.

HISTORICAL ASPECTS AND TRENDS

In 1957, two Frenchmen, Djourno and Eyries, stimulated a cochlear nerve exposed during removal of a large cholesteatoma under local anesthesia. An electrode was placed directly on the nerve and stimulated with a simple electric current. This produced an auditory sensation in the patient (1). This experiment led to the concept of direct stimulation of the auditory nerve or ganglion cells, which is the basis of cochlear implants. Consequently, a flurry of investigations ensued that led to the development of a speech processor to interface with and drive an electrode implanted in the scala tympani. The House 3M was the first commercially marketed single-channel cochlear implant. It was introduced in 1972 (2).

In 1982, the first multichannel device was developed at the University of Melbourne. It was the first of several models to be produced by Cochlear Ltd. (3). Since then, several other manufacturers have also produced and marketed multichannel cochlear implants. The competition among the various companies has driven the research and technology to produce better cochlear implants, the net result of which has been

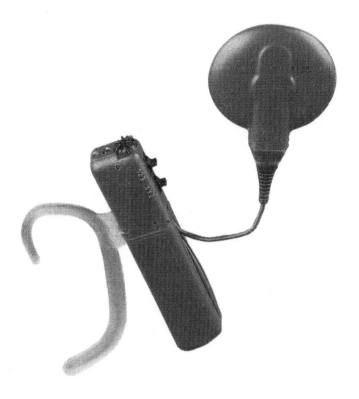

Fig. 1. The external part of a cochlear implant (MedEL TEMPO) including a microphone, a microprocessor-based speech processor, and a radio-frequency transmitting coil.

Fig. 2. The implanted portion of a cochlear implant (MedEL Combi 40+) including a radio frequency receiver coil, a microprocessor-based stimulator, and a multichannel electrode array.

Table 1
Device Implantation As of 2001, in United States and Worldwide [a]

Manufacturer (device)	Total US	Total world
Advanced Bionics (Clarion)	Not Available	~7000
Cochlear Corp. (Nucleus)	15,000	30,500
Medical Electronics (MedEl)	350	~6000

[a] Figures provided by manufacturers (personal communication, 2001).

improved electrodes, internal receivers, speech processors, and speech-coding strategies. These design changes have been particularly rapid within the past few years.

Initially, emphasis was placed on external equipment and speech-coding strategies called feature extraction strategies because their algorithm attempts to extract information from speech and encode it into the stimulus delivered by the cochlear implant. Although these earlier strategies, which proved successful during the mid-1990s, are still in use, continuous sampling strategies are available in current devices, and one device offers an analog option as well. Loudness continues to be modulated by the intensity of the electrical current in all strategies. External equipment has steadily become smaller and more versatile. Multiple programs and speech-processing strategies are now standard. Most recently, attention has once again turned toward internal electrode design, the goal being to develop internal arrays that lie closer to the modiolus and result in less insertion trauma. These electrodes run complex programming strategies with less power, opening the way for ear-level and ultimately fully implantable devices.

To date, there are more than 50,000 patients with cochlear implants worldwide, including both single-channel and multichannel models. Table 1 provides a breakdown of the numbers of cochlear implants implanted for the three major cochlear implant manufacturers offering multichannel cochlear implant systems.

COCHLEAR IMPLANT COMPONENTS AND THEIR FUNCTIONS

As indicated earlier, cochlear implants may be subdivided into five primary components, including microphone, speech processor, transmitting coil, internal receiver/stimulator, and electrode array. Cochlear implants are among the most complex devices treated in this text and we provide a brief description here.

The microphone detects sound in the listener's environment and sends that information to the speech processor. Essentially a mini-computer, the speech processor encodes the incoming sound from the microphone based on programs stored in its memory. The programs are software algorithms that encode the information provided by the microphone and provide the instructions that enable the receiver/stimulator and electrode array to stimulate the cochlea appropriately. The programming varies by device and even in the same processor for an individual. One reason for this variation is that each device uses a different algorithm to encode auditory information that, in turn, affects how electrical stimulation of the remaining spiral ganglion fibers occurs. A second reason for variation is that each patient has unique requirements for electrical stimulation.

Once the signal has been processed and encoded by the speech processor, the information is sent to the transmitting coil that is held in place over the implanted receiver/stimulator by an external magnet. The transmitting coil sends the signal provided by the speech processor to the receiver/stimulator via a radio-frequency signal. Upon reaching the receiver/stimulator, the signal is transduced into electrical pulses. Information encoded into the signal informs the receiver/stimulator which electrodes to activate and when and how to activate the electrodes, and so on. The pattern of activation of the electrodes is determined in part by the nature of the electrode array and in part by the coding strategy utilized by the programs in the speech processor. The electrical current sent through the implanted electrodes to the cochlea directly stimulates remaining auditory nerve fibers. The patterns of stimulation are conducted along the auditory nerve to the brain where interpretation as meaningful sound occurs.

CURRENTLY AVAILABLE COCHLEAR IMPLANTS IN THE UNITED STATES

The three manufacturers currently offering Food and Drug Administration (FDA)-approved and/or investigational devices in the United States are Advanced Bionics Corp., Cochlear Ltd, and Medical Electronics (MedEl) Corp. The features and functions of internal and external hardware components, as well as speech-coding strategies, are discussed here.

Clarion (Advanced Bionics)

Advanced Bionics is the manufacturer of the Clarion cochlear implant. Advanced Bionics was formed in 1994 as a spinoff of Minimed Technologies, which helped develop the University of California–San Francisco cochlear implant. The first Clarion clinical trial began in 1992 with a device offering the first reverse-telemetry electrode array focused on the cochlear modiolus, multiprogramming options, and multiple program memories. Depending on the receiver/stimulator, electrode array, or speech processor under consideration, both FDA-approved and investigational versions of the hardware are available for the Clarion cochlear implant.

Internal Components

Currently, Advanced Bionics offers three versions of the electrode array: the Enhanced Bipolar™ (spiral) electrode, and the recently developed HiFocus™ electrode array. As discussed later, the HiFocus array is available in two configurations, HiFocus I and HiFocus II.

The spiral electrode, introduced in 1991, is a precurled electrode array with 16 platinum-iridium spherical electrode contacts along 25 mm of the array that is inserted into the cochlea, utilizing a specialized insertion tool. The array is conformed to the shape of the inner ear to ensure that the electrodes maintain close proximity to the auditory nerve fibers within the modiolus. The electrodes are paired, thus enabling a fixed bipolar stimulation, and may be used with or without the Electrode Positioning System that was introduced in 1998.

The Electrode Positioning System is a flexible silicone-based polymer positioner designed for insertion alongside the electrode array in a two-stage surgical insertion process. Its purpose is to position the electrode array close to the modiolus of the cochlea, thereby enhancing the ability of the electrodes to stimulate the spiral ganglion. This system guides the electrode deep into the cochlea and prevents excess tissue formation by occupying the space immediately lateral to the electrode array.

The HiFocus I electrode array was introduced in 1999. It is precurved, but not fully spiraled like the Enhanced Bipolar electrode array. It was designed to better accommodate the Electrode Positioning System and still utilizes 16 platinum electrodes that face the modiolus of the cochlea. The electrodes are separated by dielectric partitions that focus the electric stimulation toward the modiolus and reduce electrical interaction among the electrodes. It is fully approved by the FDA.

In 2000, the HiFocus II electrode array was introduced as an investigational device. It uses a smaller positioner, directly attached to the electrode array, which reduces the electrode insertion process to a single surgical stage. The nature and functioning of the electrodes for the HiFocus II array are essentially the same as that for the HiFocus I array; however, the HiFocus II is available only via investigational trials.

Presently, two versions of the receiver/stimulator are available: the Implantable Cochlear Stimulator and the CII model. The electronics are packaged and hermetically sealed in a ceramic case that is coated with silastic. The Implantable Cochlear Stimulator, introduced in 1995 (version 1.2), was the first to implement continuous bidirectional telemetry communication between the internal and external components of the cochlear implant. Telemetry measurements enable direct monitoring of the overall status of the cochlear implant and may be made with either the handheld Portable Cochlear Implant Tester or the Computer Programming Interface normally used for programming of the speech processor. The telemetry function also provides a fail-safe system that directs the external components to shut down if aberrations in device function are detected, a feature currently offered only by the Clarion system. The circuitry of the Implantable Cochlear Stimulator is designed to support the monopolar and bipolar modes of stimulation that are utilized by the speech processor.

The CII model was released in 2001. It is the newest version of the Advanced Bionics Implantable Cochlear Stimulator and is capable of stimulation speeds faster than any other device currently on the market. The CII is designed to internally store program information that normally would be stored in the speech processor. This enables the entire cochlear implant system to operate more efficiently. The CII also contains two new diagnostic and measurement tools: Electric Field Imaging and High Resolution Neural Response Imaging. The Electric Field Imaging software enables measurement of electrode status and current delivered by the electrodes that may provide information useful for device programming. The upcoming High Resolution Neural Response Imaging software will be able to measure the response of the auditory nerve fibers to electric stimulation provided by the cochlear implant. The CII electronics package may be used with either the HiFocus I or HiFocus II electrode.

External Components

The S-Series body-worn speech processor, introduced in 1997, is compatible with the spiral, HiFocus I and HiFocus II electrodes coupled with the Implantable Cochlear Stimulator. Previous versions of the Clarion speech processor were replaced by the S-Series processor and thus are not discussed. It consists of a lightweight metal casing designed to minimize the effects of static electricity on the internal circuitry. The top of the S-Series processor has three rotary knobs that control program selection, volume level, and sensitivity setting, each of which may be adjusted independently of the others. It has the capacity to store three programs that, in turn, can implement three different coding strategies. Power for the device is supplied by a rechargeable battery pack. The overall size of the processor is 7 × 6 cm. In case of emergency, an optional battery pack is available that accepts conventional batteries. The circuitry contains an audible alarm to signal when battery power is low or the link between the external and internal components has been lost. This is especially important for caregivers of pediatric patients with this device. An accessory jack is present that allows various accessories to be utilized with the system. An aspect of the Clarion cochlear implant, which differs from the devices offered by other manufacturers, is that the headset incorporates the transmitter coil, the microphone, and the external magnet in a single unit.

The Platinum series processor is the latest release in the Clarion line of speech processors and is the smallest body-worn speech processor on the market. It measures 5 × 6 cm, including the rechargeable battery pack, which is identical to that used with the S-Series. All of the features and functions of the S-Series previously described apply to the body-worn Platinum Series processor. An ear-level version of the Platinum Series processor, in which the entire circuitry of the processor is housed in a small case behind the ear, is currently being evaluated in clinical trials.

It is anticipated that, when FDA approval has been received for the HiFocus II electrode array, the only Clarion system devices offered in North America will be the CII receiver-stimulator, HiFocus II electrode array and Platinum processor. Older generations of these components may be made available in certain circumstances but will not be the standard versions offered.

Speech-Processing Strategies

The Clarion cochlear implant system is presently capable of utilizing three FDA-approved coding strategies, including Simultaneous Analog Stimulation, Continuous Interleaved Sampling, and Paired Pulsatile Sampling, now called Multiple Pulsatile Sampling. The Simultaneous Analog Stimulation strategy is the only available speech-processing mode in which all functional electrodes are stimulated simultaneously. The Continuous Interleaved Sampling mode digitizes incoming sound from the microphone and stimulates all active channels in a sequential manner. A unique aspect of this strategy is that the order of stimulation may be varied from base to apex, apex to base, or nonsequentially. The Multiple Pulsatile Sampling strategy is a hybrid in which simultaneous and sequential stimulation are both utilized. In this mode, an envelope extraction paradigm called bin averaging is utilized and is the source of information simultaneously presented to two channels at a time. The use of bin averaging enables Multiple Pulsatile Sampling to stimulate faster than Continuous Interleaved Sampling, although not as fast as Simultaneous Analogue Stimulation.

Nucleus (Cochlear Ltd.)

Of the three cochlear implant companies offering products in the United States, Cochlear Ltd. has been on the market the longest (since 1982). Consequently, more implant recipients have received this device than any other. The current system, the Nucleus 24™, was introduced in 1997, and FDA approval in both adults and children was obtained in 1998. With the exception of the Nucleus 24 Double Array, which remains in clinical trial status in the United States, all external and internal components of the Nucleus 24 system are FDA-approved. Both a body-worn (SPrint) and an ear-level speech processor (Esprit) are available.

Internal Components

At present, Cochlear Ltd. has three electrode array configurations available: the straight electrode system, a split-array electrode designed for use in difficult insertions (e.g., ossification), and the newer Contour electrode array.

The current version of the straight electrode array, released in 2000, is the 24k electrode array. The design is based on the 22M electrode first released in 1985. Like the original array, the 24k has 22 platinum electrode bands and 10 stiffening rings. The 22 independently functioning electrodes are spaced along a 25-mm section of the array within a silicone carrier. The device has two additional electrodes that are not inserted into the cochlea. One of the extracochlear electrodes is a platinum plate attached to the receiver/stimulator unit, and the other is a ball electrode on the end of an independent lead wire, which is placed under the temporalis muscle during the surgery. The purpose of the two extracochlear electrodes is to reduce the power consumption of the cochlear implant by serving as ground electrodes. The remote grounds enable the use of a more power-efficient monopolar mode of stimulation. The device supports monopolar, bipolar, and common-ground modes of stimulation. The straight electrode is directly advanced through the cochleostomy, requiring the use of microinstruments designed for the task.

The receiver/stimulator of the 24k array is a hybrid integrated circuit contained within a hermetically sealed titanium casing coated in silicone. Also within the silicone encasement are a platinum receiver coil and titanium-coated rare earth magnet. The receiver coil functions as an antenna for the signal received from the externally positioned transmitter. The magnet is removable through a simple surgical procedure to permit the use of magnetic resonance imaging (MRI) when necessary. The 24k device, when connected to the programming interface for the Nucleus implant, can perform three types of telemetry measures, including impedance and compliance telemetry, as well as Neural Response Telemetry. This is an objective measure of auditory nerve function that may be utilized during difficult fitting sessions to estimate parameters of stimulation. Only the Nucleus system is capable of Neural Response Telemetry at this time.

The Nucleus 24 Double Array was released in 1999. It is a specially designed electrode array for patients who have ossification or other physiological abnormalities in their cochlea contraindicating the use of a single electrode array. The Double Array is split into two shorter arrays, each with 11 electrodes. Implantation requires modification of the traditional surgical approach. Two openings must be drilled, one for each of the electrode arrays, which are placed in the first and second turns of the cochlea.

The Contour, introduced in 1999, is the most recent version of the Nucleus electrode array. It is precurved, with a tapered apical portion designed to match the curvature and size of the cochlea's scalae. A stylet is embedded in the array to assist in proper positioning of the array during insertion. After the electrode array has been advanced through the cochleostomy, the stylet is withdrawn, allowing the electrode array to tightly hug the modiolus. In the Contour, the previously full-banded electrodes are now half-banded to direct stimulation toward the spiral ganglion fibers. The use and function of the two extracochlear ground electrodes have been carried over from the 24k array.

The hermetically sealed silicone-encased titanium housing of the Contour receiver/stimulator remains unchanged from that of the Nucleus 24k. However, the device is smaller and the receiver/stimulator pedestal has been reshaped into a circle for ease of placement during surgery. The magnet remains removable, allowing the device to be MRI compatible.

External Components

The current body-worn speech processor, SPrint™, was first marketed in 1997 and is compatible with both the straight and Contour internal electrode arrays. The SPrint processor has a four-program capacity and can implement three different speech-coding strategies: Spectral Peak, Continuous Interleaved Sampling, and Advanced Combination Encoder. The Spectral Peak strategy encodes selected spectral characteristics of incoming sound. Of the available 22 channels carrying the most important speech information, 6 to 8 are stimulated sequentially for each sound. It is the slowest of the three strategies, with a maximum stimulation rate of 250 Hz. In the Continuous Interleaved Sampling mode, up to 12 active channels can be used, all of which are sequentially stimulated for all incoming sounds. The Advanced Combination Encoder strategy is a hybrid that combines features of the Spectral Peak and Continuous Interleaved Sampling strategies. Up to 20 channels carrying the most important speech information are selected as in Spectral Peak and are then stimulated at various rates.

Unique to the SPrint processor is a liquid crystal display that registers information about the status of the system. Independent controls for volume level, sensitivity level and programs are present. The SPrint also offers an autosensitivity noise-reduction paradigm to assist in background noise reduction as well as a lock feature utilized primarily to keep children from altering recommended settings. The case is made of plastic and can be powered by standard batteries. The headset has a separate microphone housed behind the ear and a transmitting coil that contains a magnet. Various accessories, including FM system cables, are available for use with the SPrint processor.

In 1998, the Esprit 24™ became the first commercially available ear-level speech processor in the United States. Like the SPrint model, it is compatible with both the straight and Contour electrode arrays but differs in several ways. First, the Esprit can only utilize the Spectral Peak coding strategy. Patients preferring Continuous Interleaved Sampling or Advanced Combination Encoder strategies or patients with high stimulation requirements may be unable to wear the ear-level processor. Also, the Esprit has the capacity to store two rather than four programs. Either or both can be programmed with the autosensitivity function. A single rotary control is present and may be used as either a volume or sensitivity control but not as both within the same program. Two high-power hearing-aid batteries drive the Esprit. An updated version of the ear-level Esprit processor, the G3, was released in the summer of 2001 and implements the Continuous Interleaved Strategy and Advanced Combination Encoder strategies in addition to the Spectral Peak strategy.

MEDEL (MEDICAL ELECTRONICS)

After more than 20 years of use in Europe, the MedEl device was introduced to the US market in 1994. The current version of the internal device, the Combi 40+™, entered clinical trials in the United States in 1997. The primary speech processor, the TEMPO+, is an ear-level device. The MedEl device received FDA approval in August 2001. Although the MedEl offers both body-worn and ear-level speech processors, it is anticipated that the body-worn speech processor may be discontinued as the newer electronic circuitry of the ear-level processor supports all of the features and functions of the body-worn processor.

Internal Components

The C40+ internal electrode array has three available configurations: the standard, the compressed, and the split array. The standard version is a straight electrode array with 24 electrode contacts designed to operate as 12 pairs. Wider spacing of the electrode contacts is designed to minimize channel interaction. The Combi 40+ array utilizes an extracochlear ball electrode placed under the temporalis muscle. The standard electrode, at 31 mm, has the deepest insertion depth of any available device. The Combi 40+ array is directly advanced into the cochlea via cochleostomy.

The compressed electrode design, like the standard version, has 24 electrode contacts as well as an extracochlear ball electrode. The compressed electrode has a smaller spacing between electrode contacts, resulting in a significantly shorter electrode lead. The compressed electrode design allows the full complement of contacts to be inserted when surgical insertion is partially compromised, as in incomplete ossification.

For cases of severe ossification or cochlear malformation, the split-array electrode design is available. In this design, the 24 electrode contacts are divided into two separate electrode arrays that are implanted independently of each other. Use of the split electrode requires the drilling of two tunnels into separate turns of the cochlea and is indicated for use when insertion of either the standard or compressed electrode array is contraindicated or will result in incomplete insertion.

The receiver/stimulator of the Combi 40+ is housed in a hermetically sealed ceramic case. The circuitry permits device telemetry and measurements of voltage compliance that assists in determination of proper functioning of the internal components of the cochlear implant. Another feature of the MedEl internal device is that MRI is permissible (up to 0.2 Tesla) without removal of the internal magnet.

External Components

Med-El is attempting to phase out the CIS PRO+™ body-worn speech processor in favor of the TEMPO+, the versatile ear-level speech processor. The TEMPO+ has a modular design enabling customization of the speech processor, based on the patient's choice. This speech processor has the capacity to store three programs, a rotary dial for control of sensitivity level, an on/off switch, and a three-way switch that can be programmed with three volume levels for each program. In a sense, the TEMPO+ speech processor stores nine separate programs, three customized programs each, with three volume settings. It is powered by three high-powered hearing-aid batteries in a completely behind-the-ear format. In the future, an accessory unit that allows use of a standard battery will be available. The accessory unit is connected to the speech processor via a wire and may be worn elsewhere on the body, increasing time between battery charges. The TEMPO+ speech processor is capable of using both of the speech-coding strategies utilized by the CIS PRO+ processor. Various accessories, including Frequency Modulation systems used to increase the signal-to-noise level of a speaker's voice, can be used with the TEMPO+ processor.

Speech-Processing Strategies

The MedEl cochlear implant system offers two speech-processing strategies. The MedEl version of the Continuous Interleaved Strategy (called CIS+) utilizes a Hilbert transform for envelope detection, which is thought to improve sound quality and improve patient performance as compared to other implementations of Continuous Interleaved Strategy. The Number of Maxima (N-of-M) strategy is also a high-rate-of-stimulation strategy that utilizes some elements of the Continuous Interleaved Strategy, but stimulates only the electrodes carrying the most important speech information to be stimulated. Of these two strategies, CIS+ is used by the majority of MedEl users.

PERFORMANCE OF DIFFERENT DEVICES

Device performance is measured by having the patient listen to recorded speech, consisting of lists of words or sentences, and repeating what is heard. It is difficult to compare performance among devices, so results are presented individually by device.

Clarion

Results for the Hifocus I electrode array were presented at the sixth International Cochlear Implant Conference in Miami, Florida *(4)*. Forty three postlinguistically deafened adults with the HiFocus I electrode, who were assessed after 1 month of wearing the Clarion system, had average scores of 38% on the Consonant-Nucleus-Consonant (single-syllable) word test and 72% on the Central Institute for the Deaf sentences in quiet, without visual cues. Of these 43 subjects, 19 (44%) scored 50% or higher on the Consonant-Nucleus-Consonant word test, a noteworthy performance after 1 month of wearing a cochlear implant.

Nucleus

Results of the Nucleus 24 Contour clinical trial were presented at the seventh International Cochlear Implant Conference in Los Angeles, California *(5)*. The clinical trial results for postlinguistically deafened adults after 3 months of wearing the Nucleus 24 are presented in Table 2. The score of 59.4% on the City University of New York Sentence test was obtained with noise present during testing.

MedEl

Results for the MedEl cochlear implant system in postlinguistically deafened adults after 6 months of wearing the device are presented in Table 3. The score of 60% on the Hearing in Noise Test (sentences) was obtained with noise present during the testing *(6)*.

In summary, results for all devices are quite impressive. However, enthusiasm for these scores must be tempered with the fact that these are average scores, meaning that some individuals performed better and some performed worse than the values indicated. This cannot be overemphasized during preimplantation counseling sessions.

Table 2
Nucleus 24 Contour Results at 3 Months of Use

Test	Score
City University of New York (CUNY) Sentences in Quiet	78%
City University of New York (CUNY) Sentences in Noise (10 dBHL SNR)	59.4%
Hearing in Noise Test (HINT) Sentences in Quiet	62.5%
Consonant-Nucleus-Consonant (CNC) Single-Syllable Words	38.4%

Note: This table lists the performance scores on standardized speech tests for patients implanted with the Nucleus 24 with Contour cochlear implant.

Table 3
Results of the MedEl Device at 6 Months of Use

Test	Score
Hearing in Noise Test (HINT) Sentences in Quiet	77%
Hearing in Noise Test (HINT) Sentences in Noise (10 dBHL SNR)	60%
City University of New York (CUNY) Sentences in Quiet	89%
Consonant-Nucleus-Consonant (CNC) Single-Syllable Words	44%

Note: This table lists the performance scores on standardized speech tests for patients implanted with the Medical Electronics cochlear implant.

Also, it should be pointed out that scores for children are more difficult to obtain as well as interpret and are thus not presented. However, it has been demonstrated repeatedly that children receiving implants at a young age and, given appropriate intervention, can potentially develop normal speech and language function using a cochlear implant.

CANDIDACY, PREOPERATIVE PLANNING, AND SURGERY

Indication

Candidates for cochlear implant are adults of any age and children as young as 12 months who have been born deaf or have later lost their hearing. They must have at least a severe bilateral sensorineural hearing loss and be unable to discriminate a significant percentage of speech using appropriately fit hearing aids. The definition of a significant percentage of speech continues to change as the results of cochlear implantation improve. Currently, candidates must be able to understand no better than 40% of open-set sentences in their best-aided condition.

Criteria for cochlear implantation in children have evolved substantially in recent years. The minimum age for implantation has decreased, wheres the acceptable level of residual hearing has increased *(7)*. Cochlear implants were once restricted to children who were essentially anacoustic, whereas those with even minimal amounts of residual hearing were considered borderline, and those with any open-set word recognition were not considered candidates. Experience has shown that children with more residual hearing often perform better with implants, and those with some measurable open-set speech-recognition ability prior to implantation perform better with cochlear implants than do children without residual open-set word recognition.

Medical and radiological criteria have been expanded to include children with significant cochlear abnormalities in addition to other substantial medical conditions. Given these expansions, it is more important than ever that each case be considered individually by an experienced cochlear implant team consisting of an otolaryngologist, audiologist, rehabilitation and educational professionals, psychologists, social workers, and any others deemed necessary for a specific case (8).

Current selection criteria include age older than 1 year, severe-to-profound bilateral sensorineural hearing loss, benefit from hearing aids less than that expected from cochlear implants, no medical contraindications to general anesthesia, appropriate motivation and expectations, and family and rehabilitation support for development of oral language, speech, and hearing, especially for pediatric patients. The only absolute contraindications are agenesis of the inner ear (Michel syndrome) and absence of the cochlear nerve.

Audiological Assessment

Audiological evaluation is the primary means of determining cochlear implant candidacy. Ear-specific auditory information in both the aided and unaided conditions must be obtained. Sometimes, obtaining this information may not be possible without therapeutic intervention and repeat visits to the implant audiologists, especially in children.

Evaluation of candidacy for implantation should include assessment of the patient's general health and ability to undergo general anesthesia. A complete medical history and physical examination should be performed along with appropriate laboratory tests and imaging studies. Children and adults with all types of concurrent medical conditions are being considered for cochlear implantation, and implant surgeons often find it necessary to consult with other specialists involved in patient medical management.

Otoscopic evaluation of the tympanic membrane should be performed. On examination, special attention needs to be paid to the status of the middle ear and any congenital malformations that might be associated with abnormal facial nerve position.

In addition to medical and audiological evaluation, assessment of family expectations and support, hearing aid history, and patient motivation must all be taken into consideration. For the very young child, educational situation, primary mode(s) of communication and compliance with the therapy process are important aspects to be considered.

Computed tomography (CT) is vital in helping the surgeon assess the feasibility of implantation in some candidates. This assists in ear selection and helps the surgeon to evaluate inner ear morphology, including patency of the cochlea. Congenital cochlear abnormalities do not necessarily preclude implantation, but need to be identified preoperatively in order to assist in surgical planning, including the selection of an optimal electrode array (9).

MRI is useful when internal auditory canals are less than 1.5 mm, to demonstrate the presence of the cochlear nerve; when the CT demonstrates questionable ossification of the cochlea; and when sclerosing labyrinthitis with soft tissue obliteration may not be accurately imaged by CT scan (the T2-weighted MRI demonstrates loss of the endolymph/perilymph signal).

Surgery

Experience with cochlear implant surgery has led to surgical techniques that can simplify the procedure and avoid complications. The physical characteristics of the implantable electronic package and electrode array vary among devices, requiring device-specific surgical techniques to minimize complications.

Under general anesthesia, using routine aseptic methods and facial nerve monitoring, surgery usually takes 1.5 to 2.5 hours and patients are discharged on the day following surgery. The first postoperative visit is at 1 week to check the wound-healing status. Approximately 4 to 6 weeks later, after flap edema has resolved, fitting and mapping of the speech processor begins *(10)*.

IMPORTANCE OF PROSTHETIC DESIGN

A review of the literature reveals that many issues have been and are being investigated with respect to the bioengineering of cochlear implant prostheses. As with any device that is implanted into the body, tissue tolerance is always a serious concern, especially when the device in question is intended for long-term implantation. Hence, proper prosthetic design necessitates the use of proven biocompatible materials.

The cochlea, approximately the size of a pea when fully developed, is a very delicate organ and is thus sensitive to insertion of the electrode array. The occurrence of insertion trauma has been well documented in the literature and continues to be taken into account when new electrode designs are proposed *(11)*. When residual hearing in the implanted ear is lost following cochlear implantation, it is often secondary to insertion trauma *(12)*. This has been a consideration in recent electrode designs seeking to conserve residual hearing in the implanted ear.

Undue permanent pressure from an implanted electrode array may induce cochlear erosion and undesirable tissue ingrowth within the scalae *(13)*. As ossification alters the flow of electric current within the inner ear; it can adversely affect performance of the cochlear implant. Electrode arrays must accomplish the primary function, stimulation of the spiral ganglion fibers, while minimizing pressure on intracochlear structures. Even in cases where ossification following cochlear implantation does not occur, excessive intracochlear tissue proliferation can still occur. Aside from altering the electrical stimulation properties of the electrode, bone and tissue growth within the cochlea make explantation and reimplantation difficult.

Although the primary goal of the cochlear implant is to produce auditory sensation by direct electrical stimulation of the spiral ganglion, it is difficult to control the path of the electric current produced by the electrodes of the cochlear implant array. Sometimes the electric current produced by the electrode array stimulates structures outside of the cochlea that is referred to as extracochlear stimulation. This often produces undesirable effects, the most common of which is stimulation of the facial nerve, resulting in twitching of the facial muscles and/or pain. Electrode arrays are designed to optimally stimulate spiral ganglion fibers while minimizing extracochlear stimulation.

Although the use of electric current to stimulate the auditory nerve fibers is convenient and practical, it also has the potential to damage surrounding tissue *(14)*. Care must be exerted to minimize this risk. The voltage and amperage required for optimal stimulation should be less than the values that would damage nerve cells. As each patient will invariably present with different current requirements for each electrode, both the programming system and the speech processor must be able to safely manipulate the parameters of electric stimulation.

Previously, a significant issue in the development of the internal component of the cochlear implant has been the inability to subject an implanted patient to an MRI because the magnetic field can damage the electronic circuitry of the receiver/stimulator. Also, receiver/stimulators that utilize a magnet to attract the magnet of the transmitter coil on the surface of the skin tend to produce a localized distortion, making the MRI unreadable in the vicinity of the implanted components. Device manufacturers have taken steps to ensure that patients are able to undergo MRI testing but MRI field strength must be limited and temporary explantation of the magnet may be required.

The growing tendency to implant children at younger ages poses a special problem. The child's skull steadily grows until maturity, so adequate length must be incorporated into the electrode array, enabling it to expand with the growth of the skull. Otherwise, risk of displacement of the electrode array is high (15). Currently available devices are engineered to accommodate somatic growth.

FOLLOW-UP CONSIDERATIONS

At the University of Miami Ear Institute, patients are seen approximately 1 week postoperatively by the surgical team at which time the incision and eardrum are inspected. The surgeon looks for any indication of complications (hematoma, skin flap problem, otitis media, infection, etc.). At about 4 weeks after the surgery, the patient returns to the clinic for a routine checkup of the implant region and postoperative vestibular assessment, if warranted. If the surgical team feels that the implant region has healed appropriately by this visit, the patient is seen by the cochlear implant team for the initial stimulation. This is the first time that the external components of the cochlear implant are fitted and activated. This is when the patient first hears with the cochlear implant.

Because the patient's neurophysiological and psychological responses vary rapidly during the first few weeks of listening with the cochlear implant, follow-up programming of the speech processor is critical. Patients are seen again 1 week following the initial stimulation. If the patient is judged to be adjusting appropriately to the cochlear implant at this second programming visit, the next visit is extended to 2 weeks after the second programming. If the trend continues and the patient progresses well, the next visit is extended further. The increase in time between sessions continues as long as the patient's programming and adjustment requirements remain relatively stable until a maximum of 3 months (for children) to 6 months (for adults) between appointments is reached. From this point, it is recommended that the patient continue programming appointments biannually. It is the experience of this clinic that the majority of patients are adequately served with this timeline of follow-up and programming.

Various factors may alter this follow-up routine. Patients who choose to receive an investigational device must agree to regular testing intervals required by the clinical trial. The most common evaluation timetable for clinical trials involves audiometric and speech recognition testing at 1 month, 3 months, 6 months, and 1 year after initial stimulation of the cochlear implant. As many patients live some distance from the cochlear implant center, we usually attempt to combine the testing and device programming into a single visit.

Follow-up for children is often more difficult than for adults. The less cooperative or less experienced with sound the child is, the longer it takes to program the device. A single scheduled programming appointment may prove to be grossly insufficient to complete the programming. If this is anticipated, multiple visits on consecutive days may be scheduled to allow time to complete the programming requirements.

Postoperative complications, such as undesirable facial stimulation or pain, caused by activation of any electrodes significantly increase programming time. For these patients, many visits may be required as advanced and sometimes aggressive fitting and programming strategies are used to meet the patient's special needs. Many patients can be adequately served with a predetermined follow-up protocol, but various circumstances can arise that necessitate deviations from the protocol.

PROGRAMMING STRATEGIES OF PRACTICING CLINICIANS

It is difficult to precisely describe the strategies employed by clinicians for programming and follow-up care because each individual clinician generally adopts his or her own style. However, many clinicians seem to share the habit of easing patients into listening with their cochlear implant. Compared to the relative silence experienced prior to obtaining the implant, the sounds provided by the device are initially overwhelming for most, if not all, patients. Very few patients are able to tolerate the full dynamic range on the day of their initial stimulation and stimulation is increased at successive sessions until the full range is comfortably reached.

Programming a map for a cochlear implant involves several steps. First, the listener must confirm auditory perception. Next, the lowest level of current representing the electric correlate of auditory threshold, the "minimum response level," must be established for each active electrode. Also, for each of the same electrodes, a maximum current level representing the loudest level acceptable to the patient must be determined. Finally, it is important that loudness be essentially equal across all electrodes.

Setting the minimum response level is relatively easy. Depending on the device being programmed, this level may be set to a 100% detection rate, a 50% detection rate, or a slightly subthreshold level. To obtain the minimum response level, a patient may simply report hearing the stimulus as the current is raised and lowered until a threshold value is obtained. Patients often report hearing even when no stimulus is being presented, resulting in inaccurate thresholds. To obtain true thresholds, our clinic has the patient count the stimuli presented to an electrode. Current is raised and lowered until the softest level at which the patient can accurately count the stimulus. Setting the minimal level too high may have adverse effects, such as increasing the level of background noise or producing an audible hum in the processor, whereas setting the minimal level too low may result in the inability to hear certain soft sounds. This is more difficult when patients have little sound experience and limited language.

Setting the maximum current level is more complicated than setting the threshold levels for several reasons. What constitutes a "maximum level" varies by device. In some cases, maximum stimulation is defined as the maximum comfortable loudness level. In other cases, the recommended setting for maximum current level is the most comfortable loudness level. Correctly setting the maximum level is vital because the stimulation algorithm utilizes it as a safety feature to prevent overstimulation by uncomfortably loud sounds. When the upper levels are set too loud, speech can become distorted and certain loud sounds may be uncomfortable. However, if maximum stimulation levels are set too low, speech may be perceived as too soft, with little variation in intensity.

Setting maximum current levels is difficult because a subjective judgment of loudness by the patient is required. There is much room for error in this task, as the reference point for loudness comparison is quite subjective. Many clinicians have the patient rank the loudness of a stimulus on a scale as the stimulus is raised, stopping when the target loudness is reached. A factor that helps speech to sound smoother and more natural is when the maximum stimulation levels are equalized in loudness for all active electrodes. This requires having the patient use the same criteria used to judge the maximum level for each electrode, and comparing among electrodes. This approach calls for subjective interpretation, which introduces another source of error. If the electrodes are not properly balanced, the quality of speech may be choppy, tinny, or hollow as certain electrodes are artificially emphasized compared to others.

A procedure that attempts to reduce subjectivity in setting maximum current levels employs the acoustic stapedial reflex, which is a quantifiable neuromuscular reflex mediated via the brainstem. Intense auditory stimulation results in a contraction of the stapedius muscle. This stiffens the ossicular chain, resulting in reduced compliance of the eardrum, which can be measured with commercially available equipment. This reflex is present in many cochlear implant recipients in response to adequate electrical stimulus. There is good correlation between the level of current required to elicit the reflex and behaviorally obtained maximum stimulus levels for experienced cochlear implant users *(16)*. Maximum stimulus levels obtained via electrical reflexes are judged to be of equal loudness across electrodes by experienced cochlear implant users *(17)*. Speech recognition performance in cochlear implant users when the maximum stimulus level is set with electrical reflexes is equal to or better than when subjectively set in most cases.

SPECIAL PATIENT POPULATIONS

As might be expected, application of clinical practice varies depending on patient population. Very young hearing-impaired children differ from adults in that they depend on their implants to develop spoken language and thus require more prolonged training. Although the vast majority of pediatric cochlear implant recipients are successful, results vary. However, both congenitally deaf and postlingually deafened children have developed hearing and oral language abilities using implants, with adequate follow-up therapy.

A special subpopulation within the pediatric group includes children who are implanted when they are older after coping with hearing loss for a number of years. Older children are often challenging because they are mature enough to form opinions about the decision to obtain a cochlear implant. Children whose first language is sign may initially accept the cochlear implant but then reject it during their teen years as peer pressure and self-image exert increased influence. It must be kept in mind, however, that children implanted as adolescence approaches have become accustomed to living as deaf persons and learning to use the tremendous amount of auditory stimulation provided by the cochlear implant is difficult. This is especially true if the child has not been a consistent hearing aid user and communicates primarily through sign language. It is necessary with adolescents to determine that the child desires the implant and is not being forced to accept it by parents. The older child or preteen should clearly understand the work involved and the device limitations.

Although the cochlear implant is an effective treatment option for profound deafness, to many deaf activists it represents unwanted technology that demeans and threatens their way of life. The American deaf community sees their way of life, based on the sole use of American Sign Language to communicate, as emotionally fulfilling, promising, and independent. Organized attempts to suppress the use of cochlear implants, particularly in children, occurred throughout the 1990s, based on concerns that, if a large number of deaf children received cochlear Implants, they would become part of mainstream society, and deaf society would be diminished (18).

There are ethical questions regarding cochlear implants. They concern who should make the decision for the child and the principles on which the decision should be based. Parents or court-appointed guardians have the legal right and responsibility to exercise free informed consent on behalf of their child, assuming that their decisions are based solely on the best interests of their child. As experience continues to demonstrate both the safety and effectiveness of cochlear implants, many deaf organizations are moderating their opposition to the implantation of these devices in children.

In general, the results of cochlear implantation in the elderly have been comparable with those of younger adults (19). In our institution, patients up to age 88 years have obtained excellent results, as evaluated by both audiological and quality-of-life measures. Perioperative attention to medical and surgical details allows for safe insertion and a minimum of postoperative complications. In the elderly, cochlear implantation involves a number of unique issues that can affect patient outcomes, including age-related changes in the auditory system, prolonged duration of deafness, diminished communication ability, and co-existing medical and psychosocial problems.

Elderly patients who receive a cochlear implant present with special needs that must be addressed. Many elderly patients live alone, which creates several problems. They often have difficulty understanding and recalling the information necessary for proper operation of the device. The elderly tend to have difficulties in manual dexterity that translates to problems in changing batteries, replacing broken wires, and even in adjusting the controls of the device. In some cases, the lack of manual dexterity may contraindicate the use of the miniaturized behind-the-ear speech processors. Living alone also means that the elderly individual may not be exposed to speech on a consistent basis. Progress with the implant may be less rapid when compared to individuals living with family or friends.

Experience suggests that elderly patients may have unrealistically high expectations for the cochlear implant, expecting it to restore hearing to normal. Extensive counseling is required regarding all of these issues.

Postmeningitis Deafness and Ossified Cochlea

Bacterial meningitis is a leading cause of acquired deafness in children. Histopathological studies on human temporal bones in ears deafened by meningitis reveal a marked reduction in spiral ganglion cells and at times labyrinthitis ossificans of the cochlear scalae.

Postmeningitic patients can be electrically stimulated but they often require progressively higher stimulation levels and higher programming modes over time compared with other patients *(20)*. In addition, this population is at increased risk of extracochlear stimulation, especially facial nerve stimulation. Generally, extracochlear stimulation requires the inactivation of electrodes, decreasing maximum stimulus levels and/or the use of less optimal programming, which may result in decreased performance of the cochlear implant *(21)*.

Postmeningitic deafness may feature labyrinthine ossification but, if neural elements are present, they can be stimulated with good hearing outcome. Postmeningitic children, particularly those with ossified cochleas, may need frequent programming adjustments to maintain performance. They need close follow-up for this reason. Changes may be related to progressive cochlear ossification *(20)*.

Malformed Cochlea

Profoundly deaf children suffering from bilateral inner-ear malformations have not been implanted in many cochlear implant centers until recently because of the complexity of evaluation, surgery, and rehabilitation. Malformation of the inner ear is no longer a contraindication and the outcome is generally good after cochlear implantation *(22–24)*, but abnormal anatomy of the cochlea and facial nerve has raised concerns regarding safety and effectiveness of electrode insertion. Surgical concepts have been developed and specific solutions to problems encountered have been described *(22)*.

In malformations consisting of incomplete partition (Mondini malformation), enlarged vestibule, and dilated vestibular aqueduct observed on CTs, routine cochleostomy and insertion of the electrode are accomplished. There is an increase risk of cerebrospinal fluid leak during the cochleostomy.

In cochlear dysplasia of a more severe degree, known as common cavity deformity, the modiolus is absent and the facial nerve may run in the lateral wall of the presumptive cochlea. In cases of common cavity deformity, the electrode array may be inserted into an opening created in the lateral recess of the common cavity, roughly corresponding to the expected position of the amputated limb of the lateral semicircular canal. Insertion in this area is performed in order to avoid the abnormal facial nerve. Cochlear endoscopy can be used to confirm electrode position, following which the entire cavity is packed with muscle.

Facial nerve monitoring is used primarily for safety reasons since atypical routing and split nerves may be expected. Also, it facilitates intraoperative identification of the nerve. Monitoring is useful for positioning of the electrode array and selecting the stimulation strategy after surgery. Intraoperative X-rays should be available in cases of atypical reactions to stimulation to ensure correct electrode placement *(23)*.

Often, the electrode placement is less stable due to the lack of anatomical structures that normally provide stability to the electrode array. Management of these patients typically involves more frequent follow-up as their electrical current requirements may fluctuate.

Revision Surgery

Athough histopathological studies of insertion trauma have shown iatrogenic damage to the stria, spiral ligament, and Organ of Corti, corresponding severe loss of spiral ganglion cells did not occur. Survival of ganglion cells, which are thought to be stimulated by cochlear implants, may in part explain the overall good results of reimplantation *(25)*.

Most reimplantations have involved upgrading a single-channel implant to a multi-channel device. Several authors have commented on the safety of reimplantation of multichannel devices. Hearing outcomes have been reported to be as good as or better than with the initial cochlear implant with a satisfactory complication rate *(26)*.

EXPECTATIONS AND RECOVERY STAGES

Hearing with the cochlear implant is different from normal so the patient must relearn how to process auditory information. The period of adjustment for cochlear implant users is highly variable, ranging from almost instantaneous speech understanding to a gradual improvement over time.

Cochlear implant users encounter predictable milestones. Simply discerning the presence of sound is the first. Later, the unfamiliar sounds become recognizable. Next, the patient is able to discern the source of environmental sounds without being told. At this point, the patient is able to tell that someone is speaking but is unaware of what is being said without visual cues. Later, the user is able to discriminate a limited set of words and phrases without visual cues. In essence, this is the beginning of speech recognition. The last stage is that of speech understanding without visual cues, to the point of being able to carry on a conversation, perhaps even over the telephone. Each patient progresses at his or her own individual pace; however, some patients never attain the ability to communicate without visual cues.

Unfortunately, although predictors have been proposed, we are unable to accurately predict how a patient will perform *(27)*. A critical aspect in management is extensive and repetitive counseling regarding expectations and hopes. This is an important goal of follow-up. Some patients who attest to understanding that there are no guarantees of performance with the cochlear implant prior to surgery may quickly become disappointed and even angry following surgery when they are unable to understand speech right away. It is beneficial for prospective candidates to talk to patients who have already had this surgery.

DELAYED COMPLICATIONS

Device failure is seen in 4 to 10% of patients *(28)*. This produces considerable psychological trauma. In addition, for children, it entails decreased or absent auditory stimulation when learning to hear is critical for language development *(23)*.

There are several ways in which a cochlear implant may fail to function properly, the most common of which involves the components of the external hardware. The wires that interconnect the external components are prone to damage from wear and tear, thus causing intermittency or even total loss of communication between parts. The microphone of the cochlear implant is sensitive to moisture, which can cause device failure due to electrical short-circuiting. The telecoil may become damaged, affecting its ability to transmit information to the internal components of the cochlear implant. The speech processor itself may be damaged, affected by moisture, or fail due to defects in the circuitry that are not immediately apparent during early fitting with the device. Although interruptions in the cochlear implant function secondary to external hardware issues are annoying, they can easily be corrected once the problem is indentified.

Loss of function secondary to failure of the internal device is usually more difficult to identify and may ultimately require explantation and replacement. Malfunction in newer devices can be more easily diagnosed utilizing device telemetry. If the electrode array is the site of damage, electrode impedance measurements may provide the required information for appropriate diagnosis. If the receiver/stimulator is faulty, clinical programming systems may be unable to read telemetry from the internal components, thus suggesting device failure.

Loss of function may also result from internal device migration in which some or all of the electrode contacts migrate out of the cochlea. Generally, this occurs over time during which clinical signs, such as changes in electrode impedance, presence of extracochlear stimulation, or increases in current requirements may appear. When this problem is suspected, radiographic examination is typically diagnostic.

Device integrity checks can also detect instrument failure. Some may be performed in the clinic using evoked potential systems set up to record the electrical signal produced by the cochlear implant. Interpreting the results of device integrity checks can be difficult and requires experience. The manufacturers of these implants provide technical support to perform in-depth integrity testing with their own proprietary testing equipment.

Fortunately, complications with cochlear implantation done by experienced surgeons have been infrequent *(29)*: failure of the implant electrode (1.6% of implantations); problems requiring revision surgery (4.3%) and all other major complications such as hemorrhage, long-lasting facial palsy, meningitis, or persistent perilymph leak (5.9%), for an overall major complication rate of 10.2%. This is similar to the figure of 9.5% quoted by Hoffman and Cohen *(30)*. A minor complication, by definition, is one that can be overcome by medical or audiological management and includes wound infection, nonauditory electrode stimulation, and so on. Summerfield and Marshall report the rate of minor complications to be 24% *(31)*.

Data compiled by Hoffman and Cohen indicated that direct surgical complications occurred in less than 2% of the cases *(30)*. The most common was skin-flap-related in 1.75% of the cases. Facial nerve injury at surgery occurred in 0.58%. Postsurgical electrode migration occurred in 1.31% and facial nerve stimulation in 0.94% of cases. Complication rates in children are similar to those in adults *(28)*.

Flap Complications

Among the most frequently encountered complications are those associated with the incision and postauricular flap. They continue to be among the most common complications, although their rates fell from 5.44% in 1988 to 2.79% in 1995. Flap necrosis, flap infections, and delayed healing are the major flap complications, with 55% requiring revision surgery *(23)*.

Flap necrosis and implant exposure may require explantation of the device. However, when limited exposure of a device occurs as a result of flap breakdown, coverage may be accomplished by rotation of a local flap or by anterior rotation of the implant. One adjunctive measure advocated to increase the viability of the flap in cases of compromise is hyperbaric oxygen.

Infection of an implant is not rare. A functioning multichannel implant is a substantial financial and personal investment but, if there is evidence of systemic infection such as meningitis, it must be removed *(33)*. It should not be removed if appropriate corrective surgery with vascularized flaps and intravenous antibiotics can solve the problem safely *(32)*. If conservative options are unsuccessful, the implant should be removed. If the infection is confined to the receiver/stimulator device, one may simply remove that portion of the device and cut the electrode as it enters into the cochleostomy and then close the skin incisions. Once the infection has resolved and the flap is appropriately vascularized and healed, reimplantation is an option.

Electrode problems are also common, but most are minor. Malfunctioning electrodes may be turned off and the device programmed to take full advantage of the remaining working electrodes. The most common reasons for electrode deactivation are facial nerve stimulation, poor sound quality, and pain *(34)*. Other reasons include absence of auditory stimulation, vibration, reduced dynamic range, throat sensations, absence of loudness growth (lack of increased perceptual loudness with increasing stimulation of electrodes), or dizziness.

SUPPORT GROUPS

Over the years, various organizations and support groups have been formed (*see* the Appendix for a list of support groups). The group now known as the Cochlear Implant Association (formerly known as Cochlear Implant Club International) is widely recognized.

The Network of Educators of Children with Cochlear Implants helps train educators who have children with cochlear implants in their classrooms via inservices and workshops. The American Academy of Otolaryngology–Head and Neck Surgery offers information at http://entnet.org. Many academic centers also offer information (e.g., www.cochlearimplants.org).

Other groups that deal primarily with hearing loss but also support implant patients include Self Help for the Hard of Hearing and the Alexander Graham Bell Society. A group that advocates a rehabilitative approach that has been very successful in children with cochlear implants is Auditory-Verbal International. The contact information for these organizations is listed in the appendix to this chapter. Many states and regions have smaller organizations and support groups, for example, and they may be found by contacting the local cochlear implant center.

The World Wide Web (WWW) can be a tremendous source of information regarding the cochlear implant. Each implant company has a dedicated website that may be accessed by both professionals and the public, and at their respective URLs: Advanced Bionics (http://www.cochlearimplant.com), Medical Electronics (http://www.medel.com), and Cochlear (http://www.cochlear.com). A privately run website called "Listen-Up!" offers an enormous amount of information, including links, stories of cochlear implant experiences, as well as hints and suggestions for new cochlear implant users and may be found at http://www.listen-up.org.

Another increasingly popular source of information is the online chat room where users are able to interact in real time by typing back and forth. Some links on the websites will direct users to such chat rooms, which can be beneficial as several people can talk and discuss issues simultaneously. A caveat is in order: all websites and chat rooms are essentially open forums, meaning that anyone can present information. All information obtained via the WWW should be verified by a professional as such information may be inaccurate, either mistakenly or even deliberately. Every professional who deals with cochlear implants has a horror story or two about incorrect information obtained by a prospective patient from the WWW.

There are several effective preventive measures that should minimize the incidence of deafness. All children should receive immunizations that are currently recommended by the American Committee on Immunization Practices of the Center for Disease Control and Prevention in Atlanta, Georgia. In brief, the immunizations that are currently recommended in children are the following: Hepatitis B, Polio, Diphtheria, Pertussis, Tetanus, *Haemophilus influenzae*, Measles, Mumps, Rubella and Varicella. Vaccination against *H. influenzae* type B is recommended at 2, 4, and 12–15 months of age. Meningitis caused by this organism usually occurs before age four and can cause bilateral deafness. Immunization decreases the risk of meningitis and otitis media owing to *H. influenzae*.

Screening for hearing impairment should be performed on all neonates at high risk for hearing impairment and should take place as close to hospital discharge as possible. The optimum time is 35 gestational weeks *(35)*. Many states require screening of all newborns prior to discharge from the hospital. High-risk children not tested at birth should be screened before age three but there is insufficient evidence of accuracy to recommend routine audiological testing of all children in this age group. There is also insufficient evidence of benefit to recommend for or against hearing screening of asymptomatic children beyond the age of 3 years.

Screening is not recommended for asymptomatic adolescents or adults not exposed routinely to excessive noise. Elderly patients should be evaluated for progressive loss of hearing associated with aging, counseled regarding the availability and use of hearing aids and referred appropriately when abnormalities are detected.

Cytomegalovirus and rubella virus infections in expectant mothers can be transmitted to the fetus and cause deafness. Prenatal visits are designed to screen for a variety of complications of pregnancy while educating the patient during the pregnancy. The initial first trimester visit should include a Rubella antibody screen. Because of the theoretical risk of transmission of the live virus vaccine, patients do not receive the rubella vaccine until postpartum. Women who are known to have low or nonexistent antibody levels should be advised to avoid anyone with possible rubella infections.

Hearing loss owing to noise exposure and ototoxic drugs can be minimized or prevented. Hearing loss from noise exposure is preventable through the use of protective equipment and auditory monitoring. Avoiding hearing loss caused by ototoxic drug administration is also feasible. The most common culprits are aminoglycoside antibiotics, followed by nonsteroidal anti-inflammatory agents, diuretics, and cancer chemotherapy agents (e.g., cisplatin, nitrogen mustards, and vincristine). Patients receiving ototoxic drugs can be monitored for hearing loss via high-frequency audiometry or otoacoustic emissions, which have been shown to be highly sensitive *(36)*. Evidence of neurotoxicity should then lead to changes in drug dose or use of alternative agents.

Prevention measures for implanted patients with profound bilateral sensorineural hearing loss may seem unnecessary. However, with the increasing use of implants for patients with some residual hearing, this becomes more important. Many researchers are working to regenerate ganglion cells or hair cells. With these advances, preventing trauma to the spiral lamina or to the basilar membrane during surgery will be one of the major preventive measures in the future *(11)*.

Table 4
Patient Surveillance After Implantation of Cochlear Implant at the University of Miami Ear Institute

	Postoperative year					
	1	2	3	4	5	10
ENT visit (otologist)	2	1	1	1	1	1
Vestibular tests	1	*	*	*	*	*
Cochlear implant programming (audiologist)	12	2	2	2	2	2
Computed tomography scan	*	*	*	*	*	*

Note: This table lists the number of times each year the modality is recommended in asymptomatic patients.
*If needed, depending on otological and audiological examination.

APPENDIX: CONTACT INFORMATION FOR SUPPORT GROUPS

Alexander Graham Bell Association for the Deaf and Hard of Hearing
3417 Volta Place, NW
Washington, DC 20007-2778
202/337-5220 (Voice)
202/337-5221 (TTY)
202/337-8314 (Fax)
http://www.agbell.org/

Auditory-Verbal International
2121 Eisenhower Avenue, Suite 402
Alexandria, VA 22314
(703) 739-1049 Voice
(703) 739-0874 TDD
(703) 739-0395 Fax
E-mail: avi@auditory-verbal.org
http://www.auditory-verbal.org/Contact.htm

Cochlear Implant Association, Inc.
5335 Wisconsin Ave, NW, Suite 440
Washington, D.C. 20015-2052
Phone: (202) 895-2781
Fax: (202) 895-2782
http://www.cici.org/

Self Help for Hard of Hearing People, Inc.
7910 Woodmont Ave - Suite 1200
Bethesda, Maryland 20814
301-657-2248 Voice
301-657-2249 TTY
301-913-9413 Fax
Email: National@shhh.org
http://www.shhh.org/

University of Miami Ear Institute
www.coclearinplants.org

REFERENCES

1. Djourno A, Eyries C, Prothèse auditive par excitation électrique a distance du nerf sensoriel a l'aide d'un bobinage inclus a demeure. Presse Medicale 1957;35:14–17.
2. Niparko JK, Wilson BS. History of cochlear implants. In: Niparko JK, Kirk KI, Mellon NK, McConkey-Robbins A, Tucci DL, Wilson BS, eds. Cochlear implants: principles & practices. Philadelphia, PA: Lippincott Williams & Wilkins, 2000, pp. 103–107.
3. Clark GM. Historical perspectives. In: Clark GM, Cowan RS, Dowell RC, eds. Cochlear implantation for infants and children: advances. San Diego, CA: Singular., 1997.
4. Osberger MJ. Results from the North American HiFocus™ Electrode studies. Presented at the sixth International Cochlear Implant Conference (CI 2000), Miami, 2001. Abstract available at: www.cochlearimplant.com
5. Parkinson AJ, Arcaroli J, Staller SJ, Arndt PL, Cosgriff A, Ebinger K. The Nucleus 24 Contour Cochlear Implant System: Adult Clinical Trial Results. Presented at the seventh International Cochlear Implant Conference (CI 2001), Los Angeles, 2001. Abstract available at: www.cochlear.com
6. Results with the Med-El. Presented at Otology 2000. Birmingham, England, June 2000. Abstract available at: www.medel.com
7. Zwolen TA, Zimmerman-Phillips S, Ashbaugh CJ, et al. Cochlear implantation of children with minimal open set speech recognition skills. Ear Hear 1997;18:240–251.
8. Balkany T, Hodges AV, Goodman KW. Ethics of cochlear implantation in young children. Otolaryngol Head Neck Surg 1999;121:673–675.
9. Woolley AL, Orser AB, Lusk RP, Bahadori RS. Preoperative temporal bone computed tomography scan and its use in evaluating the pediatric cochlear implant candidate. Laryngoscope 1999;107:1100–1106.
10. Balkany TJ, Cohen NL, Gantz BJ. Surgical technique for the Clarion Cochlear Implant. Ann Otol Rhinol Laryngol 1999;108(Suppl 177):27–31.
11. Balkany T, Yang, N, Eshraghi AA. Clarion, Nucleus and Combi 40+ perimodiolar electrodes: a comparative study of scalar position and tissue damage in human temporal bones. Abstracts of oral presentations of eighth symposium of cochlear implants in children 2001 (Los Angeles), 43.
12. O'Leary MJ, Fayad J, House WF, Linthicum FH Jr. Electrode insertion trauma in cochlear implantation. Ann Otol Rhinol Laryngol 1991;100(9 Pt 1):695–699.
13. Keithley EK, Chen M, Linthicum F. Clinical diagnoses associated with histologic findings of fibrotic tissue and new bone in the inner ear. Laryngoscope 108:87–91.
14. Shannon, R. A model of safe levels for electrical stimulation. IEEE 1992;39:424–426.
15. Tucci DL, Niparko JK. Medical and surgical aspects of cochlear implantation. In: Niparko JK, Kirk KI, Mellon NK, McConkey-Robbins A, Tucci DL, Wilson BS, eds. Cochlear implants: principles & practices. Philadelphia, PA: Lippincott Williams & Wilkins, 2000, pp. 189–221.
16. Hodges AV. Electrically elicited stapedius muscle reflexes: utility in cochlear implant patients. In: Cullington H, ed. Cochlear implants: objective measures. London, UK: Whurr Publishers, 2002.
17. Hodges AV, Balkany TJ, Ruth RA, Schloffman JJ. Equal loudness balancing using electrical middle ear muscle reflexes. Presented at the International Cochlear Implant, Speech and Hearing Symposium, Melbourne, October 24–28, 1996. Abstract available by request from Annelle Hodges, PhD, University of Miami Ear Institute.
18. Balkany T, Hodges A. Misleading the deaf community about cochlear implantation in children. Ann Otol Rhinol Laryngol 1995;166(Supplement):148–149.
19. Buchman C, Fucci M, Luxford W. Cochlear implants in the geriatric population: benefits outweigh risks. Ear Nose Throat J 1999;78(7):489–494.

20. Eshraghi AA, Telischi FF, Hodges A, Balkany T. Changes over time in programming mode of cochlear implant users. Abstracts. American Academy of Otolaryngology, Head and Neck Surgery (Denver) Annual Meeting 2001. www.coclearimplants.org.
21. Kelsall D, Shallop J, Brammeier T, Prenger E. Facial nerve stimulation after Nucleus 22-Channel cochlear implantation. Am J Otol 1997;18:336–341.
22. Luntz M, Balkany TJ, Hodges AV. Surgical technique for implantation of malformed inner ear. Am J Otol 1997;18:66–67.
23. Eshraghi A, Woolley A, Balkany T. Cochlear implant. In: Wohl D. Josephson G., eds. Complications in pediatic otolaryngology. Boca Raton, FL; Taylor & Francis, 2005, pp. 501–512.
24. Graham JM, Phelps PD, Michaels L. Congenital malformations of the ear and cochlear implantation in children: review and temporal bone report of common cavity. J Laryngol Otol 2000;Supplement 25:1–14.
25. Balkany TJ, Hodges AV, Gomez-Marin O, et al. Cochlear reimplantation. Laryngoscope 1999;109:351–355.
26. Gantz BJ, Lowder MW, McCabe BF. Audiologic results following reimplantation of cochlear implants. Ann Otol Rhinol Laryngol 1989;98:12–16.
27. Waltzman S, Fisher S, Niparko J, Cohen N. Predictors of postoperative performance with cochlear implants. Ann Otol Rhinol Laryngol 1995;104 (Supp 165):15–18.
28. Luetje CM, Jackson K. Cochlear implants in children: what constitutes a complication. Otolaryngol Head Neck Surg 1997;17:243–247.
29. Proops DW, Stoddart RL, Donaldson I. Medical, surgical and audiological complications of the first 100 adult cochlear implant patients in Birmingham. J Laryngol Otol 1999;24:14–17.
30. Hoffman RA, Cohen NL. Complications of cochlear implant surgery. Ann Otol Rhinol Laryngol 1995;Suppl166:420–422.
31. Summerfield AQ, Marshall D. Cochlear implantation in the UK 1990–1994. Report by the MRC Institute of Hearing Research on the Evaluation of the National Cochlear Implant Programme. Main report. HMSO 1995:91–198.
32. Rubinstein JT, Gantz BJ, Parkinson WS. Management of cochlear implant infections. Am J Otol 1999;20:46–49.
33. Daspit CP. Meningitis as a result of a cochlear implant: case report. Otolaryngol Head Neck Surg 1991;105:115–116.
34. Stoddart RL, Cooper HR. Electrode complications in 100 adults with multichannel cochlear implants. J Laryngol Otol 1999:18–20.
35. Eshraghi A, Francois M, Narcy P. Evolution of transient evoked otoacoustic emissions in preterm newborns: a preliminary study. Int J Pediatr Otorhinolaryngol 1996;37:121–127.
36. Ress BD, Sridhar KS, Balkany TJ, Waxman GM, Stagner BB, Lonsbury-Martin BL. Effects of cis-platinum chemotherapy on otoacoustic emissions: the development of an objective screening protocol. Otolaryngol Head Neck Surg 1999;121(6):693–701.

US Counterpoint to Chapter 16

J. Gail Neely

Chapter 16 on cochlear implants is well written and comprehensive. This commentary focuses on the primary mission of this textbook: the long-term care and follow-up of patients with implanted prosthetic devices.

Because the number of implants is very small and complications are fortunately rare, trials to evaluate methods to prevent or treat problems are not very feasible. Thus, the primary mechanism of evidence gathering is through case studies in the cumulative databases of postmarket surveillance cohorts maintained by the federal government and the cochlear implant industry and by reviewing the small numbers reported in the medical literature.

RESPONSIBILITY

Patients, *clients*, and *fiduciary* are important terms in medicine and health care. Merriam-Webster's Unabridged Dictionary Online traces the etymology of these words to Latin and Middle English *(1)*. Patient refers to those suffering or in want or need. Client refers to those who are dependent or "one who has someone to lean on." Fiduciary refers to that which is founded in trust or confidence. Human communication in medicine is a complex, dynamic, bilateral process, with expectations and fiduciary responsibilities incumbent on both the physician and the patient; this is, or should be, a good example of teamwork. In the context of implanted prosthetic devices, this two-way process involves instruction and the reception of instruction. It also involves surveillance and compliance.

The company often supplies some written instructions for some of the newer prosthetic devices; however, this method may be incomplete or discontinued for older, more established prosthetics. Written instructions from the physician are valuable; however, they are easily lost, not read, or are incapable of itemizing all the possible adverse events that can occur. Nevertheless, written instructions and good record-keeping become crucially important in the unfortunate event of an adversarial relationship.

In the best of worlds, both the physician and patient trust each other to inform the other of potential or actual problems, whether or not regular return appointments are scheduled. However, perhaps the most effective safeguard for the physician and for the patient is a continued surveillance/compliance regimen when a prosthetic device is implanted, notwithstanding current managed care confounders.

EARLY PROBLEM DETECTION

Methods to avoid early complications include the use of a compressive dressing and perioperative antibiotics, avoidance of nose blowing and positive pressure ventilation for several days, avoidance of the ipsilateral eyeglass ear-piece, avoidance of trauma to the wound, keeping it clean and dry until it is well healed, good control of co-morbidities such as diabetes mellitus, and the cessation of smoking. Surveillance for problems in wound healing (such as flap necrosis, subcutaneous air, and infections) are generally accomplished by 1 month postoperatively. Patients should be instructed to return sooner if any pain, swelling, discharge, fever, headache, irritability, or stupor occur. The potential for cerebrospinal rhinorrhea or meningitis, especially in children with major congenital abnormalities of the inner ear, must be kept in mind. Flap problems once were not rare; however, now they are very unusual. Tissue coverage of the implant is imperative.

Early device failure, wound compression sores, or unexpected biocompatibility problems of the implanted component, suture materials, or the external processor may not be apparent until the wound is healed and the hookup and mapping begins. The patient and the implant team audiologist must be alert to report to the surgeon any difficulties; even minor incision line or magnet area redness, tenderness, or ulceration can be extremely important.

LATE PROBLEM DETECTION

It is important to remember that implanted devices (a) are composed of metals with potential ferromagnetic properties or are actual magnets; (b) have embedded electronics sensitive to electromagnetic effects; (c) are composed of other alloplastic materials with potential delayed biocompatibility issues; (d) may initiate tissue changes, remodeling, and/or migration; and (e) may become a nidus for infection.

The newer metals in cochlear implants are not ferromagnetic; however, they do contain an implanted magnet and an antenna coil. Strong magnetic currents, such as in magnetic resonance imaging (MRI), are contraindicated because of the probability of severe displacement of the implant (2–4). Some implants allow the removal of the magnet, with the potential risk of introducing infection or damaging the hermetic seal about the implant. Theoretically, magnetic currents or other electromagnetic forces could even then generate electrical fields via the implanted coils and damage the implant and/or the nerve. The patient must be made aware of these contraindications and treating physicians must listen to the patient.

The embedded electronics and/or the adjacent eighth nerve may be damaged by electrocautery, microwaves, and ionizing irradiation. The patient, treating surgeons, interventional psychiatrists, and dentists must be aware that monopolar cautery and similar modalities, such as defibrillators and electroconvulsive therapy, are contraindicated. Obviously, defibrillation for immediate life-threatening emergencies takes precedence over the serious damage that can be done to the eighth nerve and implant. Shielded bipolar cautery may be used, if at a distance from the implant. Care should be taken to insure that any cautery equipment used anywhere in the body is truly bipolar and shielded. For more elective procedures, the implant can be removed if absolutely necessary; however, this may create a considerable risk of damage to the cochlea or to the reimplantation pathway on that side.

Delayed biocompatibility issues are extremely rare; however, the slightest break in the hermetic seal from trauma or migration stressors may create tissue damage due to the electrical charge, short-circuiting in the device, and malfunction. Rare instances of delayed painful shocks from the implant have occurred; the cause is usually not discovered. In these situations, removal and reimplantation of the same or another type of cochlear implant generally resolves the problem. In other rare instances, the suture material used to anchor the implant has penetrated the skin over the implant, creating the potential for infection; removal of the portion of the suture penetrating the skin has solved this problem.

Tissue is dynamic, of course, and might be more accurately thought of as a fluid rather than a solid. On the other hand, alloplastic materials are rigid. The result is remodeling of tissues, including the dynamics of scar formation and contractions, about the surgical defect, implant, and under the apposed external device, allowing possible implant migration and painless, delayed overlying skin ulceration. The patient must be ever vigilant to inspect the implant area for signs of thinning or ulcerating skin or progressive or sudden implant dysfunction. If any skin problems arise, the external device must be avoided until the skin has returned to normal; then, a weaker external magnet must replace the previous one.

Infections are always a possibility from direct extension through atrophic and/or ulcerated skin, from a middle ear infection, or from embolic seeding from distant infections or dental surgery. Immediate and effective treatment of infections, especially ear infections, and prophylaxis for dental surgery is recommended. Recent concern about meningitis in post-implanted patients as a result of middle ear infection has led to the recommendation that implanted subjects be immunized against pneumoccoci.

Things to avoid fall into the following categories: (a) personal behaviors, (b) environmental and occupational hazards, and (c) particular medical and surgical diseases and treatments. Personal behaviors that might put unusual stresses on the ventilation of the middle ear or that might damage the implant directly are wise to avoid. Rough contact sports and scuba diving are some examples. Unusually stressful nose blowing, especially during a sinonasal infection, would be wise to avoid. Environmental and occupational hazards, such as proximity to potential head-trauma-prone events, electromagnetic fields, or ionizing irradiation are wise to avoid. As mentioned above, medical and surgical events to avoid are MRI, head irradiation, monopolar electrocautery, electroconvulsive therapy, defibrillation or cardioversion, and direct incisions over the implant. If these must occur, the implant surgeon should be contacted if the procedure is not being done to treat an acute life-threatening condition.

SUMMARY

Patient surveillance after cochlear implantation at Washington University School of Medicine is summarized in Table 1. This cochlear implant program is partially funded by the National Institutes of Health with the specific objective of optimizing the

Table 1
Patient Surveillance After Cochlear Implantation at Washington University School of Medicine

	Postoperative year					
Office visit to a surgeon or an audiologist	12	1	1	1	1	1
X-rays (standard 2-D or CT)	1	0	0	0	0	0
Other tests (audiometry)	12[a]	1[a]	1[a]	1[a]	1[a]	1[a]

Note. Tests used in follow-up of asymptomatic patients who have recovered from the implantation procedure. Numbers in cells indicate the number of times each year that modality is recommended. 2-D, two-dimensional; CT, computed tomography.

[a] With vestibular/balance testing, if indicated by clinical findings.

implant; as such, the audiologists in the team carefully evaluate these patients indefinitely. Each surgeon may vary this slightly; however, the principal responsibility for continued surveillance for safety and efficacy is forever in the hands of the patient.

REFERENCES

1. Merriam-Webster Unabridged Dictionary Online. http://unabridged.merriam-webster.com/ 2003, March 22.
2. Graham J, Lynch C, Weber B, Stollwerck L, Wei J, Brookes G. The magnetless Clarion cochlear implant in a patient with neurofibromatosis 2. J Laryngol Otol 1999;113:458–463.
3. Teissl C, Kremser C, Hochmair ES, Hochmair-Desoyer IJ. Magnetic resonance imaging and cochlear implants: compatibility and safety aspects. J Magn Reson Imaging 1999;9:26–38.
4. Heller JW, Brackmann DE, Tucci DL, Nyenhuis JA, Chou CK. Evaluation of MRI compatibility of the modified nucleus multichannel auditory brainstem and cochlear implants. Am J Otol 1996;17:724–729.

European Counterpoint to Chapter 16

Issam Saliba and Bernard Fraysse

INTRODUCTION

Stimulation of tissues with electricity for medical purposes is not a new idea. Two thousand years ago, Roman physicians recommended the electrical discharge of the torpedo fish for the treatment of headache and gout. Interest in electrical methods of stimulating hearing began when Volta discovered the electrolytic cell in 1790. He inserted metal rods into his own ears and connected them to a circuit that produced approximately 50 volts. When the circuit was completed, he experienced a sensation which he described as a blow to the head; it was followed a few moments later by a noise that sounded like the boiling of a viscous liquid. Volta's unpleasant experience probably discouraged others. Few attempts to investigate the phenomenon were carried out over the next 50 years, but the sensation was always momentary and lacked tonal quality. Nevertheless, it was apparent that a sense of hearing occurred with an electrical stimulation of the inner ear. A great deal of experimentation was needed to determine how this could be used to restore hearing. Many questions are still unanswered. What is the best way to "code" the complexities of sound into electrical pulse? Does it matter where the electrodes are placed inside the inner ear? What are the effects of long-term electrical inner ear stimulation on the hearing nerves and brain?

Lundberg performed the first direct stimulation of the auditory nerve in a human during an operation in 1950. His patient became aware of noise. In 1957, Djourno and Eyries implanted an electrode attached to an induction coil in the head of a deaf person (*1*). They were able to transmit a signal to the electrode via a radio antenna on the outside of the body. The person heard sounds resembling the chirping of a cricket and was also able to recognize simple words. This experiment inspired others to investigate the possibility of using implanted prostheses to enable deaf people to hear.

Progress accelerated in the 1960s. There was continued research into the electrical stimulation of the acoustic nerve. A major advance was made when investigators learned that specific auditory nerves must be stimulated with electrodes in the cochlea in order to reproduce sound. William House implanted electrodes in three patients in 1961 (*2*) and all obtained some benefit. A few years later, an array of electrodes was placed in the cochlea, with satisfactory results. More people got implants in the 1970s, and continued research led to the development of a multichannel device (*3*). By December 1984, the cochlear implant was no longer deemed experimental and was

From: *The Bionic Human: Health Promotion for People With Implanted Prosthetic Devices*
Edited by: F. E. Johnson and K. S. Virgo © Humana Press Inc., Totowa, NJ

approved by the US Food and Drug Administration for implantation into adults. In 1990, approval was extended to the use of cochlear implants in children ages 2 and older. At present, speech processors are better than ever and miniaturized so they can be incorporated into a behind-the-ear hearing aid-like device.

Cochlear implantation is a safe and effective medical procedure for individuals who are severely to profoundly deaf and who have had minimal benefit from conventional hearing aids. They are one of the most cost-effective medical interventions and can have a major long-term impact on a recipient's quality of life. Children who receive cochlear implants in early childhood are about twice as likely to be in mainstream schools as those without implants. A cochlear implant in a child saves society up to $1 million in educational and assistive benefits over the lifetime of the child. The number of implanted patients continues to increase. Performance improvement is related to improvements in the processing strategy.

PREOPERATIVE ASSESSMENT

Postlingually deafened adults, prelingually deafened adults with oral communication capability, and prelingually deafened adults with progressive deafness are candidates for a cochlear implant. Any candidate should have had at least 6 months of experience with conventional hearing aids before device insertion. A cochlear implant should be considered for a child with bilateral profound or severe deafness as soon as the benefit from a conventional hearing aid is considered insufficient following at least 6 months of experience. It should be offered as soon as possible if the hearing loss is the result of meningitis because of the risk of eventual cochlear ossification, which is a technical problem at surgery.

We employ a multidisciplinary assessment of each patient and carefully look for contraindications by audiometrical, psychiatric, and neuroradiological studies. The clinical assessment consists of questions about the previous general medical history of the patient as well as the history of his deafness, plus a clinical examination. In children, it is important to look for anomalies of the external ear as well as anomalies of the branchial arches such as cysts and cervical fistulas.

The audiometrical assessment generally starts with an analysis of subjective tests, complemented by objective testing if a cochlear implant is likely to be warranted. The subjective audiometry determines the severity of the deafness. In children, it shows the perception thresholds and, according to the child's age, the intelligibility thresholds or the extrapolation of these thresholds. Generally, a cochlear implant can be proposed if there is a limited benefit with hearing aids with a speech discrimination at 65 dB without lip reading under 30% for children and 50% for adults. All candidates should have had at least 6 months of experience, with conventional hearing aids before implantation. Different tests are given to the children, according to their linguistic level.

The objective audiometry does not require the active participation of the patient but does require the child to be quiet. In some cases it is necessary to use medication, even general anesthesia. It generally consists of the following:

- Impedance testing, which allows detection of middle ear pathology.
- Evaluation of otoacoustic emissions; if they are present, we can tell that the auditory threshold is not superior to 30 dB of auditory deprivation. However, one must be careful because they are present in central deafness with cochlear integrity.

- Evaluation of auditory brainstem response; this is important in the determination of auditory thresholds for the very young children. The principal limitation of this test is that it does not explore the low frequencies. It does not give information about residual hearing until 1000 Hz. That is why a filtered and nonfiltered click at the 1000 Hz level are used.
- A promontory test is added only for adults; in children general anesthesia is necessary for the introduction of the electrode.
- Sometimes, electrocochleography and a vestibular examination are performed.

These tests allow us to confirm the importance of the auditory deprivation. They have prognostic importance and can point to problems such as central deafness or neuropathy. Other assessments are sometimes required:

- An ophthalmological assessment with fundoscopy to rule out retinitis.
- An electrocardiogram to rule out arrhythmia.
- Timed measurement of urinary protein excretion to rule out Alport syndrome.

Multiple handicaps are not necessarily a contraindication to implantation of a cochlear prosthesis but it is important to define them to be able to assess prognosis.

Genetic testing is often proposed to the patient and the family, particularly if the etiology of deafness is unknown, as a genetic cause exists in about 50% of instances of deafness. Among these, 30% are associated with other clinical anomalies and are part of a recognized syndrome. About 70% are nonsyndromic. In 2–3% of cases, a specific genetic site is known.

Preoperative vaccination against meningitis is planned:

- Antihemophilus in children under the age of 6.
- Antipneumococcus for children and adults, all ages.

Meetings and information are part of the preoperative assessment and are important for the patients. Adult candidates are introduced to volunteers who have had cochlear implant surgery. Candidates can discuss the process of implantation and observe what results they can expect for themselves. Similarly, parents of implanted children are introduced to the families of children who are candidates. Deafness associations can also play an important educational role.

For adults, the goals of the speech and language assessment are to explore communication abilities with and without lip reading, to eliminate important cognitive disorders, to provide information about technical aspects of the cochlear implant procedure, and to explain what to expect during follow-up.

For children, assessment starts with information on cochlear implants and follow-up. Next is a review of the history of the child's deafness, his or her psychomotor development, the child's acceptance of conventional hearing aids, his or her schooling and rehabilitation program, and the communication modes the family uses with the child. Evaluation of the communication abilities of the child focuses on speech abilities, speech perception, and language understanding. Different perception tasks, according to the child's age and linguistic level, are performed:

- Ability to form words and sentences with and without lip reading and conventional hearing aids is determined.
- A videorecording of mother and child is made to document the various behaviors of the child, including production of sounds.
- The quality of the voice and the articulation of phonemes are analyzed, when possible.

Table 1
Patient Surveillance After Implantation of Cochlear Implant at Purpan Hospital–Toulouse, France

	Postoperative year					
	1	2	3	4	5	10
Office visit	4	2	2	2	2	1
X-ray (Cochlear view)	1	0	0	0	0	0
Cochlear implant test and hearing test	6	3	3	2	2	1

For adults, the psychological assessment focuses on the expectations and motivations of the adult who wishes to receive a cochlear implant. It is designed to detect psychiatric disorders and unrealistic expectations. For children, it consists of a clinical interview with the parents to analyze their position concerning their child's deafness and their expectations about the cochlear implantation. In addition, we observe how the child relates to his or her parents and to strangers, as well as the child's general behavior, particularly any pathological elements. The psychologist emphasizes that the child will remain a deaf person, even if he or she receives an implant. The necessity of rehabilitation and the key role of the parents are also stressed.

The cochlear implant is performed under general anesthesia. The extent of hair shaving is variable, according to the incision used. Surgery employs facial nerve monitoring. We have done more than 280 cochlear implants, 48% in children. In the literature, many complications related to the incision are described. These include hematoma, infections of the operative site, dehiscence of the suture line, cutaneous flap necrosis, and extrusion of the implant (5,7–15). We use a vertical s-shaped retroauricular incision that seems to give the best results (16). In all cases, the incision must follow certain rules (17). We avoid an incision parallel to the free edge of the receiver/stimulator, avoid placing the musculoperiosteal flap incision directly over the cutaneous incision, avoid a wide cutaneous and periosteal dissection, and place the implant at least 1.5 cm away from the incision. We insert the electrode in the scala tympani through the round window visualized through a posterior tympanotomy. Almost all of our patients are hospitalized for 1 day. On the first postoperative day, a cochlear view X-ray is done to detect any defect in the electrode insertion in the cochlea and the dressing is changed. The patient is discharged if all seems well. Ten days later we remove the dressing, check the wound, and make an appointment with the audiologist to start the processor programming. If there is no complication, no computed tomography scan or other tests are needed. The patient's appointment with the surgeon occurs about 10 days after surgery, 2 months after the start of programming, and every 6 months for 5 years (see Table 1). We recommend that patients refrain from any contact sport without appropriate protection such as a helmet.

REFERENCES

1. Djourno A, Eyries C. Prothèse auditive par excitation électrique à distance du nerf sensoriel à l'aide d'un bobinage inclus à demeure. Presse médicale 1957;35:14–17.

2. Niparko JK, Wilson BS. History of cochlear implant. In: Niparko JK, Kirk KI, Mellon NK, McConkey-Robbins A, Tucci DL, Wilson BS, eds. Cochlear implants: principles and practices. Philadelphia, PA: Lippincott Williams & Wilkins, 2000:103–107
3. Clark GM. Historical Perspectives. In: Clark GM, Cowan RS, Dowell RC, eds. Cochlear implantation for infants and children: advances. San Diego, CA. Singular, 1997.
4. Summerfield AQ, Marshall DH. Cochlear implantation: demand, costs and utility. Ann Otol Rhinol Laryngol Suppl 1995;166:245–248.
5. Telian SA, El-Kashlan HK, Arts HA. Minimizing wound complications in cochlear implant surgery. Am J Otol 1999;20(3):331–334.
6. Gibson WPR, Harrison HC, Prowse C. A new incision for placement of cochlear implants. J Laryngol Otol 1995;109:821–825.
7. Cohen NL, Hoffman RA, Stroschein M. Medical or surgical complications related to the Nucleus multichannel cochlear implant. Ann Otol Rhinol Laryngol 1988;97(suppl 135):8–13.
8. Harris JP, Cueva RA. Flap design for cochlear implantation: avoidance of a potential complication. Laryngoscope 1987;97(6):755–757.
9. Clark GM, Pyman BC, Bailey QR. The surgery for multiple-electrode cochlear implantation. J Laryngol Otol 1979;93:215–223.
10. Parkins W, Metzinger SE, Marks HW, Lyons GD. Management of late extrusions of cochlear implants. Am J Otol 1998;19:768–773.
11. Ishida K, Shinkawa A, Sakai M, Tamura Y, Naito A. Cause and repair of flap necrosis over cochlear implant. Am J Otol 1997;18:472–474.
12. Cohen NL, Hoffman RA. Complications of cochlear implant surgery in adults and children. Ann Otol Rhinol Laryngol 1991;100(9 Pt 1):708–711.
13. Cohen NL, Hoffman RA, Strochein M. Medical or surgical complications related to the nucleus multichannel cochlear implant. Ann Otol Rhinol Laryngol 1998;97:8–13.
14. Kempf HG, Johann K, Weber PB, Lenarz T. Complications of cochlear implant surgery in children. Am J Otol 1997;18:S62–S63.
15. Webb RL, Lehnardt E, Clark GM, Laszig R, Pyman BC, Franz BKHG. Surgical complications with the cochlear multiple-channel intracochlear Implant: experience at Hanover and Melbourne. Ann Otol Rhinol Laryngol 1991;100:131v136.
16. Hampton SM, Toner JG. Belfast Cochlear Implant Center; the surgical results of the first 100 implants. Rev Laryngol Otol Rhinol 2000;121:9–11.
17. Miyamoto RT, Robbins AM, Kirk KI, Wagner-Escobar M. Aural rehabilitation. In: Hughes G, Pensak, eds. Clinical Otology (2nd ed.). New York: Thieme Medical Publishers,1997: 395–405.

17
Ossicular Implants

Adrien A. Eshraghi, Hessam Elfiki, Steven R. Mobley, and Thomas J. Balkany

INTRODUCTION

Middle ear function requires an intact, mobile tympanic membrane, an air-containing mucosa-lined middle ear space, and a connection between the tympanic membrane and the inner-ear fluids. Restoration of function to hearing-impaired patients is often feasible with modern surgical approaches.

Custom-sculpted autologous and homologous incus prostheses have been used as ossicular implants, although difficulty in sculpting can be a problem. Homograft incus prostheses have fallen into disfavor because of their short shelf-life and the fear of transmission of infectious agents, such as HIV and hepatitis *(1)*.

Synthetic prostheses avoid the problems of tissue processing and banking and also eliminate the risk of disease transmission seen in homografts. They also eliminate the time-consuming intraoperative sculpturing of autografts during surgery and other risks associated with autografts and homografts, such as adhesion formation and absorption.

An ideal middle ear implant for ossicular reconstruction should be biocompatible, readily available, technically easy to use, and provide excellent long-term hearing *(2)*. Over the years, a variety of materials have been used. Commercially available prostheses of polyethylene, polytetrafluoroethylene (Teflon), stainless steel, titanium, gold, ceramics, and glass have enjoyed varying degrees of popularity *(3–6)*.

In 1952, Wullstein was the first to use a biomaterial in reconstructive middle ear surgery. He implanted a columella into the middle ear for the reconstruction of the ossicular chain and, although the initial hearing results were good, the implants extruded. Later, polymers, such as polyethylene, Teflon, and silicon rubber (Silastic), were used but eventually abandoned because of their high extrusion rate.

In the 1970s, developments in materials science, cell biology, and reconstructive surgery gave rise to new substances for middle ear reconstruction and a better understanding of the interaction between these implant materials and the body *(2)*. Early versions of plastipore partial ossicular replacement prostheses (PORPs) and total ossicular replacement prostheses (TORPs) were introduced by Shea in 1974 but suffered from high extrusion rates. The use of cartilage interposed over the prosthesis platform has helped reduce the extrusion rate *(7)*.

From: *The Bionic Human: Health Promotion for People With Implanted Prosthetic Devices*
Edited by: F. E. Johnson and K. S. Virgo © Humana Press Inc., Totowa, NJ

Hydroxyapatite prostheses were introduced in 1984 after Grote and others found the material to be both highly biocompatible and infrequently extruded when applied against the tympanic membrane (2). Additional advantages include their osteoconductive nature and relatively low cost. The tissue integration properties of the more recent hydroxyapatite prostheses seem to have reduced the necessity for cartilage protection, but extrusion rates remain substantial.

In the 1990s, surgeons began to use composite hydroxyapatite prostheses because of difficulty with intraoperative shaft trimming and shaping. These prostheses typically have hydroxyapatite heads attached to shafts made from other materials, including plastipore or Teflon (8).

HAPEX, introduced in 1996, is a homogeneous composite of particulate hydroxyapatite and high-density polyethylene blended in a 40:60 ratio. More recently, porous coralline (from sea coral) hydroxyapatite has been used as a head attached to a HAPEX shaft for PORPs and TORPs (9).

BIOMATERIALS

Biomaterials currently employed are of two types: polymers and ceramics. Each has its own advantages and disadvantages with regard to biocompatibility, integration capacity, and surgical application.

An implanted material is regarded as "bioinert" if the body does not react at all with the implant material. It is "biotolerated" if the local tissues regard the implant as a foreign body but do not extrude it. It is "bioactive" if the local tissues actively integrate with the implant material, leading to a firm, stable bond between them (2).

Polymers can be fabricated into porous solids, which promote integration into tissue. They are widely used in TORPs (columella between the footplate and tympanic membrane) and PORPs (between the stapes superstructure and tympanic membrane). The early porous plastic implants were "biotolerated." The surface was not favorable for tissue integration, and an interface between the tympanic membrane and the implants was necessary. Even then, a high extrusion rate was reported. In recent years, research has focused on the development of polymers with properties favoring tissue integration, with the goal of using them in reconstruction of soft tissue (2).

Ceramics comprise the other major class of materials used for ossicular implants. They are bioinert (aluminum oxide) or bioactive (glass ceramics and calcium phosphate ceramics). The bioactive materials are generally used in otology. There are several types of glass ceramic and each reacts differently with the surrounding tissues. Glass prostheses can be difficult to shape, and long-term studies have shown that resorption occurs. This led Grote (2) to study calcium phosphate ceramics. There are many types of calcium phosphate, and he focused on hydroxyapatite, which is the mineral matrix of living bone tissue. Hydroxyapatite has proved to be a bioactive material that is integrated into native bone tissue without encapsulation.

The ceramic hydroxyapatite is a calcium phosphate-polymer material that is fabricated in a dense and porous form. The dense form conducts vibratory energy well, which is valuable for ossicular implantation. The interaction with both epithelium and connective tissue is excellent, and a direct bond is formed between the material and the host tissue. The advantage of the hydroxyapatite implant for bony reconstruction is that the material is readily available. Because of standard manufacturing procedures,

Table 1
Types of Ossicular Prostheses

Partial ossicular replacement: capitulum	Connecting tympanic membrane to stapes
Total ossicular replacement prosthesis:	Connecting tympanic membrane to footplate or open oval window
Incus replacement prosthesis:	Connecting malleus to stapes capitulum
Incus stapes replacement prosthesis:	Connecting malleus to footplate or open oval window
Incudostapedial joint prosthesis:	Connecting remaining long process of incus to stapes capitulum

resorption and remodeling are controllable and predictable. Long-term clinical studies in ear surgery have validated its biocompatibility and usefulness (2,10).

TYPES OF OSSICULAR PROSTHESES

To reconstruct defects in the ossicular chain, the surgeon may employ a columella or bridge the defect in the ossicular chain. For the columella there are two possible approaches. First, when the stapes superstructure is present, a short columella interposed between the stapes head and the tympanic membrane can be used in the form of a PORP. When the stapes superstructure is missing and a mobile footplate is present, a long columella connecting the mobile footplate with the tympanic membrane can be used in the form of TORP. Bridging defects in the ossicular chain requires that the handle of the malleus be integrated into the tympanic membrane to drive the ossicular chain and restore its lever mechanism (Table 1).

For the plastipore polyethylene ossicular implants, both the TORP and the PORP prostheses are commercially available. Shea and Emmett first reported successful restoration of hearing using them in a large series of patients. Similar success in other large series has been reported by others (7,11).

Either the columella or the bridge approach can be used with hydroxyapatite prostheses (2,10). To bridge a missing incus, an incus replacement prosthesis (IRP) can be inserted between the handle of the malleus and the mobile stapes capitulum or superstructure. The IRP has a body with a depression that fits on the stapes head. In some types there is a notch that adapts to the malleus handle. When a malleus and mobile stapes footplate are present, and the incus and stapes superstructure are absent, an incus stapes replacement prosthesis (ISRP) of dense hydroxyapatite can be used as a bridge between the handle of the malleus and the mobile footplate. The ISRP consists of a shaft with a diameter of 0.6 mm and a handle (2). Figures 1 and 2 represent examples of PORP and ISRP commonly used in ossiculoplasty.

INDICATIONS FOR IMPLANTS

Audiological screening should be performed in all children with delayed speech, failed school hearing tests, or suspected syndromic hearing loss. Currently, there is insufficient evidence to justify routine audiological testing of all children. Patients with congenital abnormalities of the head and neck region are frequently at risk for associated conductive or neural hearing loss. Often, the physical abnormalities are subtle and

Fig. 1. Example of PORP (Sheehy Pop, Medtronic Xomed, Inc.)

Fig. 2. Example of ISRP (Malleus cradle total prosthesis, Medtronic Xomed, Inc.)

great attention to detail must be utilized during the physical examination of these patients. Particular attention should be focused on the auricle and periauricular area, the eyes, and extremities. Any suspicion of syndromic malformation or hearing loss should lead to prompt audiological evaluation and referral to a specialist.

Patients with recurrent otitis media, with or without effusion, should be followed closely by clinical examination and audiometric evaluation in order to detect worsening of hearing. In patients with retraction pockets of the tympanic membrane, efforts should be made to improve middle ear ventilation. Compliant patients may respond favorably to eustachian tube exercises, such as repeated daily Valsalva maneuvers. In others, transtympanic ventilation tubes may be necessary to improve ventilation of the middle ear and prevent the sequelae of recurrent otitis media or cholesteatoma. Patients with chronic otitis media should have regular otological examination to detect early signs of cholesteatoma and prevent ossicular damage.

A cholesteatoma arises from trapped squamous epithelium in a retraction pocket, tympanic membrane perforation, or the temporal bone after trauma. The epithelium continues to grow in the area of the middle ear and mastoid part of the temporal bone. This results in a mass that can cause destruction of the ossicles and, in more severe cases, lead to facial nerve, inner ear, or central nervous system complications.

In patients without contraindications to general anesthesia, cholesteatoma should be treated surgically because medical therapy is not effective. The extent of surgery is based on a careful preoperative examination, including microscopic otoscopy, and, in some cases, computed tomography (CT) scans of the temporal bones. The primary goal of surgery is removal of all squamous epithelium from the middle ear and mastoid. The secondary goal is re-establishment of hearing. Often these goals require staged operations.

Elderly patients should be evaluated by audiometry to determine the presence of hearing loss and any conductive component that may be improved surgically. Presbycusis, a form of sensorineural hearing loss, is the most common type of hearing loss in patients age 65 and over. It often responds best to hearing aids. Some elderly patients have a conductive component of their hearing loss and may be surgical candidates. Ossicular dislocation and otosclerosis are examples of surgically correctable hearing losses.

The majority of patients with a perforated eardrum are potential candidates for tympanoplasty. A conductive loss greater than 45 dB and an erosion of the long process of the incus are indications that the patient may also be a candidate for ossicular reconstruction. However, there may be only a slight retraction pocket over the long process of the incus with very little conductive component to the hearing loss. During the surgery, if the long process of the incus is discovered to be friable or stiff, reconstruction of the ossicular chain is usually required.

Patients with cholesteatoma are frequently candidates for ossicular reconstruction, as the incus and head of the malleus are often removed, even if there was only a minimal air-bone gap before surgery (12). Other candidates for the procedure may present with an intact eardrum but have a significant conductive component of their hearing loss. This may be owing to a congenital malformation of the ossicles. It may be also seen in which reconstruction was delayed to a second stage.

A history of head injury, with distortion of the eardrum or healed fracture of the bony ear canal, may indicate ossicular discontinuity and the need for reconstruction also. In general, surgery is recommended when the ear is dry and clean. Preoperative management includes history, general physical examination, and microscopic examination of both ears. Audiometric evaluation includes tuning fork testing by the surgeon and air-bone-speech audiometry.

Active infection should be eliminated, if possible. Careful comprehensive explanations are important so that fully informed consent can be obtained. Preoperative cultures and CT scan are of questionable value (9). Surgical candidacy is based on the presence or absence of active drainage, infection, and hearing loss. The goals of surgery include a dry ear, intact tympanic membrane, and improved hearing.

Hesitancy to perform ossiculoplasty in children has been attributed to higher rates of recurrent infection, eustachian tube dysfunction, and the associated difficulty controlling middle ear mucosal disease and cholesteatoma. These factors potentially result in higher rates of extrusion or failure. However, success in pediatric tympanoplasty is now routine (13). There are few differences in overall success rates between adults and children. Children differ only in that they may be more likely to require postsurgical tympanostomy tube insertion to maintain a stable ear (14).

Brackman (11) and Smyth (15) reported excellent initial results with porous polyethylene TORPs and PORPs but Smyth's results declined over time. Both studies had extrusion rates from 7% to 11%. Smyth used cartilage interposition in only 20% of his cases. He also used a stricter criterion for surgical success: closure of the air-bone gap to less than 10 dB. Brackman et al. reduced their extrusion rate from 7% to 3.5% by using cartilage interposition and by decreasing the size of the platform of the prosthesis. Jackson (3) reported poor results with porous polyethylene TORPs and PORPs. Air-bone gaps were decreased to less than 20 dB in 49% of PORPs. He used either no cartilage or homograft cartilage between the drum and prosthesis.

Goldenberg (10) compared hearing results of hydroxyapatite, homograft bone, and plastipore prostheses and did not find significant differences among them. He also reported favorable hearing results with plastipore prostheses owing to better adjustment of the prosthesis length during surgery. He reported success rates with TORPs and PORPs of 54.5 and 75.2%, respectively. In his series of 2200 cases, Portmann (16) reported that homograft prostheses yielded better hearing results than even the most biocompatible allograft prosthesis.

FOLLOW-UP

The patient is discharged on the day of surgery and the mastoid dressing is removed the next day by the patient or family. The postoperative checks are at 1 and 4 weeks. An audiogram is performed at 3 months, at which time patients may begin to perform Valsalva maneuvers, blow their noses, and swim. Patients are instructed to return again 3 months later. Usually, annual visits are scheduled thereafter, depending on initial disease (Table 2).

During follow-up visits, the history is crucial, including the status of the contralateral ear. Drainage, eustachian tube function and degree of hearing loss are key historical factors. Microscopic examination is also important. It should be performed with the patient in the supine position. The status of the eardrum and the physiology of the ear canal skin are evaluated. The possibility of recurrence of cholesteatoma or the primary disease is assessed, and a CT scan or magnetic resonance imaging (MRI) of temporal bone is ordered, if needed.

DETECTION OF COMPLICATIONS

Early in the use of these ossicular prostheses, the head of the prosthesis was placed in contact with the tympanic membrane graft and subsequently extruded in a number of cases. To lower the incidence of this complication, many surgeons began interposing a thin sheet of autograft cartilage between the head of the prosthesis and the tympanic membrane or graft. Subsequently, the extrusion rate of TORPs and PORPs dropped significantly (18).

Acute otitis media, particularly in children, places patients at high risk of implant extrusion. Directed antibiotic therapy helps prevent this complication. Frequent follow-up visits after treatment of an acute episode are warranted. It is important to remember that all children should receive immunizations that are currently recommended. Vaccination against *H. influenza* type b, in particular, can decrease the incidence of otitis media in children.

Intraoperative technical problems can result in a perilymph fistula with severe or total sensorineural hearing loss. This can be detected and managed during follow-up visits. Postoperative vertigo can occur following ossiculoplasty but generally resolves in 1 to 2 days. Vertigo that does not resolve, or is associated with more severe symptoms, can be an indication of unrecognized inner ear trauma. If the reconstruction was difficult, then the possibility of perilymph fistula and re-exploration should be considered.

If the graft is normal and the ear is healed, but there has been no improvement in hearing, the cause may be simple dislocation of the prosthesis from the stapes head or malleus. Another cause of hearing loss may be the presence of Silastic sheeting between the stapes head and the prosthesis in cases where Silastic was used as a middle ear spacer. The most common etiology of a persistent significant (>20 dB) conductive hearing loss following ossiculoplasty is that the shaft of the incus-stapes prosthesis has been dislodged from the stapedial footplate.

PROSTHESES FOR STAPES SURGERY IN OTOSCLEROSIS

Otosclerosis is a type of osteodystrophy limited to the temporal bone. It can lead to hearing loss correctible by stapes surgery. Early attempts to correct hearing loss owing to otosclerosis were made in the 19th century but were ineffective and dangerous because they resulted in several cases of fatal meningitis *(5)*.

In the 1950s, when Shea first demonstrated that the fixed stapes could be safely removed and replaced with a prosthesis, the search for the ideal replacement prosthesis began. The aim is to establish conduction of sound across the middle ear into the cochlea. The prosthesis must attach securely to the incus and fit into the narrow oval window niche in order to transmit signals to the oval window and inner ear. The biomaterial of the ideal prosthesis should be well tolerated by the patient and nontoxic to the middle and inner ear.

Proper length of the prosthesis is critical to a successful surgical outcome. If too short, it does not transmit sound to the oval window effectively. If too long, it may cause inner ear damage leading to sensorineural hearing loss, tinnitus, and vertigo *(5)*. Currently available stapes prostheses are composed of stainless steel, platinum, or polytetrafluorethylene (Teflon) *(5,19)*. Stainless steel is nonreactive with host tissue and is very well tolerated by the middle ear at both the oval widow and the incus. However, most modern prostheses are made of nonferromagnetic material so that patients can undergo MRI studies, if necessary. Platinum is an easily malleable, nonmagnetic material used to attach a wire-piston prosthesis to the incus. Because of its malleability, it can be fashioned into ribbon. Platinum, typically used with a Teflon piston shaft, is well tolerated by the incus in its ribbon form. Teflon is used often for stapes prostheses because it is well tolerated by the local tissues.

A variety of implants have been used as stapes replacement prostheses. These differ in size, shape, and weight. The mass of the implant is important because it affects the transmission of sound, particularly at high frequency. The ideal stapes prosthesis should exert no pressure on the inner ear at rest and its weight should be similar to that of the

Fig. 3. Example of prosthesis for stapes surgery (Robinson cupped piston, Medtronic Xomed, Inc.)

normal human stapes *(20)*. The main differences among stapes prostheses are at the point of interaction with host tissue. Several designs allow the prostheses to connect to the incus. Of the prostheses commonly available today, the attachment to the incus consists of either a loop that surrounds the incus or a cup into which the lenticular process fits. Most loops consist of a wire that hooks onto the incus. It is tightened to secure the prosthesis (Schuknecht, House, McGee, and Fisch prostheses). Thin stainless steel and platinum ribbon are the most common types used. Some prostheses (e.g., Causse) have a loop or ring composed of Teflon that require opening the ring prior to placement. The other common styles utilized for contact of the prosthesis to the incus are the cup and bucket-handle designs, such as Robinson (Fig. 3), Shea, and Lippy prostheses *(5)*. Most commonly, these are made of stainless steel.

Almost all prostheses used today consist of a shaft that projects into the oval window niche. It is designed with a blunt end to prevent perforation of the oval window graft. Modern shafts are composed of Teflon, stainless steel, or both. The prosthesis shaft diameter ranges from 0.3 to 0.8 mm. Most are available in a variety of lengths, but some are manufactured at a standard length that allows the surgeon to trim them to the desired length.

The main indication for the stapes prosthesis is conductive hearing loss (>20 dB air-bone gap, negative Rinne with 512 Hz tuning fork) secondary to otosclerosis with adequate cochlear reserve and good speech discrimination. In cases of bilateral otosclerosis, stapedectomy is usually performed, first on the poorer hearing ear, leaving the better hearing ear to be addressed 6 to 12 months later. The risks of potential complications are discussed at length with the patient preoperatively. All patients must

Table 2
Patient Surveillance After Ossiculoplasty at the University of Miami Ear Institute

	Postoperative year					
	1	2	3	4	5	10
ENT visit (otologist)	2	1	1	1	1	1
Audiogram (audiologist)	2	1	1	1	1	1
Computed tomography scan	*	*	*	*	*	*

Note. This table lists the number of times each year the modality is recommended in asymptomatic patients.
*If needed, depending on otological and audiological findings.

be counseled on the nonsurgical option of hearing aids. When the diagnosis is strongly supported by the clinical and audiologic examination, CT scans are not indicated.

Surgical success following primary stapedectomy, defined as less than 10 dB conductive hearing loss at 6 to 12 months following surgery, should occur in 90% of patients. In contrast, revision stapedectomy carries an increased risk for cochlear hearing loss and is successful in about 50% of patients.

Follow-up employs the same regimen, as previously described for patients undergoing ossicular chain reconstruction, with a particular emphasis on avoiding straining, heavy lifting, or strenuous physical exercise for 3 weeks. The patient should not fly during this period.

Swimming is allowed after complete healing but diving is discouraged. The risks of barotrauma from flying or diving are explained to the patient in detail preoperatively. During follow-up visits, the patient is questioned about symptoms of hearing loss, vertigo, dizziness, or imbalance. An audiogram is performed 3 months after surgery.

Patients with ossicular implants may, of course, develop an infection elsewhere in the body. Patients with certain surgically placed prosthetic devices (such as heart valves) typically receive antibiotic prophylaxis when bacteremia is likely, but we do not recommend this for our patients with ossicular prostheses. The same holds true for patients undergoing colonoscopy or dental work; we do not treat our patients any differently than patients without an implant.

Many surgical complications are first detected during follow-up. Some are self-limited; others are life-threatening. Immediate postoperative complications following stapedectomy include persistent vertigo and middle ear infections. These can be early signs of a serous labrynthitis and have the potential to result in sensorineural hearing loss. Infections can lead to meningitis, so these patients should be treated aggressively with intravenous antibiotics. Unpleasant metallic taste sensations and oral dryness can occur when the chorda tympani nerve is manipulated or divided. These symptoms nearly always disappear within 3 months and the most important therapy is reassurance from the surgeon.

Late complications include profound sensorineural hearing loss and recurrence of the conductive deafness. The incidence of profound hearing loss is less than 1% and the exact etiology is unclear. When the conductive loss recurs, it is usually due to slippage of the prosthesis off the incus or out of the oval window. Occasionally, it is owing to erosion of the incus. Secondary perilymphatic fistulas usually are caused by barotrauma from flying, scuba diving, and the like. This is exacerbated when equalization is impaired by poor eustachian tube function. Barotrauma may displace the prosthesis through the graft, resulting in leakage of perilymphatic fluid around the prosthesis shaft. Mild but persistent disequilibrium and deterioration of speech discrimination scores call for prompt surgical exploration. Sudden, marked hearing loss is a surgical emergency. High-dose steroids (e.g., prednisone at 1 mg/kg) and broad-spectrum antibiotics should be initiated, along with immediate referral to a specialist. The prosthesis should be removed and the area of fistulization repaired with perichondrial graft. Patients are instructed to avoid strenuous activity for 3 weeks and to avoid further barotrauma. Additional information about ossicular implants can be obtained through the Internet (*see* the Appendix).

Hereditary otosclerosis is an autosomal dominant condition with incomplete penetrance. A positive family history is present in about 50% of patients. Medical therapies, such as phosphorus, thyroxin and calcium phosphate, have been used with minimal success. The value of sodium fluoride in the management of active otosclerosis remains controversial. Patients with childbearing potential should be counseled that hormonal changes, such as pregnancy, have been associated with progression of otosclerosis.

APPENDIX: WEB SITES FOR FURTHER INFORMATION

http://www.Berksent.com/reconstruction.html
http://www.cholesteatoma.org/earsurge271.html
http://www.amershamhealth.com/medcyclopaedia/volume%20VI%202/ossicular%20chain.asp
http://www.earsurgery.org/surgoto.html

REFERENCES

1. Slater PW, Rizer FM, Schuring AG, Lippy WH. Practical use of total and partial ossicular replacement prostheses in ossiculoplasty. Laryngoscope 1997;107:1193–1198.
2. Grote JJ. Biocompatible materials in chronic ear surgery. In: Brachmann D, ed. Otologic surgery. Philadelphia, PA: W.B. Saunders, 1994, pp. 185–200.
3. Jackson GC, Glasscock ME, Schwaber MK, et al. Ossicular chain reconstruction: the TORP and PORP in chronic ear disease. Laryngoscope 1983;93:981–988.
4. Gjuric M, Schagerl S. Gold prostheses for ossiculoplasty. Am J Otol 1998;19:273–276.
5. Slattery III WH, House JW. Prosthesis for stapes surgery. Otolaryngol Clin North Am 1995; 28:253–264.
6. Wehrs RE. Hydroxyapatite implants for otologic surgery. Otolaryngol Clin North Am 1995; 28:273–286.
7. Sheehy JL. Tympanoplasty: cartilage and porous polyethylene. In: Brachmann D, ed. Otologic surgery. Philadelphia, PA: W.B. Saunders, 1994, pp. 179–184.
8. Pasha R, Hill III SL, Burgio DL. Evaluation of hydroxyapatite ossicular chain prosthesis. Otolaryngol Head Neck Surg 2000;123:425–429.
9. Luetje II CM. Reconstruction of the tympanic membrane and ossicular chain. In: Baily BJ, ed. Head neck surgery otolaryngology. Philadelphia, PA: Lippincott Raven, 1998, pp. 2073–2082.
10. Goldenberg RA. Hydroxyapatite ossicular replacement prostheses. Results in 157 consecutive cases. Laryngoscope 1992;102:1091–1096.
11. Brackman DE, Sheehy JL, Luxford WM. TORPs and PORPs in tympanoplasty: a review of 1042 operations. Otolaryngol Head Neck Surg 1984;92:32–37.

12. Wehrs RE. Tympanoplasty: ossicular tissue and hydroxyapatite. In: Brachmann D, ed. Otologic surgery. Philadelphia, PA: W.B. Saunders, 1994, pp. 167–178.
13. Schwetschenau EL, Isaacson G. Ossiculoplasty in young children with Applebaum incudostapedial joint prosthesis. Laryngoscope 1999;109:1621–1625.
14. Daniels RL, Rizer FM, Schuring AG, Lippy WL. Partial ossicular reconstruction in children: a review of 62 operations. Laryngoscope 1998;108:1674–1681.
15. Smyth DL. Five-year report on partial ossicular replacement prostheses and total ossicular replacement prostheses. Otolaryngol Head Neck Surg 1982;90:343–346.
16. Portmann M. Results of middle ear reconstruction surgery. Ann Acad Med Singapore 1991;20:610–613.
17. Bayazit Y, Goksu N, Beder L. Functional results of plastipore prostheses for middle ear ossicular chain reconstruction. Laryngoscope 1999;109:709–711.
18. Emmett JR. Plastipore implants in middle ear surgery. Otolaryngol Clin North Am 1995; 28:265–272.
19. House JW. Otosclerosis. In: Hughes GB, ed. Clinical otology. New York: Thieme, 1997, pp. 241–249.
20. Causse J, Gherini S, Horn KL. Surgical treatment of stapes fixation by fiberoptic argon laser stapedectomy with reconstruction of the annular ligament. Otolaryngol Clin North Am 1993;26:395–416.

US Counterpoint to Chapter 17

J. Gail Neely

Chapter 17 on ossicular implants is well written and comprehensive. This commentary focuses on the primary mission of this textbook: the long-term care and follow-up of patients with implanted prosthetic devices.

Because the numbers of patients with ossicular implants is relatively small and complications are fortunately rare, clinical trials are difficult to carry out. Useful data is contained in the cumulative databases of postmarket surveillance cohorts maintained by the government and implant industry and in series from centers of excellence reported in the medical literature.

RESPONSIBILITY

Patients, clients, and *fiduciary* are important terms in medicine and health care. Merriam-Webster's Unabridged Dictionary Online traces the etymology of these words to Latin and Middle English *(1)*. Patient refers to those suffering or in want or need. Client refers to those who are dependent or "one who has someone to lean on." Fiduciary refers to that which is founded in trust or confidence. Human communication in medicine is a complex, dynamic, bilateral process, with expectations and fiduciary responsibilities incumbent on both the physician and the patient; this is, or should be, a good example of teamwork. In the context of implanted prosthetic devices, this two-way process involves instruction and the reception of instruction. It also involves surveillance and compliance.

Some written instructions are often supplied by the company for some of the newer prosthetic devices; however, this method may be incomplete or discontinued for older, more established prosthetics. Written instructions from the physician are valuable; however, they are easily lost, not read, or are incapable of itemizing all the possible adverse events that can occur. Nevertheless, written instructions and good record-keeping become crucially important in the unfortunate event of an adversarial relationship.

In the best of worlds, both the physician and patient trust each other to inform the other of potential or actual problems, whether or not regular return appointments are scheduled. However, perhaps the most effective safeguard for the physician and for the patient is a continued surveillance/compliance regimen, beginning when a prosthetic device is implanted, notwithstanding current managed care confounders.

Follow-up recommendations for ossicular prostheses are more dependent on the type of disease being treated than the fact a prosthesis was used. It is helpful to stratify the cases into three groups: (a) those operated on for cholesteatoma, (b) those operated on for otosclerosis, and (c) those operated on for other middle ear and mastoid disease.

Those operated on for cholesteatoma and otosclerosis should perhaps be followed annually forever because cholesteatomas may recur and otosclerosis often affects both ears and can be associated with progressive sensorineural hearing loss leading to deafness. Generally, other conditions leading to ossicular reconstruction are relatively stable after 1 year and often do not need surveillance indefinitely. Each surgeon may vary this slightly. The patient bears much of the responsibility for continued surveillance and should be educated to report problems promptly.

EARLY PROBLEM DETECTION

Methods to avoid problems include the use of perioperative antibiotics and the avoidance of blowing the nose or the use of positive pressure ventilation for several days, avoidance of the ipsilateral eyeglass ear-piece, avoidance of getting the wound traumatized, dirty or wet until the scalp wound is well healed, good control of co-morbidities such as diabetes mellitus, and cessation of smoking.

Additionally, care should be taken to avoid motions that might displace the middle ear prosthesis or that might result in increased intralabyrinthine pressure in the case of stapes surgery. For example, avoidance of riding in a truck or tractor, flying, jogging, and straining to lift or for bowel movement is wise.

Surveillance for problems in wound healing of the external auditory canal, tympanic membrane, and middle ear usually require office visits at about one week, one month, and three months. Surveillance 6 and 12 months later may be all that is required to assure a stable condition. However, because the inner ear is approached in stapes surgery, being alert to the function of the inner ear in hearing and balance is important. Serous labyrinthitis (presenting as vertigo with or without sensorineural hearing loss) may require a short course of steroids. "Reparative granuloma," a very rare but considerably more serious problem, may initially present like serous labyrinthitis, but the hearing loss rapidly becomes profound and dizziness persists; this requires surgical intervention, often without recovery of hearing.

LATE PROBLEM DETECTION

After the wound has healed, hearing is dependent on how well the middle ear is aerated, how well the prosthesis is attached, and how mobile it is. Hearing is usually not tested until about three months postoperatively, unless unusually good or bad. A tuning fork examination in the office usually gives some indication of the condition of both the middle and the inner ear.

Middle ear prosthesis problems may occur at any time, but are often encountered within the first 2 years and tend to fall into four categories: (a) displacement of the prosthesis, at either end of the prosthesis; (b) fixation of the prosthesis; (c) extrusion of the prosthesis through the tympanic membrane; and (d) inner-ear complications.

Displacement of the prosthesis may occur by the effects of gravity on the mass of the prosthesis, scar contracture, excessive insufflation of air in the middle ear by a severe sneeze or nose blowing, or by erosion of the adjacent ossicle to which the prosthesis is attached. It presents as a sudden or progressive conductive hearing loss and may require replacement of the prosthesis.

Fixation of the prosthesis may occur by hyperostosis of adjacent bone, tympanosclerosis, or regrowth of otosclerosis. It presents as a rapid or progressive conductive hearing loss and may require replacement.

Extrusion of the prosthesis through the intact tympanic membrane may occur any time an alloplastic material is adjacent to the drum without an interposed layer of cartilage. However, negative pressure in the middle ear with retraction of the drum about the prosthesis is the primary cause. Evidence of a retracted drum may require middle ear ventilation with a tympanostomy tube in an attempt to avoid prosthesis extrusion. Characteristically, these present as a slowly progressive hearing loss. Interestingly, extrusion may be discovered by physical examination only without any new presenting symptoms. In this case, the prosthesis may completely extrude without changing the hearing; this is because of a slowly progressive attachment of the vibratory tympanic membrane to the capitulum of the stapes.

Delayed inner-ear problems present as sudden, progressive, or fluctuating sensorineural hearing loss with or without constant or episodic imbalance or vertigo. The cause is often a prosthesis-associated fistula from the middle ear into the inner ear through the oval window or otic capsule in the case of a recurrent progressive cholesteatoma. Inner-ear complications are more common in cases in which the stapes was removed or fenestrated in order to place a stapes replacement prosthesis. Inner-ear symptoms usually require fairly immediate medical or surgical intervention. Air in the inner ear through a fistula can create these symptoms and occasionally might be seen on a high-resolution computed tomography. In the case of otosclerosis progression associated with inner-ear dysfunction, surgical intervention is not helpful; however, sodium fluoride may retard the progression.

Many of the older middle ear prostheses, especially stapes replacement prostheses, are ferromagnetic which constitutes a contraindication to magnetic resonance imaging (MRI; 2,3). Stapes prostheses are often considered contraindications for tympanometry during audiometry. This has not been confirmed; however, very few people wish to be associated with a patient problem after impedance audiometry.

Infection of the middle ear after placement of most ossicular replacement prostheses is generally just inconvenient; however, an infection can result in poorer Eustachian tube function and the generation of potentially displacing or fixing scar. However, middle ear infection in a post-stapedectomy or stapedotomy patient may be life-threatening. Infection may follow the prosthesis into the inner ear and result in a suppurative labyrinthitis and subsequent meningitis. Because of this, it is wise not to perform a stapes operation in an ear prone to poor Eustachian tube function or recurrent otitis media. This is a primary reason to delay stapes surgery in children. Immediate otological examination and care is indicated for a patient with a stapes-operated ear and middle ear infection. Occasionally, a severe external ear infection can inflame the middle ear; therefore, external otitis in post-stapedectomy ears requires immediate attention.

THINGS TO AVOID

Things to avoid fall into the following categories: (a) personal behaviors, (b) environmental and occupational hazards, and (c) medical and surgical diseases and treatments.

Table 1
Patient Surveillance After Ossicular Implantation for Cholesteatoma/Otosclerosis at Washington University School of Medicine

	Years post-implant (within stated year)					
	1	2	3	4	5	10
Office visit to surgeon	1	1	1	1	1	1
Audiometry	1	0	0	0	0	0

Note: Tests used in follow-up of asymptomatic patients who have recovered from the implantation procedure. Numbers in cells indicate the number of times each year that modality is recommended.

Table 2
Patient Surveillance After Ossicular Implantation for Other Middle Ear and Mastoid Disease at Washington University School of Medicine

	Years post-implant (within stated year)					
	1	2	3	4	5	10
Office visit to surgeon	1	0	0	0	0	0
Audiometry	1	0	0	0	0	0

Note: Tests used in follow-up of asymptomatic patients who have recovered from the implantation procedure. Numbers in cells indicate the number of times each year that modality is recommended.

Personal behaviors that might put unusual stresses on the ventilation of the middle ear or that might damage the implant directly are wise to avoid. Rough contact sports and scuba diving *(4)* are some examples. It would be wise to avoid forceful nose blowing, especially during a sinonasal infection. Situations with the potential for causing head trauma are obviously hazardous. It is prudent to avoid strong electromagnetic fields or ionizing irradiation. MRI should be avoided if the prosthesis is ferromagnetic. Recent studies have shown that even the same prostheses from the same manufacturer may vary in their response to strong magnets; this is usually not a major problem with the newer materials.

SUMMARY

Patient surveillance after ossicular implantation at Washington University School of Medicine is summarized in Tables 1 and 2. As mentioned in the introduction, follow-up recommendations for ossicular prostheses are more dependent on the type of disease being treated than the fact a prosthesis was used. It is helpful to stratify the cases into three groups: (a) those operated for cholesteatoma, (b) those operated for otosclerosis, and (c) those operated for other middle ear and mastoid disease.

REFERENCES

1. Merriam-Webster Unabridged Dictionary Online. http://unabridged.merriam-webster.com/ 2003, March 22.
2. Williams MD, Antonelli PJ, Williams LS, Moorhead JE. Middle ear prosthesis displacement in high-strength magnetic fields. Otol Neurotol 2001;22(2):158–161.
3. Syms AJ, Petermann GW. Magnetic resonance imaging of stapes prostheses. Am J Otol 2000;21:494–498.
4. House JW, Toh EH, Perez A. Diving after stapedectomy: clinical experience and recommendations. Otolaryngol Head Neck Surg 2001;125:356–360.

European Counterpoint to Chapter 17

Goetz Geyer

RECONSTRUCTIVE MIDDLE EAR SURGERY

The principles of reparative and reconstructive microsurgery on the middle ear are identical throughout the world. The first step in chronic middle ear disease is control of the disease process. The second step is restoration of middle ear function, either immediately or after approximately 1 year, depending on the intraoperative findings. Substantial regional variation is observed as to the choice of materials employed for implantation. Thus, synthetic materials that are preferred in many Anglo-Saxon countries (including the primary authors of this chapter) have failed to become popular in Germany. In German-speaking countries, the use of synthetic materials is restricted to placement of platinum ribbon Teflon prostheses in otosclerotic surgery. Despite its biocompatibility and stability after implantation, dense hydroxyapatite has not gained acceptance because of the virtually simultaneous development of ionomeric cement implants and the advent of the currently popular titanium ossicles.

Allogenic ossicles have a limited shelf life. They have the advantage of being very inexpensive and their rate of integration in the middle ear tissue nearly equals that of autogenous ossicles *(1)*. Many authorities believe that the continued use of allogenic ossicles should be abandoned because some residual doubt persists as to the potential risk of infectious disease transmission. In the hands of a skilled otological surgeon the time spent intraoperatively for sculpting of (autogenous or allogenic) ossicles is insignificant. When the long process of the incus is placed to rest on the footplate as a columella, there is a danger that, on slight dislocation, the columella will become tightly adherent to the niche walls.

In the early 1950s, Wullstein *(2)* was the first to insert a columella made from a dental synthetic material (Palavit) between the tympanic membrane and the stapes footplate. Its early extrusion and the rejection of the Supramid implants introduced by Kley *(3)* somewhat later have been factors leading to the abandonment of synthetic materials in German-speaking countries. These failures also provide a likely explanation for the reservations of German otological surgeons toward utilizing composite synthetic materials such as Hapex for reconstructing the chronically diseased middle ear.

The biocompatibility of hydroxyapatite ceramics is undisputed. The surgical technique adopted by Grote *(4)*, for example, which is usually done as a two-stage tympanoplasty after complete healing of the ear, permits direct attachment of the prosthesis endplate to the undersurface of the drum. In this context, the osteoconductive properties of hydroxyapatite are irrelevant to ossicular reconstruction (they rather tend to have an adverse effect) and only come into play if there is no relative movement between the implant and the implant site. Reck *(5)*, using surface–active glass ceramic, has demonstrated integration of the material into the tympanic membrane. However, despite reports of excellent integration, implants may become extruded even in well-aerated middle ears owing to micromovement. The improved tissue integration properties that have been described for hydroxyapatite should have a favorable effect on the rate of extrusion. Despite this improvement, reinforcement of the drum, using a cartilage transplant, for example, is of major importance.

In contradistinction to the United States, the use of polymers as bone substitutes in the middle ear has not found wide acceptance in Germany. A number of ceramic materials (e.g., aluminum oxide ceramics) have attained only regional importance. Surface-active ceramics, which show excellent biocompatibility but lack biostability, have proved to be unsuitable for implantation in the middle ear. Composite ionomeric cement, although providing first-class biocompatibility, biostability, and workability, has been withdrawn from the market owing to problems with handling when in the liquid state. Middle ear implants made of titanium currently are a popular choice *(6,7)*.

In contrast to the concept of the primary authors of this chapter, in Germany the term columella is used exclusively to indicate a prosthesis that is inserted between the drum cover/malleus handle and the footplate. Apatite prostheses are premolded, yet require trimming to adapt them to a particular anatomic situation. This procedure, applied to the fragile hydroxyapatite ceramic, for example, is quite demanding surgically.

When performing surgery for cholesteatoma, general anesthesia may be of benefit but is not always required. After management of the pathologic process, the current standard of practice is to restore the continuity of the ossicular chain without delay. The implant material inserted in the chronically diseased ear should not be subject to degradation even if transient acute otitis media occurs. The patient is instructed that, in order to achieve improved hearing, a second surgical intervention may need to be performed. The initial implantation of a prosthesis may be delayed or, in the event of prosthesis dislocation after immediate placement, correction may be required at a later date.

Reinforcement of the drum cover with a cartilage transplant has led to improved functional and surgical results in the majority of cases for each type of prosthesis used. The anatomic results obtained after inserting partial ossicular replacement prosthesis (PORPs) are comparable for allogenic and autogenous implants *(4)*, but the functional results after use of PORP transplants (usually autogenous ossicles) are superior to those achieved with alloplastic materials. This is not surprising in that more favorable audiological results can be expected in ears with only minor ossicular destruction and the attendant better eustachian tube and middle ear function. Alloplastic implants (e.g., ionomeric cement or titanium), which usually need to be inserted in ears with extensive destruction of the ossicular chain and severe middle ear disease, yield less good clinical and functional results *(8)*. Total ossicular replacement prosthesis (TORPs) show a tendency to become firmly adherent to the adjacent vestibular walls.

After removal of the head dressing 1 to 2 days postoperatively, audiological testing of inner-ear function is done in conjunction with daily tuning fork examination and wound inspection. Approximately 3 weeks following surgery, the ear canal pack is

Table 1
Patient Follow-Up After Insertion of Ossicular Implants at the Department of Otolaryngology, Head and Neck Surgery, Municipal Hospital, Solingen

	Postoperative year					
	1	2	3	4	5	10
Office Visit (with ear microscopy)	4	1	1	1	1	1
Computed tomography or magnetic resonance imaging	*	*	*	*	*	*
Pure tone audiogram[a]	1	1	1	1	1	1

Note: The numbers in each cell represent the number of times the modality is requested during each 1-year time period.
*Obtained if recurrence of cholesteatoma is suspected or other indication arises.
[a] Using tuning forks with 512 and 1024 Hz frequencies.

removed and a pure tone audiogram obtained. Further follow-up measures are performed at the scheduled intervals (Table 1).

If dizziness occurs following a tympanoplastic procedure, loosening the pack in the ear canal will usually relieve symptoms. Persistence of the vertigo, which is frequently associated with spontaneous or positional nystagmus, is an indication for early revision to prevent lasting labyrinthine damage.

For stapedectomy/stapedotomy, platinum ribbon Teflon or titanium prostheses have replaced those made of stainless steel. Gold implants, although perfectly contoured, have not proved to be of value in stapes surgery because hearing loss after insertion can occur. The source of this is still unclear. With the advent of Teflon stapes pistons, the risks of revision surgery have diminished. Compared to the previously popular prostheses made of steel, the hazard of direct attachment or adhesion of the implant material to cutaneous labyrinthine structures has decreased. Revision surgery should preferably be undertaken with the patient under local anesthesia so that the procedure may be interrupted when indicated (e.g., if vertigo should occur). An audiological evaluation (examination of bone conduction) is done on the first day after surgery. The first threshold audiogram is obtained 1 week postoperatively, when the pack is removed from the auditory canal. Erosion of the long process of the incus constitutes one of the most frequent late complications following revision surgery. In the patients operated on by the author, slippage of the prosthesis has been distinctly uncommon.

Measures conducive to stabilization of the surgical result include perioperative antibiotic prophylaxis. In German-speaking countries, opinions concerning the need for perioperative prophylactic measures during otological surgery are conflicting. The range is wide, encompassing those who advocate antibiotic-free management (e.g., in stapedectomy/stapedotomy, a clean procedure) and proponents of 1-week antibiotic coverage (e.g., in chronic otitis media, a clean contaminated procedure). Prior to surgery, our stapedectomy/stapedotomy patients are given an antibiotic active against staphylococci (e.g., Cefuroxim) as a one-shot prophylactic measure. If exacerbation of chronic otitis media has occurred, it is recommended to institute topical treatment preoperatively using, for example, 3% hydrogen peroxide. Particularly when alloplastic middle ear prostheses are to be implanted, this should be supplemented by culture and sensitivity testing and administration of an appropriate systemic antibiotic.

In order to maintain the surgical result over the long term, the practice of inspection under otomicroscopic control during the postoperative period has become firmly established. The various procedures used are largely comparable, although slight individual variations do exist *(9,10)*. Granulation tissue, which mainly develops in cavities after radical incisions, is pared down. Depending on the individual situation, this procedure may have to be repeated several times to encourage rapid epithelialization, thus forestalling secondary infection. The otologic surgeon also removes crusts and cauterizes granulation tissue. Application of Aureodelf® ointment promotes healing by softening tenacious pack residues. Alcohol-containing gauze strips using 0.02% Dequalinium solution with alcohol are helpful in disinfecting and swabbing the operative area. Depending on the results of culture and sensitivity testing, ears showing prolonged discharge are treated with appropriate antibiotic-impregnated gauze strips. In persistently draining ears, placement of gauze strips soaked with Cefuroxim (active against *S. aureus*) and Cefsoludin sodium (active against *Pseudomonas* species, although not approved for this indication) has proved to be effective.

By obtaining a threshold audiogram immediately after pack removal, the attending otologist is able to assess middle ear function and to verify the correct position of the implant. Development of granulation tissue, if any, is caused by minor epithelial deficits as they may occur with intact canal wall procedures or elevated membranes.

Clinical scientific documentation and quality assurance measures require documentation of important events. We have developed a standardized form to record intraoperative findings *(11)*. This allows for the collection of data essential for scientific evaluation and quality assurance, irrespective of the surgeon's assessment of the situation in his operative report. Data sheets are further refined and updated as new scientific insight develops. A meaningful intraoperative videoprint, photographs or videotapes illustrate a particular intraoperative situation to the otologic specialist who is responsible for postoperative follow-up.

POSTOPERATIVE SURVEILLANCE

The intraoperative documentation is supplemented by a checklist that records otologic follow-up data in triplicate *(11)*. It emphasizes to patients the need for regular meticulous care (removal of cerumen, ablation of crusts in cases of defective meatal self-cleaning) to sustain a good operative result. The checklist also aims at ensuring regular attendance of all patients, both those dissatisfied with their operative result and others having perfectly healed ears. This facilitates a realistic assessment of all surgical results.

The otolaryngologist responsible for further follow-up forwards the original document to the hospital or, if follow-up is done in the hospital, it is filed in their records. The second copy is kept in the otolaryngologist's office. The third copy remains with the patient, together with the follow-up checklist. In close cooperation with colleagues in practice, it thus becomes possible to exchange information pertaining to the surgical results and relevant changes in otologic status. This will spare the patient the inconvenience of possibly having to travel longer distances to the clinic for follow-up visits.

In the first year, routine checks are limited to four visits and to one office visit during each of the subsequent years (Table 1). In the first few weeks after removal of the meatal pack, the situation may arise, particularly in cavities after radical surgery, that two to three visits a week are necessary until complete epithelialization is identified, and such cases do not necessarily signify a complicated course. Depending on the drainage function of the cavity and/or external meatus, and on accumulation of cerumen and epithelial proliferation, regular visits (four to five per year) to the otologist's office for aural cleaning may be required. One pure tone audiogram done at yearly intervals appears to be sufficient, a definitive hearing result being obtainable in our experience at 9 months postoperatively. After surgery for improved hearing (in otospongiotic-type disease), it is standard practice to do an audiogram immediately after surgery and a second one toward the end of the first postoperative year. For routine checks, imaging techniques and blood tests can be dispensed with.

Follow-up screening is well accepted by patients, as shown by their regular attendance in the Solingen clinic or the local otolaryngologist's office. These visits are apt to promote the patient–doctor colloquy and to convey the impression that the patient is well attended. Detailed documentation forms are currently utilized in the Solingen Ear Nose and Throat (ENT) Department and in the ENT Clinic of Würzburg University. Adoption of a similar type of documentation is recommended at congresses and seminars in otological surgery. Attempts are undertaken to achieve some degree of standardization in intraoperative documentation and postoperative follow-up, thus allowing better comparability of results among otologic surgeons.

In his follow-up evaluation, the otologist inquires about intermittent ear discharge, impaired or unstable hearing, vertigo, tinnitus, and so on. On otomicroscopic inspection, the otologist is able to assess the condition of the grafted tissue. Quite frequently, when the drum cover is transparent, the otologist can also evaluate the proper position of the implant. Further aspects warranting examination are the width of the meatal opening and epithelialization of the external auditory canal and radical cavity. The finding of a drum retraction or retraction pockets in the cavity indicates the ventilatory situation in the middle ear spaces.

A constant hearing ability suggests that prosthesis position has remained unchanged and that the implant material has good biostability. A positive Valsalva maneuver is a favorable prognostic indicator of middle ear ventilation. However, a normally aerated middle ear is observed in many instances independently of the outcome of this test. A threshold audiogram is obtained to supplement and corroborate the clinical examination.

Imaging techniques do not contribute much to evaluating the position of the prosthesis but they are indispensable when a tumor is suspected or the extent of a cholesteatoma invading the petrous apex needs to be defined in an ear with otherwise preserved hearing.

A constantly discharging ear strongly suggests an exacerbation or recurrence of chronic otologic disease. A foul-smelling exudate in particular strengthens the suspicion of new cholesteatoma growth. If, in a patient operated on for cholesteatoma, complete removal is in doubt, a second-look operation done about 18 months later contributes to eradicating residual cholesteatoma (in most cases cholesteatoma "pearls"). Definitive aural healing can then be assumed.

Independent of the implant material used, a functioning Eustachian tube and middle ear favors a good audiological outcome. To improve this desirable result, material properties, workability, design of the prosthesis endplate, contact area of the shaft to the stapes head or the footplate, low weight material for better transmission of higher frequencies, and rate of overgrowth by middle ear mucosa have to be critically tested. Titanium meets all the requirements for such a modern middle ear implant. Improvement for the integration of prosthetic devices might be achieved by coating the material surface with bioactive substances, for example, bone morphogenic protein (BMP 2) *(12)*. A socket joint between endplate and shaft of the prosthesis could postpone a possible migration of the endplate through the tympanic membrane when a slight dislocation of the implant occurs.

It is estimated that 15,000 middle ear prostheses are implanted each year in Germany. They are fabricated from various human and nonhuman materials. Comparative prospective studies are clearly desirable, but the differing surgical techniques in use make this difficult. Irrespective of the material used, the experience and skill of the operator, together with appropriate utilization of a specific material, are important determinants of outcomes *(13)*. Publication of comparative multi-institution prospective trials could provide an incentive to otologic surgeons to adopt a surgical technique or a specific material that in all probability offers optimal results.

Further information on the topic of middle ear surgery and implants (not only for experts) is available on the website of the Solingen clinic (www.klinikumsolingen.de, especially www.klinikumsolingen.de/hno). Questions sent by e-mail (geyer@klinikum solingen.de) are answered immediately. At present, no toll-free telephone numbers or support groups exist in Germany that may provide patients with information concerning middle ear problems.

REFERENCES

1. Hildmann H, Karger B, Steinbach E. Ossikeltransplantate zur Rekonstruktion der Schallübertragung im Mittelohr. Eine histologische Langzeituntersuchung. Laryngorhinootologie 1992;71:5–10.
2. Wullstein H. Die Tympanoplastik als gehörverbessernde Operation bei Otitis media chronica und ihre Resultate. Proc Fifth Internat Congress Oto-Rhino-Laryngol 1953;104.
3. Kley W. Probleme der Tympanoplastik. Z Laryng Rhinol Otol 1955;34:719–726.
4. Grote JJ. Reconstruction of the ossicular chain with hydroxyapatite prostheses. Am J Otol 1987; 396–401.
5. Reck R, Störkel S, Meyer A. Langzeitergebnisse der Tympanoplastik mit Ceravital-Prothesen im Mittelohr. Laryng Rhinol Otol 1987;66:373–376.
6. Schwager K. Titan als Gehörknöchelchenersatzmaterial. In vivo Untersuchungen im Mittelohr des Kaninchens und Untersuchungen zur Proteinadsorption am Implantatmaterial. Habilitationsschrift, Universität Würzburg, 1998.
7. Geyer G. Materialien zur Rekonstruktion des Schallleitungsapparates. HNO 1999;47:77–91.
8. Geyer G, Rocker J. Ergebnisse der Tympanoplastik Typ III mit autogenem Amboß sowie Ionomerzement und Titanimplantaten. Laryngo-Rhino-Otologie 2002;81:164–170.
9. Kley W. Nachbehandlung und Nachsorge nach hörverbessernden Operationen. HNO 1988; 36:175–180.
10. Schmelzer A, Hildmann H. Nachbehandlung nach Ohroperationen. Laryngo-Rhino-Otol 1999;78:103–106.

11. Müller J, Schön F, Joa P, Freudensprung H, Geyer G. Würzburger Ohrdokumentationsbogen (Erhebungsbogen, Nachschaubogen), Universität Würzburg 1990.
12. Jennissen HP. Accelerated and improved osteointegration of implants biocoated with bone morphogenetic protein 2 (BMP-2). Ann NY Acad Sci 2002;961:139–142.
13. Schuknecht HF, Shi SR. Surgical pathology of middle ear implants. Laryngoscope 1985;95: 249–258.

18
Vascular Prostheses

Karthikeshwar Kasirajan, Brian Matteson, John Marek, and Mark Langsfeld

INTRODUCTION

The progression of atherosclerosis is often relentless and, despite the numerous treatment options, recurrent disease is common. Vascular prosthetic implants are prone to failure, resulting in significant morbidity, cost, discomfort, and inconvenience for the patient. Surveillance protocols of vascular prostheses (grafts, stents, stent-grafts, etc.) have been developed to identify graft-threatening lesions (e.g., stenosis, aneurysmal degeneration) before the onset of graft failure. There is ample scientific evidence that vascular prosthetic surveillance is a clinically useful and cost-effective tool after most conventional vascular prostheses are placed. Percutaneous procedures have not yet been demonstrated to be superior or even equivalent to open surgical techniques, as measured by long-term results, but the minimally invasive nature of percutaneous interventions has led to significant reductions in early morbidity and mortality rates. Hence, percutaneous endovascular interventions are particularly attractive in selected patients. There are separate surveillance procedures for each type of therapy. The costs of the various procedures/methods for following these vascular patients must be considered. We recognize that reimbursement rates for follow-up tests vary among regions. Our recommendations are based on sound conservative medical practices and not on reimbursement patterns.

GENERAL CONSIDERATIONS

In many instances, the use of autogenous material (i.e., leg or arm vein) is preferred for vascular reconstruction. For patients requiring lower extremity bypass, particularly in the below-knee location, patency rates for autogenous conduits are far superior to those obtained using any prosthetic material (1,2). Occasionally, all veins have been previously harvested, necessitating the use of prosthetic grafts. In the aortic location, prosthetic implants are the first choice for reconstruction owing to the size of the aortoiliac system and the lack of a comparable autogenous conduit. Prosthetic patch material is often used for carotid surgery because it is easy to use, as durable as vein, and avoids additional incisions to harvest vein. Finally, prosthetic conduits in trauma situations, for visceral vessel and upper extremity reconstruction, and for extra-anatomic bypasses have all been shown to be effective. In 2001, it is estimated that

From: *The Bionic Human: Health Promotion for People With Implanted Prosthetic Devices*
Edited by: F. E. Johnson and K. S. Virgo © Humana Press Inc., Totowa, NJ

approximately 400,000 prosthetic grafts and patches were implanted for vascular reconstruction *(3)*. In addition, there has been an explosion in the number of "less-invasive" stents placed in both arterial and venous locations and endovascular grafts are now commonly used for repair of abdominal aneurysms.

The body's response to implantation of prosthetic graft material is well documented. It includes the complex interaction between various blood and tissue components (inflammatory mediators) and the implanted foreign body, which, under normal conditions, results in healing with eventual graft incorporation. Both proliferative and migratory mechanisms result in an ingrowth of smooth muscle cells (SMCs) and endothelial cells from the native artery to the attached graft *(4,5)*. This ingrowth of cells, however, usually only occurs several centimeters from the anastomosis. In addition, the perianastomotic endothelial cells are phenotypically altered, leading to a chronic injury state, with eventual SMC proliferation. This continued SMC proliferation (myointimal hyperplasia) can contribute to stenosis/occlusion and lead to graft failure *(6)*.

The luminal surface of the graft not lined with endothelial cells is covered with a protein-rich pseudointima composed mainly of fibrinogen, platelets, and polymorphonuclear leucocytes. Eventually, equilibrium is established between the flowing blood and the protein-rich lining of the graft. Also, in some graft materials, a transmural ingrowth of mesenchymal cells occurs. Gradually, the graft becomes incorporated by connective tissue and is securely attached to the surrounding tissue *(7,8)*.

When less-invasive techniques, such as stent placement, are used for focal stenotic lesions, they are usually preceded by balloon angioplasty. Balloon expansion of a stenosis causes a tear in the plaque and disruption of the intimal elastic lamina. Thrombus forms on the fractured plaque, with eventual ingrowth and proliferation of myofibroblasts *(9,10)*. This results in intimal thickening that can contribute to restenosis. Remodeling of the injured artery is inevitable. Stent placement prevents narrowing from remodeling, but intimal hyperplasia can lead to restenosis.

Our knowledge of the body's response to endovascular grafts placed for aneurysm repair is limited, but it appears to differ from that observed in surgically implanted grafts. The graft essentially forms a new lining for the aneurysm, and the aneurysm sac and its contents are left in place. In a study of patients treated for abdominal aneurysms, who had their endovascular grafts explanted, there was minimal tissue incorporation into the grafts despite tight fixation *(11)*. The characteristic ingrowth of endothelial and SMCs from the attachment sites was absent. Also, several of the grafts lacked any pseudointima. However, in another study of patients who had endovascular grafts placed for occlusive disease, a true neointima of the graft–artery interface was consistently present after 3 months *(12)*. Clearly, there are local environmental factors that affect the healing of endovascular grafts. We can assume that healing of these grafts is not the same as in surgically implanted devices. More research needs to be done to characterize the healing process (or lack of healing) and therefore optimize endovascular graft construction to promote graft incorporation.

Once implanted, there are several reasons why grafts may fail. Failure of a graft within 30 days of implantation is usually due to a technical error, hypercoaguable state, or inadequate outflow. Graft failure from 30 days to 2 years is usually the result of an obstructing lesion such as myointimal hyperplasia in the graft or adjacent artery *(13)*. Graft failure after 2 years is most commonly due to progression of atherosclerosis *(14)*.

Structural failure of grafts, such as infection, graft dilation, or pseudoaneurysm formation occurs in less than 10% of cases and usually occur years after implantation. Because of these multiple potential problems, graft surveillance protocols must be followed.

Patients with prosthetic implants are susceptible to development of prosthetic infection. The consequences of a graft infection can be devastating with loss of life or limb. Fortunately, they are uncommon. The majority of prosthetic infections occur from graft contamination in the perioperative period *(15–17)*. Patient factors that may play a role in the development of graft infection include bacteremia resulting from remote infection or immune compromised states. Patients requiring reoperative surgery or emergent procedures also have a higher infection rate. Careful attention to meticulous operative technique and gentle handling of tissues is important. Maintaining sterile conditions and avoidance of contact between prosthetic material and skin, as well as ligation of all vessels and lymphatics to avoid hematomas and lymph collections, are important techniques to minimize wound complications and avoid graft contamination. The perioperative use of prophylactic antibiotics is also important to minimize the chance of developing a wound infection *(18)*. Early postoperative graft infections (less than 4 months after implant) are commonly caused by more virulent organisms, such as *Staphylococcus aureus*. Patients usually present with evidence of sepsis, such as fever, leucocytosis, and local signs of wound infection. Late infections (more than 4 months after implant) are usually caused by less virulent organisms, such as *Candida* species or *Staphylococcus epidermidis (19)*. *S. epidermidis*, which is commonly found on the skin, produces an extracellular glycocalyx called slime, which protects the bacteria from host defenses and antibiotics *(20)*. These bacteria may remain dormant on the prosthetic device for years and never cause clinical problems; however, this organism is often cultured from grafts explanted for pseudoaneurysm development or graft thrombosis *(21)*. It is also the most common organism responsible for aortic graft infection, which is usually diagnosed years after implantation *(22)*.

Other mechanisms besides direct graft contamination that may cause a graft infection include bacteremia and mechanical erosion of the graft through skin or into bowel or uninary tract. Avoidance of bacteremic seeding in the perioperative period is especially important. Immunocompromised patients, or those patients with remote infection, should not receive a prosthetic graft unless absolutely necessary. Parenteral antibiotics should be administered immediately preoperatively and 24–48 hours postoperatively to avoid bacteremic contamination from sources such as intravascular catheters and Foley catheters. Culture-specific antibiotics should be used to treat any concomitant soft tissue or other remote infections, such as pneumonia, ischemic or diabetic foot infection, and so on.

As the implanted graft heals in the body, development of a pseudointima lessens the vulnerability to bacterial contamination. Nonetheless, bacterial seeding can occur even years after implantation, so aggressive treatment of local or systemic infections is mandated. Patients with prosthetic grafts, like patients with cardiac valves, are traditionally treated with antibiotics prior to dental surgery. There is little data supporting the use of antibiotic prophylaxis in patients with grafts undergoing invasive procedures, such as colonoscopy. Because the incidence of graft infection is so low, it is difficult to prove an association between bacteremia caused by dental work or cytoscopy and the development of a prosthetic graft infection. Much more important is the status of the host's

Table 1
Recommended Antibiotic Prophylaxis for Invasive Procedures

Procedure	Antibiotic
Dental: extractions, gingival procedures, implants	Amoxicillin 2 g 1 hour preprocedure (for penicillin allergy or unable to take oral:clindamycin 600 mg iv 30 minutes before)
Respiratory: tonsillectomy, surgery on respiratory mucosa, rigid bronchoscopy	Amoxicillin 2 g 1 hour preprocedure (for penicillin allergy or unable to take oral:clindamycin 600 mg iv 30 minutes before)
Esophageal: sclerotherapy of varices, stricture dilation	Amoxicillin 2 g 1 hour preprocedure (for penicillin allergy or unable to take oral:clindamycin 600 mg iv 30 minutes before)
Gastrointestinal: endoscopic retrograde pancreatography with biliary obstruction, biliary tract surgery, colon surgery or colonoscopy	Ampicillin 2 g iv plus gentamicin 1.5 mg/kg iv 30 minutes before procedure, and ampicillin 1 g iv 6 hours after
Genitourinary: prostate surgery, urethral dilation, cystocopy	Ampicillin 2 g iv plus gentamicin 1.5 mg/kg iv 30 minutes before procedure, and ampicillin 1 g iv 6 hours after

Adapted from the Sanford Guide to Antimicrobial Therapy 2002.

immune response. It is currently believed that the most effective way of preventing graft infection is avoidance of contamination at the initial implantation. Once the graft has been implanted, aggressive treatment of any infection is mandated. For patients with significant dental disease, prosthetic implementation should be done after the dental work is completed *(23)*. Because the development of a graft infection has such potentially devastating consequences, we recommend prophylactic antibiotics for any patient who has a vascular graft and is undergoing a bacteremia-producing procedure. This is especially true for endografts, as we do not yet know their long-term behavior and the body's response to their presence (Table 1).

The final mechanism for graft contamination is the erosion of a graft through the skin (from wound breakdown/infection) or into the genitourinary or gastrointestinal (GI) tract. Theoretically, adequate tissue coverage over the graft and avoidance of direct contact with the skin or bowel will prevent this complication. However, in the case of aortoduodenal fistula, some authorities believe that the aortic graft may become infected first, with later involvement of the anastomosis, development of a pseudoaneurysm, and eventual erosion into the duodenum *(24)*.

The appearance of clinical symptoms in a patient with graft infection is the result of the balance between the patient's host defenses and the virulence of the organism. The signs of an infection depend on the location of the graft and may be quite varied or subtle. Having a high index of suspicion is important in diagnosing graft infection. Overt signs of infection include fever, leucocytosis, graft exposure or graft-cutaneous sinus tract, localized perigraft abscess/cellulitis, anastomotic pseudoaneurysm, graft-enteric erosion or fistula, and graft thrombosis *(25)*. These can generally be detected by physical examination.

In patients suspected of having a graft infection, detection of fluid or air around the graft, or the presence of an anastomotic pseudoaneurysm, can help confirm the diagnosis. To this end, ultrasound in the extremities or computed tomography (CT) in the abdomen is most helpful. Endoscopy can confirm the diagnosis of a graft-enteric erosion or fistula. However, sometimes an exploratory celiotomy is necessary to make the diagnosis of aortoduodenal fistula or aortic graft infection.

Structural change and overt structural failure of prosthetic grafts have been described. Up to 20% of Dacron grafts, particularly knitted types, have been found to undergo gradual dilation (26). Structural failure, described as fiber breakdown, localized aneurysmal dilation, generalized dilation complicated by bleeding through the graft interstices, or pseudoaneurysm formation, has been estimated to occur in 0.5% to 3% of patients who have received a Dacron implant (27). The average time from implant to failure was 6.4 years in 122 reported cases in the literature (28). Because these problems usually occur more than 5 years after implant, patients with vascular grafts must be followed for life. In patients who have a history of arterial prosthetic placement in the abdomen, a careful abdominal examination should be performed to detect any pulsatile masses. In our practice, we perform a CT or duplex examination of the abdomen 5 years after aortic arterial implant. Evidence of significant graft dilation (by >20% of original graft diameter) or any other abnormality then prompts closer follow-up and possible intervention. Finally, as mentioned previously, endovascular grafts are now used frequently. Structural abnormalities of these devices have been described (29). Long-term follow-up in these patients is mandatory.

In conclusion, large numbers of patients receive prosthetic implants each year for treatment of vascular disorders. These implants may fail over time owing to one of many possible mechanisms. Close lifelong follow-up is therefore necessary in these patients. Patients are best followed by the implanting vascular surgeon. We know of no dedicated website for patients with vascular implants. However, we recommend http://www.vascularweb.org/ as it provides up-to-date information on vascular disease and related issues to patients as well as physicians. A patient may also use this website to review potential vascular disease symptoms and locate a vascular surgeon.

CAROTID REVASCULARIZATION

In the United States, approximately 500,000 people develop new strokes each year. After heart disease and cancer, cerebrovascular disease is the third leading cause of death (30). Carotid artery stenosis is responsible for about 20 to 30% of strokes (31).

Duplex scanning (B-mode ultrasound, combined with pulsed Doppler analysis of blood flow) has been shown to be the most cost-effective, safe, and accurate means of assessing extracranial cerebrovascular disease (32,33). It is an accurate screening tool in patients with cervical bruit as well as in high-risk populations such as those with peripheral vascular disease or coronary artery disease. A carotid duplex study from a reliable vascular laboratory is often used as the sole diagnostic study prior to carotid endarterectomy (34,35). After the landmark studies of the North American Symptomatic Carotid Endarterectomy Trial (36) and Asymptomatic Carotid Arteriosclerosis Study (37), carotid endarterectomy is widely practiced to prevent stroke in symptomatic and asymptomatic patients with severe carotid stenosis.

Recurrent stenosis is reported to occur in 15 to 20% of patients within 5 years following carotid endarterectomy (38,39). Neointimal hyperplasia is a reaction to most

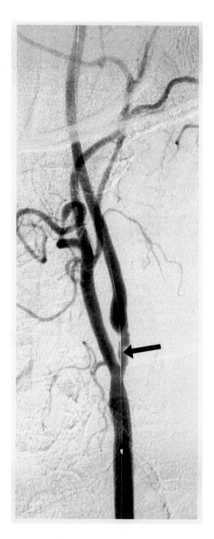

Fig. 1. Angiographic appearance of a recurrent carotid stenosis (arrow) following endarterectomy. Note the smooth, nonulcerated lesion of neointimal hyperplasia.

forms of arterial intervention, and is the commonest cause of recurrent stenosis within 2 years of the primary intervention *(40)*. The lesions of neointimal hyperplasia are smooth and uniform (Fig. 1) and lack the ulcerations typical of atherosclerosis (Fig. 2) *(41)*. Because of the benign nature of recurrent stenosis occurring within 2 years *(42)*, we reserve reintervention only for more than 80% diameter reduction in asymptomatic patients. However, symptomatic restenosis is treated if more than 50% diameter reduction is noticed. After 2 years, recurrence is usually owing to recurrent atherosclerosis, resulting in an ulcerated, soft plaque that harbors a greater embolization potential. These lesions are approached in more aggressive fashion, similar to *de novo* atherosclerosis. The recurrence rate following carotid angioplasty and stenting has not been adequately defined, because of a lack of long-term follow-up data, but the authors have encoun-

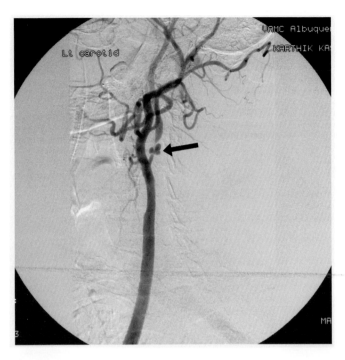

Fig. 2. Angiogram showing atherosclerotic carotid artery stenosis with an ulcerated plaque (arrow).

tered two patients with recurrent stenosis within 8 months following carotid angioplasty and stent placement (Fig. 3A,B). This suggests that careful follow-up is likely to be at least as important for these patients as for those treated by endarterectomy.

Surveillance Following Carotid Interventions

Postcarotid endarterectomy evaluation is the same whether prosthetic material is used or not. This evaluation includes a detailed history of recurrent neurological symptoms, physical examination, and duplex scanning. The neck incision should be inspected for any wound breakdown or drainage suggestive of underlying infection. At 1 month, a duplex scan is performed to provide a baseline. This is controversial in the literature, as some authors do not feel that this early scan has any significant benefit *(43)*.

Following this, we recommend yearly duplex examination and clinic visit. More frequent surveillance has not been demonstrated to decrease the incidence of postoperative transient ischemic attacks, stroke, or asymptomatic total occlusion *(44)*. Patients are asked to return at 6-month intervals if contralateral stenosis or ipsilateral disease progression to more than 50% stenosis is noted (Table 2).

Carotid angioplasty and stenting is currently done only under a clinical trial protocol. Patients are maintained on 75 mg of clopidigrel bisulfate (Plavix), once a day for 3 weeks, along with aspirin (81 mg daily for life). The protocol also requires all angioplasty and stent patients to be evaluated by an independent team of neurologists who use the National Institutes for Health Stroke Scale *(45)* at 24 hours, 1 month, and 6 months after treatment.

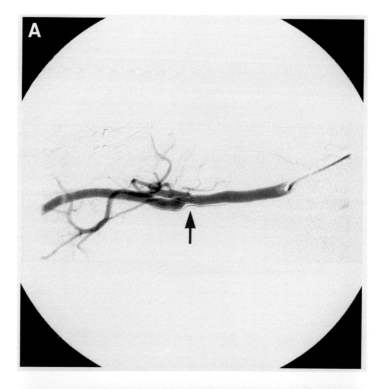

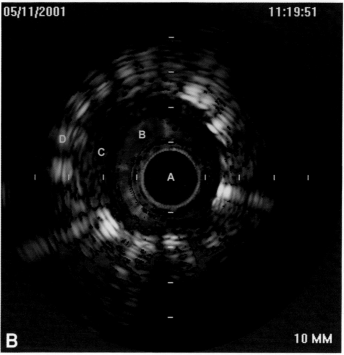

Fig. 3. (A) Angiogram showing recurrent carotid stenosis following angioplasty and stent placement. **(B)** Intravascular ultrasound (IVUS) image demonstrating the instent restenosis (A, IVUS catheter, B, blood flow channel, C, neointimal ingrowth through the stent struts, and D, stent struts).

Table 2
Surveillance Following Carotid Endarterectomy or Carotid Angioplasty/Stenting at University of New Mexico

	Postoperative year					
	1	2	3	4	5	10
Office visit	2[a]	1	1	1	1	1
Duplex scan	2[b]	1[b]	1[b]	1[b]	1[b]	1[b]

Note: The number in each cell denotes the number of times each modality is recommended during the year indicated.

[a] Office visit at 1 month and 1 year post-procedure. Physical examination and risk-factor evaluation at all office visits. Smoking cessation class, if needed, and serum cholesterol level each year.

[b] Repeat every 6 months if recurrent stenosis develops or if there is contralateral disease with >50% stenosis.

HEMODIALYSIS ACCESS

The number of patients with end-stage renal disease (ESRD) is rapidly increasing in the United States. In 1999, approximately 424,000 patients had ESRD, compared to 66,000 in 1982 *(46)*. It is estimated that 600,000 patients will be receiving dialysis by 2008. Hemodialysis access procedures have hence become one of the most commonly performed operations in the United States. Failure of hemodialysis access sites owing to thrombosis, stenosis, and low-flow states contributes to a substantial number of hospital visits and admissions, resulting in significant costs *(47)*.

The Brescia-Cimino radiocephalic fistula, first reported in 1966 *(48)*, remains the preferred method of access for patients requiring long-term hemodialysis. When direct anastomosis between artery and vein cannot be performed, the use of an intervening prosthetic material is the next best option. However, failure rates are higher with prosthetic grafts, as compared to a mature native arteriovenous fistula *(49)*. Thrombosis within 6 weeks of implant is usually caused by technical errors or mechanical problems with the conduit used. Late thrombosis commonly (after 3 months) is usually the result of neointimal hyperplasia at the site of the venous anastomosis (Fig. 4) *(50)*. The early detection and correction of graft dysfunction, especially venous stenosis, can prevent graft failure. Several retrospective studies have established that routine graft surveillance with prompt treatment of dysfunctional grafts (angioplasty or surgical revision) significantly reduces the incidence of graft thrombosis (Fig. 5A,B) *(51–53)*. Intervention before the onset of graft thrombosis results in efficient, uninterrupted dialysis, reduced need for temporary dialysis catheters, reduced rate of graft replacement or thrombectomy, fewer hospital admissions for dialysis access-related problems, and an overall improved quality of life.

Dialysis Access Surveillance

Various methods of evaluating dialysis access sites are available. The measurement of blood flow rate and intragraft pressure during dialysis is the most accurate, simple, inexpensive, and safe method of graft surveillance *(54)*. Currently, the Vascular Access

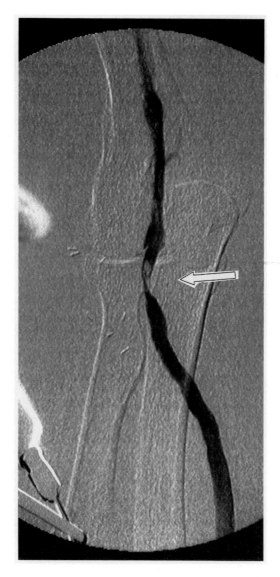

Fig. 4. Angiogram showing venous anastomotic hyperplasia causing stenosis in a dialysis graft.

Work Group of the National Kidney Foundation has recommended a multifaceted approach involving dynamic or static venous pressure measurements and routine physical examination of the graft at intervals of 1 month or less *(55)*. They have found venous pressure measurements to be the most reliable indicator of dialysis graft dysfunction and one of the simplest and most cost-effective. A 15-gage dialysis needle introduced at the venous end of the graft at flow rates of 200 to 225 mL per minute should record pressures of less than 125 to 150 mmHg. Higher pressures suggest venous outflow stenosis, but no single value is a reliable indicator of impending graft failure. Serial values with a trend demonstrating increasing pressures are more valuable *(56,57)*.

The pressure measurements should be used in conjunction with a physical examination of the graft during dialysis (Table 3). Prolonged needle hole bleeding and arm

Table 3
Surveillance of Hemodialysis Grafts

	Postoperative year					
	1	2	3	4	5	10
Physical examination[a]	12	12	12	12	12	12
Intragraft venous pressure[a]	12	12	12	12	12	12
Confirmatory contrast study[b]	12	12	12	12	12	12

Note: Number of times each modality is recommended during the year indicated.
[a] Performed at dialysis center.
[b] Performed if abnormal physical examination or abnormal venous pressures.

swelling often indicate venous outflow stenosis. A pulse to the point of stenosis often replaces the thrill present in a well-functioning graft. An abnormal thrill or bruit may also develop at the site of stenosis *(58)*. Once a flow abnormality is detected by physical examination or by pressure measurements, confirmatory angiography is then performed.

The blood flow rate during dialysis has not been found to be a sensitive or specific predictor of future graft patency. Unfortunately, average flow rates in grafts that remain functional for months or years may be the same as in grafts that fail a few days later. Other methods, such duplex sonography and recirculation measurements, are not cost-effective and have not demonstrated superiority to venous pressure measurements when used in combination with routine physical examination *(59)*.

RENOVASCULAR INTERVENTIONS

Renal artery stenosis is the etiology of hypertension in less than 5% of the hypertensive population in the United States. However, it is more prevalent (30–40% of the total) in patients with untreated diastolic blood pressure greater than 105mm Hg *(60)*. It is the cause of renal failure in about 15% of patients with ischemic nephropathy *(61)*. The growing awareness of the prevalence and the implications of atherosclerotic renovascular disease have led to increasing efforts at renal revascularization. Percutaneous angioplasty and stenting are now valuable adjuncts to open surgery. Weibull et al. reported a patency rate after surgical bypass of 93% at 2 years *(62)*. However, the morbidity and mortality rates are reported to be higher than for percutaneous interventions. This has resulted in renal angioplasty and stent placement becoming the *de facto* standard of care. The drawback of this approach is a high restenosis rate. Recent reports continue to document restenosis rates of 15 to 20% within 2 years *(63,64)*. The early diagnosis and prompt treatment of restenosis before progression to complete occlusion has produced long-term patency and clinical results equivalent to the results for stent placement of *de novo* lesions *(65)*. This is strong evidence supporting a rigorous, standardized follow-up regimen.

Surveillance After Renal Revascularization

The same surveillance protocol is followed for both successful renal artery bypass grafting (prosthetic or autogenous material) and angioplasty with stent placement. All

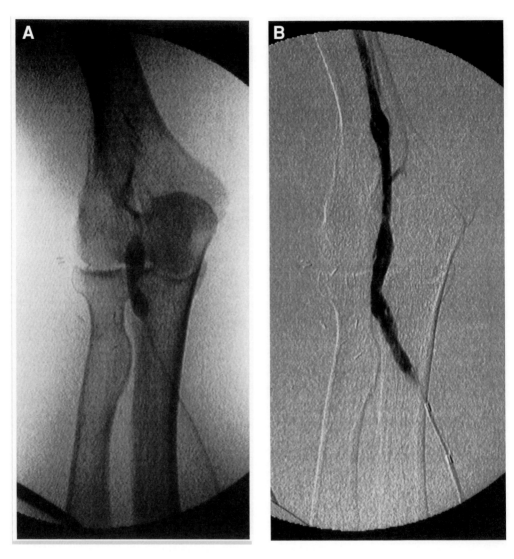

Fig. 5. (A) Balloon angioplasty of anastomotic stenosis. Note the significant waist despite high-pressure balloon inflation. **(B)** Postangioplasty venogram showing no venous stenosis in a dialysis graft.

vascular patients are maintained on 81 mg of aspirin once a day for the remainder of their lives. At each clinic visit, an assessment of the status of hypertension and antihypertensive medications is made. Blood pressure and serum creatinine and blood urea nitrogen (BUN) levels are monitored at office visits (Table 4). Renal artery duplex ultrasonography is performed at 1, 6, and 12 months, and yearly thereafter, unless abnormal. Abnormalities trigger further studies/interventions, which are beyond the scope of this book. In the hands of an experienced vascular laboratory technologist, renal duplex examinations give the clinician both anatomic and physiological information. Determination of kidney size is important to determine the progression (or stabilization) of parenchymal loss. Measurement of renal parenchymal Doppler signal resistance indi-

Table 4
Surveillance Following Renal Revascularization (Bypass or Angioplasty/stent)

	Postoperative year					
	1	2	3	4	5	10
Office visit	3[a]	1	1	1	1	1
Blood pressure	3	1	1	1	1	1
Lab studies[b]	1	1	1	1	1	1
Duplex scan[c]	3[d]	1	1	1	1	1

Note: The number in each cell denotes the number of times each modality is recommended during the year indicated.
[a] Office visits include physical examination and risk-factor assessment. In first year at 1 month, 6 months, and 1 year after procedure.
[b] Blood urea nitrogen and serum creatinine.
[c] Inconclusive or abnormal study in the apropriate clinical situation warrants angiogram.
[d] At 1, 6, and 12 months following procedure.

ces is helpful in assessing renal function. However, the main role of duplex scanning is to determine if there is significant stenosis in the main renal arteries both before and after interventions *(66)*. It is probably the most cost-effective study in following these patients. Unfortunately, it is inadequate for following accessory renal arteries and segmental branches of the main renal arteries. Should the renal duplex scan yield abnormal results, confirmatory angiogram is generally necessary.

Many patients with renal artery stenosis and renal insufficiency have associated parenchymal renal pathology owing to diabetes or hypertension. These patients often have a slow deterioration in renal function even after the renal artery stenosis has been corrected. A progressive increase in BUN and serum creatinine level with a normal renal artery duplex study indicates intrinsic renal pathology. Stabilizing the renal function by correcting renal artery stenosis is a *bona fide* benefit to the patient *(67)*. In hypertensive patients treated for renal artery stenosis, an initial improvement in blood pressure followed by a slow and partial loss of response is not uncommon, despite a normal duplex evaluation. Significant and sustained benefit is observed in only one-third of the hypertensive patients *(68)*. Inconclusive renal artery duplex examinations are best evaluated with a magnetic resonance angiogram (MRA). This is an excellent study, but it tends to overestimate stenosis, leading to false-positive results. It is also expensive and cannot be used in patients with stents containing iron, both of which limit its use in follow-up. If restenosis is identified on magnetic resonance angiography, conventional angiography is often needed.

MESENTERIC ARTERY REVASCULARIZATION

Chronic mesenteric ischemia is suggested by a constellation of clinical symptoms, including weight loss, postprandial abdominal discomfort (food fear), diarrhea, constipation, GI bleeding, and unexplained anemia. Diagnosis is based on duplex and

Table 5
Surveillance Following Mesenteric Revascularization

	Postoperative year					
	1	2	3	4	5	10
Office visit	3[a]	1	1	1	1	1
Body weight	3	1	1	1	1	1
Lab studies[b]	1	1	0	0	0	0
Duplex scan[c]	3[d]	1	1	1	1	1

Note: The number in each cell denotes the number of times each modality is recommended during the year indicated.

[a] Office visits include physical examination and risk-factor assessment. In first year at 1 month, 6 months, and 1 year after procedure.

[b] Serum albumin and transferrin.

[c] Inconclusive or abnormal study in the appropriate clinical situation warrants angiogram.

[d] At 1, 6, and 12 months following procedure.

angiographic findings. Mesenteric revascularization for acute or chronic ischemic bowel syndromes is not commonly performed. The number of times this operation is performed is unknown and the optimal postoperative surveillance strategy has not been adequately defined. Treatment is often by surgical bypass (usually prosthetic in elective cases) and less commonly by angioplasty with or without stenting *(69)*.

Surveillance After Mesenteric Bypass

The office visit is the most important part of surveillance following mesenteric revascularization. Any recurrence of the symptoms just described may indicate restenosis or disease progression in vascular segments not easily assessed by a duplex examination. A careful record of the patient's weight is a good indicator of the success of the procedure. In addition to these, we routinely check serum albumin and transferrin levels during clinic visits (Table 5). Most patients with chronic mesenteric ischemia have low levels before revascularization and an increase followed by stabilization at 6 months to 1 year is typical. Any decline in levels in the late postoperative period may indicate restenosis. Duplex ultrasound is our preferred noninvasive method of directly evaluating the mesenteric bypass *(70)* and should be performed by experienced registered vascular technologists. It involves a high degree of technical difficulty owing to respiratory motion, bowel gas, previous surgery, and bypass grafts placed within the diaphragmatic hiatus. Patients receive a clear liquid diet the day before and are given nothing by mouth for 8 hours before the test because postprandial intestinal hyperemia may falsely suggest a high-grade stenosis. Unfortunately, no validated objective criteria are available for judging stenosis of mesenteric bypass grafts. We use a 1-month baseline study to serve as a comparison for subsequent follow-up examinations. A change in arterial velocities on duplex examination or a change in patient symptoms may warrant a conventional angiogram.

INTERVENTIONS FOR AORTOILIAC OCCLUSIVE DISEASE

Aortoiliac stenosis or occlusion is a common clinical problem. Patients typically present with claudication, limb-threatening ischemia or erectile dysfunction. An aortobifemoral bypass best manages complete infrarenal aortic or iliac occlusion, but isolated aortic or iliac stenosis is more readily treated with balloon angioplasty and stent placement. Percutaneous iliac angioplasty and stenting has demonstrated patency rates nearly as good as those of aortobifemoral bypass grafting (71). Ballard et al. (72) reported far fewer complications in a percutaneous treatment group than in an open surgical bypass group (3.1 vs 22.2%; $p < 0.01$). Iliac angioplasty and stenting has therefore become a common peripheral arterial intervention in the United States. The timely diagnosis of restenosis and prompt intervention before complete thrombosis results in a 5-year secondary patency rate of 80 to 90% (73). This demonstrates that reintervention is associated with durability comparable to the primary procedure, with similar technical success and complication rates, and provides a powerful rationale for surveillance in asymptomatic patients. Recurrent symptoms in patients with either an iliac stent procedure or an aortobifemoral bypass may also be due to progression of atherosclerosis in the infrainguinal segments. Hence, surveillance should involve a method of evaluating the distal arterial blood flow pattern. This is readily done by serial measurements of the ankle-brachial index (ABI) and segmental lower extremity pressure waveform analysis.

Aortobifemoral bypass grafts have such excellent primary patency rates that routine graft surveillance has not been found to be cost-effective (74). However, late clinical reevaluation is performed to assess for the development of femoral artery anastomotic pseudoaneurysms (Fig. 6) as well as the potential complications of graft dilatation or structural failure. Should a femoral pseudoaneurysm be detected, an abdominal CT scan should be performed to evaluate the proximal anastomosis and to evaluate one important cause, namely graft infection. Typical findings include fluid around the graft and perigraft inflammation.

Most patients with aortoduodenal fistula have symptoms (unexplained anemia, malaise, abdominal pain, fever of unknown origin, and GI bleeding), but a few patients who have not bled may be detected on routine follow-up. This dreaded complication is typically observed approximately 2 to 6 years after the primary procedure, but its relative rarity (0.4 to 2.4% over a lifetime) cannot justify a routine surveillance program. The classic triad of symptoms includes GI bleeding, sepsis, and abdominal pain (Fig. 7A,B). Unfortunately, all three symptoms occur together in only 30% of patients (75). Aortic graft infection and aortoduodenal fistula, if left undiagnosed and untreated, carry very high morbidity and mortality rates.

Surveillance After Aortoiliac Angioplasty/Stenting

All patients are maintained on 81 mg of aspirin daily for the remainder of their lives for its proven benefit in reducing myocardial infarction and stroke. Antiplatelet agents have no benefit in reducing the restenosis rate but have the potential to prevent early rethrombosis. The newer antiplatelet agents or warfarin are not usually warranted, given the low incidence of rethrombosis for aortoiliac interventions. Patients are asked to

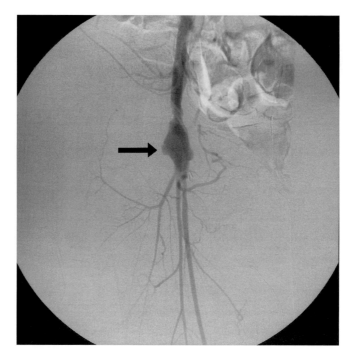

Fig. 6. Angiogram showing femoral anastomotic pseudo aneurysm (arrow), 8 years after an aortobifemoral bypass graft placement.

Table 6
Surveillance Following Iliac Percutaneous Transluminal Angioplasty/Stent

	Postoperative year					
	1	2	3	4	5	10
Office visit	3[a]	2	1	1	1	1
Ankle-brachial index[b]	3	2	1	1	1	1
Duplex scan[b]	3	2	1	1	1	1

Note: The number in each cell denotes the number in each cell denotes the number of times each modality is recommended during the year indicated.

[a] Office visits include physical examination and risk-factor assessment. In first year at 1 month, 6 months, and 1 year after procedure.

[b] At each office visit.

return promptly for recurrent symptoms such as claudication or rest pain. We routinely see patients 1 month following the procedure. At this time, an aortoiliac duplex scan and ABIs are obtained. We then routinely see patients every 6 months for 2 years as recurrent stenosis owing to neointimal hyperplasia most commonly occurs within this time *(76)* (Table 6). Patients who develop recurrent symptoms despite a normal iliac duplex scan usually have progression of distal atherosclerosis, which is readily detected

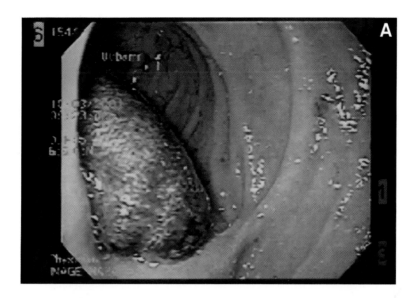
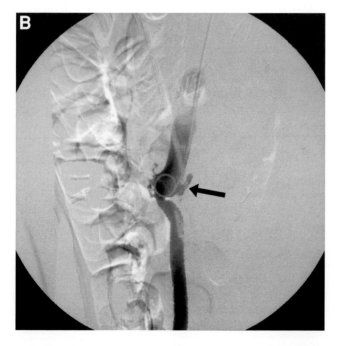

Fig. 7. (A) Upper gastrointestinal endoscopy reveals the proximal segment of an aortobifemoral bypass graft within the duodenum (aortoduodenal fistula). **(B)** Aortogram demonstrates contrast extravasation into the duodenum (arrow) in a patient with aortoduodenal fistula.

by segmental limb pressure measurements or ABIs (Fig. 8). After 2 years, patients are seen yearly and are advised to return earlier if recurrent symptoms are noted.

Patients who have had aortobifemoral bypass procedures also require routine care and surveillance (*see* Table 7). All are placed on lifelong aspirin. Patients are seen in

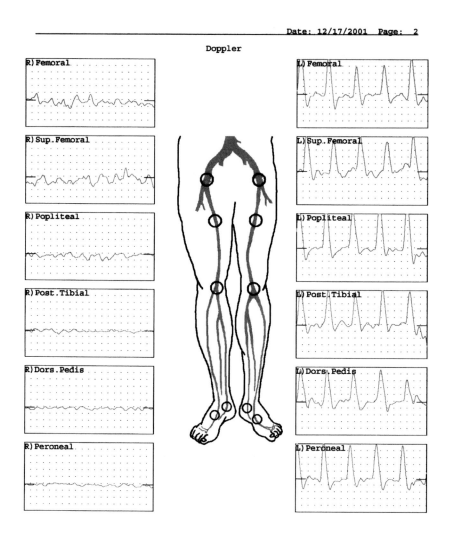

Fig. 8. Segmental limb pressures with normal triphasic waveforms on left compared to monophasic waveforms on right leg secondary to right iliac artery occlusion.

the clinic each year for the rest of their lives, or sooner, for clinical evidence of graft malfunction or for development of symptoms suggesting distal occlusive disease (Fig. 9). The authors feel the yearly follow-up is necessary because of the relentless progression of atherosclerosis in these patients, particularly those that continue to smoke. Similarly, for patients who have aortic bypass procedures for aneurysmal disease, yearly follow-up is justified to evaluate atherosclerosis progression and to detect aneurysms in the iliac, femoral, and popliteal vessels.

AORTIC ENDOGRAFTS

Abdominal aortic aneurysms were the 15th leading cause of death in the United States in 1999. In 2000, approximately 15,000 Americans died of aortic aneurysms or dissections *(77)*. Currently, about 45,000 open surgical procedures are done yearly, with 15 to 20% done on an emergency basis for leaking or ruptured aneurysms *(78)*.

Table 7
Surveillance Following Aortic Graft Placement (Open Repair) for Aneurysmal or Occlusive Disease

	Postoperative year					
	1	2	3	4	5	10
Office visit	3[a]	1	1	1	1	1
Ankle-brachial index[b]	3	1	1	1	1	1
Duplex scan[c]	0	0	0	0	1	0

Note: The number in each cell denotes the number of times each modality is recommended during the year indicated.

[a] Office visit includes physical examination and risk-factor assessment. In first year at 1, 6, and 12 months after procedure.
[b] Performed at all office visits.
[c] Performed at year 5 or if abnormal physical findings.

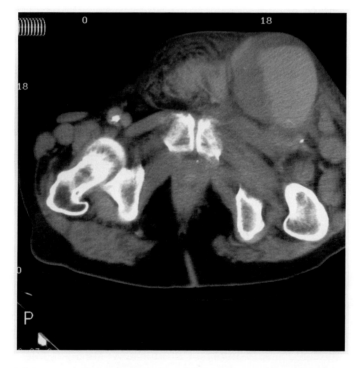

Fig. 9. Computed tomogram image of a ruptured femoral artery anastomotic pseudoaneurysm.

Because the in-hospital mortality rate is at least 50% for repair of a ruptured abdominal aortic aneurysm, it is important for the clinician to diagnose and consider elective repair of these aneurysms (79). Open elective repairs are usually associated with a 1- to 2-day stay in the intensive care unit and a total hospital stay of 5 to 10 days. The endovascular technique has helped hasten the recovery period, resulting in an earlier return to work. Currently in the United States, three devices are approved for endovascular exclusion of abdominal aneurysms. In 1999, the US Food and Drug Administration approved the

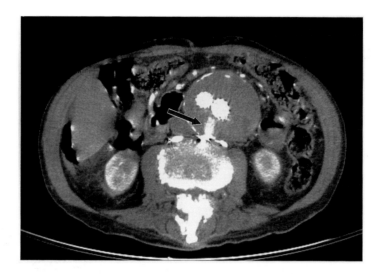

Fig. 10. CT showing type II endoleak into the sac of an "excluded" abdominal aortic aneurysm. The location of contrast within the aneurysm sac is highly indicative of lumbar artery back-flow.

Ancure (Guidant Corp., Menlo Park, CA) and the Aneurx (Medtronic, Inc., Santa Rosa, CA) devices. More recently, in late 2002, the Excluder device (W.L. Gore, Flagstaff, AZ) was approved. About 50% of all infrarenal aortic aneurysms can be excluded with these devices. Unlike the open technique of repair, long-term data on the durability of aortic endografts are unavailable *(80)*. Despite this, more than 8000 Ancure devices and 20,000 Aneurx endografts have been placed worldwide. Open surgical repair involves direct exposure and opening of the aneurysm sac, oversewing of lumbar and inferior mesenteric artery branches, and suturing of the graft to both proximal and distal sites. Hence, open surgical repair is a curative procedure. Endograft aneurysm repair may be considered a palliative procedure. It relies on thrombosis of the aneurysm sac secondary to blood flow exclusion. Endograft fixation to the aortic wall (at the proximal and distal attachment sites) is by means of friction (Aneurx, Excluder) or attachment hooks (Ancure). Branch vessels thrombose owing to limited outflow. Failure of these mechanisms to produce complete aneurysm exclusion produces a condition called "endoleak" (Fig. 10) or persistent flow into the aneurysm sac (Table 8). Endoleaks represent a failure of therapy, as a pressurized aneurysm sac has the same rupture potential as an untreated aneurysm *(81)*. Unfortunately, not all endoleaks are detected at the time of endograft placement (primary endoleak; Fig. 11). Endoleaks may develop several years following the primary procedure. This is felt to be cuased by a change in aneurysm morphology *(82)*. These later developing endoleaks may be particularly dangerous due to renewed pressurization of an atrophic aneurysm sac.

The Ancure device relies on hooks for proximal and distal attachment, whereas the Aneurx and Excluder endografts rely on graft radial force to produce a friction seal. Hence, graft migration may be less common with the Ancure system. However, the main body of the Ancure device is unsupported, potentially leading to graft kinking with shrinkage of the aneurysm sac. This phenomenon is not as common with the

Table 8
Classification of Endoleaks

Endoleak type	Definition
I	Blood enters aneurysm sac at either proximal or distal attachment site (lack of proper fixation).
II	Blood enters sac via retrograde flow through lumbar vessels or patent inferior mesenteric artery.
III	Blood enters sac through a graft fabric tear or where modular components of graft dislocate.
IV	Blood enters sac via transgraft porosity.

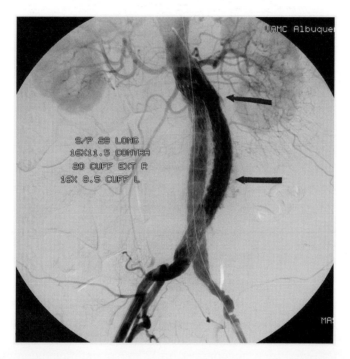

Fig. 11. Type I (proximal attachment site leak) and type IV (graft suture hole/porosity) leak seen on the completion angiogram.

Aneurx or Excluder device because they have a total body stent support, lending superior columnar strength. Besides aneurysm rupture, other late complications such as graft thrombosis (Fig. 12A,B), hook and stent fracture, graft migration, modular component dislocation, and graft material fatigue are now being reported *(80,83,84)*. A lifelong surveillance protocol is necessary to monitor these grafts *(85)*.

SURVEILLANCE AFTER PLACEMENT OF AORTIC ENDOGRAFTS

CTs and four-view plain abdominal films are the main imaging modalities for surveillance following aortic endograft placement. The noncontrast CT is performed to include native vessels 2 cm proximal and distal to the aortic endografts. This is fol-

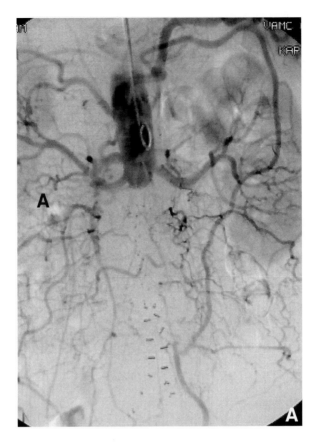

Fig. 12. (A) Complete thrombosis of an Ancure Endograft demonstrated on angiography, secondary to limb angulation with aneurysm shrinkage.

lowed by a contrast-enhanced CT scan at 3 mm intervals and continued to the femoral artery bifurcation. The contrast is administered at a rate of 4–5 mL per second for a total of 170 mL of contrast. A 10-second delay is added to the time of peak enhancement. This is followed by a noncontrast run 25 seconds after the completion of the contrast run. The delayed acquisition significantly improves the detection of low-flow endoleaks *(86)*. We currently use CT as our primary modality for surveillance following endograft placement. The CT images are easier to compare to subsequent examinations than duplex scans and less subject to operator variability. We can determine patency of the graft and its limbs, the renal arteries, and the proximal and distal native arteries from the CT images. Determination of graft migration, as well as aneurysm shrinkage or expansion, is also easily determined. Finally, endoleaks can also be visualized. Some centers now advocate measurement of aneurysm volume based on three-dimensional reconstruction of the aneurysm sac *(87)*. A baseline examination is obtained at 1 month and subsequent examinations are obtained every 6 months thereafter (Table 9). We continue CT surveillance indefinitely as the mere absence of endoleaks is no guarantee of success, and continued aneurysm expansion has been reported in this situation *(88,89)*. In addition, delayed endoleaks can occur due to graft migra-

Fig. 12. (B) Explanted Ancure endograft demonstrates the kinks in the proximal limb take-off junction.

Table 9
Surveillance After Placement of Aortic Endografts

	Postoperative year					
	1	2	3	4	5	10
Office visit[a]	3	2	1	1	1	1
Four-view plain abdominal films	3	2	1	1	1	1
Spiral computed tomography scan[b]	3	2	1	1	1	1

Note: The number in each cell denotes the number of times each modality is recommended during the year indicated.

[a] Office visits include physical examination and risk-factor assessment.

[b] Spiral computed tomography scan at 1, 6, 12, 18, and 24 months, then yearly. Frequency may need to increase if aneurysm is not shrinking or is increasing in size.

tion, proximal neck dilatation, and recanalization of lumbar (Fig. 13) or inferior mesenteric artery branch vessels (Fig. 14A,B) *(90)*. Aneurysm sac shrinkage should be considered the only reliable indicator of the success of endograft therapy. In patients with

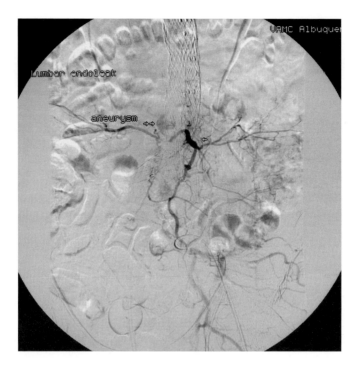

Fig. 13. Type II lumbar endoleak seen with superselective cannulation of the feeding lumbar vessel.

contrast intolerance we have used duplex ultrasonography to evaluate sac diameter and endoleaks. Although some authors have reported that duplex sonography correlates well with CT, it is subject to significant operator variability and has not gained widespread popularity *(91)*. Plain abdominal films in anteroposterior, right posterior oblique, left posterior oblique, and lateral views are also obtained at the time of CT scan. Changes in graft morphology and evidence of metal fatigue such as fracture are more readily visualized with plain abdominal films than with CT. Comparison with previously obtained images is the best indicator of potential problems with the endograft technique. As newer and better devices continue to evolve, the surveillance strategy may become less stringent. Currently, all potential endograft patients must be willing to participate in a lifetime of surveillance to assure continued aneurysm exclusion.

INFRAINGUINAL BYPASS GRAFTS

Infrainguinal arterial occlusive disease is a common problem, with claudication as the most common presenting complaint. Overall, the incidence of intermittent claudication in the United States is 1.8% in patients less than 60 years of age, 3.7% in patients 60–70 years old, and 5.2% in patients older than 70 years *(92,93)*. The rates are almost five times higher in patients with significant risk factors, such as diabetes *(94)*. However, not all patients with claudication require therapy. Seventy-five percent stabilize their walking distance to claudication with a monitored exercise program, smoking cessation, and other risk-factor modifications. Only the remaining 25% continue to deteriorate, with 6% requiring an amputation in 5 or more years *(95)*. Limb-threatening

Chapter 18 / Vascular Prostheses

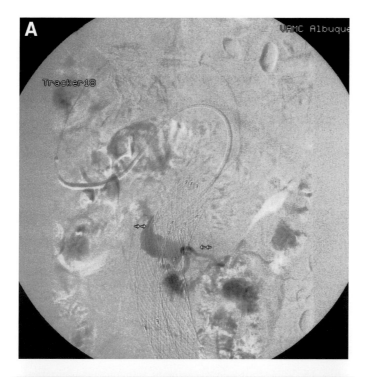

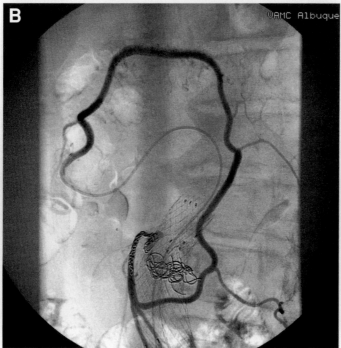

Fig. 14. (A) Angiogram showing type II inferior mesenteric artery endoleak via the large meandering mesenteric artery originating from the superior mesenteric artery. **(B)** Angiogram showing coil embolization of the endoleak. The path taken by a microcatheter from the superior mesenteric artery is well demonstrated.

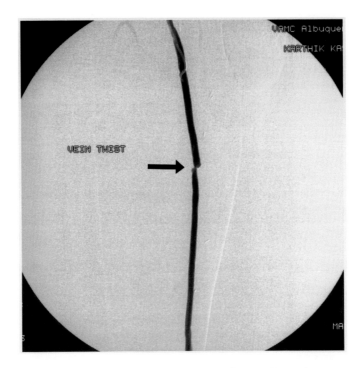

Fig. 15. Vein-graft stenosis seen on a diagnostic angiogram.

ischemia is often the result of multisegment arterial occlusion. This is usually best managed with surgical bypass grafting. Percutaneous techniques of lower extremity revascularization for limb salvage are associated with poor patency rates and are not further discussed. The role of percutaneous interventions in the treatment of claudication has not yet been defined, and at the present time must be considered experimental.

Following lower extremity revascularization, there are currently numerous clinical studies documenting the usefulness of infrainguinal vein bypass graft surveillance programs (96,97). These have been shown to improve the long-term graft patency and limb-salvage rates by as much as 10 to 15%. Since the initial report of serial angiographic surveillance by Szilagyi et al. (98), it has been adequately demonstrated that intrinsic vein graft stenosis is the most likely culprit for occlusion from 30 days to 5 years after vein graft. The vast majority are focal (Fig. 15) and easily corrected by open surgical vein patch angioplasty or percutaneous angioplasty (Fig. 16) (99,100). Prompt correction of the graft-threatening lesion returns the patency curve to that of a graft that has never developed a stenosis, that is, a 5-year patency rate of 80% (101). However, if graft thrombectomy or thrombolysis was required before identifying and treating the culprit stenosis, the 1-year patency rate is dismal (20–35%) (102,103).

Unfortunately, graft surveillance has not proven to be effective in preventing acute (<30-day) graft thrombosis. The early (<30-day) events are usually owing to poor patient selection, technical misadventures such as vein graft twist, or the use of suboptimal vein conduits (104,105). Surveillance has also not been successful in preventing failure of infrainguinal prosthetic bypasses, owing to the rare occurrence of intrinsic lesions (106). Hence, the authors do not routinely perform graft duplex scans during surveillance for infrainguinal prosthetic bypass grafts.

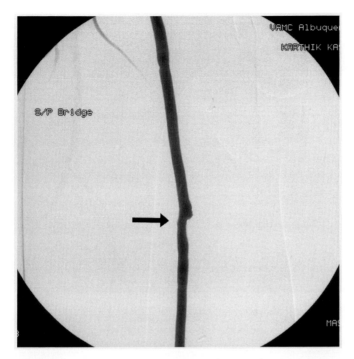

Fig. 16. Angiogram illustrating treatment of vein graft stenosis with a self-expanding stent.

Surveillance After Infrainguinal Vein Bypass Grafting

Clinic visits include a pulse examination by palpation, measurement of the ABI, and duplex assessment of the entire vein conduit, including proximal and distal anastomoses and adjacent inflow and outflow vessels. A diminished ABI (>0.15 change from a prior study) should always have an explanation such as an intrinsic graft stenosis or inflow or outflow vessel disease progression. If the culprit lesion is not identified by the duplex evaluation, a conventional angiogram or MRA should be performed. The majority of vein graft stenoses are easily diagnosed on the duplex study (Fig. 17A,B). A peak systolic velocity of greater than 300 cm per second or one that is 3.5 times that of an adjacent normal segment is highly indicative of vein graft stenosis *(107)*. These threatened grafts are further evaluated by means of a diagnostic angiogram and the culprit lesion is promptly revised. The angiogram also helps identify other potentially correctable lesions prior to reintervention.

An initial duplex scan is obtained at the time of the first clinic visit (<1 month). Subsequent clinic visits and studies are obtained at 3 months and then every 6 months for 2 years *(108)*. Annual studies are then obtained for the lifetime of the graft (Table 10). If progressive increase in the peak systolic velocity is noted in a focal segment, but does not reach the revision threshold, the studies are repeated every 2 months until the lesion resolves or progresses to the point of requiring a revision. Long-term studies have shown a small but real lifetime incidence (2–4%) of late-appearing graft stenoses, prompting lifelong follow-up *(108)*.

Patients receiving a prosthetic leg bypass are followed clinically only, with no duplex examinations. Because of the increased risk of graft failure, patients are seen in the office at 1 month, 3 months, and every 6 months for the first 2 years, then yearly there-

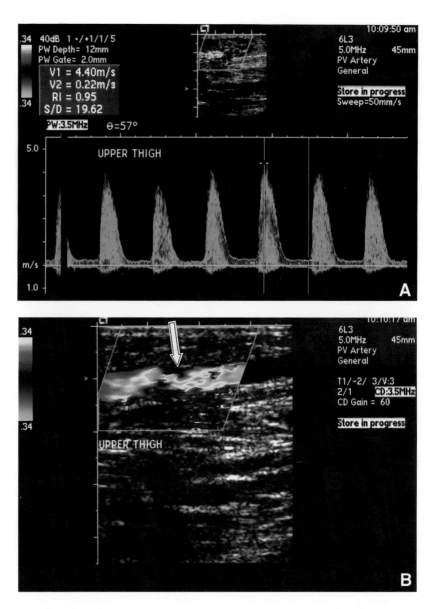

Fig. 17. (A) High peak systolic velocity in a vein graft, indicative of an intrinsic stenosis. **(B)** Duplex image of a high-grade vein graft stenosis, demonstrating high velocity flow within the stenosis.

after (see Table 11). ABIs are performed at each visit. Physical examination includes inspection of the leg for wound problems and pulsatile masses, along with pulse checks.

TRANSJUGULAR INTRAHEPATIC PORTOSYSTEMIC SHUNT

Blood from the portal vein flows through the liver sinusoids, which, in turn, drain into the hepatic veins. This is a low resistance system allowing for high flow rates. The

Table 10
Surveillance After Infrainguinal Vein Bypass Grafts

	Postoperative year					
	1	2	3	4	5	10
Office visit[a]	4	2	1	1	1	1
Ankle-brachial index[b]	4	2	1	1	1	1
Duplex scan[c]	4	2	1	1	1	1

Note. The number in each cell denotes the number of times each modality is recommended during the year indicated.

[a] In first year, at 1 month, 3 months, 6 months, and 12 months. In second year, at 18 months and 24 months.

[b] Performed at all office visits. A >0.15 drop warrants angiogram.

[c] At same time as all office visits. Abnormal scan with a focal peak systolic velocity greater than 300 cm per second and/or velocity ratio greater than 3.5 to adjacent segments, warrants angio or operative revision.

Table 11
Surveillance for Infrainguinal Prosthetic Bypass Grafts

	Postoperative year					
	1	2	3	4	5	10
Office visit[a]	4	2	1	1	1	1
Ankle-brachial index[b]	4	2	1	1	1	1

Note: Number of times each modality is recommended during the year indicated.

[a] In first year, at 1 month, 3 months, 6 months, and 12 months.

[b] Performed at all office visits. A >0.15 drop warrants angiogram.

pressure difference between the portal vein and the right atrium is usually 3 to 6 mmHg. In patients with cirrhosis, this gradient is usually more than 6 mmHg, resulting in an increase in the portal pressure to more than 11 mmHg. This is diagnostic of portal hypertension. Presently, a transjugular intrahepatic portosystemic shunt (TIPS) is reserved for Child class B and C patients, especially those best suited for liver transplant *(109)*. Surgical bypass has a higher patency rate and is the preferred method of decompression for patients with well-preserved liver function (Child class A). TIPS is unfortunately associated with poor primary patency rates owing to shunt thrombosis. Reported 1-year primary patency rates range from 20 to 66% *(110)*. However, with careful surveillance and prompt reintervention, patency rates are significantly better (72–83% at 1 year and 58–79% at 2 years) *(111)*. TIPS initiates a pronounced intimal hyperplastic response, composed mainly of myofibroblasts and collagen *(112)*, which often causes occlusion of the prosthesis. This may be overcome in the future by the use of covered stents, thus reducing the need for reintervention *(113)*. A prospective study comparing a covered stent (Viatorr Gore TIPS endoprosthesis) with the "bare"

Table 12
Etiologies for Restenosis Following Transjugular Intrahepatic Portosystemic Shunt

- Hypercoagulable states
- Intimal injury to hepatic or portal vein
- Hepatic vein stenosis
- Competitive native shunts
- Stent migration

Table 13
Criteria for Diagnosis of Transjugular Intrahepatic Portosystemic Shunt Dysfunction

- Portal vein velocity <40 cm per second
- Jet lesion within the shunt
- Change in peak shunt systolic velocity >50 cm per second
- No flow in shunt
- Reversal of flow in hepatic vein draining the shunt
- Reversal of flow in portal vein branches

Wallstent is currently being evaluated. A variety of other reasons have also been given to explain the poor patency rate of TIPS procedures (Table 12).

Surveillance for TIPS Patients

A significant stenosis is defined as a reduction in shunt diameter by greater than 50%. This is best diagnosed by a conventional venogram. However, venography as a routine screening tool is not cost-effective and subjects the patient to a significant risk of complications. Currently, duplex sonography is the test of choice for routine shunt surveillance *(114)*. The maximum velocity at the mid-shunt is commonly used to assess shunt function. Additional findings such as portal vein velocity and direction of blood flow in the portal vein and its branches are also recorded. Criteria used to assess shunt function are given in Table 13. A duplex study is obtained the day following the TIPS procedure as a baseline, and repeated at 1 month, 3 months, and every 6 months thereafter (Table 14).

SUMMARY

Prosthetic devices play a critical role in the treatment of patients with both occlusive and aneurysmal disease. Because of the progressive nature of vascular disease and multiple causes of graft failure, patients must be followed closely long term. Depending on the type of device and location of placement, different surveillance protocols have been established. It is important to follow these protocols in order to enhance graft patency and detect and treat graft complications. We recommend the website www.vascularweb.org for patients with prosthetic implants who wish to keep abreast of changes/updates in vascular care.

Table 14
Surveillance Protocol for Transjugular Intrahepatic Portosystemic Shunt

	Postoperative year					
	1	2	3	4	5	10
Office visit[a]	4	2	2	2	2	2
Duplex Scan[b]	4	2	2	2	2	2

Note: The number in each cell denotes the number of times each modality is recommended during the year indicated.
[a] In first year, at 1, 3, 6, and 12 months.
[b] Baseline duplex during hospitalization, then at 1, 3, 6, and 12 months, then every 6 months thereafter.

REFERENCES

1. Davis MG, Feeley TM, O'Mally MK, et al. Infrainguinal polytetrafluoroethylene grafts: saved limbs or wasted effort? A report on 10 years' experience. Ann Vasc Surg 1991;5: 519–524.
2. Veith FJ, Gupta SK, Ascer E, et al. Six year prospective multicenter randomized comparison of autologous saphenous vein and expanded polytetrafluoroethylene in infrainguinal arterial reconstruction. J Vasc Surg 1986;3:104–114.
3. Medical Data International, Market and Technology Report, 2001.
4. Berger K, Sauvage LR, Rao AM, et al. Healing of arterial prostheses in man: its incompleteness. Ann Surg 1972;175:118–127.
5. Sauvage LR, Berger K, Beilen LB, et al. Presence of endothelium in an axillary femoral graft of knitted Dacron with an external velour surface. Ann Surg 1975;186:749–757.
6. Ross R. Pathogenesis of atherosclerosis: an update. New Engl J Med 1986;314:488–500.
7. Clowes AW, Kohler T. Graft endothelialization: the role of angiogenic mechanisms. J Vasc Surg 1991;13:734–736.
8. Greisler HP. Vascular graft healing: interfacial phenomena. In: Greisler HP, ed. New biologic and synthetic vascular prostheses. Austin, TX: RG Landes, 1991, pp. 1–19.
9. Mondy JS, Williams JK, Adams MR, et al. Structural determinants of lumen narrowing after angioplasty in atherosclerotic nonhuman primates. J Vasc Surg 1997;26:875–883.
10. Scott NA, Cipolla GD, Ross CE, et al. Identification of a potential role for the adventitia in vascular lesion formation after balloon overstretch injury of porcine coronary arteries. Circulation 1996;93:2178–2187.
11. McArthur C, Teodorescu V, Eisen L, et al. Histopathologic analysis of endovascular stent grafts from patients with aortic aneurysms: Does healing occur? J Vasc Surg 2001;33: 733–738.
12. Marin ML, Veith FJ, Cynamon J, et al. Human transluminally placed endovascular stented grafts: Preliminary histopathologic analysis of healing grafts in aortoiliac and femoral artery occlusive disease. J Vasc Surg 1995;21:595–603.
13. Gentile AT, Mills JL, Gooden MA, et al. Identification of predictors for lower extremity vein graft stenosis. Am J Surg 1997;174:218–221.
14. DePalma RG. Atherosclerosis in vascular grafts. Atheroscler Rev 1979;6:146.
15. Bunt TJ. Synthetic vascular graft infections. I graft infections. Surgery 1983;93:733–746.
16. O'Brien T, Collin J. Prosthetic vascular graft infection. Br J Surg 1992;79:1262–1267.
17. Bandyk DF, Esses GE. Prosthetic graft infection. Surg Clin NA 1994;74:571–590.
18. van Himbeek FJ, van Knippenberg LA, Niessen MC, van Greithuysen AJ. Wound infection after arterial surgical procedures. Eur J Vasc Surg 1992;6:494–498.

19. Bandyk DF. Infection in prosthetic vascular grafts. In: Rutherford RB. Vascular surgery. Philadelphia, PA: W.B. Saunders, 2000, pp. 733–751.
20. Bergamini TM. Vascular prostheses infection caused by bacterial biofilms. Semin Vasc Surg 1990;3:101–109.
21. Vinard E, Eloy R, Descotes J, et al. Human vascular graft failure and frequency of infection. J Biomed Mater Res 1991;25:499–513.
22. Bandyk DF, Berni GA, Thiele BL, et al. Aortofemoral graft infection due to Staphylococcus epidermidis. Arch Surg 1984;119:102–108.
23. Stansby G, Byrne MT, Hamilton G. Dental infection in vascular surgical patients. Br J Surg 1994;81:1119–1120.
24. Buchbinder D, Leather R, Shah D, et al. Pathologic interactions between prosthetic aortic grafts and the gastrointestinal tract. Am J Surg 1980;140:192–198.
25. Champion MC, Sullivan SN, Coles JC, et al. Aortoenteric fistula: Incidence, presentation, recognition, and management. Ann Surg 1982;3:314–317.
26. Nunn DB, Carter MM, Donohue MT, et al. Postoperative dilation of knitted Dacron aortic bifurcation graft. J Vasc Surg 1990;12:291–297.
27. Watanabe T, Kusaba A, Kuma H, et al. Failure of Dacron arterial prostheses caused by structural defects. J Cardiovasc Surg 1983;24:95–100.
28. Nunn DB. Structural failure of Dacron arterial grafts. Sem Vasc Surg 1999;12:83–91.
29. Guidon R, Marois Y, Douville Y, et al. First generation aortic endografts: analysis of explanted Stentor devices from the Eurostar Registry. J Endovasc Ther 2000;7:105–122.
30. American Heart Association: Heart and stroke statistical update. Dallas, TX, 1997.
31. Timsit SG, Sacco RL, Mohr JP, et al. Early clinical differentiation from severe atherosclerotic stenosis and cardioembolism. Stroke 1991;23:486–491.
32. Zwiebel WJ. Spectrum analysis in carotid sonography. Ultrason Med Biol 1987;13:625–636.
33. O'Donnell TF, Erdoes L, Mackey WC, et al. Correlation of B-mode ultrasound imaging and arteriography with pathologic findings at carotid endarterectomy. Arch Surg 1985;120:443–449.
34. Strandness DE Jr. Angiography before carotid endarterectomy—no. Arch Neurol 1995;52:832–833.
35. Ricci MA. The changing role of duplex scan in the management of carotid bifurcation disease and endarterectomy. Sem Vasc Surg 1998;11:3–11.
36. North American Symptomatic Carotid Endarterectomy Trial Collaborators: Beneficial effect of carotid endarterectomy in symptomatic patients with high grade carotid stenosis. N Engl J Med 1991;325:445–453.
37. Executive Committee for the Asymptomatic Carotid Atherosclerosis Study: Endarterectomy for asymptomatic carotid stenosis. JAMA 1995;273:1421–1428.
38. Healy DA, Clowes AW, Zierler RE, et al. Immediate and long-term results of carotid endarterectomy. Stroke 1989;20:1138–1142.
39. Nicholls SC, Phillips DJ, Bergelin RO, et al. Carotid endarterectomy: relationship of outcome to early restenosis. J Vasc Surg 1985;2:375–381.
40. Clagett GP, Robinowitz M, Youkey JR, et al. Morphogenesis and clinicopathologic characteristics of recurrent carotid disease. J Vasc Surg 1986;3:10–23.
41. Kasirajan K, Cornu-Labat G, Turner JJ, et al. Electron microscopic luminal surface characteristics of carotid plaques. Vasc Surg 1997;6:769.
42. Strandness DE Jr. Screening for carotid disease and surveillance for carotid restenosis. Sem Vasc Surg 2001;14:200–205.
43. Mackey WC, Belkin M, Sindi R, et al. Routine postendarterectomy duplex surveillance. Does it prevent later stroke? J Vasc Surg 1992;16:34.
44. Ricotta JJ, O'Brien MS, DeWeese JA. Natural history of recurrent and residual stenosis after carotid endarterectomy: Implications for postoperative surveillance and surgical management. Surgery 1992;112:656–663.

45. Lai SM, Duncan PW, Keighley J. Prediction of functional outcome after stroke; comparison of the Orpington prognostic scale and the NIH stroke scale. Stroke 1998;29:1838–1842.
46. Incidence and Prevalence Data, 1999, USRDS, www.usrds.org
47. Mayers JD, Markell MS, Cohen LS, et al. Vascular access surgery for maintenance hemodialysis: variables in hospital stay. ASA 1992;10J38:113.
48. Brescia M, Cimino J, Appel K, et al. Chronic hemodialysis using venipuncture and a surgically created arteriovenous fistula. N Engl J Med 1966;275:1089–1092.
49. Harland RC. Placement of permanent vascular access devices: Surgical considerations. Adv Ren Replace Ther 1994;1:99–105.
50. Palder SB, Kirkman RL, Whittemore AD, et al. Vascular access for hemodialysis: patency rates and results of revision. Ann Surg 1985;202:235–239.
51. Cinat ME, Hopkins J, Wilson SE. A prospective evaluation of PTFE graft patency and surveillance techniques in hemodialysis access. Ann Vasc Surg 1999;13:191–198.
52. Sands JJ, Miranda CL. Prolongation of hemodialysis access survival with elective revision. Clin Nephrol 1995;44:329–333.
53. Safa AA, Valji K, Roberts AC, et al. Detection and treatment of dysfunctional hemodialysis access grafts: Effect of a surveillance program on graft patency and incidence of thrombosis. Radiology 1996;199:653–657.
54. Besarab A, Sullivan KL, Ross RP, et al. Utility of intra-access pressure monitoring in detecting and correcting venous outlet stenosis prior to thrombosis. Kidney Int 1995;47:1364–1373.
55. Schwab S, Besarab A, Beathard G, et al. NKF-DOQI clinical practice guidelines for vascular access. Am J Kidney Dis 1997;30(suppl 3):S150–S191.
56. Schwab SJ, Raymond JR, Saeed M, et al. Prevention of hemodialysis fistula thrombosis: Early detection of venous stenosis. Kidney Int 1989;36:707–711.
57. Sullivan KL, Besarab A. Hemodynamic screening and early percutaneous intervention reduce hemodialysis access thrombosis and increase graft longevity. J Vasc Interv Radiol 1997;8:163–170.
58. Trerotola SO, Scheel PJ, Powe NR, et al. Screening for access graft malfunction: comparison of physical examination with US. J Vasc Inf Radiol 1996;7:15–20.
59. Lumsden AB, MacDonald MJ, Kikeri D, et al. Cost efficacy of duplex surveillance and prophylactic angioplasty of arteriovenous ePTFE grafts. Ann Vasc Surg 1998;12:138–142.
60. Dean RH, Screening and diagnosis of renovascular hypertension. In: Novick A, ed. Renovascular Disease. Philadelphia: W.B. Saunders, 1996:225–233.
61. Rimmer JM, Gennari J. Atherosclerotic renovascular disease and progressive renal failure. Ann Intern Med 1993;118:712–719.
62. Weibull H, Bergvist D, Bergentz SE, et al. Percutaneous transluminal renal angioplasty versus surgical reconstruction of atherosclerotic renal artery stenosis: a prospective randomized study. J Vasc Surg 1993;8:841–850.
63. Harjai K, Khosla S, Shaw D, et al. Effect of gender on outcome following renal artery stent placement for renovascular hypertension. Cathet Cardiovasc Diagn 1997;42:381–386.
64. White CJ, Ramel SR, Collins TJ, et al. Renal artery stent placement: Utility in lesions difficult to treat with balloon angioplasty. J Am Coll Cardiol 1997;30:1445–1450.
65. Blum U, Krumme B, Flugel P, et al. Treatment of ostial renal artery stenosis with vascular endoprostheses after unsuccessful balloon angioplasty. N Engl J Med 1997;336:459–465.
66. Ziegler RE. Is duplex scanning the best screening test for renal artery stenosis? Sem Vasc Surg 2001;14:177–185.
67. Dean RH, Tribble RW, Hansen KJ, et al. Evolution of renal insufficiency in ischemic nephropathy. Ann Surg 1991;213:446–456.
68. Rees CR. Stents for atherosclerotic renovascular disease. JVIR 1999;10:689–705.
69. Kasirajan K, O'Hara PJ, Gray BH, et al. Chronic mesenteric ischemia: open surgery vs. percutaneous angioplasty and stenting. J Vasc Surg 2001;33:63–71.

70. Zwolak RM, Fillinger MF, Walsh DB, et al. Mesenteric and iliac duplex scanning: a validation study. J Vasc Surg 1998;27:1078–1088.
71. Henry M, Amor M, Ethevenot G, et al. Palmaz stent placement in iliac and femoropopliteal arteries: Primary and secondary patency in 310 patients with 2–4 year follow-up. Radiol 1995;197:167–174.
72. Ballard JL, Bergen JJ, Singh P, et al. Aortoiliac stent deployment versus surgical reconstruction: analysis of outcome and cost. J Vasc Surg 1998;28:94–103.
73. Vorwerk D, Guenther RW, Schurmann K, et al. Late reobstruction in iliac arterial stents: Percutaneous treatment. Radiology 1995;197:479–483.
74. Poulias GE, Doundoulakis N, Prombonas E, et al. Aortobifemoral bypass and determinants of early success and late favorable outcome: experience with 1000 consecutive cases. J Cardiovasc Surg 1992;33:664–678.
75. Connolly JE, Kwaan JHM, McCart PM, et al. Aortoenteric fistula. Ann Surg 1981;194:402–412.
76. Lafont A, Guzman LA, Whitlow PL, et al. Restenosis after experimental angioplasty: Intimal, medial, and adventitial changes associated with constrictive remodeling. Circ Res 1995;76:996–1002.
77. Deaths: Final Data for 2000. National Vital Statistics Reports 50: No. 15, Sept 2002 http://www.cdc.gov
78. Ernst CB, Rutkow IM, Cleveland RJ, et al. Vascular surgery in the United States. Report of the Joint Society for Vascular Surgery—International Society for Cardiovascular Surgery Committee on Vascular Surgical Manpower. J Vasc Surg 1987;6:611–621.
79. Johnston KW. Canadian Society for Vascular Surgery Aneurysm Study Group: ruptured abdominal aortic aneurysm: six year follow-up results of a multicenter prospective study. J Vasc Surg 1994;19:888–890.
80. Harris PL. The highs and lows of endovascular aneurysm repair: the first two years of the Eurostar registry. Ann R Coll Surg Eng 1999;81:161–165.
81. Bernhard VM, Mitchell RS, Matsumura JS, et al. Ruptured abdominal aortic aneurysm after endovascular repair. J Vasc Surg 2002;35:1155–1162.
82. Veith FJ, Baum RA, Ohki T, et al. Nature and significance of endoleaks and endotension: Summary of options expressed at an international conference. J Vasc Surg 2002;35:1029–1035.
83. Politz JK, NewmanVS, Stewart MT. Late abdominal aortic aneurysm rupture after AneuRx repair: a report of 3 cases. J Vasc Surg 2000;31:599–606.
84. Buth J. Early complications and endoleaks after endovascular abdominal aortic aneurysm repair: report of a multicenter study. J Vasc Surg 2000;31:134–146.
85. Zarins CK, White RA, Schwarten D, et al. AneuRx stent graft versus open surgical repair of abdominal aortic aneurysms: multicenter prospective clinical trial. J Vasc Surg 1999;29:292–308.
86. Glozarian J, Dussaussois L, Abada HT, et al. Helical CT of aorta after endoluminal stent-graft therapy: Value of biphasic acquisition. AJR 1998;171:329–331.
87. Singh-Ranger R, McArthur T, Corte MD, et al. The abdominal aortic aneurysm sac after endoluminal exclusion: a medium-term morphologic follow-up based on volumetric technology. J Vasc Surg 2000;31:490–500.
88. White GH, May J, Petrasek P, et al. Endotension: an explanation for continued AAA growth after successful endoluminal repair. J Endovasc Surg 1999;6:308–315.
89. Gilling-Smith GL, Martin J, Sudhundran S, et al. Freedom from endoleak after endovascular aneurysm repair does not equal treatment success. Eur J Vasc Endovasc Surg 2000;19:421–425.
90. White GH, Weiyun Y, May J, et al. Endoleak as a complication of endoluminal grafting of abdominal aortic aneurysms: classification, incidence, diagnosis, and management. J Endovasc Surg 1997;4:152–168.

91. Wolf YG, Johnson BL, Hill BB, et al. Duplex ultrasound scanning vs. computed tomographic angiography for postoperative evaluation of endovascular abdominal aortic aneurysm repair. J Vasc Surg 2000;32:1142–1148.
92. McDaniel MD, Cronenwett JL. Basic data related to the natural history of intermittent claudication. Ann Vasc Surg 1989;3:273–277.
93. Kannel WB, McGee DL. Update on some epidemiologic features of intermittent claudication: the Framingham study. J Am Geriatr Soc 1985;33:13–18.
94. Murabito JM, D'Agostino RB, Silbershatz H, Wilson PWF. Intermittent claudication: a risk profile from the Framingham Heart Study. Circulation 1997;96:44–49.
95. Imparato AM, Kim GE, Davidson T, et al. Intermittent claudication: Its natural course. Surgery 1975;78:795–801.
96. Bandyk DF, Schmitt DD, Seabrook GR, et al. Monitoring functional patency of in situ saphenous vein bypasses: The impact of a surveillance protocol and elective revision. J Vasc Surg 1989;9:286–296.
97. Erickson CA, Towne JB, Seabrook GR, et al. Ongoing surveillance of vascular laboratory surveillance is essential to maximize long term in situ saphenous vein bypass patency. J Vasc Surg 1996;23:18–26.
98. Szilagyi DE, Elliott JP, Hageman JH, et al. Biological fate of autogenous vein implants as arterial substitutes: Clinical, angiographic, and histopathologic observations in femoropopliteal operations for atherosclerosis. Ann Surg 1978;178:232.
99. O'Mara CS, Flinn WR, Gupta SK, et al. Recognition and surgical management of patent but hemodynamically failed arterial grafts. Ann Surg 1981;193:467.
100. Sanchez LA, Suggs WD, Marin ML, Veith FJ. Is percutaneous balloon angioplasty appropriate in the treatment of graft and anastomotic lesions responsible for failing vein bypasses? Am J Surg 1994;168:97–101.
101. Bergamini TM, George SM, Massey HT, et al. Intensive surveillance of femoropopliteal-tibial autogenous vein bypasses improves long term graft patency and limb salvage. Ann Surg 1995;221:507–516.
102. Sanchez L, Gupta SK, Veith FJ, et al. A ten year experience with 150 failing or threatened vein and polytetrafluoroethylene arterial bypass grafts. J Vasc Surg 1991;14:729–738.
103. Whittemore AD, Clowes AW, Couch NP, et al. Secondary femoropopliteal reconstruction. Ann Surg 1981;193:35–42.
104. Wilson YG, Davies AH, Currie IC, et al. Vein graft stenosis: Incidence and intervention. Eur J Vasc Endovasc Surg 1996;11:164–169.
105. Panetta TF, Marin ML, Veith FJ, et al. Unsuspected pre-existing saphenous vein disease: an unrecognized cause of vein bypass failure. J Vasc Surg 1991;15:102–112.
106. Collier P, Ascer E, Veith FJ, et al. Acute thrombosis of arterial grafts. In: Bergan JJ, Yao JST, eds. Vascular surgical emergencies. New York: Grune & Stratton, 1987, pp. 517–528.
107. Gupta AK, Bandyk DF, Cheanvachai D, Johnson BL. Natural history of infrainguinal vein graft stenosis relative to bypass grafting technique. J Vasc Surg 1997;25:211–225.
108. Mills, JL. Infrainguinal vein graft surveillance: How and when. Sem Vasc Surg 2001; 14:169.
109. Kerlan RK, LaBerge JM, Gordon RL, Ring EJ. TIPS: current status. AJR Am J Roentg 1995;164: 1059.
110. Martin M, Zajko AB, Orons PD, et al. Transjugular intrahepatic portosystemic shunt in the management of variceal bleeding: indications and clinical results. Surgery 1993; 114:719–727.
111. LaBerge JM, Somberg KA, Lake JR, et al. Two-year outcome following transjugular intrahepatic portosystemic shunt for variceal bleeding: results in 90 patients. Gastroenterol 1995;108:1143–1151.
112. LaBerge JM, Ferrell LD, Ring EJ, et al. Histopathologic study of stenotic and occluded transjugular intrahepatic portosystemic shunts. JVIR 1993;4:779–786.

113. Nishimine K, Saxon RR, Kichikawa K, et al. Improved transjugular intrahepatic portosystemic shunt patency with PTFE-covered stent-grafts: experimental results in swine. Radiology 1995;196:341–347.
114. Haskal ZJ, Pentecost MJ, Soulen MC, et al. Transjugular intrahepatic portosystemic shunt stenosis and revision: early and midterm results. AJR Am J Roentg 1994;163:439–444.

US Counterpoint to Chapter 18

Luis A. Sanchez

INTRODUCTION

Chapter 18 is a comprehensive review of the most commonly used vascular prostheses and the suggested long-term evaluation and follow-up based on the current available literature. In general, we agree with the authors' recommendations. They concentrate on grafts and devices commonly used in a broadly based peripheral vascular practice. They discuss in detail the topic of graft infection, which is an uncommon but extremely severe complication of vascular reconstructions. They describe in detail the etiology, frequent complications, and potential prevention techniques for prosthetic graft infections based on a wide range of published reports. We agree with their analyses and recommendations (*see* Tables 1–10). Subsequently, they review the most commonly treated areas of the vascular tree in an organized and thoughtful fashion. The Washington University vascular surgery service feels this list is appropriate and generally concurs with it.

Extracranial carotid artery disease is responsible for a significant number of cerebrovascular accidents. The aggressive treatment of these lesions for the prevention of stroke is associated with excellent short- and long-term results. The use of carotid patches (vein or prosthetic) is a perennially controversial topic among vascular surgeons. Conflicting reports suggest that patching of carotid arteries may or may not decrease the rate of restenosis after carotid endarterectomy. Most physicians agree that patching should be performed in patients with small carotid arteries (often women) and those patients undergoing secondary carotid interventions most commonly associated with restenosis. The authors suggest a comprehensive surveillance protocol for patients with extracranial carotid artery disease for continued stroke prevention. These surveillance protocols should be similarly applied to the growing patient population that has been treated with the developing technique of carotid angioplasty and stenting. This endovascular technique is rapidly becoming an important alternative for the treatment of carotid artery disease. The early data available (SAPHIRE and ARCHER trials) suggests that this technique can be applied to patients considered at high risk of perioperative complications after carotid endarterectomy, with morbidity and mortality rates comparable or even lower than those associated with carotid endarterectomy. Its role in the treatment of patients that are considered to be at low risk of complications after carotid endarterectomy is a controversial but evolving topic at this time.

From: *The Bionic Human: Health Promotion for People With Implanted Prosthetic Devices*
Edited by: F. E. Johnson and K. S. Virgo © Humana Press Inc., Totowa, NJ

Table 1
Surveillance Following Carotid Endarterectomy or Carotid Angioplasty/Stenting at Washington University

	Postoperative year					
	1	2	3	4	5	10
Office visit	3[a]	1	1	1	1	1
Duplex scan	3[b]	1[b]	1[b]	1[b]	1[b]	1[b]

Note: Number of times each modality is recommended during the year indicated.

[a] Office visit at 1 month, 6 months, and 1 year after procedure. Physical examination and risk factor evaluation at all office visits. Smoking cessation class, if needed, and serum cholesterol level each year.

[b] Duplex scan at 1 month, 6 months, and 1 year following procedure. Repeat every 6 months if recurrent stenosis develops or if there is contralateral disease with >50% stenosis.

Table 2
Surveillance of Hemodialysis Grafts at Washington University

	Postoperative year					
	1	2	3	4	5	10
Physical examination[a]	12	12	12	12	12	12
Intragraft venous pressure[a]	12	12	12	12	12	12
Confirmatory contrast study[b]	12	12	12	12	12	12

Note: Number of times each modality is recommended during the year indicated.

[a] Performed at dialysis center.

[b] Performed if abnormal physical examination or abnormal venous pressures.

Hemodialysis access procedures continue to increase as the population of patients with end-stage renal disease continues to grow and survive longer, thanks to improvements in their care. Autologous fistulas continue to be the access of choice for these patients but a significant group of patients do not have a reasonable venous conduit that can be used for hemodialysis. Prosthetic grafts are the most commonly used long-term alternative while indwelling catheters are usually used as "temporary" access. The authors discuss a sound surveillance protocol for these patients that leads to early interventions in patients with "failing" dialysis access and can avoid the repeated use of "temporary" catheters that is associated with local and systemic infections and the development of occlusive central venous lesions.

Renovascular occlusive disease is a potentially treatable etiology for hypertension and progressive renal insufficiency. Open surgical revascularizations are associated with excellent short- and long-term patency rates but are unfortunately also associated with moderate morbidity rates. Renal artery angioplasty and stenting is associated with higher recurrence rates but very limited morbidity and mortality. The authors have outlined an excellent surveillance protocol based on duplex surveillance for patients

Table 3
Surveillance Following Renal Revascularization (Bypass or Angioplasty/Stent) at Washington University

	Postoperative year					
	1	2	3	4	5	10
Office visit	3[a]	1	1	1	1	1
Blood pressure	3	1	1	1	1	1
Laboratory studies[b]	1	1	1	1	1	1
Duplex scan[c]	2[d]	1	1	1	1	1

Note: Number of times each modality is recommended during the year indicated.

[a] Office visits include physical examination and risk-factor assessment. In first year at 1 month, 6 months, and 1 year after procedure.

[b] Blood urea nitrogen and serum creatinine.

[c] Inconclusive or abnormal study in the apropriate clinical situation warrants angiogram.

[d] At 6 and 12 months followng procedure.

Table 4
Surveillance following Mesenteric Revascularization at Washington University

	Postoperative year					
	1	2	3	4	5	10
Office visit	3[a]	1	1	1	1	1
Body weight	3	1	1	1	1	1
Laboratory studies[b]	1	1	0	0	0	0
Duplex scan[c]	2[d]	1	1	1	1	1

Note: Number of times each modality is recommended during the year indicated.

[a] Office visits include physical examination and risk-factor assessment. In first year at 1 month, 6 months, and 1 year after procedure.

[b] Serum albumin and transferrin.

[c] Inconclusive or abnormal study in the appropriate clinical situation warrants angiogram.

[d] At 6 and 12 months following procedure.

undergoing renal angioplasty and stenting. A similar protocol, but perhaps not as stringent, should also be applied to patients that have undergone open surgical revascularizations because the recurrence rate is not as high as after renal angioplasty and stenting. Mesenteric revascularizations (open or endovascular) are difficult to evaluate owing to their locations and the limited experience with the noninvasive evaluations for this pathology. The authors appropriately have suggested a surveillance protocol similar to that of renal revascularizations.

Table 5
Surveillance Following Iliac Percutaneous Transluminal Angioplasty/Stent at Washington University

	Postoperative year					
	1	2	3	4	5	10
Office visit	3[a]	2	1	1	1	1
Ankle-brachial index[b]	3	2	1	1	1	1
Duplex scan[c]	2	2	1	1	1	1

Note: Number of times each modality is recommended during the year indicated.

[a] Office visits include physical examination and risk-factor assessment. In first year at 1 month, 6 months, and 1 year after procedure.
[b] At each office visit.
[c] Every 6 months for the first 2 years.

Table 6
Surveillance Following Aortic Graft Placement (Open Repair) for Aneurysmal or Occlusive Disease at Washington University

	Postoperative year					
	1	2	3	4	5	10
Office visit	3[a]	1	1	1	1	1
Ankle-brachial index[b]	3	1	1	1	1	1
Duplex scan[c]	0	0	0	0	1	0

Note: Number of times each modality is recommended during the year indicated.

[a] Office visit includes physical examination and risk-factor assessment. In first year at 1, 6, and 12 months after procedure.
[b] Performed at all office visits.
[c] Performed at year 5 or if abnormal physical findings.

Interventions for aortoiliac disease are very common and it is easier to recognize problems related to them. Clinical symptoms, physical examination, noninvasive lower extremity arterial studies such as ankle-brachial indices (ABIs), and duplex scanning, if appropriate, should be used to monitor these patients. Open surgical reconstructions, such as aortofemoral and iliofemoral bypasses, have excellent long-term patency rates. Aortoiliac endovascular interventions also have good mid-term and long-term patency rates but are more likely to develop early restenosis owing to the development of intimal hyperplasia. These patients benefit from duplex scanning of the treated arterial segment at regular intervals, as suggested by the authors. The patients that have undergone open surgical reconstructions of the aortoiliac system for occlusive or aneurysmal disease do not need to be followed as frequently as patients treated in an endovascular fashion but more frequently than they are usually followed by the treating physicians. Duplex or computed tomography scanning should be performed at 5 years (or sooner

Table 7
Surveillance for Aortic Endografts at Washington University

	Postoperative year					
	1	2	3	4	5	10
Office visit[a]	3	2	1	1	1	1
Four-view plain abdominal films	3	2	1	1	1	1
Spiral computed tomography scan[b]	3	1	1	1	1	1

Note: Number of times each modality is recommended during the year indicated.

[a] Office visits include physical examination and risk-factor assessment.

[b] Spiral computed tomography scan at 1, 6, 12 months, then yearly. Frequency may need to increase if aneurysm is not shrinking or is increasing in size, the patient develops new endoleak, or proximal or distal fixation is poor.

Table 8
Surveillance for Infrainguinal Vein Bypass Grafts at Washington University

	Postoperative year					
	1	2	3	4	5	10
Office visit[a]	4	2	1	1	1	1
Ankle-brachial index[b]	4	2	1	1	1	1
Duplex scan[c]	4	2	1	1	1	1

Note: Number of times each modality is recommended during the year indicated.

[a] In first year, at 1 month, 3 months, 6 months, 12 months. In second year, at 18 months, and 24 months.

[b] Performed at all office visits. A >0.15 drop warrants angiogram.

[c] At same time as all office visits. Abnormal scan with a focal peak systolic velocity greater than 300 cm per second and/or velocity ratio greater than 3.5 to adjacent segments, warrants further evaluation and a potential intervention if the stenosis is greater than 75%.

in some patients with associated small aneurysms being followed or progressive occlusive disease) because the most common complications of anastomotic pseudoaneurysms and progression of arterial disease are more prevalent 5 years after the original intervention. The new techniques of endovascular repair of aortoiliac aneurysms have unique problems associated with them and patients need close surveillance. Patients treated with these developing techniques and devices need surveillance to assess the reconstruction for continued exclusion of the aneurysm, development of associated endoleaks, and for structural integrity of the devices. Endoleaks are a unique complication associated with endovascular repair of aneurysms. The authors suggest that the presence of an endoleak is a "failure of therapy" but this is not completely accurate. Patients who have type I or III endoleaks require prompt treatment because they are associated with pressurization of the growing aneurysm sac. Fortunately, these leaks occur in only 1–5% of patients and most of them can be corrected with a second-

Table 9
Surveillance for Infrainguinal Prosthetic Bypass Grafts at Washington University

	Postoperative year					
	1	2	3	4	5	10
Office visit[a]	4	2	1	1	1	1
Ankle-brachial index[b]	4	2	1	1	1	1

Note: Number of times each modality is recommended during the year indicated.
[a]In first year, at 1 month, 3 months, 6 months, 12 months.
[b]Performed at all office visits. A >0.15 drop warrants further evaluation and intervention if a greater than 75% stenosis is encountered that threatens graft patency.

Table 10
Surveillance Protocol for Transjugular Intrahepatic Portosystemic Shunt at Washington University

	Postoperative year					
	1	2	3	4	5	10
Office visit[a]	4	2	2	2	2	2
Duplex Scan[b]	4	2	2	2	2	2

Note: Number of times each modality is recommended during the year indicated.
[a]In first year, at 1, 3, 6, and 12 months.
[b]Baseline duplex during hospitalization, then at 1, 3, 6, and 12 months, then every 6 months thereafter.

ary endovascular intervention. Type II endoleaks are the most prevalent ones (occuring in 5–15% of cases) but these are rarely associated with aneurysm sac pressurization and growth. In fact, most of these type II (branch) endoleaks do not require an intervention. In our experience, with more than 500 endovascular repairs, only 5% of patients had type II endoleaks that persisted for more than 6 months and 1% that required endovascular interventions for their correction owing to aneurysm enlargement. In our experience, no patient has required surgical conversion for the correction of a type II endoleak at a mean follow-up of more than 36 months. The authors suggest an appropriately comprehensive surveillance protocol for these patients that at this time should be considered a lifetime commitment.

Infrainguinal arterial occlusive disease has been most frequently treated with arterial bypasses with vein or prosthetic material. Percutaneous interventions in this arterial distribution have been associated with limited patency rates and high recurrence rates but their role is increasing with the use of a combination of evolving technologies, including thrombolytic agents, angioplasty and stenting, endovascular grafts, brachytherapy, cryotherapy, and drug-eluting stents. The authors suggest a comprehensive surveillance protocol for vein grafts using ABIs and duplex scanning. A similar surveillance protocol should be used after endovascular techniques are used for the treatment of infrainguinal arterial occlusive disease. Surveillance of patients with prosthetic grafts is not as fruitful as vein-graft surveillance and the authors do not recommend the routine use of duplex scanning. Duplex scanning can be helpful in the surveillance of these patients to evaluate proximal, distal, and anastomotic disease. In addition, low graft-flow velocities (<30 cm per second) can suggest severe distal or proximal disease that could threaten graft patency and could potentially be treated before graft failure occurs.

Transjugular intrahepatic portosystemic shunts (TIPS) are becoming the treatment of choice for a growing number of patients with portal hypertension. The early high recurrence rates are improving with better techniques and the use of covered stents. In addition, TIPS is associated with much lower morbidity and mortality rates than the surgical options for portal vein decompression. The authors suggest a comprehensive surveillance protocol based on duplex evaluations and we agree with it.

Surveillance of patients with vascular prostheses and a broad range of vascular interventions are essential. Surveillance protocols lead to improved long-term results, better patency rates, and decreased complications associated with the interventions necessary to maintain graft patency because the treatment of "failing" arterial reconstructions is associated with lower morbidity and mortality rates than treatment of occluded arterial reconstructions.

European Counterpoint to Chapter 18

S. Rao Vallabhaneni and John A. Brennan

INTRODUCTION

Vascular grafts are surgically inserted conduits used for anatomic or extra-anatomic bypass of various arterial segments. The most common indications are aneurysmal and occlusive disease. Aortic and lower limb reconstructions are common and upper limb reconstruction is somewhat rare. Large blood vessels such as aorta and iliac arteries are usually replaced with prosthetic material. Autogenous vein is the best conduit for infrainguinal reconstruction, but prosthetic conduits are used when this is not feasible. An estimate of the numbers of prosthetic arterial grafts used annually in European countries is provided in Table 1.

Dacron grafts are made of continuous multifilament polyester yarn. The handling and final characters of the conduit, to a certain extent, depend on whether the yarn is made by weaving or by knitting. Weaving involves interlacing the fabric threads in a simple over and under pattern in both the longitudinal and circumferential directions of the conduit. A tight weave results in a conduit with low porosity and high strength. This means there will be little bleeding through the graft at surgery. Tightly woven prosthetic conduits are resistant to elongation and dilatation, which is advantageous, but low compliance and a tendency to fray at the cut edges mean poorer handling characteristics. Perigraft healing is poor owing to low porosity. In a knitted graft, the yarn is oriented predominantly in one direction, usually longitudinally. The manufacturing process can vary the pore dimensions. Knitted grafts tend to have a higher porosity than woven grafts, resulting in better incorporation into surrounding tissue, better compliance and improved handling characteristics. This is at the expense of requirement for preclotting and relatively lower strength leading to long-term dilatation (1).

Some problems have been overcome by modifications to the basic knitted and woven structure. Velour surfaces can be added on either or both surfaces of the graft, which improves the handling characteristics of the graft and provides enhanced scaffolding on which healing can occur (2). Crimping of the graft imparts more flexibility, a degree of elasticity and better shape retention with bending. However, in the long term, much of the elasticity and flexibility are lost because of perigraft fibrosis. The disadvantage of the crimp is the reduced luminal diameter and an uneven inner surface with increased potential for thrombosis. In areas liable to external compression and kinking such as extra-anatomical bypasses or conduits placed across a joint, noncrimped structure with

Table 1
Estimates of the Number of Prosthetic Grafts Used Annually in Europe

Country	Population (million)	Thoracic	Abdominal	Axillo-femoral/bifemoral	Infrainguinal
United Kingdom	58.7	1170	5300	780	1500
Netherlands	15.7	675	3600	225	3100
Belgium	10.2	450	2400	150	2190
France	57.7	1200	8800	800	11,500
Germany	82.2	1500	12,000	800	9000
Italy	57.5	1100	5800	350	4050
Spain	39.7	445	3800	222	2600
Portugal	9.9	70	680	50	350
Greece	10.6	120	1020	60	750
Austria	8.1	120	680	0	725
Switzerland	7.3	190	1000	60	700
Norway	4.3	170	920	60	500
Sweden	8.9	240	1280	80	750
Denmark	5.3	200	1040	60	650
Finland	5.2	150	800	50	500
Turkey	64.5	65	340	20	525
Poland	38.7	750	4000	250	1500
Czechoslovakia	10.3	60	320	20	500
Total	495	8675	53780	4037	41,390

external support rings appears to be better than crimped structure without external support (3). Impregnation of a graft with absorbable substances such as collagen, albumen, or gelatin renders the graft impervious to leakage through the interstices without loss of the favorable handling characteristics of porous material. This sealant is completely absorbed after implantation, allowing healing and normal incorporation within the body to take place. An additional advantage of impregnation with these substances is that it allows bonding with antibiotics. Rifampicin impregnation of the graft reduces the risk of a subsequent infection following a bacterial challenge (4) and reduces the risk of a recurrent infection when replacing an infected graft (5). Reducing the thrombogenecity of prosthetic grafts has always been a priority. Experimental evidence shows that thrombogenicity of Dacron can be reduced by coating its surface with a fluoropolymer (6), a process termed *fluoropassivation*. Heparin bonding of collagen-sealed conduits has also been attempted. It is rare for reconstruction of the aorta to be undertaken with material other than Dacron in Europe.

Expanded polytetrafluoroethylene (ePTFE) is a fluorocarbon polymer. Manufacturing of the conduits is by a paste extrusion process and the porosity of the material can be controlled. The material is inelastic, impervious to blood and resistant to dilatation. Being chemically inert, highly electronegative and hydrophobic, ePTFE conduits allow suture-hole bleeding for a much longer time than any other conduit. Several modifications have been added over the years. Manufacture of thin-walled conduits with microcrimped fibrillary structure allows a degree of longitudinal stretch (stretch ePTFE) in a graft that does not stretch otherwise. Manufacture of thinner-walled conduits improves handling characteristics and imparts better conformability. External support with spirals or rings helps to prevent kinking when placed across a joint but a prospective randomized study did not show improved patency *(7)*. ePTFE continues to be the choice of most UK surgeons despite little evidence of better patency than Dacron in infrainguinal reconstruction. Thin-walled, coated Dacron grafts, however, may become equally popular in the future.

ePTFE grafts have been modified with carbon coating to reduce platelet deposition and reduce thrombogenicity. There is experimental evidence *(8)* of the effectiveness of this process and clinical trials are in progress evaluating this in infrainguinal reconstruction. Cuffs or patches of vein interposed between a prosthetic graft and native artery at the distal anastomosis have a beneficial effect on patency rates *(9)*. The most common configurations are the Miller cuff, the Taylor patch, and the St. Mary's boot. The runoff vessels at the time of graft failure also appear to be better preserved with interposition cuffs. The changes in anastomotic hemodynamics imparted by Miller cuffs *(10)* are understood to inhibit neointimal hyperplasia to improve patency. This has been exploited in designing ePTFE grafts with preshaped cuffs (Distaflo®).

Polyurethane is a polymer used to make conduits with a smooth, nonthrombogenic inner surface, a thin-walled structure with some compliance, and good handling characteristics. Despite considerable effort to develop a polyurethane graft suitable for routine use, patency rates have not been high and it is not in routine use.

Glutaraldehyde-tanned human umbilical vein is the only commercially available allograft prosthetic. Human umbilical veins are harvested, placed on mandrils for shape retention and tanned with 1% glutaraldehyde solution. Antigenic properties of biological material are lost in the process. This conduit is mainly used for infrainguinal reconstruction. Centers with a large experience with this material have reported excellent results *(11)*. However, it tends to show aneurysmal dilatation in the long term *(12)* that has discouraged wider use.

AORTIC AND SUPRAINGUINAL RECONSTRUCTION AND FOLLOW-UP OF ASYMPTOMATIC PATIENTS

Aortic aneurysm repair is performed to prevent death from rupture. For the suprarenal aortic segment a surgically interposed conduit is the mainstay of treatment. Endoluminal stent grafting, currently being widely evaluated for infrarenal aortic aneurysms, is also an option for localized fusiform aneurysms of the thoracoabdominal aorta. Occlusive arterial disease is another indication for surgical reconstruction of the aortoiliac segment. All these reconstructions are usually long-lasting. Hemodynamically critical lesions of inflow or outflow are rare in the aorta and the large lumen size and flow patterns mean neointimal hyperplasia hardly ever progresses to threaten the patency. Aortofemoral reconstructions tend to have excellent patency rates owing to the fact that profunda femoris arteries tend to provide runoff even if superficial femoral artery segments occlude due to progression of atherosclerosis.

Significant complications in long-term follow-up have been reported to be relatively uncommon *(13)* but anastomotic false aneurysms, aortoenteric fistulae, and graft sepsis

Table 2
Follow-Up of Conventional Surgery Patients
in the United Kingdom—Endovascular Aneurysm Repair Trial I[a]

	Year 1	Year 2	Year 3	Year 4
Clinic visit with assessment of adverse events and interventions	3	1	1	1
Quality-of-life questionnaire	3	1	1	1
Serum creatinine level	1	1	1	1
Contrast enhanced computed tomography of abdomen and pelvis	1	0	0	0

[a] The number in each cell represents the number of times per year each modality is required in the protocol.

have all been reported *(14)*. Femoral false aneurysm can be detected clinically but seldom causes complications when asymptomatic. Fistulae and chronic low-grade sepsis can remain occult until precipitating a catastrophe. Fortunately, their incidence is sufficiently low *(13,14)* that routine investigation is not necessary to rule them out. Good primary patency rates of aortofemoral grafts and a general lack of an effective preventive measure should an impending occlusion be detected mean that a surveillance program is of no benefit. This is in marked contrast to the view of many American surgeons, as explained in this chapter. Aortofemoral reconstructions usually require revision when they fail since it is neointimal hyperplasia at the lower anastomoses that is usually responsible and this problem cannot be effectively overcome in any other way. Endovascular treatment is not widely used for this problem in this arterial segment. Initial follow-up with a clinic visit to ensure freedom from immediate and early complications of surgery (usually wound-related) and postoperative convalescence is all that is usually practiced. Currently, in the United Kingdom, clinical trials are in progress to evaluate endovascular repair of infrarenal aortic disorders. The United Kingdom—Endovascular Aneurysm Repair Trial I involves random allocation of patients to either endovascular or conventional repair. The trial protocol requires the conventional surgery patients to undergo a structured follow-up as shown in Table 2. Routine computed tomography (CT) scan follow-up is not the standard in United Kingdom after aortic surgery. However, should the trial show any benefits of such follow-up, current practice can be expected to change.

Patients with atherosclerotic disease usually have risk factors such as hypertension, diabetes mellitus, lipid abnormalities, and so on. Long-term follow-up studies show that diseases of the cardiovascular system are the main causes of death after aortic aneurysm repair. Aneurysms of the rest of the arterial tree cause significant morbidity and mortality after infrarenal aortic aneurysm repair. Arguments have been made for liberal use of CT scans in the follow-up of these patients *(14)* because other methods are not reliable in detecting changes in renal vessels or thoracoabdominal aorta. Although this is particularly relevant in hypertensive patients who have had an abdominal aortic aneurysm repair, such follow-up is not standard practice in the United Kingdom. This is because thoracoabdominal aneurysms are usually symptomatic *(15)* and the value of screening is therefore questionable. The family practitioner should control the risk factors or detect and correct them as they surface. The most significant of the risk factors are hypertension, diabetes mellitus, smoking, lipid abnormalities, and lack of physical exercise. The low incidence of graft-related complications and the high incidence of significant cardiovascular disease in this group of patients has led to the current practice in the United Kingdom, which is to direct the follow-up at the general health of the patient rather than at the graft. Prolonged specialist follow-up is not generally necessary as long as relevant specialist services are readily accessible should problems arise. This approach also affects the management of patients with arterial prostheses who develop an infection remote from the site of the prosthesis (e.g., appendicitis in a patient with a synthetic femoropopliteal graft) or undergo a procedure known to produce transient bacteremia (e.g., colonoscopy). The decision whether to treat such patients differently from patients without a prosthetic implant is left to the primary care physician.

INFRAINGUINAL BYPASS AND FOLLOW-UP IN ASYMPTOMATIC PATIENTS

Infrainguinal bypass using a prosthetic conduit is usually done to alleviate critical limb ischemia. Cessation of function of the bypass usually precipitates a major amputation or critical ischemia and consequent severe deterioration in the quality of life. Surgeons and patients both have a vested interest in maximizing the longevity of a bypass. Most studies comparing synthetic grafts with autologous vein show synthetic grafts to be inferior to vein in femoropopliteal bypasses *(16,17)* but some show the difference only in the below-knee bypass group *(18)*. Most surgeons use a prosthetic graft only in situations in which autogenous vein is not available. However, in approximately one-third of patients, an autogenous vein is not available because of prior use for coronary or peripheral arterial bypass, severe varicose change, or too small a caliber. Some surgeons use prosthetic conduit by choice for above-knee femoropopliteal bypass *(19)*. This is with a view to save the long saphenous vein for a possible repeat reconstruction.

The reasons for bypass failure within the first 4 weeks are largely technical such as poor surgical technique, inappropriate choice of operation, and poor runoff. Anastomotic narrowing owing to neointimal hyperplasia and progression of disease in the inflow and runoff vessels is responsible for failure later on. Stenosis of the conduit, which happens frequently in vein bypass grafts *(20)*, occurs only rarely in prosthetic conduits *(21)*. Because these factors are progressive, a surveillance program might be expected to draw one's attention to correctable lesions so that secondary interventions may be planned to improve primary patency rates. In practice, however, there is conflicting evidence as to the benefit of such a surveillance program for prosthetic grafts, as measured by patency rates and amputation rates *(21–23)*. The most effective technique of surveillance and the most appropriate indication for intervention remain undetermined.

Clinical examination, symptomatology and ankle-brachial occlusion pressure indices (ABPI) can all be useful in bringing attention to possible threat to a bypass graft. But they are all somewhat subjective, may not give advance warning and by themselves cannot point to a possible remedy. Experience from vein graft studies show that

Table 3
Follow-Up of Infrainguinal Reconstruction Patients at the Royal Liverpool University Hospital, UK[a]

	Year 1	Year 2	Year 3	Year 4	Year 5	Year 10
Fasting serum lipid profile, serum creatinine, urea, electrolytes and complete blood count	2	2	2	2	2	2
Duplex ultrasound scan of the limb, including the graft	4	0	0	0	0	0
Office visit and resting ankle-brachial pressure	4	2	2	2	2	2

[a] The number in each cell represents the number of times per year each modality is obtained.

Note: There is no standardized follow-up schedule for patients following placement of arterial prostheses in other locations.

a fall by 0.1 in the ABPI pointed to 75% of threatened grafts *(24)* but this is too close to measurement error to be useful as a screening tool. Digital subtraction angiography gives detailed anatomical information regarding inflow, anastomoses, and runoff. It is probably performed in most patients undergoing secondary intervention. The cost, contrast administration, ionizing radiation, and the discomfort involved, however, make it inappropriate as a tool for routine surveillance. Duplex scanning provides a real-time B-mode ultrasonic image combined with Doppler blood flow analysis. It is noninvasive, does not involve radiation, and can provide a wealth of relevant information, making it the ideal tool for graft surveillance.

Identification of accelerated blood flow velocity with duplex scanning can be used to detect stenoses. Measurement of peak systolic velocity allows accurate estimation of the hemodynamic effect of a stenosis *(25)* and has been used effectively in prosthetic graft surveillance. Duplex scanning can also be used to assess anastomoses and lesions in inflow or runoff vessels. Another method is to compute the input impedance of the graft and the runoff vessels from the noninvasive measurement of instantaneous blood-flow velocity and pressure *(26)*. A high level of impedance points to a threat to the graft and further assessment by angiography needs to be undertaken.

The highest incidence of graft failure owing to correctable lesions is in the first year after the operation. There is little benefit in continuing surveillance duplex scanning beyond that period. At the regional vascular unit at Royal Liverpool University Hospital there is a graft surveillance program in place that encompasses synthetic grafts as well. The follow-up schedule is shown in the Table 3.

An efficacious follow-up program should achieve reduced amputation rates. There have been reports bearing conclusive proof of such benefit. A committed graft-surveillance program is expensive and increases the workload of a vascular service significantly *(27)*. It would appear as though some grafts can be rescued before they fail, but many graft failures cannot be predicted. Under these circumstances, many surgeons feel it is not justifiable to survey prosthetic grafts. The indications for intervention based on surveillance duplex scans are also somewhat subjective and not uniform.

CONCLUSION

Since the establishment of the safety and efficacy of prosthetic arterial conduits, several improvements have been added to enhance the performance of the grafts. Despite a lot of effort and expense, the ideal prosthetic conduit has not been developed yet and the search goes on. The recipients of arterial prosthetic grafts are at a significantly increased risk of morbidity and mortality from cardiac, cerebrovascular, respiratory, and metabolic diseases. In an asymptomatic patient, therefore, the goals of follow-up are mainly to control and modify the risk factors for arterial disease. The resources suggested by the primary chapter authors for those interested in further information are quite suitable and we have no further resources to suggest.

REFERENCES

1. Nunn DB, Freeman MN, Hudgkins PC. Post-operative alterations in size of dacron grafts: an ultrasonic evaluation. Ann Surg 1979;189:741–775.
2. Sauvage LR, Berger K, Wood SJ, Nakagawa Y, Mansfield PB. An external velour surface for porous arterial prosthesis. Surgery 1971;70:940–993.
3. Kennedy DA, Sauvage LR, Wood SJ. Comparison of noncrimped, externally supported (EXS) and crimped, nonsupported Dacron prosthesis for axillofemoral and above-knee femoropopliteal revascularization. Surgery 1982;92:931–936.
4. Lachapelle K, Graham AM, Symes JF. Antibacterial activity, antibiotic retention and infection resistance of a Rifampicin impregnated gelatin sealed Dacron graft. J Vasc Surg 1994; 19:675–682.
5. Goeau-Brissonniere O, Mercier F, Nicolas MH, et al. Treatment of vascular graft infection by in situ replacement with a rifampicin-bonded gelatin-sealed dacron graft. J Vasc Surg 1994;19:739–744.
6. Rhee RY, Gloviexki P, Camria RA, Miller VM. Experimental evaluation of bleeding complications, thrombogenecity and neointimal characteristics of prosthetic patch materials used for carotid angioplasty. Cardiovasc Surg 1996;4:746–752.
7. Gupta SK, Veith FJ, Kram HB. Prospective randomized comparison of ringed and non-ringed polytetrafluoroethylene femoropopliteal bypass grafts. J Vasc Surg 1991;13:163–172.
8. Tsuchida H, Cameron BL, Marcus CS, Wilson SE. Modified polytetrafluoroethylene: Indium 111-labeled platelet deposition on carbon-lined and high porosity polytetrafluoroethelyne grafts. J Vasc Surg 1992;4:643–649.
9. Stonebridge PA, Prescott RJ, Ruckley CV on behalf of the Joint Vascular Research group. Randomized trial comparing infra-inguinal PTFE bypass grafting with and without interposition cuff at the distal anastomoses. J Vasc Surg 1997;26:543–550.
10. Harris P, da Silva A, How T. Interposition vein cuffs. Eur J Vasc Endovasc Surg 1996;11: 257–259.
11. Dardik H, Wengerter K, Qin F, et al. Comparative decades of experience with glutaraldehyde-tanned human umbilical cord vein graft for lower limb revascularization: an analysis of 1275 cases. J Vasc Surg 2002;35:64–71.
12. Strobel R, Boontje AH, Van den Dunggn JJAM. Aneurysm formation in modified human umbilical vein grafts. Eur J Vasc Endovasc Surg 1996;11:417–420.
13. Johnston KW and the Canadian Society for Vascular Surgery Aneurysm Study Group. Nonruptured abdominal aortic aneurysm: six-year follow-up results from the multicentre prospective Canadian aneurysm study. J Vasc Surg 1994;20:163–170.
14. Plate G, Hollier LA, O'Brien P, Pairolero PC, Cherry KJ, Kazmier FJ. Recurrent aneurysms and late vascular complications following repair of abdominal aortic aneurysms. Arch Surg 1985;120:590–594.

15. Gilling-Smith GL, Mansfield AO. Thoraco-abdominal aortic aneurysm. Br J Surg 1995;82: 148–149.
16. Veith FJ, Gupta SK, Ascer E, et al. Six-year prospective multicentre randomized comparison of autologous saphenous vein and expanded polytetraflouroethylene grafts in infrainguinal arterial constructions. J Vasc Surg 1986;3:104–114.
17. Kumar KP, Crinnion JN, Ashley S, Case WG, Gough MJ. Vein, PTFE or Dacron for above knee femoropopliteal bypass? Int Angiol 1995;14:200.
18. Bergan JJ, Veith FJ, Bernhard VM, et al. Randomization of autogenous vein and polytetrafluoroethylene grafts in femorodistal constructions. Surgery 1982;92:921–930.
19. Michaels JA. Choice of material for above knee femoropopliteal bypass graft. Br J Surg 1989;76:7–14.
20. Tønnensen KH, Holstein P, Rørdam L, Büllow J, Helgstrand U, Dreyer M. Early results of percutaneous transluminal angioplasty (PTA) of failing below-knee bypass grafts. Eur J Vasc Endovasc Surg 1998;15:51–51.
21. Dunlop P, Sawyers RD, Naylor AR, Bell PRF, London NJM. The effect of a surveillance programme on the patency of synthetic infra-inguinal bypass grafts. Eur J Vasc Endovas Surg 1996;11:441–445.
22. Aune S, Pedersen W, Trippestad A. Surveillance of above-knee prosthetic femoro-popliteal bypass. Euro J Vasc Endovasc Surg 1998;16:509–512.
23. Golledge J, Beattie DK, Greenhalgh, Davies AH. Have the results of infrainguinal bypass improved with the widespread utilisation of postoperative surveillance? Eur J Vasc Endovasc Surg 1996;11:388–392.
24. Brennan JA, Walsh AKM, Beard JD, Bolia AA, Bell PRF. The role of simple non-invasive testing in infra-inguinal vein graft surveillance. Eur J Vasc Surg 1991;5:13–17.
25. Grigg MJ, Nicolaides AN, Wolfe JHN. Detection and grading of femoro-distal vein graft stenoses. Duplex velocity measurements compared with angiography. J Vasc Surg 1988;18: 661–666.
26. Wyatt MG, Muir RM, Tennant WG, Scott DJA, Baird RN, Horrocks M. Impedance analysis to identify the risk of femorodistal graft. J Vasc Surg 1991;13:284–293.
27. Loftus IM, Reid A, Thomson MM, London NJM, Bell PRF, Naylor AR. The increased work load required to maintain infrainguinal bypass graft patency. Eur J Vasc Endovasc Surg 1998;15:337–341.

19
Cardiac Valves

M. David Arya and Arthur J. Labovitz

INTRODUCTION

Cardiac valve replacement has had a dramatic impact on the prognosis of patients with valvular heart disease. Advances in diagnostic techniques, both interventional and surgical, have led to earlier diagnosis, more scientific selection of patients for surgery vs medical management, and increased survival of patients who undergo valve replacement. Data from the American College of Cardiology/American Heart Association (ACC/AHA) Task Force Report on valvular heart disease indicates that since the first prosthetic valve in 1960 through 1994, there were 1,957,000 valve implantations *(1)*. In the United States there are 97,000 valve replacement operations annually; outside the United States there are 149,000 annually (*see* Table 1). More than 50% of valve replacement operations in the United States involve bioprosthetic valves. The rest are mechanical valves. Outside the United States, bioprosthetic valves comprise 30% of the total and mechanical valves 70% *(2)*.

The medical literature base from which to make clinical management decisions has expanded in recent years, but gaps remain. In particular, there are few large-scale randomized trials addressing the treatment of patients with valvular disease. Clinical outcomes are dependent on operative technique, postoperative management, and referral patterns that are unique to each institution. Controlling for these factors is difficult. Each type of prosthetic valve has distinct advantages, but none is ideal for every patient.

Complications after valve replacement involve either structural deterioration or events related to the replacement of a native valve with a prosthesis. Structural complications involve value deterioration: wear, fracture, calcification, poppet escape, and leaflet tear. Nonstructural complications include the development of pannus, paravalvular leak, inappropriate sizing or positioning, valve thrombosis, embolism, bleeding complications and endocarditis. In essence, the patient exchanges a severe form of native valvular heart disease for a milder one at the time of surgery. The physician should be cognizant of the potential complications of each type of apparatus and how to diagnose them.

Early diagnostic tools available to the clinician for the assessment of prosthetic valve function included fluoroscopy, cardiac catheterization, and auscultation. Doppler echocardiography, introduced in 1979, provided a noninvasive, accurate, and reproducible way to evaluate heart function, valve area, and pressure gradient across valve

From: *The Bionic Human: Health Promotion for People With Implanted Prosthetic Devices*
Edited by: F. E. Johnson and K. S. Virgo © Humana Press Inc., Totowa, NJ

Table 1
Commonly Used Prosthetic Heart Valves

Valve type	Model	First implanted	Total implants[a]
Mechanical			
Ball-cage	Starr-Edwards	1965	200,000
Tilting disk	Björk-Shiley[b]		360,000
	Medtronic Hall	1977	178,000
	Omniscience	1978	48,000
	Monostrut	1982	94,000
Bileaflet	St. Jude	1977	580,000
	Carbomedics	1986	110,000
Bioprosthetic			
Porcine	Hancock	1970	177,000
	Hancock Modified Orifice	1978	32,000
	Carpentier Edwards (CE)	1971	400,000
	CE SupraAnnular	1982	45,000
Porcine (stentless)	Toronto Stentless	1991	5000
	Medtronic Freestyle Stentless	1992	5000
Pericardial	Carpentier Edwards	1982	35,000
Homograft	Various[c]	1962	28,000
Autologous	Pulmonary Autograft	1967	2000

[a] Approximate number of implants through 1994.
[b] The Björk-Shiley valve is no longer used. There are an estimated 36,000 patients with this valve still alive today.
[c] Includes noncommercial and Cryolife homograft valves.
Data from ref. 1.

orifices. Today, it is the primary monitoring tool used in following patients with prosthetic valves (3,4).

HISTORY OF HEART VALVE PROSTHESES

The first successful open heart surgery by Gibbon in 1953 paved the way for numerous invasive procedures, including valve replacement surgery. Currently, there are two classes of heart valve: mechanical prostheses, with rigid, manufactured occluders, and biological, or tissue valves, with flexible leaflet occluders of human or animal origin, usually mounted in a nonbiological frame.

Harken and colleagues successfully replaced the aortic valve and Starr the mitral valve with mechanical prostheses in 1960. The Starr-Edwards valve, the result of the combined work of a cardiac surgeon and a mechanical engineer, marked the beginning of the modern era of prosthetic valve replacement surgery. Problems with the durability of the silastic ball were corrected in the mid-1960s, paving the way for routine valvular replacement surgery.

Tilting disk prostheses, notably the Björk-Shiley in 1969, and bileaflet valves, such as the St. Jude in 1977, followed the Starr-Edwards caged-ball valve. It was immedi-

Chapter 19 / Cardiac Valves

ately apparent that all mechanical prosthetic valves required chronic anticoagulation to prevent catastrophic thromboembolism. Subsequent changes in the three mechanical valve types have been incremental, resulting in improved performance.

Recognition of the complications of mechanical prostheses led to the development of several types of valves prepared from biological tissues in the 1960s. Treating tissue valves with glutaraldehyde, introduced by Carpentier, paved the way for modern bioprosthetic valves *(8)*. The Hancock porcine valve, the first bioprosthetic valve, was available in 1970. Unfortunately, all bioprosthetic devices lack long-term durability. About 15 to 20% of patients require repeat valve replacement within 10 years owing to structural failure *(9)*. Similarly, the high structural failure rate of the glutaraldehyde-preserved valve constructed from bovine pericardium, developed by Ionescu, eventually led to its abandonment *(10)*.

CLASSIFICATION OF PROSTHETIC HEART VALVES

A heart valve prosthesis, in its essence, consists of an orifice through which blood flows, and an occluding device that opens and closes the orifice. The ideal valve should mimic the hemodynamic characteristics of the normal native valve. Hemodynamically, it should have no significant resistance to blood flow and transvalvular gradients similar to those of native valves. Regurgitation during valve closure should be minimal. The device should be durable and nonthrombogenic. Currently, none of the available models meets all of these characteristics so clinical decision making must weigh the advantages and disadvantages of each valve type.

Mechanical Valves

The shape of the occluder provides a simple basis for classification: caged-ball valves (Starr-Edwards), single-disc or tilting-disc (Björk-Shiley, Medtronic Hall, Omniscience) and bileaflet (St Jude, Carbomedics). A major advantage of mechanical valves is durability. The primary disadvantage is the need for long-term anticoagulation.

The caged-ball valve created by Starr and Edwards is the oldest prosthetic valve in continuous use. As its name implies, the occluder device is a ball of silicon rubber polymer that oscillates in a cage of cobalt chromium alloy (*see* Fig. 1) During opening, the ball is free to travel completely out of the orifice, thus reducing the possibility of thrombus formation. First introduced in 1960, the device underwent modifications in its initial design but has been essentially unchanged since the mid-1960s. Currently, the only caged-ball valve still in use is the Starr-Edwards 1260. Antegrade blood flow occurs around the ball with an average velocity of 2 m per second *(11)*. Turbulent blood flow owing to the location of the ball in the center of the flow stream contributes to the high thrombogenicity. The hemodynamic performance of this valve is poor, with higher transvalvular pressure gradients than observed with tilting-disc and bileaflet types.

Tilting-disc valves were designed to provide more central blood flow than the caged-ball valve. This design employs a circular disc as an occluder, which is retained by stout wire arms or closed loops that project into the orifice. The first successful tilting disc valve was introduced by Björk and Shiley in 1969. Although no longer produced, more than 360,000 such valves were implanted and an estimated 38,000 patients with these valves are still alive *(12)*. The Medtronic Hall tilting disc was introduced in 1977. It uses a guide strut that projects through a central hole in the disc (*see* Fig. 2). The opening of the disk relative to the valve annulus varies from 60 to 80°. There are two

Fig. 1. An example of the Starr-Edwards mechanical prosthesis in the closed position.

orifices for antegrade flow. The antegrade velocity across this valve averages 2 m per second. There is a small amount of normal regurgitation owing to backflow around the central strut and small gaps around the perimeter of the valve *(13)*.

Bileaflet valves are currently the most commonly used mechanical valves. Introduced by St. Jude Medical in 1977, this design uses an occluding mechanism consisting of two semicircular leaflets with small ears that pivot in butterfly-shaped recesses in the housing. The leaflets swing apart when open, creating three flow areas, two peripheral and one central (*see* Fig. 3). The opening angle of the leaflets is 75 to 90°. Each leaflet consists of a graphite core coated in pyrolytic carbon.

The St. Jude device valve has become the most widely used type. As of the year 2000, more than 1 million valves had been implanted. Other valve companies have adopted its basic design while varying elements of the housing, leaflets, and sewing ring. The Carbomedics, one such variant, is a bileaflet valve composed of pyrolytic carbon with a titanium housing that can be rotated so as to avoid interference with disc excursion by subvalvular tissue. The average antegrade flow velocity for bileaflet valves is 3 m per second in the aortic position and 1.6 m per second in the mitral position *(14)*. As with the tilting-disc models, there is a mild degree of normal regurgitation, emanating from the pivot points of the valve discs. This regurgitation is designed in part to decrease the risk of valve thrombosis.

The main advantage of mechanical valves is their durability. Actuarial survival rates range from 60 to 70% at 10 years for the Starr-Edwards valve to 85% at 9 years for the Omnicarbon valve and 94% at 10 years for the St. Jude Medical bileaflet valve *(15–19)*. Patient survival depends on factors related to each individual patient's disease and comorbid conditions. The most common types of dysfunction for mechanical valves are paravalvular leak, endocarditis, and thrombosis. Structural failure of mechanical valves is extremely rare but can have devastating complications for the patient and presents the physician with difficult treatment dilemmas. This topic is dealt with elsewhere in the book.

Fig. 2. Medtronic Hall mechanical prosthesis in fully opened position. The central strut fits through a hole in the disc. The open position creates a major and minor orifice.

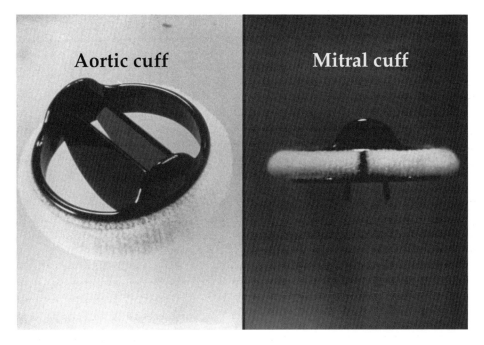

Fig. 3. St. Jude Medical bileaflet mechanical prosthesis in opened position. The two large orifices and the smaller central orifice are clearly visible.

Choosing a mechanical valve requires consideration of many factors, including the patient's age and suitability for chronic anticoagulation. Two studies at our institution are the only prospective randomized trials comparing two different prostheses (the bileaflet St. Jude and the pivoting-disk Medtronic Hall) in the mitral and aortic positions. Tateneni and colleagues studied the rest and exercise hemodynamics in 90 patients randomized to receive either St. Jude or Medtronic Hall devices in the aortic and mitral positions. They found no hemodynamic or clinical differences between the two patient groups *(85)*.

Fiore and colleagues randomized 102 patients to receive either the St. Jude Medical or Medtronic Hall prosthesis in the mitral position, with a 10-year observation period. They found no differences in late clinical performance or hemodynamic results between groups. There was no significant difference in the resting or exercise transvalvular gradient for either valve type. Ten-year actuarial survival results for the two valve types were similar at 53% (St. Jude) and 58% (Medtronic Hall) *(20)*.

Variations in the normal resting gradients of mechanical prosthetic valves may make it difficult to differentiate between normal and marginally stenotic valves. The use of exercise Doppler echocardiography may make these subtle differences clearer. A review by Dressler et al. of transvalvular gradients, under both resting and exercise conditions, helped establish what constitutes optimal prosthetic valve hemodynamics during rest and exercise. The peak instantaneous gradient for mechanical valves in the aortic position can double with exercise (35–63 mmHg) when compared with rest (18–26 mmHg). Similarly, the pressure gradient across mechanical mitral valves, which are mildly stenotic at rest (gradients 2.3–7.1 mmHg), increases considerably with exercise (gradients 5.1–16.5 mmHg) *(3)*.

All mechanical heart valves require chronic anticoagulation. Despite adequate anticoagulation, the rate of thromboembolism is 0.6% to 2.0% per patient per year for mechanical valves. Most studies have shown that the risk of thromboembolism is higher, with a mechanical valve in the mitral position *(21)*. The risk of thromboembolism is about the same for bileaflet or tilting-disc models. The 10-year actuarial freedom from thromboembolism is 72% for both the St. Jude and Medtronic Hall valves *(20)*. The use of chronic anticoagulation also entails a risk of bleeding complications, averaging approximately 1% per year *(21)*.

Biological Valves

There are a wide variety of biological valves. They include valves translocated within the same individual (autograft), harvested from a human at the time of death (homograft), or transplanted from a nonhuman animal or fashioned from animal pericardium (heterografts). All bioprosthetic valves have the advantage of nonthrombogenicity, thus obviating the need for chronic anticoagulation. Their major disadvantage is limited durability, necessitating eventual replacement.

The most frequent situation in which an autograft is used is the Ross procedure in which a patient's own pulmonic valve is autotransplanted to the aortic position. The pulmonary valve is then replaced by an aortic or pulmonary homograft or a heterograft. The procedure was first described in 1967 *(22)* and has had favorable long-term results. The surgery is more complex, however, with a higher rate of technical errors, leading to early valve failure *(23,24)*. The major long-term problem is degeneration of the

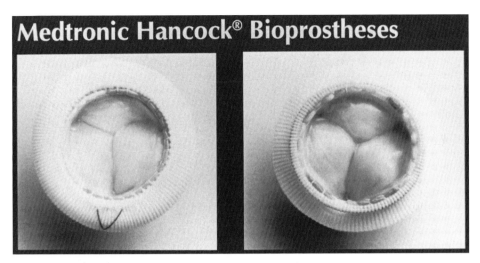

Fig. 4. Hancock bioprosthesis in the fully closed position.

right-sided homograft with need for reoperation. Progressive aortic insufficiency is observed in a small percentage of patients.

The majority of bioprosthetic valves used now are stented porcine valves, such as the Hancock and Carpentier-Edwards valves. These heterografts consist of porcine valve leaflets treated with glutaraldehyde, which sterilizes the valve tissue, destroys its antigenicity, and makes the valve more durable by stabilizing collagen cross-links. The valve leaflets are mounted on flexible stents with a sewing ring. This allows the three-dimensional structure and relationships of the valve to be maintained (*see* Figs. 4 and 5). The hemodynamic profiles of the porcine heterografts are similar to those of comparably sized mechanical prostheses. Peak blood flow is 2 to 3 m per second in the aortic position and 1.5 m per second in the mitral position. Corresponding mean transvalvular pressure gradients are 10 to 15 mmHg in the aortic position and 4 to 7 mmHg in the mitral position *(25)*. Bioprosthetic valves have almost no valvular regurgitation although echo Doppler may detect a small amount of regurgitation in up to 10% of normally functioning bioprosthetic valves *(14)*.

Long-term experience with stented bioprosthetic valves, however, has indicated that stents produce suboptimal hemodynamics. It is felt that the stent contributes to leaflet deterioration and calcification. A standardized measurement of a hemodynamic profile is the performance index. This is defined as the ratio of the effective orifice area to the area of the sewing ring. A valve with a performance index of 1 reflects optimal hemodynamics. A performance index of less than 1 indicates poorer hemodynamics. Mechanical valves have a calculated performance index of 0.6 to 0.7. In comparison, similarly sized bioprosthetic valves have performance indexes in the range of 0.3 to 0.4 *(13)*. Possible explanations for the lower score of bioprosthetic valves include the presence of semirigid stents, stiffness of the fixed leaflet tissue, and the three-dimensional structure of the leaflets *(26)*.

The stentless bioprosthetic valve (Toronto SPV, Medtronic Freestyle) was developed to improve valve hemodynamics and durability while retaining the advantages of the stented bioprosthetic valve (*see* Fig. 6). Stentless bioprosthetic valves are manufac-

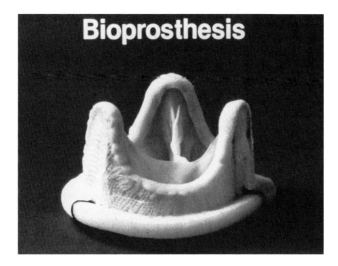

Fig. 5. Example of a stented heterograft prosthesis.

Fig. 6. Example of a Medtronic Freestyle porcine aortic prosthesis. It may be used as root replacement or it may be trimmed for subcoronary implantation.

tured from porcine valves that are processed at low pressures. The use of Dacron cloth covering (Toronto SPV) and anticalcific treatments (PSB stentless porcine valve) are some of the methods used to improve durability *(27)*.

The hemodynamics of stentless prosthetic valves appear to improve over time. The valves have only been used in the aortic position up to this time. The mean transvalvular gradient is generally about 15 mmHg at time of implantation, decreasing by 30% at 6 months follow-up *(28,29)*. Aortic regurgitation is seen in less than 5% of the patients with stentless prosthetic valves *(30)*.

Bovine pericardial valves are the third major type of bioprosthetic valve. These are made of bovine pericardium sewn into a valvular configuration on a stented frame. Bovine pericardium has a higher percentage of collagen than do porcine valves, thus providing a theoretical advantage of less shrinkage. The first commercially available bovine pericardium bioprosthetic valve was the Ionescu-Shiley. Owing to unacceptably high failure rates related to cusp dehiscence, it was taken off the market after 10 years *(10)*. Currently, the only available bovine pericardium valve in the United States is the Carpentier-Edwards Perimount aortic valve.

Although long-term anticoagulation is not required, patients with bioprosthetic valves require anticoagulation for the first 3 months while the sewing ring becomes endothelialized. The rate of thromboembolism is less than 1% per patient per year without anticoagulation *(31,32)*. As would be expected, the rate of clinically detectable bleeding in patients with bioprosthetic valves is less than half that observed with mechanical valves.

The major disadvantage relates to durability. Unlike mechanical valves, for which the structural failure rate is extremely low, structural failure is inevitable for biological valves. Degeneration, cuspal tears, fibrin and calcium deposition, and fibrosis lead to degeneration or stenosis, requiring eventual replacement. These changes can be seen as early as 3 to 5 years after implantation and by 10 years the rate of primary tissue failure averages 30% *(33)*. Several series estimate the rate of structural deterioration of porcine valves at 3.3% per patient year *(34,35)*.

Location of the prosthesis and age of the patient also play a role in durability. Jones et al. followed patients who underwent a total of 1050 aortic or mitral valve replacements with either Hancock or Carpentier-Edwards valves for an average of 10 years. Actuarial freedom from structural valve deterioration at 1- years follow-up was significantly better for valves in the aortic position (79%) than in the mitral position (63%). Furthermore, elderly patients are less likely to develop structural failure than younger patients. Freedom from reoperation at ten years for patients aged 60 or older was 92% for aortic valves and 80% for mitral valves. Freedom-from-reoperation rates at 10 years for patients under 40 were 47% (aortic) and 46 % (mitral) for those less than 40 years of age *(35)*. Accordingly, these valves are only implanted in patients less than 60 years of age under special circumstances.

MANAGEMENT OF THE PATIENT WITH A PROSTHETIC HEART VALVE

The management of patients with prosthetic heart valves is a lifelong process. The patient exchanges the disease process of valvular heart disease for that of prosthetic heart valve disease and its sequelae. The treating physicians work together to ensure that the resultant operation benefits the patient and that the potential complications from the prosthetic valve are minimized. The management focuses on routine assessment of valvular function, maintenance of anticoagulation and minimization of bleeding risk in patients requiring chronic anticoagulation therapy, prevention of infection, and early detection of structural degeneration of the prosthetic valve.

Selection of an Artificial Valve

The selection of the appropriate artificial valve is a complex decision that should involve input from the surgeon, the cardiologist, and the patient. Early and late mortality rates, prosthetic valve endocarditis rates, and reoperation rates are similar for mechanical and bioprosthetic valves in the first 5 years. There are only two randomized trials comparing mechanical and porcine valves, the Edinburgh Heart Valve Trial and the Veterans Affairs Cooperative Study on Valvular Heart Disease.

The Edinburgh study compared the durability of the Björk-Shiley (tilting disk) and the Hancock and Carpentier-Edwards bioprosthetic valves. The authors found no significant difference in the actuarial incidence of reoperation or death at 5 years. By 12 years, those patients with mechanical prostheses had a trend toward increased survival compared to those with bioprosthetic valves. The rate of reoperation was higher in patients with the porcine prosthesis (37 vs 8%). Age and location of the valve also correlated with higher rates of structural failure. Younger patients and those with bioprosthetic valves in the mitral position had higher incidences of reoperation for structural failure. Not surprisingly, the incidence of major bleeding was higher in the mechanical valve patients (18%) than those with bioprosthetic valves (7%). There was no difference in the rates of thromboembolism or endocarditis between the two valve types (32).

The Veterans Study compared survival and complication rates between two groups of men ($N = 575$) randomized to either tilting-disc (Björk-Shiley) or bioprosthetic (Hancock) valves at the time of valve replacement surgery. Both groups had a similar probability of death at 11 years of follow-up (57% for the Björk-Shiley valve; 62% for the Hancock valve). There was also a similar rate of complications (65% for the Björk-Shiley; 69% for the Hancock). There was no difference in the incidence of embolism, bleeding, and endocarditis between the two groups. As in the Edinburgh study, there was a higher incidence of structural valve failure in the bioprosthetic valve group (36% for aortic valves, 15% for mitral valves) when compared with the mechanical valve group (0% for both aortic and mitral valves) (36).

These two studies highlight the major considerations, expected durability of the valve, and the risks of anticoagulation that must be tailored to an individual patient undergoing valve replacement surgery. Mechanical valves are the prostheses of choice in younger patients and patients already on chronic anticoagulation for atrial fibrillation. Bioprosthetic valves are indicated for patients with contraindications for chronic anticoagulation, those who are noncompliant with anticoagulation treatment, and elderly patients with limited life expectancy. The decision may not always be easy.

Follow-Up Visits

The patient with a newly implanted artificial heart valve should have an outpatient evaluation within 3 to 4 weeks after discharge from the hospital. At this time, the patient's recovery should have progressed to the point that his physical abilities and functional capacity can be assessed. At that time, the ACC/AHA Guidelines recommend that the physician perform a complete history and physical and obtain an electrocardiogram, chest X-ray, two-dimensional (2-D) echocardiogram with Doppler, complete blood count, serum chemistries (creatinine, electrolytes), and , if appropriate, assessment of anticoagulation (see Table 2).

The main focus of the examination is to assess the function of the prosthesis and evaluate the patient for infection, conduction disease, valvular disorders, and coronary artery disease (CAD) (1). Any symptoms suggestive of prosthetic valvular dysfunction require immediate investigation.

Table 2
Patient Surveillance After Implantation of Prosthetic Heart Valve at Saint Louis University Hospital in the Asymptomatic Patient

	Postoperative year					
	1	2	3	4	5	10
Office visit, including history and physical	2^a	1	1	1	1	1
Complete blood count and serum electrolytes and creatinine	2^a	1	1	1	1	1
Electrocardiogram	2^a	1	1	1	1	1
Transthoracic echocardiogram	1	0^b	0^b	0^b	0^b	0^b
PT/INR (if taking chronic anticoagulation)^c	2^a	1	1	1	1	1

^{a} At 3–4 weeks postop and at 12 months.
^{b} Repeat transthoracic echocardiography is not routinely performed. Indications for echocardiography are driven by clinical findings, including new symptoms or changes in physical findings.
^{c} In patients taking warfarin. This is in addition to routine monthly monitoring by the primary medical doctor.

The first step in any office visit in a patient with prosthetic heart valve disease is a detailed history and physical examination. The importance of knowing the type and time of valve implantation cannot be overestimated. The patient may relate symptoms of fatigue or shortness of breath that may be clues to valvular dysfunction. A patient with an unexplained fever or a neurological event should be evaluated for endocarditis or thrombosis.

The physical examination is important. Findings such as elevated jugular venous pressure, peripheral edema, and pulmonary rales suggest heart failure and should prompt a more detailed evaluation of prosthetic valve function. Abnormal auscultatory findings may be the first sign of prosthetic valvular dysfunction (see Table 3).

The findings on cardiac examination vary with the type of valve. The physician should be familiar with the auscultatory characteristics of each type of prosthesis. Bioprosthetic models have very soft opening and closing sounds with no appreciable clicks and in fact may be difficult to distinguish from healthy native valves. In the aortic position, an ejection murmur is commonly heard with stented porcine valves. Mechanical valves have a variety of opening and closing clicks, depending on the valve type. The caged-ball valve has multiple sounds due to the ball moving back and forth along the stented cage. Additionally, there are distinct opening and closing clicks. Tilting disc valves have both an opening and closing click owing to the movement of the disc against the restraining struts. Bileaflet valves typically have an audible closing click, but the leaflets open silently. All mechanical valves in the aortic position have an associated systolic ejection murmur. The degree of mitral regurgitation observed with mechanical valves is rarely appreciated with auscultation.

Sequential physical examinations can prove invaluable in detecting dysfunctional valvular prostheses. The presence of an aortic diastolic murmur or a loud systolic ejection murmur is always an abnormal finding. Changes in the baseline auscultatory examination should prompt investigation. Audible aortic regurgitation generally indicates pathological valvular or paravalvular incompetence. Muffled valve clicks may represent thrombosis or pannus formation.

Table 3
**Auscultatory Findings in Patients
With Normally Functioning Prosthetic Heart Valves**

Valve type	Aortic position	Mitral position
Caged-ball (Starr-Edwards)	Sharp opening sound after S1 Systolic ejection murmur after S1 Sharp closing sound after S2 Ball rattling in cage during systole	Sharp opening sound 70–150 ms after S2 Systolic ejection murmur after closing sound Sharp closing sound at S1 Ball rattling in cage
Tilting Disk (Björk-Shiley, Medtronic Hall)	Soft opening sound after S1 Systolic ejection murmur after S1 Sharp closing sound after S2	Soft opening sound 70–150 ms after S2 Sharp closing sound at S1
Bileaflet (St. Jude)	Inaudible opening sound after S1 Systolic ejection murmur Soft closing sound after S2	Inaudible opening sound after S2 Diastolic murmur Soft closing sound at S1
Porcine Bioprosthetic (Hancock, Carpentier-Edwards)	Systolic ejection murmur	Diastolic rumble

Doppler Echocardiography

The methods used to assess prosthetic valvular function have changed over the years. Fluoroscopy allowed visualization of the prosthetic valve position and detection of dehiscence and abnormalities of the occluder. Cardiac catheterization remained the standard for the assessment of prosthetic valvular function until Doppler echocardiography became available in the mid-1980s.

The most commonly used test to assess prosthetic valve function is 2-D echocardiography with Doppler. Holen first described its utility in 1979 *(37)*. Since that time, many studies have affirmed its usefulness and it has become the procedure of choice for the evaluation of valve function *(4)*. As with native valves, it provides information about anatomical structure and valve function. A baseline echocardiogram after the surgery establishes a hemodynamic signature of the prosthetic valve with which one can compare later studies.

However, owing to metallic components in mechanical valves, the images obtained with echocardiography are limited by shadowing. This is also seen in most bioprosthetic valves owing to the stents and sewing ring. Mechanical valves, especially in the mitral position, are difficult to assess with transthoracic echocardiography due to severe reverberations from the metal in the valve. In this case, transesophageal echocardiography (TEE) is needed to obtain clear images of disc or leaflet motion and the left atrium. TEE, however, is not routinely needed and should only be performed if valvular dysfunction is suspected based on physical findings, clinical symptoms, or abnormalities on prior transthoracic echocardiography.

Stented bioprosthetic valves often allow adequate assessment of the leaflets with 2-D echocardiography, although TEE is occasionally needed. Other than thickening or increased echogenicity in the aortic annulus, stentless bioprosthetic valves, homografts, and autografts may be impossible to differentiate from native valves.

The same Doppler principles used to assess native valves apply to prostheses. Echo Doppler allows the determination of pressure gradients by use of the modified Bernoulli equation and valve area by the pressure-half-time method for mitral valves and continuity equation for aortic valves.

Doppler echocardiography can accurately measure transvalvular pressure gradients. By aligning the Doppler beam parallel to blood flow, peak and mean transvalvular pressure gradients can be measured by applying the modified Bernoulli equation: gradient = $4v_{max}^2$, where v_{max} is the maximal transvalvular velocity. One can estimate the size of a valve orifice by calculating the time it takes the initial peak pressure gradient to fall to one-half its original value (pressure half-time). The continuity equation is used during Doppler examination to calculate aortic valve area. It states that flow (F) through a conduit is a function of the instantaneous velocity (V) of the fluid and the area (A) of the conduit, $F = V \times A$. Using Doppler, the flow through the aortic valve and left ventricular outflow tract can be measured. After measuring the diameter of the left ventricular outflow tract, one can determine the area of the aortic valve. The validity of such calculations to determine mitral and aortic valve area has been proven by comparison with results obtained at cardiac catheterization *(38)*.

Doppler echocardiography reveals that all prosthetic heart valves have some degree of obstruction when compared with native valves. The velocities and gradients for prosthetic valves are higher than those of normal native valves. Determination of appropriate velocities and gradients across normal prosthetic valves depends on valve size, valve type, and flow (*see* Tables 4 and 5). Numerous studies have established the ranges of normal values for the pressure gradients and valve area of prosthetic valves measured by echocardiography.

Minor regurgitation occurs with all mechanical valves. It is important to recognize this when performing Doppler echocardiography. The amount and pattern of regurgitation, typically located at the pivot joints and along the central closure line, are unique for each type of prosthetic valve. The regurgitant volume may be up to 15 mL per beat *(39)*. The bileaflet St Jude has two to four centrally directed regurgitant jets. The tilting disc Medtronic Hall has two jet types: one through the central guiding hole and the other along the disk circumference. For the bileaflet valve, the jets are low in intensity and have minimal penetration into the atrium. With the tilting-disc valve the central jet is more prominent and longer *(40)*. Bioprosthetic valves typically have minimal regurgitation. If present, it is typically a small, centrally located jet *(41)*.

While a baseline 2-D echocardiogram with Doppler is a class I indication according to the ACC/AHA guidelines, the role of serial echocardiograms in asymptomatic patients is controversial. The frequency with which such studies should be undertaken is unknown as there are no data on which to base such a decision. Obtaining a routine serial echocardiogram at the time of annual follow-up, in the absence of any change in clinical status, is currently a class IIb indication, according to the most recent guidelines, but many feel that the need for echocardiography should be dictated by clinical circumstances such as a change in symptoms or development of a new murmur.

Table 4
Normal Prosthetic Mechanical Valve Hemodynamics

Valve type	Peak velocity (m/s) Mean ± 1 SD	Mean gradient (mmHg) Mean ± 1 SD	Valve area (cm^2) Mean (range)
Aortic			
Caged-ball			
Starr-Edwards	3.1 ± 0.5	2.4 ± 4	
Tilting disk			
Björk-Shiley	2.6 ± 0.4	14 ± 5	
Medtronic Hall	2.4 ± 0.2	14 ± 3	
Bileaflet			
St. Jude	2.5 ± 0.6	12 ± 7	
Carbomedics			
Mitral			
Caged-ball			
Starr Edwards	1.9 ± 0.4	5 ± 2	2.1 (1.2–2.5)
Tilting disk			
Björk-Shiley	1.6 ± 0.3	3 ± 2	2.4 (1.6–3.7)
Medtronic Hall	1.7 ± 0.3	3 ± 0.9	2.4 (1.5–3.9)
Bileaflet			
St. Jude	1.6 ± 0.3	3 ± 1	2.9 (1.8–4.4)
Carbomedics			

Data from refs. 26 and 38.

Table 5
Normal Bioprosthetic Valve Hemodynamics

Valve type	Peak velocity (m/s) Mean ± 1 SD	Mean gradient (mmHg) Mean ± 1 SD	Valve area (cm^2) Mean (range)
Aortic			
Hancock	2.4 ± 0.4	11 ± 2	1.8 (1.4–2.3)
Carpentier-Edwards	2.4 ± 0.5	14 ± 6	1.8 (1.2–3.1)
SPV-Toronto (stentless)	2.2 ± 0.4	3 (2–20)	— (1.8–2.3)
Aortic Homograft	1.8 ± 0.4	7 ± 3	2.2 (1.7–3.1)
Mitral			
Hancock	1.5 ± 0.3	4 ± 2	1.7 (1.3–2.7)
Carpentier-Edwards	1.8 ± 0.2	7 ± 2	2.5 (1.6–3.5)
Aortic Homograft	1.8 ± 0.4	7 ± 2	2.2 (1.4–3.0)

Data from refs. 26 and 38.

Echocardiographic evaluation should always be performed if there is any concern about valve deterioration and integrity, a new murmur, or concerns about ventricular function. Bioprosthetic valves have an increasing risk of structural deterioration at 5 to 8 years. Current guidelines recommend using changes in clinical status or physical examination as the indication for echocardiography. For patients with a history of valve thrombosis, echocardiography should be performed once a month for 6 months after the initial event. Thereafter, echocardiography to examine the valve should be performed every 6 months *(1,42)*.

Thus, routine follow-up in asymptomatic patients with prosthetic heart valves consists of annual history and physical examinations. Routine echocardiography is not indicated unless driven by individual considerations. Diagnostic testing, with echocardiography or blood work, is indicated by changes in the patient's functional status or physical exam. Other considerations, such as monitoring of chronic anticoagulation or continued assessment and treatment of concomitant conditions (such as CAD or cardiac arrhythmias), may require the physician to evaluate the patient on a more frequent basis.

CHRONIC ANTICOAGULATION: THROMBOEMBOLIC RISK AND BLEEDING RISK

Prosthetic heart valves have a risk of thromboembolic complications. These can be devastating and far outweigh the benefits of the treatment of valvular heart disease. The prevention of thrombus formation by chronic anticoagulation, however, presents another significant risk, bleeding, that may be inconvenient at least and fatal at worst.

Akins has compiled a detailed review of the complications associated with mechanical prosthetic valves *(43)*. Generating a composite linearized rate for each complication, by taking the total number of all events from available studies and dividing by the total patient-years of follow-up, he has identified the risk of bleeding and thromboembolic complications for mechanical valves in the aortic and mitral positions. Even the author, however, acknowledges the limitations of trying to determine the risk of bleeding or thromboembolism in patients with prosthetic heart valves. Reported incidences of complications depend on individual patients and are thus susceptible to change over time. Furthermore, thromboembolic and bleeding complications recorded in studies are not simply functions of the individual prosthesis because variables such as patient age, inherent hematological function, type of valvular disease, and thoroughness of the person managing the patient's anticoagulation play an important role in determining the risk of complications.

Anticoagulation

Mechanical valves require chronic anticoagulation therapy. Despite the use of warfarin, the risk of thromboembolic events in patients with mechanical valves is 1–2% per year *(21)*. The desired level of anticoagulation, as determined by the International Normalized Ratio (INR), depends on the type of mechanical valve and the location, aortic or mitral. A review by Cannegieter and colleagues determined that the incidence of thromboembolism was 0.5 per 100 patient-years for bileaflet valves, 0.7 per 100 patient-years for tilting disc valves, and 2.5 per 100 patient-years for the caged-ball valves. Valves in the aortic position had a lower incidence of thromboembolism (0.5 per 100 patient-years) than did valves in the mitral position (0.9 per 100 patient-years). Interestingly, women had a slightly higher risk of thromboembolism than men (0.8 vs 0.6 per 100 patient-years) *(44)*. Even patients with bioprosthetic valves are at risk for thromboembolic events, especially in the first few months after valve implantation.

Table 6
Anticoagulation Therapy Recommendations for Prosthetic Heart Valves

Valve type	Goal INR (range)
Mechanical Valves	
Aortic valve (bileaflet, tilting disk) and no risk factors[a]	2.5 (2.0–3.0)
Aortic valve (caged-ball) and no risk factors	3.0 (2.5–3.5)
Aortic valve and risk factor(s)	3.0 (2.5–3.5)
Mitral valve (any type) with or without risk factors	3.0 (2.5–3.5)
Bioprosthetic Valves	
First 3 months after surgery	2.5 (2.0–3.0)
Aortic valve (any type) and risk factor(s)	2.5 (2.0–3.0)
Mitral valve (any type) and risk factor(s)	3.0 (2.5–3.5)

[a] Risk factors defined as the presence of atrial fibrillation, history of prior thromboembolism, or history of hypercoagulable state. (Data from refs. 1 and 44.)

A meta-analysis of 46 observational studies on thromboembolic complications in patients with mechanical heart valves confirmed a difference in the risk of embolic events, depending on the type of mechanical valve. The incidence of thromboembolism for bileaflet valves was 50% lower and for tilting-disc valves 30% lower than for caged-ball valves (45). The desired level of anticoagulation depends on the type of mechanical valve and the position, aortic or mitral, as well as associated conditions.

The ideal INR targets, as determined from data from multiple randomized trials, ACC/AHA guidelines and the Fifth Consensus Conference on Antithrombotic Therapy, are (1) (see Table 6):

- Goal INR of 2.5 (range 2.0–3.0) for patients with a bileaflet mechanical valve in the aortic position. Left atrium must be of normal size, the patient must be in sinus rhythm, and the left ventricular ejection fraction most be normal.
- Goal INR of 3.0 (range 2.5–3.5) for patients with a tilting-disc valve in the aortic or mitral position or bileaflet valve in the mitral position. Also for patients with chronic atrial fibrillation, bileaflet mechanical valves, or caged-ball valves.
- Goal INR of 3.0 (range 2.5–3.5) in combination with low-dose aspirin (80–100 mg daily) in patients with mechanical valves and additional risk factors (atrial fibrillation, history of previous thromboembolism, left ventricular dysfunction, or history of a hypercoagulable state).

It is important to keep in mind that it is difficult to achieve optimal levels of anticoagulation within the desired INR range. Several studies have shown that actual INR vs goal INR can vary widely (46,47). These studies revealed that the INRs of patients are outside of the desired therapeutic range as much as 40% of the time. This intrapatient INR variability is a risk factor for thromboembolic events.

The addition of antiplatelet agents, notably aspirin in low doses (80–100 mg per day) should also be considered in patients with increased risk of thromboembolism (48). Such risk factors include a history of thromboembolism, atrial fibrillation, and hypercoagulable state. Some would argue that patients with severely reduced left ventricular function would also benefit although this has not been studied. The addition of aspirin has been shown to lower the risk of thromboembolism compared with warfarin therapy alone. Aspirin also decreases mortality owing to other cardiovascular diseases. The use of dipyridamole, in addition to chronic warfarin therapy, has theoretical benefits, but further studies are required to delineate its role in chronic anticoagulation therapy.

The benefits of the addition of antiplatelet agents such as aspirin are accomplished by increased risk of bleeding complications. Turpie and colleagues examined the issue of combination therapy with warfarin and aspirin in patients with mechanical valves *(49)*. They found that combination therapy reduced the risk of thromboembolic events by 77%, compared with warfarin alone, but increased the risk of all bleeding events by 5.5%. There was, however, no significant increased risk of major bleeding events for those receiving combination therapy.

Massel and Little completed a meta-analysis of the risk and benefit of adding antiplatelet therapy to standard coumadin therapy in patients with prosthetic heart valves *(50)*. They found that the addition of antiplatelet agents, primarily low-dose aspirin but also high-dose aspirin and dipyridamole, to warfarin therapy in these patients was associated with decreased risk of mortality and thromboembolic events. The relative risk reduction for both antiplatelet agents was 57% for thromboembolic events and 49% for mortality. There was an increased risk of bleeding with the use of antiplatelet agents, but the authors noted that increased risk was seen primarily in earlier trials, before the widespread use of the standardized INR, and with higher doses of aspirin (500–1000 mg) than are typically used today.

The major advantage of bioprosthetic valves is freedom from the need for chronic anticoagulation, but bioprosthetic valves are not free of all risk. The incidence of thromboembolic events has been reported as 1.7–6/100 patient-years *(51,52)*. The period of greatest risk in these patients occurs during the first 3 months after valve implantation and anticoagulation (target INR of 2.0–3.0) is recommended during this time period *(53)*. The increased risk is believed to be owing to lack of endothelialization of the sewing ring during the immediate postoperative period. Some centers use aspirin only for valves in the aortic position. Once 3 months have passed, treatment with aspirin alone is adequate *(54)*. No prospective trials have been undertaken to determine the best anticoagulation regimen for patients with bioprosthetic valves. The current anticoagulation recommendations for patients with bioprosthetic valves are *(1)* (*see* Table 6):

- Goal INR of 2.5 (range 2.0–3.0) for the first 3 months after insertion of a bioprosthetic valve in the mitral or aortic position.
- Aspirin 325 mg or low dose (80–100 mg daily) alone after 3 months in patients with bioprosthetic valves.
- Goal INR of 2.5 (range 2.0–3.0) in combination with low-dose aspirin for patients with bioprosthetic valves in the aortic position with additional risk factors (atrial fibrillation, history of previous thromboembolism, left ventricular dysfunction, or history of a hypercoagulable state).
- Goal INR 3.0 (range 2.5–3.5) in combination with low-dose aspirin for patients with bioprosthetic valves in the mitral position with additional risk factors.

As noted earlier, biological prosthetic valves are not without risk of thromboembolic events, but the risk is low. Jones and colleagues reported the 10-year actuarial rate of freedom from thrombosis and embolism in patients with bioprosthetic valves (Hancock and Carpentier-Edwards valves) to be 93% *(35)*. This rate is similar to that of patients with mechanical valves who are chronically anticoagulated with warfarin. Patients with bioprosthetic valves, however, do not have the added risk of hemorrhagic events associated with chronic anticoagulation therapy. Jones found the 10-year actuarial freedom from bleeding complications to be 98% for aortic valves and 95% for mitral valves (35). Of course, patients with other risk factors, such as chronic atrial fibrillation, previous thromboembolism, or hypercoagulable state, may require warfarin. The decision to institute chronic anticoagulation therapy in patients with decreased left ventricular function (defined as left ventricular ejection fraction <30%) has not been formally studied, although many would recommend this therapy.

The first evidence of thrombosis is often a peripheral embolus. The most common and most devastating site of embolization is the brain. The second most common site is the coronary circulation, followed by extremities, kidneys, mesenteric arteries, and retinal vessels. As noted earlier, the effectiveness of anticoagulant therapy is variable, with up to 40% of patients having subtherapeutic INR values at any given time (46,47). The suggestion of peripheral emboli by history, physical examination, laboratory tests, or imaging modalities should prompt an immediate evaluation of the valve prostheses with echocardiography and assessment of the adequacy of anticoagulation therapy.

Current ACC/AHA guidelines recommend raising the target INR range or adding aspirin in a patient who has had a definite embolic event while on adequate antithrombotic therapy. Patients taking warfarin with an INR of 2.0 to 3.0 should increase the warfarin dose to achieve an INR of 2.5 to 3.5. Aspirin (80–100 mg daily) should be added to warfarin therapy. Patients already taking low-dose aspirin should increase the dose to 325 mg daily (1).

Even patients with homografts have rates of thromboembolism similar to those with porcine or pericardial bioprosthetic valves. This may be explained by the underlying incidence of stroke and thromboembolism in the normal population. This background incidence of thromboembolism increases with age, atrial fibrillation, tobacco abuse, hypertension, and cerebrovascular disease. These risk factors are commonly seen in patients undergoing valve replacement surgery and make it difficult to draw definitive conclusions from multiple studies of thromboembolic risk and prevention.

Higher levels of anticoagulation and the addition of antiplatelet agents must be tempered with the increased risk of bleeding complications. The ideal anticoagulation regimen must be tailored for each individual patient, taking into consideration comorbid conditions that may increase the risk of bleeding complications. In addition, adequate surveillance of anticoagulation parameters must be obtained to ensure optimal levels of anticoagulant therapy.

Bleeding Complications

The use of chronic anticoagulation therapy places a patient at a constant risk for hemorrhage. Patients on chronic anticoagulation have a risk of major bleeding of less than 1.4% per year (21). The risk for bleeding with biological valves is less than half of that with mechanical valves, as one might expect, because most patients with biological valves are not anticoagulated. There have been many reports of risk factors for bleeding in anticoagulated patients. Unfortunately, the data are inconsistent.

One of the presumed risk factors for increased bleeding events is age. There has been inconsistent data on whether elderly patients are at increased risk for bleeding complications. There are many reports associating age with increased bleeding rates and many that find no such association. Similarly, there are limited clinical data supporting the association of hypertension, previous hemorrhage or previous gastrointestinal bleeding with increased risk of bleeding events in patients with prosthetic valves (45).

The addition of antiplatelet agents, such as aspirin, is clearly associated with an increased risk of bleeding events. Turpie's study comparing combination therapy of warfarin and low-dose aspirin with warfarin alone demonstrated an increased risk of events with combination therapy (35 vs 22%), but there was no significant difference in the incidence of major bleeding (12.9 vs 10.3%) (49). Laffort demonstrated that combination therapy was associated with an increased risk of gastrointestinal hemorrhage (7 vs 0%) (55). These discrepancies in the medical literature about risk factors for increased bleeding risk demonstrate the difficulty in balancing benefits and risks of hemorrhagic complications.

Anticoagulation in Patients Requiring Surgery

The recommendations for withholding anticoagulation in patients undergoing surgical procedures must be individualized owing to the dearth of clinical data establishing how anticoagulation can be safely withheld. The general practice is to discontinue oral anticoagulation approximately 3 days prior to the planned surgery. Often, the patient is treated with intravenous heparin both prior to and after the surgery. Oral anticoagulant therapy is restarted several days after the procedure and the patient remains on heparin until adequate INR values are obtained (see Table 7).

The risks of stopping warfarin are real. Stopping warfarin for 3 days has a 0.08 to 0.16% risk of an embolism, assuming an annual incidence of 10%. There are concerns that stopping and then reinstituting warfarin could result in a hypercoagulable state or that there might be a thrombotic rebound (56). Only small nonrandomized and retrospective studies have been performed to assess the optimal method for withholding anticoagulation therapy. Despite the lack of definitive findings, several recommendations can be made (1,31).

First of all, minor procedures such as skin biopsy, dental procedures, and eye surgery are associated with minimal risk of bleeding complications and do not require cessation of anticoagulation therapy. The strategies for withholding anticoagulation can be divided into low- and high-risk groups.

Low-risk patients should stop taking warfarin 48 to 72 hours prior to surgery so that the INR is less than 1.5 at the time of surgery. Reinstitution of oral anticoagulation should begin immediately postoperatively or after the control of active bleeding. If the patient takes aspirin, it should be discontinued 1 week prior to the procedure and resumed with warfarin during the postoperative period.

High-risk patients include those with history of recent thromboembolism, history of previous thromboembolism when off anticoagulation, presence of a Björk-Shiley valve, or one of the following: atrial fibrillation, hypercoagulable condition, mechanical prosthesis, and decreased left ventricular function. They should stop warfarin 72 hours prior to the surgical procedure and start intravenous heparin (target partial thromboplastin timerange of 55 to 70 seconds) when the INR is less than 2.0. Heparin should be discontinued 6 hours before the procedure and resumed 24 hours after the operation.

Table 7
Adjustment of Anticoagulation in Patients Requiring Surgical or Invasive Procedures

Type of surgery	Type of anticoagulation	Method of adjustment of anticoagulation prior to and after surgery
Minor[a]	Warfarin	No need to adjust.
	Aspirin	No need to adjust.
Major (low-risk group)[b]	Warfarin	Stop 72 hours prior to surgery. Check INR is <1.5. Resume same day after surgery or after control of bleeding. Stop 1 week prior to surgery.
	Aspirin	Resume same day after surgery or after control of bleeding.
Major (high-risk group)[b]	Warfarin	Stop 72 hours prior to surgery. Start heparin[c] when INR <2.0. Stop heparin[c] 6 hours prior to surgery. Restart heparin within 24 hours after surgery and continue until warfarin resumed and therapeutic INR obtained. Stop 1 week before surgery.
	Aspirin	Resume same day after surgery or after control of bleeding.

[a] Minor surgery includes dental procedures, eye surgery, and skin biopsies.
[b] Difference between low- and high-risk groups is that high-risk groups are those patients with recent thromboembolism, presence of atrial fibrillation, history of hypercoagulable state, left ventricular dysfunction, presence of Björk-Shiley valve, or presence of mechanical prosthetic valve.
[c] Recent studies indicate that low-molecular-weight heparins may be substituted for unfractionated intravenous heparin.
Data from refs. *1* and *44*.

The patient should remain on heparin until warfarin has been reinstituted and an INR of greater than 2.0 is obtained. Aspirin should be discontinued and resumed in the same manner as in low-risk patients.

The use of low-molecular-weight heparin is an attractive alternative to intravenous heparin because of its ease of use. Unfortunately, the medical literature does not provide enough evidence to recommend the use of this anticoagulant at this time. Vitamin K should be discouraged in patients on oral anticoagulation. Use of vitamin K to lower INR values is associated with an increased risk of a hypercoagulable state. Furthermore, it increases the time needed to reinstitute oral anticoagulation therapy with warfarin. In emergency situations, fresh frozen plasma is preferred to vitamin K therapy.

PROSTHETIC HEART VALVE DYSFUNCTION

Prosthetic valve dysfunction is a term used to encompass a wide variety of mechanical and physiological failings associated with a prosthetic valve (*see* Table 8). The failure can be structural owing to deterioration of the valve, leading to stenosis or regurgitation. It can also be nonstructural, which refers to an abnormality not intrinsic

Table 8
Potential Complications Associated With Prosthetic Heart Valves

Type of dysfunction	Aortic valve (mechanical)[a]	Aortic valve (bioprosthetic)[a]	Mitral valve (mechanical)[a]	Mitral valve (bioprosthetic)[a]
Structural	0.0	0.3–4.3	0.0	2.0–4.0
Valve thrombosis	0.0–0.2	0.0–0.2	0.2–0.8	0–0.2
Bleeding event	0.9–2.3	0.2–1.3	1.1–2.2	0.4–1.2
Endocarditis	0.4–0.5	0.2–1.3	0.2–0.7	0.0–1.0
Nonstructural dysfunction	0.4–0.8	0.05–0.2	0.4–1.4	0.0–1.5

[a] Morbidity of current cardiac valve operations expressed as % per patient-year. (Data from ref. 86.)

to the valve, which results in stenosis or regurgitation. Examples include pannus formation, paravalvular leak, or clinically significant hemolytic anemia. Other types of valvular dysfunction include endocarditis and thromboembolism.

The following sections summarize the risk, presentation, diagnosis, and treatment of these forms of prosthetic valve dysfunction.

Hemolysis

A compensated form of hemolysis occurs in most patients with mechanical prosthetic valves and can be seen in patients with bioprosthetic valves as well. Studies show that 50 to 95% of patients with mechanical valves have evidence of intravascular hemolysis but it is mild and subclinical in most cases (57,58). More severe hemolytic anemia is seen in up to 15% of patients with caged-ball and bileaflet prostheses and those with paravalvular leaks (59). The development of significant hemolytic anemia is rare in patients with bioprosthetic valves and may herald valve failure (60).

The clinical findings with hemolysis vary widely. Patients may present with jaundice, dark urine, fatigue, or heart failure. Physical examination may reveal a new regurgitant murmur. Work up for suspected anemia should include a complete blood count, reticulocyte studies, serum lactose dehydrogenase level, serum haptoglobin level, and peripheral blood smear. Patients with significant hemolysis will have sheared red blood cells (schistocytes), on peripheral blood smear, decreased serum haptoglobin levels, and elevated serum lactose dehydrogenase levels. Patients usually have a compensatory increase in reticulocytes. Iron can be lost in the urine as hemosiderin, leading to iron deficiency.

The patient should also undergo an echocardiogram to assess valve function. Severe hemolytic anemia in a patient with a mechanical valve is rare without some form of prosthetic regurgitation. The most common disturbance is a paravalvular leak. All prosthetic valves, both mechanical and prosthetic, may develop this abnormality. The risk of such leaks varies with the characteristics of the sewing ring, the surgical technique, and the quality of the tissue into which the valve was implanted. Hemolysis may also result from leakage through the valve orifice. In mechanical valves, the source of hemolysis is usually incomplete seating of the poppet. In tissue valves, hemolysis owing to a central leak may be caused by cusp retraction, perforation, or rupture. An inadequately sized aortic valve may also be associated with hemolysis.

The treatment of anemia is usually accomplished with oral iron and folate replacement. In the cases of severe anemia, associated high-output heart failure or significant valvular, or paravalvular regurgitation, replacement or repair of the prosthetic valve is indicated.

For the physician caring for an asymptomatic patient, an annual complete blood count should be performed. Additional blood work, such as serum lactose dehydrogenase or haptoglobin levels, should be dictated by the abnormalities on the blood counts or patient symptoms.

Endocarditis

Implanted prosthetic valves are at higher risk for infection than native heart valves. This is the result of abnormal flow patterns and the presence of foreign material in all valves. The risk of infection is not uniform. It is higher in the first 3 months after surgery. Afterward, the risk decreases to less than 1% annually for the life of the valve in patients taking appropriate antibiotic prophylaxis. Infection occurs with similar frequency at aortic and mitral sites. The rates of infection are similar for bioprosthetic and mechanical valves during the first 18 months after implantation, and then become higher for bioprosthetic valves *(61,63)*.

All patients with prosthetic heart valves are considered to be at high risk for endocarditis and require endocarditis prophylaxis with dental and surgical procedures (*see* Table 9). Despite the lack of controlled clinical trials to support antibiotic prophylaxis against endocarditis and the circumstantial nature of the evidence that transient bacteremia causes endocarditis, the consensus of informed opinion is that patients with prosthetic heart valves should receive antibiotic prophylaxis prior to any invasive procedure *(64,65)*.

Endocarditis involving prosthetic valves is classically categorized as either early (≤2 months after implantation) or late (>2 months after implantation). It is believed that early endocarditis is the result of infection at the time of implantation or via hematogenous spread in the immediate postoperative period. Late endocarditis is postulated to occur after the valve, sewing ring, and adjacent structures become endothelialized. Some authors have advocated lengthening the definition of early endocarditis to include the first 6 to 12 months after valve implantation.

The mortality associated with prosthetic valve endocarditis can be quite high. Mortality rates range from 30 to 80% in the early form and 20 to 40% in late endocarditis. Worse prognosis is associated with persistent bacteremia despite adequate antibiotic therapy, a new murmur, and new conduction abnormalities *(31)*.

The microbiology of prosthetic valve endocarditis is quite predictable and reflects the presumed nosocomial acquisition of infection. In the immediate post-implantation period, pathogens have direct access to the valve, ring, and anchoring sutures, because they are not endothelialized. Host proteins coat the structure and some pathogens can adhere to them. In the early period, the most common etiological agents are coagulase-negative staphylococci, particularly *Staphylococcus epidermidis*. Other agents include *Staphylococcus aureus*, Gram-negative bacilli, diptheroids, and fungi, especially *Candida* species. Of course, any bacterial or fungal species can be a cause of endocarditis.

In the late period, the etiological agents are similar to those seen in native valve endocarditis. It is postulated that the pathogens adhere to platelet-fibrin thrombi that form on the prosthesis. These agents include streptococci, *S. aureus*, enterococci, the

Table 9
Endocarditis Prophylaxis

Situation	Drug	Dosing
Dental, oral, respiratory tract, and esophageal procedure		
Standard regimen	Amoxicillin	2 g PO 1 hour before procedure.
	Ampicillin	2 g im or iv 30 minutes before procedure.
Penicillin-allergic	Clindamycin	600 mg PO 1 hour before procedure or 600 mg iv 30 minutes before procedure.
	Cefazolin or cephadroxil	2 g PO 1 hour before procedure.
	Azithromycin or clarithromycin	500 mg 1 hour before procedure.
Genitourinary or gastrointestinal procedures		
Standard regimen	Ampicillin + gentamicin	Ampicillin 2.0 g im/iv + gentamicin 1.5 mg/kg iv within 30 minutes of procedure; ampicillin 1 g im/iv or amoxicillin 1g PO 6 hours after procedure.
Penicillin-allergic	Vancomycin + gentamicin	Vancomycin 1 g iv (>1–2 hours) + gentamicin 1.5 mg/kg iv within 30 minutes of procedure.

Data from ref. *1*.

HACEK organisms (*Haemophilus* species, *Actinobacillus actinomycetemcomitans*, *Cardiobacterium hominis*, *Eikenella*, and *Kingella*), and coagulase-negative staphylococci. Endocarditis in prosthetic valves typically involves the sewing ring and is often associated with paravalvular leakage or abscesses.

Endocarditis should be suspected in any patient with a prosthetic heart valve who develops a new or changing murmur, an embolic event, petechiae, splenomegaly, or peripheral signs such as Osler's nodes or Janeway lesions. The most common presentation involves evidence of systemic embolization, seen in up to 40% of cases, and new or changing murmurs, reported in up to 50% of cases *(66)*. The clinician should have a high index of suspicion for endocarditis in any patient with prosthetic heart valve disease.

Suspicion of endocarditis should prompt the clinician to obtain multiple blood cultures and evaluate the patient's valve with echocardiography. Transthoracic echocardiography is useful for establishing the diagnosis of endocarditis and complications, such as ring abscesses, but technical difficulties are common since the materials in the valves hinder adequate ultrasonic imaging. Improved diagnostic accuracy has been obtained through the use of TEE.

The sensitivity of transthoracic echocardiography for the diagnosis of endocarditis ranges from 36 to 43%. In contrast, sensitivity for TEE ranges from 82 to 86% *(67,68)*. Additionally, TEE is more accurate at detecting the presence of valve ring abscess compared with transthoracic echocardiography. A study of 44 patients with valve abscesses found that transthoracic echocardiography detected only 28% of the lesions vs 87% detected by TEE *(69)*.

Treatment of prosthetic valve endocarditis involves antibiotic therapy directed at the etiological micro-organism as determined by blood culture. Usually repeat blood cultures are sterile after 3 to 5 days of appropriate antimicrobial therapy. After a completion of appropriate antimicrobial treatment, documented negative blood cultures should be obtained. Surgical intervention may be indicated in patients with uncontrolled infection, relapsing infection, and recurrent fever, despite negative cultures. Structural abnormalities, such as valvular incompetence or obstruction, partial valve dehiscence or the presence of an abscess, should prompt surgical intervention. The need for surgery based on the size of vegetations is controversial and must be tempered with the increased risk for embolization if a nonsurgical approach is taken.

Structural Failure of Prosthetic Valves

For mechanical valves, structural failure of a valve indicates faulty design or construction and is an unacceptable complication. On the other hand, structural deterioration with bioprosthetic valves is expected, although current designs and construction have lengthened the life expectancy of nonmechanical valves. A Veterans Administration study by Hammermeister reported in 1993 assessed the likelihood of structural failure in more than 500 patients who received mechanical or bioprosthetic heart valves. After more than 10 years of follow-up, the likelihood of structural failure with bioprosthetic valves was 15% compared with 0% for mechanical valves *(36)*.

The incidence of structural failure for mechanical valves is extremely low. The current model of the Starr-Edwards valve, with more than 200,000 implanted, has no reported fractures of the cage. The St. Jude bileaflet valve has only 12 reported structural failures, involving fractures of the disc or housing, in approximately 900,000 implantations through 1998. Manufacturers, however, do keep track of any valvular dysfunction owing to structural failures. The most infamous example of primary structural mechanical valve failure is the Björk-Shiley Convexo-Concave valve (*see* Fig. 7). A modest change in the design of this tilting-disk valve resulted in unexpected failure rates. At first, the cause of valve failure was felt to be caused by faulty welds of the strut. Further investigation, however, has determined the cause of the outlet strut fracture to be the result of overrotation of the disc, which imparts large stresses to the tip of the strut. From 1979 to 1986, 82,000 such valves were implanted in the United States and 457 cases of strut fracture were reported through 1998 *(12)*.

In bioprosthetic valves, degenerative calcification of the leaflets results in leaflet tears (and valvular regurgitation) and valvular stenosis (*see* Fig. 8). All models have a limited life expectancy. By 10 years, the rate of primary bioprosthetic valve failure is close to 30%. The Veterans Affairs Cooperative Study on Valvular Heart Disease found that, after 11 years of follow-up, the structural failure rate for bioprosthetic valves in the aortic position was 15% and in the mitral position was 36% *(34)*. Several series have shown that the percentage of bioprosthetic valves free of structural deterioration

Chapter 19 / Cardiac Valves 513

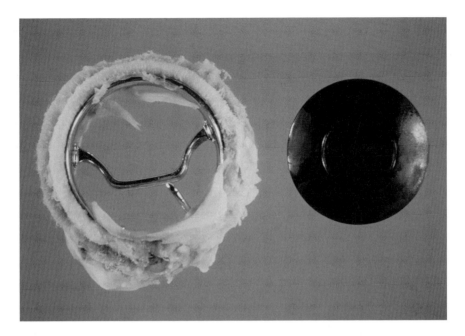

Fig. 7. This Björk-Shiley valve demonstrates the catastrophic result of strut fracture. The patient presented in shock and was found to have wide-open mitral insufficiency. One of the struts of the prosthetic mechanical valve had fractured. The disc was completely torn away from the ring and found in the iliac artery.

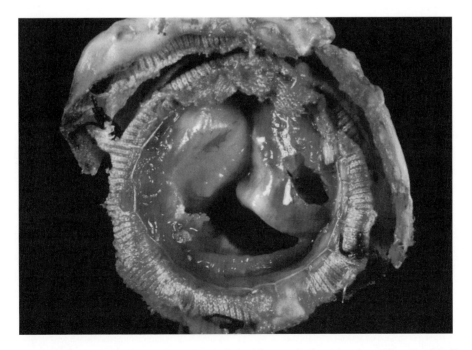

Fig. 8. The lack of long-term durability of the bioprosthetic valve is evident in this figure. This bioprosthetic valve has a tear in one of the leaflets, which led to valvular insufficiency and the need for replacement.

at 10 years after implantation is 80 to 85%. By 15 years, the percentage of valves free of structural deterioration is 25 to 65% *(33)*.

The degenerative calcification that leads to eventual valve failure is more common in younger patients. Their valves more rapidly accumulate calcium deposits for unclear reasons. A study by Burdon evaluating the structural valve deterioration rates of the Hancock bioprosthesis found that the percentage of freedom from structural deterioration varied markedly, depending on the age of the patient *(70)*. Patients under 50 had a 50 to 67% structural failure rate at 10 years, but patients over 60 had failure rates of 7 to 27% over the same time period. Several other studies have shown similar results. The rate of deterioration is also higher in valves in the mitral position.

Modern pericardial tissue valves, such as the Carpentier-Edwards Perimount, have rates of structural deterioration similar to that of porcine prosthetic valves. An exception to this is the Inosecu-Shiley valve. This valve, which was the first pericardial valve produced, had dramatic failure rates with a rate of deterioration approaching 50% at 10 years *(71)*. The valve was subsequently withdrawn from the market. Recent advances, including low-pressure preparation and anticalcification processes, may increase durability of tissue valves, particularly the newer stentless variety.

Nonstructural Dysfunction of Prosthetic Valves

Nonstructural dysfunction refers to abnormalities not intrinsic to the valve that nonetheless result in compromised valve integrity and deterioration of valve function. Such abnormalities include pannus formation, paravalvular leak, patient–prosthetic mismatch, endocarditis, and thrombosis.

Pannus Formation

Pannus formation refers to fibrous tissue ingrowth that compromises valvular motion. It is less common than thrombus formation as a cause of prosthetic valve obstruction. A study of 112 patients who underwent replacement of mechanical valves owing to obstruction found pannus to be the cause in 11% of the cases. Thrombus was the cause in 77% of the cases and the combination of thrombus and pannus formation was the cause in the remaining 12% of cases *(72)*.

The presence of pannus can cause either prosthetic valvular stenosis or regurgitation. Stenosis is owing to the restriction of the disk motion. In tissue valves, the pannus causes obstruction of leaflet motion. Regurgitation involving pannus formation is caused by the failure of the disk, ball, or bileaflet occluder to completely close.

Echocardiography, particularly transesophageal, plays a central role in detection of pannus. Because of its fibrous composition, pannus is echogenic and firmly adherent to the valve apparatus. It typically involves a small dense mass that does not extend beyond the prosthesis *(73)*. Thrombus is typically more mobile and echogenic. In some cases, it may be difficult to differentiate pannus from thrombus. It is important to make this differentiation as the treatments are different. Thrombus is often amenable to thrombolytic therapy, but reoperation is the only effective therapy for pannus formation.

Thrombosis

The most common cause of valve obstruction is thrombus formation. Most commonly, thrombus develops in the setting of inadequate anticoagulation. The risk varies with valve location. The incidence may be as high as 13% in the first year in patients with prostheses in the tricuspid position or a little as 0.2 to 6% per patient-year in those with aortic or mitral valves *(74,75)*.

As with pannus formation, the presence of thrombus formation can lead to valvular stenosis or regurgitation. Similarly, the diagnostic method of choice is echocardiography, usually TEE. Characteristics of thrombus are large size, low density, and extension beyond the valve ring. Differentiation between thrombus and pannus can also be predicted by the history of inadequate anticoagulation therapy *(73)*.

Surgical replacement of the valve is indicated in cases where valve function is severely compromised due to thrombus. Fibrinolytic therapy, however, may be considered in certain patients. This typically involves streptokinase or urokinase, followed by intravenous heparin and institution of oral anticoagulation with warfarin therapy. An extensive review of 200 published studies of thrombolytic therapy for left-sided prosthetic valve thrombosis demonstrated a success rate of 82%. Complications included thromboembolism (12%), stroke (5–10%), death (6%), major bleeding (5%), and recurrent thrombosis (11%) *(42)*.

Patients considered to be candidates for thrombolytic therapy include those with functional class III/IV who are at high risk for surgery and those with contraindications to surgery. Patients with functional class I/II and no contraindications to surgery should undergo operative replacement of the valve. This is because of the low operative mortality when compared with the 12 to 17% risk of embolism posed by thrombolytic therapy *(76)*.

Small clots may be treated with intravenous heparin and addition of oral anticoagulation. Patients whose thrombus does not respond to anticoagulation or whose clot is large and mobile should undergo operation. Regardless of the treatment of the thrombus, target INR levels in patients with a single episode of thrombosis are higher than the typical goals of warfarin therapy. All patients should have a target INR of 3.0 (range 2.5–3.5) after treatment of thrombosis. The addition of aspirin (81 mg daily) is strongly encouraged, unless otherwise contraindicated. Surveillance with monthly echocardiography for the first 6 months is recommended. Thereafter, echocardiography should be performed every 6 months *(42)*.

Paravalvular Leak

This form of regurgitation is defined by leakage of blood around the valve and is not caused by incompetence of the valve obstructing mechanism (leaflets, disk, or valve tissue). It is the result of dehiscence of the valve ring secondary to endocarditis with abscess formation, broken sutures, or technical difficulties at time of implantation. Patients often present with signs and symptoms consistent with heart failure. Hemolysis is common and should immediately raise the clinician's suspicion for paravalvular leak. Diagnosis is accomplished with echocardiography. TEE is usually needed to thoroughly examine the valve ring. Severe dehiscence of the valve may appear as laminar flow, making it difficult to differentiate from valvular regurgitation.

Treatment of paravalvular leakage is dependent on the amount of regurgitation. Unlike the small physiological amounts of transvalvular regurgitation seen with mechanical valves and trace amounts of regurgitation seen in bioprosthetic valves, any amount of paravalvular leakage is abnormal. Marked hemolysis or hemodynamic compromise requires surgical intervention with valve replacement or repair as indicated.

Patient–Prosthesis Mismatch

Patient–prosthesis mismatch describes inappropriate sizing of the replacement valve, primarily in the aortic position. Rahimtoola has defined mismatch as occurring when the prosthetic valve area is less than that of the normal native valve *(82)*. Such a definition means that all mechanical and bioprosthetic valves represent mismatch. However, the outcome of such a mismatch is of no consequence when it is mild or moderate. Severe patient–prosthesis mismatch, on the other hand, can result in a patient who is hemodynamically and symptomatically worse after valve replacement *(83)*.

Replacing a native valve with a valve that too small results in functional valvular stenosis even though the prosthetic valve is functioning properly. Smaller valve sizes, more typically seen in female patients, have higher peak velocities and mean gradients and may have hemodynamic parameters almost indistinguishable from valve obstruction. There appears to be no performance advantage of one prosthesis over another in this setting. A study at our institution compared the St. Jude Medical and Medtronic Hall valves for aortic valve replacement in the small aortic root. There was no difference between prostheses in their performance in smaller annuli *(84)*. Such small valves should be reserved for patients with a body surface area of less than or up to 1.7 m^2 in order to avoid hemodynamic and functional compromise. Normally, regression of left ventricular hypertrophy of 12–44% has been reported after aortic valve replacement. However, incomplete regression of left ventricular hypertrophy often results when aortic valve prostheses of 21 mm or less are implanted *(77,84)*.

Other

Although the aforementioned are common complications, there are additional, albeit less common, complications. These are usually the result of dehiscence of paravalvular structures, with resultant pseudoaneurysms, fistulas, or abscesses. All are rare and are the result of infection or mechanical destabilization of the structure involved.

Pseudoaneurysms may form as a result of aortic or mitral valve replacement. With the aortic valve, dehiscence may occur at the aortic anastomosis or coronary anastomosis. The result is the formation of a pseudoaneurysm, hematoma, or abscess. Patients may present with a variety of symptoms, including fatigue, dyspnea, fevers, or they may be asymptomatic. TEE can determine the size and extent of the pseudoaneurysm with a high degree of accuracy. San Roman et al. demonstrated that TEE detected 90% of abscesses and 100% of pseudoaneurysms and fistulas *(78)*. Computed tomography or magnetic resonance imaging may also be diagnostic, but they lack the dynamic information obtained via echocardiography.

Mitral valve replacement may also be complicated by the formation of left ventricular pseudoaneurysm. Several series have determined the incidence to be 0.5 to 2.0% *(79,80)*. Commonly, the pseudoaneurysm originates at the posterior annular suture site and ruptures at the posterior base of the left ventricle. Fatal hemodynamic collapse can occur. In some cases, a thin-walled pseudoaneurysm forms. Patients may be asymptomatic or present with chest pain or congestive heart failure. A new systolic murmur may be heard. The diagnosis is best confirmed with echocardiography. Surgical intervention to repair the aneurysm (and the valve, if indicated) is the treatment of choice. Improved surgical techniques of valve implantation has resulted in a reduced incidence of pseudoaneurysm formation.

Another less common complication is the development of a fistula after valve replacement surgery. They are rare and are the result of dehiscence of the valve ring or abscess formation with rupture. Communications between the right and left ventricle, aorta and either ventricle, or aorta and either atrium have been reported. The diagnostic modality of choice is echocardiography and the treatment of choice is surgical repair.

EVALUATION AND TREATMENT OF CORONARY HEART DISEASE IN PATIENTS WITH PROSTHETIC VALVES

The prevalence of CAD depends on well-known clinical risk factors, including age and gender of the patient, smoking history, and so on. In patients undergoing valve replacement, these same clinical variables are used to determine the likelihood of CAD. The prevalence of CAD in patients who undergo aortic valve replacement for severe aortic stenosis is 33% *(81)*. Similarly, patients with mitral valve replacement for mitral regurgitation have a high incidence of CAD. In fact, CAD is frequently the cause of mitral regurgitation.

Patients with CAD often undergo bypass grafting at the same time as valve surgery and the clinician must be aware of the requirement for continued surveillance of the ischemic heart disease as well as the patient's prosthetic valve. Symptoms of fatigue, heart failure, or chest pain may be attributable to a change in the patient's coronary status and not owing to valve complications. Follow-up with appropriate study modalities, such as stress testing, is indicated on an individual basis.

Role of Trials

There are few large-scale multicenter trials addressing the treatment and management of patients with prosthetic heart valves. Much of the research on prosthetic heart valves has focused on the incidence of complications and the prevalence of structural deterioration. There have been few studies that compare different types of mechanical or prosthetic valves and these are warranted.

There is also a dearth of clinical trials concerning follow-up. Recommendations for the frequency of blood tests and noninvasive studies, such as echocardiography, are based on guidelines published by the ACC/AHA Task Force Report on Valvular Heart Disease. However, the Committee on Management of Patients With Valvular Heart Disease did not reach consensus on the issue of noninvasive testing. The appropriate frequency of echocardiographic evaluation of patients with prosthetic heart valve disease is unknown. A minority of the task force recommended annual echocardiographic examination for asymptomatic patients, whereas the majority felt assessment with echocardiography should be driven by clinical changes in the patient. There are also no guidelines for the appropriate timing of serial echocardiography in asymptomatic patients with bioprosthetic valves for whom there is increased risk for structural deterioration five to ten years after valve implantation. A randomized trial to evaluate the role of annual or routine echocardiography in asymptomatic patients needs to be undertaken *(1)*.

Support Groups

Patients with prosthetic heart valves may be anxious or concerned because of the presence of a manmade device in such an important organ. The benefit of support and information that can be provided to patients should not be underestimated or underappreciated. There are numerous support groups through which a patient may contact patients with similar concerns and exchange information. The advent of the World Wide Web has opened the door to such groups and a few are listed (*see* the Appendix).

APPENDIX: RESOURCES FOR PHYSICIANS AND PATIENTS

Manufacturer/Health Resource	Address	Phone number, Web address
Baxter Healthcare Corp	Cardiovascular Group 17221 Red Hill Ave Irvine, CA 92614	1-800-424-3278 www.baxtercvg.com
Medtronic, Inc.	8299 Central Ave. NE Spring Lake Park, MN 55432	1-800-328-2518 www.medtronic.com
St. Jude Medical, Inc.	One Lillehei Plaza St. Paul, MN 55117	1-800-544-1664 www.sjm.com
WebMD		www.webmd.com

REFERENCES

1. Bonow RO, Carabello B, de Leon AC Jr, et al. American College of Cardiology/American Heart Association Guidelines for the Management of Patients with Valvular Heart Disease. J Am Coll Card 1998;32:1486-1588.
2. Unpublished data from industry.
3. Dressler FA, Labovitz, AJ. Exercise evaluation of prosthetic heart valves by Doppler Echocardiography: comparison with catheterization studies. Echocardiography 1992;9: 235–241.
4. Labovitz AJ. Assessment of prosthetic heart valve function by Doppler echocardiography: a decade of experience. Circulation 1989;80:707–709.
5. Gibbon JH Jr. The application of a mechanical heart and lung apparatus to cardiac surgery. 1954;37:Minn Med 171.
6. Harken DE, Soroff MS, Taylor WJ, et al. Partial and complete prosthesis in aortic insufficiency. J Thorac Cardiovasc Surg 1960;40:744.
7. Starr A, Edward ML. Mitral replacement: clinical experience with a ball-valve prosthesis. Ann Surg 1961;154:726.
8. Carpentier A. From valvular xenograft to valvular bioprosthesis (1965–1977). Med Instrum 1977;11:98.
9. Spencer FC, Galloway AC, Colvin SB. Acquired disease of the mitral valve. In: Sabiston DC ed. Surgery of the chest. New York: W.B. Saunders, 1995, pp. 1673–1684.
10. Gabbay S, Factor SM, Strom J, Becker R, Frate RW. Sudden death due to cuspal dehiscence of the Ionescu-Shiley valve in the mitral position. J Thorac Cardiovasc Surg 1982;84: 313–314.
11. Alton ME, Paserski TJ, Orsinelli DA, Eaton GM, Pearson AC. Comparison of transthoracic and transesophageal echocardiography in evaluation of 47 Starr-Edwards prosthetic valves. J Am Coll Card 1992;20:1503–1511.
12. Wieting DW, Eberhardt AC, Ruel H, Breznock EM, Schreck SG, Chandler JG. Strut fracture mechanisms of the Björk-Shiley Covexo-Concave heart valve. J Heart Valve Dis 1999: 8:206–217.
13. Yoganathan AP, Heinrich RS, Fontaine AA. Fluid dynamics of prosthetic valves. In Otto CM ed. The practice of clinical echocardiography. Philadelphia, PA: W.B. Saunders, 1997, pp. 773–796.

14. Reisner SA, Meltzer RS. Normal values of prosthetic valve Doppler echocardiographic parameters: a review. J Am Soc Echocardiograph 1988;1:201–210.
15. Thevenet A, Albat B. Long-term follow up of 292 patients after valve replacement with Omnicarbon prosthetic valve. J Heart Valve Dis 1995;4:634–639.
16. Nitter-Hauge S, Abdelnoor M, Svennevig JL. Fifteen year experience with Medtronic-Hall valve prosthesis: a follow up study of 1104 consecutive patients. Circulation 1996;94(suppl II):105–108.
17. Tatoulis J, Chaiyaroj S, Smith JA. Aortic valve replacement in patients 50 years old or younger with the St. Jude Medical valve: 14 year experience. J Heart Valve Dis 1996;5: 491–497.
18. Godje OL, Fischlein T, Adelhard K, Nollert G, Klinner W, Reichart B. Thirty-year results of Starr-Edwards prostheses in the aortic and mitral position. Ann Thorac Surg 1997;63: 613–619.
19. Orszulak TA, Schaff HV, Puga FJ, et al. Event status of the Starr-Edwards aortic valve to 20 years: a benchmark for comparison. Ann Thorac Surg 1997;63:620–626.
20. Fiore AC, Barner HB, Swartz MT, et al. Mitral valve replacement: randomized trial of St. Jude and Medtronic Hall prostheses. Ann Thorac Surg 1998;66:707–713.
21. Cannegeister SC, Rosendaal FR, Briet E. Thromboembolic and bleeding complications in patients with mechanical heart valves. Circulation 1994;89:635–641.
22. Ross DN. Replacement of aortic and mitral valves with pulmonary autograft. Lancet 1967;2:956–958.
23. Chambers JC, Somerville J, Stone S, Ross DN. Pulmonary autograft procedure for aortic valve disease: long-term results of the pioneer series. Circulation 1997;96:2206–2214.
24. Ross DN. Reflections on the pulmonary autograft. J Heart Valve Dis 1993;2:363–364.
25. Nottestad SY, Zabalalgoitia M. Echocardiographic receognition and quantitation of prosthetic valve dysfunction. In: Otto CM, ed. The practice of clinical echocardiography. Philadelphia, PA: W.B. Saunders, 1997, pp. 797–820.
26. Otto CM, ed. Prosthetic valves. In: Valvular heart disease. Philadelphia, PA: W.B. Saunders, 1999, pp. 380–416.
27. Del Rizzo DF, Goldman BS, David TE. Aortic valve replacement with a stentless porcine bioprosthesis: multicenter trial. Canadian Investigators of the Toronto SPV Valve Trial. Can J Cardiol 1995;11;597–603.
28. Mohr FW, Walther T, Baryalei M, et al. The Toronto SPV bioprosthesis: one-year results in 100 patients. Ann Thorac Surg 1995;60:171–175.
29. Vrandecic MP, Gontijo BF, Fantini FA, et al. The new stentless aortic valve: clinical results from the first 100 patients. Cardiovasc Surg 1994;2:407–414.
30. Goldman BS, David TE, Del Rizzo DF, Sever J, Bos J. Stentless porcine bioprosthesis for aortic valve replacement. J Cardiovasc Surg 1994;35(Suppl I):105–110.
31. Vongpatanasin W, Hillis LD, Lange RA. Prosthetic heart valves. N Engl J Med 1996;335: 407–416.
32. Bloomfield P, Wheatley DJ, Prescott RJ, Miller HC. Twelve-year comparison of a Björk-Shiley mechanical heart valve with porcine bioprostheses. N Engl J Med 1991;324: 573–579.
33. Grunkemeier GL, Li H, Naftel DC, Starr A, Rahimtoola SH. Long-term performance of heart valve prostheses. Curr Probl Cardiol 2000;25:78–154.
34. Fann JI, Miller DC, Moore KA, et al. Twenty-year clinical experience with porcine bioprostheses. Ann Thorac Surg 1996;62:1301–1311.
35. Jones EL, Weintraub WS, Craver JM, et al. Ten-year experience with the porcine bioprosthetic valve: interrelationship of valve survival and patient survival in 1,050 valve replacements. Ann Thorac Surg 1990;49:370–383.
36. Hammermeister KE, Sethi GK, Henderson WG, Oprian C, Kim T, Rahimtoola SH. A comparison of outcomes in men 11 years after heart-valve replacement with a mechanical valve

or bioprosthesis. Veterans Affairs Cooperative Study on Valvular Heart Disease. N Engl J Med 1993;328:1289–1296.
37. Holen J, Simonsen S, Froysaker T. An ultrasound Doppler technique for the noninvasive determination of the pressure gradient in the Björk-Shiley mitral valve. Circulation 1979; 59:436–442.
38. Labovitz AJ, Williams GA. Doppler Echocardiography: the quantitative approach. Philadelphia, PA: Lea and Febiger, 1992
39. Baumgartner H, Czer L, DeRobertis M, et al. Normal regurgitation in mechanical valve prostheses: impact on Doppler studies. Amer Coll Cardiol Learning Center Highlights 1992;8:1.
40. Flachskampf FA, O'Shea JP, Griffin BP, et al. Patterns of normal transvalvular regurgitation in mechanical valve prostheses. J Am Coll Cardiol 1991;18:1493.
41. Mohr-Kahaly S, Kupferwasser I, Erbel R, et al. Regurgitant flow in apparently normal valve prostheses: improved detection and semiquantitative analysis by transesophageal two-dimensional color-coded Doppler echocardiography. J Am Soc Echocardiogr 1990;3:187.
42. Lengyel M, Fuster V, Keltai M, et al. Guidelines for management of left-sided prosthetic thrombosis: a role for thrombolytic therapy. J Am Coll Cardiol 1997;30:1521–1526.
43. Akins CW. Results with mechanical cardiac valvular prostheses. Ann Thorac Surg 1995; 60:1836–1844.
44. Cannegieter SC, Rosendaal FR, Wintzen AR, Van der Meer FJ, Vandebroucke JP, Briet E. Optimal oral anticoagulant therapy in patients with mechanical heart valves. N Engl J Med 1995;333:11–17.
45. Cannegieter SC, Torn M, Rosendaal FR. Oral anticoagulant treatment in patients with mechanical heart valves: how to reduce the risk of thromboembolic and bleeding complications. J Int Med 1999;245:369–374.
46. Connolly SJ, Laupacis A, Gent M, Roberts RS, Cairns JA, Joyner C. Canadian Atrial Fibrillation Anticoagulation Study. J Am Coll Cardiol 1991;18:349–355.
47. Tiede DJ, Nishimura RA, Gastineau DA, Mullany CJ, Orszulak TA, Schaff HV. Modern management of prosthetic valve anticoagulation. Mayo Clin Proc 1998;73:665–680.
48. Cappelleri JC, Fiore LD, Brophy MT, Deykin D, Lau J. Efficacy and safety of combined anticoagulant and antiplatelet therapy versus anticoagulant monotherapy after mechanical heart valve replacement: a metaanalysis. Am Heart J 1995;130:547–552.
49. Turpie AGG, Gent J, Laupacis A, et al. A comparison of aspirin with placebo in patients treated with warfarin after heart valve replacement. N Engl J Med 1993;329:524–529.
50. Massel D, Little SH. Risks and benefits of adding anti-platelet therapy to warfarin among patients with prosthetic heart valves: a meta-analysis. J Am Coll Cardiol 2001;37:569–578.
51. Pumphrey CV, Fuster V, Cheseboro JH. Systemic thromboembolism in valvular heart disease and prosthetic heart valves. Mod Concepts Cardiovasc Dis 1982;51:131.
52. Edmunds LH. Thromboembolic complications of current cardiac valvular prostheses. Ann Thorac Surg 1982;34:96.
53. Heras M, Chesebro JH, Fuster V, et at. High risk of thromboemboli early after bioprosthetic cardiac valve replacement. J Am Coll Cardiol 1995;25:1111–1119.
54. Stein PD, Alpert JS, Copeland J, Dalen JE, Goldman S, Turple AG. Antithrombotic therapy in patients with mechanical and biological prosthetic heart valves. Chest 1996;108(Suppl 4):371S–379S.
55. Laffort P, Roudaut R, Roques X, et al. Early and long-term (one-year) effects of the association of aspirin and oral anticoagulant on thrombi and morbidity after replacement of the mitral valve with the St. Jude medical prosthesis: a clinical and transesophageal echocardiographic study. J Am Coll Cardiol 2000;35:739.
56. Genewein U, Haeberli A, Straub PW, Beer JH. Rebound after cessation of oral anticoagulant therapy: the biochemical evidence. Br J Haematol 1996;92:479–485.

57. Skoularigis J, Essop MR, Skudicky D, Middlemost SJ, Sareli P. Frequency and severity of intravascular hemolysis after left-sided cardiac valve replacement with Medtronic-Hall and St.Jude prostheses and influence of prosthetic type, position, size and number. Am J Cardiol 1993;71:587–591.
58. Barmada H, Starr A. Clinical hemolysis with the St. Jude heart valve without paravalvular leak. Med Prog Technol 1994;20:191–194.
59. Ismeno G, Renzulli A, Carozza A, et al. Intravascular hemolysis after mitral and aortic valve replacement with different types of mechanical prostheses. Int J Cardiol 1999;69:179.
60. Enzenauer RJ, Berrenberg JL, Cassell PR Jr. Microangiopathic hemolytic anemia as the initial manifestation of porcine valve failure. South Med J 1990;83:912.
61. Arvay A, Lengyel M. Incidence and risk factors of prosthetic valve endocarditis. Eur J Cardiothorac Surg 1988;2:340
62. Grover FL, Cohen DJ, Oprian C, et al. Determinants of the occurrence of and survival from prosthetic valve endocarditis. J Thorac Cardiovasc Surg 1994;108:207.
63. Rutledge R, Kim BJ, Applebaum RE. Actuarial analysis of the risk of prosthetic valve endocarditis in 1598 patients with mechanical and bioprosthetic valves. Arch Surg 1985; 120:469.
64. Durack DT, Beeson PB. Experimental bacterial endocarditis, I: colonization of a sterile vegetation. Br J Exp Pathol 1972;53:44–49.
65. Van der Meer JT, Van Wijk W, Thompson J, Vandenbroucke JP, Valkenburg HA, Michel MF. Efficacy of antibiotic prophylaxis for prevention of native-valve endocarditis. Lancet 1992;339;135–139.
66. Ben Ismail M, Hannachi N, Abid F, et al. Prosthetic valve endocarditis. A survey. Br Heart J 1987;58:72.
67. Daniel WG, Mugge A, Grote J, et al. Comparison of transthoracic and transesophageal echocardiography for detection of abnormalities of prosthetic and bioprosthetic valves in the mitral and aortic positions. Am J Cardiol 1993;71:210.
68. Alton ME, Pasierski TJ, Orsinelli DA, et al. Comparison of transthoracic and transesophageal echocardiography in evaluation of 47 Starr-Edwards prosthetic valves. J Am Coll Cardiol 1992;20:1503.
69. Daniel WG, Mugge A, Martin RP. Improvement in the diagnosis of abscesses associated with endocarditis by transesophageal echocardiography. N Engl J Med 1991;324:795.
70. Burdon TA, Miller DC, Oyer PE, et al. Durability of porcine valves at fifteen years in a representative North American patient population. J Thorac Cardiovasc Surg 1992;33: 526–533.
71. Masters RG, Walley Vm, Pipe AL, Keon WJ. Long-term experience with the Ionescu-Shiley pericardial valve. Ann Thorac Surg 1995;60:S288—S291.
72. Deviri E, Sareli P, Wisenbaugh T, et al. Obstruction of mechanical heart prosthesis: clinical aspects and surgical management. J Am Coll Cardiol 1991;17:646.
73. Barbetseas J, Nagueh SF, Pitsavos C, et al. Differentiating thrombus from pannus formation in obstructed mechanical prosthetic valves: an evaluation of clinical, transthoracic, and transesophageal echocardiographic parameters. J Am Coll Cardiol 1998;32:1410.
74. Thorburn CW, Morgan JJ, Shanahan MX, Chang VP. Long-term results of tricuspid valve replacement and the problem of prosthetic valve thrombosis. Am J Cardiol 1983;51: 1128–1132.
75. Roudaut R, Labbe T, Lorient Roudaut MF, et al. Mechanical cardiac valve thrombosis: is fibrinolysis justified? Circulation 1992;86(Suppl II):II-8–15.
76. Birdi I, Angelini GD, Bryan AJ. Thrombolytic therapy for left-sided prosthetic heart valve thrombosis. J Heart Valve Dis 1995;4:154–159.
77. Dolan MS, Castello R, St Vrain JA, et al. Relationship of left ventricular hypertrophy regression and outflow turbulence. Circulation 1996;94(Suppl I):721.

78. San Roman JA, Vilacosta I, Sarria C, et al. Clinical course, microbiologic profile, and diagnosis of periannular complications in prosthetic valve endocarditis. Am J Cardiol 1999;83:1075.
79. Roberts WC, Morrow AG. Late postoperative pathological findings after cardiac valve replacement. Circulation 1967;35(Suppl I):48–62.
80. Karlson KJ, Ashraf MM, Berger RL. Rupture of left ventricle following mitral valve replacement. Ann Thorac Surg 1988;46:590–597.
81. Iung B, Drissi MF, Michel PL, et al. Prognosis of valve replacement for aortic stenosis with or without coexisting coronary heart disease: a comparative study. J Heart Valve Dis 1993;2:430–439.
82. Rahimtoola SH. The problem of valve prosthesis–patient mismatch. Circ 1978;58:20–24.
83. Barner HB, Labovitz AJ, Fiore AC. Prosthetic valves for the small aortic root. J Card Surg 1994;9(Suppl):154–157.
84. Fiore AC, Swartz M, Grunkemeier G, et al. Valve replacement in the small aortic annulus: prospective randomized trial of St. Jude with Medtronic Hall. Eur J Cardiothoracic Surg 1997;11:485–492.
85. Tatineni S, Barner HB, Pearson AC, Halbe D, Woodruff R, Labovitz AJ. Rest and exercise evaluation of St. Jude Medical and Medtronic Hall prostheses. Influence of primary lesion, valvular type, valvular size, and left ventricular function. Circulation 1989;80(Suppl 3, Pt 1):116–123.
86. Edmunds LH. Evolution of prosthetic heart valves. Am Heart J 141(5): 849–855.

US Counterpoint to Chapter 19

Kristine J. Guleserian and Marc R. Moon

Chapter 19 provides a broad review of prosthetic heart valves, beginning with a historical perspective of early valve development and attempts at surgical replacement to current valve characterization and their postoperative management regimens. The practice at Washington University is essentially the same (*see* Table 1). The authors of Chapter 19 nicely outline the types of commonly used prosthetic valves and differences in hemodynamics among them. They recommend specific postoperative anticoagulation and endocarditis prophylaxis practices, with which we agree. Recommendations provided in the text are based on the 1998 Guidelines for the Management of Patients with Valvular Heart Disease from the American College of Cardiology and American Heart Association. Online resources for both the physician and the patient are cited in the Appendix to the chapter; these are excellent sources of additional information.

Like the authors of Chapter 19, we base valve selection on patient age, life expectancy, and suitability for long-term anticoagulation. However, in recent years, we have tended to favor bioprosthetic rather than mechanical valves in a progressively younger population, noting that the risks of long-term anticoagulation are not insignificant. In a recent review presented at the 83rd meeting of the American Association for Thoracic Surgery in May 2003, Ikonomidis and associates reported their 20-year experience with the St. Jude bileaflet mechanical valve in the aortic (478 patients) and mitral (359 patients) position (1). All patients received long-term anticoagulation therapy with warfarin. Their cumulative incidence of bleeding complications at 10 years and 20 years for aortic valve patients was 20% ± 2% and 25% ± 3%, respectively, compared to 13% ± 2% and 23% ± 4% for mitral valve patients. Furthermore, their cumulative incidence of thromboembolic complications at 10 years and 20 years for aortic valve patients was 14% ± 2% and 21% ± 3%, respectively, compared to 20% ± 3% and 29% ± 4% for mitral valve patients.

In an attempt to elucidate the risk of reoperative aortic valve replacement, Gaudiani and colleagues analyzed their series of 1213 patients undergoing aortic valve replacement at the 52nd annual Scientific Session of the American College of Cardiology (2). In their series, including 134 (11%) reoperations, the operative mortality rate was not different between primary aortic valve replacement (4.1%) and reoperative aortic valve

Table 1
**Patient Surveillance After Implantation
of a Prosthetic Heart Valve at Washington University Hospital**

	Postoperative year					
	1	2	3	4	5	10
Office visit, including history and physical	2^a	1	1	1	1	1
Complete blood count and serum electrolytes and creatinine	2^a	1	1	1	1	1
Electrocardiogram	2^a	1	1	1	1	1
Transthoracic echocardiogram	1	0^b	0^b	0^b	0^b	0^b
PT/INR (if taking chronic anticoagulation)^c	2^a	1	1	1	1	1

^{a} At 3 to 4 weeks postoperative and at 12 months.
^{b} Repeat transthoracic echocardiography is not routinely performed. Indications for echocardiography are driven by clinical findings, including new symptoms or changes in physical findings.
^{c} In patients taking warfarin. This is in addition to routine monitoring by the primary medical doctor.

replacement (3.1%) ($p > 0.89$). Their conclusion was that the risk of reoperative aortic valve replacement was lower than the published risks of long-term anticoagulation and that "any adult who requires aortic valve replacement may be well advised to consider tissue prostheses." We currently feel that, in experienced hands, the risk of reoperative aortic valve replacement is similar to, or only slightly higher than, that of primary valve replacement. In addition, modern advances in valve preparation technology (anticalcification treatments, low-pressure fixation) may improve longevity for biologic prostheses in the aortic position. We currently recommend bioprosthetic replacement for patients 60 years or greater but are not uncomfortable inserting tissue valves in even younger patients who want to avoid long-term anticoagulation for lifestyle reasons. With regard to valve choice for mitral valve replacement, liberal use of mitral valve repair in most centers has made this less of an issue in recent years. Based on data from the Society of Thoracic Surgeons National Database, in 1993 valve repair represented only 26% of the 4790 procedures performed on the mitral valve nationwide, rising to 48% of the 6251 mitral procedures performed in 2002 *(3)*. When the valve absolutely cannot be repaired (severe rheumatic involvement of the chordae and leaflets, ischemia with disruption of the papillary muscle), we currently recommend bioprosthetic valves for patients 65 years old or greater.

Patients with endocarditis represent an important subset of patients undergoing valve replacement in that their initial operative risk is higher and long-term survival rate is lower than patients with noninfective valvular pathology. In 2001, we reported on 306 patients undergoing valve replacement for native valve endocarditis (209 patients) or prosthetic valve endocarditis (97 patients) with either a bioprosthetic (221 patients), mechanical (65 patients), or homograft (20 patients) valve *(4)*. Overall, there was no difference in the long-term survival rate between patients who underwent bioprosthetic (51% ± 4% at 10-year) vs mechanical (50% ± 8%) ($p > 0.27$) valve replacement. For patients less than or equal to 60 years of age, the long-term survival rate was similar in those who received a mechanical (61% ± 9% at 10 years) vs bioprosthetic (58% ± 4% at 10 years) valve ($p > 0.29$). However, for patients older than 60 years of age, the long-term survival rate tended to be lower in patients who received a mechanical valve (18%

± 12% at 10 years) compared to those who received a bioprosthetic valve (31% ± 6% at 10 years) ($p = 0.08$). Importantly, the long-term survival rate was superior for patients with native valve endocarditis (44% ± 5% at 20 years) compared with those with prosthetic valve endocarditis (16% ± 7% at 20 years) ($p < 0.003$).

For all patients, the freedom-from-reoperation rate was relatively high in those with mechanical valves (74% ± 9% at 10 years and 15 years) but started to decline steeply by the year 10 for bioprosthetic valve recipients (56% ± 5% at 10 years, 22% ± 6% at 15 years). In younger patients (60 years of age), the freedom-from-reoperation rate was lower with bioprosthetic (64% ± 4% at 10 years, 48% ± 4% at 15 years) compared with mechanical (19% ± 8% at 10 years, 23% ± 9% at 15 years) valve replacement. For patients older than 60 years of age, the freedom-from-reoperation rate was acceptable with either a mechanical (100% ± 0% at 10 years) or a bioprosthetic (91% ± 6% at 10 years, 91% ± 5% at 15 years) valve. From these data, we concluded that, for patients with endocarditis, mechanical prostheses are appropriate for patients with native valve endocarditis who are less than 60 years of age, have no contraindication to long-term anticoagulation, and have a life expectancy that is otherwise not limited by other major medical problems. Bioprosthetic valves, on the other hand, are best for patients older than 60 years of age with either native valve endocarditis or prosthetic valve endocarditis. Bioprosthetic valves are also acceptable for selected younger patients with prosthetic valve endocarditis who have limited life expectancy (e.g., coronary artery disease, left ventricular dysfunction, end-stage renal failure) or a history of intravenous drug use.

We no longer routinely heparinize patients immediately following mechanical or bioprosthetic valve replacement unless the patient is in persistent atrial fibrillation (>24 hours) or the initiation of warfarin therapy has to be postponed greater than 3 to 4 days (e.g., while awaiting pacemaker placement). We feel that these patients are generally not prone to clot formation in the immediate postoperative period owing to an underlying coagulopathy consequent to cardiopulmonary bypass.

Regarding postoperative anticoagulation, we agree with the authors of Chapter 19 and favor warfarin therapy for mechanical prostheses and, in general, adhere to the American College of Cardiology/American Heart Association guidelines for target international normalized ratio. Based on these same guidelines, the authors also recommend anticoagulation with warfarin for the first 3 months following bioprosthetic valve replacement in either the aortic or mitral position. Here we differ slightly: we currently utilize antiplatelet therapy alone (325 mg of aspirin daily) for our patients undergoing biological aortic valve replacement who are in normal sinus rhythm at the time of hospital discharge and have no other compelling indication for anticoagulation with warfarin.

Few studies focus solely on the issue of anticoagulation following bioprosthetic valve replacement, and the decision to anticoagulate with warfarin for the first 3 months postoperatively has generally been based on circumstantial evidence suggesting that upward of 50% of all thromboembolic episodes occur during the first 6 weeks (5). No data exists, however, to suggest that warfarin therapy during this period decreases the thromboembolic risk. Moinuddeen and co-authors at Yale carried out a retrospective study over a 10-year period (1987 to 1996) to determine whether anticoagulation with heparin and coumadin in the early postoperative period following bioprosthetic aortic valve replacement was beneficial in preventing cerebral ischemic events (6). The anticoagulation group included 109 patients who received heparin followed by warfarin for 3 months (prothrombin time, 20 to 25 seconds). The no-anticoagulation group included 76 patients who received no heparin or warfarin postoperatively. Aspirin was given to most patients in the anticoagulation group (91%) and in those undergoing concomitant coronary bypass grafting in the no-anticoagulation group (37%). Mean follow-up duration was 47 ± 26 months and 59 ± 31 months, respectively, for the anticoagulation and no-anticoagulation groups. Moinuddeen reported no statistically significant difference in the incidence of ischemic cerebral events between the two groups at 24 hours (4.6% for the anticoagulation group vs 3.9% for the no-anticoagulation group), from 24 hours to 3 months (2.8 vs 2.6%), or long-term (>3 months: 11 vs 11.8%) ($p > 0.10$ for all). Furthermore, there was no difference in the incidence of cerebral ischemia with the addition of aspirin (13% with aspirin vs 10% without aspirin). Finally, using Kaplan-Meier analyses, they found no statistically significant difference in the stroke-free survival rate or the long-term survival rate between the anticoagulation and no-anticoagulation groups. The authors concluded that early anticoagulation offered no advantage in the prevention of early ischemic cerebral events following biologic aortic valve replacement and that long-term valve function and survival were not adversely impacted by withholding early anticoagulation. They did, however, appropriately point out that, in their series, there was no disadvantage to the use of early anticoagulation (no increase in bleeding complications or prolonged hospital stay) and that its use is not inherently dangerous.

Based on the findings of Moinuddeen and co-workers (6), and our own clinical impressions, although admittedly anecdotal, we employ anticoagulation with heparin and warfarin only selectively following bioprosthetic aortic valve replacement. Unfortunately, no such data exists regarding warfarin anticoagulation following bioprosthetic replacement of the mitral valve; therefore, we generally recommend 2 to 3 months of empiric warfarin therapy. However, if the patient has any contraindication to full anticoagulation, we feel quite comfortable treating these patients with antiplatelet therapy alone.

APPENDIX: RESOURCES FOR PHYSICIANS AND PATIENTS

Manufacturer/ health resource	Address	Phone number, Web address
Baxter Healthcare Corp	Cardiovascular Group 17221 Red Hill Ave Irvine, CA 92614	1-800-424-3278 www.baxtercvg.com
Medtronic, Inc	8299 Central Ave. NE Spring Lake Park, MN 55432	1-800-328-2518 www.medtronic.com
St. Jude Medical, Inc.	One Lillehei Plaza St. Paul, MN 55117	1-800-544-1664 www.sjm.com
WebMD		wwww.webmd.com

REFERENCES

1. Ikonomidis JS, Kratz JM, Crumbley AJ, et al. Twenty-year experience with the St. Jude Medical mechanical valve prosthesis. J Thorac Cardiovasc Surg 2003;126:2022–2031.

2. Gaudiani VA, Castro LJ, Grunkemeier GJ, Fisher AL, Wu Y, Aziz S. The risks and benefits of reoperation aortic valve replacement. J Am Coll Cardiol 2003;41(Suppl A):516A.
3. STS National Database Fall 2002 Executive Summary (p. 22). The Society of Thoracic Surgeons. Available at: http://www.sts.org. Accessed June 6, 2003.
4. Moon MR, Miller DC, Moore KA, et al. Treatment of endocarditis with valve replacement. The question of tissue versus mechanical prosthesis. Ann Thorac Surg 2001;71:1164–1171.
5. Oyer PE, Stinson EB, Reitz BA, Miller DC, Rossiter SJ, Shumway NE. Long-term evaluation of the porcine xenograft bioprosthesis. J Thorac Cardiovasc Surg 1979;78:343–350.
6. Moinuddeen K, Quin J, Shaw R, Dewar M, Tellides G, Kopf G, Elefteriades J. Anticoagulation is unnecessary after biological aortic valve replacement. Circulation 98 1998;(Suppl II):II-95–II-99.

European Counterpoint to Chapter 19

Alan J. Bryan and Gianni D. Angelini

In the United Kingdom approximately 7000 heart valve replacement operations are performed each year *(1)*. We are fortunate in the United Kingdom to have established methods of data collection, both for numbers and early outcomes for patients undergoing valve surgery *(1)*, and for late follow-up of patients who have undergone heart valve replacement *(2)*. The first of these is the UK Cardiac Surgical Register, which has now evolved into the National Adult Cardiac Surgical Database *(1)*. Since 1977, this has collected basic data with respect to heart valve replacement and early outcome. This data shows that there was a steady but small increase in the number of valve replacement operations in the United Kingdom from around 5000 procedures in 1977 to around 7000 procedures in 2000. There was a steady increase in the mean age of patients undergoing valve replacements and, in the latter half of the 1990s, this was in the range of 64–65 years. The immediate hospital mortality for commonly performed first-time valve replacements in 1999–2000 in the United Kingdom was 3.5% for aortic valve replacement, 6.7% for mitral valve replacement, and 10% for aortic and mitral valve replacement *(1)*.

The second source of information is the UK Heart Valve Registry, which collects data on early and late outcome. It has now been in operation for 15 years *(2)*. During this period, there have been marked increases in the numbers of patients of 70 years and older undergoing aortic and mitral valve replacement and a marked decrease in the number of younger patients undergoing mitral valve replacement *(2)*. This reflects the decline in the incidence of rheumatic etiology in the United Kingdom as an indication for mitral valve replacement, as well as the increasing predominance of mitral valve repair as a preferred option for degenerative mitral valve disease. The overall 30-day mortality rate for heart valve replacement fell to 5.6% in 2000; this was the fourth consecutive year in which a fall was observed *(2)*. This data must be seen as encouraging in the context of an increasingly elderly surgical population.

CURRENT PRACTICE

In the United Kingdom, in 1990, 75% of all prosthetic heart valves implanted were mechanical. In 2000, only 56% were mechanical valves *(2)*. This steady decline is predominantly a reflection of the increasing age of the surgical population, perhaps coupled with enhanced durability of the current range of biological prostheses. Of the mechanical valves, bileaflet pyrolytic carbon valves now dominate the market. About 92% of mechanical valves implanted are now of this design, with only small numbers of single disk and ball and cage valves *(2)*. This has been a consistent trend in recent years, which is likely to persist unless the design of mechanical heart valves advances significantly. Forty-four percent of the valves implanted in the United Kingdom in 2000 were bioprostheses. Within this group, 55% of the valves were porcine bioprostheses and 45% were bovine pericardial prostheses *(2)*. Overall, 84% of bioprosthetic valves were stented and 16% were stentless. Around 80 to 100 homograft valves (1%) are implanted in the United Kingdom each year. Eighty-eight percent of bioprosthetic valves and 64% of mechanical valves were implanted in the aortic position. This presumably reflects the reduced durability and the perceived limited benefits of bioprostheses in the mitral position. In comparison to the population in the United States undergoing heart valve replacement, it remains likely that European patients will remain somewhat younger but, contrary to the 70/30 split between mechanical and bioprostheses outside the United States described in Chapter 19, the contemporary data from the United Kingdom suggests that the 56/44 split is now very close to that observed across the Atlantic.

The overall steady but modest increase in the incidence of heart valve replacement over 15 years does conceal quite major changes in the patient population. The principal change has been the decline in the rheumatic valvular heart disease population leading to fewer young patients needing double and triple valve surgery. At the same time, more older people have needed single valve replacements for degenerative or ischemic etiologies. In the aortic position, the patients are increasingly aged and, as a consequence, bioprostheses predominate *(2)*. In the mitral position there remain concerns about the durability of biological valves, and avoidance of warfarin anticoagulation is often not possible in the presence of atrial fibrillation. The implantation of mechanical valves, therefore, tends to predominate.

With respect to selection of particular valve prostheses, obviously the mechanical valve manufacturers each claim superiority with respect to hemodynamic performance and thromboembolic risk. On occasion there are anecdotal reports of poor performance or clusters of valve thromboses *(3)*. The single randomized prospective trial of two bileaflet prostheses has found no difference in outcome at 5 years, and overall intuitively it seems unlikely that there will be much difference in clinical outcome *(4)*. In the National Health Service, cost is a major issue and, to this end, the United Kingdom is currently in the process of agreeing on a national pricing structure for heart valves in an attempt to purchase heart valve replacements at a lower cost.

The use of bioprostheses has been stimulated both by the increasing age of the patients and the recognition that modern stented bovine pericardial valves are demonstrating excellent durability, good hemodynamics, and ease of implantation *(5)*. The use of stentless bioprostheses, homografts, and pulmonary autograft operations is likely to remain limited by the increased technical complexity of the operative procedure and doubts about whether the more complete left ventricular regression and improved hemodynamics observed with stentless valves does actually translate into clinical benefit.

Table 1
**Patient Surveillance After Implantation
of Prosthetic Heart Valve at Bristol Heart Institute**

	Postoperative year					
	1	2	3	4	5	10
Office visit	3[a]	1	1	1	1	1
Chest X-ray	3[a]	0	0	0	0	0

Note: The number in each cell is the number of times the modality is recommended during that year. Anticoagulation is managed by the family doctor or local cardiologist, in general. Other tests, including echocardiogram and blood tests, are done only as clinically indicated.

[a]At 6 weeks, 6 months, and 1 year postoperatively.

FOLLOW-UP AFTER MECHANICAL HEART VALVE REPLACEMENT

In general, the follow-up protocols for patients after mechanical heart valve replacement are similar to those described by our North American colleagues. The widespread use of intraoperative transesophageal echocardiography (TEE) means that immediate confirmation of satisfactory valve function is accepted as adequate evidence. In general, in the United Kingdom, the history and physical examination are only supplemented by special investigations, such as transthoracic echocardiography, where clinical suspicions exist with respect to valve function. If early progress is satisfactory, then annual follow-up is commenced thereafter either by the referring cardiologist or family doctor (Table 1). For patients with uneventful recovery after mechanical aortic valve replacement it is not at all unusual for follow-up to be carried out by an experienced family practitioner. Problems like prosthetic valve endocarditis rarely occur coincident with scheduled annual outpatient appointments. In general, the globalization of health care means that many physicians and surgeons in the United Kingdom will accept the American College of Cardiology/American Heart Association guidelines as the best model for practice in patients after heart valve replacement *(6)*. Special investigations like transthoracic echocardiography tend to be reserved for patients with unsatisfactory clinical progress or where special concerns about valve dysfunction exist. For instance, annual echo examination of patients 10 years after bioprosthetic replacement is much more likely and useful as the failure patterns of particular valves becomes obvious.

TEE is used liberally in those patients where accurate delineation of complex problems is required and, in particular, for the imaging of mitral valve prostheses when clinical suspicion of a paravalvar leak or prosthetic valve endocarditis has been raised by transthoracic echocardiography *(7)*.

MANAGEMENT OF INFECTIONS AND PREVENTION OF PROSTHETIC VALVE ENDOCARDITIS

Any remote infection has the potential to produce a bacteremia and, as a consequence, result in prosthetic valve endocarditis. Any such infection should be treated in the conventional manner with the appropriate antibiotics. Difficulty can arise in patients with systemic sepsis with confirmed bacteremia from an identifiable remote site who have a prosthetic valve *in situ* in determining whether the valve is infected. In this situation, careful echocardiographic evaluation, usually with TEE, represents the most sensitive diagnostic modality *(7)*. Vegetations, abscess cavities, and valve dysfunction are clear indicators of prosthetic valve endocarditis with an attendant increased operative risk. Biological replacement devices, such as homografts, may offer a better chance to eradicate infection in this situation.

Prevention of prosthetic valve infection is preferable to treatment and the United Kingdom guidelines are widely publicized in the British National Formulary (www.bnf.org) and these guidelines are updated regularly by a working part of the British Society of Antimicrobial Therapy *(8)*. Essentially, antibiotic prophylaxis is indicated for all dental and other invasive procedures. Detailed information is on the website but is beyond the scope of this chapter.

MANAGEMENT OF ANTICOAGULANTS

In general, the management of anticoagulation in patients after mechanical heart valve replacement is carried out in the community either by family practitioners or in anticoagulant clinics managed from district general hospitals in any particular area.

There is agreement among cardiologists and surgeons with respect to the guidelines for anticoagulation, as laid out in Chapter 19. The concept of prosthesis-specific anticoagulation is accepted as a concept promoted by Butchart in the United Kingdom *(9)*. The importance of this has been highlighted by the evidence recently presented to suggest that good anticoagulant control is associated with improved outcome after heart valve replacement *(10)*. The problems in applying these standards in the United Kingdom revolve around organizational factors. Tertiary cardiac surgery/cardiology centres are few and far between and cover large geographical areas. The management of patients with heart valve replacements in the community means that the standard British Society of Hematology guidelines for mechanical heart valve anticoagulation indicate a target international normalized ration (INR) of 3.5 *(11)*. In Europe there is no consensus, and other recommended ranges of INR are 2.0–3.0 *(12)* and 2.5–3.5 *(13)*. This is inappropriately high for many patients with modern mechanical bileaflet prostheses but concerns may exist in the community that differentiating among prosthetic heart valves may be imperfect and therefore the INR range reflects some patients who may have more thrombogenic valves. The encouraging data from recent reports of patient-monitored anticoagulation suggest that this is likely to be an important aspect of the future care of these patients in developed countries *(14)*.

The changing etiology of valvular disease and demography of the patient population suggests that there will be a proportionately older population more suited to bioprosthetic replacements. In addition, the increasing application of reparative techniques to the mitral valve indicates that valve replacement will become less common. There will be an increased focus on prosthesis-specific anticoagulation and patient-directed anticoagulation, particularly if further evidence is produced linking stability of anticoagulation with better outcome. This could all change with radical new designs with new biomaterials but the relative stagnation with respect to new valve replacement devices, make this, in our view, unlikely. For those readers interested in further information, we endorse the sources cited in Chapter 19.

REFERENCES

1. Keogh BE, Kinsman R. National Adult Cardiac Surgical Database Report 1999–2000. The Society of Cardiothoracic Surgeons of Great Britain and Ireland.
2. The United Kingdom Heart Valve Registry Report 2000.
3. Rosengart TK, O'Hara M, Long SJ, et al. Outcome analysis of 245 CarboMedics and St. Jude valves implanted at the same institution. Ann Thorac Surg 1998;66: 1684–1691.
4. Lim KH, Caputo M, Ascione R, Wild J, West R, Angelini GD, Bryan AJ. A prospective randomised comparison of CarboMedics and St Jude medical bileaflet mechanical heart valve prostheses. J Thorac Cardiovasc Surg 2002;123:21–32.
5. Banbury MK, Cosgrove DM, Thomas JD, et al. Haemodynamic stability during 17 years of the Carpentier-Edwards aortic pericardial bioprosthesis. Ann Thorac Surg 2002;73: 1460–1465.
6. ACC/AHA Guidelines for the management of patients with valvular heart disease. Executive Summary. Circulation 1998;98:1949–1984.
7. Lengyel M. The impact of transoesophageal echocardiography on the management of prosthetic valve endocarditis: experience of 312 cases and review of the literature. J Heart Valve Disease 1997;6:204–211.
8. Freeman R, Hall RJC. Infective endocarditis. In: Julian DG, Camm AJ, Fox KM, Hall RJC Poole-Wilson P, eds. Diseases of the heart (2nd ed.). Philadelphia, PA: WB Saunders, 1996, pp. 888–910.
9. Butchart EG. Prosthesis specific and patient specific anticoagulation. In: Butchart EG, Bodnar E, eds. Current issues in heart valve disease: thrombosis, embolism and bleeding. London: London ICR Publishers, 1992, pp. 293–317.
10. Butchart EG, Payne N, Li H-H, Buchan K, Mandana K, Grunkemeier GL. Better anticoagulation control improves survival after valve replacement. J Thorac Cardiovasc Surg 2002; 123:715–723.
11. British Society for Haematology. British Committee for Standards in Haematology, Haemostasis and Thrombus Task Force. Guidelines on oral anticoagulation. J Clin Pathol 1990;43:177–183.
12. Gohlke-Barwolf C, Acar J, Oakley C, et al. Guidelines for prevention of thromboembolic events in valvular heart disease. Study group of the working group on valvular heart disease of the European Society of Cardiology (2nd ed.). Eur Heart J 1995;16:1320–1330.
13. Stein P, Alpert J, Copeland J, Dalen J, Turpie A. Antithrombotic therapy in patients with mechanical and bioprosthetic heart valves. Chest 1995;108:371S–379S.
14. Sidhu P, O'Kane HO. Self-managed anticoagulation: results from a two-year prospective randomised trial with heart valve patients. Ann Thorac Surg 2001;72:1523–1527.

20
Intravascular Filters and Stents

Lazar J. Greenfield and Mary C. Proctor

INTRODUCTION

In 1953, Sven Ivar Seldinger solved the problem of simple vascular access and opened the door to the current era of endovascular diagnostics and therapeutics *(1)*. Twenty years later, Greenfield introduced the pulmonary embolectomy catheter and vena caval filter for treatment of pulmonary embolism *(2)*. The introduction of intravascular stents by Palmaz in 1985 marked the most recent advance in the progress toward less invasive treatment for vascular disease *(1)*.

These early investigators kept careful records of patients undergoing these new procedures and have frequently reported outcomes *(3–5)*. As vena caval filters and vascular stents have become standard care, a need for routine surveillance procedures has developed. This chapter deals with the routine evaluation of vena caval filters and stents used in the venous system. We do not address arterial or cardiac stents.

VENA CAVAL FILTERS

Vena caval filters are used to prevent pulmonary embolism in patients with a high risk for pulmonary embolism and a contraindication to or complication of anticoagulation. They are also occasionally used in conjunction with anticoagulation for patients at very high risk. More recently, filters are being used prophylactically in patients without thromboembolic disease but who are at high risk of developing pulmonary embolism. In order to determine appropriate surveillance for patients with vena caval filters, it is important to understand the risk for adverse events and the liabilities of each design. Corrosion resistance, toxicity, thrombo-resistance, and device/wall interaction must be evaluated.

Data regarding vena caval filters extends back over 30 years to the experience with the Mobin-Uddin umbrella. Review of patient outcomes demonstrated a high rate of vena caval occlusion and migration of the umbrella to the right heart. These unacceptable outcomes led to the development of the Greenfield vena caval filter in the 1970s. This device is intended for permanent placement in the inferior vena cava (IVC). It is manufactured of type 316 stainless steel or beta titanium. The titanium version has greater resistance to fatigue and less corrosion relative to stainless steel. The filter is composed of six limbs affixed to a central apex creating a cone. Each limb is attached to the caval wall by small hooks. The geometry affords it the unique ability to maintain blood flow within the vena cava despite potentially significant volume reduction by entrapped thrombus *(6)*.

Since 1990, four additional filters have been approved by the Food and Drug Administration for marketing in the United States. The VenaTech filter *(7,8)* is a cone-shaped device of flat wire manufactured of Elgiloy, a variant of 316 stainless steel (SS). In an attempt to center and stabilize the filter within the vena cava, it is equipped with six stabilizing legs that attach to the IVC wall. Because there seems to be a direct association between the area of metal exposed to blood flow and the amount of initial thrombus formation and neointimal thickening, this device has had a higher risk for adverse outcomes.

The Simon Nitinol filter is a cone-shaped device topped by a petal-shaped second trapping level. It is attached to the IVC by six tiny L-shaped hooks. Manufactured of nitinol, a blend of nickel and titanium, it has unique properties of shape memory and superelasticity *(9,10)*. These facilitate insertion using a very low-profile delivery system. There is a potential for toxicity from nickel as 0.1% of the population is allergic to nickel and about 0.1% of them will react to nitinol. A second potential problem with this device is related to the upper trapping level. Because of the shape memory of nitinol, the upper level opens to a fixed dimension regardless of the size of the vena cava. When oversized with respect to the cava, local tissue remodeling and inflammatory hyperplastic changes can take place in the vessel wall. Finally, because of the amount of metal in contact with flowing blood, especially between the two trapping levels, there is an increased risk of thrombosis.

The Bird's Nest filter is the only one of the five filters that does not use a cone shape *(11,12)*. It is free-formed as a wire nest within the IVC at the time of placement. It is manufactured of 304 SS and affixed to the IVC by four hooks located at the four ends of the wire. This design also involves a larger amount of metal in the blood stream and because the wires are in very close proximity, there has been an increased incidence of caval thrombosis.

The newest vena caval filter, the "Trap-ease," is marketed by the Cordis Division of Johnson and Johnson and has been available for 3 years. Little data regarding its performance exist. The device is manufactured of nitinol broad wire and is shaped like two cones end to end. One of its striking features is the high radial force exerted when it is deployed. Several potential liabilities exist, including the risk of caval thrombosis owing to the type and amount of metal, damage to the endothelium of the IVC caused by the force exerted against the caval wall and the other factors listed for nitinol devices.

Questions regarding the need for ongoing anticoagulation of filter patients are frequently raised. We have not found this to be necessary. Decisions regarding long-term anticoagulation should be based on the patient's underlying thromboembolic disease. Those with recurrent thromboembolism with no contraindication to anticoagulation may benefit from extended therapy but it is not required to maintain the patency of the filter.

A second issue is the limitation on physical activity. In our experience, there is no need to abstain from any form of physical activity because of the presence of a Greenfield filter. We have experience with professional athletes and even a Navy fighter jet pilot who is subjected to extreme gravitational forces. There have been no clinical or structural problems. Once a filter has been placed and the underlying condition has

Table 1
Surveillance After Implantation of Vena Caval Filter

	Postoperative year					
	1	2	3	4	5	10
Office visit	1	1	1	1	1	1
Abdominal X-ray (anterior-posterior, lateral and oblique)	1	1	1	1	1	1
Ultrasound scan (inferior vena cava and lower extremities)	1	1	1	1	1	1
Prothrombin time (international normalized ratio)	1^a	1^a	1^a	1^a	1^a	1^a
Other (computed tomography of abdomen)	1^b	1^b	1^b	1^b	1^b	1^b

^{a} Only if anticoagulated.
^{b} Only if suspicion of penetration or occlusion exists.

resolved, there are no maintenance requirements for the filter. Because of the number of filters that are placed without sufficient patient education, we maintain a website (http://www.greenfieldfilter.com) that provides general filter information and answers to specific questions raised by patients.

The purpose of surveillance is to monitor the proper function of these permanently placed devices to provide appropriate intervention for problems identified early and to provide ongoing evaluation and treatment of the underlying venous disease (Table 1). The primary use of these data is for direct patient care. Secondary use is acquisition of information regarding the performance of these devices over time. We currently maintain data for more than 2500 filter placements spanning 30 years. Periodically, we review this information to provide guidance regarding patient outcomes and publish the findings (13–16). These reports and similar studies with other filters have been responsible for growing use of this treatment.

More than 25,000 filters are placed each year with more than 500,000 currently in use. Prior to 1987, filters were placed by surgeons. With the introduction of percutaneous devices, interventional radiologists also became involved. Patient follow-up varied widely from biannual examinations to specific follow-up. Annual surveillance has been our practice during the past 30 years. A consensus conference was held in 1998 to establish reporting standards for filter studies. This conference included both surgeons and radiologists. The need for annual evaluation with objective testing was one of the points of agreement (17). Our routine surveillance for the asymptomatic patient includes three areas: clinical history and examination; duplex ultrasound of the IVC and lower extremities; and an abdominal plain film in three projections, anterior-posterior, lateral and oblique.

The focus of the clinical examination is related to progression or recurrence of thromboembolic disease and its treatment. Patients are questioned about the concurrent use of anticoagulants and any adverse effects. The need for hospitalization or surgical procedures is ascertained and patients are questioned explicitly regarding any occurrence of pulmonary embolism. The lower extremities are examined for signs of chronic venous insufficiency.

The duplex examination includes both lower extremities, the filter access site and the inferior vena cava. It provides real-time images of the vessel wall and lumen, allowing identification of acute or chronic thrombosis. Particular attention is paid to the site of prior thrombosis to document the degree of resolution. We find new acute thrombosis in 1 to 2% of patients screened. The filter insertion site is investigated to determine its long-term patency. The incidence of venous occlusion is about 8% with the majority being asymptomatic. Patients with acute or subacute thrombosis are evaluated and treated, whereas those with chronic disease are referred for further evaluation. The ultrasound examination of the IVC incorporates both duplex images and Doppler signals. The majority of patients can be scanned successfully with adequate preparation. They are asked to fast from midnight the evening before and to have the scan completed as early in the morning as possible to limit interference from bowel gas. The Doppler signal is used to locate the renal arteries and to detect blood flow above and below the filter. With duplex, it is usually possible to image the filter in either the transverse or longitudinal plane. If the filter is imaged, the presence or absence of thrombus can be determined. In addition to bowel gas, abdominal scar tissue or body habitus can make it difficult to obtain images or detect a Doppler signal resulting in a suboptimal study. However, failure to image the filter and detect Doppler signals in an optimal study may indicate caval occlusion. This can happen in the presence of massive thromboembolism or as a result of caval contraction or intimal hyperplasia. This finding should be confirmed with a contrast computed tomography (CT) or magnetic resonance (MR) venography to determine the cause. The majority of patients with documented caval occlusion are asymptomatic as it develops slowly, allowing time for collateral channels to develop.

A few patients will have thrombus in the filter that extends well beyond the apex. In these cases, a second filter may be required proximal to the first to provide continued protection from pulmonary embolism. Alternatively, lytic therapy may be indicated if the thrombus is fresh and does not extend more than 5 cm above the apex.

Because the ultrasound scan is a relatively costly procedure, if we are unable to perform a successful test for 2 consecutive years, we delete that part of the examination but continue to scan the lower extremities. Patients who had prophylactic filters placed in the absence of thromboembolism are only scanned at the first follow-up unless new thrombus is detected, which occurs in 16% of these patients. Those who are free of thrombosis return only for an abdominal X-ray.

At the first surveillance visit, the abdominal films are obtained in the interventional radiology department. By using the same equipment and imaging protocols, new films are comparable to the post placement films. The position of the filter is evaluated with respect to a spinal landmark to determine whether any longitudinal movement has taken place. The widest distance between two opposing filter legs is measured and used as a surrogate for potential penetration of the IVC by a filter leg. When the change is greater than 7 mm, a CT scan is recommended to determine potential involvement of adjacent structures. Reduction of the diameter suggests caval stenosis or obstruction and requires CT or MR imaging. Oblique views are used to evaluate the distribution or presence of possible crossed filter limbs. Subsequent X-rays may be obtained in other imaging facilities as further changes are readily discernible. Plain films are satisfactory to document movement and filter base expansion but the specific degree of filter tilting or distribution of the limbs can only be done by venacavography and contrast CT, respectively.

Maintaining a record of routine surveillance examination results is extremely important in determining expected outcomes for a wide range of patients. Indications for filter use have expanded during the past 10 years. Devices are being placed in younger patients with a life expectancy of 50 years or more. Because there is no evidence regarding performance that approaches that time frame, it is essential to evaluate these patients prospectively and gather data for the future.

ENDOVASCULAR STENTS

Vascular stents are metallic devices that are placed percutaneously to treat abnormal vessel geometry in a variety of situations, including persistent filling defects, tortuosity, increased resistance, dissections, or intimal flaps *(18)*. They were originally designed for use in the coronary arteries but are currently being used in many other arteries and in limited venous applications, including superior vena caval syndrome, subclavian vein obstruction, and following thrombolytic treatment of iliofemoral or other large vein thrombosis.

Dotter first conceived the use of a rigid scaffold to maintain the patency of compromised vessels. Three basic types have been developed: self-expanding, balloon expanding, and thermal expanding. The Palmaz stent was the earliest balloon expandable stent. An angiographic balloon is inflated within the stent until it reaches the desired diameter expansion. The balloon is then deflated and removed. The self-expanding devices are constrained during insertion and, when properly positioned, the retaining covering is released. Thermal-expanding stents use the unique properties of nitinol described previously. These devices are more flexible and useful at sites of flexion. However, it is more difficult to deploy them at an exact location as the stent is dependent on temperature increase to resume its desired size and shape *(18,19)*. Three of the materials used to fabricate IVC filters are also used in stents: stainless steel, titanium, and nitinol. The properties of these metals also affect blood and tissue reactions and include corrosion resistance, toxicity, thrombogenicity, and wall incorporation.

Because stents are permanent devices, resistance to corrosion and subsequent metal fatigue are significant concerns. It may be more difficult to detect a stent fracture than a fracture of a filter as the wires are closely knit and the shape integrity may be maintained despite the break. The clinical significance of such a fracture is unknown. Other deformations or multiwire fractures may have greater clinical importance as the device may no longer provide the intended support. Stents can also be crushed but this should be clearly visible on radiographs. This is more likely to occur when the stent is positioned at a point of flexure or when adjacent to bony compression sites such as the thoracic outlet. Metal toxicity may be an issue with the nitinol stents owing to the prevalence of patients with a sensitivity to nickel.

Thrombogenicity is a concern with all devices because occlusion of a stent can have serious implications and require subsequent interventions to reestablish patency. The rate of blood flow through the vessel is associated with the risk of thrombosis as is the type of material and the surface preparation of the metal. Smooth, highly polished surfaces are more resistant to thrombus formation. As it is easier to prevent an occlusion than to treat it, a routine surveillance plan is warranted. Duplex scans can be used to identify slowing of flow or reduction of the luminal area that requires intervention. The final characteristic is wall healing. In general, the greater the amount of metal in contact with the vessel wall, the greater the proliferative response. Stents vary greatly with respect to the surface area and the radial force exerted. Over the initial 8 weeks, the process of vessel wall incorporation takes place. During this time, the stent becomes enveloped in a platelet-rich covering that matures to fibrin strands. Within 3 to 4 weeks, a neointimal layer replaces the thrombotic covering, which finally acquires an endothelial covering. During the final 4 weeks, the neointima resorbs, leaving an extracellular matrix. If the neointima continues to thicken, occlusion can take place. Some manufacturers are studying the role of special coatings to prevent aggressive neointimal growth. The administration of anticoagulants during the procedure also may be useful.

The high radial force of the stent may cause remodeling of the vessel wall, resulting in a thinning of the media because of mechanical compression or altered flow patterns. Devices with a more rigid configuration that apply higher radial force may be more prone to these complications (20). In the most extreme form, stents can erode through the vessel wall and cause damage to adjacent structures. Knowing the characteristics of the location and the type of stent that has been used may help to focus the surveillance testing.

Stents are often used as an alternative to more invasive surgical procedures. Therefore, the frequency and method of evaluation is dictated by the pre-existing pathology. We are unaware of any surveillance guidelines that have been developed exclusively for stents. The most common sites for venous stent placement are the iliac and the subclavian vein and often they are placed in conjunction with lytic therapy for deep vein thrombosis (21). Because the flow of blood is significantly slower in these locations compared to arterial sites, it is important to assess the patency of the vessel. Physical examination of the extremity and Doppler-duplex ultrasound, MR, or intravascular ultrasound can be used to verify patency.

The role of venography and duplex ultrasound are well established but the utility of intravascular ultrasound has gained importance in the evaluation both prior to and after stent placement. It facilitates assessment of the luminal size and contour of the implantation site. It is superior to angiography in detecting incomplete stent opening. It is also capable of characterizing luminal morphology and transmural anatomy (22).

The mechanical stability of the stent can be determined by plain films or CT (23). This may be difficult to interpret with those stents that are less radio-opaque. In order to fully evaluate stents that pass through the thoracic outlet, it may be necessary to move the extremity through a series of provocative maneuvers to determine whether it is being compressed (24). It is inadvisable to place a stent in this location without prior correction of the bony abnormality.

Nazarian et al. reported outcomes for 56 patients treated with venous stents. The primary and secondary patency rates at 1 year were 50 and 81% (25) in this large series. Patency rates for stents used in the intrahepatic IVC have been 100%. However, the mean length of follow-up was 1 month (26).

When stents are used in conjunction with balloon dilation of chronic iliac vein obstruction, 1-year primary and assisted patency rages for 94 patients were 82 and 91%, respectively (27). Miller reported primary clinical success in 19 or 23 patients when Wall stents were used to treat malignant superior vena cava obstruction (28). The

Table 2
Surveillance After Implantation of Venous Stents

	Postoperative year					
	1	2	3	4	5	10
Office visit	4[a]	1	1	1	1	1
X-ray of stent site	1	1	1	1	1	1
Ultrasound scan of stent site	4[a]	1	1	1	1	1
Prothrombin time (international normalized ratio)	1[b]	1[b]	1[b]	1[b]	1[b]	1[b]
Other (intravenous ultrasound or venogram)	1[c]	1[c]	1[c]	1[c]	1[c]	1[c]

[a] At 1, 3, 6, and 12 months.
[b] Only if anticoagulated.
[c] If transcutaneous ultrasound shows occlusion.

group at Stanford has the largest single institution experience and was able to resolve obstruction in 97% of cases. Their long-term follow-up demonstrated a primary patency rate of 90% and secondary patency rate of 95% *(29)*.

The use of venous stents is a very recent treatment method and few series report more than 12-month outcomes. The true value of this intervention will not be known for 5 to 10 years, when long-term data become available. However, the initial results are impressive, especially for conditions such as chronic iliac occlusion that has few acceptable alternative treatments. Patients' medical records should note the presence of IVC filters or stents. Several authors have reported problems with guidewires being caught in filters that were dislocated when aggressive attempts were made to free the wire. This happens when endovascular procedures are attempted following filter placement without adequate visualization of the vessels *(30,31)*. In very rare instances, infection of vascular stents has been reported *(32)*. *Staphylococcus aureus* is usually bacterium identified, causing a necrotizing angitis at the stent that can become systemic. Treatment requires intravenous antibiotics and supportive care. Failure to identify this potential site of infection may delay appropriate care.

There are no standardized follow-up guidelines but a regimen of 1-, 3-, 6-, and 12-month evaluations followed by yearly visits appears reasonable (Table 2). The primary methods of evaluation should be physical examination, duplex ultrasound, and plain films in addition to any tests specific to the underlying condition. Documentation should include patency, mechanical stability, and the need for secondary interventions. These data should be reported by physicians who place large numbers of stents to guide others in making reasonable risk–benefit assessments.

IVC filters and vascular stents are just two of a series of permanent implants made possible by developments in endovascular procedures. Unlike angioplasty or thrombolysis, the patient is left with a permanent metallic device that is subject to various complications at different periods of time. When signs of device failure are found early, appropriate intervention may be provided, preventing more serious complications. Physicians who implant such devices are responsible either to provide ongoing care or to arrange for appropriate follow-up with another physician. In this way, long-term optimal patient outcomes can be assured.

REFERENCES

1. Fogarty TJ, Biswas A. Evolution of endovascular therapy: diagnostics and therapeutics. In: White RA, Fogarty TJ, eds. Peripheral endovascular interventions. New York: Springer-Verlag, 1999, pp. 3–10.
2. Greenfield LJ, Kimmell GO, McCurdy WC. Transvenous removal of pulmonary emboli by vacuum-cup catheter technique. J Surg Res 1969;9:347–352.
3. Greenfield LJ, Zocco J, Wilk J, Schroeder T, Elkins R. Clinical experience with the Kim-Ray Greenfield vena caval filter. Ann Surg 1977;185:692–698.
4. Greenfield LJ, Peyton R, Crute S, Barnes RW. Greenfield vena caval filter experience: Late results in 156 patients. Arch Surg 1981;116:1451–1455.
5. Greenfield LJ, Proctor MC, Cho KJ, et al. Extended evaluation of the titanium Greenfield vena caval filter. J Vasc Surg 1994;20:458–465.
6. Greenfield LJ, McCurdy JR, Brown PP, Elkins R. A new intracaval filter permitting continued flow and resolution of emboli. Surgery 1973;73:599–606.
7. Crochet DP, Stora O, Ferry D, et al. Vena Tech-LGM filter: long-term results of a prospective study. Radiology 1993;188:857–860.
8. Crochet DP, Brunel P, Trogrlic S, Grossetete R, Auget JL, Dary C. Long-term follow-up of vena tech-LGM filter: predictors and frequency of caval occlusion. JVIR 1999;10:137–142.
9. Brown R. Simon Nitinol filter clincal study. 1991. Anonymous. Unpublished.
10. Kim D, Grassi CJ, Simon M, Kleshinski S, Siegel J, Porter DH. Simon Nitinol filter clinical trial: Final results. RSNA 1991.
11. Firkin A, Walters N, Thomson K, Atkinson N. Inferior vena cava "bird's nest" filters—2-year follow-up. Aust Radiol 1992;36:286–288.
12. Roehm J, Gianturco C, Barth M, Wright K. Percutaneous transcatheter filter for the inferior vena cava: A new device for treatment of patients with pulmonary embolism. Radiology 1984;150:255–257.
13. Greenfield LJ, Proctor MC, Saluja A. Clinical results of Greenfield filter use in patients with cancer. Cardiovasc Surg 1997;5:145–149.
14. Greenfield LJ, Proctor MC. Suprarenal filter placement. J Vasc Surg 1998;28:432–438.
15. Greenfield LJ, Cho KJ, Proctor MC, Sobel M, Shah S, Wingo J. Late results of suprarenal Greenfield vena cava filter placement. Arch Surg 1992;127:969–973.
16. Greenfield LJ, Proctor MC. Twenty-year clinical experience with the Greenfield filter. Cardiovasc Surg 1995;3:199–205.
17. Bonn J, Cho KJ, Cipolle M, et al. Recommended reporting standards for vena caval filter placement and patient follow-up. J Vasc Surg 1999;30:573–579.
18. Deithrich EB. Endovascular suite design. In White RA, Fogarty TJ, eds. Peripheral endovascular interventions. New York: Springer-Verlag, 1999, pp. 133–142.
19. Johnston KW. Value and limitations of percutaneous transluminal angioplasty, stents, and lytic agents for aortoiliac disease. In: Veith FJ, ed. Current critical problems in vascular surgery (vol. 7). St. Louis: Quality Medical Publishing, 1996, pp. 275–282.
20. Back MR, White RA. Biomaterials: considerations for endovascular devices. In: White RA, Fogarty TJ, eds. Peripheral endovascular interventions. New York: Springer-Verlag, 1999, pp. 219–246.
21. Semba CP, Dake MD. Iliofemoral deep venous thrombosis: Aggressive therapy with catheter-directed thrombolysis. Radiology 1994;191:487–494.
22. Cavaye DM, Diethrich EB, Santiago OJ, Kopchok GE, Laas TE, White RA. Intravascular ultrasound imaging: an essential component of angioplasty assessment and vascular stent deployment. Int Angiol 1993;12(3):214–220.
23. Phipp LH, Scott JA, Kessel D, Robertson I. Subclavian stents and stent-grafts: cause for concern? J Endovasc Surg 1999;6:223–226.

24. Hall LD, Murray JD, Boswell GE. Venous stent placement as an adjunct to the staged, multimodal treatment of Paget–Schroetter syndrome. JVIR 1995;6:565–570.
25. Nazarian GK, Bjarnason H, Dietz CA, Bernadas CA, Hunter DW. Iliofemoral venous stenosis: effectiveness of treatment with metallic endovascular stents. Radiology 1996;200: 193–199.
26. Fletcher WS, Lakin PC, Pommier RF, Wilmarth T. Results of treatment of inferior vena cava syndrome with expandable metallic stents. Arch Surg 1998;133:935–938.
27. Neglen P, Raju S. Balloon dilation and stenting of chronic iliac vein obstruction: technical aspects and early clincal outcome. J Endovasc Ther 2000;7:79–91.
28. Miller JH, McBride K, Little F, Price A. Malignant superior vena cava obstruction: stent placement via the subclavian route. Cardiovasc Intervent Radiol 2000;23:155–158.
29. Bergan JJ, Thorpe PE. Endovenous Surgery. In White RA, Fogarty TJ, eds. Peripheral endovascular interventions. New York: Springer-Verlag, 1999, pp. 515–529.
30. Andrews RT, Geschwind JFH, Savader SJ, Venbrux AC. Entrapment of J-tip guidewires by Venatech and stainless steel Greenfield vena cava filters during central venous catheter placement: percutaneous management in four patients. Cardiovasc Intervent Radiol 1998; 21:424–428.
31. Dardik A, Campbell KA, Yeo CJ, Lipsett PA. Vena cava filter ensnarement and delayed migration: an unusual series of cases. J Vasc Surg 1997;26:869–874.
32. Latham JA, Irvine A. Infection of endovascular stents: an uncommon but important complication. Cardiovasc Surg 1999;7:179–182.

US Counterpoint to Chapter 20

David M. Hovsepian

Chapter 20 provides a brief overview of two very different types of implants used in the venous system; namely, inferior vena cava (IVC) filters and stents. Remarkably, there are still no guidelines regarding the proper indications for use, choice of device, or follow-up protocols, especially for IVC filters. The authors offer their own unique views, and it is for the reader to decide whether enough scientific proof is given in evidence to justify adoption of the concepts presented therein. For many types of implanted medical devices discussed in this book, there were many patient support groups, telephone hotlines, websites, and so on, available as resources for those interested in additional information. We know of no such sources, other than device manufacturers and the US Food and Drug Administration, covering caval filters and stents at present.

We are the fortunate recipients of Dr. Greenfield's many contributions to science during his three decades of experience in the development of his own IVC filter. His is one among a number of commercially now available in the United States, each with its own unique design features and applicability. When making a choice between filters, cost differences are not insignificant.

Caval filtration is a delicate balance between effective thrombus-trapping and physiological invisibility. There must be enough surface area to capture significant emboli, but with minimal alteration of IVC blood flow, thrombogenicity, and foreign body reaction. A permanent IVC filter must attach itself securely and resist becoming dislodged by normal physical activity, Valsalva maneuver, or external forces, such as the Heimlich maneuver-like "quad cough" used for pulmonary toilet in paralyzed patients *(1–3)*. No one filter possesses all of these desirable features. Fortunately, all of the filters in current use have been found to be effective in reducing the risk of life-threatening pulmonary embolism, with an acceptably low morbidity.

In Chapter 20, the descriptions of the non-Greenfield-type filters presented are brief and focus on negative attributes. To be fair, one should also point out that there have been published reports of serious complications following Greenfield filter placement. These include tilting, which can lead to incomplete thrombus-trapping *(4)*; perforation of the IVC and adjacent aorta, bowel, ureter, and vertebral body *(5–10)*; recurrent pulmonary embolism or caval thrombosis *(11–15)*; strut fracture and/or structural failure

(16–18); and migration, both caudal and cephalad, with reports of resultant pericardial tamponade, arrhythmia, and/or death *(19,20)*.

The Greenfield filter has continued to evolve since its inception. The material has changed from stainless steel to titanium and back again to stainless steel. The anchoring hook design has been modified. The size of the introducer has been reduced and the delivery system is once again over-the-wire. Functional alterations may accompany each of these technological innovations. The latest version, a 12-Fr stainless steel model, appears a worthy successor, but long-term results from the original 24-Fr stainless steel filter cannot automatically be extrapolated to any of the subsequent iterations.

With respect to follow-up, there is no consensus. A joint conference held in 1998 with both vascular surgeons and radiologists in attendance found little difference among the IVC filters in current use. The panel made the recommendation that "patients with permanently implanted devices deserve routine follow-up"*(21)*, including both clinical evaluation and objective testing, but they did not specify the interval(s) at which this follow-up should be obtained.

A related topic concerns the patient with a caval filter or stent who develops an infection elsewhere or who is about to undergo a procedure such as bowel surgery that is likely to produce transient bacteremia. Does such a patient deserve different antibiotics, higher doses of antibiotics, or a longer course of antibiotics than a similar patient without an intravascular filter or stent? We are aware of no data supporting modification of treatment to prevent infection of the implanted prosthetic device and thus do not alter our antibiotic therapy in these patients.

There are two contrasting issues at stake with regard to follow-up, namely the cost of surveillance, often referred to as monitoring, and the incidence of morbid complications. In Dr. Greenfield's experience, the incidence of caval occlusion is about 8%, with the majority being asymptomatic. Based on this, one could argue that the costs of physician and clinic time, ultrasound examinations, and abdominal X-rays are not justified. Following up on a potential problem with more intensive imaging, such as computed tomography (CT) or magnetic resonance imaging (MRI), may only increase costs, with little actual benefit. The most cost-effective strategy has yet to be determined but it may be prudent to evaluate only those patients who are symptomatic.

It would be interesting to learn how often Dr. Greenfield's diligent follow-up led to an intervention. How many times does such surveillance detect a problem that can be treated? Immediately after filter insertion, one might still be able to intervene for crossed legs, migration, or symptomatic caval occlusion. However, once the filter has become incorporated into the caval wall, approximately 4 to 6 weeks later, it is unlikely to move on its own or be moved intentionally. Any thrombus that has formed in or around the filter during that period will have become organized and adherent.

Delayed occlusion of the IVC can occur after sudden trapping of a large volume of thrombus—the extreme situation of the job that it was intended to do. However, that thrombus, which originated in the lower extremities, is often older and more organized. Although one might be tempted to try thrombolysis in this setting, the likelihood of a long-term positive outcome is poor. So, there are very few filter-related complications in which one is able to intervene effectively.

Table 1
Washington University Regimen
of Follow-Up After Vena Cava Filter Insertion

	Postoperative year					
	1	2	3	4	5	10
Office visit	0[a]	0[a]	0[a]	0[a]	0[a]	0[a]
Lab tests	0[a]	0[a]	0[a]	0[a]	0[a]	0[a]
X-rays	0[a]	0[a]	0[a]	0[a]	0[a]	0[a]
MR/CT	0[a]	0[a]	0[a]	0[a]	0[a]	0[a]
Other tests	0[a]	0[a]	0[a]	0[a]	0[a]	0[a]

[a] Only obtained as clinical symptoms or physical findings warrant. MR/CT: magnetic resonance/computed tomography.

Most physicians perform the minimal follow-up of caval filters that is considered prudent. Our hospital is a large university medical center and patients are often elderly and quite ill, frequently as the result of malignancy. Monitoring in this latter group of patients is unlikely to have a beneficial effect on long-term outcome. Our standard follow-up routine is to visit patients the morning after filter placement to evaluate the access site and lower extremities. Beyond that, we rely on the patients or their referring physicians to contact us if they perceive that a problem has developed (Table 1).

It would not be unreasonable to schedule a 30-day office visit and obtain an abdominal X-ray at that time, depending on patient logistics, for that is still within the time frame in which one might find a problem that is amenable to treatment. In the absence of hard scientific evidence that such a program is cost-effective, however, it is understandable that few physicians are willing to shoulder the logistical and time demands of implementation.

The follow-up of endoluminal stents in the venous system is another matter. Unlike patients who receive filters, stents are typically placed for treatment of occlusive disease. The most frequent indication is a complication related to central venous access. Stenting has been shown to be more effective than angioplasty alone for venous obstruction (22–24), but reintervention to maintain patency is not uncommon, especially in dialysis patients (25,26).

For stents placed in the superior vena cava or IVC, the indication is almost always malignancy. Therefore, the need for long-term follow-up may not be as important as for patients with benign disease. With few exceptions, stent-related problems in cancer patients are owing to progression of underlying disease and most patients remain asymptomatic unless the occlusion happens rapidly or there are insufficient collaterals. Monitoring of the superior vena cava or IVC using ultrasound is technically challenging and often impossible. Most radiologists would view ultrasound follow-up of these areas as a laborious, unnecessary expense.

In the case of venous stents for benign disease, one could argue in favor of a limited follow-up regimen, such as an office visit and duplex ultrasound obtained at annual intervals, to look for primary or secondary signs of venous obstruction. However, most central venous problems cannot be diagnosed by ultrasound, and both CT and MRI are

Table 2
Washington University Regimen of Follow-Up After Venous Stent Placement

	Postoperative year					
	1	2	3	4	5	10
Office visit	0[a]	0[a]	0[a]	0[a]	0[a]	0[a]
Lab tests	0[a]	0[a]	0[a]	0[a]	0[a]	0[a]
X-rays	0[a]	0[a]	0[a]	0[a]	0[a]	0[a]
MR/CT	0[a]	0[a]	0[a]	0[a]	0[a]	0[a]
Other tests	0[a]	0[a]	0[a]	0[a]	0[a]	0[a]

[a] Only obtained as clinical symptoms or physical findings warrant. MR/CT: magnetic resonance/computed tomography.

unlikely to provide a clear picture of what is happening inside the stent. Specifically, the principal problem with CT is an artifact that is caused by the metal that produces streaks that crisscross and obscure the image, regardless of stent type. For MRI, stainless steel stents create a local distortion of the magnetic field that blacks out a region around and including the stent. Nitinol stents, although nonferromagnetic, also pose problems for MR imaging, particularly of the lumen. Venography is the gold standard for follow-up of any venous implant and may be the best first step to diagnosis, with the advantage that intervention may be possible at the same time.

Therefore, in contrast to the cost- and labor-intensive approach presented in Chapter 20 for the follow-up of vena cava filters and venous stents, my fellow interventional radiologists at Washington University and I take a minimalist approach (Table 2). Most thrombotic problems associated with filters go unnoticed. If symptoms arise, then the appropriate workup can be instituted and tempered by whether or not the knowledge likely to be gained can be acted upon. For patients who receive venous stents, the strategy has been to investigate and treat on an *ad hoc* basis, usually with venography as the initial imaging modality and treatment is often immediate. Clinical follow-up and monitoring with ultrasound may be conceptually appealing but there is little evidence to support this strategy except in unusual circumstances.

REFERENCES

1. Balshi JD, Cantelmo NL, Menzoian JO. Complications of caval interruption by Greenfield filter in quadriplegics. J Vasc Surg 1989;9:558–562.
2. Kinney TB, Rose SC, Valji K, Oglevie SB, Roberts AC. Does cervical spinal cord injury induce a higher incidence of complications after prophylactic Greenfield inferior vena cava filter usage? J Vasc Interv Radiol 1996;7:907–915.
3. Greenfield LJ. Does cervical spinal cord injury induce a higher incidence of complications after prophylactic Greenfield filter usage? J Vasc Interv Radiol 1997;8:719–720.
4. Katsamouris AA, Waltman AC, Delichatsios MA, Athanasoulis CA. Inferior vena cava filters: in vitro comparison of clot trapping and flow dynamics. Radiology 1988;166:361–366.
5. Teitelbaum GP, Jones DL, van Breda A, et al. Vena caval filter splaying: potential complication of use of the titanium Greenfield filter. Radiology 1989;173:809–814.

6. Dabbagh A, Chakfe N, Kretz JG, et al. Late complication of a Greenfield filter associating caudal migration and perforation of the abdominal aorta by a ruptured strut. J Vasc Surg 1995;22:182–187.
7. Howerton RM, Watkins M, Feldman L. Late arterial hemorrhage secondary to a Greenfield filter requiring operative intervention. Surgery 1991;109:265–268.
8. Lok SY, Adkins J, Asch M. Caval perforation by a Greenfield filter resulting in small-bowel volvulus. J Vasc Interv Radiol 1996;7:95–97.
9. Berger BD, Jafri SZ, Konczalski M. Symptomatic hydronephrosis caused by inferior vena cava penetration by a Greenfield filter. J Vasc Interv Radiol 1996;7:99–101.
10. Kim D, Porter DH, Siegel JB, Simon M. Perforation of the inferior vena cava with aortic and vertebral penetration by a suprarenal Greenfield filter. Radiology 1989;172:721–723.
11. Athanasoulis CA, Kaufman JA, Halpern EF, Waltman AC, Geller SC, Fan CM. Inferior vena caval filters: review of a 26-year single-center clinical experience. Radiology 2000; 216:54–66.
12. Ferris EJ, McCowan TC, Carver DK, McFarland DR. Percutaneous inferior vena caval filters: follow-up of seven designs in 320 patients. Radiology 1993;188:851–856.
13. Richenbacher WE, Atnip RG, Campbell DB, Waldhausen JA. Recurrent pulmonary embolism after inferior vena caval interruption with a Greenfield filter. World J Surg 1989;13: 623–628.
14. McAuley CE, Webster MW, Jarrett F, Hirsch SA, Steed DL. The Greenfield intracaval filter as a source of recurrent pulmonary thromboembolism. Surgery 1984;96:574–547.
15. Braun TI, Goldberg SK. An unusual thromboembolic complication of a Greenfield vena caval filter. Chest 1985;87:127–129.
16. Lang W, Schweiger H, Fietkau R, Hofmann-Preiss K. Spontaneous disruption of two Greenfield vena caval filters. Radiology 1990;174:445–446.
17. Plaus WJ, Hermann G. Structural failure of a Greenfield filter. Surgery 1988;103:662–664.
18. Bury TF, Barman AA. Strut fracture after Greenfield filter placement. J Cardiovasc Surg (Torino) 1991;32:384–386.
19. Lahey SJ, Meyer LP, Karchmer AW, et al. Misplaced caval filter and subsequent pericardial tamponade. Ann Thorac Surg 1991;51:299–300.
20. Becker DM, Philbrick JT, Selby JB. Inferior vena cava filters. Indications, safety, effectiveness. Arch Intern Med 1992;152:1985–1994.
21. Greenfield LJ, Rutherford RB. Recommended reporting standards for vena caval filter placement and patient follow-up. Vena Caval Filter Consensus Conference. J Vasc Interv Radiol 1999;10:1013–1019.
22. Nazarian GK, Bjarnason H, Dietz CA, Jr., Bernadas CA, Hunter DW. Iliofemoral venous stenoses: effectiveness of treatment with metallic endovascular stents. Radiology 1996;200: 193–199.
23. Nazarian GK, Austin WR, Wegryn SA, et al. Venous recanalization by metallic stents after failure of balloon angioplasty or surgery: four-year experience. Cardiovasc Intervent Radiol 1996;19:227–233.
24. Gross CM, Kramer J, Waigand J, et al. Stent implantation in patients with superior vena cava syndrome. Am J Roentgenol 1997;169:429–432.
25. Gray RJ, Horton KM, Dolmatch BL, et al. Use of Wallstents for hemodialysis access-related venous stenoses and occlusions untreatable with balloon angioplasty. Radiology 1995;195:479–484.
26. Vesely TM, Hovsepian DM, Pilgram TK, Coyne DW, Shenoy S. Upper extremity central venous obstruction in hemodialysis patients: treatment with Wallstents. Radiology 1997; 204:343–348.

European Counterpoint to Chapter 20

A. E. Healey and Derek A. Gould

There exists little significant evidence on which to base the use of inferior vena cava (IVC) filters *(1–3; see* Table 1) and, at present, there are no UK or European regulatory guidelines for their use. They are therefore used on a case-by-case basis with moderate variation between centers. The Royal Liverpool University Hospital Interventional Radiology Department places about 12 IVC filters annually although a neighboring center with a much smaller population catchment area inserts at least twice as many each year. Even in the same unit there is a wide range of use between individual radiologists and referring clinicians. The incidence of use of IVC filters in the United Kingdom seems much less than in the United States. Although we can find no comparative data in the current literature, the incidence of filter use in the United States in recent years is well documented with 30,000 to 40,000 filters inserted per year *(4)*.

The indications for use of IVC filters in the United Kingdom include the following *(5,6)*:

- Recent lower limb deep venous thrombosis (DVT) in or above the femoropopliteal region where there is a contraindication to anticoagulation or where complications of anticoagulation require their discontinuation .
- Recurrent pulmonary emboli despite hematologically effective levels of anticoagulation *(8)*.
- Chronic pulmonary hypertension and reduced cardiac reserve.

Relative indications are as follow:

- Free-floating thrombi demonstrated on imaging of the pelvic veins or IVC.
- Previous DVT and planned orthopedic or pelvic surgery.
- Surgical therapy for pulmonary embolism.
- Recurrent septic pulmonary emboli or pulmonary emboli in cancer patients *(9)*.
- Some cases of polytrauma *(10–14)*.

It must be stressed that regulatory differences have allowed a much broader range of products to be used in the United Kingdom and Europe than in the United States. In the United Kingdom, the use of filters can be divided into those destined for permanent placement (permanent filters) and those placed for the short term (temporary or retrievable filters).

Temporary filters retain an external communication for removal and may be of value in short-term protection from pulmonary embolism *(15)*. Problems with these devices

From: *The Bionic Human: Health Promotion for People With Implanted Prosthetic Devices*
Edited by: F. E. Johnson and K. S. Virgo © Humana Press Inc., Totowa, NJ

**Table 1
Definitions**

Filter:	Implantable, biocompatible, vascular device that has the ability to trap migrating particles of clot, preventing their passage to the lungs.
Guide-wire:	Biocompatible metallic or composite wire used in the introduction of catheters into vessels or organs.
Catheter:	Plastic tube that is passed into a vessel or organ to inject opaque dye or other material, or to perform a therapeutic procedure.
Stent:	Supportive metal/alloy structure designed to support the internal structure of a blood vessel, usually where this is narrowed, thereby increasing the diameter of the lumen.
Stent graft:	Stent that is coated with an impervious layer of biocompatible fabric.
Balloon-mounted stent:	Stent that is deployed around an expandable balloon.
Self-expanding stent:	Stent that is composed of a spring or "shape memory" material that expands following withdrawal of a covering membrane.

have, however, included thrombosis owing to the filter, particularly at the time of removal, when there may be an associated risk of pulmonary embolization. Under such circumstances, adjunctive thrombolysis and perhaps a secondary, permanent or retrievable filter may be required to allow removal of the temporary filter.

Short-term placement of *retrievable* filters can be considered and is favoured in our current practice *(2)*. Removal of the Gunther Tulip (Cook, Copenhagen, Netherlands) is recommended at up to 14 days *(16)*, although removal has been attempted beyond this by some workers. The mechanism of removal involves a snare to secure a small hook and collapse the filter into a sheath. If the filter remains *in situ* longer than the recommended, there is the inherent risk of endothelialization with incorporation of the fixing legs of the filter into the structure of the IVC itself *(16–18)*. We have experience of this with failed attempted removal at 42 days. Removal of the filter after endothelialization carries a risk of laceration of the IVC or failed retrieval. There are reports that regular repositioning of a temporary filter can extend its total implantation time *(19)*. Some of the more recent filter designs appear to have addressed this problem (e.g., Trap-Eze, Cordis, Roden, Netherlands; ALN, ALN Implants, Chirurgicaux, Ghisonaccia, France), allowing removal at up to 90 days using a retrieval device *(20–22)*. Retrieval of the Recovery nitinol filter (Bard Peripheral vascular, Tempe, AZ) has been reported up to 134 days after insertion *(23)*. Retrievable devices such as the Gunther Tulip filter will act as an excellent permanent device should they be left *in situ* intentionally or through retrieval failure *(24)*.

Filter construction using titanium alloys (e.g., Gunther Tulip, Cook, Copenhagen, Netherlands) carries an additional benefit in a lack of paramagnetic effects during magnetic resonance imaging.

Chapter 20 focused on *permanent* filter placement. It may well be that there is some overuse of these devices without a firm evidence base for long-term benefit *(2,25)*. As yet, the perfect IVC filter does not exist, although clearly there are varying advantages, as illustrated here, with some permanent filters having excellent clot trapping abilities, for example the Birds Nest filter (Cook, Copenhagen, Netherlands), with up to 98% of patients protected from pulmonary emboli *(26)*. None, however, has all the criteria required to be considered optimal. Features that the ideal filter system might possess include the following *(27)*:

- Trapping of all significant (i.e., clinically relevant) emboli.
- Nonthrombogenic, 100% caval patency.
- No migration.
- No perforation of the caval wall.
- High durability, fatigue resistance, biocompatibility.
- Low profile and easy insertion.
- Retrievability.

As the chapter indicates, the potential for thrombotic complications of IVC filters is in part related to the surface area of the filter *(28,29)*. The Birds Nest filter has a high surface area and, predictably, an IVC thrombosis rate of 4.7% in one series *(27)*. A balance has to be struck in filter design between the requirement of sufficient structural elements to catch free clot and the risk of IVC thrombosis *(28,30)*. Thrombosis centered around the filter itself occurs in up to 19% of cases *(31)*. At the same time, it may be the case that many instances of IVC occlusion are the result, not of *in situ* thrombosis, but of efficient clot trapping, the functional intention of the device. The issue of long-term anticoagulation is a moot one and, although this is recommended by some authors *(32)*, it is often contraindicated by the patient's condition and indeed may be the underlying reason for the requirement of filter placement. Although infection in relation to an IVC filter has been reported *(33)*, antibiotic prophylaxis during periods of risk of bacteremia is not generally advocated in the United Kingdom and is not part of our routine practice. Neoplastic involvement by tumor emboli has also been reported *(25)*.

The complications of migration *(34–37)* and IVC perforation *(38,39)* represent another design dichotomy. A filter that is resistant to migration through effective engagement of fixation hooks is more likely to result in epithelial trauma to the IVC, particularly during implantation and, if retrievable, during removal. Even when distant migration has occurred, there may be no significant sequelae. Removal of a migrated filter can often be undertaken surgically or by the interventional radiologist *(40,41)*.

While follow-up for patients with IVC filters is not usual in the United Kingdom, the placement of a filter, particularly a permanent one, requires caution in any future venous catheter procedure where guide wire entrapment is a real and well-described risk *(42–46)* and can result in filter displacement *(47)*. A clear record, in the patient's case notes, of the presence of an IVC filter is therefore mandatory.

Some of these issues may be addressed with development of novel designs of drug-eluting, permanent implantable devices, as used in arterial stenting *(48)*, or temporary devices specifically designed for short-term use.

The lack of any consensus for follow-up of implanted filters (Table 2) may in part be related to the fact that the majority of permanently implanted IVC filters in the United Kingdom are used in patients with terminal illness *(49)*. For those not implanted in the terminally ill, monitoring in the community by primary health care practitioners would be the norm, although with no recommended schedule. Some of these patients are also regularly reviewed in an anticoagulation clinic, although those with a contraindication

Table 2
Patient Surveillance After Device Implantation at the Royal Liverpool University Hospital (IVC Filters)

	Postoperative year					
	1	2	3	4	5	10
Office visit	0	0	0	0	0	0
Plain radiographs	0	0	0	0	0	0
Computed tomography	0	0	0	0	0	0
Ultrasound	0	0	0	0	0	0

Note: Prior to discharge from hospital and within 2 weeks of implantation the decision to remove a retrievable device is made. This grid concerns nonremovable filters. Follow-up by referring clinicians is performed on a case-by-case basis. There is no scheduled follow-up of patients by the radiologist involved in the device implantation.

to anticoagulants will not be reviewed. If problems develop, then review would be undertaken by a hematologist, occasionally with radiological input. The investigations performed would be tailored to the individual patients and their presenting symptoms.

VENOUS STENTS

The difference in practice between the United Kingdom and the United States appears marked in the use of central venous stents (Table 1; *50*). This probably relates to the prevalence of mediastinal fibrosis in the United States secondary to histoplasmosis infection, a rarely encountered problem in the United Kingdom *(51,52)*. A significant application of venous stents in the United Kingdom is in relief of the distressing symptoms of superior vena cava obstruction that is usually due to malignant compression or malignant invasion *(53,54)*. Accordingly, superior vena caval stents are by essence almost always used palliatively for symptomatic relief; as yet there is no firm evidence base for their use in benign superior vena cava obstruction. Nonetheless, some workers have demonstrated value in benign central vein obstruction *(55)*. Owing to the prevalence of advanced malignant disease in many of these cases, the requirement for long-term follow-up is generally not an issue.

Central venous stenting is also carried out for thoracic outlet syndrome *(56)*. The use of venous stent placement in the United Kingdom in relief of symptoms secondary to central vein stenosis following iatrogenic vessel injury in central venous access line placement has been advocated by some workers *(57)*, although others have shown disappointing long-term results *(58)*. Better results are obtained in the few cases in the literature of stenting for venous stenosis secondary to surgery for congenital cardiac malformations *(59–62*; Table 3).

The disappointing patency of venous stents in relation to the upper limb in the authors' experience suggests little evidence for primary placement, and we have found this to be particularly the case at locations where compression can occur between the clavicle and the first rib: stent compression in this way has been well recognized elsewhere *(63–65)*. Stents in the brachiocephalic and subclavian veins therefore have a high rate of occlusion and fracture and primary placement cannot be recommended in this region

Table 3
Patient Surveillance
at the Royal Liverpool University Hospital (Venous Stents)

	\multicolumn{6}{c}{Postoperative year}					
	1	2	3	4	5	10
Office visit	0	0	0	0	0	0
Plain radiographs	0	0	0	0	0	0
Computed tomography	0	0	0	0	0	0
Ultrasound	0	0	0	0	0	0

Note: Follow-up by referring clinicians in their routine clinics is performed but this is on a case-by-case basis. There is no scheduled follow-up of patients by the radiologist involved in the device implantation.

though is indicated in some cases of failed balloon angioplasty of central venous stenosis, particularly where there is a need to maintain function in dialysis fistulae.

In the lower limb, however, venous stents have greater long-term patency and, for example, are valuable in management of venous compression in May-Turner syndrome *(66,67)* and strictures in Klippel-Trenaunay syndrome *(68)*. Stent placement may be of value as a part of the management of recurrent or acute iliofemoral thrombosis *(69,70)*. Following venous thrombolysis or mechanical thrombectomy, the demonstration of underlying common or external iliac stenosis is a developing indication for stent placement, in view of the high restenosis rate in this region following balloon dilation alone. Venous claudication, insufficiency, and chronic obstruction are also areas of increased use for venous stenting *(71–73)*. It is therefore clear that treatment with venous stenting can be recommended in some cases of benign venous obstruction *(74)*.

Intracranial use of venous stents for conditions such as benign raised intracranial pressure, intracranial venous hypertension, and vein of Galen aneurysm formation is now being advocated by some authors *(75–77)*.

The development of transjugular intrahepatic portosystemic shunts (TIPS), initially by Rosch in dogs, and first performed with stents in humans using the Palmaz stent, has become a highly significant therapeutic mode in bleeding esophageal varices owing to portal hypertension in hepatic cirrhosis. Ascites is another relative indication for this procedure that requires an approach to the hepatic veins from the internal jugular vein in the neck. A curved Colapinto needle is then directed into the portal vein and a stent placed to maintain this communicating channel and allow decompression from the portal vein into the right atrium.

Stent placement in portal venous obstruction *(78–80)* and Budd-Chiari syndrome *(81,82)* have been valuable in these difficult management situations *(83)*. Venous stents have also been used following liver transplantation for venous complications (Table 4; *84*).

Stenosis in the venous limb of dialysis fistulae or grafts may be a source of significant impairment of satisfactory hemofiltration. Where balloon angioplasty of such stenoses is unsuccessful, the use of stents may occasionally produce excellent results *(85,86)*. The use of balloon-expandable stents in this situation carries the major risk of compression from trivial external pressure with thrombosis and occlusion of the fis-

Table 4
Patient Surveillance at Royal Liverpool University Hospital (Transjugular Intrahepatic Portosystemic Shunts)

	Postoperative year					
	1	2	3	4	5	10
Office visit	0	0	0	0	0	0
Plain radiographs	0	0	0	0	0	0
Computed tomography	0	0	0	0	0	0
Ultrasound	4	4	4	4	4	4

Note: Follow-up by referring clinicians is performed on a case-by-case judgment. The follow-up of patients by the radiologist involves ultrasound examination of the stent every 3 months.

tula. Therefore, stents in this location are generally self-expanding, such as the Wallstent (Boston Scientific, Natick, MA) *(87)*. The stent position may be marked on the skin surface by tattoo to reduce the risk of strut fracture by inadvertent cannulation of the stented segment during hemodialysis. An alternative to the use of stents in dialysis fistula stenoses that are resistant to balloon angioplasty is the use of a cutting balloon. Nonetheless, stent placement remains of value where balloon techniques fail and indeed has an application to one of the significant complications of dialysis fistula angioplasty. Balloon angioplasty of dialysis fistulae carries a risk of destruction of the fistula by thrombosis or rupture. Where rupture of an arteriovenous fistula has occurred during balloon angioplasty, self-expanding stents such as the Wallstent have been shown to be effective in maintaining patency while preventing further hemorrhage *(88)*. The use of an uncovered stent to achieve hemostasis in this way is interesting, although the mechanism may be related to recompression of disrupted layers of vessel wall, thereby resealing the vessel. Clearly, this may not always be achieved and surgery or stent grafting (*see* Table 1) may be necessary.

Infection has rarely been reported with venous stents *(89)*. Although there is no evidence base for a recommendation, it would seem reasonable to advocate antibiotic prophylaxis during implantation of such permanently implantable devices and during periods of risk of bacteremia. The evidence for single-dose antibiotic prophylaxis is doubtful *(90)*. While we administer antibiotics prophylactically in some cases during stent implantation in our institution, this is certainly not uniform practice.

Although there is no recommended long-term follow-up schedule in the United Kingdom in central venous stent placement (Table 3), follow-up by the clinical team principally involved in the patient's management will generally be sufficient to monitor the development of any recurrent problems. Re-referral to interventional radiology is the norm if the patient develops recurrent or new symptoms. Duplex ultrasound is valuable in the investigation of problems with an existing stent. If this is inconclusive, or symptoms are intermittent or posture related, venography may prove to be of greater value.

COMMENT

The long-term follow-up of these patients in the United States is to be commended. We suspect the lack of long-term follow-up of patients with venous implants in the

United Kingdom is related to differing pathologies and the predominance of palliative applications in the United Kingdom. Financial constraints and the restrictions of a public healthcare system in the United Kingdom also undoubtedly have an impact. We also suspect that the way radiologists have traditionally tended to work in the United Kingdom, with, generally, a dearth of exposure to the follow-up of patients, contributes to these differences in practice. Follow-up after interventional procedures is generally the brief of the clinical team caring for the patient, with radiology input invited perhaps after identification of a complication. We suspect that there is a strong case for greater involvement of radiologists in the management and clinical follow-up of their patients and there is certainly a wide recognition of this; it is hoped that radiological practice will slowly change in this respect. At the least there must be a record of any existing IVC filter within the patient's records in view of potential complications, not least during venous catheterization procedures such as central venous catheter placement. The adoption of a more proactive stance in the United Kingdom could aid early identification of complications such as infection or migration that can occur in venous implants. Many of these cases do however have a relatively limited prognosis and the suggested approach described for monitoring in the United States may be seen to be overly consumptive of resources in the relatively financially stringent UK service.

The subject of arterial stents and stent grafts is broad and beyond the scope of a mere commentary. Very wide applications have developed in management of obstructive arterial disease of the limbs, renal, carotid and visceral territories using balloon and self-expandable devices. The development of stent grafts has extended interventional radiology into the management of aneurysms, aortic dissection, and vascular rupture, thereby greatly broadening the applications of these techniques. Currently in the United Kingdom, the National Institute for Clinical Excellence *(91)* recommends that evidence on safety and efficacy does not yet support use of endovascular repair of abdominal aortic aneurysms without special arrangements for consent and for audit or research.

In the United Kingdom, the follow-up of stent grafts for treatment of abdominal aortic aneurysms with endovascular aneurysm repair (EVAR) is probably the most comprehensive of any interventional radiological procedure in the United Kingdom *(92)*. Within the current EVAR trial, which compares stent-graft abdominal aortic aneurysm repair with conventional surgery or best medical therapy, follow-up involves clinical assessment and Doppler ultrasound at 1, 3, and 6 months and then annually. Contrast enhanced computed tomography (CT) assessment is conducted at 3, 6, 12, 18, and 24 months, and then annually. This regimen may indicate a theoretical optimum follow-up for current and new vascular devices. Unquestionably, however, this carries a heavy radiation burden from the CT studies, although these might be reduced by use of magnetic resonance or ultrasound. We suspect that financial, regulatory, and medicolegal concerns will determine the future shape of implant surveillance.

Mention must also be made of the vast range of implantable embolization material for use in both the arterial and venous vascular systems. It is hoped that a future edition of this work will cover arterial stent and stent-graft applications, as well as embolization materials and devices, in greater detail. For those interested in further information, a list of resourcesis provided in the Appendix.

APPENDIX: SPECIALIST ORGANIZATIONS IN INTERVENTIONAL RADIOLOGY

Professional organizations	Web address	Contact name	Address	Telephone no.	Fax no.	E-mail address
Royal College of Radiologists	www.rcr.ac.uk	Damion Clarke Brettles Ltd. Public Relations Brettles Yard 16-18 Cambridge Road Stansted, Mountfitchet Essex CM 24 8BZ	The Royal College of Radiologists 38 Portland Place London W1B 1JQ	+44(0) 20 7636 4432	+44(0) 20 7323 3100	enquiries@rcr.ac.uk
British Society of Interventional Radiologists	http://www.BSIR.org	Lavinia Gittins	4 Verne Drive Ampthill Bedford MK45 2PS	+44(0) 1525 403026	+44(0) 1525 751384	office@bsir.org administrator@bsir.org
Vascular Surgical Society	http://www.vssgbi.org	Jeanette Roby	Vascular Surgical Society Office The Royal College of Surgeons of England 35-43 Lincoln's Inn Fields London WC2A 3PE	+44(0) 20 7973 0306	+44(0) 20 7430 9235	Administrator@vssgbi.org
Device manufacturers						
Cook, Inc.	http://www.cookgroup.com		P.O. Box 489 Bloomington, IN 47402-0489 USA	812-339-2235 Toll Free 800-457-4500	800-554-8335 Toll Free	
Boston Scientific	http://www.bsci.com		Boston Scientific Corp. Hdqtrs One Boston Scientific Place Natick, MA 01760-1537	888-272-1001		

Bard	http://www.crbard.com	C.R. Bard, Inc. 730 Central Avenue Murray Hill, NJ 07974 USA	908-277-8000	908-277-8240
St. Jude Medical	http://www.sjm.com	St. Jude Medical, Inc. One Lillehei Plaza St. Paul, MN 55117-9983 USA	Toll Free 800-328-9634 651-483-2000 Voice mail 800-444-4069	651-482-8318 Telex 298453
Medtronic	http://www.medtronic.com	Medtronic World Headquarters 710 Medtronic Parkway Mail Stop: L100 Minneapolis, MN 55432-5604	763-514-4000 or 763-574-4000	763-514-4879 or 763-574-4879

REFERENCES

1. Girard P, Stern JB, Parent F. Medical literature and vena cava filters: so far so weak. Chest 2002;122:963–967.
2. Decousus H, Leizorovicz A, Parent F, et al. A clinical trial of vena caval filters in the prevention of pulmonary embolism in patients with proximal deep vein thrombosis. N Engl J Med 1998;338:409–415.
3. Ferris EJ, McgowanTC, Carver DK, McFarland DR. Percutaneous inferior vena caval filters: follow-up of seven designs in 320 patients. Radiology 1993;188:851–856.
4. Magnant JG, Walsh DB, Juravsky LI, Cronenwett JL. Current use of inferior vena cava filters. J Vasc Surg 1992:16:701–706.
5. Gassi CJ, Swan TL, Cardella JF, et al. Quality improvement guidelines for percutaneous permanent inferior vena cava filter placement for the prevention of pulmonary embolism. SCVIR standards of practice committee. J Vasc Interv Radiol 2001;12:137–141.
6. Becker DM. Inferior vena cava filters: Indications, safety, effectiveness. Arch Intern Med 1992;152:1985–1994.
7. Hynek K, Spalova I, Spatenka J, Mates M. Possibilities of using vena cava filters in pregnant women with thromboembolism. Sb Lek 2002;103:451–454.
8. Reekers JA, Harmsen H, Hoogeveen YL, Gunther RW. Vena cava filter devices. In: Oudkerk M, Van Beek EJR, Ten Cate JW, eds. Diagnosis and treatment of pulmonary embolism. Malden, MA: Blackwell Science, 1999, pp. 330–349.
9. Schwarz RE, Marrero AM, Conlon KC, Burt M. Inferior vena cava filters in cancer patients: indications and outcome. J Clin Oncol 1996;14:652–657.
10. Sharp RP, Gupta R, Gracies V, et al. Incidence and natural history of below-knee deep venous thrombosis in high–risk trauma patients. J Trauma Inj Infect Crit Care 2002;53/6: 1048–1052.
11. Khansarinia S, Dennis JW, Veldenz, et al. Prophylactic greenfeild filter placement in selected high-risk trauma patients. J Vasc Surg 1995;22:231–236.
12. Langhan EM, Miller RS, Cassey WJ, et al. prophylactic inferior vena cava filters in trauma patients at high risk: follow-up examination and risk/benefit assessment. J Vasc Surg 1999; 30:484–490.
13. Carlin AM, Tyburski JG, Wilson RF, et al. Prophylactic and therapeutic inferior vena cava filters to prevent emboli in trauma patients. Arch Surg 2002;137:521–527.
14. Greenfield LJ, Proctor MC, Michaels AJ, et al. Prophylactic vena caval filters in trauma: the rest of the story. J Vasc Surg 2000;32:490–497.
15. Offner PJ, Hawkes A, Madayag R, Seale F, Maines C. The role of temporary inferior vena cava filters in critically ill surgical patients. Arch Surg 2003;138:591–594.
16. Millward SF, Bhargava A, Aquino J Jr., et al. Gunther Tulip filter: preliminary clinical experience with retrieval. J Vasc Intervent Radiol 2000;11:75–82.
17. Neuerburg JM, Gunther RW, Vorwerk D, et al. Results of a multicenter study of the retrievable tulip vena cava filter: Early clinical experience. Cardiovasc Intervent Radiol 1997;20:10–16.
18. Millward SF, Olivia VL, Stuart SD, et al. Gunther tulip retrievable vena cava filter: results from the registry of the Canadian Interventional Radiology Association. J Vasc Intervent Radiol 2001;12:1053–1058.
19. Tay K, Martin ML, Fry PD, et al. Repeated gunther tulip inferior vena cava filter repositioning to prolong implantation time J Vasc Intervent Radiol 2002;104:509–512.
20. Asch MR. Initial experience in humans with a new retrievable vena cava filter. Radiology 2002;225:835–844.
21. Nutting C, Coldwell D. Use of Trap-ease device as a temporary caval filter. J Vasc Interv Radiol 2001;12:991–993.
22. Rousseau H, Pierre P, Otal P, et al. The 6-F nitinol Trap-ease inferior vena cava: results of a prospective multicenter trial. J Vasc Interv Radiol 2001;12:299–304.

23. Asch MR. initial experience in humans with a new retrievable inferior vena cava filter. Radiology 2002;225:835–844.
24. Matsuura JH. Retrievable vena cava filters. Endovascular Today 2003; May June:24–26.
25. Neeman Z, Auerbach A, Wood BJ. Metastatic involvement of a retrieved inferior vena cava filter. J Vasc Interv Radiol 2003;14:1585.
26. Nicholson AA, Ettles DF, Paddon AJ, Dyet JF. Long-term follow-up of the Bird's Nest IVC Filter. Clin Rad 1999;54:759–764.
27. Kinney TB. Update on inferior vena cava filters. J Vasc Interv Radiol 2003;14: 425–440.
28. Thomas JH, Cornell KM, Siegel EL, et al. Vena cava occlusion after bird's nest filter placement. Am J Surg 1998;176:595–600.
29. Athanasoulis CA, Kaufman JA, Halpern EF, Waltman AC, et al. Inferior vena caval filters: review of a 26 single centre clinical experience. Radiology 2000:216:54–66.
30. Tardy B, Page Y, Zeni F, et al. Acute thrombosis of a vena cava filter with a clot above the fiter. Chest 1994;106:1607–1609.
31. Becker DM. Inferior vena cava filters: Indications, safety, effectiveness. Arch Internal Med 1992;152: 1985–1994.
32. Gomes MPV, Kaplan K, Deitcher SR. Patients with inferior vena cava filters should receive chronic thromboprophylaxis. Med Clin North Am 2003;87:1189–1203.
33. Lin M, Bien Soo T, Chung Horn L. Successful Retrieval of Infected Günther Tulip IVC Filter J Vasc Interv Radiol 2000;11:1341–1343.
34. Raghavan S, Akhtar A, Bastani B. Migration of inferior vena cava filter into renal hilum. Nephron 2002;91:333–335.
35. Porcellini M, Stassano P, Musumeci A, Bracale G. Intracardiac migration of nitinol TrapEase vena cava filter and paradoxical embolism. Eur J Cardiothorac Surg 2002;22: 460–461.
36. Friedell ML, Goldenkranz RJ, Parsonnet V, et al. Migration of a Greenfield filter to the pulmonary artery: a case report. J Vasc Surg 1986;3:929–931.
37. Freezor RJ, Huber TS, Welborn BM, et al. Duodenal perforation with an inferior vena cava filter: an unusual cause of abdominal pain. J Vasc Surg 2002;35:1010–1012.
38. Mohan CR, Hoballah JJ, Sharp WJ, et al. Comparative efficacy and complications of vena cava filters. J Vasc Inter Radiol 1995;21:235–246.
39. Taylor FC, Awh MH, Kahn CE, et al. Vena Tech vena cava filter; experience and early follow-up. J Vasc Inter Radiol 1991;2:435–440.
40. Deutsch L-S, Percutaneous removal of intracardiac Greenfield vena cava filter. Am J Roentgenol 1988;151:677–679.
41. Arjomand H, Surabhi S, Wolf NM. Right ventricular foreign body: percutaneous transvenous retrieval of a Greenfield filter from the right ventricle—a case report. Angiology 2003; 54:109–13.
42. Ellis PK, Deutsch LS, Kidney DD. Interventional radiological retrieval of a guide-wire entrapped in a Greenfield filter—Treatment of an avoidable complication of central venous access procedure. Clin Rad 2000;55:238–239.
43. Goldberg ME. Entrapment of an exchange wire by an inferior vena caval filter: a technique for removal. Anesthesia & Analgesia 2003;96:1235–1236.
44. Amesbury S, Vargish T, Hall J. An unusual complication of central venous catheterization. Chest 1994;105:905–907.
45. Kaufman JA, Thomas JW, Geller SC, et al. Guide-wire entrapment by inferior vena caval filters: invitro evaluation. Radiology 1996;198:71–76.
46. Loesberg A, Taylor FC, Awh MH. Dislodgement of inferior vena caval filters during "blind" insertion of central venous catheters. Am J Roentgenol 1993;161:637–638.
47. Marelich GP, Tharratt RS. Greenfield inferior vena cava filter dislodged during central venous catheter placement. Chest 1994;106:957–959.
48. Duda SH, Poerner TC, Wiesinger B, et al. Drug-eluting stents: potential applications for peripheral arterial occlusive disease. J Vasc Interv Radiol 2003;14:291–301.

49. Jarrett BP, Dougherty MJ, Calligaro KD. Inferior vena cava filters in malignant disease. J Vasc Surg 2002;36:704–707.
50. Irving JD, Dondelinger RF, Reidy JF, et al. Gianturco self-expanding stents: clinical experience in the vena cava and large veins. Cardiovasc Intervent Radiol 1992;15:328–333.
51. Jirat S, Varejka P, Chochola M, et al. Interventional management of the inferior vena cava syndrome. Cas Lek Cesk 2002;141:773–775.
52. Vorwerk D, Guenther RW, Wendt G, et al. Iliocaval stenosis and iliac venous thrombosis in retroperitoneal fibrosis: percutaneous treatment by use of hydrodynamic thrombectomy and stenting. Cardiovasc Intervent Radiol 1996;19:40–42.
53. Wilson E, Lyn E, Lynn A, Khan S. Radiological stenting provides effective palliation in malignant central venous obstruction. Clin Oncol 2002;14:228–232.
54. Brady TM, Swischuk JL, Castaneda F, et al. Central venous thrombolysis/PTA/stenting. Semin Intervent Radiol 2000;17:121–138.
55. Petersen BD, Uchida BT. Long-term results of treatment of benign central venous obstructions unrelated to dialysis with expandable Z stents. J Vasc Interv Radiol 1999;10:757–766.
56. Ibrahim IM, Lipman SP, Alasio T, et al. Percutaneous venous stenting for thoracic outlet syndrome. Vasc Surg 1996;30:407–412.
57. Peters S, Beath SV, Puntis WL, John P. Superior vena cava thrombosis causing respiratory obstruction successfully resolved by stenting in a small bowel transplant candidate. Ach Dis Child 2000;83:163–164.
58. Thony F, Moro D, Ferretti G, et al. Percutaneous treatment of superior vena cava obstruction. Rev Mal Respir 1999;16:731–743.
59. Khasnis A, Dalvi B. Stenting for SVC obstruction in an infant operated for total pulmonary venous return. Indian Heart J 2001;53:214–217.
60. Miura T, Sano T, Hikaru. Intravascular stenting of systemic venous baffle stenosis after corrective surgery for double outlet right ventricle with left isomerism. Heart 1999;81:218–220.
61. Rosenthal E, Qureshi SA. Stenting of systemic venous pathways after atrial repair for complete transposition. Heart 1998;79:211–212.
62. Abdulhamed JM, Alyousef SA, Mullins C. Endovascular stent placement for pulmonary venous obstruction after mustard operation for transposition of the great arteries. Heart 1996;75:210–212.
63. Rosenfield K, Schainfeld R, Pieczek A, Haley L, Isner JM. Restenosis of endovascular stents from stent compression. J Am Coll Cardiol 1997;29:328–338.
64. Mathur A, Dorros G, Iyer SS, Vitek JJ, Yadav SS. Roubin GS. Palmaz stent compression in patients following carotid artery stenting. Cathet Cardiovasc Diagn 1997;41:137–140.
65. Windecker S, Maier W, Eberli FR, Meier B, Hess OM. Mechanical compression of coronary artery stents: potential hazard for patients undergoing cardiopulmonary resuscitation. Catheter Cardiovasc Interv 2000;51:464–467.
66. Nazarian GK, Bjarnason H, Dietz CA, Bernadas CA, Hunter DW. Iliofemoral venous stenoses: effectiveness of treatment with metallic endovascular stents. Radiology 1996;200: 193–199.
67. Heijmen RH, Bollen TL, Duyndam DAC, Overtoom TTC, et al. Endovascular stenting in May Turner syndrome. J Cardiovasc Surg 2001;42:83–87.
68. Stone DH, Adelman M, Rosen RJ, et al. A unique approach in the management of vena caval thrombosis in a patient with Klippel-Trenaunay syndrome. J Vasc Surg 1997;26:155–159.
69. Fearon WF, Semba CP. Iliofemoral venous thrombosis treated by catheter-directed thrombolysis, angioplasty and endoluminal stenting. West J Med 1998;168:277–279.
70. Semba CP, Dake MD. Catheter-directed thrombolysis for iliofemoral venous thrombosis. Semin Vasc Surg 1996;9:26–33.
71. Neglen P, Thrasher TL, Raju S. Venous outflow obstruction: An underestimated contributor to chronic venous disease. J Vasc Surg 2003;38:879–885.

72. Neglen P, Raju S. Balloon dilation and stenting of chronic iliac vein obstruction: Technical aspects and early clinical outcome. J Endovasc Ther 2000;7:79–91.
73. Blatter W, Blatter IK. Relief of obstructive pelvic symptoms with endoluminal stenting. J Vasc Surg 1999;29:484–488.
74. Wohlgemuth WA, Weber H Loeprecht H, et al. PTA and stenting of benign venous stenoses in the pelvis: Long-term results. Cardiovasc Intervent Radiol 2000;23:9–16.
75. Higgins JNP, Owler BK, Cousins C, Pickard JD. Venous sinus stenting for refractory benign intracranial hypertension. Lancet 2002;359:228–230.
76. Brew S, Taylor W, Reddington A. Stenting of a venous stenosis in vein of Galen aneurysm malformation: a case report. Intervent Neuroradiol 2001;7:237–240.
77. Malek AM, Higahida RT, Balousek PA, et al. Endovascular recanalization with balloon angioplasty and stenting of an occluded occipital sinus for treatment of intracranial venous hypertension: technical case report. Neurosurgery 1999;44:896–901.
78. Yamakado K, Nakatsuka A, Tanaka N, Fujii A, Terada N, Takeda K. Malignant portal venous obstructions treated by stent placement: significant factors affecting patency. J Vasc Interv Radiol 2001;12:1407–1415.
79. Yamakado K, Nakatsuka A, Tanaka N, et al. Portal venous stent placement in patients with pancreatic and biliary neoplasms invading portal veins and causing portal hypertension: initial experience. Radiology 2001;220:150–156.
80. Hiraoka K, Kondo S, Ambo Y, et al. Portal venous dilation and stenting for bleeding jejunal varices: Report of two cases. Surg Today 2001;31:1008–1011.
81. Witte AMC, Kool LJS, Veenendaal R, et al. Heptic vein stenting for Budd-Chiari syndrome. Am J Gastroenterol 1997;92:498–501.
82. Zhang CQ, FU LN, Xu L, et al. Long-term effect of stent placement in 115 patients with Budd-Chiari syndrome. World J Gastroenterol 2003;9:2587–2591.
83. Ferguson JM, Palmer K, Garden OJ, Redhead DN. Transhepatic venous angioplasty and stenting: a treatment option in bleeding from gastric varices secondary to pancreatic carcinoma. HPB Surg 1997;10:173–175.
84. Weeks SM, Gerber DA, Jaques PF, et al. Primary Gianturco stent placement for inferior vena cava abnormalities following liver transplants. J Vasc Interv Radiol 2000;11:177–187.
85. Quinn SF, Schuman ES, Demlow TA, et al. Percutaneous transluminal angioplasty versus endovascular stent placement in the treatment of venous stenoses in patients undergoing hemodialysis: intermediate results. J Vasc Interv Radiol 1995;6:851–855.
86. Vorwerk D, Guenther RW, Mann H, et al. Venous stenosis and occlusion in hemodialysis shunts: follow-up results of stent placement in 65 patients. Radiology 1995;195:140–146.
87. Gunther RW, Vorwerk D, Bohndorf K, et al. Venous stenoses in dialysis shunts: treatment with self-expanding metallic stents. Radiology 1989;170:401–405.
88. Welber A, Schur I, Sofocleous CT, Cooper SG, Patel RI, Peck SH. Endovascular stent placement for angioplasty-induced venous rupture related to the treatment of hemodialysis grafts. J Vas Interv Radiol 1999;10:547–551.
89. Bukhari RH, Muck PE, Schlueter FJ, et al. Bilateral renal artery stent infection and pseudoaneurysm formation. J Vasc Interv Radiol 2000;11:337–341.
90. Deibert P, Schwarz S, Olschewski M, et al. Risk factors and prevention of early infection after implantation or revision of transjugular intrahepatic portosystemic shunts: results of a randomised study. Dig Dis Sci 1998;43(8):1708–1713.
91. National Institute for Clinical Excellence. www.nice.org.uk.
92. Gould DA, Edwards RD, McWilliams RG, et al. Graft distortion after endovascular repair of abdominal aortic aneurysm: association with sac morphology and mid term complications. Cardiovasc Intervent Radiol 2000;23:5.

21
Vascular Access Devices

Christopher N. Compton and John H. Raaf

HISTORY

Sir Christopher Wren, the eminent British architect, was perhaps the first to devise a method for delivery of intravenous therapy. In 1657, he constructed a cannula from the quill of a bird feather and used it to inject drugs into the vein of a dog. The first demonstration of *central* venous catheterization is attributed to Werner Forssman, a courageous German physician and physiologist. In 1929, Forssman inserted a 4-Fr ureteral catheter a distance of 65 cm into his own antecubital vein, then walked to his X-ray department and documented radiographically the position of the catheter tip in his right atrium *(1,2)*. Forssman was awarded the Nobel Prize in Physiology or Medicine in 1956, with Drs. André Cournand and Dickinson Richards, for laying the foundation of angiography and cardiac catheterization. In 1949, Duffy described a technique for central venous catheterization via the external jugular vein *(3)*. Soon after, Aubaniac introduced subclavian vein access using an infraclavicular approach *(4)*. Seldinger, in 1953, contributed a method of percutaneous vascular catheterization over a guidewire *(5)*.

Currently used venous access devices (VADs) can be categorized according to the maximum safe indwelling time (Table 1). Short-term devices are left in place up to 2 or 3 weeks, intermediate-term devices 2 weeks to 6 months, and long-term devices months to years. Central venous catheters (CVCs) were first made of stiff polyethylene plastic and were used only as short-term devices. Short-term catheters now are usually made of polyurethane, while intermediate- and long-term ones most often are constructed of silastic.

The introduction of total parenteral nutrition *(6)*, hemodialysis *(7)*, and intensive chemotherapy created a need for long-term CVCs. Broviac *(8)*, attempting to provide access for parenteral nutrition, and Hickman *(9)*, working with bone marrow transplant patients, introduced the silastic catheters that bear their names, both designed for prolonged use. These catheters are tunneled, cuffed, and less traumatic to vessel endothelium than those made from stiffer plastic materials.

Current indications for use of central VADs are presented in Table 2. Over the past two decades these indications have expanded greatly and now more than 5 million CVCs of all types are placed annually in the United States *(10)*.

From: *The Bionic Human: Health Promotion for People With Implanted Prosthetic Devices*
Edited by: F. E. Johnson and K. S. Virgo © Humana Press Inc., Totowa, NJ

Table 1
Types of Central Venous Catheters

Short-term
    Polyurethane CVCs— single-lumen and multilumen
Intermediate-term
    PICC lines
    Non-cuffed, thin-walled silastic CVCs—Centrasil, Intrasil
    Groshong catheter
    P.A.S. Port
Long-term
    Single-lumen, cuffed silastic CVCs—Broviac, Hickman
    Multi-lumen, cuffed silastic CVCs—Quinton-Raaf dual-lumen and triple-lumen
        —Quinton-Raaf PermCath
    Subcutaneously implanted ports—Mediport, Port-a-Cath

CVC, central venous catheter; PICC, peripherally inserted central catheter.

Table 2
Indications and Uses of CVCs

Short-term CVCs
    Intravenous fluids
    Blood products
    Intravenous medications
    Repeated blood sampling
    Hemodynamic monitoring

Intermediate- and long-term CVCs
    Intravenous fluids
    Blood products
    Intravenous medications (e.g., antibiotics, analgesics, antiemetics)
    Repeated blood sampling
    Chemotherapy
    Total parenteral nutrition
    Hemodialysis
    Plasmapheresis
    Bone marrow transplantation

CVC, central venous catheter.

SHORT-TERM, PERCUTANEOUS, NON-TUNNELED CENTRAL VENOUS CATHETERS

Short-term CVCs are available with either single or multiple lumens. The Arrow (Arrow International, Inc., Reading, PA) 7-Fr triple lumen, 20 cm catheter (Fig. 1) is an example of a multilumen, short-term CVC that is frequently used in critically ill patients. Multilumen catheters can be used for simultaneous administration of medications that are incompatible when mixed. Swan-Ganz catheters are multilumen short-term CVCs that are used in an intensive care setting. They allow hemodynamic

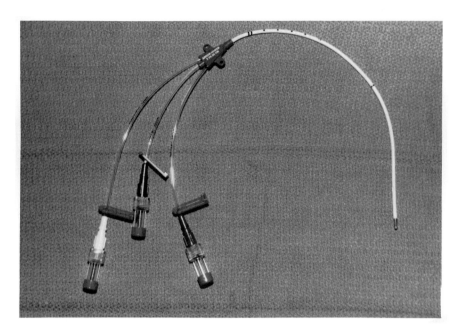

Fig. 1. A short-term, polyurethane, 7-Fr triple-lumen 20 cm central venous catheter (Arrow International). (From ref. *11*.)

monitoring by measuring central venous pressure, pulmonary artery pressure, pulmonary capillary wedge pressure, and cardiac output. Typically, a Swan-Ganz catheter has five lumens. Two lumens lead to the right atrium and one each to a thermistor, the pulmonary artery, and the balloon at the tip that is wedged peripherally in the pulmonary artery system. These catheters are usually left in place for several days. A description of the special methods for their sterile insertion and maintenance is beyond the scope of this chapter.

Short-term CVCs are currently manufactured from relatively stiff polyurethane plastic. The limited duration of their use is the result of the risk of infection and the potential damage to the vascular endothelium that occasionally results in vessel wall perforation *(12)*. The catheter tip should lie well down in the superior vena cava (SVC) at the right atrial—SVC (RA-SVC) junction, not high at the junction of the brachiocephalic veins. If the tip is too high, especially if the catheter has been inserted from the left side, the end of the catheter is likely to lie against the wall of the SVC and result in resistance to blood withdrawal through the catheter (often termed "withdrawal occlusion"). In this position, it may also cause damage to the vein wall.

Advantages of short-term, percutaneous, non-tunneled catheters are that they can be placed and removed easily at the bedside and they can be readily changed over a guidewire. However, routine replacement of short-term CVCs over a guidewire, for example every 3 days, does not reduce the incidence of infection. In a randomized trial, Cobb *(13)* found that routine catheter exchange over a wire actually increases the risk of bacteremia.

Disadvantages of short-term catheters are that they are less secure and have a higher incidence of catheter-associated infection than those that are tunneled and have a

Dacron cuff. Short-term catheters require daily dressing changes and strict exit site care. They are generally left in place at most 2 or 3 weeks. In this chapter, we do not address the management of patients who receive multiple short-term catheters over an extended period of time.

INTERMEDIATE-TERM CENTRAL VENOUS ACCESS DEVICES

If single-lumen central venous access is needed for more than 2 weeks, but not more than 2 or 3 months, then an intermediate-term CVC may be appropriate. These include the peripherally inserted central catheter (PICC) and the Centrasil catheter (Baxter Travenol, Deerfield, IL). The Groshong catheter (Davol Inc., Cranston, RI) and the peripherally placed P.A.S. Port (Pharmacia Deltec, Inc., St. Paul, MN) share features with long-term catheters. They are, in our experience, more likely to be used for an intermediate length of time so we include them in the intermediate-term CVC category.

PICC lines are 16- or 18-gage percutaneously inserted CVCs placed into the basilic or cephalic vein near the antecubital fossa, with the tip in the SVC. They are made of silastic or polyurethane, and guidewires are usually used to facilitate insertion, typically through a 14- or 15-gage peelaway introducer cannula. Meticulous handling of PICC lines during insertion is observed to avoid deposition of talc on their surface, which could induce phlebitis.

The Centrasil is a small bore, 16-gage silastic catheter that is noncuffed and non-tunneled and is usually inserted via the subclavian vein. It must be sutured to the skin at the exit site to prevent dislodgement. Broadwater et al. *(14)* and Raad et al. *(15)* have described its extensive use at M.D. Anderson Cancer Center in patients receiving chemotherapy. This type of catheter can be used for weeks to months. Because the Centrasil is thin-walled, it is more easily damaged and collapses more readily than the Broviac or Hickman. Attention by a skilled intravenous nursing team is required to maintain the skin suture, declot the catheter, or replace it because of damage.

The Groshong catheter is similar to the Broviac because it is constructed of silastic and is cuffed. It has a closed distal tip with a slit valve that is designed to open only on infusion or aspiration. Because of the closed nature of tip, some who use the Groshong catheter feel it is unnecessary to flush it with heparin; rather, they flush with saline alone. Our experience has been that, despite the slit valve, blood can still enter the catheter tip and result in clot formation within the catheter. Pasquale et al. *(16)* compared Groshong to Hickman catheters in patients receiving chemotherapy. They found no difference in septic complications but withdrawal occlusion occurred significantly less frequently with Hickman catheters. Warner et al. *(17)* conducted a randomized, prospective trial of dual-lumen Hickman catheters vs dual-lumen Groshong catheters in children with cancer. They also found withdrawal occlusion to be less frequent with Hickman catheters.

The P.A.S. Port is a small titanium port with a silastic septum 6.6 mm in diameter, connected to a polyurethane catheter with a 1 mm internal diameter lumen. The catheter is introduced by cutdown into a forearm vein near the antecubital crease, and the port is implanted through the same incision into a subcutaneous pocket on the volar surface of the forearm. Some authors have found the P.A.S. Port to be useful in patients with cystic fibrosis because it avoids implantation of a port on the chest wall. Chest percussion therapy is therefore unimpeded. However, a significant risk of brachial and axillary vein thrombophlebitis exists with use of the P.A.S. Port *(18)*, plus a high incidence of withdrawal occlusion, and these problems have limited its application in our practice. The remainder of this chapter focuses on long-term devices and we do not discuss further the care of patients with intermediate-term devices.

LONG-TERM, SILASTIC, TUNNELED, CUFFED RIGHT ATRIAL CATHETERS

Long-term right atrial catheters are available in single or multilumen designs, all of which are silastic, cuffed, and tunneled. The Broviac catheter has a 1.0-mm internal diameter and the Hickman a 1.6-mm internal diameter. The Quinton-Raaf dual-lumen and triple-lumen silastic catheters *(11,19)* (Fig. 2) provide long-term access in patients requiring intensive intravenous support. The Quinton-Raaf PermCath *(21,22)* (which has two large 2-mm diameter lumens and was originally described in the literature as the Quinton-Raaf HemoCath *[19]*) is a long-term CVC (Fig. 3) that provides central venous access for hemodialysis, plasmapheresis, and support of bone marrow transplantation. Lazarus et al. *(24)* demonstrated the efficacy of this large-bore, dual-lumen silastic catheter in patients undergoing autologous peripheral blood progenitor cell transplantation. They placed either a Quinton-Raaf PermCath or a similar Bard-Hickman hemodialysis/apheresis dual-lumen catheter in each patient and used it both for collection of progenitor cells by leukapheresis and for subsequent administration of high-dose chemotherapy, reinfusion of hematopoietic cells, and intensive supportive care.

The tip of any long-term silastic catheter should, in our opinion, lie in the right atrium just below the RA–SVC junction. All of these long-term silastic catheters are tunneled, separating the venipuncture site from the skin exit site. The Dacron cuff provides security from displacement and (likely) infection, as a result of tissue ingrowth into the cuff. The advantages of tunneled catheters over non-tunneled, temporary catheters are their lower risk of infection or dislodgement, and their long-term viability. A disadvantage is that they require more anesthesia for placement, usually as intravenous sedation for adults and general anesthesia for children. Long-term catheters often require a cutdown on the Dacron cuff for removal, and they are more expensive than non-tunneled catheters.

LONG-TERM SUBCUTANEOUS VENOUS ACCESS PORTS

Implanted venous access ports (Fig. 4), which have no external components, were originally popularized by Niederhuber et al. *(25)*. They consist of a plastic, stainless steel, or titanium chamber connected to a Broviac or Hickman-size silastic right atrial catheter. These ports are placed in a subcutaneous pocket on the anterior, infraclavicular chest wall and are secured by one or more sutures to the pectoralis fascia or, in obese patients, to the fatty posterior surface of the port pocket. Because the device is entirely internal, there is no need for a protective dressing once the incision has healed, thus providing an excellent cosmetic result. As with long-term right atrial catheters having an external hub, we usually use the right central veins (and therefore the right anterior chest wall for implantation of the port) because of the more direct route to the right atrium. Occasionally, however, a right-handed patient will request placement of a port on the left chest wall rather than the right to avoid trauma to the port during firing of a rifle or shotgun.

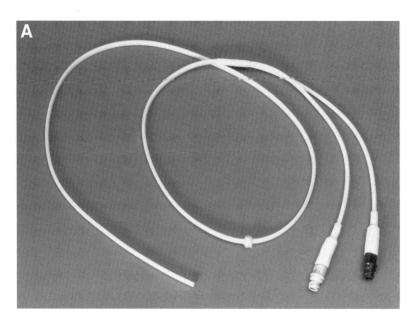

Fig. 2. (A) Raaf 2.2 mm dual-lumen right atrial catheter (Quinton Instrument Co.). This long-term silastic catheter has two 1.0-mm (Broviac size) lumens. It can be introduced through a 10.5 Fr peel-away sheath. **(B)** Long-term, silastic venous access catheters, all cuffed, with external hub(s). Left to right: Broviac, Hickman, Raaf 2.2 mm, and Raaf 3.2 mm right atrial catheters. (From ref. *20*.)

Dual ports having two chambers are available and allow administration of two simultaneous but separate intravenous infusions, but the complication rate (catheter thrombosis and infection) for these bulky devices has been excessive in our experience so we use them infrequently.

To access a port, one inserts a Huber needle through the skin and port septum into the chamber. The needle is typically attached to a segment of plastic connector tubing.

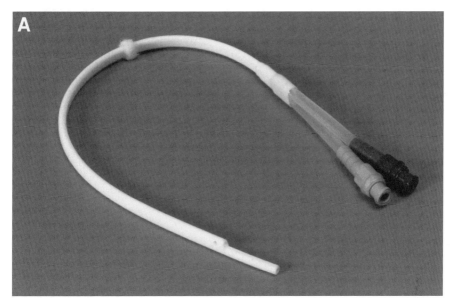

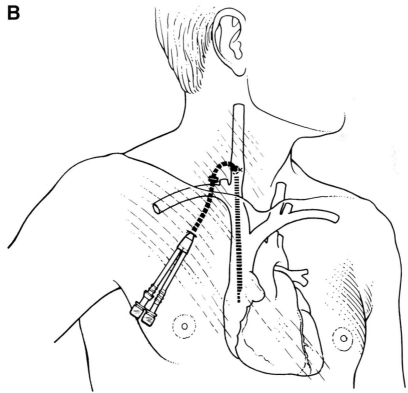

Fig. 3. (A) Dual-lumen Quinton-Raaf PermCath right atrial catheter. The PermCath is a large-bore, staggered tip catheter that can be used for hemodialysis, plasmapheresis, or bone marrow transplantation. (From ref. *20.*) **(B)** To avoid obstruction of flow through the catheter owing to catheter tip malposition, the surgeon should confirm by fluoroscopy that the PermCath tip is precisely positioned "free floating" in the right atrium. (From ref. *23.*)

Fig. 4. Implanted venous access ports are available in a variety of materials, sizes, and silastic septum diameters. (From ref. 26.)

The Huber needle design (with a bent tip) was thought to decrease the chance of damage by coring the silastic septum by the needle. However, a study of this question showed that coring occurs with equal but very rare frequency using both conventional and Huber needles (27).

Many patients tolerate insertion of a needle through the skin overlying their port without much discomfort, but some adults and most children do not. A degree of dermal analgesia can be provided to these patients by applying EMLA cream (lidocaine/prilocaine—ASTRA Pharmaceutical Products, Westborough, MA) to the skin over the port prior to each needle insertion.

Infusion of fluids or blood products via an implanted port may be slower than through a percutaneous catheter because all of the infused fluid must pass through a 19- or 22-gage Huber needle. Subcutaneous ports do not require any restriction on patient activity; both showering and swimming are permitted after the port pocket incision has healed. Subcutaneous ports should be flushed with heparin-saline at least once a month when not in use, compared to twice weekly for externalized silastic catheters (28). Ports require an operative procedure for removal. They are more difficult to access than external ports, and malposition of the inserted Huber needle can lead to injection of chemotherapeutic agents or other medications into the port pocket. Drug extravasation following such an injection outside the port chamber, or because of separation of the catheter from the port owing to an insecure locking ring, can result in subcutaneous tissue necrosis, skin breakdown, or port-pocket infection. A logical strategy for managing the extravasation of chemotherapeutic drugs that are vesicants or local irritants has been described by Larson (29,30).

Several authors have concluded that implanted ports are associated with a lower rate of infection compared with external right atrial catheters *(31–33)*. However, two prospective, randomized trials comparing infectious complications for ports vs long-term silastic external catheters showed no significant difference *(34,35)*. Each device has its own advantages; therefore, patient and staff preferences often become important factors in deciding which device to use in an individual patient.

IMPLANTABLE INFUSION PUMPS

The implantable infusion pump is another type of vascular access device that can deliver intravenous or intra-arterial drugs, in both outpatients and inpatients. Two such devices are currently used, the Infusaid 400 (Infusaid Corp., Norwood, MA) and the Medtronic Synchromed.

Over the past 20 years there have been many reported attempts to treat tumors in the liver, most commonly metastases from colorectal cancer, by intra-arterial chemotherapy using an infusion pump. The pump is implanted in a subcutaneous pocket over the abdomen and the catheter tip is secured in a retrograde direction in the gastroduodenal artery, with the tip just at the origin of the gastroduodenal artery from the hepatic artery. A perfusion scan can document catheter position and function prior to beginning intra-arterial infusion. The pump reservoir should never be allowed to empty completely because lack of flow through the catheter is likely to result in catheter thrombosis and occlusion. A predetermined refill schedule based on flow rate and reservoir capacity prevents clotting. Refilling and accessing the pumps is performed with a Huber needle that is passed percutaneously through the pump reservoir septum, whose location is determined by palpation or by a template placed on the skin *(36)*.

Implantable pumps have also been used for intraperitoneal insulin infusion in diabetics, spinal delivery of opioids for chronic pain, and intravascular delivery of heparin in patients with recurrent thromboembolism *(37)*. Complications of pumps are the same as for implanted ports, including catheter migration and occlusion, vessel thrombosis, and infection. As in the case of implanted venous access ports, the attached catheter can retract back into the device pocket, leading to drug extravasation into soft tissue. If not securely fixed to its subcutaneous pocket or underlying fascia, the pump can invert, making cannulation through the silastic septum impossible *(38)*. Inversion of the pump should be suspected if the pump chamber cannot be cannulated with a Huber needle.

CVC MAINTENANCE AND SURVEILLANCE FOR COMPLICATIONS

Rather than assessing the status and function of CVCs on a regularly scheduled basis, it is customary to examine the patient and his or her intravenous device and the surrounding tissue each time it is used (Tables 3 and 4). If the device is not currently being used for drug infusion, it is usually flushed with heparin-saline (to prevent clotting) and assessed for infection or other complication, according to a "minimum maintenance" schedule, as outlined in Tables 3 and 4. This involves observation and flushing on a daily basis for short-term devices, twice a week for long-term silastic external catheters, and once a month for implanted ports. A nurse can do this in the office, or in an outpatient setting. In some cases, a family member can be trained to do these procedures *(39)*. Some flexibility should be allowed related to the distance the patient lives from the physician's office or hospital, the length of time the device has been in place,

Table 3
Long-Term CVC Maintenance and Surveillance
After Implantation: Tunneled Silastic Catheters With an External Hub[a]

	Postoperative year				
	1	2	3	4	5
Minimum number of "visits"[b] per year	104	104	104	104	104

[a]Tunneled silastic catheters with an external hub are rarely used longer than 2 years.

[b]Catheter maintenance ("a visit") including examination and cleansing of the implantation site, and a heparin-saline flush, is performed *each time* the device is used, or a minimum of twice a week for the lifetime of the device. This flushing and inspection routine is usually performed in the office, or at home by an oncology or home health nurse, or sometimes by the patient or a trained family member.

When inspecting the device and insertion site, one looks for evidence of infection, catheter occlusion, catheter breakage or leak, vein thrombosis, or other complication. *The routine does not vary month-to-month, or year-to-year, and is continued for the entire lifetime of the device.* Other tests (e.g., X-rays, ultrasound, venogram, blood culture, or white blood count) are ordered only when inspection of the device and patient suggests that a complication has occurred.

Table 4
Long-Term CVC Maintenance and Surveillance
After Implantation: Implanted Ports With an Attached Silastic Catheter[a]

	Postoperative year				
	1	2	3	4	5
Minimum number of "visits"[b] per year	12	12	12	12	12

[a]Implanted ports with an attached silastic catheter are rarely used longer than two years.

[b]Port maintenance ("a visit") including examination of the implantation site and a heparin-saline flush, is performed *each time* the device is used, or a minimum of once a month for the lifetime of the device. This flushing and inspection routine is usually performed in the office, or at home by an oncology or home health nurse, or sometimes by the patient or a trained family member. When examining the patient and insertion site, one looks for evidence of infection, catheter occlusion, catheter breakage or leak, vein thrombosis, or other complication. *The routine does not vary month-to-month, or year-to-year, and is continued for the entire lifetime of the device.* Other tests (e.g., X-rays, ultrasound, venogram, blood culture, or white blood count) are ordered only when inspection of the device and patient suggests that a complication has occurred.

and the skills and reliability of the patient and family. A discussion of CVC care protocols used in Great Britain has been published in the *British Journal of Haematology (40)*.

It is the authors' opinion that patients should direct the majority of their questions regarding CVC care to their physicians and nurses. Additional useful information (but also some misinformation) may be derived from patient support groups. As far as we are aware, no such groups focus solely on CVC care. However, many support groups organized for individuals with specific illnesses such as cancer or gastrointestinal conditions (e.g., short bowel syndrome) exist in local communities and on the Internet.

Variations in minimal maintenance techniques and guidelines exist from institution to institution, as well as variations in permitted patient activity and protection of exter-

Table 5
Complications of Venous Access Devices

Immediate
    Arterial injury
    Hemorrhage, hematoma
    Nerve (e.g., brachial plexus) injury
    Pneumothorax, hemothorax
Immediate or delayed
    Air embolism
    Cardiac arrhythmias
    Catheter malposition
    Hydrothorax
Delayed
    Arteriovenous fistula
    Catheter damage—separation, leak, breakage, fragment embolism
    Catheter migration
    Catheter "pinch-off" under clavicle
    Catheter-associated sepsis—exit site, tunnel, port pocket, bacteremia
    Catheter thrombosis
    Drug extravasation into soft tissue
    Endocardial wall damage
    Needle or catheter dislodgement
    Withdrawal occlusion (one-way catheter) owing to fibrin sheath or flap or clot
    Pericardial perforation, tamponade
    Port pocket seroma, infection
    Port or pump inversion
    Skin erosion, necrosis, slough
    Superior vena cava syndrome
    Tricuspid valve occlusion by clot
    Tumor metastasis in subcutaneous catheter tunnel
    Twiddler's syndrome
    Vein erosion, perforation (e.g., superior vena cava)
    Vein thrombosis, phlebitis

Adapted from refs. *11* and *41*.

nal CVCs during bathing and showering. For example, in stable, non-neutropenic patients, we have found that, once a long-term external CVC has become fixed to the subcutaneous tissue by its Dacron cuff and any exit site suture has been removed, routine dressings may no longer be required, and exposure of the catheter and exit site to shower water can be permitted. Obviously the hub must be securely capped, and meticulous aseptic technique should be observed each time the catheter is flushed with heparin-saline, to avoid infection.

The purpose of CVC surveillance is to detect complications related to these devices as early as possible so that appropriate treatment can be given and loss of the device avoided. Table 5 presents a summary of early and late CVC complications. In the remainder of this chapter, we discuss the recognition, and to some extent management, of four common CVC complications: infection, catheter occlusion, catheter damage or breakage, and vein thrombosis. As an introduction to this discussion, we wll first summarize the results of a survey of Ohio surgeons who implant long-term VADs because the results provide insights into common practice and areas of agreement and disagreement in vascular access surgery.

SURVEY OF OHIO SURGEONS WHO IMPLANT VENOUS ACCESS DEVICES

We mailed a questionnaire regarding placement and use of silastic long-term right atrial catheter and ports to 600 general and vascular surgeons in Ohio who are fellows or associate fellows of the American College of Surgeons. There were 186 responses, a 31% response rate. The results (42) are presented in Table 6. Most long-term devices placed by these surgeons are inserted in the operating room under local anesthesia with intravenous sedation. About two-thirds of the surgeons routinely give perioperative antibiotics, usually a cephalosporin. There is a low threshold for infusing platelets preoperatively and a somewhat higher one for giving fresh frozen plasma.

The majority of surgeons use intraoperative fluoroscopy for optimal positioning of the catheter tip. At the time of the survey, few used Doppler ultrasound to study the venous anatomy preoperatively. Opinions varied regarding the optimal position of the tip of a silastic right atrial catheter. Approximately one-third felt that either the right atrium or the SVC was satisfactory. About half prefer the SVC. Some reasons for these opinions are listed under question 13. More than 90% of the surgeons place single-lumen or multilumen silastic right atrial catheters into the subclavian vein, using a peel-away sheath, as the preferred site and method of placement. Large-lumen silastic right atrial catheters, such as the PermCath, are also placed by two-thirds of the surgeons into the subclavian vein using a peel-away sheath. Others use the jugular vein, employing a peel-away sheath or direct cutdown.

Answers given to question 17 reflect the lack of consensus as to whether multilumen silastic catheters are more frequently associated with thrombotic or septic complications than single-lumen ones. About half the surgeons thought that implanted ports had about the same complication rate as silastic catheters with an external hub. Regarding desired mechanical features of a venous access port, there was a preference for a large septum, detachable catheter, and low cost. Eighty-three percent of surgeons felt that a bent tip Huber needle was less damaging to a silastic port septum than a regular straight hypodermic needle. There was relatively little experience with the P.A.S. Port or the Groshong slit-tip silastic CVC.

Ninety-three percent of surgeons did not feel that unexplained fever was an indication for mandatory removal of a silastic right atrial catheter. About half thought that such indications include fever and bacteremia, fever not responsive to antibiotics, or a positive culture from blood drawn through the access device. Approximately two-thirds would remove such a device on the strong recommendation of an infectious disease consultant. A significant number of surgeons had patients who experienced serious complications from long-term VADs, as listed in question 26, demonstrating that this is not "minor surgery."

Table 6
Survey of 183 Ohio Surgeons' Practices With Regard to Placement of Long-Term Vascular Access Devices

1. Where in the hospital do you usually place Broviac/Hickman/Mediport venous access devices?

	Percent
Operating room	87
Outpatient treatment room	2
Bedside	1
Different locations	10

2. Anesthetic most often used in placing silastic right atrial catheters/ports

	Percent
Local	15
Local and intravenous sedation	78
General	7

3. Perioperative antibiotics used in patients receiving Broviac/Hickman catheters?

	Percent
Routinely	58
Only if patient immunocompromised	16
Almost never	21
Never	6

4. Perioperative antibiotics used in patients receiving implanted ports (e.g., Mediport)

	Percent
Routinely	67
Only if patient immunocompromised	12
Almost never	15
Never	6

5. If you use a perioperative antibiotic, which do you use?

	Percent
A cephalosporin	91
Vancomycin	8
Another drug	1

6. Venous anatomy studied preoperatively by Doppler ultrasound before placement of a silastic right atrial catheter?

	Percent
Routinely	1
Selectively	15
Almost never	35
Never	49

7. When do you give platelets prior to placement of a silastic right atrial catheter?

	Percent
If platelet count <50,000	43
If platelet count <20,000	30
If platelet count <10,000	8
Almost never	12
Never	6

(continued)

**Table 6 *(Continued)*
Survey of 183 Ohio Surgeons' Practices With Regard
to Placement of Long-Term Vascular Access Devices**

8. When do you infuse fresh frozen plasma preoperatively?

	Percent
For any abnormality in PT or PTT	5
Only for a marked abnormality	79
Almost never	13
Never	3

9. How do you confirm the correct position of the catheter tip?

	Percent
Intraoperative fluoroscopy	60
Fluoroscopy and chest X-ray	12
Intraoperative chest X-ray	7
Postoperative chest X-ray	21

10. What is the optimum position of the tip of a silastic right atrial catheter?

	Percent
Right atrium, for best catheter function	14
SVC, to avoid arrhythmias and cardiac trauma	49
Right atrium and SVC both satisfactory	37

11. If you feel strongly about the catheter tip position, why? (individual responses given)

- Less "positional" if in RA; functions better
- More thrombotic problems & more prone to back out if tip in SVC
- RA or RA-SVC junction allows for growth
- A close friend died of RA thrombus and massive embolism produced by RA catheter placement
- To avoid risks of arrhythmia and atrial damage, I place the tip at the SVC/RA junction
- There are pros and cons, so I accept both

12. Preferred method to place a single-lumen silastic right atrial catheter

	Percent
Peel-away sheath into subclavian vein	91
Peel-away sheath into neck vein	1
Open cutdown to cephalic vein	6
Open cutdown to a neck vein	2
Axillary vein cutdown	1

13. Preferred method to place a double or triple-lumen silastic right atrial catheter

	Percent
Peel-away sheath into subclavian vein	90
Peel-away sheath into a neck vein	2
Open cutdown to a neck vein	6
Cephalic or axillary vein cutdown	2

(continued)

Table 6 (Continued)
Survey of 183 Ohio Surgeons' Practices With Regard to Placement of Long-Term Vascular Access Devices

14. Preferred method to place a very large silastic right atrial catheter (PermCath) for hemodialysis or apheresis

	Percent
Peel-away sheath into subclavian vein	59
Peel-away sheath into a neck vein	11
Open cutdown to a neck vein	11
Cephalic or axillary vein cutdown	2
Not answered	17

15. Are multilumen silastic right atrial catheters associated more frequently with thrombotic or septic complications than single-lumen ones?

	Percent
Yes	41
No	38
No opinion	20

16. Do multilumen silastic right atrial catheters with round lumens have fewer complications than those with double-D or wedge-shaped lumens?

	Percent
Yes	6
No	19
No opinion	75

17. Compared to Broviac/Hickman silastic catheters with an external hub, do implanted ports have in their subsequent use:

	Percent
Overall fewer complications	44
More frequent complications	6
About the same complication rate	50

18. In patients who need a single lumen long-term venous access device for mainly outpatient use, do you prefer:

	Percent
A Broviac/Hickman catheter with an external hub	12
A Groshong catheter with an external hub	10
An implanted port	49
Usually defer to the referring physician's and patient's preferences	28

19. Opinions about the mechanical features of a venous access port

	Percent responding	
	Important	Not important
No metal	17	83
Titanium ok, but not stainless steel	30	70
Large septum	59	41
Detachable catheter	58	42
Lowest possible cost	75	25

(continued)

Table 6 (Continued)
Survey of 183 Ohio Surgeons' Practices With Regard to Placement of Long-Term Vascular Access Devices

20. Is a bent tip Huber needle less damaging to a silastic Mediport septum than a regular straight hypodermic needle?

	Percent
Yes	83
No	17

21. Opinion regarding the Pharmacia P.A.S. Port peripheral arm-implanted port

	Percent
Functions as well or better than ports placed on the chest wall (Mediport)	1
Functions less well and with a greater complication rate than the Mediport	11
No difference in function	6
No experience with the P.A.S port	82

22. Opinion regarding the Groshong slit-tip silastic CVC

	Percent			
	Yes	No	No difference	No opinion
More convenient than B/H catheters (no heparin)	34	10	13	43
Higher complication rate than B/H catheters	14	19	20	47

23. What is a mandatory indication for removal of a silastic right atrial catheter?

	Percent responding who answered yes
Unexplained fever	7
Fever and bacteremia	45
Fever not responsive to antibiotics	52
Positive blood culture from blood drawn through the access device	49
Strong ID consult recommendation	60

24. Have you had at least one patient with the following significant complication?

	Percent responding yes
Death from catheter-related sepsis	10
Broken catheter fragment (± catheter embolism)	25
Drug leakage with tissue inflammation or loss	33
Hemorrhage with return to operating room	13
Hemothorax (chest tube or thoracotomy required)	26

25. Should silastic long-term right atrial catheters (Broviac/Hickman/Mediport) be placed by nonsurgeons?

	Percent
Never	82
In selected patients	16
In most patients	1

26. How do you feel about performing venous access surgery?

	Percent
I like the challenge	38
Not very rewarding, but I do it as a service	59
I avoid it if possible	2

PT, prothrombin time; PTT, partial thromboplastin time; SVC, superior vena cava; RA, right atrial; CVC, central venous catheter; B/H, Broviac/Hickman; ID, infectious disease.

CVC SURVEILLANCE: CONTROVERSIES AND POTENTIAL FUTURE STUDIES

Catheter-Associated Infection

Routine assessment of an indwelling CVC involves careful observation of the exit site of percutaneous catheters or the implantation site of intravenous ports or pumps. One looks for redness or purulence that might suggest infection at the exit opening or within the tunnel. One also inspects any exit site suture for intactness because an insecure suture can allow a catheter to "fall out" and be lost. The suture can be safely removed from cuffed catheters once tissue ingrowth into the cuff has occurred. This may take several weeks in steroid-treated or irradiated patients.

The skin over a pump or port should be observed for inflammation, edema, or impending extrusion owing to erosion by the underlying device. Redness of the skin over a port suggests possible port-pocket infection or extravasation of medication into the port pocket owing to a catheter or septal leak, or infusion outside the port resulting from Huber needle malposition.

Infection is the most common complication of CVCs; at least 400,000 cases of catheter-associated infection occur annually *(10)*. Catheter-related infection is a controversial and complex subject, about which hundreds of papers have been written. In his exhaustive review, Greene *(43)* estimates the incidence of catheter-related infection for short-term CVCs to be between 3 and 7%, compared to 1 to 2% for Hickman or Broviac catheters. Controversy exists regarding what exactly constitutes a catheter-related infection, particularly in neutropenic patients. The following definitions of the various types of catheter-related infection are based on those proposed by Greene:

1. *Exit site infection*: Purulence from the catheter exit site, or less than 2 cm of inflammation around the exit site.
2. *Tunnel infection*: Greater than 2 cm of inflammation extending proximal from the catheter exit site of tunneled catheters.
3. *Port-pocket abscess or cellulitis*: Fluctuance around the subcutaneous implanted port with evidence of surrounding inflammation overlying the port.
4. *Catheter-related bacteremia or fungemia*:
   a. A positive blood culture drawn from the catheter with clinical manifestations of sepsis and no apparent source for the infection other than the catheter, or no suspected translocation of organisms from bowel to bloodstream in the febrile neutropenic patient.
   b. A 5- to 10-fold or greater increase in colony-forming units (CFUs) or organisms per milliliter of blood obtained through the device compared with simultaneous peripheral blood cultures.
   c. More than 1000 CFUs or organisms obtained through the catheter in the absence of peripheral blood cultures.
   d. A catheter tip culture of greater than 15 CFUs when the device is removed specifically for suspected catheter-related infection.
5. *Septic thrombophlebitis:* Evidence of a venous thrombus associated with an indwelling catheter and positive blood cultures with clinical manifestations of sepsis.
6. *Infusate-related bacteremia:* The presence of the same pathogen in blood and infusate without identification of alternative sites of infection.
7. *Catheter colonization:* Less than 15 CFUs of an organism on the removed catheter tip.

Methods of treatment of catheter-related infection are even more vigorously debated than the question of what constitutes such an infection; a discussion of this topic is beyond the scope of this chapter. One can find excellent descriptions of the manage-

Table 7
Indications for Central Venous Catheter Removal

a. Therapy has finished; catheter is no longer needed.
b. Progressive exit-site or tunnel inflammation (increasing erythema, tenderness, or purulence) despite adequate trial of saline soaks, local application of povidone-iodine ointment, and systemic antibiotics. A positive *Pseudomonas* culture from the tunnel or exit site, or a systemic *Candida* infection, suggest more rapid catheter removal.
c. Fever associated with bacteremia without an obvious source, which does not clear after a trial of intravenous antibiotics given through the catheter, or if such a patient develops hypotension thought to be the result of sepsis.
d. Chills or hypotension that reproducibly follow irrigation of the catheter (must be distinguished from heparin allergy).
e. Evidence of endocarditis, or a septic pulmonary infarct.

Adapted from ref. *11*.

ment of the various types of CVC-associated infection in the published BCSH Guidelines *(40)*, and in reviews by Oppenheim *(44)* and Greene *(43)*. Factors to be considered in deciding on treatment include the type and location of the infection as discussed above, the organism and its susceptibility to specific antibiotics, and the immune status of the patient. There is now a consensus that most CVC-related infections do not require immediate catheter removal. Commonly accepted reasons for removing long-term CVCs are listed in Table 7.

Prevention of CVC-related infections is a rapidly developing field, as summarized by Mermel *(45)*, with many opportunities for clinical trials. Standard aseptic techniques, of course, are critical during catheter insertion and maintenance. There is little evidence to support the use of perioperative antibiotics for long-term CVC placement, though our survey of practicing surgeons shows this is often done. Al-Sibai et al. *(46)* claim that, in their nonrandomized study of 200 Hickman catheter insertions, the number of exit site infections was reduced by use of prophylactic antibiotics. Many of their patients were immunosuppressed. They now routinely use vancomycin perioperatively. Other authors, including Mermel, caution that such use will result in the emergence of antimicrobial resistance, and they argue strongly against this practice. More randomized studies are needed.

The recent development of antimicrobial-coated or -impregnated catheters and cuffs also provides the opportunity for future research. Mermel, Stolz, and Maki *(47)* have demonstrated the surface antimicrobial activity of heparin-benzalkonium chloride bonded Swan-Ganz pulmonary artery catheters. The VitaCuff (Vitaphore Corp.), made of biodegradable collagen that contains bactericidal silver, was designed for subcutaneous implantation around short-term catheters near the exit site. Studies of whether this type of silver-impregnated cuff can reduce the incidence of catheter-related infection have not been conclusive. Its general use is not yet recommended *(45)*, particularly not for silastic catheters that already have a Dacron cuff positioned in mid-tunnel *(48)*.

Animal studies demonstrating the efficacy of catheters coated with antimicrobials in reducing catheter colonization and infection will no doubt lead to their greater clinical evaluation and possible application in the future *(43,45)*. Such devices include CVCs coated with chlorhexidine-silver sulphadiazine, cefazolin, or the combination of minocycline and rifampin.

Catheter Occlusion

At each use of a CVC, the ease of infusion and withdrawal should be noted. A patent, unobstructed catheter should permit free infusion and withdrawal. X-rays, venograms, or ultrasound scans are not, therefore, routinely performed in patients with nonoccluded CVCs. Partial or complete occlusion (often, initially, withdrawal occlusion alone) may result from one of the following: catheter tip malposition against a vein wall, catheter kinking, catheter "pinch-off," partial catheter intraluminal thrombosis, or occlusion at the tip by thrombus or a fibrin sheath.

Occlusion caused by catheter tip malposition can occur shortly after initial placement, particularly when insertion has been via the left subclavian vein and the tip lies perpendicular to the SVC wall, usually high in the SVC. Peterson et al. *(49)* elegantly demonstrated that this type of catheter malfunction is statistically correlated with position of the tip greater than 4 cm superior to the RA-SVC junction. We strongly agree with their recommendation that, on insertion, the tip should be placed at the RA-SVC junction, or just inside the RA, yet we know from our survey of Ohio surgeons that this is not standard practice.

Silastic catheter occlusion at a later time may be owing to migration of the tip from an ideal central position to a more peripheral one, perhaps at the time of straining or coughing. Repeated flushing with heparin-saline, hydrating the patient, or repositioning the patient may relieve this type of catheter occlusion caused by simple tip malposition. If these maneuvers do not relieve the occlusion, a chest X-ray should be obtained to look for uncorrected tip malposition or catheter kinking. Several methods have been proposed to correct catheter malposition that does not respond to simple maneuvers. These include redirection using an angiographic guidewire and pigtail catheter *(50)*, or a 2 Fr Fogarty balloon-tipped catheter *(51)*. If these techniques are unsuccessful, surgical replacement of the CVC may be required.

Another cause of occlusion is "pinch-off" of the catheter as it passes through the narrow angle between the clavicle and the first rib, as originally described by Aitken and Minton *(52)*. Krutchen et al. *(53)* argue, based on radiological studies, that pinch-off is not actually due to catheter compression between the first rib and clavicle, but rather entrapment in the subclavius musculocostoclavicular ligament complex. They feel that this can be avoided by employing a fluoroscopically guided puncture technique to enter the subclavian vein lateral to the first rib. Hinke et al. *(54)* detail the radiological features of catheter compression due to pinch-off. They recommend prompt removal of all catheters exhibiting occlusion and radiological evidence of pinch-off to avoid the possibility of catheter fracture and embolization.

If tip malposition, or catheter kinking or pinch-off, does not seem by chest X-ray to be the cause of occlusion, then a contrast study through the catheter may reveal a thrombus or fibrin sheath *(55,56)* at the tip. If a fibrin sheath is present, or if the catheter is partially occluded by thrombus, fibrinolysis can be attempted. Until recently, the most frequently used fibrinolytic drug was urokinase (Abbokinase, Abbott Labs). Injected into the catheter was 0.4–0.6 mL of a 5000 U/mL solution, and the catheter was clamped for 30 minutes *(57,58)*. This was followed by unclamping the catheter and aspirating the contents. If bolus urokinase infusion was unsuccessful, a 6-hour infusion of 250,000 U of urokinase could be attempted *(59)*. Urokinase, an effective and safe drug, was unfortunately removed from the market in 1999 and is not available at this time. An alternate thrombolytic agent is streptokinase *(57,60)*, but it has been associated with anaphylactic reactions. Therefore, we now use tissue plasminogen activator (t-PA; Ateplase; Genentech, South San Francisco, CA) to clear intraluminal catheter thrombus. A protocol for using t-PA for this purpose has been published by Hadaway *(61)*.

Finally, CVC occlusion may be the result of mixing and precipitation of incompatible medications within the catheter lumen. Agents that have been used to restore patency owing to specific drug precipitation include hydrochloric acid, sodium bicarbonate, and 70% ethanol *(62)*.

Catheter Damage/Breakage

As mentioned earlier, catheter pinch-off at the costoclavicular angle leads, in a few patients, to catheter breakage and embolization *(63–65)*. Patients in whom this occurs are often asymptomatic but, if the CVC is being used for infusions, they may experience pain near the clavicle resulting from drug extravasation, if the sheared, nonembolized end of the catheter has slipped out of the vein. Obviously, this part of the device should be removed, but there may be no compelling reason to attempt radiological removal of a small fragment of silastic catheter lodged in the right atrium or ventricle in an asymptomatic patient. Nonetheless, these patients are often referred for such a radiological procedure.

Damage or transection of a silastic right atrial catheter having an external hub may occur during a dressing change. Obviously, sharp instruments such as scissors should be used only with great care when removing an old dressing. If the external segment of such a catheter is damaged or divided, the catheter should be clamped with a noncrushing instrument just on the patient's side of the point of damage and a repair kit (obtained from the catheter manufacturer) used under sterile conditions to replace the hub(s) and damaged segment. Catheters that develop a leak in the segment that is subcutaneous should be replaced with a new catheter. Similarly, a subcutaneous port and attached catheter that develops a leak should be replaced to avoid extravasation of drugs into soft tissue.

Venous Thrombosis and Stenosis

No aspect of vascular access surgery is more controversial than catheter-related venous thrombosis—its detection, treatment, and possible prevention. CVC-related venous thrombosis may manifest as swelling of the ipsilateral arm or face, or pain in these areas, but often it is clinically silent due to collaterals that form around the stenosis. Yet the incidence of catheter-associated venous thrombosis and vessel stenosis appears to be high. Cimochowski et al. *(66)* found that in patients with temporary dialysis catheters placed via the subclavian vein, central vein thrombosis or stenosis detected by venogram occurred in 50% of patients. In contrast, all of their patients receiving internal jugular catheters were free of venogram abnormalities.

Subclavian vein stenosis is a particularly serious complication in dialysis patients because the arm on the side of the stenosis is then no longer a feasible site for future dialysis access. A fistula or graft placed on that side will likely lead to arm swelling. Accordingly, current Dialysis Outcome Quality Initiative guidelines recommend that subclavian vein access be used only if the jugular vein is not available (67).

Catheter-related central vein thrombosis is very dependent on the catheter material. Welch et al. (68) demonstrated in a dog model that polyethylene catheters are much more likely to cause central vein thrombosis than silastic catheters. With polyethylene, thrombophlebitis occurred early, by day 2, and it was not associated with bacterial growth.

Catheter-induced venous thrombosis, most often seen in chronically ill patients receiving chemotherapy (69) or hemodialysis, is a secondary process due to vessel trauma from the catheter or from the drugs infused through it. The underlying process is inflammatory, with gradual but limited thrombosis. Brismar et al. (70) studied catheter-related venous thrombosis by catheter phlebography and found two types: (a) sleeve thrombus, that is, thrombus that coats the catheter surface, often its entire intravascular length, and sometimes has a floating component that extends beyond the tip into the SVC or right atrium; and (b) mural veno-occlusive thrombosis, which usually occurs at the point of catheter insertion into the vein. Mural veno-occlusive thrombosis is most often limited to the subclavian-axillary vein. However, mural veno-occlusive thrombus can also form near the catheter tip in the SVC and will occasionally give rise to SVC syndrome.

The relationship of SVC thrombosis to the position of the catheter tip was studied by Puel et al. (71), who found a strong correlation between catheter-related SVC syndrome and position of the catheter tip high in the SVC, with insertion from the left side. These authors believe that, with the tip in this position, there is a higher chance of trauma to the vessel wall and also of irritation by infused drugs delivered near the vessel wall at the site of chronic microtrauma. Woodyard et al. (72) have shown that another way SVC syndrome can occur is by catheter placement through a segment of SVC that has been previously strictured, for example by radiation therapy.

Rarely, a large thrombus can form in the right atrium at the tip of a silastic catheter, as illustrated in case reports by Oh et al. (73) and Chakravarthy et al. (74). A thrombus of this type has the potential to become infected (causing fever), obstruct the tricuspid valve, or break free and cause pulmonary embolism. There is no standard treatment for these large, infrequently seen, RA thrombi. Management may include observation, catheter removal, anticoagulation, thrombolysis, endovascular thrombectomy, surgery with right atriotomy, or a combination of these interventions.

Symptomatic pulmonary emboli from catheter-related vessel thrombosis are rare. The frequency of all catheter-related pulmonary emboli, symptomatic or not, is not known, although it is generally thought to be low (75). A sleeve thrombus appears more likely to embolize than mural veno-occlusive thrombus (70).

Secondary, catheter-induced thrombophlebitis is different from primary, effort-induced subclavian-axillary vein thrombosis (SAVT, also known as Paget-von Schrötter Syndrome). Primary SAVT typically occurs in the dominant arm of a young male (average 30 years old) and presents as painful swelling with venous engorgement. It is associated with repetitive use of the arm, often work- or sport-related. In primary SAVT, there is usually an underlying anatomic defect such as abnormal anterior scalene muscle; abnormal subclavius muscle; congenital band; abnormal pectoralis minor muscle; aberrant accessory phrenic nerve; abnormal ribs, clavicle or costoclavicular ligament; or a congenital web or malformed valve (Robert B. Rutherford, personal communication). Effective treatment of SAVT is catheter-directed thrombolysis followed by positional venogram and appropriate surgery, such as first rib resection, directed toward the underlying anatomic defect *(76)*.

In secondary, catheter-induced venous thrombosis, severe swelling and pain are rare, and treatment is generally conservative: catheter removal in many patients, arm elevation, and anticoagulation or thrombolytic treatment in selected patients. Thrombectomy, angioplasty, or venous reconstruction are almost never done for catheter-induced venous thrombosis.

Our approach, if symptoms or signs of CVC-related venous thrombosis occur, is first to consider whether the patient really needs the catheter. If not, the catheter should be removed. If venous access through the device is still required, we obtain a venogram to document the size and location of the thrombus. Heparin infusion is initiated through the CVC at 1000 U per hour and continued until coumadin, begun simultaneously, has caused the patient's prothrombin time to reach therapeutic range. Coumadin is given for 3 months. Similar strategies have been described by Whitman *(75)* and by Moore et al. *(77)*. Drakos et al. *(78)* reported their use of low-molecular-weight heparin (Enoxaparin, Rhone Poulenc Rorer, Antony, France) to treat Hickman catheter-related venous thrombosis in thrombocytopenic patients undergoing bone marrow transplantation. No hemorrhagic complications were observed.

Thrombolytic therapy may be indicated in some patients if there are other sites of venous thrombosis, if symptoms are severe, or if the thrombus enlarges on heparin therapy. Fraschini et al. *(79)* describe a method for infusion of urokinase through a catheter positioned at or within the clot. After discontinuation of urokinase, their patients received heparin for 5 to 7 days, followed by warfarin for 2 to 3 months in some patients. These authors observed complete lysis of 25 of 30 thrombi that were directly infused with urokinase.

Experience with both anticoagulation and fibrinolytic therapy for catheter-induced venous thrombosis has been reported by Haire et al. *(80)* and by Seigel et al. *(81)*. There is currently no standard protocol for this problem, particularly because urokinase is no longer commercially available. Fortunately, most patients do well if one simply removes their catheter.

There is increasing interest in giving prophylactic anticoagulant or thrombolytic drugs to at least some patients with long-term CVCs, to prevent catheter or vessel thrombosis. Bern et al. *(82)* randomly assigned 82 patients to receive 1 mg of warfarin orally or nothing, starting 3 days before catheter insertion and continuing for 90 days. This low dose does not alter coagulation parameters. Venography was performed in all patients at 90 days after catheter placement, or earlier if symptoms of thrombosis occurred. The incidence of venous thrombosis was 9.5% in those receiving warfarin (all with a thrombus in this group were symptomatic) and 37.5% in those who did not (33% with thrombosis were asymptomatic). The difference was highly significant ($p < 0.001$). Based on this study, some institutions are now routinely using prophylactic low-dose warfarin in patients with CVCs. However, Whitman *(75)* pointed out that, in the Bern study, catheter tip position and side of insertion were not controlled and very high (brachiocephalic vein) tip position was considered acceptable. Therefore, we feel the results should be confirmed and greater experience obtained with this approach, particularly regarding the question of appropriate patient selection, prior to widespread application of low-dose warfarin prophylaxis.

CONCLUSION

The large numbers of CVCs now being used (particularly silastic long-term devices) and the numerous unresolved questions regarding best management, suggest it is time to conduct well-designed clinical trials to resolve existing controversies. Surveys such as the one presented in this chapter regarding opinions of Ohio surgeons can highlight current practices. Such practices can then be compared to recent studies as summarized in literature reviews on vascular access such as those by Whitman *(75)*, Namyslowski and Patel *(83)*, and Blum *(84)*.

Questions that could be addressed by future clinical trials are, for example: What is the benefit, if any, of perioperative antibiotics? What degree of thrombocytopenia requires perioperative platelet infusion? Should intraoperative ultrasound be used routinely or selectively *(85)*? Should the subclavian vein be avoided altogether if the internal jugular is available? What is the best and safest location for the catheter tip, and is this different for silastic vs the stiffer short-term catheters? Do new catheter materials (e.g., heparin-, antiseptic-, and antibiotic-bonded) offer significant benefits and are they cost-effective? Are some patients prone to catheter or venous thrombosis? If so, how can we identify them? Are there safe and effective methods (anticoagulation or fibrinolysis) that can be used to treat or prevent thrombotic complications? Finally, what constitutes an ideal vascular access team? To optimize patient care, particularly in larger hospitals, should members of a specialized vascular access team or service be responsible for placement and management of intermediate- and long-term CVCs?

REFERENCES

1. Forssman W. Experiments on myself: memoirs of a surgeon in Germany. New York: St. Martin's Press, 1974.
2. Kalso E. A short history of central venous catheterization. Acta Anaesth Scand 1985;81:7–10.
3. Duffy BJ. The clinical use of polyethylene tubing for intravenous therapy. Ann Surg 1949; 130:929–936.
4. Aubaniac R. L'injection intraveineuse sous-claviculaire: avantages et technique. Presse Med 1952;60:1456.
5. Seldinger SI. Catheter replacement of the needle in percutaneous arteriography. Acta Radiol [Diagn] (Stockh) 1953;39:368–376.
6. Dudrick SJ, Wilmore DW, Yars HM, Rhoads JE. Long-term total parenteral nutrition with growth, development, and positive nitrogen balance. Surgery 1968;64:134–142.
7. Quinton WE, Dillard D, Scribner BH. Cannulation of blood vessels for prolonged hemodialysis. Trans Am Soc Artif Intern Organs 1960;6:104–108.
8. Broviac JW, Cole JJ, Scribner BH. A silicone rubber atrial catheter for prolonged parenteral alimentation. Surg Gynecol Obstet 1973;136:602–606.
9. Hickman RO, Buckner CD, Clift RA, et al. A modified right atrial catheter for access to the venous system in marrow transplant recipients. Surg Gynecol Obstet 1979;148:871–875.
10. Raad II, Darouiche RO. Catheter-related septicemia: risk reduction. Infect Med 1996;13: 807–812, 815–816, 823.
11. Raaf JH. Administration of chemotherapeutic agents: techniques and controversies. Supportive Care in Cancer 1994;2:335–346.

12. Ellis LM, Vogel SB, Copeland EM. Central venous catheter vascular erosions: diagnosis and clinical course. Ann Surg 1989;209:475–478.
13. Cobb DK, High KP, Sawyer RG, et al. A controlled trial of scheduled replacement of central venous and pulmonary-artery catheters. N Engl J Med 1992;327:1062–1068.
14. Broadwater JR, Henderson MA, Bell JL, et al. Outpatient percutaneous central venous access in cancer patients. Am J Surg 1990;160:676–680.
15. Raad I, Davis S, Becker M, et al. Low infection rate and long durability of non-tunneled silastic catheters. Arch Intern Med 1993;153:1791–1796.
16. Pasquale MD, Campbell JM, Magnant CM. Groshong versus Hickman catheters. Surg Gynecol Obstet 1992;174:408–410.
17. Warner BW, Haygood MM, Davies SL, Hennies GA. A randomized, prospective trial of standard Hickman compared with Groshong central venous catheters in pediatric oncology patients. J Am Coll Surg 1996;183:140–144.
18. Salem RR, Ward BA, Ravikumar TS. A new peripherally implanted subcutaneous permanent central venous access device for patients requiring chemotherapy. J Clin Oncol 1993;11:2181–2185.
19. Raaf JH. Results from use of 826 vascular access devices in cancer patients. Cancer 1985;55:1312–1321.
20. Raaf JH, Heil D, Rollins DL. Vascular access, pumps, and infusion. In: McKenna RJ, Murphy GP eds. Cancer surgery. Philadelphia, PA: J. B. Lippincott, 1994, pp. 47–62.
21. Schwab SJ, Buller GL, McCann RL, et al. Prospective evaluation of a Dacron cuffed hemodialysis catheter for prolonged use. Am J Kidney Dis 1988;11:166–169.
22. Moss AH, McLaughlin MM, Lempert KD, Holley JL. Use of a silicone catheter with a Dacron cuff for dialysis short-term vascular access. Am J Kidney Dis 1988;12:492–498.
23. Raaf JH. Vascular access, catheter technology and infusion pumps. In: AR Moossa, SC Schimpff, Robson MC eds, Comprehensive textbook of oncology (2nd ed.). Baltimore, MD: Williams & Wilkins, 1991, pp. 583–589.
24. Lazarus HM, Trehan S, Miller R, Fox RM, Creger RJ, Raaf JH. A multi-purpose central venous catheter for both collection and transplantation of hematopoietic progenitor cells. Bone Marrow Transplant 2000;25:779–785.
25. Niederhuber JE, Ensminger W, Gyves JW, et al. Totally implanted venous and arterial access system to replace external catheters in cancer treatment. Surgery 1982;92:706–712.
26. Raaf JH, Vinson D. Vascular access: central venous catheters, Section XII. Vascular surgery. In: Levine BA, Copeland EM III, Howard RJ, Sugerman H, Warshaw AL eds. Current practice of surgery. London: Churchill Livingstone, 1993, pp. 1–11.
27. Raaf J, Comella S, Wyzgala M. No difference in silastic coring from venous access ports by Huber vs standard needles. ASCO Proceedings 1989;8:331.
28. Shapiro CL. Central venous access catheters. Surg Oncol Clin N Am 1995;4:443–451.
29. Larson DL. Treatment of tissue extravasation by antitumor agents. Cancer 1982;49:1796–1799.
30. Rudolph R, Larson DL. Etiology and treatment of chemotherapeutic agent extravasation injuries: a review. J Clin Oncol 1987;5:1116–1126.
31. Groeger JS, Lucas AB, Thaler HT, et al. Infectious morbidity associated with long-term use of venous access devices in patients with cancer. Ann Intern Med 1993;119:1168–1174.
32. Lokich JJ, Bothe A, Benotti P, Moore C. Complications and management of implanted venous access catheters. J Clin Oncol 1985;3:710–717.
33. Ross MN, Haase GM, Poole MA, et al. Comparison of totally implanted reservoirs with external catheters as venous access devices in pediatric oncologic patients. Surg Gynecol Obstet 1988;167:141–144.
34. Mirro J Jr, Rao BN, Stokes DC, et al. A prospective study of Hickman/Broviac catheters and implantable ports in pediatric oncology patients. J Clin Oncol 1989;7:214–222.

35. Mueller BU, Skelton J, Callender DPE, et al. A prospective randomized trial comparing the infectious and noninfectious complications of an externalized catheter versus a subcutaneously implanted device in cancer patients. J Clin Oncol 1992;10:1943–1948.
36. Johnson GB. Nursing care of patients with implanted pumps. Nurs Clin North Am 1993;28:873–883.
37. Buchwald H, Rohde TD, Schneider PD, et al. Long-term, continuous intravenous heparin administration by implantable infusion pump in ambulatory patients with recurrent venous thrombosis. Surgery 1980;88:507–516.
38. Seeger J, Woodcock TM, Richardson JD. Complications of implantable chemotherapy pump: a case of pump inversion. Cancer 1985;56:2428–2429.
39. Cole D. Selection and management of central venous access devices in the home setting. J Intraven Nurs 1999;22:315–319.
40. BCSH Guidelines on the insertion and management of central venous lines. Br J Haematol 1997;98:1041–1047.
41. Gorman RC, Buzby GP. Difficult access problems. Surg Oncol Clin N Am 1995;4:453–472.
42. Data previously presented ("Controversies in the use of silastic catheters for venous access") at the annual meeting of the Ohio Chapter, American College of Surgeons, Cincinnati, OH, May 1996.
43. Greene JN. Catheter-related complications of cancer therapy. Infect Dis Clin North Am 1996;10:255–295.
44. Oppenheim, BA. Optimal management of central venous catheter-related infections: what is the evidence? J Infection 2000;40:26–30.
45. Mermel LA. Prevention of intravascular catheter-related infections. Ann Intern Med 2000;132:391–402.
46. Al-Sibai MB, Harder EJ, Faskin RW, et al. The value of prophylactic antibiotics during the insertion of long-term indwelling silastic right atrial catheters in cancer patients. Cancer 1987;60:1891–1895.
47. Mermel LA, Stolz SM, Maki DG. Surface antimicrobial activity of heparin-bonded and antiseptic-impregnated vascular catheters. J Infect Dis 1993;167:920–924.
48. Groeger JS, Lucas AB, Coit D, et al. A prospective, randomized evaluation of the effect of silver impregnated subcutaneous cuffs for preventing tunneled chronic venous access catheter infections in cancer patients. Ann Surg 1993;218:206–210.
49. Peterson J, Delaney JH, Brakstad MT, et al. Silicone venous access devices positioned with their tips high in the superior vena cava are more likely to malfunction. Am J Surg 1999;178:38–41.
50. Walker TG, Geller SC, Waltman AC, et al. A simple technique for redirection of malpositioned Broviac or Hickman catheters. Surg Gynecol Obstet 1988;167:246–248.
51. Schaefer C, Geelhoed GW. Redirection of misplaced central venous catheters. Arch Surg 1980;115:789–791.
52. Aitken DR, Minton JP. The "pinch-off" sign: a warning of impending problems with permanent subclavian catheters. Am J Surg 1984;148:633–636.
53. Krutchen AE, Bjarnason H, Stackhouse DJ, et al. The mechanisms of positional dysfunction of subclavian venous catheters. Radiology 1996; 200:159–163.
54. Hinke DH, Zandt-Stastny DA, Goodman LR, et al. Pinch-off syndrome: a complication of implantable subclavian venous access devices. Radiology 1990;177:353–356.
55. Hoshal VL, Ause RG, Hoskins PA. Fibrin sleeve formation on indwelling subclavian central venous catheters. Arch Surg 1971;102:353–358.
56. Tschirhart JM, Rao MK. Mechanism and management of persistent withdrawal occlusion. Am J Surg 1988;54:326–328.
57. Hurtubise MR, Bottino JC, Lawson M, McCredie KB. Restoring patency of occluded central venous catheters. Arch Surg 1980;115:212–213.

58. Lawson M, Bottino JC, Hurtubise MR, McCredie KB. The use of urokinase to restore the patency of occluded central venous catheters. Am J Intraven Ther & Clin Nutrit 1982;29–32.
59. Haire WD, Lieberman RP. Thrombosed central venous catheters: restoring function with a 6-hour urokinase infusion after failure of bolus urokinase. JPEN 1992;16:129–32.
60. Rubin RN. Local installation of small doses of streptokinase for treatment of thrombotic occlusions of long-term access catheters. J Clin Oncol 1983;1:572–573.
61. Hadaway LC. Managing vascular access device occlusions, part 1. Nursing 2000 1999;30:20.
62. Krzywda EA. Predisposing factors, prevention, and management of central venous catheter occlusions. J Intraven Nurs 1999;22:S11–S17.
63. Klotz HP, Schopke W, Kohler A, et al. Catheter fracture: a rare complication of totally implantable subclavian venous access devices. J Surg Oncol 1996;62:222–225.
64. Noyen J, Hoorntje J, de Langen Z, et al. Spontaneous fracture of the catheter of a totally implantable venous access port: case report of a rare complication. J Clin Oncol 1987;5:1295–1299.
65. Prager D, Hertzberg RW. Spontaneous intravenous catheter fracture and embolization from an implanted venous access port and analysis by scanning electron microscopy. Cancer 1987;60:270–273.
66. Cimochowski GE, Worley E, Rutherford WE, et al. Superiority of the internal jugular over the subclavian access for temporary dialysis. Nephron 1990;54:154–161.
67. National Kidney Foundation—Dialysis Outcomes Quality Initiative clinical practice guidelines for vascular access. Am J Kidney Dis 1997;30:5150–5191.
68. Welch GW, McKeel Jr DW, Silverstein P, Walker HL. The role of catheter composition in the development of thrombophlebitis. Surg Gynecol Obstet 1974;138:421–424.
69. Lokich JJ, Becker B. Subclavian vein thrombosis in patients treated with infusion chemotherapy for advanced malignancy. Cancer 1983;52:1586–1589.
70. Brismar B, Hårdstedt C, Jacobson S. Diagnosis of thrombosis by catheter phlebography after prolonged central venous catheterization. Ann Surg 1981;194:779–783.
71. Puel V, Caudry M, Le Métayer P, et al. Superior vena cava thrombosis related to catheter malposition in cancer chemotherapy given through implanted ports. Cancer 1993;72:2248–2252.
72. Woodyard TC, Mellinger JD, Vann KG, Nissenbaum J. Acute superior vena cava syndrome after central venous catheter placement. Cancer 1993;71:2621–2623.
73. Oh WK, Lee BH, Sweitzer NK. Right atrial thrombus associated with a central venous catheter in a patient with metastatic adrenocortical carcinoma. J Clin Oncol 2000;18:2638–2639.
74. Chakravarthy A, Edwards WD, Fleming R. Fatal tricuspid valve obstruction due to a large infected thrombus attached to a Hickman catheter. JAMA 1987;257:801–803.
75. Whitman ED. Complications associated with the use of central venous access devices. Curr Prob Surg 1996;33:309–378.
76. Rauwerda JA, Bakker FC, van den Broek TAA, Dwars BJ. Spontaneous subclavian vein thrombosis: A successful combined approach of local thrombolytic therapy followed by first-rib resection. Surgery 1988;103:477–480.
77. Moore C, Heffernan I, Sheldon S. Diagnosis and management of subclavian vein thrombosis. J Infusional Chemotherapy 1992;2:151–152.
78. Drakos PE, Nagler A, Or R, et al. Low molecular weight heparin for Hickman catheter-induced thrombosis in thrombocytopenic patients undergoing bone marrow transplantation. Cancer 1992;70:1895–1898.
79. Fraschini G, Jadeja J, Lawson M, et al. Local infusion of urokinase for the lysis of thrombosis associated with permanent central venous catheters in cancer patients. J Clin Oncol 1987;5:672–678.
80. Haire WD, Lieberman RP, Edney J, et al. Hickman catheter-induced thoracic vein thrombosis: Frequency and long-term sequelae in patients receiving high-dose chemotherapy and marrow transplantation. Cancer 1990;66:900–908.

81. Seigel EL, Jew AC, Delcore R, et al. Thrombolytic therapy for catheter-related thrombosis. Am J Surg 1993;166:716–719.
82. Bern MM, Lokich JJ, Wallach SR, et al. Very low doses of warfarin can prevent thrombosis in central venous catheters. Ann Intern Med 1990;112:423–428.
83. Namyslowski J, Patel NH. Central venous access: a new task for interventional radiologists. Cardiovasc Intervent Radiol 1999;22:355–368.
84. Blum AS. The role of the interventional radiologist in central venous access. J Intravenous Nurs 1999;22:S32–S39
85. Fry WR, Clagett GC, O'Rourke PT. Ultrasound-guided central venous access. Arch Surg 1999;134:738–741.

US Counterpoint to Chapter 21

Eric D. Whitman

VASCULAR ACCESS DEVICES

Chapter 19 provides a well-researched review of current practices and devices for providing secure venous access for a variety of medical therapies. In addition, the chapter includes the results of a survey the authors performed of general and vascular surgeons in Ohio. This survey provides a unique perspective on the prevailing practice patterns of the general surgeons in the community who probably insert the majority of venous access devices in that geographic area. I found this survey and its results quite interesting because they reinforce my opinions on the general pattern of venous access device usage in the United States.

However, I feel it is important to include in this counterpoint focused and decisive opinions and recommendations on what I consider a key aspect of medical care related to venous access. I believe these recommendations are essential to the goals of this text, as I understand them.

The first and perhaps most important treatment decision about vascular access devices to be made is the *choice* of a particular venous access device. This choice has multiple ramifications. It is important to remember that vascular access devices (VADs) *enable* therapy. Therefore, they must be appropriate for the intended therapy without burdening the patient and/or caregiver with unnecessary issues of device maintenance or cost. The access device industry has generally divided its products into short- or long-term devices. Long-term VADs are typically constructed of thermoplastic polymers (that soften when warmed by the body's temperature) such as silicone or various polyurethanes. Most "long-term" devices are designed to either be tunneled from the venous entry site to a remote catheter exit site, with a fibrous cuff positioned at or near the exit site *(1)*, or as an infusion port where the catheter is attached to a subcutaneously implanted reservoir *(2)*. Tunneled catheters exit the skin and require significant nursing care at the exit site. Infusion ports are completely covered by the skin and therefore (when not being used) require little nursing care. When an access needle is placed through the skin into an infusion port and left in place, as might be done for continuous, several-day chemotherapy infusion, the nursing care issues become very similar to those of the tunneled catheter devices. Devices generally designed for short-term use are not tunneled and not completely implanted subcutaneously. Originally, they were also constructed of inherently different polymers. However, there are now

From: *The Bionic Human: Health Promotion for People With Implanted Prosthetic Devices*
Edited by: F. E. Johnson and K. S. Virgo © Humana Press Inc., Totowa, NJ

several short-term products on the market made of silicone or polyurethanes identical to the long-term devices. Short-term devices, because they have an external component, also require significant nursing care but are less expensive and may be inserted outside of a typical operating room or procedure room environment *(3,4)*. In general, all available venous access devices may be used for currently approved therapies without limitations. Although there are substantial differences between the levels of nursing care required for various types of VADs, this nursing care should be generally available in any community in the United States.

The VAD selected must be consistent with the *standards* of the local community. If a newer type of device is implanted without the knowledge or approval of the treating clinicians, this may cause unnecessary and avoidable problems for both the patient and the treating clinician. For example, if a new type of device is brought to market that requires substantially different nursing care patterns, the penetration of the necessary nursing education to utilize that device may not be adequate within the treating community to advocate use of the new product. Over the past 15 years, there have been multiple examples of VADs that required such different nursing care procedures that they were overcome by a tidal wave of resistance within the nursing community. In the late 1980s, an infusion port that had a side-entry septum instead of the top-entry septum seen on every other product was released. The marketing concept was to enable the access needle to slide into the septum flush with the skin. The reality in the marketplace was that nurses continued to try to access the device from the top down because they could not tell that this was a different device from its appearance under the skin. This led to a high level of frustration by both the patients and the nurses and this product is no longer on the market. Other devices, also employing an inherently different design that demanded alterations in standard medical or nursing care, have not reached nearly the market penetration originally projected *(5)*.

Similarly, there are little if any randomized, prospective data comparing relative outcomes based on VAD choice. Any assertions by authors that one type of device is superior, particularly in terms of complication rates, to another are inherently flawed and biased. If there is a true difference in risk of complications between devices, it is more likely to be secondary to treatment patterns and patient selection *(6–10)*. Even so-called short-term catheters have been used at large institutions for chronic or long-term treatments *(11)*. Although there may be a theoretical disadvantage in using catheters not necessarily designed for long-term implantation, these catheters are also less expensive and can be removed/replaced (if necessary) with less hassle. For example, an infusion port is generally considered by many physicians to be intrinsically resistant to infection because they are completely implanted beneath the skin and therefore in theory are easier to keep sterile. Unfortunately, infusion ports do become infected and these infections may become life-threatening because they ultimately present as a localized abscess (known as a port-pocket infection [12]), often causing systemic signs and symptoms that may even lead to septic shock. Does a port's purported relative resistance to infection lead to otherwise avoidable delay in diagnosing and treating a port-related infection? This question has not been addressed in the literature.

In patients with implanted vascular access devices, infections may also occur in sites distant from the catheter. In these patients, the clinician may consider treating the infection longer, using higher doses of antibiotics, or using different antibiotics. We find no support in the literature for this practice.

Overall, the choice of an appropriate VAD is based on projected duration of use, prevailing local care patterns, and patient and physician preference. Long-term therapies, such as multicycle chemotherapy and chronic total parenteral nutrition, should usually be enabled by one of the more durable-design devices, such as a tunneled catheter or an infusion port. There are multiple models from multiple manufacturers of each device and little to distinguish them. In some institutions, typical care patterns dictate the use of VADs that globally are considered "short term." The ongoing success of these clinical programs (11), and their continued use of short-term devices, suggests again that device success and longevity may not necessarily be related (in current market conditions and with currently marketed devices) to device design. Finally, the prevailing standard of care in some clinical practices is to match up device A with disease/treatment B, mostly on an arbitrary basis. The apparent success of this decision model is likely the result of the familiarity of the regional nursing support staff with the device choice(s) and their ability to maximize the performance and longevity of the "routine" choice.

In summary, I must disagree with the authors' firm recommendations regarding VAD choice. For any prolonged infusion therapy in patients with poor peripheral venous access or with agents that require central venous infusion, either a so-called short-term or long-term design catheter may be appropriate. The choice is best reserved for the treating physician who has the experience and the nursing support staff to best guide that clinical decision.

In the future, there may be new VADs with a clear and undeniable clinical advantage to their use. For example, there are now several antibiotic-treated or coated short-term central VADs on the market. Randomized prospective studies have clearly shown that the incidence of infection with these devices is lower than with nontreated control devices (13,14).

However, these catheters are not currently available in traditional long-term designs such as tunneled catheters or infusion ports. In addition, the randomized trials only looked at patients receiving intensive, very short-term, therapy, typically in an intensive care unit where the risk of catheter infection is extremely high. The long-term efficacy of these treated catheters to prevent infection in patients receiving chronic infusion therapy is unknown. Other ongoing research, which has not yet resulted in the commercial release of a modified VAD, has looked at the prevention of catheter-associated thrombus. The rationale for this approach is the observation that catheter infection is often related to catheter-associated venous thrombosis, even though the venous thrombosis is often if not usually asymptomatic (15). Preclinical studies have suggested that catheters coated with thrombus or components of thrombus are much more likely to become infected in a laboratory setting than uncoated, clean, catheters (16). Based on these results, it is possible that prevention of catheter-related venous thrombosis may actually also prevent catheter-related infection.

I have summarized our current recommended practice in the accompanying table (Table 1). There is no known clinical advantage to routine surveillance testing for any VAD. The only routine element to the chronic care of a VAD is flushing each lumen of the device with the indicated solution, as specified by the device's manufacturer. Oth-

Table 1
Patient Surveillance After Implantation of Venous Access Devices at the Washington University Medical Center, St. Louis, Missouri

	Postoperative year					
	1	2	3	4	5	10
Office visit	0[a]	0[a]	0[a]	0[a]	0[a]	0[a]

[a] Office visits, X-rays, blood tests, scans, and so on, are obtained as indicated by clinical findings only. Device flushing is carried out according to the manufacturer's instructions.

erwise, any other scans, radiographs, or tests are reserved for indicated clinical situations (17). Specifically, malfunctioning catheters should be investigated with a standard chest radiograph. This will confirm that the catheter remains in the appropriate position and has not fractured. If a catheter ceases to function, with a normal chest radiograph, then additional testing should be performed to include venous Doppler studies of the subclavian and internal jugular veins and, potentially, a venogram to further investigate the possibility of a venous thrombosis. Blood tests and local (swab) or systemic cultures should only be performed when there is a clinically apparent infection and the catheter device is considered a potential source of the infection. Finally, new onset of pain near the catheter during medication infusion should be investigated with a catheter venogram to eliminate the possibility of drug extravasation from catheter fracture.

REFERENCES

1. Hickman RO, Buckner CD, Clift RA, et al. A modified right atrial catheter for access to the venous system in marrow transplant recipients. Surg Gynecol Obstet 1979;148:871–875.
2. Niederhuber JE, Ensminger W, Gyves JW, et al. Totally implanted venous and arterial access system to replace external catheters in cancer treatment. Surgery 1982;92:706–712.
3. McCready D, Broadwater R, Ross M, et al. A case–control comparison of durability and cost between implanted reservoir and percutaneous catheters in cancer patients. J Surg Res 1991;51:377–381.
4. Raad I, Davis S, Becker M, et al. Low infection rate and long durability of nontunneled silastic catheters. A safe and cost-effective alternative for long-term venous access. Arch Int Med 1993;153:1791–1796.
5. Andrews JC, Walker-Andrews SC, Ensminger WD. Long-term central venous access with a peripherally placed subcutaneous infusion port: initial results. Radiology 1990;176:45–47.
6. Christensen ML, Hancock ML, Gattuso J, et al. Parenteral nutrition associated with increased infection rate in children with cancer. Cancer 1993;72:2732–2738.
7. Mirro J, Jr., Rao BN, Stokes DC, et al. A prospective study of Hickman/Broviac catheters and implantable ports in pediatric oncology patients. J Clin Oncol 1989;7:214–222.
8. Timsit J, Sebille V, Farkas J, et al. Effect of subcutaneous tunneling on internal jugular catheter-related sepsis in critically ill patients: a prospective randomized multicenter study. JAMA 1996;276:1416–1420.
9. Wagman LD, Kirkemo A, Johnston MR. Venous access: a prospective, randomized study of the Hickman catheter. Surgery 1984;95:303–308.

10. Whitman ED. A neural network to predict prospectively the risk of central venous access device infection. Surg Forum 1996;XLVII:630–632.
11. Broadwater JR, Henderson MA, Bell JL, et al. Outpatient percutaneous central venous access in cancer patients. Am J Surg 1990;160:676–680.
12. Whitman ED. Complications associated with the use of central venous access devices. Curr Prob Surg 1996;33:309–378.
13. Maki DG, Stolz SM, Wheeler S, et al. Prevention of central venous catheter-related bloodstream infection by use of an antiseptic-impregnated catheter. Ann Intern Med 1997;127:257–266.
14. Raad I, Darouiche R, Dupuis J, et al. Central venous catheters coated with minocycline and rifampin for the prevention of catheter-related colonization and bloodstream infections. Ann Intern Med 1997;127:267–274.
15. Raad I, Luna M, Khalil SM, et al. The relationship between the thrombotic and infectious complications of central venous catheters. JAMA 1994;271:1014–1016.
16. Vaudaux P, Pittet D, Haeberli A, et al. Fibronectin is more active than fibrin or fibrinogen in promoting Staphylococcus aureus adherence to inserted intravascular catheters. J Infect Dis 1993;167:633–641.
17. Whitman ED. Vascular Access for Cancer. In: Norton JA, Bollinger RR, Chang AE, et al., eds. Surgery: basic science and clinical evidence. New York: Springer, 2001, pp. 1795–1821.

European Counterpoint to Chapter 21

Matthias Lorenz, Carsten N. Gutt, and Stefan Heinrich

BACKGROUND

Since their introduction into clinical use, long-term infusion devices have become very popular, especially in cancer patients. They provide convenient venous access and are suitable for outpatient use *(1)*. These systems cause fewer infectious and thrombotic complications than non-tunneled infusion systems but they are not complication free. Because they are used for chronically ill patients (e.g., cancer patients), device-associated complications lead to treatment interruptions and carry a risk of disease progression. For this reason, the main goal of follow-up is maintenance of the device by avoiding complications. If a complication does occur, appropriate treatment should be given as soon as possible.

Several long-term infusion devices have been developed. Although they differ somewhat, the implantation techniques are all similar. The best known tunneled central lines in Europe are Hickman™ (Broviac) and Groshong™ catheters as well as the PermCath™ long-term dialysis catheters. They all have an extracorporeal part so they can be directly connected to infusion sets. These are generally used for parenteral nutrition, stem-cell transplantation, chemotherapy or dialysis. Other devices lie completely subcutaneously. The catheter is connected to a subcutaneous infusion chamber that is punctured by a particular needle for infusions. The main indication is systemic chemotherapy, but long-term intravenous antibiotic therapy, repeated transfusions, parenteral nutrition (and chronic fluid substitution), and chronic pain therapy are other indications *(2)*. Intra-arterial infusion devices are mainly employed for regional chemotherapy for primary or secondary hepatic tumors. The catheter can either be connected to implantable infusion pumps or port systems. If ports are used, an extracorporeal pump can be connected to the subcutaneous port chamber for continuous treatment. Occasionally, implantable pumps are also used for intrathecal delivery of narcotics for patients with chronic pain, insulin infusion for brittle diabetics, and so on.

Long-term infusion catheters can be introduced via any peripheral vein, even into the saphenous, femoral or gonadal veins *(3)*. However, the insertion into the superior vena cava (SVC) via the subclavian or jugular vein is the preferred route. Two techniques have been described for device implantation. The most common employs percutaneous puncture of the subclavian or internal jugular vein. The catheter is inserted by Seldinger technique through a peel-off introducer and can easily be advanced into

From: *The Bionic Human: Health Promotion for People With Implanted Prosthetic Devices*
Edited by: F. E. Johnson and K. S. Virgo © Humana Press Inc., Totowa, NJ

the SVC. We prefer the cutdown technique, whenever possible, utilizing the cephalic vein, through which the catheter is inserted into the SVC. A puncture with the risk of a pneumothorax is not necessary, and the catheter is easy to introduce into the subclavian vein. The catheter is fixed in place by one ligature.

Dependent on the infusion system, a subcutaneous pocket on the pectoral muscle or a subcutaneous tunnel is then prepared for a port chamber or a central line, respectively. The central line is fixed by skin stitches, which can usually be removed after 2 weeks. To achieve the best cosmetic result, the incision for the port chamber is closed by an absorbable intracutaneous running suture. The optimal position of the catheter tip is below T3, which is demonstrated by fluoroscopy at the end of the implantation. This position has been shown to cause the fewest catheter-associated complications such as thrombosis and displacement (3–5). In any case, damage or narrowing of the catheter must be avoided. The cutdown technique has some advantages compared to the puncture technique and it does not take more time in experienced hands. Pneumothoraxes occur in 1.2% of punctures and never during the cutdown technique (4). Interestingly, the risk for a pneumothorax is not reduced by puncturing the jugular instead of the subclavian vein (6). We routinely perform both implantation and removal of these long-term infusion devices under local anesthesia on an outpatient basis. Sedation or general anesthesia is reserved for special indications.

As described by Compton and Raaf, intra-arterial pumps are used for regional chemotherapy, mainly for liver metastases from colorectal cancer and hepatocellular carcinoma. The catheter is implanted under general anesthesia, according to the technique reported by Watkins (7). If not previously performed, all patients undergo cholecystectomy to avoid chemical cholecystitis. The catheter is inserted into the gastroduodenal artery and carefully advanced to its origin from the common hepatic artery. All branches of the hepatic artery distal to the origin of the gastroduodenal artery (i.e., right gastric artery) and up to 2 cm proximal to the gastroduodenal artery must be ligated to avoid chemoperfusion of the stomach or small bowel. The correct perfusion of the liver is then proven by Wood's lamp after fluorescein injection. Finally, the catheter is connected to the heparin-filled infusion device, which can either be a port chamber or a pump. Port systems are subcutaneously placed onto the right costal arch, and the infusion pumps are implanted into a subcutaneous pocket in the right lower abdominal wall.

As an alternative, percutaneous port implantation into the hepatic artery has been reported during recent years (8,9). The subclavian artery is punctured under ultrasound guidance and the catheter is advanced into the gastroduodenal artery. A side hole of this catheter is placed into the hepatic artery and the gastroduodenal artery is embolized with coils for catheter fixation. These results have only been reported in small series. Complication rates are comparable to the open technique but neither laparotomy nor general anesthesia are necessary. Further trials are needed to investigate which technique should be standard in the future.

PREVENTION AND TREATMENT OF COMPLICATIONS

Complications of long-term infusion devices are listed in the chapter by Compton and Raaf and mainly include wound infection (at the exit sites of tunneled catheters), device infection, catheter occlusion, venous thrombosis, and catheter damage (10,11).

Special attention should be paid to the implantation procedure because it is a crucial step in avoiding complications.

To avoid catheter-associated infection, sterile operative technique is important. Because a foreign body is always at higher risk for contamination, perioperative antibiotic prophylaxis appears to be reasonable. However, as Compton and Raaf point out, there are few data about the value of such prophylaxis and the Ohio survey reflects this uncertainty. We always perform perioperative antibiotic prophylaxis. In addition to the nonrandomized trial of Al-Sibai (12), the rationale for this routine is supported by a randomized trial in patients with hematological malignancies showing that antibiotic prophylaxis significantly decreased the catheter-related sepsis rate after Hickman catheter implantation (13). Also, large series on other foreign-body implantations, such as mesh repair for inguinal hernias, demonstrated the beneficial effect of antibiotic prophylaxis (14).

To avoid infection during catheter use, puncture of the port membrane and connection of infusions to the tunneled catheters should be performed under sterile conditions, including desinfection of the region of interest and use of sterile gloves. After the puncture of a port chamber, the needle should be covered by a sterile dressing until its removal. At each visit, the skin around the infusion device should be observed closely to detect any infection early. The rate of line infections may be decreased and the infection-free period extended by the addition of antibiotics to the flush solution, as suggested by a recent randomized trial in pediatric patients (15). Generally, early use of prophylactic broad-spectrum antibiotics for severe bacterial infections elsewhere in the body appears reasonable to avoid catheter colonization. Although we are aware that there is no scientific confirmation of this policy, we strongly recommend it.

As Compton and Raaf point out in Chapter 21, several protocols for the treatment of catheter infection have been reported, and it is up to the judgment of the physician which protocol is used. Although not all infected catheters need to be explanted immediately (16), the threshold for removing a catheter should be low, especially if an infection does not respond to antibiotic treatment or recurs.

Catheters can be pinched in such a way that they eventually break. This is a rare but potentially dangerous complication because the catheter tip can embolize (2). It seems to be caused by a sharp angle of the catheter over the first rib owing to the puncture technique and has never been described after the cutdown technique. The only treatment for catheter damage is the removal of the device, unless the damage is in a part of the system that can be exchanged. Removal of an intravascular fragment after catheter breakage usually requires an invasive procedure. Usually this is done by an interventional radiologist. For this reason, any damage must be avoided and careful implantation of the catheter (see above) and use are mandatory.

Catheter occlusion is a serious complication and may be caused by catheter breakage, clotting, displacement, or vascular thrombosis. In order to prevent catheter occlusion, the devices should always be flushed and locked with heparin at the end of each infusion. During treatment pauses, the catheters should be flushed every 4 weeks. If the injection causes pain, chest X-ray or contrast injection through the device should be performed prior to any manipulation. Lysis, as described in Chapter 21, is certainly the method of choice for clots. We usually treat thrombotic occlusion of the infusion catheter or the catheter bearing vessel by instilling 50,000 E urokinase + heparin (1 mL)

Table 1
**Technical Complications
of Continous Intra-Arterial Chemotherapy of the Liver**[a]

	Port (n = 110)	Pump (n = 70)
Catheter-associated		
Occlusion	28	3
Dislocation	15	7
Leakage	9	—
Arterial thrombosis	5	2
Device-associated		
Pump failure	—	6
Local infection	2	4
Rupture of the membrane	2	—
Fluid collection	1	3
Puncture not possible	4	2
Skin necrosis	—	5

[a] Regional chemotherapy was either given through a subcutaneously implanted pump or port with connection to an extracorporeal pump (20).

into the infusion device. After 24 hours, the infusion device is carefully flushed with saline. High-pressure irrigation of occluded catheters should be avoided but may be performed by experienced clinicians, if chemical lysis is not effective. An angiogram is necessary after this maneuver because catheter rupture or displacement is possible.

As described earlier, the position of the intravenous catheter tip has a direct influence on the complication rate. The risk for thrombosis is increased if the catheter is inserted from the left side and if the tip is positioned above the T3-level (4,17). We place the tip below T3 but try not to advance the tip into the right atrium because the catheter may cause cardiac arrhythmias in this position.

As Compton and Raaf mentioned, 1 mg prophylactic oral warfarin per day significantly reduces the rate of venous thrombosis (18). In addition, low-molecular-weight heparin (LMWH) significantly decreased the rate of thrombotic events in one randomized trial from 62 to 6% (19). This trial was stopped early because of the unexpectedly high difference between the groups. Currently, an international randomized double-blind trial is ongoing, in which the value of LMWH is again being investigated after implantation of long-term infusion devices.

The causes of venous thrombosis and the various treatment options have been described in Chapter 21. In asymptomatic patients, the removal of the device is usually sufficient. However, progression of the thrombosis must be avoided and symptomatic thromboses need to be treated. We recommend the same treatment as for noncatheter-related deep vein thrombosis. Whether anticoagulation should be achieved by coumadin or by LMWH has not been investigated and is therefore at the preference of the treating physician. A regional lysis therapy can be added (e.g., see earlier) but the decision to use it should be made case by case. In addition, the catheter usually has to be removed.

We have recently evaluated our experience with 180 arterial catheters (Table 1) (20). Because of the constant flow, infusion pumps (e.g., Infusaid) have a lower com-

Table 2
Recommendations for Follow-Up of Port Catheters That Are Not in Use (From Three European Centers)[a]: Practice at Frankfurt, Germany

	Postoperative year				
	1	2	3	4	5
Office visit	12	12	12	12	12

[a] Tunneled long-term infusion devices such as Perm-Caths™, Groshong™, or Hickman™ catheters are very rarely used in these institutions. If so, they are removed as early as possible. For this reason, no follow-up scheme is available, and these patients would be followed-up on an individual basis (certainly more often than patients with ports).

Note: No X-rays, blood tests, or other examination are ordered unless clinical findings warrant them.

Unused ports are usually not left in place for more than 1 year. However, if the patient does not want the catheter to be removed, this follow-up scheme is used.

At each visit, the skin is examined for infection and the patient is interviewed for any complaints, such as pain, fever, swelling, etc. The follow-up examinations are performed in the oncology outpatient clinic, usually during a follow-up examination for the underlying disease.

After flushing the catheter with 20 mL of saline, 10 mL of heparin is instilled and the infusion set is removed.

plication rate than arterial ports and enable a longer infusional treatment. In accordance with the recommendations for the use of intravenous infusion devices, any intraarterial infusion system should be flushed regularly with heparin. Implanted pumps are filled with heparinized saline during treatment pauses. An angiogram via the infusion device should be performed to investigate complaints of the patient and at 3-month intervals to detect catheter migration early and avoid misperfusion. Whether the routine use of LMWH may reduce the rate of arterial thrombosis has not been investigated yet.

SUMMARY

Long-term infusion devices have revolutionized the management of patients with chronic infusional treatment, by enabling outpatient treatment and providing continuous vascular access for both blood sampling and infusions, but device survival is limited by technical complications and device failure.

In summary, asymptomatic patients do not require a particular treatment or follow-up after implantation of long-term infusion devices if they are being actively used. A summary of the follow-up protocols of the three contributing European centers for patients with implanted devices that are not being used is given in Tables 2–4. However, certain precautions and prophylactic measures (sterile handling, regular flushing) should be performed to avoid complications and extend the catheter survival. If complications occur, sufficient treatment must be started early. The majority of recommen-

Table 3
Recommendations for Follow-Up of Port Catheters That Are Not in Use (From Three European Centers)[a]: Practice at Heidelberg, Germany

	Postoperative year				
	1	2	3	4	5
Office visit	6	6	6	6	6

[a] Tunneled long-term infusion devices such as PermCaths™, Groshong™, or Hickman™ catheters are very rarely used in these institutions. If so, they are removed as early as possible. For this reason, no follow-up scheme is available, and these patients would be followed-up on an individual basis (certainly more often than patients with ports).

Note: No X-rays, blood tests, or other examination are ordered unless clinical findings warrant them.

Unused ports are usually not left in place for more than a year. However, if the patient does not want the catheter to be removed, this follow-up scheme is used.

At each visit, the skin is examined for infection and the patient is interviewed for any complaints, such as pain, fever, swelling, etc. The follow-up examinations are performed in the oncology outpatient clinic, usually during a follow-up examination for the underlying disease.

After flushing the catheter with 20 mL of saline, 10 mL of heparin is instilled and the infusion set is removed.

Table 4
Recommendations for Follow-Up of Port Catheters That Are Not in Use (From Three European Centers)[a]: Practice at Zurich, Switzerland

	Postoperative year				
	1	2	3	4	5
Office Visit	4	4	4	4	4

[a] Tunneled long-term infusion devices such as PermCaths™, Groshong™, or Hickman™ catheters are very rarely used in these institutions. If so, they are removed as early as possible. For this reason, no follow-up scheme is available, and these patients would be followed-up on an individual basis (certainly more often than patients with ports).

Note: No X-rays, blood tests, or other examination are ordered unless clinical findings warrant them.

(Continued)

**Table 4 (Continued)
Recommendations for Follow-Up
of Port Catheters That Are Not in Use
(From Three European Centers)[a]:
Practice at Zurich, Switzerland**

Unused ports are usually not left in place for more than a year. However, if the patient does not want the catheter to be removed, this follow-up scheme is used.

At each visit, the skin is examined for infection and the patient is interviewed for any complaints, such as pain, fever, swelling, etc. The follow-up examinations are performed in the oncology outpatient clinic, usually during a follow-up examination for the underlying disease.

After flushing the catheter with 20 mL of saline, 10 mL of heparin is instilled and the infusion set is removed.

dations for prophylaxis and treatment of complications are supported by poor quality evidence. We fully agree with Compton and Raaf about questions that need to be addressed in future trials. Because more than 500,000 device implantations per year are carried out in the United States alone, these trials are important and appear to be feasible *(1)*. For those interested in further information, we believe the resources cited by Compton and Raaf are excellent. Additional resources are listed in the Appendix.

APPENDIX: RESOURCES WITH ADDITIONAL INFORMATION ABOUT VASCULAR ACCESS DEVICES

Deutsche Krebsgesellschaft e. V.
Hanauer Landstraße 194
60314 Frankfurt am Main
Telephone: +49 (0) 69-63-00-96-0
Fax: +49 (0) 69-63-00-96-66
E-mail: http://www.krebsgesellschaft.de

Krebsinformationsdienst KID
Deutsche Krebsforschungszentrum
Im Neuenheimer Feld 280
Telephone: +49 (0) 62-21-41-01-21
Fax: +49 (0) 62-21-40-18-06
E-mail: www.krebsinformation.de

Pharmacia AG
Lagerstrasse 14
8600 Dubendorf
Switzerland
Telephone: +49 (1) 802-82-00
Fax: +49 (1) 802-82-99
E-mail: http://www.pharmacia.ch

Fresenius Kabi Deutschland GmbH
Else-Kroner-Strasse 1
61352 Bad Homburg
Germany
Telephone: +49 (0) 6172 – 68 60
Fax: +49 (0) 6172 – 68-62
E-mail: www.fresenius-kabi.de

REFERENCES

1. Freytes CO. Vascular access devices problems revisited: the multinational association of supportive care in cancer (MASCC) experience. Support Care Cancer 1998;6:13–19.
2. Kock HJ, Pietsch M, Krause U, Wilke H, Eigler FW. Implantable vascular access systems: experience in 1500 patients with totally implanted central venous port systems. World J Surg 1998;22:12–16.
3. DiCarlo I, Cordio S, La Greca G, et al. Totally implantable venous access devices implanted surgically. Arch Surg 2001;136:1050–1053.
4. Eastridge BJ, Lefor AT. Complications of indwelling venous access devices in cancer patients. J Clin Oncol 1995;13:233–238.
5. Cohn DE, Mutch DG, Rader JS, Farrell M, Awantang R, Herzog TJ. Factors predicting subcutaneous implanted central venous port function: the relationship between catheter tip location and port failure in patients with gynecologic malignancies. Gynecol Oncol 2001;83:533–536.
6. Ruesch S, Walder B, Tramer MR. Complications of central venous catheters: internal jugular versus subclavian access—a systematic review. Crit Care Med 2002;30:454–460.
7. Watkins E, Khazei AM, Nahra KS. Surgical basis for arterial infusion chemotherapy of disseminated carcinoma of the liver. Surg Gynecol Obstet 1970;130:581–605.
8. Wacker F, Boese-Landgraf J, Wagner A, Albrecht D, Wolf K, Fobbe F. Minimally invasive catheter implantation for regional chemotherapy of the liver: a new percutaneous transsubclavian approach. Cardiovasc Intervent Radiol 1997;20:128–132.
9. Yoshikawa M, Ebara M, Nakano T, Minoyama A, Sugiura N, Ohto M. Percutaneous transaxillary catheter insertion for hepatic artery infusion chemotherapy. Am J Roentgenol 1992;158:885–886.
10. Lorenz M, Hottenrott C, Seufert RM, Encke A. Long-term experience with totally subcutaneously implanted infusion chambers used as permanent central venous access. Langenbeck's Arch Surg 1988;373:302–309.
11. Ballarini C, Intra M, Pisani Ceretti A, et al. Complications of subcutaneous infusion port in the general oncology population. Oncology 1999;56:97–102.
12. Al-Sibai MB, Harder EJ, Faskin RW, Johnson GW, Padmos MA. The value of prophylactic antibiotics during the insertion of long-term indwelling silastic right atrial catheters in cancer patients. Cancer 1987;60:1891–1895.
13. Lim SH, Smith MP, Machin SJ, Goldstone AH. A prospective randomized study of prophylactic teicoplanin to prevent early Hickman catheter-related sepsis in patients receiving intense chemotherapy for haematological malignancies. Eur J Haematol 1993;51:10–13.
14. Yerdel MA, Akin EB, Dolalan S, et al. Effect of single-dose prophylactic ampicillin and sulbactam on wound infection after tension-free inguinal hernia repair with polypropylene mesh: the randomized, double-blind, prospective trial. Ann Surg 2001;233:26–33.
15. Henrickson KJ, Axtell RA, Hoover SM, et al. Prevention of central venous catheter-related infections and thrombotic events in immunocompromized children by the use of vancomycin/ciprofloxacin/heparin flush solution: a randomized, multicenter, double-blind trial. J Clin Oncol 2000;18:1269–1278.
16. Longuet P, Douard MCA, G., Molina JM, Benoit C, Leport C. Venous access port-related bacteremia in patients with acquired immunodeficiency syndrome or cancer: the reservoir as a diagnostic and therapeutic tool. CID 2001;32:1776–1783.
17. Gould JR, Carloss HW, Skinner WL. Groshong catheter-associated subclavian venous thrombosis. Am J Med 1993;95:419–423.
18. Bern MM, Lokich JJ, Wallach SR, et al. Very low doses of warfarin can prevent thrombosis in central venous catheters. A randomized prospective trial. Ann Intern Med 1990;112:423–428.

19. Monreal M, Alastrue A, Rull M, et al. Upper extremity deep venous thrombosis in cancer patients with venous access devices—prophylaxis with a low molecular weight heparin (Fragmin). Thromb Haemost 1996;75:251–253.
20. Heinrich S, Petrowsky H, Schwinnen I, et al. Technical complications of continuous intra-arterial chemotherapy with 5-fluorodeoxyuridine and 5-fluorouracil for colorectal liver metastases. Surgery 2003;133:40–48.

22
Osseointegrated Dental Implants

Steven E. Eckert and Sreenivas Koka

INTRODUCTION

Edentulism, the condition where some or all teeth are missing, is treated by the use of dental prostheses. Conventional prostheses are satisfactory for most patients. Unacceptable results are usually the result of unfavorable anatomy. Anatomic improvements are possible through procedures such as the vestibular extension or vestibuloplasty that create a broader area for denture coverage *(1)*. The recognition that this approach provides only moderate improvement in stability while having less positive effects on retention and support of dental prostheses has led to the development of more definitive methods.

Implant-supported dental prostheses are designed to provide better results than with conventional dentures *(2–4)*. Implants are placed beneath the oral mucosa and provide retention, support, and stability for prostheses. Endosseous implants extend into the underlying bone (Fig. 1). Subperiosteal implants do not extend into the bone but instead rest directly on the bone beneath the periosteum *(5–10)*.

Dental implants may exist as "closed systems" where there is no direct communication with the oral cavity or as "open systems" where a portion of the implant passes from bone through mucosa into the oral cavity. Although closed implant systems carry fewer biological risks because they do not communicate with the harsh oral environment, they provide less direct patient benefit owing to a lack of mechanical denture retention. Examples of closed implant systems are the subperiosteal magnet or the hydroxyapatite ridge augmentations described in the 1980s *(11–14)* (Fig. 2). Unfortunately, the soft tissue used to cover the implant is often placed under tension, making it prone to dehiscence. This results in the loss of a portion of the implanted structure in the case of hydroxyapatite implant or the loss of the entire submucosal magnet.

Open implant systems have become the mainstay of modern dental implant practices. Because the implant projects through the soft tissue, the protruding portion of the implant can be used to provide some or all of the retention, support, and stability to the dental prosthesis (Fig. 3). Unfortunately, the mechanical advantages of an open implant system carry with them the biological toll associated with communication with the oral flora *(15–24)*. Because the dental implant passes into the oral cavity, the natural mucosal barrier is lost. Although teeth also pass through the tissue, there is a biological connection between the gum tissue and the tooth that does not exist with the implant. The

From: *The Bionic Human: Health Promotion for People With Implanted Prosthetic Devices*
Edited by: F. E. Johnson and K. S. Virgo © Humana Press Inc., Totowa, NJ

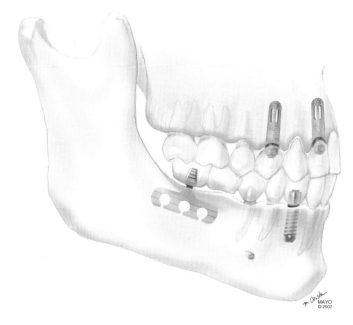

Fig. 1. Endosseous implants are placed within the bone and pass through the soft tissue to support the dental prosthesis.

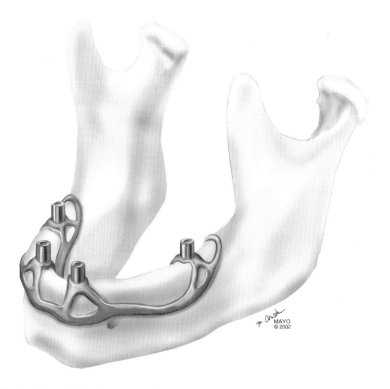

Fig. 2. Subperiosteal implants rest directly on the bone beneath that gum tissue. These implants do not project into the bone itself but do pass through the tissue into the oral cavity to support the prosthesis.

Chapter 22 / Osseointegrated Dental Implants

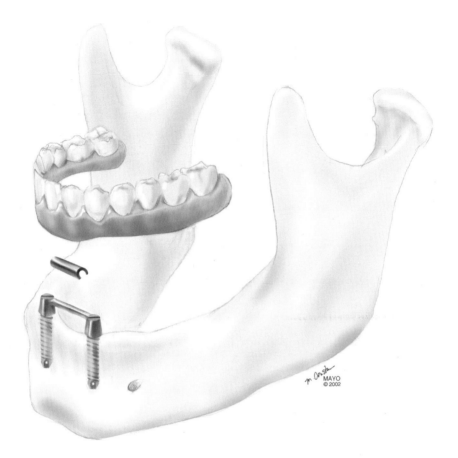

Fig. 3. The implants are connected by a rigid metal bar that allows the denture to be clipped into place. Retention of the denture is supplied by the clip.

implant exposes the underlying tissue to oral bacteria. They can cause infection within the gum tissue or the supporting bone. Although host resistance to infection is variable, modern dental implants are associated with surprisingly few infections.

In the early 1950s, Branemark and colleagues *(25–29)*, a Swedish orthopedic group, –began investigation of wound-healing. They recognized that, under a specific set of circumstances, threaded titanium implants could form an apparent union to underlying bone. The term *osseointegration* was eventually used to describe this apparent union of living bone with alloplastic implant. It was also observed that transmucosal or transdermal placement did not adversely affect the process (Fig. 4) *(27,28,30–35)*.

Early animal experimentation led to human studies that demonstrated similar results *(2,26,29,36)*. Specific steps were followed to achieve osseointegration. The osteotomy site must be meticulously prepared. The bone must be kept below 47°C during the drilling procedure because higher temperatures damage bone and prevent osseointegration. A sterile implant with an acceptable oxide surface layer is critical for bone adaptation. It is also important for osseointegration that the implant be immediately mechanically stable. To achieve this, a threaded implant design is used to engage the surrounding bone. In order to maintain the implant in an undisturbed state, the soft

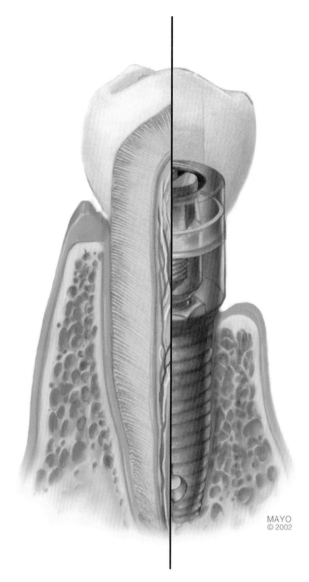

Fig. 4. Endosseous implants are placed into the bone and function like the natural tooth root. Unlike natural teeth, the mucosal connection to the implant is one of close adaptation rather than biological connection.

tissue above the implant is closed primarily. Limiting or eliminating the use of a transitional dental prosthesis further improves the chances for osseointegration.

After it was shown that osseointegration of titanium-threaded implants could be successful, it was determined that other materials, other surfaces, and other designs could also be successful *(37–42)*. Branemark described predictably high long-term survival rates of the osseointegrated implant *(28,29,43,44)*. Careful review of the dental implant literature suggests that some modifications of the Branemark design may have adverse long-term consequences. Failure after functional loading is unacceptably high with cylindrical titanium plasma sprayed implants *(45–51)* and hydroxyapatite-coated implants *(52)*. Unfortunately, long-term studies comparing different implant designs, materials, and surface coatings under a variety of clinical scenarios are currently lacking. This creates the potential for misinterpretation of literature describing various implants. Although there may be specific clinical scenarios that make one design, material, or surface prone to failure, they have not been clearly defined. The clinician is faced with the arduous task of making decisions when some important evidence is not yet available.

Implant complications generally fall into one of two main categories. Biological complications are those in which the body responds adversely to the implant. These include soft-tissue swelling or infection and bone loss. The resulting loss of osseointegration causes loss of the implant. Mechanical complications are those in which the problem is breakage of the implant or the prosthesis that is supported by the implant *(16,53–57)*. Some of these can be repaired but fracture of the implant requires implant removal and replacement.

Clearly, it is possible to create a biological failure through mechanical means. Excessive force can exceed the limits of bone strength with the resulting failure of the implant. Fortunately, this is not common and for this reason it is not discussed in this chapter.

IMPLANT-SUPPORTED PROSTHESES

In the United States, it is estimated that more than 30 million people are missing all of the teeth (edentulism) in at least one jaw. The number of people missing at least one tooth (partial edentulism) approaches 200 million in this country *(58–61)*. Restorations using dental implants may be indicated in many of these patients. Although dental implants have only recently become a mainstream approach to tooth replacement, the number of implants placed in the United States each year now exceeds 500,000. In the future, it is likely that this number will continue to increase *(62–73)*.

Patient demand is strong. Implants offer absolute prosthesis stability while avoiding damage to any of the remaining natural teeth. The modern endosseous implant is a predictable, long-lasting method of securing a dental prosthesis *(43,74–76)*. In most patients, the implants will last for the rest of their lives.

Many different types of dental implants are currently available in the marketplace. Some designs and materials have been tested for more than 35 years, whereas others are relatively new *(see* the Appendix). Experience at the Mayo Clinic has been primarily limited to the use of threaded-titanium implants *(77,78)*. The design of this implant has undergone few fundamental changes since its development in the mid-1960s. The prosthetic components, however, have been modified to create designs for virtually all clinical applications. Use of a standard implant design with modifiable prosthetic components has allowed the clinicians at Mayo to develop a treatment protocol and a maintenance program that is consistent, predictable, and effective.

With a 10-year survival rate of 90–95%, dental implants are quite reliable *(77)*. Perhaps the most severe complication is the failure of an implant to maintain its connection with the bone, resulting in progressive loosening. Often, this is not noticed by the patient and can only be recognized by the clinician. There is usually no pain associated with the early phases of loosening of the implant unless the situation progresses to an acute infection of the soft tissue surrounding the implant. Fortunately, this is very rare.

Some authors have described degeneration of the bone that supports an implant as the result of poor oral hygiene *(79–85)* or excessive force transmission to the implant during chewing or habitual jaw movements *(86,87)*. These problems are uncommon at the Mayo Clinic. It is possible that some implants are more subject to plaque-induced bone loss than are implants of different materials or designs. Problems owing to excessive force have probably been avoided through meticulous management of occlusion during the reconstructive phases of treatment. Designs that incorporate a "weak link"— in the form of a material or a retaining screw that breaks under heavy load—allow the prosthesis to loosen before excessive chewing forces cause bone destruction. We believe that patient education and prosthesis design combine to protect the implant–bone interface from the ravages of excessive forces.

When "saucer-like" bone loss does occur, it is frequently associated with fracture of the implant body, a situation that can only be rectified by removal and replacement of the implant *(88,89)*. Saucer defects in the bone are observed on dental radiographs. Prior to bone loss, the clinician may observe loosening of the dental prosthesis, perhaps owing to the use of a "weak-link" design. Because repeated prosthesis loosening is an indicator of overloading forces that may lead to implant fracture, it is an indication for prompt intervention *(54,90)*. This is one of the reasons follow-up evaluations are provided for implant patients. Patients are seen at specific intervals to evaluate prosthesis movement before bone loss develops.

Therapeutic efforts should address the cause: typically excessive force transmission owing to improper design, inadequate adjustment of the occlusion, or adverse habitual behavior patterns on the part of the patient. Although all can be treated, tooth grinding and/or clenching is particularly difficult to correct *(91–95)*. The dentist is likely to use an approach that addresses the effects of the behavior while accepting the activity itself. Placing a protective guard between the teeth during sleeping hours protects the teeth, jaws, and implants from heavy grinding forces.

Soft–tissue inflammation may be caused by dental plaque or by loosening of the prosthesis *(54,79–85)*. Prosthesis loosening, as discussed previously, requires the intervention of the dentist to tighten the offending loose screw or to replace broken components. Plaque control, however, is a joint effort on the part of the patient and the dental restorative team. Plaque control begins with an understanding of the techniques for plaque removal on the part of the patient. With most implant-supported prostheses, plaque is removed through routine brushing and flossing (Fig. 5). Normally this is the most effective way to remove dental plaque and requires no special techniques or equipment. There are, however, some exceptions. These occur when the implant-supported prosthesis is either very close to the underlying soft tissue or when there is a large amount of space between the prosthesis and the tissue. Both situations differ from what is seen with conventional crowns, bridges, and dentures. The difference occurs because the implant may be positioned differently than the natural tooth roots, causing the dentist to alter the prosthesis design to accommodate the implant position.

In the maxillary anterior region of the oral cavity it is often necessary, for esthetic or functional reasons, to place the prosthesis very close to soft tissue, making the use of dental floss difficult or impossible. In those situations, the use of an oral irrigating device (Water Pic®) is recommended as a supplement to standard tooth brushing. Conversely, in the edentulous mandible there are rarely esthetic or phonetic concerns. The

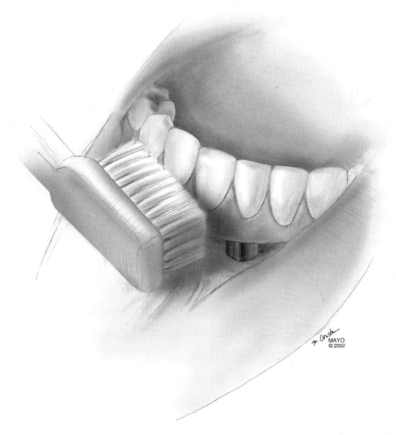

Fig. 5. Toothbrush is used to remove plaque from the implant-supported prosthesis.

lower lip simply does not move enough to expose the under side of the mandibular prosthesis and there are no sounds in the English language that require tongue positions below the incisal edge of the mandibular incisor teeth. Because these concerns do not exist, many clinicians have elected to create space beneath a mandibular prosthesis to allow access for cleaning (Fig. 6). Because the prosthesis does not move on its implant supports, the space poses no problem with food being crushed between the tissue and the prosthesis. Food debris and plaque must, however, be removed from beneath the prosthesis. In many instances, this is accomplished by toothbrush and toothpaste but there are situations where a thick cotton yarn must be threaded beneath the prosthesis and around the implant supporting posts to keep the prosthesis clean (Fig. 7). Interproximal brushes and end-tufted brushes may also be useful (Fig. 8). Because this cleaning technique is somewhat foreign to the patient, instructions are given and reinforced at the early prosthesis re-evaluation appointments.

Despite the best efforts of the patient, there are times when plaque and calculus-tartar accumulate on the dental prostheses. Removal of calculus from an implant-supported prosthesis requires the intervention of a dental hygienist. Because titanium is relatively soft and subject to scratching, and because a scratched surface may be more plaque-retentive, roughening of the titanium surfaces should be avoided. Gold-plated

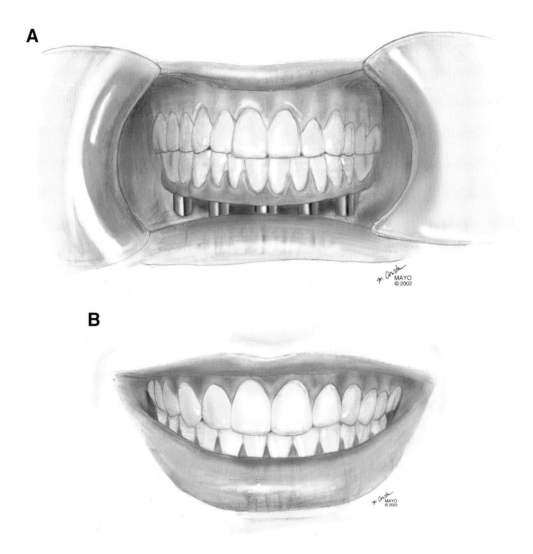

Fig. 6. In the lower jaw, space beneath the implant-supported prosthesis (**A**) causes no cosmetic or speech concerns because the resting lips (**B**) entirely cover the implant.

dental scalers or plastic scalers are used to remove calculus from the titanium components (96). The materials that make up the prosthesis, however, are alloys that may be handled conventionally. Recommendations for routine prophylaxis when commercially pure threaded-titanium implants are used should be individualized. If the patient is routinely removing plaque and avoiding the accumulation of calculus, prophylaxis is performed to remove stain from the prosthesis. For some patients, such prophylactic stain removal may be needed only rarely. Other patients have difficulty in removing plaque and must, therefore, be seen more frequently for prophylaxis.

Routine checks of the patient who uses an implant-supported prosthesis should be performed. In those patients who experience few problems with plaque control, these checkup appointments may be required annually. More frequent evaluations are needed for patients with a propensity for prosthesis loosening and for those who are unable to

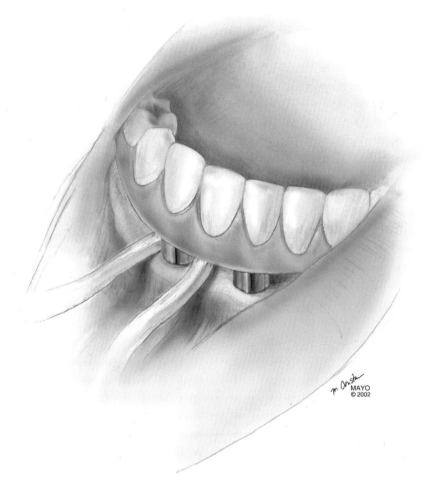

Fig. 7. Cotton yarn is used to remove plaque from the implants.

remove dental plaque well. Fortunately, the complication rate for implant–supported prostheses is normally well below 5% per year *(16,57,97–101)*. Because the complications are most frequently mechanical in nature, rather than biological, patients can assist in evaluating the need for office visits by periodically checking their prostheses for stability *(54,77)*.

TYPICAL CLINICAL RETURN VISIT

Once the dental prosthesis is fabricated, the patient requires follow-up evaluations. At each of these visits, certain factors must be checked to ensure the continued successful function of the implant–supported prostheses. These are as follows:

- Adequacy of occlusion
- Stability of prosthesis
- Health of soft tissue
- Bone health
- Appearance of prosthesis

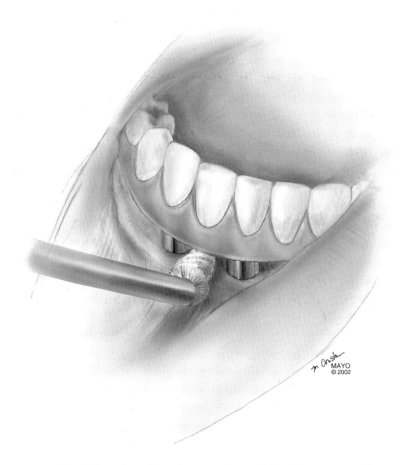

Fig. 8. Interdental brushes may be used for plaque removal.

Evaluation of the occlusion is both subjective and objective. The patient provides the subjective evaluation by responding to questions related to comfort and evenness of the bite. In order to objectively confirm the patient's perception of a comfortable occlusion, the dentist places very thin paper-like materials between the teeth to confirm that the implant-supported prosthesis is holding the material as natural teeth do. It is very important that the implant-supported prosthesis contact no more forcibly than the natural teeth. Following confirmation of an adequate bite, the prosthesis is checked for mobility. This is a relatively simple evaluation in which the dentist tries to move the prosthesis. This can be done with simple finger pressure or the dentist may use instruments to try to "wiggle" it while carefully observing for signs of movement.

The overall appearance of the prosthesis is evaluated. Material fracture or discoloration may indicate the need for repair or remaking of the prosthesis. The soft tissue surrounding an implant-supported prosthesis should be healthy. It should be pink and firm to the touch. It should not bleed easily with light pressure. There should be no signs of infection such as swelling, tenderness, redness, or drainage. The implant-supported prosthesis and the remaining natural teeth should be evaluated for dental plaque and calculus. If hard deposits of calculus are present the patient will require a dental pro-

Table 1
Patient Surveillance After Placement of Implant-Supported Prosthesis at Mayo Clinic

	Postoperative year					
	1	2	3	4	5	10
Office visit	4[a]	1	1	1	1	1
Radiographs[b]	1	—	—	—	—	—

[a] At 1, 3, 6, and 12 months following insertion of the prosthesis. This includes patient education, evaluation of home care, and dental prophylaxis.
[b] Radiographs are not routinely made after the first year of clinical service unless there are clinical signs and symptoms of deterioration.

phylaxis while the presence of plaque alone suggests a need for improved home care. Bone health is evaluated in a number of ways. Dental radiographs are made at the time of implant prosthesis insertion and at the first annual recheck appointment but are not needed at every annual recheck from that point forward unless clinical signs and symptoms warrant them. The dentist must use sound judgment regarding the making of new radiographs. The general rule is that in the absence of soft tissue changes and in the absence of repeated prosthesis loosening, radiographs should be made sparingly.

Ongoing maintenance of an implant-supported prosthesis is a cooperative effort on the part of the patient and the dental professional team. The patient should be diligent in evaluating the comfort and appearance of the prosthesis and the surrounding tissue. If changes occur, the dentist should be notified and corrective measures taken. The dentist must evaluate the prosthesis at intervals, not to exceed 1 year, to ensure long-term success. There may be times when the implant-supported prosthesis is removed simply to confirm that all of the components of the prosthesis are in good working order. Because these prostheses use mechanical connections to the implant, these connections can deteriorate and require attention. With experience, the dentist can determine when prophylactic removal of the implant-supported prosthesis is indicated.

With proper attention, dental implant-supported restorations should provide reliable, long-lasting service. The care described in this chapter, combined with diligent home care on the part of the patient, help to ensure patient satisfaction.

APPENDIX: PARTIAL LIST OF IMPLANT MANUFACTURERS AND SUPPORT GROUPS[a]

Name	Type of group	Web address	Phone no.
American Dental Association	Professional organization	www.ada.org	312-440-2500
Academy of Osseointegration	Professional organization	www.osseo.org	847-439-1919
American College of Prosthodontists	Professional organization		312-573-1260
American Academy of Periodontology	Professional organization	www.perio.org	1-800-282-4867

American Academy of Implant Dentistry	Professional organization	www.aaid-implant.org	312-335-1550
American Academy of Oral and Maxillofacial Surgeons	Professional organization	www.aaoms.org	847-678-6200
International Congress of Oral Implantologists	Professional organization	www.dentalimplants.com	800-442-0525
Nobel Biocare	Implant manufacturing company	www.nobelbiocare.com	800-322-5001
Straumann	Implant manufacturing company	www.straumann.com/home	800-524-6752
Dental Implant Support Group	Support group	www.groups.yahoo.com/group/implantsupport/	
"The Dental Implant Support Group" (United Kingdom)	Support group	www.dental-implant-support-group.org.uk	

[a] Partial list of implant manufacturers and support groups with URL and telephone numbers is supplied. Due to the limited number of adverse long-term complications, support groups for implant patients are relatively rare.

REFERENCES

1. Froschl T, Kerscher A. The optimal vestibuloplasty in preprosthetic surgery of the mandible. J Craniomaxillofac Surg 1997;25:85–90.
2. Laney WR, Tolman DE, Keller EE, Desjardins RP, Van Roekel NB, Branemark PI. Dental implants: tissue-integrated prosthesis utilizing the osseointegration concept. Mayo Clin Proc 1986;61:91–97.
3. Laney WR. Selecting edentulous patients for tissue-integrated prostheses. Int J Oral Maxillofac Implants 1986;1:129–138.
4. Eckert SE, Laney WR. Patient evaluation and prosthodontic treatment planning for osseointegrated implants. Dent Clin North Am 1989;33:599–618.
5. Benson D. Tissue considerations for subperiosteal implants. Implantologist 1978;1:59–69.
6. Bodine RL, Yanase RT, Bodine A. Forty years of experience with subperiosteal implant dentures in 41 edentulous patients. J Prosthet Dent 1996;75:33–44.
7. Garefis PN. Complete mandibular subperiosteal implants for edentulous mandibles. J Prosthet Dent 1978;39:670–677.
8. Homoly P. Subperiosteal implants fill niche when ridge is too thin. Dentist 1989;67:25.
9. James RA, Lozada JL, Truitt PH, Foust BE, Jovanovic SA. Subperiosteal implants. CDA Journal 1988;16:10–14.
10. Linkow LI. Titanium subperiosteal implants. J Oral Implantol 1989;15:29–33, 36–40.
11. Kent JN, Homsy CA, Gross BD, Hinds EC. Pilot studies of a porous implant in dentistry and oral surgery. J Oral Surg 1972;30:608–615.
12. Leak DL, Kent JN, LaVelle W. New method for alveolar ridge augmentation. J Calif Dent Assoc 1973;1:50–52.
13. Quinn JH, Kent JN, Hunter RG, Schaffer CM. Preservation of the alveolar ridge with hydroxylapatite tooth root substitutes. J Am Dent Assoc 1985;110:189–193.
14. Jennings DE. Treatment of the mandibular compromised ridge: a literature review. J Prosthet Dent 1989;61:575–579.
15. Ellen RP. Microbial colonization of the peri-implant environment and its relevance to long-term success of osseointegrated implants. Int J Prosthodont 1998;11:433–441.
16. Esposito M, Hirsch JM, Lekholm U, Thomsen P. Biological factors contributing to failures of osseointegrated oral implants. (II). Etiopathogenesis. Eur J Oral Sci 1998;106:721–764.
17. Quirynen M, Listgarten MA. Distribution of bacterial morphotypes around natural teeth and titanium implants ad modum Branemark. Clin Oral Implants Res 1990;1:8–12.

18. Alcoforado GA, Rams TE, Feik D, Slots J. Microbial aspects of failing osseointegrated dental implants in humans. J Parodontol 1991;10:11–18.
19. Bauman GR, Mills M, Rapley JW, Hallmon WW. Plaque-induced inflammation around implants. Int J Oral Maxillofac Implants 1992;7:330–337.
20. Edgerton M, Lo SE, Scannapieco FA. Experimental salivary pellicles formed on titanium surfaces mediate adhesion of streptococci. Int J Oral Maxillofac Implants 1996;11:443–449.
21. Koka S, Razzoog ME, Bloem TJ, Syed S. Microbial colonization of dental implants in partially edentulous subjects. J Prosthet Dent 1993;70:141–144.
22. Leonhardt A, Adolfsson B, Lekholm U, Wikstrom M, Dahlen G. A longitudinal microbiological study on osseointegrated titanium implants in partially edentulous patients. Clin Oral Implants Res 1993;4:113–120.
23. Rams TE, Roberts TW, Feik D, Molzan AK, Slots J. Clinical and microbiological findings on newly inserted hydroxyapatite-coated and pure titanium human dental implants. Clin Oral Implants Res 1991;2:121–127.
24. Apse P, Ellen RP, Overall CM, Zarb GA. Microbiota and crevicular fluid collagenase activity in the osseointegrated dental implant sulcus: a comparison of sites in edentulous and partially edentulous patients. J Periodontal Res 1989;24:96–105.
25. Branemark PI, Adell R, Breine U, Hansson BO, Lindstrom J, Ohlsson A. Intra-osseous anchorage of dental prostheses. I. Experimental studies. Scand J Plast Reconstr Surg 1969; 3:81–100.
26. Adell R, Hansson BO, Branemark PI, Breine U. Intra-osseous anchorage of dental prostheses. II. Review of clinical approaches. Scand J Plast Reconstr Surg 1970;4:19–34.
27. Branemark PI, Lindstrom J, Hallen O, Breine U, Jeppson PH, Ohman A. Reconstruction of the defective mandible. Scand J Plast Reconstr Surg 1975;9:116–128.
28. Branemark PI, Hansson BO, Adell R, et al. Osseointegrated implants in the treatment of the edentulous jaw. Experience from a 10-year period. Scand J Plast Reconstr Surg Supplementum 1977;16:1–132.
29. Adell R, Lekholm U, Rockler B, Branemark PI. A 15-year study of osseointegrated implants in the treatment of the edentulous jaw. Int J Oral Surg 1981;10:387–416.
30. Albrektsson T, Branemark PI, Hansson HA, Lindstrom J. Osseointegrated titanium implants. Requirements for ensuring a long-lasting, direct bone-to-implant anchorage in man. Acta Orthop Scand 1981;52:155–170.
31. Tjellstrom A, Lindstrom J, Hallen O, Albrektsson T, Branemark PI. Osseointegrated titanium implants in the temporal bone. A clinical study on bone-anchored hearing aids. Am J Otol 1981;2:304–310.
32. Tjellstrom A, Rosenhall U, Lindstrom J, Hallen O, Albrektsson T, Branemark PI. Five-year experience with skin-penetrating bone-anchored implants in the temporal bone. Acta Otolaryngol 1983;95:568–575.
33. Tjellstrom A, Yontchev E, Lindstrom J, Branemark PI. Five years' experience with bone-anchored auricular prostheses. Otolaryngology - Head Neck Surg 1985;93:366–372.
34. Adell R, Lekholm U, Rockler B, et al. Marginal tissue reactions at osseointegrated titanium fixtures (I). A 3-year longitudinal prospective study. Int J Oral Maxillofac Surg 1986; 15:39–52.
35. Parel SM, Holt GR, Branemark PI, Tjellstrom A. Osseointegration and facial prosthetics. Int J Oral Maxillofac Implants 1986;1:27–29.
36. van Steenberghe D, Branemark PI, Quirynen M, De Mars G, Naert I. The rehabilitation of oral defects by osseointegrated implants. J Clin Periodontol 1991;18:488–493.
37. Gammage DD, Bowman AE, Meffert RM, Cassingham RJ, Davenport WA. Histologic and scanning electron micrographic comparison of the osseous interface in loaded IMZ and Integral implants. Int J Periodontics Restorative Dent 1990;10:124–135.
38. Block MS, Gardiner D, Kent JN, Misiek DJ, Finger IM, Guerra L. Hydroxyapatite-coated cylindrical implants in the posterior mandible: 10-year observations. Int J Oral Maxillofac Implants 1996;11:626–633.

39. Gammage DD, Bowman AE, Meffert RM, Cassingham RJ, Davenport WA. Histologic and scanning electron micrographic comparison of the osseous interface in loaded IMZ and Integral implants. Int J Periodontics Restorative Dent 1990;10:124–135.
40. Block MS, Finger IM, Fontenot MG, Kent JN. Loaded hydroxylapatite-coated and grit-blasted titanium implants in dogs. Int J Oral Maxillofac Implants 1989;4:219–25.
41. Kirsch A. The two-phase implantation method using IMZ intramobile cylinder implants. J Oral Implantol 1983;11:197–210.
42. Kirsch A. Plasma-sprayed titanium-I.M.Z. implant. J Oral Implantol 1986;12:494–497.
43. Adell R, Eriksson B, Lekholm U, Branemark PI, Jemt T. Long-term follow-up study of osseointegrated implants in the treatment of totally edentulous jaws. Inter J Oral Maxillofac Implants 1990;5:347–359.
44. Branemark PI, Adell R, Albrektsson T, Lekholm U, Lundkvist S, Rockler B. Osseointegrated titanium fixtures in the treatment of edentulousness. Biomaterials 1983;4:25–28.
45. el Askary AS, Meffert RM, Griffin T. Why do dental implants fail? Part I. Implant Dent 1999;8:173–185.
46. el Askary AS, Meffert RM, Griffin T. Why do dental implants fail? Part II. Implant Dent 1999;8:265–277.
47. Gammage DD, Bowman AE, Meffert RM. Clinical management of failing dental implants: four case reports. J Oral Implantol 1989;15:124–131.
48. Meffert RM. Repairing the ailing implant. J Gt Houst Dent Soc 1991;63:3–4.
49. Meffert RM. How to treat ailing and failing implants. Implant Dent 1992;1:25–33.
50. Meffert RM. Treatment of the ailing, failing implant. J Calif Dent Assoc 1992;20:42–45.
51. Meffert RM. What is peri-implantitis and how do we prevent and treat it? J Mich Dent Assoc 1992;74:32–3, 36–39.
52. Johnson BW. HA-coated dental implants: long-term consequences. J Calif Dent Assoc 1992;20:33–41.
53. Esposito M, Hirsch JM, Lekholm U, Thomsen P. Biological factors contributing to failures of osseointegrated oral implants. (I). Success criteria and epidemiology. Euro J Oral Sci 1998;106:527–551.
54. Tolman DE, Laney WR. Tissue-integrated prosthesis complications. Int J Oral Maxillofac Implants 1992;7:477–484.
55. Goodacre CJ, Kan JY, Rungcharassaeng K. Clinical complications of osseointegrated implants. J Prosthet Dent 1999;81:537–552.
56. McGlumphy E, Larsen P, Peterson L. Etiology of implant complications: anecdotal reports vs. prospective clinical trials. Compendium 1993;Suppl:544–548.
57. Balshi TJ. Opportunity to prevent or resolve implant complications. Implant Soc 1990;1:7–9, 15.
58. Marcus SE, Drury TF, Brown LJ, Zion GR. Tooth retention and tooth loss in the permanent dentition of adults: United States, 1988–1991. J Dent Res 1996;75:684–695.
59. MacEntee MI. The prevalence of edentulism and diseases related to dentures—a literature review. J Oral Rehabil 1985;12:195–207.
60. Douglass CW. Implications of demographic and dental disease changes for the financing of geriatric dental services. Health Matrix 1988;6:13–19.
61. Douglass CW, Gammon MD, Atwood DA. Need and effective demand for prosthodontic treatment. A report: Part one. Oral Health 1988;78:11–7, 21–23.
62. Landesman HM. Dental implants in the predoctoral curriculum. J Calif Dent Assoc 1992;20:58–59.
63. Schnitman PA. Dental implants. State of the art, state of the science. Int J Technol Assess Health Care 1990;6:528–544.
64. Stillman N, Douglass CW. The developing market for dental implants. J Am Dent Assoc 1993;124:51–56.

65. Arbree NS, Chapman RJ. Implant education programs in North American dental schools. J Dent Educ 1991;55:378–380.
66. Bavitz JB. Dental implantology in U.S. dental schools. J Dent Educ 1990;54:205–206.
67. Bell FA, Hendricson WD. A problem-based course in dental implantology. J Dent Educ 1993;57:687–695.
68. Berge TI. Public awareness, information sources and evaluation of oral implant treatment in Norway. Clin Oral Implants Res 2000;11:401–408.
69. Henry PJ. Educational perspectives in implant prosthodontics. Aust Prosthodont J 1993;7: 51–55.
70. Judy KW. Dental implants: the need for expanded educational commitments. N Y State Dent J 1986;52:7–8.
71. Payant L, Williams JE, Zwemer JD. Survey of dental implant practice. J Oral Implantol 1994;20:50–58.
72. Schnitman PA. Education in implant dentistry. J Am Dent Assoc 1990;121:330, 332.
73. Simons AM, Bell FA, Beirne OR, McGlumphy EA. Undergraduate education in implant dentistry. Implant Dent 1995;4:40–43.
74. Zarb GA, Lewis DW. Dental implants and decision making. J Dent Educ 1992;56:863–872.
75. Naert I, Koutsikakis G, Duyck J, Quirynen M, Jacobs R, van Steenberghe D. Biologic outcome of single-implant restorations as tooth replacements: a long-term follow-up study. Clin Implant Dent Relat Res 2000;2:209–218.
76. Naert IE, Duyck JA, Hosny MM, Quirynen M, van Steenberghe D. Freestanding and tooth-implant connected prostheses in the treatment of partially edentulous patients Part II: An up to 15-years radiographic evaluation. Clin Oral Implants Res 2001;12:245–251.
77. Eckert SE, Wollan PC. Retrospective review of 1170 endosseous implants placed in partially edentulous jaws. J Prosthet Dent 1998;79:415–421.
78. Keller EE, Desjardins RP, Tolman DE, Laney WR, Van Roekel NB. Reconstruction of the severely resorbed mandibular ridge using the tissue-integrated prosthesis. Int J Oral Maxillofac Implants 1986;1:101–109.
79. Lindquist LW, Carlsson GE, Jemt T. Association between marginal bone loss around osseointegrated mandibular implants and smoking habits: a 10-year follow-up study. J Dent Res 1997;76:1667–1674.
80. Soehren SE. Similarities between the development and treatment of plaque-induced peri-implantitis and periodontitis. J Mich Dent Assoc 1996;78:32–36.
81. Ciancio SG, Lauciello F, Shibly O, Vitello M, Mather M. The effect of an antiseptic mouth rinse on implant maintenance: plaque and peri-implant gingival tissues. J Periodontol 1995;66:962–965.
82. Gatewood RR, Cobb CM, Killoy WJ. Microbial colonization on natural tooth structure compared with smooth and plasma-sprayed dental implant surfaces. Clin Oral Implants Res 1993;4:53–64.
83. George K, Zafiropoulos GG, Murat Y, Hubertus S, Nisengard RJ. Clinical and microbiological status of osseointegrated implants. J Periodontol 1994;65:766–770.
84. Pontoriero R, Tonelli MP, Carnevale G, Mombelli A, Nyman SR, Lang NP. Experimentally induced peri-implant mucositis. A clinical study in humans. Clin Oral Implants Res 1994;5:254–259.
85. Schou S, Holmstrup P, Hjorting Hansen E, Lang NP. Plaque-induced marginal tissue reactions of osseointegrated oral implants: a review of the literature. Clin Oral Implants Res 1992;3:149–161.
86. Isidor F. Loss of osseointegration caused by occlusal load of oral implants. A clinical and radiographic study in monkeys. Clin Oral Implants Res 1996;7:143–152.
87. Miyata T, Kobayashi Y, Araki H, Motomura Y, Shin K. The influence of controlled occlusal overload on peri-implant tissue: a histologic study in monkeys. Int J Oral Maxillofac Implants 1998;13:677–683.

88. Rangert B, Krogh PH, Langer B, Van Roekel N. Bending overload and implant fracture: a retrospective clinical analysis. Int J Oral Maxillofac Implants 1995;10:326–334.
89. Eckert SE, Meraw SJ, Cal E, Ow RK. Analysis of incidence and associated factors with fractured implants: a retrospective study. Int J Oral Maxillofac Implants 2000;15:662–667.
90. Parein AM, Eckert SE, Wollan PC, Keller EE. Implant reconstruction in the posterior mandible: a long-term retrospective study. J Prosthet Dent 1997;78:34–42.
91. Clark GT, Beemstervoer P, Rugh JD. The treatment of nocturnal bruxism using contingent EMG feedback with an arousal task. Behav Res Ther 1981;19:451–455.
92. Etzel KR, Stockstill JW, Rugh JD, Fisher JG. Tryptophan supplementation for nocturnal bruxism: report of negative results. J Craniomandib Disord 1991;5:115–120.
93. Solberg WK, Clark GT, Rugh JD. Nocturnal electromyographic evaluation of bruxism patients undergoing short term splint therapy. J Oral Rehabil 1975;2:215–223.
94. Rugh JD, Johnson RW. Temporal analysis of nocturnal bruxism during EMG feedback. J Periodontol 1981;52:263–265.
95. Rugh JD. Psychological stress in orofacial neuromuscular problems. Int Dent J 1981;31: 202–205.
96. Rapley JW, Swan RH, Hallmon WW, Mills MP. The surface characteristics produced by various oral hygiene instruments and materials on titanium implant abutments. Int J Oral Maxillofac Implants 1990;5:47–52.
97. Berman CL. Curbing implant complications. J Am Dent Assoc 1998;129:1666.
98. McGlumphy EA, Larsen PE, Peterson LJ. Etiology of implant complications: anecdotal reports vs. prospective clinical trials. Compendium 1993:S544–8;quiz S565–6.
99. Watson CJ, Tinsley D, Sharma S. Implant complications and failures: the single-tooth restoration. Dent Update 2000;27:35–38, 40, 42.
100. Watson CJ, Tinsley D, Sharma S. Implant complications and failures: the complete overdenture. Dent Update 2001;28:234–238, 240.
101. Esposito M, Hirsch J, Lekholm U, Thomsen P. Differential diagnosis and treatment strategies for biologic complications and failing oral implants: a review of the literature. "Evolution of prosthetic heart valves." Int J Oral Maxillofac Implants 1999;14:473–490.

US Counterpoint to Chapter 22

G. E. Ghali and John N. Kent

Endosseous dental implants are used to support fixed and removable prostheses for totally and partially edentulous patients. Branemark and colleagues demonstrated that titanium screw implants could support fixed prostheses in completely edentulous arches, with 84% implant retention in the maxilla and 93% implant survival in the mandible over 15 years *(1,2)*. They provide a rigid connection of implant to bone, with nearly complete lateral immobility *(3)*. Osseointegrated dental implants have revolutionized the prosthetic rehabilitation of edentulous or partially edentulous patients *(4–6)*. The clinical use of these implants, with modifications through biological and materials research, has expanded markedly over the past two decades. Most academic centers in the United States place endosseous dental implants and the teaching of this reconstructive modality is considered standard in all oral and maxillofacial surgery training programs.

Implant-supported dental prostheses (closed system applications with no direct communication into the oral cavity such as magnets and hydroxyapatite ridge augmentation) are uncommon today because of the success of endosseous dental implants. Subperiosteal implants, which are not endosseous in nature, extend into the oral cavity and are not frequently used. Clinical follow-up parameters of this system should be included in the endosseous open systems because the implant passes through both bone and mucosa. Perhaps the greatest test of any implanted device in the body is the permucosal or percutaneous device, which also penetrates bone. Our clinical observations on these devices show that both biological and biomechanical challenges generate few failures, oftentimes fewer than some nonpenetrating implant devices. Follow-up assesses both biological (soft- and hard-tissue health) and biomechanical (restorative dental status) issues because the submucosal-bony environment is dramatically affected by the patient's own health and hygiene status and the restoration is influenced by the environment of the oral cavity as well as occlusal forces.

Early endosseous implants were bullet-shaped or cylindrical in nature with the implant–bone interface consisting of plasma-sprayed titanium or hydroxyapatite-covered titanium. The Mayo Clinic authors state that failure after functioning is unacceptably high with these devices. We question whether the failure may be secondary to the inherent learning curves associated with placement and restoration. More recent designs feature threaded implants that outperform cylindrical versions. Surgical and restorative techniques have also improved and we now have a fairly clear understanding of the conditions required for a successful implantation. Our clinical experiences at both campuses of the Louisiana State University support these statements. The authors nicely classify implant complications into two categories: biological factors involving the soft and hard tissues and mechanical problems associated with restorative dental prostheses placed on the implant. The long-term incidence of failure in both categories beyond a 10-year follow-up are only now being elucidated.

When looking for failure etiology, the soft-tissue mucosal barrier conditions (e.g., presence or absence of fixed keratinized tissues), location of the implant device within the oral cavity (e.g., anterior vs posterior or maxillary vs mandibular), relationship of the mucosa to the prostheses, and/or systemic health disorders adversely affecting the supporting bone or soft tissues are all areas of concern. Restorative techniques vary widely, according to the preference of the dentist, ranging from free-standing devices connected to the remaining natural teeth to endosseous implants supporting removable or fixed denture prostheses. Prospective studies will help to isolate factors leading to failures in the several categories of implant usage: single implants, multiple free-standing implants, multiple implants connected to natural teeth, implants supporting removable dentures, and implants supporting nonremovable extended bridgework. In the short term, biomechanical failures associated with improper restorative designs and/or adverse occlusal forces can be fairly well controlled to generate highly satisfactory follow-up statistics. The long-term biomechanical failures associated with either restorative material failure and/or uncontrollable forces via the occlusion are only now surfacing after 10 to 15 years. Retrospective analysis of these failures will help define factors necessary for successful long-term (>20 years) retention of restorations on dental implants.

Failure associated with biological conditions such as soft-tissue inflammation from dental plaque, lack of fixed keratinized mucosa, and/or lack of professional hygiene are the most common reasons for implant loosening and failure. Both the surgeon and the restorative dentist should understand the relationship between the soft-tissue and implant-supported prostheses. Aesthetic concerns in certain cases may lead to earlier implant failure; patients need to be aware of this sacrifice for aesthetic gain. Furthermore, implant success rates tend to vary based on location within the oral cavity. Anterior (mandibular as well as maxillary) implants are in more dense bone and have longer survival than posterior implants.

Our current practice of long-term care and follow-up of patients with dental implants includes four postoperative office visits (1 week, 1 month, 6 months, 1 year) during the first year following implantation. Then, based on the patient's clinical status, office visits are tapered to yearly clinical and radiographic evaluations. In addition to these office visits, professional dental hygiene cleaning or prophylaxis is mandatory every six months without time limitation. Radiographic evaluation may include panoramic and/or periapical films. Clinical exams performed on a yearly basis beginning in year two should include: gingival bleeding index, plaque and calculus index, a description and measurement of attached gingiva, and radiographic assessment of vertical bone level (Table 1). Chapter 22 lists assessment of adequacy of occlusion, stability of prosthesis, health of soft tissues, bone health, and appearance of prostheses as evaluation

Table 1
Patient Surveillance After Implantation
of Endosseous Dental Implants at Louisiana State University

	Postoperative year					
	1	2	3	4	5	10
Office visit	4[a]	1	1	1	1	1
Panoramic and/or periapical dental X-rays	2	1	1	1	1	1
Professional dental hygiene cleanings	2	2	2	2	2	2

[a] Number of times the modality is requested during each 1-year time period.

criteria at the time of each office visit. We agree that these are certainly important in the qualitative description of the implant. However, the criteria described here provide more meaningful objective data to support clinical impressions.

The components of the yearly office visit examination (with prostheses removed) include the following:

Gingival bleeding index *(7)*
0 — tissue color normal; no bleeding on probing
1 — tissue color normal to slightly erythematous; no bleeding on probing
2 — tissue color red; bleeding on probing
3 — tissue color markedly red/edematous; bleeds on finger pressure or spontaneously

Plaque and calculus index *(8)*
0 — no plaque/no calculus
1 — plaque can be scraped off but is not visible to the clinician, or supragingival calculus extends no more than 1 mm below free gingival margin
2 — visible plaque within gingival crevice or on the tooth and gingival margin, or subgingival calculus extends more than 1 mm into the device, or moderate amounts of supragingival and subgingival calculus
3 — heavy accumulation of plaque within the crevice or on the tooth and gingival margin or heavy accumulation of supra- and subgingival calculus

The attached gingiva should be graded as present or absent on the midlabial and midlingual implant surfaces. The width of the attached gingiva can be determined by measuring (in millimeters) the width of the keratinized gingiva with a periodontal probe for both the tooth and the implant and subtracting the probe depth for that location. The level of the attached gingiva should be noted by using standardized landmarks, including the location of the shoulder of the abutment on the implants and the margin of the crowns or coping on the teeth, and measuring from those landmarks to the free gingival margins. Vertical bone levels should also be noted using custom fabricated film holders; this permits comparisons to the baseline exposure of these radiographs immediately after the implants have been exposed and restored. If possible, the restorative prostheses should be removed and appropriate radiographs taken to determine any changes in vertical bone height.

Reconstructive oral and maxillofacial implant surgery encompasses the use of implants to rehabilitate and restore form and function to the edentulous or partially edentulous jaws using fixed and removable prostheses. They may also be utilized to assist in the stabilization of prostheses that replace missing facial parts such as the nose, eyes, and ears. Oral implant reconstruction enables patients to regain normal mastication, speech, and deglutition; resolves pain, gagging, and dysfunction from conventional removable prostheses. Advances in implant science, biomaterials, and biotechnology, together with a better understanding of the biology of osseointegration about the bone–implant interface, and biomechanics, have resulted in improved outcomes and expanded applications for dental implants.

For those seeking further information, the Appendix provides useful resources.

APPENDIX: SOURCES OF FURTHER INFORMATION ON OSSEOINTEGRATED DENTAL IMPLANTS

Best Web site overall:
Academy of Osseointegration
85 West Algonquin Road
Suite 550
Arlington Heights, IL 60005-4425 USA
Tel: 847-439-1919
Fax: 947-439-1569
Web site: www.osseo.org

Best implant sites (manufacturers):
Nobel Biocare USA, Inc.
22715 Savi Ranch Parkway
Yorba Linda, CA 92887 USA
Tel: 800-993-8100
Fax: 714-998-9236
Web site: www.nobelbiocare.com
3i Implant Innovations
4555 Riverside Drive
Palm Beach Gardens, FL 33410
Tel: 800-342-5454
Fax: 561-776-1272
Web site: www.3i-online.com

Best textbooks:
Block MS, Kent JN, eds. *Endosseous Implants for Maxillofacial Reconstruction.* Philadelphia, PA: W.B. Saunders; 1995.
Block MS, Kent JN, Guerra L, eds. *Implants in Dentistry.* Philadelphia, PA: W.B. Saunders; 1997.
Block MS. *Color Atlas of Dental Implant Surgery.* Philadelphia, PA: W.B. Saunders; 2001.

REFERENCES

1. Adell R, Lekholm U, Rockler B, Branemark PI. A 15 year study of osseointegrated implants in the treatment of the edentulous jaw. Int J Oral Surg 1981;10:387–416.
2. Albrektsson T, Zarb G, Worthington P, Eriksson AR. The long-term efficacy of currently used dental implants: a review and proposed criteria of success. Int J Oral Maxillofacial Implants 1986;1:11–25.
3. Schnitman P. Discussion section. J Oral Implantol 1986;12:460–470.

4. Rangert B, Gunne J, Glantz P-O, Svenson A. Vertical load distribution on a three unit prostheses supported by a natural tooth and a single Branemark implant. Clin Oral Impl Res 1995;6:40–46.
5. Higuchi KW, Folmer T, Kultje C. Implant survival rates in partially edentulous patients: a 3-year prospective multicenter study. J Oral Maxillofac Surg 1995;53:264–268.
6. Naert I, Quirynen M, van Steenberghe D, Darius P. A six-year prosthodontic study of 509 consecutively treated implants for the treatment of partial edentulism. J Prosthet Dent 1992; 67:236–245.
7. McKinney R, Koth D, Steflic D. The single crystal sapphire endosseous dental implant II. Two year results of clinical animal trials. J Oral Implantol 1983;14:619.
8. Gettleman L, Schnitman PA, Kalis P, et al. Clinical evaluation criteria of tooth implant success. J Oral Implantol 1978;8:12–28.

European Counterpoint to Chapter 22

Anthony J. Summerwill and John I. Cawood

INTRODUCTION

The purpose of clinical guidelines is to improve the effectiveness and efficiency of clinical care through the identification of good clinical practice and desired clinical outcomes *(1)*. Although guidelines exist for the assessment of patients and subsequent provision of implant-supported restorations, there is a paucity of published information on long-term follow-up regimens to maintain these restorations. The only published guidelines available in the United Kingdom were produced by a joint working group consisting of members of the British Association of Oral and Maxillofacial Surgery and The British Society for the Study of Prosthetic Dentistry *(2)*.

Currently, endosteal dental implants are the nearest equivalent to replacing the natural tooth root that are available to manage patients with missing teeth as a result of disease, trauma, or developmental abnormalities. Dental implants have been used to replace missing teeth for many years, in many cases with limited success. It was not until the interfacial behavior between commercially pure titanium and bone was reported that the true potential of implant dentistry could be realized. The direct structural and functional connection between ordered, living bone and the surface of a load-carrying implant was described as osseointegration *(3)*. In Chapter 22, Eckert and Koka provide a detailed summary of the development of modern dental implants and the clinical processes to optimize success. We would agree with the authors that, owing to the lack of well-designed long-term studies, the literature is prone to misinterpretation. The statement about the unacceptably high failure rates for cylindrical titanium plasma spray implants exemplifies this. The citations referred to deal predominantly with the already failing implant. Our experiences in Liverpool with the IMZ system do not confirm these findings *(4)*, and this is supported by other authors *(5–7)*. Implants with roughened surfaces generally have a smooth machined collar that is exposed to the oral environment. It is only when a pathological pocket develops around the implant that the roughened surface becomes exposed to potential bacterial colonisation. Roughened surfaces may prevent the natural removal of bacterial plaque and prevent a larger area for colonization. In addition, oral hygiene measures may be less effective than on a smooth surface *(8)*.

The primary role of a dental implant is to support and retain a prosthesis that artificially replaces the missing tooth/teeth. The prosthesis can either be fixed onto the implant using a luting cement or screw, or it can be removable. Maintenance requirements differ between individual patients based largely on the treatment modality that is dependent on the clinician and the ability of the patient to maintain a good standard of oral hygiene.

THE PERI-IMPLANT ENVIRONMENT

As described in Chapter 22, the implant protrudes through the overlying oral soft tissue into the mouth in a similar manner to the tooth that it is replacing. The structure of the soft tissue surrounding the neck of the implant is in many ways similar to the tissues surrounding a natural tooth *(9,10)*. However, unlike the situation found with the natural dentition, implants are not designed to interact with the soft tissues as a tooth interacts with the periodontium. The main difference is the manner in which the peri-implant connective tissues interface with the titanium implant surface *(9)*. As there is no cementum on the surface of an endosseous implant or fiber insertion into its surface, the peri-implant seal is reinforced by a network of collagen fibers. Investigators have demonstrated fibers running parallel to a machined implant surface *(11)* and also a "circular ligament" running parallel around nonsubmerged titanium screws *(12,13)*.

Breakdown of the peri-implant seal can occur. In the 1980s, the term peri-implantitis was introduced to describe a destructive inflammatory process affecting the soft and hard tissues surrounding the implant, leading to the formation of a peri-implant pocket and loss of the surrounding bone *(14)*. A peri-implantitis bony defect usually assumes the shape of a saucer around the implant and is well demarcated *(15)*. In contrast to this, peri-implant mucositis is a self-limiting reversible soft-tissue inflammatory reaction without any loss of supporting bone, usually as a result of a localized irritant such as a loose prosthetic superstructure or plaque accumulation. This localized reaction has all of the characteristics of gingivitis found in dentate individuals *(16)*.

PERI-IMPLANTITIS AND BACTERIAL PLAQUE

The specific role of bacteria in the development of peri-implant infection was observed in the 1980s following the microscopic examination of samples taken from around a range of different implant systems. The findings from these studies suggested that peri-implantitis is a site-specific disease process with a range of microorganisms similar to those encountered in chronic periodontal disease affecting natural teeth *(17–19)*. However, most radiographic reports of peri-implantitis show a defect involving the whole surface of the implant with a lack of site specificity *(20)*.

In successful two-stage implant systems in edentulous subjects, the bacteria colonizing the surface are similar to those found in plaque on nondiseased teeth and, in most cases, this remains stable *(21,22)*. There is, however, a clear relationship between the presence of bacterial plaque on the surface of a dental implant and the development of peri-implant inflammatory changes *(16,23–25)*. However, because the development of infections around implants appears to parallel that around teeth, it is logical to apply the same, or at least similar, clinical parameters when monitoring periodontitis sites and peri-implantitis sites *(26,27)*.

MECHANICAL COMPLICATIONS

The forces of occlusion from either function or parafunction are transmitted through the implant into the underlying bone. Unlike teeth, dental implants have no periodontal ligament. Overload of the implant can result in either mechanical complications of the implant components (28) or marginal bone loss. As part of our pretreatment assessment in Liverpool, a parafunctional history is undertaken and caution is exercized in accepting patients with uncontrolled bruxism (1). In all aspects of advanced prosthodontics, care is taken to control the static and dynamic occlusal interrelationships. We adopt an approach like that at the Mayo Clinic to prevent overload of the supporting tissues by building a safety valve into the design of the superstructure. In doing so, the serious complications of marginal bone loss and catastrophic implant failure are minimized. Extreme care is also exercised to ensure passive fit of the framework. Ill-fitting frameworks and excessive biting forces have been suggested to be the primary cause of implant failure (29). The majority of fractures occur in the posterior region where a bending overload occurs as a result of excessive occlusal forces. The presence of a distal cantilevered extension on the superstructure exacerbates the problem (30,31).

Other mechanical complications include screw loosening, screw fractures, framework/resin/veneer material fractures, implant prosthesis fractures, opposing prosthesis fractures and retention problems with removable overdentures (32). Follow-up is essential to identify mechanical failure and to ascertain the cause of failure. Early treatment is essential to prevent propagation of the complication leading to more serious problems.

IMPLANT-SUPPORTED PROSTHESES IN THE UNITED KINGDOM

It is estimated that approximately 45% of people over age 65 and 5% of the 35 to 64-year age bracket are fully edentulous. The over age 65 segment has lost on average more than 10 of 28 teeth and those in the 55- to 64-year age bracket have lost an average of 9 of 28 teeth (33). There are a variety of predisposing factors that have resulted in an increase in the use of dental implants. These include the following:

- Increasing age of the population.
- Leisure behavior resulting in the loss of teeth.
- Increased preference for the use of implants to replace missing teeth.
- Documented success and predictability of implant-supported restorations.

The UK market is the smallest market in Europe in terms of penetration rate. In 1998, the market volume for dental implants was 17,000 but, by the year 2003, the market totaled 32,600 implants at an annual growth rate above 10% (33).

MAINTENANCE OF DENTAL IMPLANTS

To maintain healthy tissues around oral implants it is important to institute an effective preventive regimen (34). As mentioned earlier in the chapter, maintenance requirements differ among individual patients. In Liverpool we usually review the patient at 1, 3, 6, and 12 months after insertion of the prosthesis (Table 1). Following this, patients are generally reviewed on an annual basis with hygiene appointments at 3 or 6 months, based on individual patient needs. This is similar to the pattern of review adopted by other UK centers (35) and replicates the pattern of review at the Mayo

Table 1
**Patient Surveillance After Placement
of Implant-Supported Prostheses at Liverpool Dental Hospital**

	Postoperative year					
	1	2	3	4	5	10
Office visit	4[a]	1	1	1	1	1
Peri-implant recordings[b]	0	1	1	1	1	1
Radiographs	1	0[c]	0[c]	0[c]	0[c]	0[c]

[a] Visits at 1, 3, 6, and 12 months following insertion. These visits include evaluation of oral hygiene and reinforcement of oral hygiene instruction and also dental prophylaxis with the hygienist.
[b] Includes evaluation of plaque index, probing depth, and bleeding index.
[c] Radiographic examination is not routinely carried out after year 1 without evidence of clinical signs and symptoms.

Clinic. The clinical outcome measures that we look at are broadly the same as those described in Chapter 22, namely:

- prosthetic factors
- soft-tissue factors
- bone factors
- oral hygiene factors

Prosthetic Factors

In Liverpool the majority of maintenance complications are related to problems with the overlying prosthesis *(36)*. A detailed examination of all prosthodontic factors is an integral part of any maintenance program. Examination of the occlusion is carried out by placing articulating paper between the occluding surfaces of the teeth. When the prosthesis is opposed by natural teeth, particular attention is paid to evidence of accelerated wear on the prosthesis that might indicate a parafunctional activity or premature contact. Removable prostheses are checked for retention and stability. The active retentive components are checked for signs of wear and tear and replaced, if necessary. If the prosthesis has combined support from both the implant and the oral soft tissues, it is important to confirm the adaptation of the fitting surface onto the edentulous ridge. Selective resorption of the ridge can potentially increase the load being transmitted through the implant.

We do not routinely remove all fixed restorations during follow-up appointments unless there is a problem that needs addressing. Screw-retained restorations are checked to ensure that the screw remains tight. A brief examination of the soft tissue around the implant can reveal loosening of the abutment and the development of a peri-implant mucositis. If this is the case, the abutment is removed and the implant irrigated with a chlorhexidine wash before reseating. The prosthodontist should be aware that abutment loosening is more likely to occur in conjunction with a parafunctional activity and occlusal contacts should be checked in all excursive pathways. Localized marginal hyperplasia can also occur with ill-fitting cemented restorations. This can be caused by the presence of the marginal gap between the restoration and the abutment that acts as a plaque retentive factor, but it may also be caused by excess cement located in the peri-implant pocket.

Soft-Tissue Factors

The mucosa surrounding the implant should be free of superficial inflammation. The transmucosal part of the implant may either emerge through keratinized mucosa (masticatory mucosa) or, in cases where there has been advanced resorption of the alveolar ridge, through nonkeratinized mucosa. The two are quite different in appearance with the nonkeratinized mucosa appearing red and more mobile. Some studies suggest that implants inserted into nonkeratinized, mobile mucosa are more susceptible to peri-implantitis (37). A clinical investigation on the influence of masticatory mucosa in both edentulous and partially dentate implant subjects failed to support this (38).

Given the importance of bacterial dental plaque as an etiological agent in the development of mucositis and peri-implantitis (39,40) we carry out a modified plaque index (41). We also look at the bleeding response to probing around the implant using a modified sulcus bleeding index (14). The absence of bleeding on probing may be a useful indicator of the stability of peri-implant attachment loss (42).

Absence of mobility is an important factor for continued success of a dental implant. If an implant is clinically visible, this indicates a complete failure of osseointegration but many implants may have some degree of clinical attachment loss without showing obvious signs of mobility. To interpret low degrees of mobility there are several electronic devices available. Presently, these devices have a role in academic research but their benefit in routine clinical follow-up remains equivocal.

Bone Factors

The radiograph remains the most important source of information for determining the amount of peri-implant bone (43). Radiographs are used to check the fit of the prosthesis at the time of insertion and at the 1-year review appointment. We would agree wholeheartedly with the comments of the team from the Mayo Clinic that follow-up radiographs should be used sparingly and only in cases where there is evidence of peri-implant complications diagnosed from a thorough clinical examination. If progressive bone loss is evident, the clinician must judge whether this has been caused by bacterial-induced inflammation or occlusal factors. In many cases it may be multifactorial.

Oral Hygiene Factors

Oral hygiene factors need to be addressed prior to any active implant treatment. If plaque control is inadequate on the remaining dentition or dentures preoperatively, the introduction of a dental implant is ill-advised. If oral hygiene measures are adequate, then most implant-supported prostheses should be maintainable without regular professional intervention, as discussed in Chapter 22. The patient should be able to maintain a state of health equivalent to that which exists around the natural dentition. This can be achieved using conventional toothbrushing techniques supplemented with dental floss in single-tooth restorations. There is currently no evidence to suggest that powered or sonic toothbrushes offer any advantages over manual toothbrushing (34).

As the authors from the Mayo Clinic explain, restorations in the aesthetic zone of the maxilla can present additional difficulties in terms of oral hygiene maintenance.

Where possible, we try to overcome this by sanitary prosthetic design or the use of removable prostheses to prevent soft-tissue encroachment with a fixed restoration. If this is not possible, then mechanical oral hygiene aids such as irrigation devices may be employed. Similarly, if the position of the implant is less than ideal, then compromises may have to be made in the design and contour of the overlying prosthesis. If this leads to the development of stagnation areas and the localized accumulation of dental plaque, modification of the oral hygiene technique may need to be considered on an individual basis.

SUMMARY

The follow-up regimens for dental prostheses supported by endosteal implants are based on accepted prosthodontic and periodontal principles. The focus is on creating a biologically favorable environment by planning, design and patient education to optimize the outcome of treatment. In this regard, our principles of care and maintenance closely resemble those presented in Chapter 22.

REFERENCES

1. Royal College of Surgeons of England, Faculty of Dental Surgery, National Clinical Guidelines 1997. Guidelines for selecting appropriate patients to receive treatment with dental implants: priorities for the NHS.
2. Guidelines on standards for the treatment of patients using endosseous dental implants. Supplement to the British Dental Journal, March 1995.
3. Albrektsson T, Branemark PI, Hansson HA, Lindstom J. Osseointegrated titanium implants. Requirements for ensuring a long-lasting, direct bone-to-implant anchorage in man. Acta Orthop Scand 1981;52:155–170.
4. Chan MF, Johnston C, Howell RA, Cawood JI. Prosthetic management of the atrophic mandible using endosseous implants and overdentures: a six year review. Br Dent J 1995; 179:329–337.
5. Esposito M, Hirsch J-M, Leckholm U, Thomsen P. Failure patterns of four osseointegrated oral implant systems. J Mat Sci Mater Med 1997;8:843–847.
6. Boerrigter EM, Van Oort RP, Raghoebar GM, Stegenga B, Schoen PJ, Boering G. A controlled clinical trial of implant-retained mandibular overdentures: clinical aspects. J Oral Rehab 1997;24:182–190.
7. Fugazzotto PA, Wheeler SL, Lindsay JA. Success and failure rates of cylinder implants in type IV bone. J Periodontol 1993;64:1085–1087.
8. Esposito M, Hirsch J-M, Lekholm U, Thomsen P. Biological factors contributing to failures of osseointegrated oral implants (II). Etiopathogenesis. Eur J Oral Sci 1998;106:721–764.
9. Listgarten MA, Lang NP, Schroeder HE, Schroeder A. Periodontal tissues and their counterparts around endosseous implants. Clin Oral Implants Res 1991;2:1–19.
10. Cochran DL, Hermann JS, Schenk RK, Higginbottom FL, Buser D. Biologic width around titanium implants. A histometric analysis of the implanto–gingival junction around unloaded and loaded nonsubmerged implants in the canine mandible. J Periodontol 1997; 68:186–198.
11. Berglundh T, Linde J, Ericsson I, Marinello CP, Lijenberg B, Thomsen P. The soft tissue barrier at implants and teeth. Clin Oral Implants Res 1991;2:81–90.
12. Ruggieri A, Franchi M, Marini N, Trisi P, Piatelli A. Supracrestal circular collagen fiber network around osseointegrated nonsubmerged titanium implants. Clin Oral Implants Res 1992;3:169–175.
13. Ruggieri A, Franchi M, Trisi P, Piatelli A. Histologic and ultrastructural findings of gingival crevicular ligament surrounding osseointegrated nonsubmerged loaded titanium implants. Int J Oral Maxillofac Implants 1994;9:36–43.

14. Mombelli A, Van Oosten MAC, Schürch E, Lang NP. The microbiota associated with successful or failing osseointegrated titanium implants. Oral Microbiol Immunol 1987:2: 145–151.
15. Mombelli A. Microbiology and anti-microbial therapy of failing implants. Periodontology 2000 2002;28:177.
16. Berglundh T, Lindhe J, Marinello C, Ericsson I, Liljenberg B. Soft tissue reaction to de novo plaque formation on implants and teeth. An experimental study in the dog. Clin Oral Implants Res 1992;3:1–8.
17. Mombelli A, Van Oosten MAC, Schürch E, Lang NP. The microbiota associated with successful or failing osseointegrated titanium implants. Oral Microbiol Immunol 1987:2: 145–151.
18. Rams TE, Link CC. Microbiology of failing dental implants in humans: electron microscopic observations. J Oral Implantol 1983:11:93–100.
19. Rams TE, Roberts TW, Tatum H Jr, Keyes PH. The subgingival microbial flora associated with human dental implants. J Prosthet Dent 1984:51:529–534.
20. Tonetti MS. Risk factors for osseointegration. Periodontology 2000;17:55–62.
21. Mombelli A, Mericske-Stern R. Microbiological features of stable osseointegrated implants used as abutments for overdentures. Clin Oral Implants Res 1990:1:1–7.
22. Becker W, Becker BE, Newman MG, Nyman S. Clinical and microbiologic findings that may contribute to dental implant failure. Int J Oral Maxillofac Surg 1990:5:31–38.
23. Ericsson I, Berglundh T, Marinello C, Liljenberg B, Lindhe J. Long-standing plaque and gingivitis at implants and teeth in the dog. Clin Oral Implants Res 1992;3:99–103
24. Ericsson I, Persson LG, Berglundh T, Marinello CP, Lindhe J, Klinge B. Different types of inflammatory reactions in peri-implant soft tissues. J Clin Periodontol 1995;22:255–261.
25. Pontoriero R, Tonelli MP, Carnevale G, Mombelli A, Nyman SR, Lang NP. Experimentally induced peri-implant mucositis. Clin Oral Implants Res 1994;5:254–259.
26. Lang NP, Brägger U, Walther B, Beamer B, Kornman KS. Ligature induced peri-implant infection in cynomolgus monkeys. 1. Clinical and radiographic findings. Clin Oral Impl Res 1993;4:2–11.
27. Lang NP, Mombelli A, Brägger U, Hämmerle CH. Monitoring disease around dental implants during supportive periodontal treatment. Periodontology 2000, 1996;12:60–68.
28. Schwarz MS. Mechanical complications of dental implants. Clin Oral Implants Res 2000; 11(suppl):156–158.
29. Lekholm U, van Steenberghe D, Herrmann I, et al. Osseointegrated implants in the treatment of partially edentulous jaws: a prospective 5-year multicenter study. Int J Oral Maxillofac Implants 1994;9:627–635.
30. Rangert B, Krogh P, Langer B, Van Roekel N. Bending overload and implant failure: a retrospective clinical analysis. Int J Oral Maxillofac Implants 1995;10:326–334.
31. Balshi T. An analysis and management of fractured implants: a clinical report. Int J Oral Maxillofac Implants 1996;11:660–666.
32. Goodacre CJ, Kan JYK, Rungcharassaeng. Clinical complications of osseointegrated implants. J Prosthet Dent 1999;81:536–552.
33. Datamonitor Plc. European Dental Implants and Bone Augmentation 1999. Published: Datamonitor Plc., 106 Baker Street, London.
34. Esposito M, Worhtington HV, Coulthard P, Jokstad A. Interventions for replacing missing teeth: maintaining and re-establishing healthy tissues around dental implants (Cochrane Review). 2002 Cochrane Library, Issue 4 (ISSN 1464-780X).
35. Palmer R, Palmer P, Howe L. Dental implants: complications and maintenance. Br Dent J 1999;187:653–658.
36. Chan MF, Johnston C, Howell RA. A retrospective study of the maintenance requirements associated with implant stabilized mandibular overdentures. Eur J Prosthodont Restor Dent 1996;4:39–43.

37. Warrer K, Buser D, Lang N, Karring T. Plaque induced peri-implantitis in the presence or absence of keratinized mucosa. An experimental study in monkeys. Clin Oral Implants Res 1995;6:131–138.
38. Wennström J, Bengazi F, Leckholm U. The influence of masticatory mucosa on the peri-implant soft tissue condition. Clin Oral Implants Res 1994;5:1–8.
39. Mombelli A, Lang NP. Antimicrobial treatment of peri-implant infections. Clin Oral Implants Res 1992;3:162–168.
40. Mombelli A, Marxer M, Gaberthüel T, Grunder U, Lang NP. The microbiota of osseointegrated implants in patients with a history of periodontal disease. J Clin Periodontol 1995;22:124–130.
41. Mombelli A, Van Oosten MAC, Schürch E, Lang NP. The microbiota associated with successful or failing osseointegrated titanium implants. Oral Microbiol Immunol 1987;2: 145–151.
42. Jepsen S, Rühling A, Jepsen K, Ohlenbusch B, Albers H. Progressive peri-implantitis. Incidence and prediction of attachment loss. Clin Oral Implants Res 1996;7:133–142.
43. Verhoeven JW, Cune MS, de Putter C. Reliability of some clinical parameters of evaluation in implant dentistry. J Oral Rehab 2000;27:211–216.

23
Cardiac Pacemakers

Preben Bjerregaard and Amr El-Shafei

INDICATIONS FOR PACEMAKER THERAPY AND NEED FOR FOLLOW-UP

The first artificial pacemaker was implanted in Stockholm, Sweden, in 1958, by Elmquist and Senning *(1)*. Today, more than 1 million pacemakers have been implanted worldwide, mostly in older people *(2)*. At least 90,000 of the approximately 115,000 pacemakers implanted in theUnited States in 2004 were implanted in people 60 or older. The median age for males who receive their first pacemaker today is approximately 75 and for females 78. The indications have widened and the number of pacemakers per million inhabitants in the United States who receive their first pacemaker each year increased from approximately 200 in 1985 to more than 400 in 1998 *(3)*. Their impact on symptoms and mortality has been so overwhelming that no randomized trials have been necessary to prove their efficacy. The indications for pacemaker placement include specific electrocardiographic (ECG) abnormalities and symptoms. The latest American College of Cardiology/American Heart Association indications were published in 1998 *(4)*. Roughly 90% are implanted for sinus node dysfunction or atrioventricular (AV) block. Patients with a normal electrical system, such as patients with carotid sinus hypersensitivity or hypertrophic cardiomyopathy, may occasionally be candidates. Dual-site atrial pacing has been used to prevent atrial fibrillation and biventricular pacing may be useful to synchronize contraction of the right and the left ventricles in patients with severe left ventricular dysfunction. This has made implantation of pacemakers technically more challenging and increased the likelihood of lead dislodgment and programming difficulties. Syncope is the most frequent symptom prior to pacemaker implantation (40% of cases), followed by dizzy spells (25%),and symptomatic bradycardia (20%). Devices are becoming smaller. Most pacemakers today weigh less than 30 g and are less than 8 mm thick. They are usually implanted in the infraclavicular area with the leads entering the venous system either by subclavian puncture or via cutdown to the cephalic vein.

After the patient has left the hospital with a well-functioning pacemaker system, follow-up is lifelong and most commonly performed in a specialized pacemaker clinic *(5)*. The specific clinical goals for such a clinic include monitoring pulse generator and lead function, optimizing the clinical effectiveness of pacing by making suitable pacing system adjustments as required, maximizing pulse-generator longevity and pacing system function, minimizing pacemaker-related symptoms and complications, antici-

From: *The Bionic Human: Health Promotion for People With Implanted Prosthetic Devices*
Edited by: F. E. Johnson and K. S. Virgo © Humana Press Inc., Totowa, NJ

pating the need for elective replacement of pulse generators and leads, educating and reassuring patients and their families, addressing other clinical problems in a timely fashion as they arise, keeping records of lead and pulse-generator performance including anomalous behavior or malfunction, and maintaining contact with patients to address possible device recalls or advisories. Between January 1990 and December 2000, 408,500 pacemakers were subject to recalls or safety alerts *(6)*.

The two basic methods of patient follow-up are direct evaluation and transtelephonic monitoring. Both can be applied in a pacemaker follow-up center or private office but transtelephonic monitoring is often performed as an independent physician-based or proprietary service. In pediatric and geriatric care, where travel may be an issue, the adequacy of family resources or facilities has to be taken into consideration. Direct evaluation in which every known test can be applied is indispensable under certain circumstances. The observations and measurements made during a complete direct evaluation are summarized in Table 1. Remote monitoring uses fewer tests but is very useful because patients do not need to travel to a follow-up center.

At the North American Society of Pacing and Electrophysiology (NASPE) policy conference held in San Diego, California in 1993, the following were recommended as the minimum requirements for every direct evaluation service *(5)*:

1. At least one pacemaker-programming device appropriate for every pulse-generator model commonly encountered by the service should be available, with easy access to electronic programmers that are frequently needed and manufacturers' representatives to employ them optimally.
2. A digital interval timer (or its computer equivalent) for measuring various rates and durations of patient parameters and device parameters.
3. An ECG strip chart recorder for 12-lead ECG recording.
4. An ECG display monitor.
5. Pacemaker magnets appropriate for use with all pulse generators normally encountered.
6. A crash cart equipped with an external defibrillator.

It was suggested that the service should have ready access to the patient's clinical and pacemaker programming history, the results of past evaluations, and the operating specifications of pulse generators and leads. Free-running and magnet-mode rhythm strips (or equivalent computer files of digitized signals) should be maintained for reference.

The importance of having qualified staff capable of providing the full scope of clinical and administrative services was stressed. Industry-employed allied professionals are being used more frequently in such clinics. In order to define the appropriate role for these company representatives, a policy statement was issued recently after a meeting between the NASPE officials and representatives from major pacemaker companies *(7)*. It emphasized that these technicians should provide this specialized evaluation with the physician nearby.

Transtelephonic monitoring has two components: a transmitter and a receiver. The transmitter is used by the patient to convert ECG signals into an audible tone for transmission over conventional telephone lines, together with encoded results of stimulus-duration measurements and marker artifacts produced whenever a stimulus is detected by the device. The receiving device in the monitoring facility regenerates the ECG signals for plotting (and in some cases computerized storage) and prints out the mea-

Table 1
Observations and Measurements Made During a Complete Direct Evaluation

History and physical examination
    Special attention to the operative site
    Consideration of symptoms

Pacing rate
    Free running
    With magnet

Stimulus characteristics, A and V
    Amplitude
    Duration
    Waveshape

Stimulation thresholds, A and V, documented by rhythm strip

Sensing thresholds, A and V

Maneuvers for detecting muscle potential sensing by a unipolar electrode system, with temporary reprogramming to a triggered mode if appropriate

Tests for retrograde ventriculoatrial conduction in dual chamber systems

Testing for potential AV cross-talk in dual chamber systems

Adaptive rate functions (rate modulation characteristics)
    Response to onset of exercise
    Response to cessation of exercise

Evaluation and documentation of interaction with spontaneous rhythms
    Rhythm strip, free running
    Rhythm strip, with magnet
    Assessment of underlying spontaneous rhythm
    Intracardiac electrogram telemetry
    Timing diagram telemetry
    Device evaluation, including interrogation (data telemetry) of the implanted pulse generator
    Patient and family education
    Information based on the capabilities of the pacemaker and the electrocardiographic context in which it operates.

AV, atrioventricular.

surement results. Telephone contact between the patient and the monitoring service affords prompt patient-initiated contact when pacemaker-related symptoms occur. Because transmissions can be conducted using any standard telephone, there are no geographical constraints. A symptomatic patient is often more inclined to initiate a telephone contact rather than to schedule a direct evaluation visit, and that contact may be useful for triage if loss of capture or presyncope are present. Remote monitoring may also provide evidence of serious intermittent problems, such as failure to capture. If a problem is detected in this way, subsequent direct evaluation should be performed to confirm it whenever feasible. Temporary measures, such as magnet placement to eliminate pacemaker inhibition owing to oversensing, or to interrupt re-entrant tachycardia, may be required immediately, however. There are certain inevitable limitations in the use of transtelephonic monitoring. The tracings may be poor, or even uninterpretable, as a result of unsupervised electrode placement, motion artifacts, tele-

Table 2
1984 Medicare Guidelines for Cardiac Pacemaker Follow-Up: Transtelephonic Monitoring

Pacemaker type	Months after implantation	Scheduled TM contacts
Pacemakers that meet the criteria[a]		
Single-chamber	≤1	Every 2 weeks
	2–48	Every 12 weeks
	49–72	Every 8 weeks
	≥73	Every 4 weeks
Dual-chamber	≤1	Every 2 weeks
	2–30	Every 12 weeks
	31–48	Every 8 weeks
	≥49	Every 4 weeks
Pacemakers that do not meet the criteria		
Single-chamber	≤1	Every 2 weeks
	2–36	Every 8 weeks
	≥37	Every 4 weeks
Dual-chamber	≤1	Every 2 weeks
	2–6	Every 4 weeks
	7–36	Every 8 weeks
	≥37	Every 4 weeks

[a] Pacemakers that have demonstrated better than 90% longevity at 5 years, whose output voltage decreases less than 50% over at least 3 months, and whose magnet mode (asynchronous) rate decreases less than 20% or 5 pulses per minute over the same period. In the absence of "official" statistics on pulse-generator longevity, the second half of this table is more broadly applicable in current practice.

TM, transtelephonic monitoring.

phone line interference, inappropriate choice of ECG leads, and so on. It is, however, a reliable method of identifying changes in the pacing rate and, therefore, a very reliable way to check the pacemaker battery status.

THE CLINIC VISIT

Direct evaluation in the follow-up clinic is usually performed 3 months after a new implantation and then every 6 months for dual-chamber devices and every 12 months for single-chamber devices. Transtelephonic monitoring at Saint Louis University does not follow 1984 Medicare guidelines (Table 2) as they are out of date. Current pacemakers are very reliable. Our practice is summarized in Table 3 and is our modifications of current recommendations of the Heart Rhythm Society. The schedules and procedures used can, however, be modified as required in order to provide adequate protection for patients with various risk factors. The following issues should be taken into account:

1. Degree of pacemaker dependency.
2. Actions to be taken if a device problem is made known by recall or advisory.
3. Severity of underlying heart disease.
4. Special considerations of pediatric pacing that may create a need for routine ambulatory ECG recording and exercise testing.
5. Implants with epicardial electrodes.
6. Pending application of cardioversion, defibrillation, electrocautery, or other procedures that can affect pacing-system operation and integrity.

Table 3
**Patient Surveillance After Implantation
of Pacemaker at Saint Louis University Hospital**[a]

	Year 1	Year 2	Year 3	Year 4	Year 5	Year 6
Office visit	3	2	2	2	2	2
Transtelephonic monitoring	1	2	2	4	4	10

[a] Number of visits recommended per year for asymptomatic patients.

7. High stimulation thresholds requiring unusually high output parameter setting or institution of drug therapy that might affect the threshold of stimulation.
8. Undersensing or interference sensing problems not completely resolved by reprogramming.
9. Concurrent use of an implanted cardioverter defibrillator or other implanted electrical device separate from the pacemaker.

When clinically indicated for optimizing patient's system function or for troubleshooting, other tests may also be warranted, such as chest X-ray (for evaluation of electrode position or verification of conductor fracture), ambulatory ECG recording (for identification of possible problems or for assistance in optimizing pacing function in transient situations), and Doppler echocardiography (for assistance in adjusting rate and AV interval for hemodynamic optimization).

During every clinical encounter with a pacemaker patient, history and physical examination is important. *Palpitations* are usually unrelated to the pacemaker. They are more often related to the underlying heart disease but this is a situation where an event recorder may be very useful in order to get an exact diagnosis of the patient's problems. Sometimes this is the result of oversensitive pacemaker rate response. Occasionally, a dual-chamber pacemaker tracks fast atrial tachycardias incorrectly, which can lead to palpitations. *Pacemaker-mediated tachycardias* can be sustained by dual-chamber pacemakers whenever there is retrograde ventricular–atrial conduction. Many of these problems can be solved by reprogramming the pacemaker and many newer pacemakers have special features capable of preventing such events (e.g., mode switching to prevent fast tracking of atrial arrhythmias). *Weakness and fatigue* are also common in pacemaker patients. They are generally related to the underlying disease but sometimes are the result of inappropriate programmed rate-response parameter or inappropriate AV delay in a dual-chamber pacemaker. *Pacemaker syndrome* refers to a conglomerate of symptoms and signs present in patients in whom the atrial rhythm is dissynchronous with the ventricular rate or those in whom, despite the presence of AV synchrony, inappropriately timed atrial contraction does not precede the paced ventricular event. It occurs mainly in patients with single-chamber ventricular pacemaker systems and sinus rhythm but patients with dual-chamber pacemakers can, under certain circumstances, also develop relative AV dissynchrony in which spontaneous P-waves do not fall in optimal relationship to paced ventricular complexes. Both a suboptimal cardiac output and activation of atrial baroreceptors owing to stretch, leading to reflex peripheral vasodilatation and hypotension, contribute to the weakness and other symptoms of cerebral hypoperfusion seen in this syndrome. The awareness by neck vein pulsations, which can be uncomfortable, usually is caused by cannon waves. *Shortness of breath, orthopnea, or paroxysmal nocturnal dyspnea* can be part of a pacemaker syndrome or underlying cardiac disease. *Hiccoughs* are rare. They can occur intermittently owing to direct stimulation of the diaphragm by a ventricular pacing lead. Phrenic nerve stimulation by an atrial pacing lead has also been reported. Patients may complain of *twitching of muscles* near the pacemaker. This is usually a sign of insulation break and needs further assessment. *Presyncope or dizziness* can be related to failure to capture or inappropriate inhibition of pacing and always require meticulous assessment of both pacemaker and lead system.

During the physical examination of a pacemaker patient, it is especially important to examine the pacemaker pocket for any sign of infection, hematoma, or impending skin perforation. Signs of infection usually occur within the first 6 months after implantation or replacement of a pacemaker. If infection is suspected, the wound should be irrigated and cultures performed. To obtain cultures, the skin is cleaned and painted with antiseptic solution and a needle is directed to the pulse generator, taking care to avoid indwelling leads. The fluid collected should be cultured for both aerobic and anaerobic organisms. Detection of pocket infection mandates immediate antibiotic therapy. Explanation of the pulse generator and leads is usually required. If infections develop at a site distant from the pacemaker site, it should receive standard care without any special attention to the pacemaker. Also, there is no indication for subacute bacterial endocarditis prophylaxis in patients with pacemakers. The pulse generator should normally be slightly moveable and the overlying skin nonedematous and of normal color and temperature. Manipulation of the pocket and its contents is usually performed at each visit. The jugular venous pulsations should be examined for the presence of cannon A-waves. Swelling of the ipsilateral extremity, shoulder, neck, or face is usually an indication of venous obstruction. Subclavian vein thrombosis around the pacemaker wires occurs in up to 44% of patients with transvenous pacemaker system *(8)*. Symptomatic venous thrombus requires anticoagulation, usually with heparin, followed by Coumadin. Thrombolysis has also been reported to give good results *(9)*, and balloon angioplasty can be used to restore lumen patency *(10)*. Cardiac auscultation may occasionally reveal a pacemaker sound after delivery of the stimulus output. This can be caused by intercostal muscle or diaphragmatic contraction or to a symptomless myocardial penetration. The first heart sound may be variable because of the changing relationship of atrial to ventricular systole in patients with single-chamber ventricular pacing systems and a difference in paced or sensed AV intervals in patients with dual-chamber pacing systems. The second heart sound is often paradoxically split owing to right ventricular depolarization occurring in advance of left ventricular depolarization, mimicking left bundle branch block. Other auscultatory findings include systolic nonejection clicks possibly resulting from slapping of the lead against the tricuspid valve and systolic murmurs caused by hemodynamically insignificant tricuspid and mitral valve regurgitation; these are benign and not unexpected. If a pericardial friction rub is present early after system implantation, myocardial penetration or perforation should be suspected and evaluated. Visible or palpable contractions of the pectoral muscle or diaphragm occurring in the patient should be noted because they might signify insulation defect, lead fracture, or an excessively high programmed output. Evaluation of inhibition of triggering by electrical activity from striated muscles in close proximity to the pacemaker is assessed by raising the arm close to the pacemaker and pressing the hands together. These maneuvers should be performed under ECG monitor. An important part of pacemaker follow-up is detection of impending battery failure. The elective replacement time for a pacemaker is signaled by a change in pacemaker function known as the elective replacement indicator. This alerts the clinician that the device should be replaced within a few months. It is usually triggered when the available battery voltage, measured at a specific point in the pacing cycle, falls below a preset value (2.2–2.4 volt). It may take the form of a rate change and/or an automatic change to a simpler pacing mode, for instance from a dual-chamber mode to a single-chamber mode or from a rate-responsive mode to a non-rate-responsive mode. Real-time telemetry of the actual battery voltage and battery impedance (higher impedance with depletion) is very helpful. If the battery voltage drops below 2.1 volt, the electronic components may no longer be able to operate and this point is commonly referred to as end of life. This may be associated with sensing failure, complete cessation of pacing, or grossly irregular function. The pulse generator should obviously be replaced well before this. Pacemaker manufacturers have taken a conservative approach when establishing the available battery voltage level that triggers the elective replacement indicator. In most cases, pacemakers may continue for approximately 6 months at abnormal parameters before the end of life is reached. Most pacemakers have a voltage monitoring circuit that changes the pacing rate when the battery voltage decreases. In some pulse generators, the rate decreases abruptly; in others the rate changes gradually and may take several months to complete. It is important that these changes are known for each pacemaker because they are very pacemaker-specific.

During follow-up, it is also important to evaluate the integrity of the pacing lead system. A *break in lead insulation* will allow current to flow from the lead through the break. If this current flow is near muscle, the muscle may be stimulated. In addition, insufficient current may be available at the lead tip to polarize the myocardium, leading to failure to capture. An ECG can display increased pacing artifact amplitude, reflecting current flow directly into body tissue. Insulation break has also important implications for appropriate sensing of the intracardiac signal. If the leads cannot conduct, the sensing signal has poor fidelity and undersensing will result. Although lead insulation failure may be suspected on clinical or ECG grounds, validation is best achieved through telemetry of impedance and battery current drain information. As resistance to current flow declines in the presence of insulation break, telemetry impedance is low and current drain from the battery increases. Until the patient's system can be revised, increasing the pulse-generator output is a reasonable temporary measure. In some systems, lead polarity programming can also be useful.

A break in the lead wire or loose connections between the lead and the pulse generator increases resistance to current flow. The current reaching the lead tips may be insufficient to stimulate cardiac muscle, resulting in noncapture. The condition is confirmed by a documented increase in lead impedance during pulse-generator interrogation. Particular maneuvers, such as arm stretching or change in body position, may be required to elicit the problem and the patient should be specifically asked about symptoms during these maneuvers. The treatment is replacement of the lead.

Following history and physical examination during clinical follow-up, the pacemaker is interrogated and its function tested using an electronic programmer that communicates with the device by electromagnetic signaling. The details are beyond the scope of this chapter but current pacemakers are furnished with automated diagnostic maneuvers used to evaluate the patient's intrinsic cardiac profile and the pacemaker's pacing and rate-response operation. They include rate histograms, which automatically collects long- and short-term heart rate and percent-paced data and also indicate rate profile that automatically collects sensor rate data for the programmed rate-responsive mode. High-rate event counters automatically collect basic data on high-rate episodes. The clinician can select detailed data of an electrogram with a rate trend. Upon interrogation by the programmer, the pacemaker reports permanent programmed parameter settings, along with battery status and device identification. During each visit, an ECG rhythm strip is obtained during brief inhibition of the pacemaker in order to assess the underlying rhythm. The amplitudes of P-waves and QRS complexes are measured, followed by an assessment of the amount of energy needed to stimulate the heart. This is called the threshold for pacing. The pacemaker is usually programmed at a voltage output twice the threshold, or pulse duration three times the threshold, in order to assure an appropriate safety margin for pacing.

Most pacemakers today are so-called rate-responsive pacemakers, which means that they respond to physiological demand for an increase in heart rate. Occasionally, it is necessary to assess this function as well. The approach most widely used is that of having the patient walk in the clinic corridors at self-determined casual and brisk paces while the rate response is monitored. Although it is difficult to establish rigid guidelines for desired rate increments, we normally expect a 10–15 beats per minute (bpm) increment over the baseline during casual walking. If this is not achieved, the pacemaker is reprogrammed to another rate-response setting and the exercise repeated. The patient is then asked to undertake a brisk walk based on the patient's own definition of brisk. The rate increment desired with a brisk walk varies substantially from patient to patient. A rate increment of 30–40 bpm may be desirable in a young, active patient and a lower increment may be desirable in an older, less vigorous patient or one with coronary artery disease (CAD). Occasionally, even a clinic visit does not provide enough information and a 30-day event recorder or 24-hour Holter monitor may be valuable. Fine-tuning of the AV interval may require transthoracic echocardiography.

The clinic visit is also an opportunity to evaluate the patient's underlying heart condition and discuss preventive measures. Many disorders affecting the conducting system are, to some extent, preventable or at least amenable to slowing of the rate of progression. The most important is CAD. Reducing risk factors such as smoking, hypercholesterolemia, and hypertension can prolong life and improve quality of life. Re-evaluation of a patient's underlying heart disease during clinical follow-up may reveal progression that necessitates reprogramming of the pacemaker. In certain instances, replacement of a pacemaker with a more sophisticated device may be indicated. It is especially important to attack new-onset atrial fibrillation because it carries the risk of thromboembolic complications. Atrial fibrillation also decreases cardiac performance, causing palpitations, dizziness, or even syncope. New or worsening congestive heart failure in a patient with pacemaker always raises the question of whether different programming of the pacemaker would be of any benefit. Occasionally, a simple increase in the minimum heart rate may provide some benefit. It is important to program the AV interval that provides the highest cardiac output. Biventricular pacing with resynchronization of the two ventricles is no longer experimental and may become a standard form of pacing in the future. For example, in patients with dilated cardiomyopathy and left bundle branch block morphology with very wide QRS complexes, pacing of the left ventricle via a pacing catheter in the coronary sinus and simultaneous pacing of the right ventricle can improve cardiac performance *(11)*.

Angina pectoris can result when the heart rate is too high in patients with a pacemaker and significant CAD. This can often be prevented by reprogramming. A stress test is often valuable in this situation to select the correct rate.

Because ventricular pacing alters ventricular activation and repolarization, the diagnosis of acute myocardial infarction (MI) and/or ischemia during pacing is often difficult. The QRS complex during right ventricular pacing resembles that of spontaneous left bundle branch block. This means that the diagnosis of MI can be suspected during ventricular pacing by applying the criteria used in complete left bundle branch block *(12)*. The patient artifact, particularly during unipolar pacing, frequently distorts the initial part of the QRS complex and makes Q-waves difficult to see. An extensive anterior wall MI may cause an initial Q-wave in Leads I, AVL, and V5 to V6 producing a stimulus qR pattern and an initial r in V-1 producing stimulus rS pattern. Although the sensitivity of this combined finding is low, the specificity approaches 100%. A similar pattern can, however, be seen in patients with dilated cardiomyopathy. An underlying inferior MI can result in the inscription of a QR, or Qr complex in leads II, III, and AVF. Again, the sensitivity is low but specificity somewhat better *(13)*. The Cabrera sign, which is a late notching of the ascending limb of the QRS complex in the left precordial leads, can be present during an anterior wall MI. Its development during ventricular pacing usually suggests the presence of an extensive anterior MI *(14)*. The sensitivity of this sign is low but its specificity has been high in the few patients where it has been examined. *Cardiac memory* is characterized by persistent but reversible T-wave changes that can occur in a variety of clinical scenarios, among them following ventricular pacing. Characteristically, the T-wave becomes negative in normally conducted beats where the QRS complex was negative during ventricular pacing. The underlying mechanism for both short- and long-term cardiac memory has not been completely characterized and do not seem to be related to myocardial ischemia. It is therefore important not to use these findings in the diagnosis of CAD.

Certain metabolic factors and clinical states can cause an increase in myocardial pacing threshold (the minimum output energy of the pulse-generator stimulus that is needed to depolarize myocardial tissue). Especially important are acidosis, alkalosis, severe hyperkalemia, and severe hypocalcemia, which can increase the threshold by up to 80%. Acute MI and myxedema can cause the ventricular threshold for pacing to go up. Under these circumstances, a programmed low voltage may become insufficient to produce consistent pacing. Patients with renal failure or diabetes (in whom there is a continuing risk for electrolyte abnormalities or abnormal glucose metabolism), should therefore be provided with a larger safety margin of pacemaker output voltage than usually recommended.

ENVIRONMENTAL EFFECTS ON PATIENTS WITH PACEMAKERS

Whenever a patient with a pacemaker is within a field of electromagnetic energy, there is a potential for interference with the function of the pacemaker, but most of the evidence of clinically significant events occurring under real-world conditions are anecdotal and difficult to confirm. Recently, the Practice Guideline Committee of the NASPE published a practical guide that addresses some of these issues *(15)*. *Home appliances* have been implicated in pacemaker interference, but radios, televisions (and remote controls), electric blankets, electric shavers, ham radios, heating pads, metal detectors, microwave ovens have not been shown to cause damage to pacemakers, changes in pacing rates, or total inhibition of pacemaker output in general. Power-generating equipment, arc-welding equipment, and powerful magnets would, however, be able to cause pacemaker inhibition if a patient with a pacemaker came near such equipment. *Cellular telephones* have the potential to interfere with pacemaker function, but to do so the cellular telephone would need to be closer than 10 cm to the pacemaker pocket and switched on. No instances of clinically significant interference have been observed when the cellular telephone was held at a patient's ipsilateral ear. Analog cellular telephones are much less likely than digital devices to interfere with pacing system function. *Airport security gates* do normally not affect the function of a pacemaker, but the metal detecting signal may be activated. In contrast, hand-held metal detectors contain magnets and therefore can markedly affect a pacemaker. It is therefore recommended that patients present their pacemaker identification card to security personnel prior to screening at airports. *Anti-shoplifting gates* or electronic article surveillance systems also produce electromagnetic interference. The acoustomagnetic and the magnetic audio frequency gates produce strong enough fields to cause asynchronous pacing, oversensing, or extrasystoles. These interactions are generally transient, but can be symptomatic. Pacemaker patients should be educated to minimize their exposure to such systems by walking normally through airport gates and avoiding direct contact with anti-shoplifting gates.

Pacemaker patients often encounter electro-mechanical interference in the hospital. Some general guidelines are applicable. The pacemaker should be interrogated by a specialist prior to possible exposure to electro-mechanical interference in order to determine whether the patient is pacemaker-dependent, to determine the rate and stability of the intrinsic cardiac rhythm and to program off rate-adaptive capabilities, because the latter may affect pacemaker behavior if the patient becomes agitated during the procedure. ECG monitoring is mandatory during any procedure where significant electro-mechanical interference is a possibility. After the procedure, the pacemaker is again interrogated and programming changes made if necessary.

Electrocautery interferes with the function of a pacemaker unless special precautionary measures are taken. If unipolar cautery is used, the indifferent electrode should be positioned as close to the surgical field and as far away from the pacing system as possible. Good contact of the indifferent cautery electrode should be maintained throughout the procedure and electrocautery use should be kept to a minimum. Bipolar cautery only interferes with a pacemaker if used in very close proximity to the device and therefore is usually not a problem. If pacemaker function is disturbed by electrocautery, placing a magnet over the pacemaker will assure continuous function by converting the device to a noninhibited (asynchronous fixed rate) mode. *Transthoracic cardioversion* can lead to changes in the programming of a pacemaker and a temporary increase in the threshold for pacing. It is recommended that surface cardioverting electrodes be positioned in the anterior posterior position more than 5 cm from the pacemaker. If a patient is pacer-dependent, transcutaneous pacing should be available. Also the pacemaker output should be maximized during cardioversion. *Diagnostic radiation* appears to have no effect on pacemakers, however, *therapeutic radiation* may damage the circuitry of a pacemaker and should be avoided if possible. Even if outside the direct radiation field, the pacemaker should be shielded. After radiation therapy is complete, the device should be tested frequently for several months. *Magnetic resonance imaging* is usually considered to be absolutely contraindicated in patients with pacemakers, because rapid cardiac pacing and total inhibition of output can occur. Even patients with residual pacemaker leads after a generator has been removed are at risk because of the "antenna" effect of the electrode system. This may lead to rapid stimulation of the heart induced by the alternating magnetic field and rapid radiofrequency pulses emitted during the scan. *Extracorporeal shock-wave lithotripsy,* which is sometimes used for conditions such as kidney stones, produces hydraulic shocks that could damage a pacemaker. To decrease that possibility, the distance between the lithotriptor focal point and pacemaker should be maximized. Many pacemaker patients have undergone this procedure without any complications. *Dental equipment* does not appear to affect pacemaker function and *electroconvulsive therapy* appears to be safe with respect to pacemaker function as well. *Short-wave or microwave diathermy* may result in pacemaker inhibition and permanent pulse generator damage and should not be used in pacemaker patients.

Questions regarding pacemakers can always be directed to the Heart Rhythm Society (1400 K Street, N.W., Suite 500, Washington, DC 20005. Phone: 202-464-3400, Fax: 202-464-3401, Web site address: http://www.hrsonline.org).

Trials evaluating ways to prevent atrial fibrillation or improve hemodynamic performance in patients with heart failure using pacemakers have been promising enough to warrant larger multicenter trials. Trials focusing on prevention of ventricular tachycardia by pacing would be of great interest and should be done. The unique features of storing capabilities in newer pacemakers have opened up opportunities to detect asymptomatic arrhythmias such as paroxysmal atrial fibrillation, which could have great clinical significance for prevention of strokes.

Finally, it would be of great interest and economically important to perform multicenter trials to define whether our current follow-up practice for patients with pacemakers is the most appropriate. By improvement in pacemaker technology, it is very likely that less frequent follow-up would be adequate and money-saving and at the same time make it less inconvenient to have a pacemaker.

REFERENCES

1. Senning A. Physiologic P-wave cardiac stimulator. *J Thorac Cardiovasc Surg* 1959;38: 639–643.
2. Lamas GA, Pashos CL, Normand SLT, McNeil B. Permanent pacemaker selection and subsequent survival in elderly medicare pacemaker patients. *Circulation* 1995;91:1063–1069.
3. Daley WR, Kaczmarek RG. The epidemiology of cardiac pacemakers in the older US population. *J Am Geriatr Soc* 1998;8:1016–1019.
4. Gregoratos G, Cheitlin MD, et al. ACC/AHA guidelines for implantation of cardiac pacemakers and antiarrhythmia devices. *JACC* 1998;31:1175–1209.
5. Bernstein AD, Irwin ME, Parsonnet V, et al. Report of the NASPE policy conference on antibradycardia pacemaker follow-up: effectiveness, needs, and resources. *PACE* 1994;17: 1714–1729.

6. Maisel WH, Sweeney MO, Stevenson MG, Ellison KE, Epstein LM. Recalls and safety alerts involving pacemakers and implantable cardioverter-defibrillator generators. *JAMA* 2001;286:793–799.
7. Hayes JJ, Juknavorian R, Maloney JD. The role(s) of the industry employed allied professional. *PACE* 2001;24:398–399.
8. Lee ME, Chaux A. Unusual complications of endocardial pacing. *J Thorac Cardiovasc Surg* 1980;80:934–940.
9. Bradof J, Sands MJ, Lakin PC. Symptomatic venous thrombosis of the upper extremity complicating permanent transvenous pacing. Reversal with streptokinase infusion. *Am Heart J* 1982;104:1112–1113.
10. Yakirevich V, Alagem D, Papo J. Fibrotic stenosis of the superior vena cava with widespread thrombotic occlusion of its major contributaries: an unusual complication of transvenous cardiac pacing. *J Thorac Cardiovasc Surg* 1983;85:632–636.
11. Lupi G, Brignole M, Oddone D, Bollini R, Menozzi C, Bottoni N. Effects of left ventricular pacing on cardiac performance and on quality of life in patients with drug refractory heart failure. *Am J Cardiol* 2000;86:1267–1270.
12. Barold SS, Levine PA, Ovsyshcher IE. The paced 12-lead electrocardiogram should no longer be neglected in pacemaker follow-up. *PACE* 2001;24:1455–1458.
13. Barold SS, Falkoff MD, Ong LS. Electrocardiographic diagnosis of myocardial infarction during ventricular pacing. *Cardiol Clin* 1987;5:403–414.
14. Barold SS, Ong LS, Banner RL. Diagnosis of inferior wall myocardial infarction during right ventricular apical pacing. *Chest* 1976;69:232–240.
15. Goldschlager N, Epstein A, Friedman P, Gang E, Krol R, Olshansky B. Environmental and drug effects on patients with pacemakers and implantable cardioverter/defibrillators. *Arch Intern Med* 2001;161:649–655.

US Counterpoint to Chapter 23

Marye J. Gleva

INTRODUCTION

Chapter 23 is a thorough and succinct review of current management practice for patients with an implantable pacemaker. Implantable cardiac devices can be divided into those that have algorithms to treat or address rhythm disorders and/or cardiac timing and those that monitor pressures. Cardiac rhythm management devices include pacemakers, implantable cardiac defibrillators, cardiac resynchronization devices, and implantable loop recorders. The implantable hemodynamic monitor, an implantable device that has been studied for measuring intracardiac pressures, is currently not approved in the United States. The authors' recommendations on cardiac pacemaker use and follow-up are based on the published guidelines from experts at the American College of Cardiology, the American Heart Association, the North American Society of Pacing and Electrophysiology (1), as well as their own recognized extensive clinical experience. The chapter is applicable to both physicians and other health care providers who plan to care for these patients.

Chapter 23 begins with a global review of pacemaker implantation indications and application in the United States. As noted by the authors, the majority of cardiac rhythm devices implanted to date are for patients with symptomatic sinus node and atrioventricular (AV) node dysfunction. Newer devices, based on cardiac pacemaker principles, have been developed for the treatment of medically refractory congestive heart failure and for the treatment and prevention of atrial fibrillation. As the incidence of arrhythmias and congestive heart failure (CHF) symptoms increases with age, the use of these devices is expected to increase.

Pacemaker therapy is considered permanent, and the authors note that follow-up is lifelong unless the patient undergoes cardiac transplantation. The goals of the follow-up visit are multiple. These include assessment of device function and the treatment and prevention of current and future cardiac disease. Device treatment for CHF symptoms is now possible. This new form of pacemaker therapy, termed biventricular pacing or cardiac resynchronization therapy, has been approved for certain patients with medically refractory CHF. It is available in both pacemaker and implantable defibrillator forms. Biventricular pacing involves placing a third endocardial pacing lead into a branch of the coronary sinus to pace the left ventricle. Although mortality benefit has

From: *The Bionic Human: Health Promotion for People With Implanted Prosthetic Devices*
Edited by: F. E. Johnson and K. S. Virgo © Humana Press Inc., Totowa, NJ

Table 1
Patient Surveillance After Implantation of Cardiac Pacemakers at Washington University Year After Device Implantation

	Postoperative year					
	1	2	3	4	5	10
Physician office visit	1	1	1	1	1	2
Pacemaker interrogation in physician's office by registered nurse	2	2	2	2	2	2
Chest X-ray	0	0	0	0	1	1
Transtelephonic monitoring	6	6	6	6	6	12

not yet been shown, biventricular-pacing therapy has resulted in modest improvements in New York Heart Association (NYHA) class, 6-minute walk distance, quality of life, and a decrease in the number of hospital admissions for exacerbations of heart failure. Patients considered eligible for this therapy have NYHA class III symptoms despite optimized medical therapy, a wide baseline QRS (>120 ms in duration), and left ventricular enlargement *(2)*.

As pointed out by the chapter authors, standard endocardial pacing in the right ventricle results in a paced QRS morphology that has left bundle branch block configuration. This abnormal electrical activation sequence can result in asynchronous electro-mechanical activation of the ventricles and can exacerbate symptoms of CHF. Patients with atrial fibrillation and CHF who have undergone AV node ablation and pacemaker implantation may be particularly prone to heart failure exacerbations. A recent report focused on 20 such patients. Investigators from Emory University demonstrated a statistically significant improvement in left ventricular ejection fraction during biventricular resynchronization therapy *(3)*. Although these results have not been confirmed in a prospective randomized trial, the preliminary data is suggestive of a benefit.

There are no published guidelines for follow-up in patients treated with resynchronization therapy. As modest numbers of patients currently have biventricular devices, it is reasonable to address some of the unique issues associated with it. The authors acknowledge the challenges in lead placement; coronary venous anatomy can vary widely between individuals. Coronary sinus lead dislodgments have been reported in up to 10% of new biventricular pacing systems. Branches of the coronary sinus considered more optimal for biventricular pacing may be diaphragmatic in location or stimulation at these resynchronization sites can simultaneously stimulate the phrenic nerve. Diaphragmatic stimulation is more prevalent in these patients, compared to patients with standard pacing systems, and is seen early and late. Ventricular activation timing is unique to resynchronization therapy and has required individual optimization during outpatient visits. This is done using echocardiography, similar to what is done for AV interval optimization. Most patients do not have underlying bradycardia so concerns about failure to capture, for example, are less important. Table 1 shows the patient surveillance approach at Washington University.

Patients with implantable cardiac devices, like any patient, experience bacterial infections. Infections fall into one of three general categories: the device site, the blood, or remote from the device site. There are no randomized clinical trials comparing standard courses of antibiotics with prolonged courses for cardiac device patients with remote infections. S*taphylococcus aureus* infections of the device site with or without bacteremia and staphlococcal bacteremia present two special situations and are discussed here.

The management of infected cardiac devices is particularly problematic. It is well recognized that pacemaker and implantable defibrillator system infections are a result of either skin flora contamination or hematogenous seeding. The practice at Washington University differs somewhat from that of the authors. In many cases, cultures taken from the pacemaker pocket are not diagnostic microbiologically, especially if those patients have been treated with a prior course of oral or intravenous antibiotics. Our current practice is to culture any available drainage and avoid introduction of any needle into the pocket. If *Staphylococcus aureus* is recovered from the blood or wound, complete removal of all implanted hardware is required for infection eradication. Conservative therapy with partial removal and pocket debridement is best used as a bridge to full device explantation.

In a retrospective review of cardiac device-related infections seen at the Cleveland Clinic *(4)*, erythema and pain over the pocket were the most common presenting symptoms. Relapse of infection occurred in 1 of 117 patients in whom all device hardware was removed and in three of six patients who did not have complete removal of all hardware. The mortality rate in this case series was 8%. Investigators from Duke University published a prospective evaluation of treatment for *Staphlococcus aureus* bacteremia in 33 patients with cardiac devices *(5)*. The patients in whom the device was not removed had a threefold increase in mortality and a twofold increase in death or treatment failure. The Duke investigators recommend the "systematic extraction of cardiac devices among most patients who develop *Staphlococcus aureus* bacteremia both with and without clear device involvement, as long as skilled transvenous extraction is available to the patient."

Internet-based device follow-up has been developed (Medtronic CareLink™ Network). Currently, the technology is approved for patients with an implanted cardioverter defibrillators (North American Society of Pacing and Electrophysiology Late Breaking Clinical Trials, May 2002). Patients are provided with a monitor to transmit data to a secure website. Health care professionals can then access this information. If this technology is approved and extended to patients with an implantable pacemaker, it may decrease the number of routine follow-up clinic visits in asymptomatic patients.

In conclusion, implanted cardiac devices are quite common. Newer indications for cardiac device-based therapy will increase the number of patients and disease processes treated.

REFERENCES

1. Gregoratos G, Cheitlin MD, Conil A, et al. ACC/AHA Guidelines for implantation of cardiac pacemakers and antiarrhythmia devices. J Am Coll Cardiol 1998;31:1175–1209.
2. Cazeau S, Leclercq C, Lavergne T, et al. Effects of multisite biventricular pacing in patients with heart failure and intraventricular conduction delay. N Engl J Med 2001;334:873–880.
3. Leon AR, Greenberg JM, Kanuru N, et al. Cardiac resynchronization in patients with congestive heart failure and chronic atrial fibrillation. Effect of upgrading to biventricular pacing after chronic right ventricular pacing. J Am Coll Cardiol 2002;39:1258–1263.

4. Chua JD, Wilkoff BL, Lee I, et al. Diagnosis and management of infections involving implantable electrophysiologic cardiac devices. Ann Int Med 2000;133:604–608.
5. Chamis AL, Peterson GE, Cabell CH, et al. *Staphylococcus aureus bacteremia* in patients with permanent pacemakers or implantable cardioverter-defibrillators. Circulation 2001; 104:1029–1033.

European Counterpoint to Chapter 23

Derek T. Connelly

At the end of the 1980s, there were major differences in pacemaker implant rates between comparable industrialized countries. Although the pacemaker implant rate in the United States was 359 implants per million population per year, the rate in the United Kingdom was 148 per million, the rate in Japan was even lower at 89 per million, and the implant rate in some European countries was considerably higher than in the United States (Belgium: 538 implants per million per year) *(1)*. Internationally recognized guidelines *(2)* for pacemaker implantation and the dissemination of research findings have led to the situation where there are now many similarities and few differences between European and North American practices in the treatment of bradycardias by cardiac pacing *(3)*. The rates of pacemaker implantation still vary substantially among countries, however. The implant rate seems to be a function not simply of national prosperity but of the percentage of the gross domestic product each nation spends on health care and of the number of cardiologists per million population (and, more particularly, the number of centres that implant pacemakers). In the United Kingdom, the current pacemaker implant rate is approximately 300 per million population per year *(4)*; for most other western European countries, the figure is closer to 450 per million per year. There is now agreement among UK cardiologists that an implant rate of 450 per million per year is an appropriate level of pacemaker implantation, based on current pacemaker indications *(5)*.

Data for the United Kingdom and the Republic of Ireland for the year 2001 *(4)* confirm that the indications for pacing (both in terms of symptoms and electrocardiograph [ECG] diagnosis) are similar to those in the United States. Forty-two percent of patients presented with syncope, 32% with dizzy spells, 13% with bradycardia with less severe symptoms, 6% with heart failure, and 7% with other or no symptoms. Twenty-eight percent of implants were for complete heart block, 14% were for lesser degrees of atrioventricular (AV) block, 27% for sinus node dysfunction, 20% for atrial fibrillation with slow ventricular rate, and 11% for other reasons.

The implant technique is of course the same worldwide. Patients are either kept in the hospital for 24 hours after the procedure or, in some centers, discharged home the same day. For those kept in the hospital overnight, it is customary to perform posteroanterior and lateral chest radiographs prior to discharge from the hospital to check the

From: *The Bionic Human: Health Promotion for People With Implanted Prosthetic Devices*
Edited by: F. E. Johnson and K. S. Virgo © Humana Press Inc., Totowa, NJ

Table 1
Equipment for Pacemaker Follow-Up Clinic

Mandatory

Resuscitation equipment
Multichannel/ECG recorder
Magnet
Relevant range of pacemaker programmers
Manuals for all relevant pacemakers and programmers
Electronic device for measurement of stimulus-stimulus intervals and pulse duration
Contact telephone numbers for all relevant manufacturers
File of technical notes from relevant government department and notices from manufacturers
Access to X-ray facilities, exercise testing, ambulatory electrocardiography
Twenty-four-hour telephone-answering facilities manned by competent staff
Access to temporary pacing facilities
Facilities to hospitalize patients as emergencies at any time

Recommended

Electrocardiograph screen monitoring
Equipment for pulse waveform analysis
Computer for patient database
External pacemaker for chest wall stimulation
Reference books
Access to tilt table testing and electrophysiological testing
Proper maintenance and calibration of all equipment

Adapted from ref. 6.

positions of the pacemaker lead(s) and to exclude the presence of a pneumothorax (which is a complication of subclavian puncture in approximately 1% of cases). However, when pacemaker lead implantation is routinely performed via the cephalic vein, there is no risk of pneumothorax, and patients can usually be discharged home safely on the day of implant without a chest X-ray.

FOLLOW-UP

The goals of the pacemaker follow-up clinic are as stated in the accompanying chapter. Guidelines on pacemaker follow-up have been published for the British Cardiac Society by a working party of the British Pacing and Electrophysiology Group (BPEG) (6). The aims of the pacemaker clinic are to optimize the pacemaker system to the patient's needs, to identify system abnormalities and complications, to predict the end of life of the pulse generator, to provide patient support and education, to accumulate a database on pacing systems, to train and educate medical and paramedical staff, and to provide cardiological follow-up where appropriate. The equipment needed at a pacemaker clinic is listed in Table 1.

In Europe, there is considerably less reliance on transtelephonic monitoring of pacemakers than in the United States. The reasons for this are complex, but probably the most pressing reason is the greater distances involved in the United States. Many patients may have traveled far beyond their home territory to have a pacemaker implanted, and transtelephonic monitoring helps minimize the burden of long journeys for pacemaker follow-up. However, most US patients with pacemakers have much more frequent follow-up checks (either transtelephonic or office-based) than is customary in the United Kingdom. The reasons for this may be partly cultural, partly traditional, partly related to physician reimbursement, and partly a reflection of different levels of litigation in different countries. Whereas transtelephonic monitoring works well for most patients, problems in transmission or reception of data may occur, and there have even been reports of "Munchausen syndrome by telemetry" (7), in which a patient deliberately tried to deceive the physician in the transmission of ECG data.

In the United Kingdom, most pacemaker follow-up clinics are run by clinical cardiac scientific officers, with a physician either in attendance or immediately available to provide advice and troubleshoot problems (6). High standards of training are expected for cardiac technicians undertaking pacemaker follow-up, and many technicians undertake examinations of competence in this field, as assessed by the BPEG. In addition, an increasing number of cardiac technicians are voluntarily seeking a higher level of certification in the speciality by taking the examination of special competence in pacing and defibrillation which is set by the North American Society of Pacing and Electrophysiology (NASPE, Heart Rhythm Association). It is unusual for a representative from a pacemaker company to be routinely present at pacemaker implants or pacemaker follow-up visits in the United Kingdom. This is in marked contrast to the situation in the United States, where company representatives are present at two-thirds of implants and a higher proportion in nonacademic centers (3). However, on the rare occasion that a problem is found that cannot be solved by the hospital's technical or medical staff, the technical advice of the manufacturer may be sought.

The frequency of follow-up checks performed at European pacing centers probably differs considerably from one hospital to another. For this and other reasons, a quality improvement initiative has been developed by the three largest implanting centers in Europe: the Cardiothoracic Centre, Liverpool, UK, University Hospital, Lund, Sweden, and the Medical Clinic at University of Mainz, Germany (Table 2). This initiative, the European Quality Improvement Program in Pacing (EQUIPP), has led to harmonization of pacemaker follow-up procedures in these major European centers. Our agreed follow-up protocol consists of a first visit at 2 months following implant and subsequent visits at 12 monthly intervals until 4 years post-implant. After 4 years, visits become more frequent as the battery begins to run down. Our routine is to see patients at 6 monthly intervals from 48 months onward, until the pacemaker needs to be replaced. In some situations, follow-up visits may be scheduled more frequently after 4 to 6 years, depending on the patient's underlying cardiac rhythm, the type of pacemaker, and the manufacturer's specific recommendations for particular models.

MODE SELECTION AND PROGRAMMING

The need for programming or optimization of rate response should be assessed at the clinic visit. Rate-responsive pacemakers can increase the pacing rate in response to a physical or physiological signal (such as movement or vibration of the pacemaker can, or changes in respiratory ventilation) other than the patient's own heart rate. These devices are usually more expensive than those that do not have this feature, and not all patients require a rate-responsive pacemaker. Most patients who require a pacemaker

Table 2
Patient Surveillance
After Implantation of a Pacemaker[a]

	Postoperative year					
	1	2	3	4	5	10
Hospital visit[b]	2	1	1	1	2	2

[a] Number of visits recommended per year for asymptomatic patients; Consensus developed by the European Quality Improvement Program in Pacing (EQUIPP) program: Drs. RG Charles and DT Connelly, Cardiothoracic Centre–Liverpool, UK, Dr. J Brandt, University Hospital, Lund, Sweden, and Dr. B Novak, Medical Clinic, University Mainz, Germany.

[b] Hospital visits: at 2 months, then 12, 24, 36, and 48 months following implant.

From 48 months: 6 monthly visits until elective replacement indicator reached.

Depending on advice from manufacturers, some patients are monitored at 3-month intervals after 48 months.

Note: No transtelephonic monitoring.

are elderly (the mean age at implant is 74.8 years) and many lead a relatively sedentary life. National guidelines are available for tailoring the pacemaker prescription to the patient's symptoms and arrhythmia *(8)*, although these guidelines are now more than 10 years old and in need of updating on the basis of recently published clinical trials on pacemaker selection. In the United Kingdom and Ireland in 2001, approximately 44% of pacemaker implants have rate responsiveness and 56% do not. This contrasts with the situation in the United States, where implantation of the more expensive rate-responsive devices is performed more liberally. In 1997, 90% of pacemakers implanted in the United States were rate-responsive devices I

Although most physicians in Europe and the United States tend to tailor the pacemaker prescription to the patient's physiological needs, there is actually very little evidence that "physiological pacing" results in either improved long-term outcomes or better symptom relief than simple ventricular pacing. The Canadian Trial of Physiological Pacing (CTOPP) *(9)* demonstrated that physiological pacing reduces the incidence of atrial fibrillation compared to ventricular pacing, but at the expense of a higher complication rate. However, subgroup analysis showed that patients who were pacemaker-dependent (i.e., patients who had persistent rather than intermittent bradycardia) had more to gain from physiological pacing *(10)*. This study recruited patients with both sinus node disease and AV block. A similar study in the United States, the Mode Selection Trial (MOST) *(11)*, recruited patients with sinus node disease only and showed a reduction in the incidence of atrial arrhythmias with atrial-based ("physiological") pacing but no significant difference in quality of life or major cardiovascular end points. Recently the results of the UK Pacing and Cardiovascular Events study (UK-PACE) *(12)* have been reported. This study recruited elderly patients with AV block and, once again, physiological pacing was not shown to improve quality of life or prevent major cardiovascular events. A meta-analysis of these studies is planned

(13), and our recommendations regarding pacemaker mode selection may need to be modified in the light of the evidence from these clinical trials.

NEW DEVELOPMENTS

Biventricular pacing has become a routine treatment for patients with symptomatic heart failure despite optimal medical therapy who have poor left ventricular function and evidence of dyssynchronous left ventricular contractility (as assessed either by a wide QRS complex or by echocardiographic indices of delayed left ventricular contraction). Recent trials *(14)* have shown that biventricular pacing in such patients can not only improve symptoms and exercise tolerance but also prevent progression of heart failure and reduce the need for hospitalization and intravenous diuretic therapy. There may even be cost advantages in initiating this therapy at a relatively early phase. More recent studies have shown that this subgroup of patients with heart failure, poor left ventricular function and wide QRS complexes on the ECG are also at particularly high risk of sudden death *(15)*, and are also likely to benefit from an implantable device with defibrillator backup *(16)*.

REFERENCES

1. Parsonnet V, Bernstein AD. The 1989 world survey of cardiac pacing. Pacing Clin Electrophysiol 1991;14:2073–2076
2. Gregoratos G, Abrams J, Epstein AE, et al. ACC/AHA/NASPE 2002 guideline update for implantation of cardiac pacemakers and antiarrhythmia devices: summary article. Circulation 2002;106:2145–2161.
3. Bernstein AD, Parsonnet V. Survey of cardiac pacing and implanted defibrillator practice patterns in the United States in 1997. Pacing Clin Electrophysiol 2001;24:842–855.
4. Cunningham AD, Cunningham MW, Rickards AF. United Kingdom National Pacemaker and ICD Database Annual Report 2001. London, UK: National Pacemaker Database/Central Cardiac Audit Database, Royal Brompton Hospital. Available online at www.ccad.org.uk.
5. Hall R, Moore R, Camm J, et al. Fifth report on the provision of services for patients with coronary heart disease. Heart 2002;88 (suppl III):iii1–iii59.
6. Sutton R. Guidelines for pacemaker follow up. Report of a British Pacing and Electrophysiology Group (BPEG). Heart 1996;76:458–460
7. Hamdan M, Hart K, Fitzpatrick A. Munchausen's syndrome by telemetry. Am Heart J 1996; 131:1046–1047.
8. Clarke M, Sutton R, Ward D, et al. Recommendations for pacemaker prescription for symptomatic bradycardia. Report of a working party of the British Pacing and Electrophysiology Group. Br Heart J 1991;66:185–191.
9. Connolly SJ, Kerr CR, Gent M, et al. Effects of physiologic pacing versus ventricular pacing on the risk of stroke and death due to cardiovascular causes. Canadian Trial of Physiologic Pacing Investigators. N Engl J Med 2000;342:1385–1391
10. Tang AS, Roberts RS, Kerr C, et al. Relationship between pacemaker dependency and the effect of pacing mode on cardiovascular outcomes. Circulation 2001;103:3081–3085.
11. Lamas GA, Lee KL, Sweeney MO, et al. Ventricular pacing or dual-chamber pacing for sinus-node dysfunction. N Engl J Med 2002;346:1854–1862.
12. Toff WD, Skehan JD, De Bono DP, Camm AJ. The United Kingdom pacing and cardiovascular events (UKPACE) trial. United Kingdom Pacing and Cardiovascular Events. Heart 1997;78:221–223.

13. Montanez A, Hennekens CH, Zebede J, Lamas GA. Pacemaker mode selection: the evidence from randomized trials. Pacing Clin Electrophysiol 2003;26:1270–1282.
14. Abraham WT, Fisher WG, Smith AL, et al. Cardiac resynchronization in chronic heart failure. N Engl J Med 2002;346:1845–1853.
15. Moss AJ, Zareba W, Hall WJ, et al. Prophylactic implantation of a defibrillator in patients with myocardial infarction and reduced ejection fraction. N Engl J Med 2002;346: 877–883.
16. Bristow MR, Feldman AM, Saxon LA. Heart failure management using implantable devices for ventricular resynchronization: Comparison of Medical Therapy, Pacing, and Defibrillation in Chronic Heart Failure (COMPANION) trial. J Card Fail 2000;6:276–285.

24
Joint Prostheses and Internal Fixation Devices

Thomas J. Otto and Coles E. L'Hommedieu

INTRODUCTION

Orthopedics, as a specialty, is traced back to 1741, when Nicholas Andre coined the term "orthopaedic" from the Greek words *orthos* (meaning straight, or free from deformity) and *paedia* (meaning child). His book, entitled *L'Orthopaedie*, describes the recognition of deformity and ways in which these deformities could be prevented *(1)*. No invasive surgical procedures were described for treatment of musculoskeletal disease. There were external remedies, various ingested remedies, and some descriptions of manipulations. It was not until the work of Lister, in the 19th century, and the understanding of the importance of asepsis, that orthopedic implants were accepted as a technique in the management of orthopedic diseases.

Orthopedic biomaterials can be divided into two major categories (Fig. 1). First are implants used to perform joint arthroplasty or joint reconstruction known as joint prostheses. The second category consists of all other implanted material used in the treatment of injuries and disorders of the musculoskeletal system. This includes pins, plates, screws, rods, and suture material. It also includes bone substitutes, primarily used to stimulate new bone formation or to act as scaffolds to direct new bone formation. The monitoring and the long-term management of patients with total joint implants tend to be distinctively different than in patients with any of the commonly implanted fracture fixation devices. Orthopedic surgeons who perform joint arthroplasty are keenly aware of the many problems these implants can pose. Although each patient is monitored according to his or her specific set of circumstances, at our institution we have found it useful to set standards unique to each group.

Joint prostheses carry with them the potential for catastrophic complications. Their implantation usually requires the removal of some significant segment of diseased articular and juxta-articular tissue. They are intended for lifelong implantation and are subject to enormous repetitive stresses from body movement and weight-bearing.

In 1890, Gluck was credited with inserting an ivory ball into the neck of the femur *(2)*. This is believed by many to be the first attempt at joint arthroplasty. It was not until the recognition of aseptic technique that widespread investigation into surgery and artificial implants began. In the 1930s, Smith-Petersen reported a variety of implantable materials, including glass, Pyrex, and Bakelite, all of which proved to be too brittle and structurally insufficient for repetitive weight-bearing *(3)*. Smith-Petersen later used a

From: *The Bionic Human: Health Promotion for People With Implanted Prosthetic Devices*
Edited by: F. E. Johnson and K. S. Virgo © Humana Press Inc., Totowa, NJ

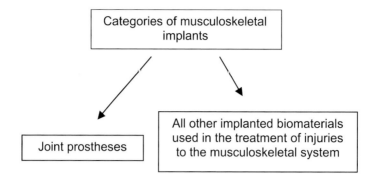

Fig. 1. Categories of musculoskeletal implants.

Fig. 2. Mold arthroplasty, circa 1940s through 1960s.

much stronger metal alloy, Vitallium, in the form of a cup called a "mold arthroplasty." His clinical results were considered very encouraging for the time (Fig. 2).

Many of the improvements in joint arthroplasty design were developed to treat end-stage arthritis. The French physicians Judet and Judet described the use of an acrylic ball used for femoral head replacement. Later versions of this prosthesis employed Vitallium instead of acrylic *(4)*. In efforts to treat increasing numbers of patients with hip arthritis and hip fractures, many biomechanical alterations and modifications of early prosthetic designs took place in the late 1940s and early 1950s. In 1950, Moore added an intramedullary stem to a Vitallium femoral head prosthesis *(5)*. The original fenestrated curved stem was later replaced with a straight I-beam design. The unipolar prostheses of Moore (Fig. 3) and Fred Thompson have been used for roughly 50 years with great success in managing the difficult problem of hip fracture in the elderly *(6)*.

The term "endoprosthesis" is used to describe a biomaterial implanted as a replacement part for some segment of bone that is removed. Total joint arthroplasty has developed out of a need to better manage advanced end-stage arthritides. Both total hip and total knee joint replacement have evolved over the past 40 years, largely owing to the efforts of Sir John Charnley in England. He is credited with first using the polymer methylmethacrylate to rigidly fix the acetabular and femoral components in position

Fig. 3. Moore-type hemiarthroplasty, still in use today.

(7,8). Methylmethacrylate had been used as a cranial bone substitute and had been studied in animal models *(9,10)*. The use of this rapidly curing polymer intraoperatively to achieve rigid prosthetic fixation greatly improved the longevity and functionality of hip arthroplasties. One recent study has shown 25-year prosthetic survivorship rates of 77.5–86.5% *(11)*. Charnley is also credited with using high-density polyethylene as a socket material to articulate against a chrome-cobalt alloy femoral head *(12)*. He coined the term "low-friction arthroplasty" to describe this interface.

Chrome-cobalt against polyethylene is still the benchmark by which other artificial weight-bearing surfaces are compared (Fig. 4). Orthopedic surgeons agree that the clinical results of the first generation total joint arthroplasties, as described by Charnley, have been excellent. The number of people with degenerative joint diseases needing joint arthroplasty is increasing. Despite their generally excellent longevity, there is a need to continue to try to improve upon the long-term results. This has led to considerable in vitro and in vivo experimentation with various designs and materials. Hydroxyapatite, ceramics, and various metal alloys are currently used in joint arthroplasty. Much research is currently aimed at prevention of arthritis and devising nonsurgical ways to repair or reverse arthritis. Until these goals are achieved, joint replacement will remain the mainstay in management of end-stage arthritis.

Fig. 4. Total hip arthroplasty, chrome-cobalt alloy against polyethylene bearing surface.

Implantable devices for the treatment of fractures and musculoskeletal deformity have evolved somewhat independently of the joint prostheses. Sporadic reports of the use of metal implants appeared in the early 1900s. Sir William Arbuthnot Lane was the first to report the use of silver wire to repair fractures (13). Widespread use of metallic implants for treatment of fractures did not occur until the 1930s when Smith-Petersen of Boston described an intramedullary metallic nail to transfix and stabilize intertrochanteric hip fractures (3). Since then, the use of various biomaterials for management of orthopedic conditions has escalated. There have been whole texts devoted to orthopedic biomaterials, their design, biomechanical characteristics, host responses, and complications (14). There have been numerous classifications of orthopedic biomaterials, usually related to specific material characteristics. In this chapter, we have divided these materials into arthroplasty and nonarthroplasty groups, largely because post-implantation patterns care for these patient groups differ from one another.

Partial and total joint arthroplasty has become an increasingly common surgical procedure (Fig. 5). US Medicare data collected between 1988 and 1997 shows that the combined incidence of hip, knee, and shoulder total joint arthroplasty doubled during this period and is now roughly 1 per 100 Medicare patients per year (15).

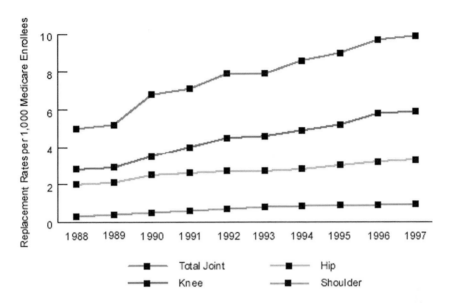

Fig. 5. The rise in joint replacement surgery in the 1990s. (From ref. *15*.)

In the United States, there are currently more than 500,000 joint prostheses placed yearly in either the knee or hip alone *(16)*. Data collected in the United States indicate about 5 million partial, total, and revision arthroplasties were performed in the 1990s. Of those, roughly 132,000 were primary joint prostheses placed in the shoulder. There were an estimated 2.3 million primary total or partial hip prostheses placed between 1990 and 1999. There were 2.1 million primary total knee prostheses placed during the same time span. The remainder were total elbow, total ankle, total wrist, and a very small number of metacarpophalangeal or interphalangeal prostheses. Revision of a prior joint prosthesis accounted for nearly 500,000 procedures over those 10 years, nearly all of which occurred at the hip or knee *(16)*. Implantation of prosthetic devices for fracture fixation, such as screws, plates, wires, and intramedullary nails is also common. During 1996, the last year of both inpatient and outpatient data collection, 655,600 procedures involving internal fixation were performed in the United States *(16)*.

Although little data exists regarding the total number of prosthetic joints in living patients, many studies have been performed to evaluate survivorship of implants, suggesting the prevalence of total joint prostheses. Ohzawa et al. reviewed patient survival following total knee replacement and found a nearly 65% 9-year survival rate in 165 patients, with the osteoarthritis patients showing no increase in mortality compared to the normal population *(17)*. A study by Tayot et al. reported a 69% patient survival rate in 376 total knee replacement recipients at a mean follow-up duration of 11.5 years, whereas Back et al. reported an 87% patient survival rate in 369 patients at a mean of 9 years follow-up for total knee arthroplasty *(18,19)*.

Hamadouche et al. studied 45 patients following total hip arthroplasty and reported an 89% patient survival rate at 8.5 years *(20)*. Another recent study by Urban et al. evaluated 64 patients who were 21 years status post total hip arthroplasty and found a 33% patient survival rate *(21)*.

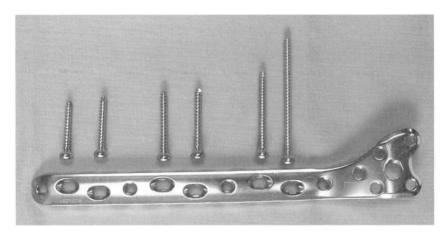

Fig. 6. Stainless steel plate and screws for fracture fixation.

Given this data and the steadily increasing rate of total joint arthroplasty, one can estimate that more than 10 million patients in the United States will be living with total joint prostheses by 2010 with a similar number of patients having retained hardware from internal skeletal fixation.

TYPES OF ORTHOPEDIC BIOMATERIALS

The types of orthopedic biomaterials are classified into the following groups: metals, polymers, ceramics, biodegradables and composites. The mechanical properties of each biomaterial strongly influences how it can be utilized in orthopedics.

Implants for the management of fractures of the skeleton are almost always made of stainless steel or titanium alloys. Although there are more than 50 commercial-grade stainless steels recognized by the American Society for Testing Materials (ASTM), only ASTM 55-66 are approved for human implantation *(22)*. These are commonly known as 316 and 316 low-carbon vacuum-molded (LVM) stainless steel. Stainless steel is an iron-based alloy that has excellent qualities for fracture management. It resists loads in bending and tension. It is the most common material utilized for fracture plates and screws (Fig. 6). Theoretical drawbacks are the potential for corrosion, most notably crevice corrosion. This leads to material degradation and metal ion release, mostly nickel ions. A small fraction of patients have a nickel allergy, which can cause clinical symptoms *(23)*. Detection of symptoms related to nickel allergy is one goal of follow-up.

Titanium alloys are used for both fracture fixation devices and total joint prostheses (Fig. 7). The modulus of elasticity of titanium alloys is closer to that of human cortical bone than that of stainless steel. This may be advantageous for healing of some fractures because a fixation device that is too rigid can cause a relative loss of bone or osteopenia owing to stress shielding *(24)*. Titanium is also less susceptible to corrosive wear than is stainless steel. At our institution, most intramedullary rod fixation utilizes titanium alloy products and most plate and screw fixation utilizes stainless steel products.

Most manufacturing companies recommend the routine removal of stainless and titanium alloy fixation hardware after fracture healing. Because of the extremely low

Fig. 7. Titanium alloy intramedullary device for fracture fixation.

incidence of harmful systemic effects, the value of routine removal remains controversial (25,26). Orthopedic surgeons must weigh the risks of further surgery against the benefits of removal.

More recently, orthopedic surgeons and implant manufacturers have shown interest in biodegradable devices (27). There are obvious theoretical advantages to biodegradable materials for fracture fixation. These materials must remain structurally sound long enough to allow adequate bone and soft-tissue healing. At the same time, they must be resorbed and degraded without local or systemic effects. Currently, polyglycolic acid and polylactic acid products are available for implantation. These substances have been used for many years as absorbable suture material. They have limited indications for use as small pins, screws, and soft-tissue anchors.

TOTAL JOINT ARTHROPLASTY

Extensive clinical experience with a variety of metals and basic research with mechanical stress simulators has led to the current state of the art in usage of total joint arthroplasty materials. Currently, the materials used in joint arthroplasty are metal alloys, polymers, and ceramics.

Metal prosthetics for joint replacement are under considerable loads over extended periods of time. During the normal gait cycle, stresses up to 10 times body weight must be endured by these devices (24). Fatigue and wear were common early problems seen in the 1970s (28). Although stainless steel is an excellent material for fracture fixation, it does not have the ability to withstand repetitive long-term weight-bearing loads. Changes in design and materials have evolved in efforts to reduce the incidence of prosthetic failure. The majority of metals currently used in joint arthroplasty are either titanium-based alloys or chrome-cobalt and molybdenum-based alloys. The absence of nickel from these alloys makes allergy problems very rare. Nickel allergy is an uncommon problem but it can be disastrous for certain patients (23). Today, gross structural failure of joint implants is unusual and is most frequently related to factors other than the material properties of the metal (29–31).

Long-term complications related to wear at the weight-bearing surface have received much attention in recent years. Polyethylene wear debris has been known for many years to have deleterious local effects, which can lead to loosening and prosthetic failure (32). Figure 8 shows the soft-tissue reaction to excessive polyethylene wear and subsequent metal particulate wear in a total knee arthroplasty. This wear often leads to prosthetic loosening and frequently requires revision of all components. Concerns about this have led to changes in polyethylene manufacturing processes, including new molding and sterilization methods (33). Ceramics, such as alumina and zirconia, have been extensively tested and are currently available as weight-bearing surfaces for both hip

Fig. 8. Intraoperative findings in a total knee arthroplasty with failure of the polyethylene and loosening of the components.

and knee prostheses *(21)*. Their major advantage is a coefficient of friction that is considerably lower than that of metal on polyethylene *(34)*. Ceramic-on-ceramic and ceramic-on-polyethylene interfaces are commercially available. All-metal articulations are being investigated as well and are available for implantation *(35)*. Laboratory and clinical studies have not yet demonstrated conclusively which weight-bearing surfaces are best. Chrome-cobalt against high-density polyethylene remains the most common interface implanted at our institution (Fig. 9).

Another area of intense study is the bone–prosthesis interface. The rapidly curing polymer polymethylmethacrylate (often referred to as "bone cement") has been used since the inception of joint arthroplasty to solidly bond the prosthesis. Its major advantage is immediate and rigid fixation of the prosthesis. Concerns regarding its long-term success have led to the fabrication of prostheses with surfaces designed to promote osseointegration, so-called "bone ingrowth surfaces." These implants require adequate host bone to achieve long-term solid fixation. One such surface is a new product called Trabecular Metal. This biomaterial is elemental tantalum that is formed into a porous trabecular pattern. Eighty percent of the material is porous, allowing bone ingrowth to occur. Trabecular Metal more closely mimics the physical structure of cancellous bone and has shown significantly increased bone ingrowth and resistance to shear stresses when compared to traditional bone ingrowth surfaces *(36)*. The use of such implants may not be reasonable in older patients with osteopenic skeletal tissue, as the ability of such bone to rigidly fix the prosthesis is greatly decreased *(35)*.

The last major category of orthopedic biomaterials is the group known as composites. These are not biocompatible and are used primarily for external support such as casts, splints, and external fixators. Carbon fiber, fiberglass, and plaster of Paris are examples. These materials are beyond the scope of this text.

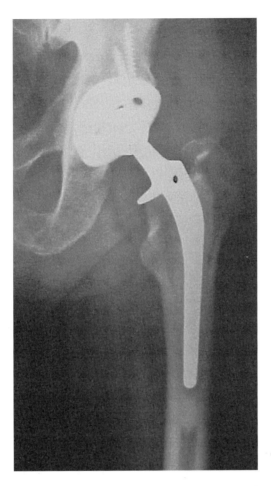

Fig. 9. Plain X-ray image of left total hip arthroplasty.

DETECTION OF FAILURE OF FRACTURE FIXATION DEVICES

Failure of implants that are used in the management of fractures is unusual when appropriate biomechanical principles are followed during their insertion *(37)*. Inadequate fixation can be associated with implant loosening or fracture. If normal fracture healing is delayed, implant fatigue may occur leading to bending or fracture of the metallic device. Local or systemic diseases, such as infection at the fracture site or metabolic imbalances, can delay healing and ultimately lead to implant failure. Implant failure in the presence of a healed fracture is very unusual *(38)*. Implant failure that occurs after the fracture has had a sufficient time for healing usually indicates problems with delayed union or nonunion. Late failure of spinal instrumentation, for example, may be the only evidence to support a diagnosis of pseudoarthrosis. Late failure of fixation devices requires the physician to consider both the management of the delayed union and the replacement of the broken fixation device. The routine follow-up of patients with fracture fixation devices allows the physician to monitor the progress of fracture healing and address any cases that may show early signs of failure of fixation.

Table 1
Routine Office Follow-Up Grid After Total Joint Arthroplasty. Patient Surveillance After Implantation of Total Joint Prosthesis at Saint Louis University.

	Postoperative year					
	1	2	3	4	5	10
Office visit	4[a]	1	1	1	1	1
AP and lateral plain X-ray of involved joint	4[a]	1	1	1	1	1

[a]At 2 weeks, 6 months, and 1 year.

Failure of total joint implants can be very difficult to diagnose in some instances. Failure can be to the result of either aseptic or septic etiologies. Early failure may be due to inadequate fixation or infection. Late failure (more than 1 to 2 years after implantation) may still be caused by infection but may also be simply owing to wear of the component surfaces.

In many instances, a patient's history can suggest not only failure of the implant but also the cause of failure. Failure is in most cases associated with symptoms of pain. The time of onset, location, and quality of pain are important to note when formulating a differential diagnosis. Acute onset of pain suggests fracture or other mechanical failure of a component, whereas progressive pain may indicate loosening of a component or infection.

After a careful history and physical examination, the most important tool in detecting implant failure is X-ray imaging (Fig. 9). It is routine and customary for orthopedists to utilize plain X-rays when assessing the musculoskeletal system. They are sometimes complemented with other imaging studies such as computed tomography (CT) scan, magnetic resonance imaging (MRI), and bone scintography but cannot be replaced. Our follow-up schedule specifies plain films of the involved prosthesis at practically every office visit (Table 1).

If prosthetic failure is suspected, plain X-rays of the affected area in two projections (typically anterior–posterior and lateral views) should be obtained. The majority of prosthetic failures are detected with this relatively simple study. A recent study showed preoperative radiographs to have a sensitivity of 94% and specificity of 100% for detecting loosening of an acetabular component (39).

Other imaging tools can be helpful when plain X-rays are inconclusive. In combination with plain films, technetium-99 bone scans are very sensitive in detecting implant loosening (40). One study evaluated bone scan use in total hip arthroplasty and showed 95% accuracy in detecting acetabular component loosening and 89% accuracy in detecting femoral component loosening (41). Typically, loosening is associated with an increase in bone metabolism, which leads to an increased uptake of technetium on bone imaging. Technetium scans rarely suggest the cause of loosening so further studies, such as arthrography and joint aspiration, may be required. This is addressed in more detail in the following section.

In the case of total joint implants, detection of failure is both challenging and important. The deleterious effects of implant failure in joint arthroplasty cannot be overstated. Loosening cause dby osteolysis, or resorption of bone tissue around the prosthesis, can be a progressively destructive process if unchecked. Routine follow-up plain films are used to evaluate this process. Revision of total joint prostheses associated with significant loss of bone tissue is a formidable procedure and results are less favorable than the primary arthroplasty *(42)*. In the case of a loose prosthesis, it is important to distinguish between aseptic and septic loosening. When infection is present the clinical and operative strategy is much different than for aseptic loosening. This is discussed in more detail in the next section.

DETECTION OF PROSTHETIC INFECTION

Accurate diagnosis of infection within and around the musculoskeletal system is very important. The majority of microorganisms that orthopedists encounter in the United States are bacterial *(43)*. Mycobacterial, fungal, and parasitic infections are seldom encountered but need to be considered in the differential diagnosis. It is important not only to know which bacteria are present and what the antibiotic sensitivities are, but where the infection is located, whether it is acute or chronic, and whether it is associated with implant loosening or failure. Other factors that play an important role in formulating a rational treatment plan are related to general health concerns. The appropriate treatment cannot be instituted without a thorough understanding of the extent of infection. While infection remains uncommon in total joint prostheses, routine follow-up may allow for the identification of early radiographic and clinical signs of a joint infection.

Joint prosthesis infections and extra-articular prosthesis infections are managed differently. Most infections following insertion of implants for uncomplicated, closed injuries to the skeleton occur acutely. The mechanism is thought to be primarily intra-operative contamination, this despite the routine use of prophylactic antibiotics and appropriate sterile technique *(44)*. This topic is also discussed in Chapter 10. Early diagnosis is important. The sooner treatment is instituted, the better patients respond. Usually these infections are not difficult to diagnose. The typical symptoms of increased pain, swelling, erythema, fever and chills occur within the first several months of injury *(45)*. The most common organisms are *staphylococcus* species and *streptococcus* species. They are usually isolated in the laboratory without difficulty. Late infection around hardware used to treat musculoskeletal disorders is rare but can occur. In most cases this is thought to result from hematogenous spread *(46)*.

The principles that need to be followed for a reasonable chance of eradicating infection are as follows:

1. Perform early aggressive debridement of all dead or infected tissue.
2. When treating an acute fracture-site infection in the presence of rigid fracture fixation, leave the fixation device in place. Fracture instability is one of the factors believed to contribute to the host's inability to defend against infection. When fixation is inadequate, other methods to stabilize the fracture, such as external fixation devices, may need to be used.
3. Culture appropriate specimens to identify the pathogen. Superficial swabs of a draining wound are notoriously inaccurate in isolating the true pathogen. Surgical cleansing and subsequent removal of specimens from the fracture site tends to be more accurate. This is followed by administration of the appropriate antibiotic. Many patients are already on antibiotics at the time of tissue debridement. This can make isolation of an organism impossible. Occasionally, organisms that cannot be cultured may be recognized on micro-

scopic exam or by serological testing. In our institution, complicated infections such as these require the expertise of infectious disease specialists.
4. Aggressive soft-tissue management and closure. This occasionally requires the expertise and assistance of plastic or microvascular surgeons.

Infections that occur following total joint arthroplasty can be much more difficult to diagnose. These infections also are classified as early or late. Late infections can also be either acute or chronic. The lifetime risk of infection for primary total joint arthroplasty is about 1 to 2% *(43)*. Although this is consistent with the risk associated with most major surgical procedures, infection in and around a total joint is a very difficult problem for both patient and physician. Patients are understandably frustrated, whereas the orthopedic surgeon tends to feel some measure of responsibility and remorse. In the case of acute postoperative infections following total joint arthroplasty, the diagnosis is usually based upon history, physical examination and blood tests, such as complete blood count (CBC), erythrocyte sedimentation rate (ESR), and C-reactive protein (CRP) *(47)*. Loosening of a previously well-fixed prosthesis within 1 year of implantation is suggestive of infection. It is the late infection that tends to present with symptoms that have occurred over a period of weeks to months in which the diagnosis is difficult to make. Most patients complain of pain but it may not just be related to weight-bearing or stresses applied to the prosthesis. Many times, patients complain of night pain or a deep, constant aching pain in the involved region of the body. Older debilitated patients may only present with malaise, weight loss, or unwillingness to bear weight. Classic signs of infection, such as tenderness, erythema, fever, and soft-tissue swelling, are usually not present. This may lead the unwary practitioner to dismiss infection as part of the differential diagnosis. Most patients with late-infected arthroplasties have evidence of prosthetic loosening on plain X-rays. Late cases of loosening may be related to wear rather than infection but, at our institution, patients who present with a painful prosthesis more than 1 year after implantation are presumed to have infection until proven otherwise. The usual management of a painful loose uninfected prosthesis is a revision or exchange arthroplasty. In the presence of undiagnosed infection, primary exchange is usually associated with recurrent infection, failure of the prosthesis and possible systemic infection for the patient. The morbidity and mortality from these complications can be high *(48)*. Patients suspected of infection have plain X-rays of the affected joint, CBC, ESR, serum CRP level, concomitant technetium-99 bone scan and indium leukocyte scan, and joint aspiration with culture. When there remains doubt, intraoperative biopsy and extensive tissue cultures are performed *(47)*.

The CBC typically demonstrates an elevated white blood cell count with a left shift when the infection is acute. In late or chronic infections, the CBC and the differential leukocyte count may be normal but normochromic anemia may be present.

The ESR is typically elevated up to levels of 50–60 mm per hour following both significant musculoskeletal trauma and total joint arthroplasty procedures. In the case of uncomplicated joint arthroplasty, it usually returns to normal levels within 4 to 6 months *(49)*.

In late acute infections, the ESR can be expected to be elevated. When there is evidence of systemic symptoms, it may be elevated above 100 mm per hour. However, in the setting of a late chronic indolent total joint infection, the ESR may be normal.

The serum CRP, some believe, may be a better indicator of active infection. It rises rapidly following surgery but returns to baseline much more rapidly than the ESR *(50)*.

It is reliably elevated in cases of acute active infection but, in the patient with a chronic low-grade prosthetic infection, it may be normal or minimally elevated.

The technetium-99 bone scan is used to evaluate blood flow and bone cellular activity *(51)*. It is very sensitive, but not very specific, in detecting increases in bone metabolism *(52)*. In a recent evaluation of infected total hip prostheses, the accuracy of technetium-99m bone scan was 79%, with a sensitivity of 83% and a specificity of 79% *(53)*. It is, for example, sensitive in detecting a loose joint arthroplasty, which is associated with increases in osteoblast and osteoclast activity at the prosthetic bone interface. It does not reliably differentiate between aseptic and septic loosening but very intense uptake usually indicates infection *(51)*.

The indium leukocyte scan, which, at our institution, is performed at the same time as the technetium-99 bone scan for these evaluations, is much more sensitive in detecting infection. When used to detect joint infection, its reported sensitivity is 85% and its specificity is 75% *(54)*.

In cases of suspected infection, joint aspiration utilizing meticulous sterile technique is performed (Fig. 10). In some joints, this is performed with fluoroscopic guidance and control. Minimizing the chance for contamination is imperative. Intraoperative biopsy and culture is the last resort. When the diagnosis is suspected but still not confirmed at the time of surgical exploration, biopsy of appropriate tissue samples should be submitted for culture, frozen section, and permanent histological examination. The pathologist should be alerted preoperatively to the importance of identifying the pathogen so that special stains can be used, if necessary. Aerobic, anaerobic, fungal, and mycobacterial cultures are also routine. If frozen section examination shows more than five polymorphonuclear cells per high-power field, then infection should be suspected *(55)*. Diagnostic tests based on bacterial DNA identification will probably become more common in the future.

Once a diagnosis of infection is made, the specific management will depend on many factors. The following guidelines are used at our institution. In patients who present with an acute postoperative infection and who have a well-fixed prosthesis, incision and thorough debridement, combined with appropriate intravenous antibiotic treatment for 6 weeks, can be attempted without removal of the implant. The same can be attempted for patients with an acute infection of a previously healed well-functioning joint arthroplasty. In patients with a chronically infected joint, the prosthesis is removed, the joint is aggressively debrided, an antibiotic-impregnated spacer is inserted, and appropriate intravenous antibiotic therapy is initiated. This is followed by delayed reimplantation of a new joint prosthesis, usually 3 to 6 months later. Patients with very low-grade infections caused by bacteria that are sensitive to relatively nontoxic antibiotics and whose prosthesis is clinically stable and functional can occasionally be treated successfully with long-term suppressive oral antibiotics *(56)*. When infections are not responsive to these therapies, permanent resection of the arthroplasty, joint fusion, or, in extreme cases, amputation may be required *(57,58)*.

MANAGEMENT OF THE CONDITION FOR WHICH THE IMPLANT WAS INSERTED

Outside of routine follow-up and postoperative care, it is rarely necessary to pay special attention to the disorder for which the prosthesis was placed. For patients with retained fracture-fixation devices, the initiating event is usually an isolated instance of

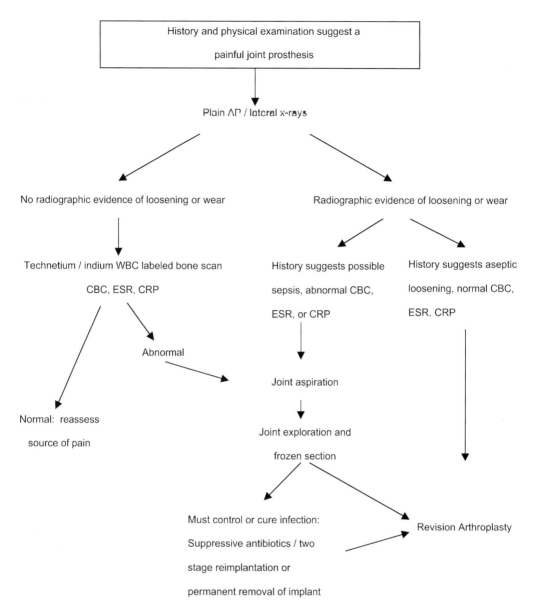

Fig. 10. Evaluation of the painful arthroplasty.

low- or high-energy trauma. In instances where the fracture is the result of cancer or metabolic disease, the underlying disease is almost always under treatment at the time of injury and cancer specialists deliver general medical care.

Prosthetic joint replacements are done primarily to manage degenerative disorders of joints. Although patients with rheumatological diseases can still have bouts of active synovitis following total joint arthroplasty, the destructive character of the disease seems to be halted. Routine radiographic examination is done more to observe for complications related to the implant rather than to follow disease progression. Management of the underlying disorder is usually well established prior to surgery and the patient's primary care physician or rheumatologist generally has a well thought-out plan for the long-term management of the disorder.

Detection of Other Conditions of Medical Significance

Allergic reactions to orthopedic implants have been sporadically described for years *(59)*. They are thought to occur in response to metal ions released by implants through corrosive processes. Cutaneous symptoms of rash and urticaria have been reported most commonly with nickel-containing alloys. The nickel content of 316 L stainless steel is 10–17% and the nickel content of chrome cobalt alloys is 2.5% *(60)*. It has been postulated that allergic reactions to implant materials may cause or contribute to loosening of joint implants owing to a local reaction *(61)*. Conversely, it has been suggested that loosening of the implant may contribute to ionization of metals and thereby increase the risk of allergy. Well-documented cases of such reactions to any of the materials approved for implantation in orthopedic surgery are rare. The utility of testing for allergies to the substances contained in orthopedic implants is controversial *(59)*. In our opinion, patients who have unexplained symptoms consistent with allergic reaction following implantation of an orthopedic device should undergo evaluation and testing by an allergist. Most implant manufacturers have available samples of the individual metals from their prostheses to use in such testing. In the event the implant under suspicion can be removed without sacrificing structural integrity (e.g., a fracture-fixation device in a patient with a healed fracture) it should be removed. In the case of a joint arthroplasty, treatment may require revision to a nonallergenic implant. This is a very significant undertaking, involving real risks, and should be considered only as a last resort.

At our institution, patients who are suspected of having metal allergies currently undergo a battery of commercially available skin tests. Since beginning joint arthroplasty procedures in 1969, there has been only one documented case of metal hypersensitivity that required removal of a joint prosthesis at Saint Louis University.

PREVENTIVE MEASURES

Patients can take measures to avoid infection of their orthopedic prostheses. Dental cleaning and surgical or other procedures on the pulmonary or genitourinary system can lead to transient bacteremia. The level of risk of bacteremia varies considerably among procedures. Although any episode of transient bacteremia has the potential to infect an implant, in fact it occurs infrequently. Excluding total joint prostheses, the risk of secondary infection of other orthopedic implants (plates, screws, wires, etc.) from bacteremia is very small. The risk of secondary infection of a total joint arthroplasty from bacteremia is not well established. Therapeutic management of known infections is well accepted, but the value of prophylactic antibiotics during procedures associated with a risk of bacteremia is less established. Because of the catastrophic nature of these infections, our institution recommends aggressive antibiotic management of known infections (both close to and distant from an implant) and prophylactic antibiotic coverage for dental or surgical procedures that can be associated with transient bacteremia, including dental cleaning procedures.

The American Academy of Orthopaedic Surgery, in conjunction with the American Dental Association, has published an advisory statement clarifying which patients should be treated with routine antibiotic prophylaxis at the time of dental work. Tables 2–4 outline which patients are considered at risk for infection, which procedures are

Table 2
Patients at Increased Risk for Prosthetic Joint Infection Following Dental Procedures

A. Immunocompromised/immunosupressed patients
   Inflammatory arthropathies: rheumatoid arthritis, systemic lupus erythematosis
   Disease, drug or radiation-induced immunosuppression

B. Other patients
   Insulin-dependent (type 1) diabetes
   First 2 years following joint placement
   Previous prosthetic joint infections
   Malnourishment
   Hemophilia

Source: The American Academy of Orthopaedic Surgeons, joint advisory statement from the American Dental Association and The American Academy of Orthopaedic Surgeons.

Table 3
Dental Procedures Considered to Pose Higher and Lower Risk for Causing Prosthetic Joint Infection

Higher risk[a]

   Dental extractions
   Periodontal procedures including surgery, subgingival placement of antibiotic fibers/
      strips, scaling and root planing, probing, recall maintenance
   Dental implant placement and reimplantation of avulsed teeth
   Endodontic (root canal) instrumentation or surgery only beyond the apex Initial placement
      of orthodontic bands but not brackets
   Intraligamentary local anesthetic injections
   Prophylactic cleaning of teeth or implants where bleeding is anticipated

Lower risk[b]

   Restorative dentistry[c] (operative and prosthodontic) with/without retraction cord[d]
   Local anesthetic injections (nonintraligamentary)
   Intracanal endodontic treatment; post-placement and buildup
   Placement of rubber dam
   Postoperative suture removal
   Placement of removable prosthodontic/orthodontic appliances
   Taking of oral impressions
   Fluoride treatments
   Taking of oral radiographs
   Orthodontic appliance adjustment

[a] Prophylaxis should be considered for patients with total joint replacement who meet the criteria in Table 2. No other patients with orthopedic implants should be considered for antibiotic prophylaxis prior to dental treatment/procedures.
[b] Prophylaxis not indicated.
[c] This includes restoration of carious (decayed) or missing teeth.
[d] Clinical judgment may indicate antibiotic use in selected circumstances that may create significant bleeding. (From The American Academy of Orthopaedic Surgeons, joint advisory statement of the American Dental Association and The American Academy of Orthopedic Surgeons, incidence stratification of bacteremic dental procedures.)

Table 4
**Recommended Prophylactic
Antibiotic Dosing for Patients Undergoing Dental Procedures**[a]

Patients not allergic to penicillin: cephalexin, cephradine or amoxicillin: 2 g orally 1 hour prior to dental procedure.

Patients not allergic to penicillin and unable to take oral medications: cefazolin 1 g or ampicillin 2 g intramuscularly/intravenously 1 hour prior to the procedure.

Patients allergic to penicillin: clindamycin: 600 mg orally 1 hour prior to the dental procedure.

Patients allergic to penicillin and unable to take oral medications: clindamycin 600 mg intramuscularly/intravenously 1 hour prior to the procedure.

[a] No second doses are recommended for any of these dosing regimens. (Source: The American Academy of Orthopaedic Surgeons, suggested antibiotic prophylaxis regimens.)

---

more likely to cause bacteremia, and which antibiotic regimens are recommended. There are other protocols for prophylaxis and these are discussed at some length in Chapter 10. We commonly recommend oral cefadroxil (500 mg the day before, the day of, and the day after dental work is performed) as an alternative for dental cleansing manipulations. Until there are studies to show it is ineffective, we will continue to recommend this regimen to our total joint patients.

Along with infection, venous thromboembolism remains a major cause of perioperative morbidity and mortality following total joint arthroplasty and hip fracture fixation. It is the leading cause of re-admission to the hospital following total hip arthroplasty (28). The incidence of deep venous thrombosis following total knee arthroplasty is estimated at up to 80%, whereas the incidence of deep venous thrombosis (DVT) following total hip arthroplasty is estimated at 50% (62,63). DVT following femoral neck or intertrochanteric femur fracture in the elderly is also common (64,65).

DVTs are often characterized as proximal or distal, referring to their relationship to the venous trifurcation in the calf. Proximal thombosis is more likely to result in pulmonary embolism (38). The risk of death from pulmonary embolism has resulted in the adoption of routine anticoagulation prophylaxis following total knee and total hip arthroplasty as well as fixation of hip fractures in the elderly. Warfarin has become the agent most commonly used at our institution to prevent venous thromboembolic disease with the ultimate goal of minimizing the incidence of fatal pulmonary embolism. Warfarin has been shown to significantly reduce proximal DVT formation following lower extremity arthroplasty and has subsequently reduced the postoperative mortality rate attributed to pulmonary embolism to 0.05% from previous rates of 1.71% without postoperative anticoagulation (35).

At our institution, patients are given a daily dose of 5 mg of warfarin orally beginning the evening of surgery. The prothrombin time (PT) and international normalization ratio (INR) are monitored daily while the patient is hospitalized. The goal of therapy is an INR of 2.0 to 2.5. The oral warfarin dose is modified daily to maintain this INR range. Patients have biweekly PT/INR values drawn to assess anticoagulation following discharge from the hospital. Asymptomatic patients are typically anticoagulated for 6 weeks following surgery.

Other anticoagulation regimens have been developed to reduce the risk of hemorrhage associated with warfarin use as well as the cost of multiple laboratory studies to regulate the warfarin dose. A recent prospective, multi-institutional trial with more than 2700 patients showed 153 bleeding complications in more than 2000 patient-years (7.6 per 100 patient-years) *(66)*. Aspirin, subcutaneous heparin, subcutaneous low-molecular-weight (LMW) fractionated heparin, and mechanical modalities such as pneumatic foot and calf pumps are all being used in an effort to decrease postoperative DVT. This is currently an area of active debate but, to date, warfarin remains the standard of care in orthopedic postoperative anticoagulation. On Suspicion of DVT (calf pain and/or tenderness, asymmetric lower extremity edema or swelling, etc.), we perform color flow Doppler ultrasound evaluation of both lower extremities *(67)*. Patients with symptoms of pulmonary embolism are evaluated with high-resolution spiral CT scan of the chest *(68)*. Patients who have DVT or pulmonary embolism are started on intravenous heparin therapy. Some centers are using outpatient LMW heparins as an alternative to inpatient therapy *(69)*. The goal of heparin therapy is a partial thromboplastin time of 60–90 seconds. These patients are then treated with warfarin for a prolonged time period based on the diagnosis: 6 weeks for distal DVT, 3 months for proximal DVT, and 6 months for pulmonary embolism *(70)*.

Treatment of osteoporosis also deserves some attention. This is a major contributing cause of fractures in the elderly, particularly those of the hip and wrist. The rate of hip fractures continues to rise in this country as the number of elderly people increases. The cost is enormous. Recent data estimates the overall cost of treating proximal femur fractures at $8 billion per year *(71)*. The morbidity is great and aggressive medical management of osteoporosis seems warranted. Simple items, such as ambulatory aids and patient education regarding safety measures, are useful but frequently overlooked. In younger patients with trauma, appropriate preventative education is the most important measure.

POSTOPERATIVE FOLLOW-UP

Postoperative follow-up for orthopedic patients varies widely, based on the type of procedure performed and implant received. The major difference in follow-up exists between patients with prosthetic joints and those with fracture fixation. Patients with prosthetic joint replacements are followed on a routine basis for the life of their prosthesis, with more frequent visits in the immediate postoperative course (*see* Table 1). Patients with fracture fixation are typically followed until fracture healing has taken place and then on an as-needed basis. As mentioned previously, occasionally fracture-fixation hardware must be removed. In these cases, the patient is followed postoperatively until the wound has healed.

A thorough review of the literature regarding scheduled postoperative follow-up yielded no prospective studies as to the benefits or drawbacks of any particular postoperative visit schedule. However, many of the leading authors have recommended particular follow-up schedules. One such author recommends initial routine follow-up at 3 to 6 weeks postoperatively, 3 months postoperatively, and yearly thereafter *(72)*. The Mayo Clinic standardized knee evaluation protocol calls for evaluation at 2 to 3 months, 1 year, 2 years, 5 years, and every subsequent 5 years postoperatively *(73)*. Our current recommendations are included in Table 1.

Table 5
Routine Office Follow-Up After Skeletal Trauma: Patient Surveillance After Implantation of Fracture-Fixation Fevice at Saint Louis University

	Postoperative year					
	1	2	3	4	5	10
Office visit	4[a]	0	0	0	0	0
Anterior–posterior and lateral plain X-ray of involved joint	4[a]	0	0	0	0	0

[a]At 2 weeks, 6 weeks, 6 months, and 1 year.

The postoperative follow-up of patients treated for fractures and other traumatic injuries is even more variable and highly individualized to account for the location of the fracture, severity of the injury, type of fixation, and patient characteristics (age, co-morbidities, social situation, etc.). We have included a generalized follow-up protocol for skeletal trauma in Table 5.

Orthopedic implants have only been available for a few decades. Rapid progress has been made without multi-institutional prospective studies, but such studies are needed. They are difficult to carry out for several reasons. Inherent patient factors are quite variable and clearly influence host response to the various biomaterials. Patients must be followed for long periods of time in order to ascertain failure of a total joint implant. Studies of sufficient duration and scope are just now being published in the orthopedic literature. These studies should allow for the rational and evidence-based selection and refinement of the most appropriate treatment methods for total joint prostheses and fracture fixation.

Adequate controlled clinical studies comparing and/or rating joint prostheses have proved difficult to achieve for several reasons. Inherent patient factors such as obesity, diabetes, cigaret smoking, steroid use, age at implantation, patient activity level, and many others are quite numerous and clearly influence the host response to the various biomaterials. Patients must be followed for long periods of time in order to ascertain failure rates of any particular total joint implant. For this reason, there are no adequately controlled studies comparing outcomes among prostheses.

WHAT HEALTH MAINTENANCE MEASURES ARE WARRANTED?

There are a number of health maintenance measures that can have an impact related to the use of implanted biomaterials in orthopedic surgery. Educational measures are currently being developed and utilized by orthopedic surgeons to reduce the risk of injury that occurs secondary to recreational trauma, work-related trauma, and trauma caused by motor vehicle accidents. The American Academy of Orthopaedic Surgery has a wealth of information available at its website and in print regarding specific injury prevention programs, including reducing back injuries, sports injuries, and osteoporosis-related fractures.

Hip fractures account for more than 350,000 hospital admissions each year in the United States (74). Indications are that this number will continue to rise. Education and awareness programs dealing with the elderly patient and hip fractures may ultimately help reduce the numbers of people incurring this devastating injury each year. Better insight into the management of osteoporosis would presumably reduce the incidence of hip fractures. Smoking cessation programs should be recommended to all patients, particularly the elderly. Smoking has been shown to significantly increase the risk of osteoporosis and of resultant hip fractures in the elderly *(75)*. Alcoholism, although not typically associated with musculoskeletal disease, also increases both the risk of osteoporotic fracture and the complications associated with fracture. A recent retrospective study of nearly 900,000 Medicare patients showed that patients with alcoholism are more than two and a half times as likely to suffer a hip fracture when compared to randomly matched nonalcoholics. Of those who do suffer hip fractures, the patients with alcoholism also showed a significantly higher mortality rate at 1 year *(76)*.

Measures to educate the population as a whole may ultimately have a better impact in overall orthopedic patient care than on an individual basis. The effects of smoking have typically already had a toll on patients before their injury or skeletal disease presents itself.

Obesity is often cited as a risk factor for fracture and a contraindication for joint replacement and other orthopedic injury. Although obesity exposes patients to a host of cardiovascular and other health risks, it does not appear to increase the risk of fracture, particularly in the elderly. A recent prospective study of postmenopausal fractures showed a significantly lower risk of fracture in the obese *(77)*. Obesity may pose some intraoperative technical issues for the orthopedic surgeon, but data do not support the belief that obese patients do worse with joint prostheses or wear them out more quickly. Recent studies show no difference in results between obese and nonobese patients at 1-year or 6-year follow-up, as judged by scoring systems, patient satisfaction, or implant survival rates *(78,79)*.

Many patients are taking a more active role in their own health care and seeking out supplementary information regarding their musculoskeletal disease. There are a large number of support groups available to patients who have either had total joint arthroplasty or suffer from one of the conditions frequently leading to prosthetic joint replacement. The groups listed in the Appendix have a large amount of patient-oriented information available either by mail or directly from their Internet sites.

APPENDIX: SUPPORT GROUPS AND PATIENT INFORMATION FOR MUSCULOSKELETAL DISEASES

American Academy of Orthopedic Surgery
6300 North River Road
Rosemont, IL 60018-4262
Phone: (847)-823-7186 or (800)-346-AAOS
Fax: (847)-823-8125
www.aaos.org

American College of Rheumatology
1800 Century Place,
Suite 250
Atlanta, GA 30345
Phone: (404) 633-3777
Fax: (404) 633-1870
www.rheumatology.org

(continued)

APPENDIX: SUPPORT GROUPS AND PATIENT INFORMATION FOR MUSCULOSKELETAL DISEASES (CONTINUED)

Arthritis Foundation
P.O. Box 7669
Atlanta, GA 30357-0669
Phone: (800)-283-7800
www.arthritis.org

Centers for Disease Control and Prevention
1600 Clifton Rd.
Atlanta, GA 30333
Phone: (404)-639-3311
www.cdc.gov

US Food and Drug Administration
5600 Fishers Lane, Rockville MD 20857-0001
Phone: (888)-INFO-FDA (888-463-6332)
www.fda.org

National Institute of Arthritis and Musculoskeletal and Skin Diseases Information Clearinghouse

National Institutes of Health
1 AMS Circle
Bethesda, Maryland 20892-3675
Phone: (301)-495-4484 or (877)-22-NIAMS (toll free)
TTY: (301)-565-2966
Fax: (301)-718-6366
www.niams.nih.gov

REFERENCES

1. Lyons AS, Petrucelli RJ. Medicine: an illustrated history. New York: Harry N. Abrams, Inc., 1978.
2. Fischer LP, Planchamp W, Fischer B, Chauvin F. The first total hip prostheses in man (1890– 1960). Hist Sci Med 2000;34:57–70.
3. Smith-Peterson MN. Anthroplasty of the hip: a new method. J Bone Joint Surg 1939;21: 430–431.
4. Judet J, Judet R, Lagrange J, Dunoyer J. Resection reconstruction of the hip. Arthroplasty with acrylic prosthesis. Edinburgh, UK: E & S Livingston, 1954.
5. Moore AT. Metal hip joint: A new self-locking Vitallium prosthesis. South Med J 1952;45: 1015.
6. Thompson, FR: Two and a half years' experience with a Vitallium intramedulary hip prosthesis. J Bone Joint Surg Am 1954;36A:489–500.
7. Charnley J. Acrylic cement in orthopedic surgery. Edinburgh, UK: E&S Livingston, 1970.
8. Charnley J, Follacci FM, Hammond BT. The long-term reaction of bone to self-curing acrylic cement. J Bone Joint Surg Br 1968;50:822–829.
9. Miller GR Tenzel RR. Orbital fracture repair with methylmethacrylate implants. Am J Ophthalmol 1969;68:717–719.
10. Wiltse LL, Hall RH, Stenehjem, JC. Experimental studies regarding the possible use of self-curing acrylic in orthopedic surgery. J Bone Joint Surg Am 1957;39A:961–972.
11. Berry DJ, Harmsen WS, Cabanela ME, Morrey BF. Twenty-five-year survivorship of two thousand consecutive primary Charnley total hip replacements: factors affecting survivorship of acetabular and femoral components. J Bone Joint Surg Am 2002;84-A(2):171–177.

12. Charnley J. Total hip replacement by low-friction arthroplasty. Clin Orthop 1970;72:7–21.
13. Lane, WA. The operative treatment of fractures. London: Medical Publishing, 1914.
14. Mears DC. Materials and orthopaedic surgery. Baltimore, MD: Williams & Wilkins, 1979.
15. Weinstein, J. The Dartmouth atlas of musculoskeletal health care. Chicago, IL: AHA Press, 2000.
16. National Centers for Health Statistics, National Hospital Discharge Survey. Data extracted and analyzed by the American Academy of Orthopaedic Surgeons, Dept. of Research and Scientific Affairs. Washington DC.
17. Ohzawa S, Takahara Y, Furumatsu T, Inoue H. Patient survival after total knee arthroplasty. Acta Med Okayama 2001;55:295–299.
18. Tayot O, Ait Si Selmi T, Neyret P. Results at 11.5 years of a series of 376 posterior stabilized HLS1 total knee replacements. Survivorship analysis, and risk factors for failure. Knee 2001;8:195–205.
19. Back DL, Cannon SR, Hilton A, Bankes MJ, Briggs TW. The Kinemax total knee arthroplasty. Nine years' experience. J Bone Joint Surg Br 2001;83:359–363.
20. Hamadouche M, Kerboull L, Meunier A, Courpied JP, Kerboull M. Total hip arthroplasty for the treatment of ankylosed hips: a five to twenty-one-year follow-up study. J Bone Joint Surg Am 2001;83-A:992–998.
21. Urban JA, Garvin KL, Boese CK, et al. Ceramic-on-polyethylene bearing surfaces in total hip arthroplasty. Seventeen to twenty-one-year results. J Bone Joint Surg Am 2001;83-A: 1688–1694.
22. Buckwalter JA, Einhorn TA, Simon SR. Orthopaedic basic science (2nd ed.). Rosemont, IL: American Academy of Orthopaedic Surgery, 2000.
23. Elves MW, Wilson JN, Scales JT, Kemp HB. Incidence of metal sensitivity in patients with total joint replacements. Br Med J 1975;4:376–378.
24. Mow VC, Hayes WC. Basic orthopedic biomechanics (2nd ed.). Philadelphia, PA: Lippincott-Raven, 1997.
25. Bostman O, Pihlajamaki H. Routine implant removal after fracture surgery: a potentially reducible consumer of hospital resources in trauma units. J Trauma 1996;41:846–849.
26. Brown OL, Dirschl DR, Obremskey WT. Incidence of hardware-related pain and its effect on functional outcomes after open reduction and internal fixation of ankle fractures. J Orthop Trauma 2001;15:271–274.
27. Bozic KJ, Perez LE, Wilson DR, Fitzgibbons PG, Jupiter JB. Mechanical testing of bioresorbable implants for use in metacarpal fracture fixation. J Hand Surg Am 2001;26:755–761.
28. Seagroatt V, Tan HS, Goldacre M, Bulstrode C, Nugent I, Gill L. Elective total hip replacement: incidence, emergency readmission rate, and postoperative mortality. BMJ 1991;303(6815):1431–1435.
29. Heck DA, Partridge CM, Reuben JD, Lanzer WL, Lewis CG, Keating EM. Prosthetic component failures in hip arthroplasty surgery. J Arthroplasty 1995;10:575–580.
30. Palmer SH, Morrison PJ, Ross AC. Early catastrophic tibial component wear after unicompartmental knee arthroplasty. Clin Orthop 1998;350:143–148.
31. Van Audekercke R, Martens M, Mulier JC, Stuyck J. Experimental study on internal fixation in femoral neck fractures. Clin Orthop 1979;141:203–212.
32. Koval KJ. Orthopedic knowledge update (7th ed.). Rosemont, IL: American Academy of Orthopaedic Surgery, 2002.
33. Kurtz SM, Muratoglu OK, Evans M, Edidin AA. Advances in the processing, sterilization, and crosslinking of ultra-high molecular weight polyethylene for total joint arthroplasty. Biomaterials 1999;20:1659–1688.
34. Patel AM, Spector M. Tribological evaluation of oxidized zirconium using an articular cartilage counterface: a novel material for potential use in hemiarthroplasty. Biomaterials 1997;18:441–447.

35. Pellici PM, Tria AJ, Garvin KL, editors. Orthopaedic knowledge update: hip and knee reconstruction (2nd ed.). Rosemont, IL: American Academy of Orthopaedic Surgery, 2000.
36. Bobyn JD, Stackpool GJ, Hacking SA, Tanzer M, Krygier JJ. Characteristics of bone ingrowth and interface mechanics of a new porous tantalum biomaterial. J Bone Joint Surg Br 1999;81:907–914.
37. Robinson CM, Adams CI, Craig M, Doward W, Clarke MC, Auld J. Implant-related fractures of the femur following hip fracture surgery. J Bone Joint Surg Am 2002;84-A(7): 1116–1122.
38. Beaty, JH. Orthopaedic knowledge update (6th ed.). Rosemont, IL: American Academy of Orthopaedic Surgery, 1999.
39. Udomkiat P, Wan Z, Dorr LD. Comparison of preoperative radiographs and intraoperative findings of fixation of hemispheric porous-coated sockets. J Bone Joint Surg Am 2001; 83-A(12):1865–70.
40. Hunter JC, Hattner RS, Murray WR, Genant HK. Loosening of the total knee arthroplasty: detection by radionuclide bone scanning. AJR Am J Roentgenol 1980;135:131–136.
41. Lieberman JR, Huo MH, Schneider R, Salvati EA, Rodi S. Evaluation of painful hip arthroplasties. Are technetium bone scans necessary? J Bone Joint Surg Br 1993;75:475–478.
42. Stromberg CN, Herberts P, Palmertz B. Cemented revision hip arthroplasty. A multicenter 5-9-year study of 204 first revisions for loosening. Acta Orthop Scand 1992;63:111–119.
43. Hanssen AD, Rand JA. Evaluation and treatment of infection at the site of a total hip or knee arthroplasty. Instr Course Lect 1999;48:111–122.
44. Ritter MA. Operating room environment. Clin Orthop 1999;369:103–109.
45. Robbins GM, Masri BA, Garbuz DS, Duncan CP. Evaluation of pain in patients with apparently solidly fixed total hip arthroplasty components. J Am Acad Orthop Surg 2002; 10:86–94.
46. Nasser S. The incidence of sepsis after total hip replacement arthroplasty. Semin Arthroplasty 1994;5:153–159.
47. Moore AT, Bohlman HR. Metal hip joint: a case report. J Boint Surg 1943;25A:688–692.
48. Hanssen AD. Managing the infected knee: as good as it gets. J Arthroplasty 2002;17(4 Suppl 1):98–101.
49. Kolstad K, Levander H. Inflammatory laboratory tests after joint replacement surgery. Ups J Med Sci 1995;100:243–248.
50. Moreschini O, Greggi G, Giordano MC, Nocente M, Margheritini F. Postoperative physiopathological analysis of inflammatory parameters in patients undergoing hip or knee arthroplasty. Int J Tissue React 2001;23:151–154.
51. Williams ED, Tregonning RJ, Hurley PJ. 99Tcm-diphosphonate scanning as an aid to diagnosis of infection in total hip joint replacements. Br J Radiol 1977;50:562–566.
52. Schneider R, Gruen D, Brause B. Diagnosis of infected joint prostheses. Semin Arthroplasty 1995;6:167–175.
53. Itasaka T, Kawai A, Sato T, Mitani S, Inoue H. Diagnosis of infection after total hip arthroplasty. J Orthop Sci 2001;6:320–326.
54. Hakki S, Harwood SJ, Morrissey MA, Camblin JG, Laven DL, Webster WB Jr. Comparative study of monoclonal antibody scan in diagnosing orthopaedic infection. Clin Orthop 1997;335:275–285.
55. Munjal S, Phillips MJ, Krackow KA. Revision total knee arthroplasty: planning, controversies, and management—infection. Instr Course Lect 2001;50:367–377.
56. Windsor RE. Management of total knee arthroplasty infection. Orthop Clin North Am 1991; 22:531–538.
57. Fenelon GC, Von Foerster G, Engelbrecht E. Disarticulation of the hip as a result of failed arthroplasty. A series of 11 cases. J Bone Joint Surg Br 1980;62-B:441–446.
58. Kostuik J, Alexander D. Arthrodesis for failed arthroplasty of the hip. Clin Orthop 1984; 188:173–182.

59. Hallab N, Merritt K, Jacobs JJ. Metal sensitivity in patients with orthopaedic implants. J Bone Joint Surg Am 2001;83-A:428–436.
60. Simon SR, ed. Orthopaedic basic science. Rosemont, IL: American Academy of Orthopaedic Surgeons, 1994.
61. Wooley PH, Nasser S, Fitzgerald RH. The immune response to implant materials in humans. Clin Orthop 1996;326:63–70.
62. Clagett GP, Anderson FA Jr, Heit J, Levine MN, Wheeler HB. Prevention of venous thromboembolism. Chest 1995;108(4 Suppl):312S–334S.
63. Cohen SH, Ehrlich GE, Kauffman MS, Cope C. Thrombophlebitis following knee surgery. J Bone Joint Surg Am 1973;55:106–112.
64. Castle ME, Orinion EA. Prophylactic anticoagulation in fractures. J Bone Joint Surg Am 1970;52:521–528.
65. Culver D, Crawford JS, Gardiner JH, Wiley AM. Venous thrombosis after fractures of the upper end of the femur. A study of incidence and site. J Bone Joint Surg Br 1970;52:61–69.
66. Palareti G, Leali N, Coccheri S, Poggi M, Manotti C, D'Angelo A, Pengo V, Erba N, Moia M, Ciavarella N, Devoto G, Berrettini M, Musolesi S. Bleeding complications of oral anticoagulant treatment: an inception-cohort, prospective collaborative study (ISCOAT). Italian Study on Complications of Oral Anticoagulant Therapy. Lancet 1996;348:423–428.
67. Leutz DW, Stauffer ES. Color duplex Doppler ultrasound scanning for detection of deep venous thrombosis in total knee and hip arthroplasty patients. Incidence, location, and diagnostic accuracy compared with ascending venography. J Arthroplasty 1994;9:543–548.
68. Remy-Jardin M, Remy J. Spiral CT angiography of the pulmonary circulation. Radiology 1999;212:615–636.
69. Levine M, Gent M, Hirsh J, et al. A comparison of low-molecular-weight heparin administered primarily at home with unfractionated heparin administered in the hospital for proximal deep-vein thrombosis. N Engl J Med 1996;334:677–681.
70. Deitcher SR, Carman TL. Deep venous thrombosis and pulmonary embolism. Curr Treat Options Cardiovasc Med 2002;4:223–238
71. Kellam JF, Fisher TJ, Tornetta III P, Bosse MJ, Harris MB. Orthopaedic knowledge update: trauma (2nd ed.). Rosemont, IL: American Academy of Orthopaedic Surgery, 2000.
72. Callaghan JJ, Rosenberg AG, Rubash HE. The adult hip. Philadelphia, PA: Lippincott-Raven, 1998.
73. Morrey BF, ed. Total joint replacement arthroplasty. New York: Churchill-Livingstone Inc., 1991.
74. Rodrigues J, Sattin RW, Waxweiler RJ. Incidence of hip fractures, United States, 1970–83. Am J Prev Med 1989;5:175–181.
75. Hollenbach KA, Barrett-Connor E, Edelstein SL, Holbrook T. Cigarette smoking and bone mineral density in older men and women. Am J Public Health 1993;83:1265–1270.
76. Yuan Z, Dawson N, Cooper GS, Einstadter D, Cebul R, Rimm AA Effects of alcohol-related disease on hip fracture and mortality: a retrospective cohort study of hospitalized Medicare beneficiaries. Am J Public Health 2001;91:1089–1093.
77. Kato I, Toniolo P, Zeleniuch-Jacquotte A, et al. Diet, smoking and anthropometric indices and postmenopausal bone fractures: a prospective study. Int J Epidemiol 2000;29:85–92.
78. Deshmukh RG, Hayes JH, Pinder IM. Does body weight influence outcome after total knee arthroplasty? A 1-year analysis. J Arthroplasty 2002;17:315–319.
79. Spicer, DD, Pomeroy, DL, Badenhausen, WE, Jr, et al. Body mass index as a predictor of outcome in total knee replacement. Int Ortho 2001;25:246–249.

US Counterpoint to Chapter 24

James A. Keeney and John C. Clohisy

We extend our commendation to Drs. Otto and L'Hommedieu for their thorough review in Chapter 24 of orthopedic biomaterials technology and the management of patients with implanted orthopaedic devices. Our practice does not differ significantly from theirs. Improvements in aseptic technique, availability of antibiotic prophylaxis, and enhancements in both internal fixation devices and joint prostheses have revolutionized orthopedic care since the 1950s, improving the quality of life for millions of patients across all orthopedic subspecialty areas.

In total hip arthroplasty, the work of Sir John Charnley remains the gold standard (1). The wide variation in available implants for total hip and total knee replacement surgery, however, attests not only to the progress in biomaterials technology but also to the release of some implants prior to proven long-term efficacy. The recognition of high rates of late aseptic loosening of cemented acetabular components has led to an increased utilization of uncemented acetabular fixation (2). Improvements in femoral stem geometry and circumferential coating of uncemented femoral components have decreased the rate of mechanical failure of the femoral component and femoral loosening, respectively. Cemented femoral fixation has been enhanced by third generation improvements in the cement mantle (centrifugation, retrograde insertion and pressurization) (3,4). It is not clear whether intended improvements in cemented femoral stems made since the 1980s have translated into superior clinical outcomes. Textured, precoated stems were developed to provide enhanced bonding with cement at insertion but may create increased cement debris once debonding occurs (5,6). Early failure of some of these "improved" cemented designs has led some surgeons to return to the use of highly polished stems as advocated by Charnley in his initial arthroplasty work. Studies have demonstrated that, although polished stems may debond from the cement mantle, this does not correlate with mechanical or clinical failure (7).

The recognition of the biological response to particulate polyethylene wear debris as the cause of osteolysis (8) and as the major long-term problem with total joint arthroplasty (9) has led to improvements in polyethylene processing and renewed interest in alternative bearing surfaces. Enhanced polyethylene crosslinking and compression molding of polyethylene inserts have improved wear characteristics. Emphasis has also been placed on reducing "back-side wear" (10) by improving the locking mechanisms

From: *The Bionic Human: Health Promotion for People With Implanted Prosthetic Devices*
Edited by: F. E. Johnson and K. S. Virgo © Humana Press Inc., Totowa, NJ

between metal and polyethylene components for both total hip and total knee arthroplasty.

Although our practice continues to use cobalt chromium surfaces against highly crosslinked polyethylene, we recognize that other centers are using alternative bearing surfaces with at least good short- and medium-term results *(11)*. Ceramics provide the theoretical advantage of decreased wear debris, although the mechanism of failures in the absence of wear has not been elucidated *(11,12)*. The significance, or insignificance, of elevated serum and urinary metal ions in patients with metal on metal articulations likewise has not been fully explained *(13)*. We are waiting for longer term follow-up results before incorporating these potentially promising alternatives into our clinical practice.

In addition to advances in implant design and the development of "low-wear" biomaterials, the biology of implant particle-induced osteolysis continues to be investigated. The identification of critical biologic factors that mediate particle-driven osteoclast activation and periprosthetic osteolysis may result in therapeutic strategies to block these processes at the cellular and molecular levels. Since the late 1990s, essential molecules have been targeted in experimental systems. Tumor necrosis factor-α has been identified as a pro-osteoclastogenic cytokine essential in particle-induced intracellular signaling *(14)* and particle-stimulated osteolysis in vivo *(15)*. Another molecule, receptor activator of nuclear factor (NF)-$\kappa\beta$ ligand and its nuclear transcription factor, NF-$\kappa\beta$, have also been shown to be essential for particle-stimulated osteoclast formation in vitro *(16,17)*. The pharmacological regulation of these and other processes may play a significant role in targeted inhibition of periprosthetic osteolysis in the future.

Surveillance of implants used in fracture care differs from those used in total joint arthroplasty. Fractures are followed at regular intervals until fracture union has been obtained. Nonsurgical or operative interventions can be undertaken as needed to enhance fracture healing if delayed union or nonunion is evident. Failure of hardware typically occurs from fatigue (related to delay in fracture healing rather than mechanical failure) unless the particular fracture geometry relies excessively on the fixation device for support. In such cases, restricted weight-bearing or augmentation of stability with a cast, brace, or other external support should be considered until adequate fracture healing is noted on radiographs.

Table 1 summarizes our recommendations for the management of patients with orthopedic fracture-fixation devices. It should be emphasized that the purpose of radiographic evaluation is to monitor the healing process more than the status of the orthopedic implant. Because the time to union varies by location of the fracture, final X-rays of fracture healing may be as early as 6 weeks and as long as 1 year following injury or surgery. Once fracture healing is complete, the implanted orthopedic hardware is no longer primarily involved in load-bearing and risk of spontaneous fracture or other complication is low. Routine surveillance with X-rays following fracture healing to monitor the status of the hardware is not cost-effective and will not likely be reimbursed under current health care guidelines in the United States.

It is recognized that smooth pins inserted for maintaining fracture reduction are at risk of migration *(18,19)* and they are routinely removed. Pediatric patients with retained hardware following fracture union may have elective hardware removal

Table 1
Routine Office Follow-Up After Implantation of Fracture-Fixation Device: Patient Surveillance at Washington University

	Postoperative year					
	1	2	3	4	5	10
Office visit	3[a]	0	0	0	0	0
Anterior–posterior and lateral plain X-ray of involved bone	3[a]	0	0	0	0	0

[a]At 2 weeks, 6 weeks, and 3 months.

If the fracture is healed and the patient has normal functional recovery, further office visits and imaging tests are not required. Additional office visits and imaging tests may be recommended until functional recovery is attained.

Table 2
Routine Office Follow-Up After Total Joint Arthroplasty: Patient Surveillance at Washington University

	Postoperative year					
	1	2	3	4	5	10
Office visit	4[a]	1	1	1	1[b]	1
Anterior–posterior and lateral plain X-ray of involved joint	3[c]	1	1	1	1[b]	1

[a]At 6 weeks, 3 months, 6 months, and 1 year.
[b]We recommend follow-up at 7 years using the same modalities as at 5 years follow-up for high demand, young patients (<50 years old).
[c]At 6 weeks, 6 months, and 1 year.

because appositional bone growth could encase the hardware and complicate future elective removal if it were to become necessary. Theoretically, a rigid metal plate could also function as a "stress riser" in this active subset of patients and increase the risk of subsequent fracture at either end of the plate. Patients with anticipated reconstructive surgery may have staged implant removal, especially where cortical deficiencies could be expected to negatively influence cement pressurization at the time of joint replacement surgery.

Our routine surveillance for joint arthroplasty implants is summarized in Table 2. In general, patients are seen at 6 weeks, 3, 6, and 12 months for clinical evaluation following the initial procedure. Radiographs are obtained at 6 weeks, 6 and 12 months and subsequent annual follow-up. The examination at 3 months includes a clinical assessment of range of motion for knee replacements and functional assessment of total hip replacement patients who are allowed to discontinue joint dislocation precautions after the 3-month visit. Ongoing radiographic surveillance is directed to identify early symptomatic loosening, polyethylene wear, or progressive asymptomatic osteolysis. It has been recognized that accelerated polyethylene wear may occur in patients with improved functional scores and higher activity levels in patients under the age of 50 (20). It must be emphasized that implant-associated osteolysis is frequently a clinically silent disease that progresses over time and can result in massive periprosthetic bone loss and/or catastrophic implant failure. If this occurs, revision surgery to reconstruct the affected joint can be very complex. Therefore, we strongly recommend continued implant surveillance over the patient's lifetime. This enables early detection and surgical treatment (21,22) that should prevent catastrophic implant failure.

Findings on radiographs suggestive of acetabular loosening include circumferential radiolucency around the acetabular component, change in cup position, defects in the acetabular cement mantle, if present, or inferior/ischial lysis, suggesting cup migration. Circumferential radiolucency around femoral stem or cement mantle, cement fracture, subsidence of the stem, change of stem position in varus/valgus, or development of a pedestal below the implant without evidence of bony ingrowth are all suggestive or femoral component loosening. Progressive osteolysis around stable femoral or acetabular components associated with asymmetrical wear of the polyethylene is also an indication for revision surgery.

Prevention of infection following surgery begins with adherence to principles of aseptic surgical technique. Our operating rooms are also equipped with horizontal laminar airflow and ventilated suits are utilized for all joint replacement procedures. Patients receive Gram-positive antibiotic coverage (Ancef or Vancomycin) given 30 minutes prior to incision and continuing for 48 hours postoperatively. We also incorporate tobramycin into polymethylmethacrylate (PMMA)cement for both primary and revision procedures.

Management of periprosthetic infections has been delineated by others (23–26). Signs of deep wound infection include fever, pain with joint motion, purulent wound drainage, and diffuse incisional erythema. Focal erythema around the incision may be associated with a sterile suture abscess, but the threshold for intervention should be low as the potential for adverse outcome in delayed treatment of a periprosthetic infection is significant (26). In our practice, patients presenting with an early deep wound infection (within 3 weeks of surgery) may be considered for liner exchange, wound irrigation and debridement followed by parenteral antibiotics for 6 weeks. Infectious disease specialist consultation is routinely obtained for optimization of antibiotic therapy.

Patients presenting with deep wound infections between 3 and 12 months following surgery are typically treated with a two-stage reconstruction. All components are removed and intraoperative cultures are taken from multiple sites. Irrigation and debridement is performed. PMMA spacers impregnated with tobramycin are placed for total knee arthroplasty for stability, local antibiotic delivery, and facilitation of exposure at time of reimplantation surgery. Antibiotic spacers are not typically used for hip reconstruction surgery, as our impression has been that instability following re-implantation is less likely with contracture of the soft-tissue envelope.

Patients with late hematogenous infection, typically more than 24 months following surgery, may be treated with component retention if the patient presents within the first 24 to 48 hours following the onset of symptoms. Delay in presentation leads to less predictable salvage with liner exchange and debridement. The patient's overall medical condition and condition of the soft-tissue envelope also play in to the decision to retain components instead of performing a two-stage revision.

Aseptic loosening is more common following total joint replacement than septic loosening. A 1% per year failure rate from aseptic loosening is expected over the lifetime of a total hip or total knee implant. Septic loosening is reported to occur at 1% incidence within the initial year and up to 2% for the lifetime of the prosthesis. Nonetheless, all patients presenting with presumed aseptic loosening are evaluated to rule out sepsis prior to revision surgery. Clinical history suggestive of infection includes failure of pain relief following initial surgery, fever (>101.5°F) in the immediate perioperative period, management with oral or parenteral antibiotics beyond the initial 48 hours postoperatively, and history of prolonged wound drainage following surgery.

We evaluate all patients with suspected aseptic loosening with serological studies: complete blood count (CBC), erthryocyte sedimentation rate (ESR), and C-reactive protein (CRP). Patients with elevated ESR or CRP undergo aspiration (in the office for knee, in the radiology suite for hips). Tissue is sent at the time of revision surgery for patients with suspected aseptic loosening. Presence of greater than five white blood cells per high-power field suggests infection and treatment is undertaken as outlined above. Technetium and indium scans are not frequently obtained in our practice.

As noted in Chapter 24, the American Dental Association and the American Academy of Orthopedic Surgeons have published recommendations for prophylactic antibiotics around dental procedures. This risk may also be considered for other instrumented procedures (e.g., cystoscopy, colonoscopy). It is our general practice to provide antibiotic prophylaxis for all of these procedures due to the potential debilitation from a periprosthetic infection.

In general, patients are encouraged to consult with their primary care physician for evaluation of infection at sites distant to their prosthetic joint. Early recognition and treatment of bacterial infections (e.g., pneumonia, urinary tract infection, cellulitis, dental abscess, etc.) is recommended to prevent the development of sepsis and the potential for seeding the prosthetic joint. Antibiotic choice and duration of treatment is dictated by the primary infection source and not typically influenced otherwise by the presence of a total joint replacement.

Like Drs. Otto and L'Hommedieu, our practice is to use coumadin for deep venous thrombosis (DVT) prophylaxis. We also use mechanical compression devices during the hospitalization and thigh-high anti-thrombotic stockings for 6 weeks following reconstructive surgery. Although this is our practice, we recognize that several options for DVT prevention have been supported by the literature.

Future improvements in biomaterials science are likely to be coupled with modification of biological processes responding to the presence of implanted biomaterials. Hopefully, this will yield a net result of prolonged prosthesis retention and sustained independent patient function. Although there will still be a place for revision surgery, the frequently expressed goal of a "permanent" joint replacement may be attainable. Readers interested in further information should consult the sources cited in Chapter 24. We believe they are the best available starting points for patients with questions.

REFERENCES

1. Callaghan JJ, Albright JC, Goetz DD, Olejniczak JP, Johnston RC. Charnley total hip arthroplasty with cement. Minimum twenty-five year follow up. J Bone Joint Surg Am 2000;82:487.
2. Clohisy JC, Harris WH. Matched-pair analysis of cemented and cementless acetabular reconstruction in primary total hip arthroplasty. J Arthroplasty 2001;16:697–705.

3. Clohisy JC, Harris WH. Primary hybrid total hip replacement performed with insertion of the acetabular component without cement and a precoat femoral component with cement. An average ten-year follow-up study. J Bone Joint Surg Am 1999;81:247–255.
4. Oishi CS, Walker RH, Colwell CW. The femoral component in total hip arthroplasty. Six to eight year follow up of one hundred consecutive patients after use of a third generation cementing technique. J Bone Joint Surg Am 1994;76:1130–1136.
5. Collis DK, Mohler CG. Comparison of clinical outcomes in total hip arthroplasty using rough and polished cemented stems with essentially the same geometry. J Bone Joint Surg Am 2002;84:586–592.
6. Sporer SM, Callaghan JJ, Olejniczak JP, Goetz DD, Johnston RC. The effects of surface roughness and polymethylmethacrylate precoating on the radiographic and clinical results of the Iowa hip prosthesis. A study of patients less than fifty years old. J Bone Joint Surg Am 1999;81:481–492.
7. Berry DJ, Harmsen WS, Ilstrup DM. The natural history of debonding of the femoral component from the cement and its effect on long term survival of Charnley total hip replacements. J Bone Joint Surg Am 1998;80:715–721.
8. Maloney WJ, Smith RL. Periprosthetic osteolysis in total hip arthroplasty: the role of particulate wear debris. J Bone Joint Surg Am 1995;77:1448–1461.
9. Harris WH. The problem is osteolysis. Clin Orthop 1995;311:46–53.
10. Mikulak SA, Mahoney OM, DelaRosa MA, Schmalzried TP. Loosening and osteolysis with the Press-Fit Condylar posterior cruciate-substituting total knee replacement. J Bone Joint Surg Am 2001;83:398.
11. D'Antonio J, Capello W, Manley M, Bierbaum B. New experience with alumina on alumina ceramic bearings for total hip arthroplasty. J Arthroplasty 2002;17:390–397.
12. Hamadouche M, Boutin P, Daussange J, Bolarden ME, Sedel L. Alumina on alumina total hip arthroplasty. A minimum 18.5 year study. J Bone Joint Surg Am 2002;84:69–77.
13. Savarino L, Granchi G, Cenni E, et al. Ion release in patients with metal on metal hip bearings in total joint replacement: a comparison with metal on polyethylene bearings. J Biomed Mater Res 2002;63:467–474.
14. Merkel KD, Erdmann JM, McHugh KP, Abu-Amer Y, Ross FP, Teitelbaum SL. Tumor necrosis factor-alpha mediates orthopaedic implant osteolysis. Am J Pathol 1999;154:203–210.
15. Schwarz EM, Lu AP, Goater JJ, et al. Tumor necrosis factor-alpha/nuclear transcription factor-kappa B signaling in periprosthetic osteolysis. J Orthop Res 2000;18:472–480.
16. Clohisy JC, Hirayama T, Frazier E, Han S, Abu-Amer Y. NF-KB signaling blockade abolishes implant particle-induced osteoclastogenesis. J Orthop Res 2004;22:13–20.
17. Clohisy, JC, Frazier E, Hirayama T, Abu-Amer Y. RANKL is an essential cytokine mediator of polymethylmethacrylate particle-induced osteoclastogenesis. J Orthop Res 2003;21:202–212.
18. Goodset JR, Pahl AC, Glaspy JN, Schapira MM. Kirschner wire embolization to the heart: an unusual cause of pericardial tamponade. Chest 1999;115:291–293.
19. Stahl S, Schwartz O. Complications of K-wire fixation of fractures and dislocations in the hand and wrist. Arch Orthop Trauma Surg 2001;121:527–530.
20. Crowther JD, Lachiewicz P. Survival and polyethylene wear of porous coated acetabular components in patients less than 50 years old: results at 9 to 14 years. JBJSA 2002;84A:729–735.
21. Maloney WJ, Herzwurm P, Paprosky W, Rubash HE, Engh CA. Treatment of pelvic osteolysis associated with a stable acetabular component inserted without cement as part of a total hip replacement. J Bone Joint Surg Am 1997;79:1628–1634.
22. Maloney WJ, Paprosky W, Engh CA, Rubash HE. Treatment of pelvic osteolysis. Clin Orthop 2001;393:78–84.

23. Segawa H, Tsukayama DT, Kyle RF, Becker DA, Gustilo RB. Infection after total knee arthroplasty. A retrospective study of the treatment of eighty-one infections. J Bone Joint Surg Am 1999;81:1434–1445.
24. Tsukayama DT, Estrada R, Gustilo RB. Infection after total hip arthroplasty. A study of the treatment of one hundred and six infections. J Bone Joint Surg 1996;78:512–523.
25. Hanssen AD, Rand JA. Instructional Course Lectures, the American Academy of Orthopaedic Surgeons-Evaluation and treatment of infection at the site of a total hip or knee arthroplasty. J Bone Joint Surg Am 1998;80:910–922.
26. Spangehl MJ, Younger SE, Masri BA, Duncan CP. Instructional Course Lectures, the American Academy of Orthopaedic Surgeons-Diagnosis of infection following total hip arthroplasty. J Bone Joint Surg Am 1997;79:1578–1588.

European Counterpoint to Chapter 24

Michael P. Manning

TOTAL JOINT ARTHROPLASTY

Drs. Otto and L'Hommedieu are to be commended for their account in Chapter 24 of the development of orthopedic implants. There are, however, a few points where I would disagree with their interpretation of history. They state that Charnley used a chrome-cobalt and high-density polyethylene combination and coined the term "low-friction arthroplasty" to describe the combination. In fact, Charnley used a stainless steel stem against a high-density polyethylene socket and the term "low-friction arthroplasty" is a contraction of the term "low-frictional torque arthroplasty." The term does not describe the friction of the bearing surfaces but rather the low-turning moment on the outer surface of the socket consequent on the significant difference in diameter between the outer surface and the bearing surface arising as a result of the small femoral head *(1)*. Charnley hoped that low frictional torque would reduce the force on the socket cement–bone interface and thereby reduce loosening rates. Stainless steel is still employed in the Charnley implant, the latest stems being made of a trademarked high nitrogen cold-worked stainless steel called Ortron 90 (De Puy, Leeds, UK). Just as in the United States, British surgeons strive to improve the long-term results of joint replacement by employing cementless devices, alternate bearing surfaces (ceramic and metal on metal), and by revisiting the concept of surface replacement.

There were 44,048 total hip replacements and 35,351 total knee replacements performed in the United Kingdom in 2000. The estimated cost of these procedures was £405 million (approximately US $648 million) *(2)*. The follow-up of well-functioning implants has often been sporadic, with many patients being discharged at about 1 year postoperatively *(3)*. Primary hip replacements generally start to fail between 5 and 10 years, many more failing after 10 years *(4)*.

The British Orthopaedic Association has recently published a statement of "best practice" in which it was concluded that patients should be followed up clinically and radiologically in the long term, the minimum required being an antero-posterior and lateral X-ray of the hip at 5 years and every 5 years thereafter *(5)*. Should problems of wear or loosening become apparent, follow-up must be performed more frequently and revision considered. The aim of long-term follow-up is to detect aseptic loosening and polyethylene wear at an early stage and intervene before severe bone loss occurs. Symp-

Table 1
Follow-Up After Primary Total Hip Replacement:
Patient Surveillance at Whiston Hospital, UK

	Postoperative year					
	1	2	3	4	5	10[a]
Office visit	3	1	1	1	1	1
X-rays (antero-posterior)	2	1	1	1	1	1
X-rays (lateral)	0	0	1	1	1	1
Other tests	0	0	0	0	0	0

[a]And every 5 years thereafter unless problems develop.

Table 2
Follow-Up After Primary Total Knee Replacement:
Patient Surveillance at Whiston Hospital, UK

	Postoperative year					
	1	2	3	4	5	10[a]
Office visit	3	1	1	1	1	1
X-rays (antero-posterior and lateral)	2	1	1	1	1	1
Other tests	0	0	0	0	0	0

[a]And every 5 years thereafter unless problems develop.

Table 3
Follow-Up Revision Hip and Knee Replacements:
Patient Surveillance at Whiston Hospital, UK

	Postoperative year					
	1	2	3	4	5	10[a]
Office visit	3	1	0	1	0	1
X-rays (antero-posterior and lateral)	2	1	0	1	0	1
Other tests	0	0	0	0	0	0

[a]And every 2 years thereafter unless problems develop.

toms of socket wear and loosening in particular are often silent initially and significant radiological features can predate symptoms by up to 2 years (6).

There is no reason to suggest that knee replacements should behave any differently and, although there is as yet no official published consensus, clearly long-term follow-up is desirable. Our practice with revision hip and knee replacements is to follow patients indefinitely at two yearly intervals. Again, if untoward features are detected on follow-up X-rays, review must be performed more frequently and repeat revision surgery considered. Tables 1–3 suggest follow-up regimens. The number of shoulder, elbow, ankle, and other joint replacements is small, but exact figures for the number of these procedures performed are not available. Again, there is nothing to suggest that these implants behave differently to hip replacements, and long-term follow-up seems reasonable. Using the methodology employed in Chapter 24, the number of people in the United Kingdom living with a total joint replacement is projected to be about 700,000 by the year 2010. The Department of Health and the Welsh Assembly Government have set up the National Joint Registry to create a database of hip and knee arthroplasties performed in England and Wales. The Registry became operational on April 1, 2003. Professional bodies encourage surgeons to enter data following primary and revision procedures. In time, it is hoped that good and bad implant devices will be identified at an earlier stage than previously and that surgical practice will be enhanced generally by the identification of best practice.

TOTAL JOINT ARTHROPLASTY FAILURE

Current practice in the United Kingdom is similar to that advocated in Chapter 24 in that a careful history and examination followed by simple X-ray imaging is the mainstay of the assessment of possible implant failure, just as it forms the basis of follow-up of well-functioning implants (5,6). The management of wear, aseptic loosening, and prosthetic infection is very similar in the United States and the United Kingdom. It is, however, worthy of note that one-stage revision, as opposed to two-stage revision, for deep infection has proved successful in more than 91% of patients in one series (7) and continues to be practiced where the patient's frailty might preclude two-stage surgery and where the infecting organism is known.

Infection developing elsewhere in the body, and the potential effect of that infection on a well-functioning total joint replacement, was reviewed by Ainscow and Denham (8). They showed that patients with rheumatoid arthritis and skin ulceration were particularly prone to develop hematogenous infection of a previously satisfactory total joint replacement. They prospectively reviewed 1000 patients who had received 1112 total joint replacements, (984 total hip replacements and 128 total knee replacements). Over a 6-year period, three patients developed hematogenous infection and subsequent failure of their total joint replacement. Two of them had rheumatoid arthritis and all three had suffered skin ulceration (the first had infected skin over a rheumatoid nodule, the second had septicaemia from varicose ulcers, and the third had ulceration over an infected bunion). In contrast, 228 patients had developed urinary tract infection, respiratory infection, or multiple infections at various sites; some had also undergone dental or surgical procedures but none had developed hematogenous infection of their total joint replacements. This study suggests that prophylactic antibiotics are not routinely required to cover "clean" dental and surgical procedures, and that, with the exception of infected skin lesions, which require prompt treatment, infections occurring elsewhere in an otherwise fit patient with a satisfactory total joint replacement do not require specific treatment over and above that which would be required to treat the infection in a patient without a total joint replacement.

IMPLANTS FOR FRACTURE FIXATION

The practice of fracture fixation is very similar in the United Kingdom and the United States. By far the most common fractures presenting for surgical treatment involve the proximal femur in elderly patients. These occur at a rate of 120 per 100,000 population

Table 4
Follow-Up of Internally Fixed Fractures[a] **(Exact Details May Vary Depending On Specific Fracture): Patient Surveillance at Whiston Hospital, UK**

	Postoperative year					
	1	2	3	4	5	10
Office visit	3	1[b]	0	0	0	0
X-rays (antero-posterior and lateral)[c]	3	1	0	0	0	0
Other tests	0	0	0	0	0	0

[a] Exact details may vary depending on specific fracture.
[b] If the fracture is slow to heal, or if the patient is being considered for removal of fixation devices.
[c] In diaphyseal fractures, X-rays should show the joint above and the joint below the fracture site.

per year, making a total of about 60,000 proximal femur fractures per year in the United Kingdom *(9)*. Approximately half of this group is treated by internal fixation, usually a sliding screw-plate device. This group of elderly patients suffers a high mortality rate as a consequence of their fracture and co-morbidities, the 1-year mortality rate being at least 20% *(10)*. For this reason, and because the sliding screw-plate device has such an enviably low rate of implant failure, these patients are followed up only if the implant positioning was suboptimal or the fracture configuration was one that suggested an increased risk of failure. Examples of such fractures would include subtrochanteric fractures or displaced subcapital fractures in the younger patient that had been reduced and internally fixed.

The follow-up of other instrumented fractures depends on the age of the patient, the adequacy of recovery from injury and the anatomical site. Skeletally immature patients are followed up to ensure that they do not suffer growth arrest (generally after physeal injuries), overgrowth (particularly after femoral shaft fractures), or other deformity. Patients from any age group are followed up at least until their fracture has healed and they confirm that they are untroubled by residual metal work. Patients with displaced intra-articular fractures are followed up for several years to ensure that arthritis does not develop and to consider early intervention if it does. Drs. Otto and L'Hommedieu consider it unnecessary to routinely remove asymptomatic fracture fixation devices, which accords with current UK practice. A very general scheme of follow-up is provided in Table 4.

THROMBOEMBOLIC PROPHYLAXIS

Historically, deep vein thrombosis (DVT) and pulmonary embolus were considered to be extremely common complications of total hip replacements *(11)*. Recently, however, evidence has been presented to suggest that the prevalence of fatal pulmonary embolism is far lower than quoted in historical publications, even in the absence of chemical prophylaxis *(12)*.

Although chemical prophylaxis (heparin, low-molecular-weight heparin (LMWH) and warfarin) reduces the rate of radiologically confirmed DVT *(13)*, there is no good evidence that the rate of fatal pulmonary emboli is reduced by these maneuvers.

Because of this, and because of the known risks of warfarin, particularly in the elderly, it is generally avoided as a prophylactic agent. The British Orthopaedic Association guide to best practice (5) does not regard the use of chemical prophylaxis as mandatory, but LMWHs are widely used in both elective and emergency orthopedic surgery, as are intermittent foot- and calf-compression devices.

WHAT HEALTH MEASURES ARE WARRANTED?
Arthritis and Total Joint Replacements

Although common sense suggests that obesity predisposes to the premature failure of total joint replacements, until recently data to support this suspicion has been lacking. Vazquez-Vela Johnson et al. have recently supplied evidence showing that, in their hands, obese males 60 years or younger undergoing total knee replacement fared particularly badly, with a 10-year device survival rate of 35.7% (14). In contrast, nonobese osteoarthritic females 60 years or older enjoyed a 99.4% 10-year device-survival rate. The numbers involved in this study were small and the age and activity levels of the young obese males may have contributed to the high failure rate. However, nonobese males under 60 did significantly better with a 10-year device-survival rate of 92.9%. This suggests that, at least in this subgroup, obesity is a significant risk factor for early failure, and that obese young males considering total knee replacement should be encouraged to lose weight before surgery and to avoid obesity subsequently.

Fractures and Fixation

The cost of treating osteoporotic fractures in the United Kingdom is estimated to be $3 billion per year (9). The National Osteoporosis Society disseminates information to patients and health care workers, and campaigns for the treatment of osteoporosis and the prevention of fractures. A comprehensive booklet, "Primary Care Strategy for Osteoporosis and Falls," can be downloaded from the National Osteoporosis Society website. The increasing awareness of osteoporosis and its risks has led to the development of protocols in most hospitals so that patients presenting with insufficiency fractures are identified and appropriate treatment commenced. Similarly patients with risk factors, for example early oophorectomy, hypogonadism, steroid therapy, etc., are also identified, investigated to assess bone density and bone biochemistry and treated accordingly.

For those desiring more information, useful references are listed in the Appendix.

APPENDIX: SOURCES OF ADDITIONAL INFORMATION

Arthritis Research Campaign (formerly the Arthritis and Rheumatism Council for Research)
Copeman House
St. Mary's Court
St. Mary's Gate
Chesterfield
Derbyshire
S41 7TD
Tel: 0870 850 5000 or
+44 (0) 1246 558033
Fax: +44 (0) 1246 558007
URL: http://www.arc.org.uk

(continued)

APPENDIX: SOURCES OF ADDITIONAL INFORMATION (CONTINUED)

National Osteoporosis Society
Camerton
Bath
BA2 0PJ
Tel: 01761 471771
Fax: 01761 471104
URL: http://www.nos.org.uk

The British Orthopaedic Association
35-43 Lincoln's Inn Fields
London
WC2A 3PN
Tel: (020 7) 405 6507
Fax: (020 7) 831 2676
URL: http://www.boa.ac.uk

The National Joint Registry
NJR Centre
329 Harwell
Didcot
Oxon
OX11 0QJ
Tel: 01235 433433
Fax: 01235 433961
URL: http://www.njrcentre.org.uk

REFERENCES

1. Charnley J. Low friction arthroplasty of the hip. Theory and practice. Berlin, Heidelburg, New York: Springer, 1979.
2. Arthritis: the big picture. The Arthritis Research Campaign, with statistics from the ARC Epidemiology Unit. Derbyshire, UK.
3. Best AJ, Fender D, Harper WM, McCaskie AW, Oliver K, Gregg PJ. Current practice in primary total hip replacement: results from the National Hip Replacement Outcome Project. Ann R Coll Surg Eng 1998;80:350–355.
4. Malchau H, Herberts P, Ahnfelt L. Prognosis of total hip replacement in Sweden. Follow-up of 92,675 operations performed in 1978–1990. Act Orthop Scand 1993;64:497–506.
5. Total hip replacement: A guide to best practice. The British Orthopaedic Association, London, UK, October 1999.
6. Hodgkinson JP, Shelley P, Wroblewski BM. The correlation between the roentgenographic appearances and operative findings at the bone cement junction of the sockets in Charnley low friction arthroplasties. Clin Orthop 1977;127:123–132.
7. Wroblewski BM. One-stage revision of infected cemented total hip arthroplasty. Clin Orthop 1986;211:103-107.
8. Ainscow DAP, Denham RA. The risk of haematogenous infection in total joint replacement. J Bone Joint Surg (Br) 1984;66-B:580–582.
9. Torgeson DJ, Iglesis CP, Reid DM. The economics of fracture prevention. In: Barlow E, Francis D, Miles R, eds. The effective management of osteoporosis. Aesculapius Medical Press; 2001, pp. 111–121.
10. Cooper C, Atkinson EJ, Jacobsen SJ, O'Fallon WM, Melton LJ. Population-based study of survival after osteoporotic fracture. Am J Epidemiol 1993;137:1001–1005.

11. Johnson R, Green JR, Charnley J. Pulmonary embolism and its prophylaxis following the Charnley total hip replacement. Clin Orthop 1977;127:123–132.
12. Warwick D, Williams MH, Bannister GC. Death and thromboembolic disease after total hip replacement. A series of 1162 cases with no routine prophylaxis. J Bone Joint Surg (Br) 1995;77B:6–10.
13. Thromboemboli, Risk Factors Consensus Group. Risk of and prophylaxis for venous thromboembolism in hospital patients. BMJ 1992;305:567–574.
14. Vazquez-Vela Johnson G, Worland RL, Keenan J, Norambuena N. Patient demographics as a predictor of the ten-year survival rate in primary total knee replacement. J Bone Joint Surg (Br) 2003;85B:52–56.

Index

A

Abdominal aortic aneurysm,
    devices for endovascular exclusion, 455–457
    endoleak classification, 456, 457, 477, 478
    epidemiology, 454
    mortality, 454, 455
    surveillance after endograft placement, 457–460, 477, 478, 482–484
ABI, see Ankle-brachial index
Agency for Healthcare Research and Quality (AHRQ), databases, 121, 122
AHRQ, see Agency for Healthcare Research and Quality
Ankle-brachial index (ABI),
    aortoiliac occlusive disease surveillance after angioplasty and stenting, 451–453
    infrainguinal vein bypass grafting surveillance, 463, 464
AMS Sphincter 800™, 313, 317, 324
Antibiotics,
    artificial urethral sphincters, 322
    breast implant surgery, 244–246
    coatings, 163
    hernia mesh repair, 274, 285
    penile prostheses, 300, 301, 307
    prophylaxis,
        Europe, 187, 188, 199, 200
        heart valve prostheses, 511
        indications, 169, 170
        orthopedic implants, 669–671
        perioperative antibiotics, 179–181
        principles, 162, 164
    vascular prostheses, 440
Anticoagulation therapy, see Warfarin
Aortic aneurysm, see Abdominal aortic aneurysm
Aortoiliac occlusive disease,
    clinical presentation, 451
    surveillance after angioplasty and stenting, 451–454, 476
    treatment options, 451

Artificial heart,
    candidate abundance, 110
    costs, 109, 110
    cultural meaning of heart, 109
    historical perspective, 108
Artificial urethral sphincters (AUS),
    AMS Sphincter 800™, 313, 317, 324
    design, 317, 319, 324, 325
    failure,
        detection, 316
        rates, 328
        revisions, 326, 327
    follow-up costs, 37
    frequency of implantation, 313
    historical perspective, 313, 324
    indications, 316, 325
    infection, 165, 166, 322
    patient selection, 322
    postoperative care, 322
    prospects for study, 319, 329, 330
    resources, 330
    surgery, 313, 315, 327, 328
    surveillance, 315–319, 323, 328, 329
    urinary incontinence,
        etiology, 325
        support groups, 320
        treatment options, 324
Aspirin,
    total joint arthroplasty patient management, 672
    warfarin combination therapy for heart valve prosthesis, 504, 507, 526
AUS, see Artificial urethral sphincters

B

Bayes' rule, decision tree modeling, 209, 210
Biofilm, formation and modeling, 151–153, 159, 160, 187
Biomaterials,
    biofilm formation and modeling, 151–153
    ceramics, 137
    histopathology of explanted prostheses,

breast implants, 148
dental implants, 147
ophthalmic implants, 147, 148
orthopedic implants, 148, 149
vascular prostheses, 148
mechanical properties,
shear modulus, 139
strain rate, 139
stress, 139
stress relaxation time, 142
tissues, 141, 142
viscoelasticity,
definition, 139, 140
Maxwell model, 140
Maxwell-Weichart model, 140
Voigt model, 140
Voigt-Kelvin model, 140
Young's modulus of elasticity, 138
metals,
reactive oxygen species formation, 136, 137
redox potentials, 136, 137
types, 136
orthopedic implants, 656, 657, 660–662, 680
ossicular implants, 413–415, 420, 434
overview and applications, 133, 134
prospects in prosthetic design, 153
synthetic polymers,
configurations, 134, 135
physical properties, 136
tissue reaction,
acute vs chronic, 142, 150, 151
coagulation cascade, 144, 145
complement system, 145, 146
inflammatory mediators, 145, 147
kinin system, 142, 143
Biomaterials Access Assurance Act, 106
Bionic, definition, 3
Biventricular pacing, 645, 646, 653
Björk-Shiley Convexo-Concave heart valve,
fracture rates, 97, 98
withdrawal from market, 98
Food and Drug Administration response, 99
notification of physicians and patients, 99, 100
litigation, 100
legislative response, 101, 103
data deficiency, 101, 112
follow-up scheme evaluation,
overview, 206
quantification, 222, 223
sensitivity analysis, 223
structuring of problem, 221, 222
utilities and evaluation, 223
state-transition model for prognosis,
multistate life table, 215, 217, 218
overview, 206, 221
sensitivity analysis, 220, 221
simple life table, 214, 215
structuring of model, 218
transition probabilities, 218, 219
utilities and evaluation, 219, 220
Breast implants, *see also* Silicone breast implants,
breastfeeding, 239, 240, 247
capsular contracture,
classification, 238, 239
epidemiology, 244, 264, 265
infection, 245, 246, 265, 266
risk factors, 244, 245, 265
complication rate, 246, 247, 256
decision tree for rupture diagnosis,
Bayes' rule, 209, 210
decision tree construction, 206, 207
overview, 206, 214
probabilities, 208, 209
sensitivity analysis, 212
structuring of decision tree, 207, 208
utilities and evaluation, 210, 212
designs, 235, 236
endoscopy, 244, 245
failure,
bleed, 241
causes, 252
diagnosis, 241, 242
leak, 241
rate, 233, 234
rupture, 241, 242, 266
follow-up costs, 36
frequency of implantation,
United Kingdom, 259
United States, 164, 237
histopathology of explanted prostheses, 148
historical perspective, 92–94, 231–234
incision approaches, 236, 255
infection organisms and prevention, 164, 165, 196, 197, 263, 264
magnetic resonance imaging, 240, 241, 243, 244, 256, 257, 266, 267
mammography, 239–243, 263
Novagold™ implants, 261
palpable masses, 240, 241

PIP hydrogel® prostheses, 260
placement, 236, 237, 255, 256
polyurethane carcinogenicity, 234
postoperative care, 238, 239
preoperative care, 237–239
resource groups, 5, 7
soya bean oil-filled implants, 260
support groups, 248, 249
surface texture, 233
surveillance, 239, 256, 257
toxic shock syndrome risks, 246
ultrasound, 243
United Kingdom experience,
  augmentation, 261, 252
  mastectomy reconstruction, 262, 263
  resources, 267, 268

C

Cardiac pacemakers,
  biventricular pacing, 645, 646, 653
  environmental interferences, 641–643
  follow-up,
    annual visits, 646
    coronary artery disease patients, 640, 641
    costs, 41
    direct evaluation components, 634–641, 650, 651
    equipment, 650
    frequency of visits, 651
    importance, 633, 634
    Internet-based follow-up, 647
    lead inspection, 639
    physical examination, 637–639
    programming, 639, 640, 651–653
    telephonic monitoring, 634–636, 650, 651
  frequency of implantation, 175, 649
  frequency of implantation, 633
  indications, 633, 645
  infection, 175, 176, 646, 647
  surgery, 649, 650
Carotid stenosis,
  recurrence after endarterectomy or stenting, 441–443
  screening, 441
  surveillance following revascularization, 443, 445, 473
CBA, see Cost–benefit analysis
CDRH, see Center for Devices and Radiological Health
CEA, see Cost-effectiveness analysis

Center for Devices and Radiological Health (CDRH), 10, 61
Centers for Medicare and Medicaid Services (CMS), databases, 120, 121
Central venous catheters, see Vascular access devices
Cerebrospinal fluid shunts,
  clinical trials and prospects for study, 372, 373
  design,
    flow-regulating shunts, 343, 344
    materials, 340, 344
    parts, 340
    valves, 340–342
  follow-up,
    annual surveillance, 350, 359, 360, 366–369
    computed tomography, 348–351, 361, 366–369
    costs, 37
    first year, 348–350, 366–368
  frequency of implantation, 164, 363, 364
  historical perspective, 334, 335, 365, 366
  hydrocephalus,
    epidemiology, 335, 336, 363
    history of study, 333, 334
    pathophysiology, 336–338, 364, 365
    prevention, 371
    resources, 353, 361, 373
  hydrodynamics,
    flow, 339
    modeling in shunt design, 339, 340
    pressure, 338, 339
    resistance, 339
    viscosity, 339
  indications, 359
  infection,
    organisms, prevention, and treatment, 174, 175, 188–190, 352, 353, 360, 369–371
  malfunction,
    classification, 352
    failure and complication rate, 367–369
  pregnancy, 360
  restrictions on patients, 350, 351
  surgery, 344–348, 361
  types, 341–344, 359, 360
Chlorhexidine-silver sulfadiazine, catheter impregnation, 3
Clarion cochlear implant, see Cochlear Implants
  external components, 383, 384
  internal components, 382, 383

       performance, 388
       speech processing strategies, 384
   Clinical practice guidelines,
       application, 53, 54
       limitations, 53, 54
   CMS, *see* Centers for Medicare and Medicaid
           Services
   Coagulation, pathways, 144, 145
   Cochlear implants,
       Clarion cochlear implant,
           external components, 383, 384
           internal components, 382, 383
           performance, 388
           speech processing strategies, 384
       complications,
           flap complications, 398, 399
           infection, 398, 399, 406
           prevention, 406
           rates, 397, 398
       components and functions, 106, 379, 381, 382
       deaf community attitudes, 107, 108
       deafness,
           prevention, 400
           support groups, 399, 401
       design, 391, 392
       eligibility, 106
       expectations and recovery, 397
       follow-up,
           annual visits, 401, 407
           children, 393
           costs, 38
           early problem detection, 405
           late problem detection, 405, 406
           postoperative, 392, 393
           responsibility, 404, 405
       frequency of implantation, 381
       historical perspective, 379, 381, 408, 409
       infection organisms and prevention, 178,
               198, 200
       magnetic resonance imaging
               compatibility, 392
       MedEl cochlear implant,
           external components, 387, 388
           internal components, 387
           performance, 388, 389
           speech processing strategies, 388
       Nucleus 24™ cochlear implant,
           external components, 386
           internal components, 385, 386
           performance, 388
       outcomes, 106, 107

       patient selection and audiological
               assessment, 389, 390, 409–411
       programming strategies, 393, 394
       special populations,
           bacterial meningitis and ossified
               cochlea, 395, 396
           children, 393–395
           elderly, 395
           malformed cochlea, 396
           revision surgery, 396, 397
       surgery, 391, 411
   Complement system, components, 145
   Computed tomography (CT),
       abdominal aortic aneurysm endografts,
               457–460
       cerebrospinal fluid shunts, 348–351, 361
       vena cava filters, 543–545
   Computer modeling,
       advantages and limitations, 205
       cost-effectiveness modeling, 223, 224
       decision tree for breast implant rupture
               diagnosis,
           Bayes' rule, 209, 210
           decision tree construction, 206, 207
           overview, 206, 214
           probabilities, 208, 209
           sensitivity analysis, 212
           structuring of decision tree, 207, 208
           utilities and evaluation, 210, 212
       follow-up scheme evaluation,
           overview, 206
           quantification, 222, 223
           sensitivity analysis, 223
           structuring of problem, 221, 222
           utilities and evaluation, 223
       state-transition model for Björk-Shiley
               Convexo-Concave valve prognosis,
           multistate life table, 215, 217, 218
           overview, 206, 221
           sensitivity analysis, 220, 221
           simple life table, 214, 215
           structuring of model, 218
           transition probabilities, 218, 219
           utilities and evaluation, 219, 220
       uncertainty, 226
       verification, 224–226
   Convexo-Concave heart valve, *see* Björk-
           Shiley Convexo-Concave heart valve
   Cost–benefit analysis (CBA), device
           surveillance testing,

Index

cost-effectiveness analysis comparison, 25, 26
valuation of life, 26, 27
Cost-effectiveness analysis (CEA), device surveillance testing,
computer modeling, 223, 224
cost–benefit analysis comparison, 25, 26
overview, 25
utility analysis comparison, 27, 28
Coumadin, see Warfarin
C-reactive protein (CRP),
cerebrospinal fluid shunt infection detection, 370
orthopedic implant infection detection, 666, 667, 683
CRP, see C-reactive protein
CT, see Computed tomography
Current Bibliographies in Medicine, 11

D

Databases, medical devices,
epidemiology study databases,
academic databases of public data, 124, 125
Canada government databases, 123
private health care databases, 123, 124
United States government databases,
Agency for Healthcare Research and Quality, 121, 122
Centers for Medicare and Medicaid Services, 120, 121
complex sampling consequences, 122, 123
prospects,
availability, 102, 103
privacy concerns, 127, 128
technological sophistication, 127
requirements, 101, 102
Decision tree, see Computer modeling
Deep venous thrombosis (DVT), total joint arthroplasty patients, 671, 683, 689, 690
Dental implants,
advantages over dentures, 603
cleaning, 608–611
complications,
biological vs mechanical, 607, 620, 626
bone degeneration, 608
rate, 611
soft tissue inflammation, 608, 620
endossous vs subperiosteal implants, 603, 604
failure rates, 607, 619, 620, 624
follow-up,
annual visits, 613, 621, 627
bone factors, 628
checklist, 611–613, 621
costs, 41
oral hygiene factors, 628, 629
prosthetic factors, 627, 628
soft tissue factors, 628
frequency of implantation, 176, 607, 626
histopathology of explanted prostheses, 147
indications, 603, 621, 622, 624
infection organisms and prevention, 176
osseointegration, 605, 606
peri-implant environment, 625
plaque, 608–611, 625
resources, 613, 614, 622
types, 606, 607, 619, 620
Diagnostic test performance, see Surveillance test performance
Doppler echocardiography, heart valve prostheses, 500–503
DuPont, termination of polyethylene sales for medical devices, 105, 106
DVT, see Deep venous thrombosis

E

Elasticity, see Biomaterials
Endocarditis, see Heart valve prostheses
Endoscopy, breast implants, 244, 245
Epidemiology,
data,
academic databases of public data, 124, 125
Canada government databases, 123
database limitations for epidemiology studies, 119, 120
device-specific study sources,
academic studies, 118
manufacturer-sponsored studies, 118
privacy concerns, 127, 128
private health care databases, 123, 124
registries, 117, 118
surveys,
market research, 117
national, 116, 117
technological sophistication, 127
types, 115, 116
United States government databases,

Agency for Healthcare Research and
Quality, 121, 122
Centers for Medicare and Medicaid
Services, 120, 121
complex sampling consequences,
122, 123
Food and Drug Administration research,
78, 79
meta-analysis, 125
scope of implanted device studies, 115
surveillance test evaluation, 14
technology assessment, 125–127
Erectile dysfunction, *see* Penile prostheses
Evidence-based medicine, deficiencies in
implanted prosthetic devices, 47, 48
Expanded polytetrafluoroethylene, grafts,
481, 482

F

FBGCs, *see* Foreign body giant cells
FDA, *see* Food and Drug Administration
Follow-up costs,
breast implants, 36
cardiac pacemakers, 41
cerebrospinal fluid shunts, 37
cochlear implants, 38
cost analysis of testing,
cost–benefit analysis,
cost-effectiveness analysis
comparison, 25, 26
valuation of life, 26, 27
cost-effectiveness analysis, 25
overview, 22, 23
perspective of analysis, 23
product function description, 23–25
sensitivity analysis, 29
utility analysis,
cost-effectiveness analysis
comparison, 27, 28
quality-adjusted life years, 29
utility values for health states, 28, 29
data sources and methodology, 34, 35, 43
dental implants, 41
frequency of screening, 17, 33
heart valve prostheses, 39
hernia prostheses, 36
intravascular filters and stents, 40
joint prostheses, 41
Medicare-allowed charges, 42, 43
ossicular implants, 38
penile prostheses, 37
urethral sphincters, 37

variability, 4, 43
vascular access devices, 40
vascular prostheses, 38, 39
Follow-up, *see* Surveillance
Food and Drug Administration (FDA),
access to information, 82, 83
Björk-Shiley Convexo-Concave heart
valve response, 99
Center for Devices and Radiological
Health, 10, 61
clinical trials in device approval,
bioresearch monitoring program, 72, 73
challenges, 70, 71
Investigational Device Exemption, 71, 72
contacts, 10, 11
custom device exemption, 84, 85
freedom of information, 11, 12, 83
functions, 61–63
global harmonization of regulation, 83
legislative authorization, 61–63
postmarket enforcement,
field inspections, 82
recall authority, 81, 82
tracking, 80, 81
postmarket oversight,
overview, 73
surveillance and risk assessment,
adverse event/problem reporting,
74–76
epidemiological research, 78, 79
implant retrieval, failure analysis,
device reliability, 76, 77
mandated studies, 77, 78
premarket review of medical devices,
applications,
class III products, 67–70
510k premarket notification, 66, 67
humanitarian device exemption
applications, 68, 69
Premarket Approval Application,
63, 67, 68
Product Development Protocol, 69, 70
medical device classification,
class I, 65
class II, 65, 66
class III, 66
Medical Device Advisory
Committee, 66
regulatory controls,
general controls, 64
special controls, 64, 65

Systematic Technology Assessment of Medical Products, 83, 84
Foreign body giant cells (FBGCs),
  distribution calculation, 151
  histopathology of explanted prostheses, 148–150

G

Genetic testing, 57
GHTF, see Global Harmonization Task Force
Global Harmonization Task Force (GHTF), mission, 83
Grafts, see Vascular prostheses

H

Health care costs,
  managed care, 91, 92
  professional vs consumer models of health care, 112
  rationing of health care, 57, 58, 89, 90, 110, 111
  trends, 90
  worldwide market for implanted devices, 115
Heart, artificial, see Artificial heart
Heart valve prostheses, see also Björk-Shiley Convexo-Concave heart valve,
  anticoagulation therapy,
    aspirin combination therapy, 504, 507, 526
    bioprosthetic valves, 505, 525
    bleeding complications, 506, 507, 523
    heparin indications, 507, 508, 525
    International Normalized Ratio targets, 503, 504, 531
    surgery patient management, 507, 508
    thromboembolism incidence, 503, 525, 526
    thrombosis sites, 506
  classification,
    biological valves, 494–497, 529
    mechanical valves, 491, 494, 529
    table of types, 490
  complications,
    endocarditis, 510–512
    fistula, 517
    hemolysis, 509, 510
    overview, 489
    pannus formation, 514
    paravalvular leak, 515
    patient–prosthesis mismatch, 516
    pseudoaneurysm, 516
    structural failure, 512, 514
    thrombosis, 514, 515
  coronary artery disease patients, 517
  endocarditis effects on outcome, 524
  follow-up,
    annual visits, 499, 524, 530
    costs, 39
    Doppler echocardiography, 500–503
    physical examination, 499
    postoperative, 498
    transesophageal echocardiography, 530
  frequency of implantation, 96, 489, 528
  historical perspective, 96, 97, 490, 491
  infection organisms and prevention, 160, 167–170, 197, 198, 510–512, 530, 531
  prospects for study, 517
  requirements, 97
  selection of valve, 497, 498, 523–525
  support groups and resources, 517, 518, 526
  United Kingdom experience, 528–532
Hemodialysis access,
  end-stage renal disease epidemiology, 445
  graft limitations versus native arteriovenous fistula, 445
  surveillance of grafts, 445–447, 474
Heparin,
  heart valve prosthesis indications, 507, 508, 525
  total joint arthroplasty patient management, 672, 689, 690
  vascular access device indications, 596, 597
Hernia mesh repair,
  advantages, 277
  European experience, 284, 285
  failure, 274
  follow-up costs, 36
  frequency, 271
  hiatal hernia repair, 281, 282
  historical perspective, 271
  incisional hernia repair, 279, 280
  infection, 194, 195, 274, 285
  inguinal hernia repair, 277–279
  materials, 271–273, 285
  placement, 273, 274
  preventive care, 275, 286
  United States experience, 277–279
Hip implants, see Orthopedic implants
Histopathology, explanted prostheses,
  breast implants, 148
  dental implants, 147
  heart valve prostheses, 148

ophthalmic implants, 147, 148
orthopedic implants, 148, 149
Hydrocephalus, *see* Cerebrospinal fluid shunts

I

ICDs, *see* Implantable cardioverter defibrillators
IDE, *see* Investigational Device Exemption
Implantable cardioverter defibrillators (ICDs),
    frequency of implantation, 175
    infection organisms and prevention, 175, 176
Infection,
    artificial urethral sphincters, 165, 166
    biofilm formation and modeling, 151–153, 159, 160, 187
    breast implants,
        capsular contracture, 245, 246, 265, 266
        organisms and prevention, 164, 165, 196, 197, 263, 264
    cardiac pacemakers, 175, 176, 646, 647
    central venous catheters, 577–579, 595
    cerebrospinal fluid shunts, 174, 175, 188–190, 352, 353, 360, 369–371
    clinical presentation, 160
    cochlear implants, 398, 399, 406
    heart valve prostheses, 160, 167–170, 197, 198, 510–512, 530, 531
    hernia mesh repair, 194, 195, 274, 285
    orthopedic implants, 170–172, 192–194, 665–671, 682, 683, 688
    ossicular implants, 431, 432
    patient factors, 188
    penile prostheses, 160, 295, 296, 300, 301, 305, 307, 308
    prevention,
        antibiotic prophylaxis,
            Europe, 187, 188, 199, 200
            indications, 169, 170
            perioperative antibiotics, 179–181
            principles, 162, 164
        breast implants, 164, 165, 196, 197
        cardiac pacemakers, 175, 176
        cerebrospinal fluid shunts, 174, 175, 188–190
        cochlear implants, 178, 198, 200
        dental implants, 176
        heart valve prostheses, 160, 167–170, 197, 198
        hernia mesh repair, 194, 195
        implantable cardioverter defibrillators, 175, 176
        intraoperative measures, 161–163

intravascular filters and stents, 176–178
ophthalmic implants, 167
orthopedic implants, 170–172, 192–194
penile implants, 160, 165–167, 190, 191
postoperative measures, 163, 164
preoperative measures, 161
vascular prostheses, 172–174, 195, 196
prosthesis factors, 186, 187
Infective endocarditis, *see* Heart valve prostheses
Inferior vena cava filters, *see* Vena cava filters
Inflammation, histopathology of explanted prostheses, 145, 147
Infrainguinal vein bypass grafting, *see* Intermittent claudication
INR, *see* International Normalized Ratio
Intermittent claudication,
    epidemiology, 460
    surveillance after infrainguinal vein bypass grafting, 462–464, 478, 479, 484, 485
    treatment options, 460, 462
Internal fixation devices, *see* Orthopedic implants
International Normalized Ratio (INR), heart valve prostheses anticoagulation targets, 503, 504, 531
Intravascular filters and stents,
    endovascular stents,
        failure, 537
        infection, 552
        patency, 538, 539
        surveillance, 538, 539, 550–553
        thrombogenicity, 537, 538
        types, 537
        United Kingdom experience, 550–552
    follow-up costs, 40
    historical perspective, 533
    infection organisms and prevention, 176–178
    resources, 554, 555
    vena cava filters,
        anticoagulation therapy, 534
        devices, 533, 534, 542, 543
        frequency of implantation, 535
        ideal criteria, 542, 549
        imaging, 543–545
        indications, 533, 547
        patient education, 534, 535
        surveillance, 535–537, 543–545, 549, 550
        temporary placement, 547, 548

Index

thrombosis, 534, 549
Investigational Device Exemption (IDE), 71, 72

J

Joint prostheses, *see* Orthopedic implants

K

Kinin system, overview, 142, 143

L

Lens implants, *see* Ophthalmic implants
Life tables, state-transition modeling,
  multistate life table, 215, 217, 218
  simple life table, 214, 215
Likelihood ratio, diagnostic test performance evaluation, 17, 18
Litigation,
  Björk-Shiley Convexo-Concave heart valve, 100
  breast implants, *see* Silicone breast implants
  fear impact on clinical decisions, 54, 55
  Proplast hip implant, 105
  scientific testimony and jury reaction, 95

M

Magnetic resonance imaging (MRI),
  breast implants, 240, 241, 243, 244, 256, 257, 266, 267
  cochlear implant compatibility, 392
  vena cava filters, 543–545
Mammography, breast implant recipients, 239–243, 263
Managed care, principles, 91, 92
Materials, *see* Biomaterials
MedEl cochlear implant,
  external components, 387, 388
  internal components, 387
  performance, 388, 389
  speech processing strategies, 388
Medical device, definition, 2
Medical Devices Register, 6
Mesenteric artery revascularization,
  surveillance after bypass, 450, 475
  treatment options, 450
Meta-analysis, applications, 53, 125
Mondor's disease, breast implant risks, 247
MRI, *see* Magnetic resonance imaging

N

Notification, physicians and patients of implant problems, 99, 100
Novagold™ breast implants, 261
Nucleus 24™ cochlear implant,
  external components, 386
  internal components, 385, 386
  performance, 388

O

Ophthalmic implants,
  follow-up charges, 41
  histopathology of explanted prostheses, 147, 148
  infection organisms and prevention, 167
Orthopedic implants,
  allergy, 669
  biomaterials, 656, 657, 660–662, 680
  classification, 655, 656
  disease management, 667, 668
  fixation device failure detection, 663–665, 688, 689
  frequency of implantation, 170, 658–660, 686
  histopathology of explanted prostheses, 148, 149
  historical perspective, 655–657, 679, 686
  infection, 170–172, 192–194, 665–671, 682, 683, 688
  obesity and outcomes, 674, 690
  osteoporosis prevention, 672–674, 690
  Proplast hip implant,
    Food and Drug Administration approval, 104
    litigation, 105
    Teflon failure, 104
  resources and support groups, 674, 675, 690, 691
  surveillance, 672, 673, 680–683, 686–688
  total joint arthroplasty,
    bone–prosthesis interface, 662, 679, 680
    complications, 661, 688
    deep venous thrombosis, 671, 683, 689, 690
Ossicular implants,
  follow-up,
    annual visits, 418, 421, 431
    complication detection, 418, 419
    costs, 38
    early problem detection, 425
    late problem detection, 425, 426
    postoperative, 418, 430–434
    responsibility, 424, 425
  historical perspective, 413, 414, 419, 420
  indications, 415–418
  infection prevention, 431, 432
  materials, 413–415, 420, 434

postoperative care, 430
precautions for patients, 425, 427
resources, 422, 434
stapes surgery in otosclerosis, 419–422
surgery principles, 429, 430
synthetic prosthesis advantages, 413, 429
types, 415
Osteoporosis, prevention, 672–674, 690

P

Pacemakers, *see* Cardiac pacemakers
Patient input, clinical decisions after prosthesis implantation, 51
Penile prostheses,
 complications, 308, 309
 costs, 303, 304
 design and operation, 290–292, 300, 305, 310
 erectile dysfunction,
  epidemiology, 289, 290
  evaluation, 293
  prevention, 297
  resources, 297, 298
  treatment options, 292, 293, 305, 306
 European resources, 310
 failure, 294, 295, 303, 308, 309
 follow-up costs, 37
 frequency of implantation, 164, 304, 305
 historical perspective, 290, 294, 302, 303
 indications, 303, 304
 infection organisms and prevention, 160, 165–167, 190, 191, 295, 296, 300, 301, 305, 307, 308
 postoperative care, 294, 307
 preoperative care, 300, 301, 306
 satisfaction rate, 292, 293
 sensitivity outcomes, 292
 surgery, 293, 294, 304, 306, 307
 surveillance, 294, 301, 309, 310
 vendors, 298
 weight gain and penile length, 296
Polyurethane, carcinogenicity, 234
Predictive value, diagnostic test performance, 16
Proplast hip implant,
 Food and Drug Administration approval, 104
 litigation, 105
 Teflon failure, 104
Prostate cancer, survivor groups, 298

R

Rationing, health care, 57, 58, 89, 90, 110, 111
Receiver operating characteristic (ROC) curve, diagnostic test performance evaluation, 18, 19
Renal artery stenosis,
 epidemiology, 447
 surveillance after revascularization, 447–449, 474, 475
Rifampicin, graft impregnation, 481
ROC curve, *see* Receiver operating characteristic curve

S

Selective serotonin reuptake inhibitors, sexual dysfunction, 289
Sensitivity, diagnostic test performance, 15
Silicone breast implants, *see also* Breast implants,
 historical perspective, 92–94, 232–234
 litigation history, 4, 94
 return to market, 95, 96
 silicone properties, 4
 United Kingdom, 259, 260
 withdrawal from market, 94, 95, 234, 235
Smooth muscle cell, vascular graft response, 438
Specificity, diagnostic test performance, 15
STAMP, *see* Systematic Technology Assessment of Medical Products
State-transition model, *see* Computer modeling
Stents, *see* Intravascular filters and stents; Vascular prostheses
Surgery, historical perspective, 1, 3
Surveillance, *see also* specific implants,
 computer modeling for follow-up scheme evaluation,
  overview, 206
  quantification, 222, 223
  sensitivity analysis, 223
  structuring of problem, 221, 222
  utilities and evaluation, 223
 costs, *see* Follow-up costs
 duration, factors affecting, 55
 primary prevention of disease, 55–58
 secondary benefits, 55
Surveillance test performance,
 cost analysis, *see* Follow-up costs
 cost–benefit analysis,

Index

cost-effectiveness analysis
  comparison, 25, 26
  valuation of life, 26, 27
cost-effectiveness analysis, 25
  overview, 22, 23
  perspective of analysis, 23
  product function description, 23–25
  sensitivity analysis, 29
  utility analysis,
    cost-effectiveness analysis
      comparison, 27, 28
    quality-adjusted life years, 29
    utility values for health states, 28, 29
diagnostic test characteristics,
  reliability, 16, 17
  validity, 14–16, 51, 52
  yield, 17
epidemiological principles for evaluation, 14
frequency of screening, 17, 33
importance of study, 13
likelihood ratios in test performance
  evaluation, 17, 18
receiver operating characteristic curve
  analysis, 18, 19
screening program requirements, 18
testing threshold determination, 20, 21
treatment threshold determination, 19, 20
Systematic Technology Assessment of
  Medical Products (STAMP), overview,
  83, 84

T

Temporomandibular joint (TMJ),
  anatomy, 103
  disease management, 103
  implant,
    requirements, 103
    Vitek implant, 103, 104, 111
TIPS, see Transjugular intrahepatic
  portosystemic shunt
TMJ, see Temporomandibular joint
Toxic shock syndrome (TSS), breast implant
  risks, 246
Transjugular intrahepatic portosystemic
  shunt (TIPS),
  indications, 464, 465
  restenosis, 465, 466
  surveillance, 466, 467, 479
Treatment threshold, determination, 19, 20
TSS, see Toxic shock syndrome

U

Ultrasound,
  breast implants, 243
  Doppler echocardiography of heart valve
    prostheses, 500–503
Uncertainty, computer models, 226
Urethral sphincters, see Artificial urethral
  sphincters
Urinary incontinence, see Artificial urethral
  sphincters
Utility analysis, device surveillance testing,
  cost-effectiveness analysis comparison,
    27, 28
  utility values for health states, 28, 29
  quality-adjusted life years, 29
Utility-sensitive decisions, attributes, 50, 51

V

VADs, see Vascular access devices
Value, concept in outcomes research and
  economic analysis, 49
Valve prostheses, see Heart valve prostheses
Vascular access devices (VADs),
  arterial catheter complications and
    prevention, 596, 597
  central venous catheters,
    maintenance, 569–571
    removal indications, 578
    surveillance,
      damage, 580, 595
      infection, 577–579, 595
      occlusion, 579, 580, 595, 596
      purpose, 571, 572
      thrombosis and stenosis, 580–583, 596
  follow-up,
    annual visits, 590, 591
    costs, 40
    European experience, 597–599
    prospects for study, 583
    strategies, 43
  historical perspective, 561
  indications, 561, 562, 594
  resources, 599
  survey of Ohio surgeon practices, 572–576
  types,
    intermediate-term devices, 564, 565
    long-term devices,
      infusion pumps, 569
      right atrial catheters, 565
      venous access ports, 565, 566, 568, 569

overview, 561, 562, 593, 594
selection factors, 588–590
short-term, percutaneous, non-tunneled central venous catheters, 562–564
Vascular prostheses,
  abdominal aortic aneurysm,
    devices for endovascular exclusion, 455–457
    endoleak classification, 456, 457, 477, 478
    epidemiology, 454
    mortality, 454, 455
    surveillance after endograft placement, 457–460, 477, 478, 482–484
  antibiotic prophylaxis, 440
  aortoiliac occlusive disease,
    clinical presentation, 451
    surveillance after angioplasty and stenting, 451–454, 476
    treatment options, 451
  carotid stenosis,
    recurrence after endarterectomy or stenting, 441–443
    screening, 441
    surveillance following revascularization, 443, 445, 473
  costs, 437
  failure causes, 438–441
  follow-up,
    costs, 38, 39
    strategies, 43
  frequency of use, 438, 481
  hemodialysis access,
    end-stage renal disease epidemiology, 445
    graft limitations versus native arteriovenous fistula, 445
    surveillance of grafts, 445–447, 474
  histopathology of explanted prostheses, 148
  infection, 172–174, 195, 196, 439–441
  intermittent claudication,
    epidemiology, 460
    surveillance after infrainguinal vein bypass grafting, 462–464, 478, 479, 484, 485
    treatment options, 460, 462
  materials and design, 480–482
  mesenteric artery revascularization,
    surveillance after bypass, 450, 475
    treatment options, 450
  renal artery stenosis,
    epidemiology, 447
    surveillance after revascularization, 447–449, 474, 475
    tissue response, 438
  transjugular intrahepatic portosystemic shunt,
    indications, 464, 465
    restenosis, 465, 466
    surveillance, 466, 467, 479
Vena cava filters,
  anticoagulation therapy, 534
  devices, 533, 534, 542, 543
  frequency of implantation, 535
  ideal criteria, 542, 549
  imaging, 543–545
  indications, 533, 547
  patient education, 534, 535
  surveillance, 535–537, 543–545, 549, 550
  temporary placement, 547, 548
  thrombosis, 534, 549
Verification, computer models, 224–226
Viscoelasticity,
  definition, 139, 140
  Maxwell model, 140
  Maxwell-Weichart model, 140
  Voigt model, 140
  Voigt-Kelvin model, 140

W

Warfarin,
  heart valve prostheses anticoagulation,
    aspirin combination therapy, 504, 507, 526
    bioprosthetic valves, 505, 525
    bleeding complications, 506, 507, 523
    International Normalized Ratio targets, 503, 504, 531
    surgery patient management, 507, 508
    thromboembolism incidence, 503, 525, 526
    thrombosis sites, 506
  total joint arthroplasty patient management, 671, 672, 689, 690

X

X-ray, internal fixation device failure detection, 664, 680, 687

Y

Young's modulus of elasticity, 138